Salt Tectonics:

A Global Perspective

Based on the Hedberg International Research Conference
Bath, U.K., September 1993

Edited by

M. P. A. Jackson

D. G. Roberts

S. Snelson

AAPG Memoir 65

Published by
The American Association of Petroleum Geologists
Tulsa, Oklahoma, U.S.A.
Printed in the U.S.A.

Association Editor: Kevin T. Biddle
Science Director: Richard Steinmetz
Publications Manager: Kenneth M. Wolgemuth
Special Projects Editor: Anne H. Thomas
Production: Kathy and Dana Walker, Editorial Technologies, Renton, WA

This and other AAPG publications are available from

AAPG Bookstore
P.O. Box 979
Tulsa, OK 74101-0979
U.S.A.
Tel (918) 584-2555
  or (800) 364-AAPG (U.S.A.—book orders only)
Fax (918) 560-2652
  or (800) 898-2274 (U.S.A.—book orders only)
  or e-mail: bookstore@aapg.org

Australian Mineral Foundation
AMF Bookshop
63 Conyngham Street
Glenside, South Australia 5065
Australia
Tel (08) 379-0444
Fax (08) 379-4634

Geological Society Publishing House
Unit 7, Brassmill
Enterprise Centre
Brassmill Lane
Bath BA1 3JN
United Kingdom
Tel 0225-445046
Fax 0225-442836

Canadian Society of Petroleum Geologists
#505, 206 7th Avenue S.W.
Calgary, Alberta T2P 0W7
Canada
Tel (403) 264-5610

AAPG
Wishes to thank the following
for their generous contributions
to

Salt Tectonics:
A Global Perspective

BP Exploration Inc.

Geological Survey of Canada

Hans Ramberg Tectonic Laboratory,
University of Uppsala

Rijks Geologische Dienst

Shell E&P Technology Company

Shell U.K. Exploration and Production

Contributions are applied against the production
costs of publication, thus directly reducing the
book's purchase price and making the volume
available to a greater audience.

Acknowledgments

We thank all the far-flung attendees, who converged on Bath for the Hedberg International Research Conference, for contributing so much to the success of the meeting. We also express our appreciation to all the institutions and companies that supported the attendees, provided time for research, and released vital data—especially the major new regional syntheses in this book. We thank British Petroleum for generously donating funds that subsidized the costs of some speakers from academia who otherwise could not have attended the meeting. We are grateful to Debbi Boonstra, who handled the registration and logistics for the conference, and Bob Millspaugh, who represented AAPG at Bath. The conference was opened and closed by D. G. Roberts and sessions were chaired by M. P. Coward, F. A. Diegel, A. D. Gibbs, S. G. Henry, R. G. Hickman, J. R. Hossack, M. P. A. Jackson, J. Letouzey, K. R. McClay, D. G. Roberts, S. Snelson, P. Szatmari, C. J. Talbot, and S. Wu.

To all the authors that contributed to this book, thank you for the countless hours spent toiling on manuscripts while stretched by other demands and for your efforts to comply with the editing and reviewing schedule. Each chapter was reviewed by another author and by "outside" experts in certain aspects of salt tectonics. Accordingly, we would like to thank the following reviewers for invaluable assistance in the onerous task of reviewing the manuscripts.

D. J. Anastasio	H. Koyi
C. J. Banks	S. E. Laubach
R. T. Buffler	L. M. Liro
S. C. Cameron	D. B. McGuinness
N. L. Carter	R. G. Martin
P. R. Cobbold	T. Nelson
M. P. Coward	K. T. Nilsen
J. C. deBremaecker	S. Nybakken
I. Davison	F. J. Peel
P. W. Dickerson	S. C. Reeve
F. A. Diegel	M. G. Rowan
G. Eisenstadt	H. Schmeling
R. Evans	D. D. Schultz-Ela
S. G. Henry	D. C. Schuster
J. R. Hossack	S. J. Seni
W. M. House	C. J. Talbot
M. R. Hudec	P. Tauvers
J. L. Jackson	B. C. Vendeville

The final compilation of this book was carried out with the invaluable help of the following individuals at the Bureau of Economic Geology, The University of Texas at Austin: Jeannette Miether for stylistic copyediting; Margaret Evans and Joel Lardon for helping with computer graphics; and Hongxing Ge for general assistance.

Finally, it has been a rewarding and pleasurable experience to collaborate with AAPG's highly professional production staff, especially Anne Thomas and Kathy Walker.

Martin Jackson

David Roberts

Sig Snelson

Foreword

Thirty years ago, a symposium on "Diapiric and Related Structures" formed part of the 1965 AAPG Annual Meeting in New Orleans. The resulting collection of papers was published as the classic Memoir 8, *Diapirism and Diapirs,* edited by J. Braunstein and G. D. O'Brien. Until now, this was the last AAPG book that focused on salt diapirs, although many papers continued to be published in the *Bulletin.* Since 1965, the data base of salt tectonics has evolved at an accelerating rate based on field observation, seismic interpretation, and advanced structural modeling. These findings and the recognition of the exploration importance of salt tectonics in more than 80 basins worldwide indicated that another global attempt to unravel the complexities of salt tectonics was long overdue.

In August 1990, AAPG's Science Director, Gary Howell, first suggested to one of us the idea of holding a Hedberg International Research Conference on "Salt Tectonics" with worldwide scope. After a formal meeting with him in May 1991, the three compiling editors accepted the task of convening the symposium, which was held in Bath, U.K., on September 13–17, 1993.

The timing of this conference was propitious. The preceding four years had seen an explosive expansion in papers dealing with salt tectonics. In particular, this field of study had just passed through a major conceptual breakthrough based on three-dimensional seismic data and innovative kinematic and dynamic modeling of salt tectonics. The objective of the conference was to highlight and share these advances in understanding the structural geology and tectonics of salt structures at seismic scales in the context of hydrocarbon exploration and development, including classic areas having abundant data as well as lesser known basins characterized by newly discovered structural styles. Of special interest were newer aspects of salt tectonics that were unknown at the symposium 30 years ago. These included controls on the shape and structural evolution of salt structures, especially the style and rate of sedimentation; the influence of faulting in extensional, contractional, transtensional, transpressional, and inversion regimes; emplacement of allochthonous salt sheets and canopies and their subsequent segmentation and redistribution; the creation of salt welds and fault welds; and the vernerable but previously neglected field of contractional salt tectonics.

The program of the Bath conference was the largest and most comprehensive ever dealing with salt tectonics. Some 46 papers were presented orally and another 34 displayed as posters interspersed with discussion sessions that continued well beyond the day's proceedings. Invitations to the 75 attendees were deliberately skewed to attract a majority of contributions from industry to ensure release of abundant new geological and geophysical data of high quality. As a result, the papers spanned a broad range, covering state-of-the-art seismic imaging, restoration techniques, and mechanical modeling followed by examination of salt tectonics (and some shale tectonics) in the following regions: Gulf of Mexico, Arctic Canada, Barents Sea, North Sea, Spain, Mediterranean Sea, Algeria, Tunisia, Red Sea, Iran, Pakistan, Kazakhstan, Caribbean, Angola, Gabon, and Brazil. The conference was notable for the wealth of new data and novel ideas offered through formal presentations and informal discussions among attendees. During the conference, many attendees expressed a wish to see a fuller collection of papers published rather than the volume of extended abstracts available only to attendees. Accordingly, they were asked to respond to a subsequent questionnaire in which they strongly reconfirmed their earlier wish to see publication of a collection of full-length papers as an AAPG Memoir.

This volume is the result. To ensure prompt publication of an affordable book, the length was restricted to 21 chapters (coincidentally the same number of papers as in Memoir 8) out of the 80 papers presented at the conference. Not all the presenters were able to contribute full-length papers, but nevertheless we were able to be highly selective in choosing papers that highlighted key advances in salt tectonics. Our choice preserves the wide geographic coverage of the conference and includes both detailed regional syntheses of classic areas, such as the Gulf of Mexico and the North Sea, as well as preliminary investigations of equally fascinating but less-explored regions of salt tectonics, such as the deep-water Santos Basin (Brazil), offshore Yemen (Red Sea), Parry Islands (Arctic Canada), and the Nordkapp Basin (Barents Sea). We hope that the investigations of salt tectonics presented in this volume will serve as classic examples of a wide range of structural styles involving evaporites. We also offer these examples as case studies that can be applied as guides to petroleum and mineral exploration in other salt basins around the world. With this broad scope, this book is entitled *Salt Tectonics: A Global Perspective.*

The organization of the book has a structure similar to that of the Bath conference. A historical review chapter is followed by four chapters on section balancing and modeling (later chapters also incorporate modeling research, but their focus is more geographic). The next 16 chapters are organized along geographic lines in loose order of their current hydrocarbon production, from the Gulf of Mexico, where mature exploration has been invigorated by the subsalt play, to remote reaches of the Arctic Ocean, where exploration is in its infancy.

Martin Jackson
David Roberts
Sig Snelson

Contents

About the Editors

Born and educated in Zimbabwe, **Martin Jackson** received his Bachelor's degrees in geology from the University of London in 1968 and 1969. His research career began by investigating lunar structures and processes from Lunar Orbiter images. After two years as a mineral exploration geologist for Cominco in southern Africa, he joined the Precambrian Research Unit at the University of Cape Town, which awarded him a Ph.D. in Geology in 1976 for work on a previously unmapped 10,000 km^2 Precambrian terrane of high-grade metamorphic and igneous rocks in the Namib Desert of Namibia. He spent the next four years teaching in the Department of Geology at the University of Natal, South Africa, from where he carried out field-based research in Swaziland and Zululand. His African experiences led to co-authorship of the textbook *Crustal Evolution of Southern Africa: 3.8 Billion Years of Earth History* (1982).

Martin joined the Texas Bureau of Economic Geology in 1980, where he is currently a Senior Research Scientist. After a desperate Texas-wide search for something vaguely familiar, he was attracted to salt structures as low-temperature analogs of high-grade gneisses and has been fascinated by salt tectonics around the world ever since. In 1984, he was Visiting Scientist at Uppsala University, Sweden, where he first started modeling salt diapirism. He then led an international team investigating the geology and dynamics of salt diapir emplacement in central Iran, which was published as GSA Memoir 177 (1990). In 1988, he established the industry-funded Applied Geodynamics Laboratory at the Bureau. Martin received the J. C. Sproule Memorial Award and the George C. Matson Award from AAPG and the Macgregor Medal from the University of Zimbabwe. He lectured in 1983–84 in AAPG's Structural Geology School, was an AAPG Distinguished Lecturer in 1991–92, and served two terms as Associate Editor for the *AAPG Bulletin* and the *GSA Bulletin*. His work has been reprinted in six languages, none of which he can speak.

David G. Roberts graduated from the University of Manchester in 1965 and commenced his career mapping volcanoes in the West Indies. He subsequently joined the National Institute of Oceanography (later the Institute of Oceanographic Sciences) developing major research on the structure, tectonics, and stratigraphy of passive margins, and he led many geoscience cruises to the Atlantic and Indian oceans. He served as Co-Chief Scientist on two Deep Sea Drilling Project Legs and was chairman of the JOIDES Passive Margin Panel. He was awarded a Doctorate of Science (D.Sc.) in 1979 for his published work on continental margin geology. He joined BP in 1981 as Head of the Basin Analysis Group and became Deputy Chief Geologist International in 1985. In 1988, he transferred to the USA as Chief Geologist of BP America. Since 1991, he has been based in London as a worldwide exploration consultant.

Sig Snelson received his Ph.D. from the University of Washington in 1957, and after completing his dissertation on the Ruby–East Humboldt Range in northeastern Nevada, he carried out research in the Austrian Alps under a Fulbright Grant.

Sig began his 33-year career with Shell Oil Company in 1958 by carrying out pioneering field mapping in northern Alaska (pre–Prudhoe Bay). He was involved in numerous field-based research studies in California, Oregon, and Washington, and in 1972 he received the AAPG Levorsen Memorial Award for a paper on thin-skinned deformation in the Appalachians. Following assignments on Central and South American basins, Sig supervised for many years the Global Geology research section at Shell Development Company. Here he was involved in long-term studies of worldwide tectonic regimes and the development of computer applications to solve problems in sequence stratigraphy, basin modeling, structural analysis, and plate tectonics. In 1980, he co-authored a paper with A. W. Bally on the application of plate tectonics principles to basin classification. In 1981 he was named as a Fellow in the AAAS for his work on the application of plate tectonics to petroleum exploration. From 1987 to his retirement in 1991, Sig served as Exploration Consultant at Shell and focused on geological studies in the deep-water Gulf of Mexico. In 1993, he was co-recipient (with D. M. Worrall) of the GSA's Structure and Tectonics Division Best Paper Award for their publication on salt tectonics in the Gulf of Mexico. In 1994, he co-authored several papers (with L. R. Russell) on the Rio Grande rift for both AAPG and GSA volumes.

Sig has served on numerous committees, including the U.S. Geodynamics Committee, the Passive Margin Drilling Panel for IPOD, the International Geology Committee of the National Research Council, the Grants-in-Aid subcommittee of the AAPG Research Committee, and the GSA Penrose Conference Committee, serving as its chairman in 1990–91. He has convened symposia on worldwide thrust belts (AAPG) and on deep-water hydrocarbon exploration (Offshore Technology Conference) and has served as an Associate Editor of *Marine and Petroleum Geology* and the *AAPG Bulletin.* He now resides on Orcas Island in Washington State.

Authors' Addresses

Editors

Dr. Martin P. A. Jackson
Bureau of Economic Geology
The University of Texas at Austin
University Station, Box X,
Austin, Texas 78713
U.S.A.
phone: 512 475-9548
fax: 512 471-9800
e-mail: jacksonm@begv.beg.utexas.edu

Dr. David G. Roberts
BP Exploration Operating Company Ltd.
Uxbridge One, 1 Harefield Road
Uxbridge, Middlesex UB8 1PD
U.K.
phone: 44 1895 877000
fax: 44 1895 877569

Dr. Sig Snelson
P.O. Box 1179
Terrill Beach Road
Eastsound, Washington 98245
U.S.A.
phone: 360 376 4976
fax: 360 376 5354
e-mail: ssnelson@pacificrim.net
or sigS206376@aol.com

Contact Authors

Dr. Peter R. Cobbold
Géosciences-Rennes
Université de Rennes
35042 Rennes Cedex
FRANCE
phone: 33 99 286096
fax: 33 99 286780
e-mail: peterr.cobbold@univ-rennes1.fr

Prof. Michael P. Coward
Geology Department
Imperial College
London University
London SW7 2AZ
U.K.
phone: 44 71 589-5111
fax: 44 71 225-8544
e-mail: m.coward@ic.ac.uk

Dr. Menno J. de Ruig
Shell UK ExPro
Shell-Mex House
Strand, London WC2R ODX
U.K.
phone: 44 71 2574007
fax: 44 71 2574892

Dr. Fred Diegel
Shell Offshore, Inc.
701 Poydras St.
New Orleans, LA 70139
U.S.A.
phone: 504-588-7219
fax: 504-588-7493
e-mail: diegel@shellus.com

Dr. Ray Fletcher [contact]
Dr. Michael R. Hudec
Exxon Production Research Co.
P.O. Box 2189
Houston, TX 77252-2189
U.S.A.
phone: 713 965-7877
fax: 713 965-7305
e-mail: michael.r.hudec@exxon.sprint.com

Dr. J. Christopher Harrison
Institute of Sedimentary and Petroleum Geology
Geological Survey of Canada
3303 33rd Street NW
Calgary, Alberta T2L 2A7
CANADA
phone: 403 292-7137
fax: 403 292-5377
e-mail: charrison@gsc.emr.ca

Mr. Richard C. Heaton [contact]
Dr. Martin P. A. Jackson
Bureau of Economic Geology
The University of Texas at Austin
University Station, Box X,
Austin, Texas 78713,
U.S.A.

Dr. Robert J. Hooper
Conoco, Inc.
Geoscience Resources
P.O. Box 2197
Houston, TX 77252-2197
U.S.A.
phone: 713 293-6755
fax: 713 293-3833
home: 713 579-3213

Dr. Jake R. Hossack
BP Exploration Operating Company Ltd.
Uxbridge One, 1 Harefield Road
Uxbridge, Middlesex UB8 1PD
U.K.
phone: 44 1895 877547
fax: 44 1895 877877
e-mail: hossacjr@txcap.hou.xwh.bp.com

Dr. Hemin Koyi
Hans Ramberg Tectonic Laboratory
Institutionen för geovetenskap
Villavagen 16
S-752 36 Uppsala
SWEDEN
phone: 46 18 18 25 63
fax: 46 18 18 25 91
e-mail: hemin.koyi@geo.uu.se

Dr. Jean Letouzey
Institut Français du Pétrole
Geology and Geochemistry Department
1 et 4 avenue de Bois-Préau
BP 311
92 Rueil-Malmaison Cedex 92506
FRANCE
phone: 33 1 47526925
fax: 33 1 47527067

Mr. Louis M. Liro
Texaco Central Exploration Department
4800 Fournace Place
Bellaire, Texas 77401
U.S.A.
phone: 713 432-6808
fax: 713 661-7463
e-mail: lirolm@texaco.com

Dr. Webster U. Mohriak
Petróleo Brasileiro S.A. /SUTEX/DITEC
Av. Chile 65 13° Andar, Room 1305
Rio de Janeiro 20.031-170
BRAZIL
phone: 55 21 534-3696
fax: 55 21 534-1076

Mr. Kåre T. Nilsen [contact]
Dr. Bruno C. Vendeville
Bureau of Economic Geology
The University of Texas at Austin
Box X, University Station
Austin, TX 78713
U.S.A.
phone: 512 471-7721
fax: 512 471-0140
e-mail: vendevillb@begv.beg.utexas.edu

Dr. Frank J. Peel
e-mail: peelfj@txpcap.hou.xwh.bp.com
[contact]
Dr. Christopher J. Travis
BP Exploration
P.O. Box 4587
200 Westlake Park Blvd.
Houston, TX 77210-4587
U.S.A.
phone: 713 560-3815
e-mail: traviscj@txpcap.hou.xwh.bp.com

Mr. Gijs Remmelts
Rijks Geologische Dienst
10 Richard Holkade
Postbus 157
2000 AD Haarlem
THE NETHERLANDS
phone: 31 23 300 236 / 300 336
fax: 31 23 367540

Dr. Mark. G. Rowan
Department of Geological Sciences
Campus Box 250
University of Colorado
Boulder, Colorado 80309-0250
U.S.A.
phone: 303 492-8141
fax: 303 492-2606
e-mail: mark@lolita.colorado.edu

Ms. Maura Sans
Dept. de Geologia Dinàmica, Geofísica i Paleontologia
Universitat de Barcelona
Zona Universitaria de Pedralbes
08028 Barcelona
SPAIN
phone: 34 3 402 13 76
fax: 34 3 402 13 40
e-mail: maura@natura.geo.ub.es

Dr. David C. Schuster
(Independent)
9105 Columbia Road
Olmsted Falls, OH 44138-2426
U.S.A.
phone: 216 235 4115
fax: 216 235 2229
e-mail: daveschust@aol.com

Prof. Christopher J. Talbot
Hans Ramberg Tectonic Laboratory
Institutionen för geovetenskap
Villavagen 16
S-752 36 Uppsala
SWEDEN
phone: 46 18 18 25 70
fax: 46 18 18 25 91
e-mail: christopher.talbot@geo.uu.se

Jackson, M. P. A., 1995, Retrospective salt tectonics, *in* M. P. A. Jackson, D. G.
Roberts, and S. Snelson, eds., Salt tectonics: a global perspective: AAPG
Memoir 65, p. 1–28.

Chapter 1

Retrospective Salt Tectonics

M. P. A. Jackson

Bureau of Economic Geology
The University of Texas at Austin
Austin, Texas
U.S.A.

*The truth is that whoever touches this enticing sub-
ject...is bound to indulge freely in speculation. The problem
is so broad, the factors involved are so numerous, and the
work to be done with regard to salt structures is so great
that we cannot...[restrict our speculation to the narrow]
limits of exact knowledge.*

—Everett DeGolyer, 1925

*Although this is no place in which to describe the adven-
tures of a petroleum geologist it may, perhaps, be said that
the carrying out of the geological work referred to was greatly hampered owing to much of the time being spent as a
prisoner in the hands of Italian, Turk and Arab.*

—Arthur Wade, mapping salt domes,
Red Sea coast of Arabia, 1912

Abstract

The conceptual breakthroughs in understanding salt tectonics can be recognized by reviewing the history of
salt tectonics, which divides naturally into three parts: the pioneering era, the fluid era, and the brittle era.

The *pioneering era* (1856–1933) featured the search for a general hypothesis of salt diapirism, initially dominated by bizarre, erroneous notions of igneous activity, residual islands, *in situ* crystallization, osmotic pressures,
and expansive crystallization. Gradually data from oil exploration constrained speculation. The effects of buoyancy versus orogeny were debated, contact relations were characterized, salt glaciers were discovered, and the
concepts of downbuilding and differential loading were proposed as diapiric mechanisms.

The *fluid era* (1933–~1989) was dominated by the view that salt tectonics resulted from Rayleigh-Taylor instabilities in which a dense fluid overburden having negligible yield strength sinks into a less dense fluid salt layer,
displacing it upward. Density contrasts, viscosity contrasts, and dominant wavelengths were emphasized,
whereas strength and faulting of the overburden were ignored. During this era, palinspastic reconstructions
were attempted; salt upwelling below thin overburdens was recognized; internal structures of mined diapirs
were discovered; peripheral sinks, turtle structures, and diapir families were comprehended; flow laws for dry
salt were formulated; and contractional belts on divergent margins and allochthonous salt sheets were recognized. The 1970s revealed the basic driving force of salt allochthons, intrasalt minibasins, finite strains in diapirs,
the possibility of thermal convection in salt, direct measurement of salt glacial flow stimulated by rainfall, and
the internal structure of convecting evaporites and salt glaciers. The 1980s revealed salt rollers, subtle traps, flow
laws for damp salt, salt canopies, and mushroom diapirs. Modeling explored effects of regional stresses on
domal faults, spoke circulation, and combined Rayleigh-Taylor instability and thermal convection. By this time,
the awesome implications of increased reservoirs below allochthonous salt sheets had stimulated a renaissance
in salt tectonic research.

Blossoming about 1989, the *brittle era* is actually rooted in the 1947 discovery that a diapir stops rising if its
roof becomes too thick. Such a notion was heretical in the fluid era. Stimulated by sandbox experiments and
computerized reconstructions of Gulf Coast diapirs and surrounding faults, the onset of the brittle era yielded
regional detachments and evacuation surfaces (salt welds and fault welds) along vanished salt allochthons, raft
tectonics, shallow spreading, and segmentation of salt sheets. The early 1990s revealed rules of section balancing for salt tectonics, salt flats and salt ramps, reactive piercement as a diapiric initiator resulting from tectonic
differential loading, cryptic thin-skinned extension, influence of sedimentation rate on the geometry of passive
diapirs and extrusions, the importance of critical overburden thickness to the viability of active diapirs, fault-
segmented sheets, counter-regional fault systems, subsiding diapirs, extensional turtle structure anticlines, and
mock turtle structures.

PREAMBLE

The Hedberg International Research Conference, held in September 1993, was attended by a diverse group from 15 countries. The range of salt basins described was equally wide. Moreover, the lexicon of salt tectonics was evolving rapidly. For all these reasons, a review of terminology was needed so that all attendees could attach the same meaning to the same word. In addition, to facilitate scientific progress at the conference, the basic conceptual framework of salt tectonics needed to be reviewed. By identifying the main supports of the conceptual structure, as well as the components that were weak or missing, new breakthroughs during the conference were more likely to be recognized.

Accordingly, an opening presentation was designed to review both terminology and concepts. The oral review had two parts: (1) a retrospective contemplation of the past and (2) a prospective look ahead at unsolved problems. An attempt to look into the future can be a stimulating oral presentation, for views ahead are always thought provoking (e.g., Talbot, 1992). However, conjectures about the future become swiftly outdated in a rapidly evolving research field, and new problems continually add to such a list. Thus, this chapter deals strictly with the conceptual evolution of salt tectonics up to early 1993, before abstracts for the Hedberg conference were submitted. Chapters in this volume present much of the progress since that time.

The literature of salt tectonics is vast. A bibliography of diapirism and diapirs compiled by Braunstein and O'Brien (1968) contained about 1800 entries; since 1968, this number may have doubled. Any review attempting to document most of these contributions would be intolerably weighty and tedious to read. The approach here is to highlight the minority of papers that seem to have wielded major influence on salt tectonics by elucidating or catalyzing major conceptual breakthroughs. Many of these achievements were impelled or permitted by technological breakthroughs, such as seismic processing and modeling techniques. However, to prevent this review from sinking under its own weight, this technological aspect is all but ignored. Dates quoted are those of the final emergence of an idea as a published paper (or abstract if no paper followed soon after). For papers originating from academia, publication is typically delayed by 1–3 years. For papers from industry, the lag time to publication may be 3–5 years, although delays of 10 years or more are not uncommon.

Readers wishing to explore the early literature are referred to the reviews by DeGolyer (1925), Rios (1948), Lotze (1957), and Braunstein and O'Brien (1968), whose conveniently spaced bibliographies provided invaluable signposts to papers previously unknown to me.

Despite the moderating and broadening efforts of those called upon to review this paper and my own efforts, it remains a personal view of salt tectonics. As such, the review is incomplete, the selection of papers is biased, and the weight assigned to various ideas is subjective.

THREE ERAS OF SALT TECTONICS

From the perspective of the 1990s, three eras of greatly disparate length encapsulate progress in understanding salt tectonics. The *pioneering era* (1856–1933) featured the search for a general hypothesis for salt diapirism. The *fluid era* (1934–~1989) overwhelmingly saw salt tectonics as analogous to the overturn of two fluids with initially inverted densities. The *brittle era* (~1989–present) treats the overburden as a strong, brittle encasement whose deformation controls the style of salt tectonics.

Parallel with this scientific evolution was the shift in focus from salt domes to *salt tectonics*. Salt domes were scrutinized in early decades because they are the most widespread near-surface expression of salt tectonics. As the subsurface was increasingly probed by seismic tools and wells, the picture widened to salt tectonics as a whole, comprising salt source layer, salt structures of all types, underlying basement, and overlying overburden, commonly stretched or compressed by regionally imposed lateral forces. This paper includes under the term *salt diapir* any structurally discordant body of salt, regardless of its emplacement mechanism.

THE PIONEERING ERA (1856–1933)

Salt domes have been known and mined for millennia for their prized reserves of preservative "white gold" in places such as the arid coasts of the Persian Gulf and Red Sea. Solution mining of rock salt began about 3000 years ago in Poland and was probably also widespread elsewhere, given the simple technology required to extract this precious commodity. Starting 700 years ago, a vast cathedral, comprising 200 km of galleries and headings, was carved from salt in the bottom of the still-producing Wieliczka mine in southern Poland.

However, in the scientific sense, the pioneering era of salt tectonics opened with the first recorded discovery of a salt dome in the geologic literature. In 1856, Ville described a salt mountain, Ran el Melah, exposed near Djelfa in the Saharan Atlas of Algeria (Figure 1). He recognized "geyser-like" forceful emplacement of the salt. Soon afterward, the first subsurface salt dome was discovered in Louisiana when a brine well at Petite Anse, one of the coastal Five Islands, struck salt in 1860 (Thomassy, 1863).

The study of salt tectonics became established among the hundreds of salt domes exposed in central Europe, especially in the Carpathians. As long ago as 1871, salt bodies here were known to be intrusions emplaced discordantly against their country rocks (Figure 2) (Posepny, 1871). However, it was another 40 years before Mrazec (1907) coined the term *diapir* for folds cored by piercing salt. Even at the dawn of the pioneering era, Posepny (1871) recorded two other characteristics of salt domes: (1) angular unconformities in flanking strata and (2) similar style folds in their interiors (e.g., Vizakna, near Salzburg, Austria).

Figure 1—The first salt diapir described in the geologic literature (Ville, 1856). The panorama shows the southern contact of the 100-m-high "Ran el Melah" (Rocher de Sel de Djelfa), a 1.4-km-wide plug of Triassic salt in the northern fringe of the Saharan Atlas range, Algeria. The diapir's contact is marked by the white line. North of the diapir (not shown) is the megabreccia residue of a wasted salt glacier.

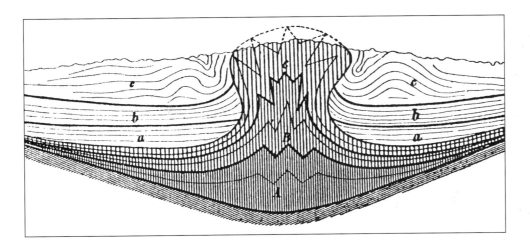

Figure 2—Posepny's (1871) schematic cross section captures some fundamental characteristics of salt diapirs: inside the diapir, increasing conformity of evaporite layers toward the diapiric contact; outside the diapir, marked discordance of most flanking strata, onlap of deep strata against the diapir, and an overturned collar around the shoulders of the diapir.

Unconstrained Speculation

Geologic understanding was first hindered, then driven, by economics. Before abundant hydrocarbons were known to be associated with salt domes, fragmentary knowledge about them came entirely from rare, fortuitous exposures. Because of this dearth of data, speculation ran wild and no hypothesis was too bizarre to omit. "As is always the case when exact information is too scarce to limit imagination seriously, widely different theories of origin were promulgated. Because the chief test for excellence in theory is that it shall not violate known fact, it is not surprising that a wide variety of theories should have been put forward" (DeGolyer, 1925, p. 835). Working on an edifice built by a century of intensive research on salt tectonics, it is easy for us today to dismiss many of these hypotheses, which were necessarily based on slender evidence. However, field data were scarce, hard won, and frequently dangerous to collect, as exemplified by the opening quote from Wade (1931).

Choffat (1882) proposed the term *tiphonique* (after mythological Typhon) to describe intrusions of saliferous and gypsiferous marl in Portugal. Choffat's (1882) section across the valley of the Serra del Rei recorded the overturned collar of sediments surrounding the diapir, but Veatch (1899, 1902) seems to have been the first to recognize doming of surrounding strata as an inherent characteristic of most salt domes.

Initially, the most prevalent view of their origin was that salt domes represented residual outlying islands of salt surrounded by the deposits of younger seas (e.g., Lockett, 1871; Hildgard, 1872). Others suggested that salt domes originated by local folding (Kennedy, 1892; Harris and Veatch, 1899) or plutonic emplacement (Lerche, 1893).

The spectacular blowout of hydrocarbons from the cap rock of Spindletop, Texas, in 1901 not only initiated mass production of petroleum but also provided a powerful economic incentive to obtain subsurface data. The quantity of available information exploded. Within a mere

Erosional Differential Loading

Differential loading refers to lateral variations in the thickness, density, or strength of overburden strata above salt. Three types of differential loading would eventually be recognized. The first of these was that due to erosional removal of overburden (Harrison, 1927). The most spectacular example cited by Harrison was the deep, meandering canyon cut by the Colorado River in Utah (Figure 6). Erosional unloading of the floor of the canyon locally decreased the pressure on the salt. The weight of the canyon walls on the underlying Pennsylvanian Paradox Salt squeezed salt sideways into the low-pressure zone below the floor of the canyon, where salt bulged up to form a meandering anticline. In several places, evaporites broke through the erosionally thinned overburden to emerge as gypsum diapirs.

Salt Fountains and Salt Glaciers

Steady-state diapirism requires a balance between the supply of salt from below and the loss of salt at the surface. Escher and Kuenen (1929) inferred that the Romanian diapirs must still be rising in order to remain as mounds of rock salt exposed to a moist climate with rainfall of 80–90 cm per year. Apparently unknown to them, geologists in the Middle East were directly observing the spectacular effects of an oversupply of salt. Lees (1927) and De Böckh et al. (1929) described salt plugs in the Zagros of southern Persia that emitted salt extruding downhill (Figure 7); they called these *salt glaciers* because they resembled ice glaciers. Harrison (1931) documented the salt plugs and their glaciers in considerable detail and speculated on the processes of emplacement. He seems to have been the first to deduce that the salt plugs emerged as early as the Late Cretaceous because these strata include conglomerates containing clasts of distinctive Hormuz igneous rocks. Noting that the results of deformation experiments were incompatible with the existence of salt glaciers, Harrison presciently speculated that flow could be aided by traces of water in the salt. Wade (1931, p. 358) went further: "Heavy rains are rare in the ranges [of Zeit somewhere by the northern Red Sea], but at such times the flow [of gypsum glaciers] is noticeably quick-

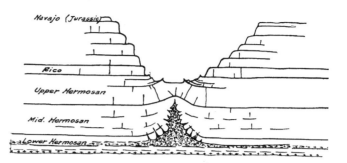

Figure 6—Upwelling of Pennsylvanian Paradox evaporites (patterned) by erosionally induced differential loading below the Colorado Canyon, Utah (from Harrison, 1927).

ened." Wade even attempted to measure the flow rate of glacial gypsum, and Bailey (1931) suggested doing the same for salt glaciers by staking them. He also originated the metaphor of extruding salt diapirs as "slow-motion fountains."

Depositional Differential Loading

Depositional differential loading is the second type of differential loading. Bailey (1931) was apparently one of the first to propose it as a mechanism for triggering diapirism in regions of "orogenic tranquillity." Subsequently, Rettger (1935) modeled the effects of a wedge-shaped differential load above a mobile substrate (Figure 8). The substrate was squeezed laterally and welled up along the periphery of the deltaic differential load.

Downbuilding

Until the early 1930s, the consensus was that salt diapirs rise discordantly, piercing thick, previously deposited piles of sedimentary rock. This penetration requires upward displacement or stoping of a huge volume of country rock. In modern parlance, such diapirs have a severe room problem. Wade (1931) was skeptical

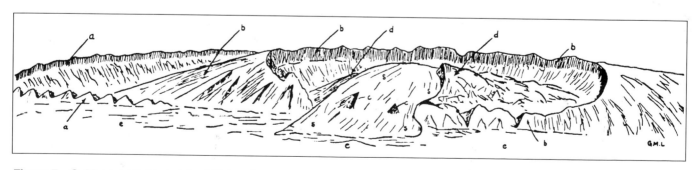

Figure 7—G. M. Lees' field sketch of Kuh-e-Anguru salt plug, Iran, visited in the early 1920s. Along with an accompanying photograph (not shown) of Kuh-e-Namak (Dashti), this was the first illustration of an identified salt glacier (De Böckh et al., 1929). The glacier overflows Cretaceous–Oligocene carbonates; two Eocene inliers (c) project through the flow.

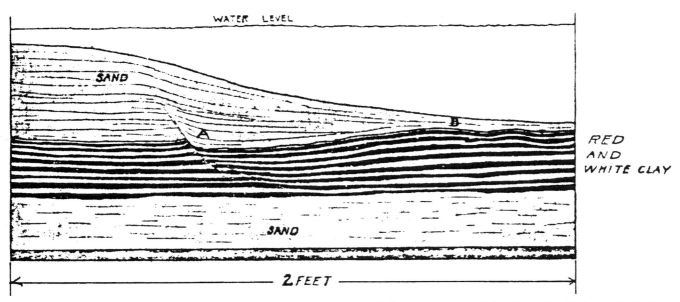

WATER LEVEL

SAND

RED
AND
WHITE CLAY

SAND

|← ——————— 2 FEET ——————— →|

Figure 8—Sedimentary differential loading demonstrated by a model of a subaqueous delta of sand, which results in the squeezing of a mobile substratum of laminated clay from beneath A to B (abridged from Rettger, 1935).

that "putty nails" could be driven through wooden boards. However, surprisingly few geologists seem to have been bothered by the severe mechanical difficulties required for diapiric piercement through great thicknesses of preexisting strong overburden. At this point, a fundamental conceptual breakthrough was made.

Previously, diapirs were thought to have grown upward from a relatively deep, static base. Barton (1933) imaginatively proposed that a diapir actually grew downward from a relatively shallow, static crest, a process he called *downbuilding* (Figure 9). Barton envisaged a dome growing syndepositionally. The diapir's crest remains at or near the surface of sedimentation while its base sinks together with the surrounding, subsiding strata. Because the diapir's crest is continually emergent, no country rocks are displaced or lifted. The diapir has no room problem even though its contacts are discordant to the surrounding strata. By downbuilding, a diapir can appear to penetrate or pierce many kilometers of strong overburden because the latter is deposited only along the flanks of the diapir (disregarding any ephemeral veneer of sediments over the crest). To support his idea, Barton (1933) discussed geologic observations and mechanics. Some of the mechanical reasoning in this substantial paper is debatable (for a discussion, see Jackson et al., 1988), but the essential validity of the downbuilding hypothesis was so apparent that it was soon overwhelmingly accepted by American geologists. It diffused to Europe far more slowly, possibly because it was incompatible with the hypothesis of regional contraction, for which the evidence was strong in Europe.

Thus, by the early 1930s, far-ranging intellectual exploration had recognized many important facets of salt tectonics. In the next era, some of these hypotheses became unfashionable in the shadow of a commanding and durable new hypothesis.

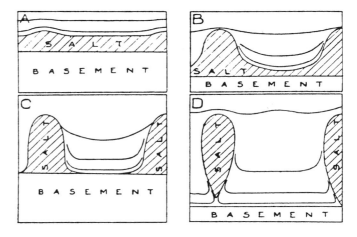

Figure 9—Barton's (1933) concept of diapiric downbuilding.

THE FLUID ERA (1933–~1989)

Not many earth scientists have been able to develop a hypothesis that dominated their subdiscipline for 55 years and also laid the foundations of a theory that eventually supplanted their own. This double feat of inaugurating the *fluid era* and laying the foundation for the succeeding brittle era was achieved by Nettleton (1934) and Nettleton and Elkins (1947). In his investigation of the fluid mechanics of salt domes, Nettleton (1934) assumed that when scaled down, salt and its overburden could be represented by two viscous fluids of negligible strength (oil and syrup). A transparent cylinder containing a stable arrangement of dense fluid beneath less dense fluid was inverted, and the overturned fluids slowly returned to equilibrium as the denser fluid sank (Figure 10).

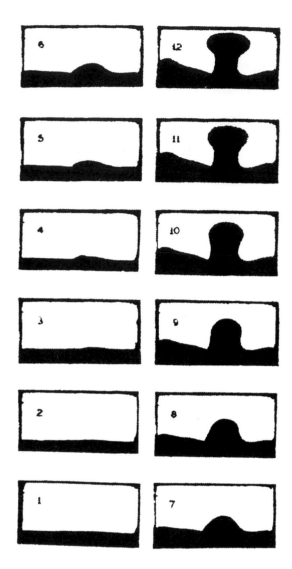

Figure 10—Dawn of the fluid era. Nettleton's (1934) gravitational overturn of a fluid-fluid system initially comprising corn syrup (white) overlying less dense crude oil (black) (abridged from Nettleton, 1934).

Nettleton (1934) demonstrated that gravity alone could generate diapir-like shapes and a surrounding peripheral sink from an undeformed source layer. Eventual evacuation of buoyant "salt" from the peripheral sink could cut off the supply of "salt" to the base of the diapir. Nettleton's basic hypothesis and assumptions prevailed for the next 55 years, even though he himself was to question them only 13 years later. Nettleton's modeling approach was widely applied, especially by Hans Ramberg. Ramberg's prodigious and elegant research, which began in the 1960s and was summarized in his books (1967, 1981), propelled the fluid approach ever higher. His work broke new ground on almost every facet of gravity tectonics, but it is probably of greater relevance to crustal tectonics than to salt tectonics.

The fluid hypothesis was widely applied because it was simple and because it focused on the progressive change in shape of the salt source layer and diapir. Had it been more explicit about structures in the overburden, the hypothesis would not have been so durable.

Palinspastic Restorations

Exactly when the first palinspastic reconstruction was published is uncertain. An early example of a detailed restoration is by Rios (1948) (Figure 11). He graphically removed the contractional overprint of Alpine orogeny to reveal the Oligocene form of a basin floored by Triassic Keuper evaporites. His restoration shows how the deepest part of the original basin became the most uplifted part during subsequent regional contraction—a process termed *inversion* since 1978 (Coward and Stewart, 1995; Letouzey et al., 1995).

Décollement Styles and Downward Salt Bulges

By the 1930s, geologists were familiar with the idea that deformation in the cover could be very different from that in the underlying basement if separated by evaporites. For example, de Cizancourt (1934) distinguished between Romanian style thick-skinned contraction and North African style thin-skinned contraction, in which the base of the evaporite was roughly planar over a nondeforming basement but the top of the salt was strongly deformed by incorporation within folds and thrust faults in the deformed cover. He also proposed that diapirs could be laterally squeezed until they expressed all the salt from within them.

Later, in Persia, the geometry of a new type of decoupling both above and below the evaporites became clear. Elaborating on earlier observations by Lees (1938), O'Brien (1957) described Miocene salt that decoupled upper cover from lower cover in such a way that salt flowed into overlying anticlines and underlying synclines. This segregation formed salt bulges laterally separated by much thinner salt. Locally, the salt bulges became diapiric and broke through their thrusted carapace.

Peripheral Sinks

Elements of Stille's (1925) idea of salt tectonics driven by orogeny still lingered in Germany in the 1950s. For example, Richter-Bernburg and Schott (1959) interpreted vortex structures within the salt as indicating the sudden release of compressional forces during eruptive emplacement of salt intrusions during orogenic phases.

At the same time, this orogenic linkage was being eclipsed by methodical interpretations of seismic data from the rejuvenated exploration of oil and gas in Germany. Trusheim (1957, 1960) saw most German salt structures as resulting from autonomous, isostatic rise of salt, which he termed *halokinesis*. This idea was old, but his technique of analysis was novel. His was the first sys-

S

N

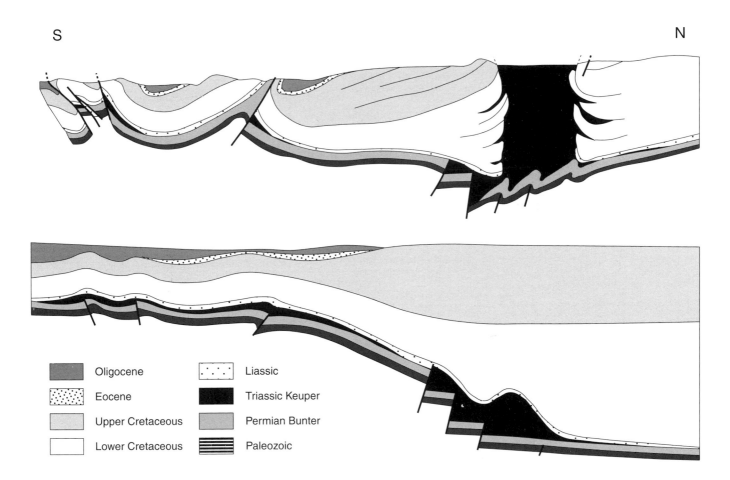

	Oligocene		Liassic
	Eocene		Triassic Keuper
	Upper Cretaceous		Permian Bunter
	Lower Cretaceous		Paleozoic

Figure 11—An early example of detailed palinspastic restoration of salt tectonics in the Pyrenees, without vertical exaggeration (redrawn from Rios, 1948). A basin floored by Triassic Keuper evaporites (bottom) became inverted during Alpine contraction (top). Another interpretation might show the diapir growing passively throughout the Cretaceous, periodically extruding to form the lateral flanges depicted, and more normal faults in the restored section because of null points along the faults in the upper section.

tematic attempt to infer the history of salt flow from the sedimentary record of the surrounding strata (Figure 12). Trusheim (1957, 1960) used lateral thickness changes in strata deposited during salt flow to infer when underlying salt was flowing laterally—an interpretive technique that survives today. He recognized the distinctive geometry of primary peripheral sinks, which form around pillows, and secondary peripheral sinks, which form around diapirs. From these characteristics, he was able to chart the structural evolution of salt diapirs entirely from the sedimentary record (Figure 12). He also recognized the structural inversion ("transformation of structural relief") that accompanies the transformation of a pillow into a diapir and forms turtle structure anticlines.

Trusheim's hypothesis provided geologists with a rational basis for interpreting salt structures, and his method was unquestioningly applied for several decades. In retrospect, two of his assumptions look weak: that diapirs must necessarily evolve through a preceding pillow stage and that a pillow forms solely by buoyancy halokinesis, even where the overburden is thick and largely uniform (compare Coward and Stewart, 1995).

Trusheim (1960) also recognized that extensional structures could be overprinted by contractional structures, which he attributed to collapse of salt structures. Today, basement-involved inversion would be a more popular explanation of the contractional reactivation.

Internal Structures of Diapirs

Because German salt diapirs have been exploited largely for scarce potash rather than rock salt, mapping their internal structure and stratigraphy was essential. Conversely, U.S. Gulf Coast salt diapirs were mined largely for rock salt. Their internal structures were thus long neglected except for safety-related features such as fractures and gas pockets. The Hans Cloos school in Germany had developed mapping techniques for studying the igneous and deformational structures of plutons. Balk (1949, 1953) transferred these techniques to the Grand Saline dome (Texas) and Jefferson Island dome (Louisiana). Mapping of mined cavern roofs between rock pillars revealed steeply plunging folds with steeply dipping curved axial surfaces. These were the *curtain folds*

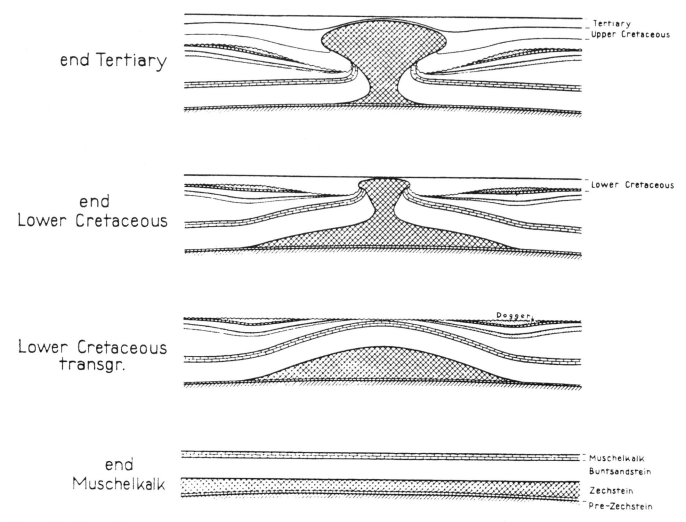

Figure 12—The sedimentary record begins to be systematically deciphered by Trusheim (1960). A flat-lying sequence containing Permian Zechstein salt (bottom) evolves into a pillow surrounded by primary peripheral sinks (Dogger time), then into a diapir and adjoining secondary peripheral sinks (top of Lower Cretaceous), and finally into a postdiapir stage overlain by Tertiary peripheral sinks (top). Adjoining turtle structure anticlines are formed as the flanks of the primary sink subside over the flanks of the deflating pillow.

described in Germany by Stier (1915) and simulated by the experiments of Escher and Kuenen (1929). A few exposures had closed folds like eyes; these were subsequently attributed to *constrictional sheath folds* (Talbot and Jackson, 1987). Balk (1949, 1953) also noted that the bedding defined by disseminated anhydrite curved into parallelism with the external contact of the diapirs and that folds tightened toward the contact. Both observations pointed to what would eventually be called an *external shear zone* (Kupfer, 1976) in the outermost parts of the diapir.

Balk also advanced knowledge on smaller scale structures in these domes. Fabric studies showed that disseminated anhydrite and spindle-shaped aggregates of anhydrite defined a strong mineral lineation throughout most of the mine in the Grand Saline dome (Balk, 1949). The lineation was axial to the steeply plunging curtain folds, all within 20° of the vertical. Halite grains were less

strongly lineated because they had recrystallized. Balk also stressed the rarity of macroscopic fluid inclusions, fractures, faults, and foreign inclusions.

Early Initiation of Salt Upwelling

By the 1960s, views on the mechanical properties of rock salt that were derived from experiments began to conflict with real-world observations. Experiments on the dislocation creep of dry salt (e.g., Handin and Hager, 1958) had indicated that salt could only flow at temperatures above 205°C, corresponding to a burial depth of more than 7 km (Gussow, 1968). This belief conflicted with (1) the long domal growth histories deduced by Trusheim (1960), and (2) the concept of downbuilding (Barton, 1933). More tellingly, salt glacial flow could only be reconciled with the experimental data if salt extrusion was rapid and red-hot (Gussow, 1968). Yet, many salt

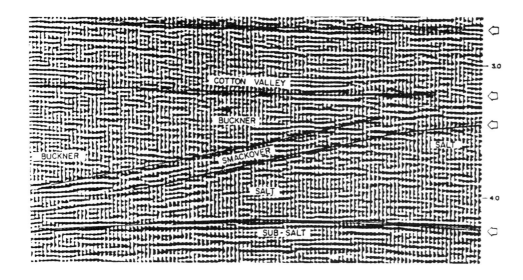

Figure 13—Seismic evidence of upwelling of Jurassic Louann Salt below a thin overburden of Smackover carbonates (central Mississippi). Buckner reflectors onlap against domed Smackover reflectors on the flanks of the 1.8-km-high pillow (abridged from Rosenkrans and Marr, 1967).

glaciers in Iran were obviously still flowing at present-day surface temperatures (Kent, 1958, 1966). Also, seismic data (Figure 13) indicated early upwelling and flow of salt at shallow depths. In the Gulf Coast interior basins, salt flowed under overburdens ~350 m thick (Rosenkrans and Marr, 1967; Hughes, 1968) and in the North Sea Basin below overburdens ~610 m thick (Brunstrom and Walmsley, 1969). This evidence took the form of onlaps, truncations, and lateral thickness changes against dome flanks; reefs were predicted and discovered on dome crests.

Salt Nappe

By the late 1960s, two distinct concepts in salt tectonics were well established but perceived to be unrelated. First, the fold and thrust belts in Europe, North Africa, and Iran, for example, were known to have involved large tectonic translations over evaporites. Second, allochthonous salt masses were known to exist (e.g., Lotze, 1934) but were generally referred to as "overhangs" or "buried extrusions" and regarded as merely local anomalies in a world of vertical salt diapirism.

In 1969, these concepts began to be linked. The horizontal component of salt tectonics was brought to the foreground for the first time with new data from the Sigsbee Scarp—a lobate, arcuate scarp separating the continental rise from the continental slope of the northern Gulf of Mexico. The scarp was first interpreted by Ewing and Antoine (1966) as a deformation front: salt welled up vertically at the base of the continental slope because of basinward flow of salt confined within the autochthonous layer. Loading by landward sediments caused this flow, and continued episodes of loading caused new salt walls to rise seaward of older ones.

After Ewing and Antoine (1966), increasingly daring concepts evolved to explain the scarp's origin. Amery (1969) first glimpsed salt's underworld by identifying an allochthonous salt tongue below the Sigsbee scarp (Figure 14). The salt itself was dimly imaged and recognized mainly as a high-velocity anomaly, which pulled up underlying reflectors of strata overridden by the salt tongue. This velocity effect was augmented by the water wedge above the overlying scarp. Amery (1969, 1978), too, thought that salt was flowing basinward, mainly within the autochthonous layer but locally as a 7-km-wide lateral tongue below the scarp. He also favored an extrusive origin for the tongue in which a thin (200–300 m) veneer of sediments overlay denser salt. The mechanics of salt drive remained obscure; these authors favored an ill-defined blend of gravity gliding and sedimentary differential loading.

Humphris (1978), in turn, emphasized the importance of the differential load applied by the prograding continental margin in driving salt laterally (Figure 15). He advocated two new concepts. First, lateral extrusion of salt occurred for tens of kilometers across younger strata (rather than mere lateral flow of salt within an autochthonous layer), which required equally large overthrusting of the overburden, like a giant thrust nappe riding on salt. Second, salt structures landward of the salt nappe were merely remnants of the salt nappe left after salt was largely displaced basinward to the front of the salt sheet.

Contractional Belts on Divergent Margins

Reyer (1888) hypothesized that gravity gliding alone over a weaker layer could cause thin-skinned contraction downdip of an extensional zone. Similar fold and thrust belts were produced by modeling (Nettleton and Elkins, 1947). Such contractional belts were eventually discovered in deep water on divergent continental slopes such as the Gulf of Mexico, first over a Paleogene shale décollement in the Mexican Ridges (Bryant et al., 1968; Garrison and Martin, 1973) then over autochthonous Jurassic salt in the Perdido foldbelt (Blickwede and Queffelec, 1988).

Finite Strain in Diapirs

Structures within diapirs were further quantified by strain modeling in the 1970s. These studies were aimed primarily at understanding the internal structures of

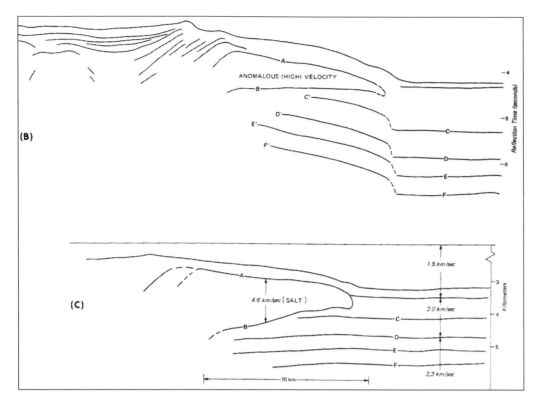

Figure 14—The first known illustration of allochthonous salt (from Amery, 1969). The original sparker profile (too indistinct to reproduce here) across the Sigsbee scarp, Gulf of Mexico, was interpreted in (B) time and (C) depth to illustrate a wedge of extrusive allochthonous salt over flat-lying reflectors.

gneiss domes, but they provided the first strain maps within model diapirs. Fletcher (1972) investigated the internal strains of gneiss domes and anticlines by analytical modeling. Dixon (1975) measured the three-dimensional finite strains within centrifuged diapiric walls. He showed that in the crest of the diapir, initial vertical stretching was followed by vertical flattening. Other parts of his diapirs also underwent what appeared to be polyphase deformation as the diapir evolved through a single overturn. This partly explains the notoriously complex folding observed in salt mines.

Thermal Convection and Salt Tectonics

All hypotheses of salt diapirism invoked surrounding clastic rocks until Talbot (1978) proposed that the high thermal expansion of halite would enable salt diapirs to rise as thermal plumes through salt. Because the thermal expansion of halite in a geothermal gradient is greater than its elastic compressibility (bulk modulus) due to confining pressure, hot, deep salt would expand, lose density, and rise. Cooler, denser salt sinking from the upper salt contact would replace it. Such a convecting system could operate even without overburden, although thermal effects could be enhanced by an insulating overburden. Talbot (1978) speculated that a salt layer more than 2 km thick in the Danakil Depression of Ethiopia along the East African rift system could be convectively overturning as a result of the high geothermal gradient.

Thermal convection was invoked on a smaller scale for evaporites in Boulby potash mine, England. Talbot et al. (1982) convincingly explained hundreds of recumbent lobes of sylvinite as the products of increasingly thorough convective stirring on two scales, roughly 100 m and 400 m wide in axial section. The evaporites were estimated to have deformed under temperatures as high as 170–390°C. The recumbent sylvinite lobes were later intruded by sill-like sheets of rock salt.

Salt transmits heat more efficiently by conduction than by thermal convection due to expansion. Diapirs act as heat chimneys for conduction, locally cooling their deep surroundings and locally heating their shallow surroundings. Such conductive perturbations could be expected to shift oil and gas windows in flanking strata up or down. However, such perturbations are difficult to model because thermal convection in surrounding pore fluids typically swamps the thermal effects that might be attributed to conductive salt.

Flow Rates and Salt Budget of a Salt Glacier

In the late 1970s, unique fieldwork was also being carried out on twin salt glaciers in southwestern Iran. Earlier, Harrison (1930) and Kent (1958, 1966) had provided qualitative geologic evidence that some Iranian salt glaciers were still flowing and wasting at air temperatures of only 10–45°C. However, a real understanding of the process of extrusion took another decade. Wenkert (1979) calculated the flow rates of five Iranian salt glaciers by assuming a steady-state balance between glacial flow and dissolution. Calculated flow rates for dry salt were much slower than the extrusive rates he had calculated, so Wenkert (1979) suggested that moisture provided by rainfall allowed the salt to flow by intergranular liquid diffusion at the calculated faster rate.

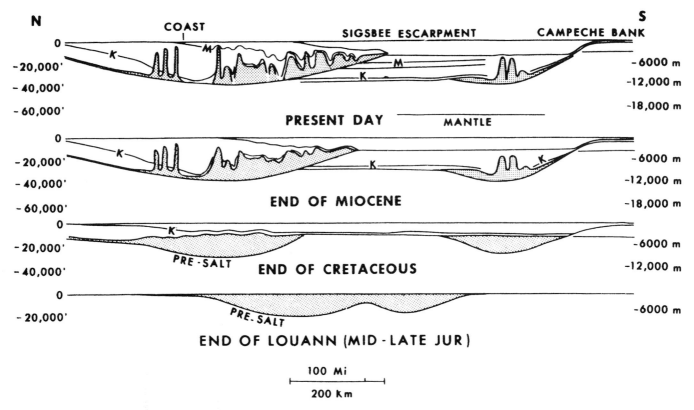

Figure 15—Sedimentary differential loading as the driving mechanism for the allochthonous Sigsbee salt nappe, Gulf of Mexico (from Humphris, 1978).

Nearly 50 years after direct measurement of salt glaciers was suggested by Bailey (1931), Talbot and Rogers (1980) finally established a rate of glacial flow by repeatedly surveying markers painted on the northern glacier of Kuh-e-Namak (Dashti, Iran). Their monitoring program was aborted by the Iranian revolution. However, incomplete data indicated that in the dry season the glacier oscillated back and forth diurnally as it heated and cooled. Conversely, during brief periods after rainfall, the glacier flowed downhill as much as 0.5 m per day. This gain was later slightly reduced as the glacier (or "namakier") dried and shrank.

Talbot and Jarvis (1984) meticulously investigated the salt budget of the Kuh-e-Namak extrusive dome and calculated that it approximated a 1-km-high, parabolic viscous fountain rising at almost 17 cm/year and spreading extrusively under its own weight at a rate of 2 m/year. These rates were some 40 to 80 times faster than existing estimates for the rise of salt domes and suggested that water had a marked softening effect that enhanced the flow of salt. This evidence of rapid flow would prove to be especially significant to the origin of allochthonous salt sheets.

Internal Structures of Salt Glaciers

Talbot's (1979, 1981) thorough work on salt glaciers included an investigation of their mesoscopic and microscopic structure. He found that the Kuh-e-Namak salt glacier was a nappe complex of recumbent sheath folds bounded by subhorizontal shear zones (Figure 16). Approaching a 20-m-high bedrock obstruction, the flowing salt decelerates and thickens, forming asymmetric flow folds and crenulations as streamlines are deflected across the color bands. Over the obstruction, glacial flow accelerates and tightens these folds into isoclinal sheath folds. The process of folding and tearing is cyclically repeated as the glacier encounters each of about 15 bedrock scarps. Farther downstream, the flowing salt becomes sufficiently weakened by intense mylonitization for the glacier to undulate over obstructions without forming sheath folds.

Perturbation of Doming Stresses by Regional Stresses

As the roof of a rising diapir arches upward, it stretches by normal faulting. Partly this fault pattern reflects the shape of the diapir. The physical models of Link (1930) and Parker and McDowell (1955) confirmed that subradial faults form above diapiric plugs, whereas subparallel faults formed above diapiric walls. In both plugs and walls, most faults were curved rather than straight. Other authors had also speculated whether the regional stress pattern could control the orientation of domal faults (Balk, 1936; Cloos, 1968). However, it was Withjack and Scheiner (1982) who first systematically investigated the

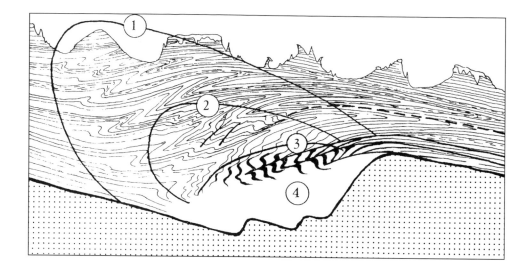

Figure 16—Glacial structures created by nonuniform flow of salt (left to right) over an obstructing bedrock step in Kuh-e-Namak's northern salt glacier, Dashti, Iran. Structural zones are regenerated over each bedrock step in the upper glacier: (1) flow folds, (2) crenulations, (3) salt veins, and (4) static salt (from Talbot, 1981).

combined effects of diapiric shape and regional stresses during gentle doming by clay and analytical modeling. Without regional strain, only normal faults developed over diapirs. In their models of regional extension, all normal faults were roughly perpendicular to the extension direction; strike-slip faults trending 60° from the extension direction formed in restricted sectors near the periphery of circular domes. During regional shortening, normal faults developed subparallel to the shortening direction, and both strike-slip faults (30° from the shortening direction) and reverse faults (perpendicular to the shortening direction) formed in restricted sectors on the periphery of the domes.

Salt Rollers

In the 1980s, extensional salt tectonics began to be revealed in more detail. For about three previous decades (e.g., Quarles, 1953), the lower footwalls of some normal faults were known to comprise ridges of mobile salt or shale. These diapiric ridges were thought to have initiated the faults above them. However, these diapirs remained unnamed until Bally (1981) featured seismic illustrations of them, labeled "salt rollers"; he did not comment on the origin of the term, which remains obscure. *Salt rollers* are low-amplitude, asymmetric salt structures comprising two flanks: one flank in conformable stratigraphic contact with the overburden and the other in normal-faulted contact with the overburden. Salt rollers are now thought to form entirely by regional extension, typically thin skinned and gravity driven. (In this review, "regional" deformation denotes a scale larger than a single diapir, regardless of whether the basement is involved.)

Subtle Traps

Salt domes had long been known to create structural traps over their crests and against their flanks. By the late 1960s, structural traps were well-established targets and had been classified (e.g., Halbouty, 1967). Seni and Jackson (1983a) focused on an equally wide range of subtle, facies-related traps associated with halokinesis; some of these subtle halokinetic traps were as much as 20 km from the actual diapir. In synthesizing and analyzing the sedimentologic and tectonic history of all 16 shallow salt domes in the East Texas Basin, they found that subtle facies changes characterized certain depositional systems at certain stages of dome growth. Applying statistical techniques to volumetric data from about 2000 wells, Seni and Jackson (1983b) were able to quantify crudely the individual and collective growth histories, gross and net salt flow rates, and strain rates of diapirs. Maximum growth rates were greatly in excess of regional aggradation rates, implying considerable outflow and dissolution of salt at the surface.

Flow Law for Damp Salt

Water had long been known to soften salt by the Joffee effect (Kleinhanns, 1914), but until the 1980s, experimental rock mechanics had focused on the rheology of dry rock salt, known to deform by dislocation creep (Carter and Hansen, 1983). Stimulated by Talbot's correlation of glacial flow with rainfall, Urai and co-workers investigated the effect of trace amounts of water on the evaporites bischofite, carnallite, and halite (Urai, 1983, 1985; Spiers et al., 1986; Urai et al., 1986). They found that damp evaporites could deform by solution-transfer creep under much lower stresses and much more rapidly than the dislocation creep characteristic of dry evaporites. This explained the extraordinarily high flow rates of damp glacial salt.

Mushroom Diapirs

The term *mushroom-shaped* has long been loosely applied to diapirs shaped like light bulbs, despite the fact that actual mushrooms are not shaped like bulbs. Truly mushroom-shaped diapirs have a bulb fringed by one or more pendant peripheral lobes. These lobes are common-

(a)

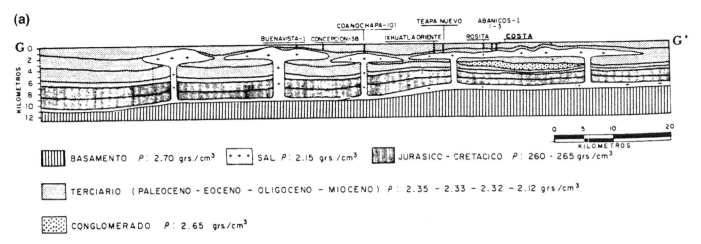

(b)

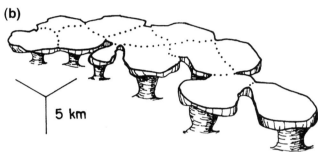

Figure 17—(a) Coalesced salt bodies were first deduced as two-dome structures on the basis of gravity modeling along the coastal Salina Basin, southwest Gulf of Mexico (from Correa Perez and Gutierrez y Acosta, 1983). (b) Other salt bodies were then mapped as a twelve-dome coalesced structure called a salt canopy in the Great Kavir of central Iran (from Jackson et al., 1987).

ly contained entirely within the diapir (internal mushroom), but in rare cases may enclose infolded country rock (external mushroom). Mushroom diapirs were first numerically modeled in two dimensions by Daly (1967). By the 1980s, mushroom diapirs had been physically modeled in three dimensions (Jackson and Talbot, 1989a) and found to be identifiable by crescentic folds in horizontal section and by downward-facing folds in vertical section. These features were recognized in at least 24 external mushroom diapirs of the Great Kavir in central Iran (Jackson and Talbot, 1989a; Jackson et al., 1990). In a minority of these external mushroom domes, the peripheral flanges rolled inward and upward to form vortex structures. Mining data from several German diapirs were also reinterpreted to indicate internal mushroom and vortex structures (Jackson and Talbot, 1989a).

Salt Canopies

Laterally spreading salt sheets have the potential to eventually coalesce with their neighbors. Curiously, though, such structures were not described or even predicted until the 1980s. Correa Perez and Gutierrez y Acosta (1983) referred to a laterally intrusive sheet of salt coalesced from two feeders in the Salina Basin of coastal Mexico (Figure 17a). They speculated that the allochthonous salt could be continuous throughout a wider area offshore. In the Great Kavir of central Iran, an exposed, coalesced cluster of 12 salt domes was independently mapped and described as a salt canopy (Figure 17b) (Jackson and Cornelius, 1985; Jackson et al., 1987, 1990; Jackson and Talbot, 1989b). A cellular structure resulting

from the incorporation of two distinct evaporite units indicates that this salt canopy is a composite structure. Salt canopies of vast extent comprising probably hundreds of feeder stalks are now known to characterize the Gulf of Mexico (e.g., see Diegel et al., 1995; Peel et al., 1995; Rowan, 1995; Schuster, 1995).

Rayleigh-Taylor Acme

The fluid era initiated by Nettleton (1934) drew to a close in the late 1980s even as modeling of Rayleigh-Taylor instability (sinking of denser fluid into less dense fluid) reached new heights of sophistication.

Two topics were investigated by finite-element modeling in two dimensions. Schmeling (1987) systematically investigated how the final wavelength and geometry of fluid upwellings could be influenced by initial irregularities in the interface between the overturning fluids. Earlier modelers tended to unquestioningly assume that the spacing of diapirs could be predicted solely from viscosities, thicknesses, densities, and boundary conditions. Schmeling showed that the final wavelength could be up to four times larger than predicted, depending on the spacing of the initial irregularities. Schmeling (1988) investigated the overturn of fluids whose densities differed owing to both composition and temperature. Various combinations of fluids and temperatures could stabilize after a single overturn, or repeatedly overturn in the same directions, or even repeatedly overturn in reversed directions each time.

Talbot et al. (1991) investigated the three-dimensional patterns of Rayleigh-Taylor overturn in detail. Combining physical modeling and geologic observations, they showed that where not perturbed by edge effects or lateral surface forces, fluids overturn in a complex spoke pattern. Two sets of laterally and vertically interlocking spokes carry fluids of different densities along the top and bottom boundaries into the nearest rising or sinking plume. Each plume flows to the opposite boundary from where it started, then expands into a broad bulb between spokes of the other fluid. This geometry implies that below dense fluid overburdens, less dense fluid diapirs rise from the triple junctions of deep polygonal ridges in their source layer.

THE BRITTLE ERA (~1989 TO PRESENT)

About 1989, the scientific tide changed. Salt tectonics began to be increasingly approached as a system involving a strong, brittle, fractured overburden rather than a weak, fluid one. Experiments with fluid overburdens have continued into the 1990s, but they now seem to be relevant only for understanding crust at or below the brittle-ductile transition. Because petroleum exploration focuses on brittle crust in and above the oil and gas generation windows, a new set of salt tectonics models has appeared. No single conceptual development triggered this paradigm shift from the fluid era to the brittle era, so there is no precise starting date. However, two breakthroughs seem to have been catalysts: (1) revitalizing an old modeling approach and (2) developing a new tool for structural restoration.

Physical Modeling Using Brittle Overburdens

The roots of this approach are slender but deep. Concerned that Nettleton's (1934, 1943) fluid-fluid models were not reproducing the pervasive faulting seen around Gulf Coast salt domes, Nettleton and Elkins (1947) simulated brittle overburden with granular materials overlying a viscous substratum representing salt. In contrast to models of fluid-fluid systems, where the rise of diapirs is accelerated by thicker overburdens, they found that the rise of diapirs was inhibited if an overburden of finite strength exceeded a certain critical thickness. Hubbert (1951) provided the theoretical imprimatur for simulating brittle deformation of rock with dry sand—as he had for Nettleton's fluid-fluid approach (Hubbert, 1937).

In the most heroic undertaking of geologic modeling ever published, Parker and McDowell (1955) constructed about 800 physical models of diapirs piercing both granular and liquid overburdens. They confirmed that diapiric growth could be terminated much more easily by additional sedimentation than by exhausting the source layer. Bishop (1978) argued on theoretical grounds that the strength of the overburden and sedimentation rates and styles played an all-important role in salt tectonics, a view that would receive greater support today.

The modeling approach of using dry sand over a viscous fluid was revitalized by using paraffin wax or silicone to simulate salt (McGill and Stromquist, 1979; Vendeville et al., 1987; Vendeville and Cobbold, 1987, 1988; Vendeville, 1988, 1989; Cobbold et al., 1989). This modeling focused on extension over a "salt" substratum and produced a far more realistic range of "salt" structures than previously accomplished (Figure 18). Vendeville and Cobbold (1988) discovered that a wide range of structural styles could result simply from varying the aggradation rate. Salt rollers were shown to be the *response* of salt flow to extension rather than the *cause* of the extension above them. Vendeville (1988) also demonstrated how monoclinal flexure of a salt-based sequence above a basement normal fault formed a graben in the overburden some distance away from the basement fault.

Computerized Reconstruction

A contrasting impetus for the brittle era was the introduction of computerized palinspastic reconstructions and section balancing to salt tectonics. Although balanced section construction did not originate in the eastern Canadian Rockies, it was first widely applied there (Douglas, 1950; Bally et al., 1966). This provenance ensured that, as the technique propagated in the 1970s and 1980s, it remained rooted in the contractional style of fold and thrust belts. Then Gibbs (1983) adapted the conventions of balancing to extensional terranes, and Worrall and Snelson (1989) adapted them to salt tectonics (Figure 19). Their structural reconstructions in the Gulf of Mexico were accurately carried out using proprietary software and depth-converted seismic data without vertical exaggeration. Strata could be progressively decompacted, backstripped, unfolded, and unfaulted. The widths of gaps and overlaps between fault blocks rigorously constrained the restoration.

Worrall and Snelson's (1989) restorations indicated the following: (1) emplacement of many restored domes was accompanied by faulting; (2) downbuilding was a dominant mechanism of dome growth; and (3) broad, flat-topped salt structures were covered by thin, bathyal shale veneers that during progressive subsidence and steepening became shale sheaths over the progressively narrowing, subsiding passive domes.

Hossack and McGuinness (1990) highlighted important ambiguities in section balancing that are peculiar to salt tectonics. Both extension and salt withdrawal are capable of lowering overburden strata below their regional elevation. Thus, accommodation space for new sediments can be created either by extension or by salt withdrawal. Because of this ambiguity, distinguishing extension from withdrawal is nontrivial in section balancing. An analogous ambiguity results because both contraction and some types of salt emplacement raise overlying strata above their regional elevation (Hossack, 1995).

Contraction in the deep-water foldbelts was accurately estimated to be much less than updip extension of equivalent age. This discrepancy suggested a role for dis-

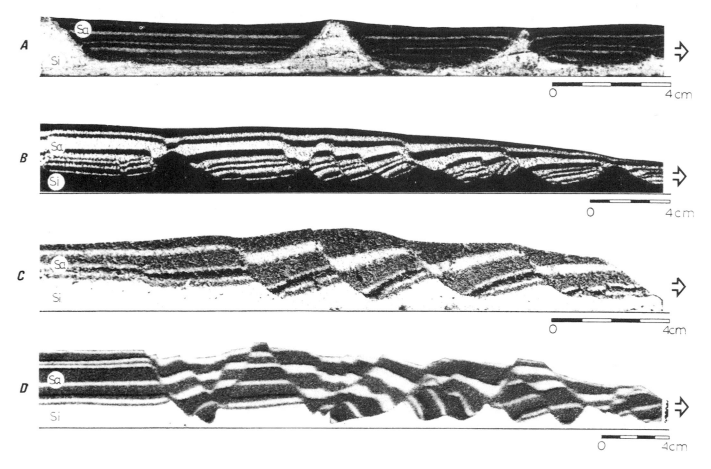

Figure 18—Dawn of the brittle era. Several structural styles were simulated during gravity gliding experiments by variable sedimentation rates of dry sand overburdens and silicone source layers by Vendeville and Cobbold (1987).

placed allochthonous salt in accommodating updip extension (Jackson and Cramez, 1989; Worrall and Snelson, 1989; Diegel and Cook, 1990; Hossack and McGuinness, 1990; Hossack, 1995; Peel et al., 1995).

Major Detachments along Vanished Salt

Worrall and Snelson (1989) stressed that growth faults were not merely surficial basinward slumps produced by gravity gliding. Rather, the growth faults extend and their hanging wall strata expand by displacing salt during regional-scale gravity spreading. Extension provides lateral accommodation space, and displacement of salt or mobile shale at depth creates vertical accommodation space. Some major growth faults terminate downward against laterally pinching-out salt (Figure 20). Restorations indicated that the salt was formerly thicker in many places. This finding led to a radical concept: many major extensional detachments in the Gulf Coast Basin follow vanished sheets of displaced allochthonous salt. Although presented speculatively, this hypothesis is now widely accepted and supported by newer seismic data (see Diegel et al., 1995; Peel et al., 1995; Rowan, 1995; Schuster, 1995).

Segmentation of Salt Sheets

Humphris (1978) envisaged that the updip parts of salt sheets could be segmented by sedimentary loading. Much of the salt displaced by this loading moved basinward and supplied salt for the downdip leading margin of the allochthonous salt mass (Figure 15).

Worrall and Snelson (1989) documented the vast size of the Sigsbee salt nappe on the Louisiana slope. Deltaic systems loaded the rear of the nappe to produce minibasins, some of which displaced all the salt below them. Strata at the base of the minibasins commonly rest directly on an evacuated salt surface underlain by significantly older Cenozoic strata, creating a sedimentary hiatus representing the time interval when allochthonous salt was at or just below the sea floor. Worrall and Snelson (1989) also identified a "Louisiana-style" (including easternmost Texas) of growth faulting: short, arcuate faults dip both landward and basinward and are spatially associated with abundant shallow salt structures of irregular shape. They attributed this structural style to uneven (in time and space) loading by irregular, shifting deltaic systems. This irregular load could only inefficiently displace salt basinward, unlike the sweeping efficiency of long, sub-

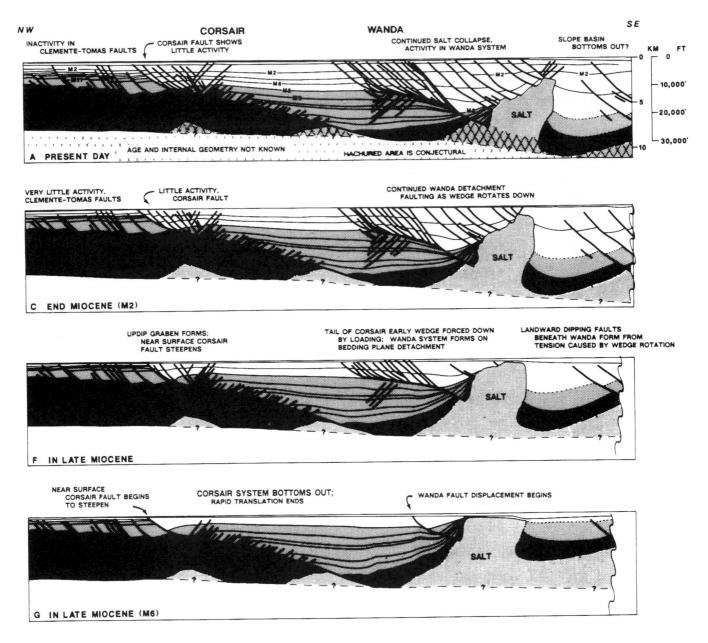

Figure 19—Another catalyst for the brittle era. The first computerized restorations in salt tectonics, as applied to the Wanda and Corsair fault systems in the northern Gulf of Mexico (from Worrall and Snelson, 1989).

parallel "Texas-style" faults dipping basinward. Sumner et al. (1990) described the circular to elliptical minibasins full of ponded sediments enclosed by the arcuate Louisiana-style growth faults. These minibasins have formed largely by salt withdrawal rather than by regional extension.

Raft Tectonics, Salt Welds, and Fault Welds

Burollet (1975) had postulated that the Angolan margin extended by gravity gliding over a thin layer of lubricating salt (Figure 21, top). As it stretched, the overburden broke into diverging raft-like blocks (*radeaux*) separated by widening grabens or half-grabens. These fault-bounded depocenters rapidly filled with younger sediments. Some of the rafts were thought to have been originally separated by diapiric salt walls. The walls subsided below deepening depocenters until only remnants were left (Figure 21, top). Burollet also mentioned a salt scar (*cicatrice salifère*). This was a residual smear of salt left by diapiric subsidence. The scar formed along the subvertical boundary between a gliding block and the adjoining fault-bounded depocenter.

Such smears of salt began to be recognized elsewhere. Worrall and Snelson (1989) noted "salt evacuation surfaces" in the Gulf of Mexico. Jackson and Cramez (1989)

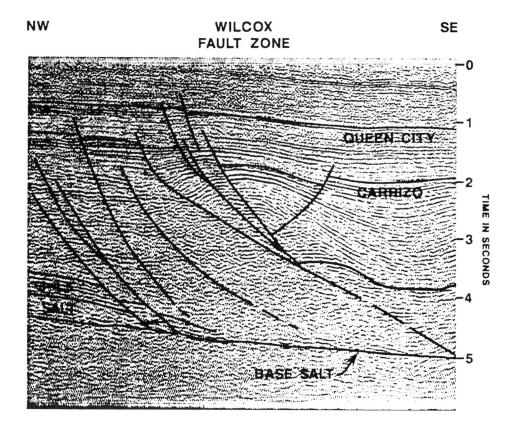

NW   **WILCOX**   **SE**
                        FAULT ZONE

QUEEN CITY

CARRIZO

BASE SALT

TIME IN SECONDS

Figure 20—Major detachments, such as the Wilcox fault zone, created accommodation space because salt withdrew from beneath them, culminating in complete loss of salt. This led to the concept that many major detachments in the Gulf of Mexico follow vanished sheets of autochthonous or allochthonous salt (from Worrall and Snelson, 1989).

surveyed a wide range of enigmatic residual structures they called "salt welds," introducing the symbol of paired dots to denote it on sections and maps (Figure 21, below). A *salt weld* joins strata originally separated by depleted or vanished salt, consisting of thin salt or brecciated insoluble residue. A weld is typically recognizable by structural discordance or local inversion above it. Salt welds can be divided into primary (formed by removal of autochthonous salt), secondary (formed by removal of steep-sided diapirs, such as Burollet's salt scar), or tertiary (formed by removal of gently dipping allochthonous salt). Hossack and McGuinness (1990) extended the concept to *fault welds,* equivalent to a salt welds along which there has been significant shear.

Examples of salt welds are common in raft tectonics. Duval et al. (1992) demonstrated that raft tectonics in the Kwanza Basin took place in two phases: the first formed small, asymmetric, rotated rafts when the overburden was only a few hundred meters thick; the second formed large, nonrotated glide blocks, which themselves comprise smaller rafts. Duval et al. (1992) also compared the roles of contraction, allochthonous salt emplacement, and sea floor spreading in creating the space necessary for thin-skinned extension.

Shallow Emplacement of Salt Sills

Broad lateral flanges of salt in Algeria were interpreted as intrusive salt sills by Ehrmann (1922) (Figure 22). Much later, analysis of high-quality seismic data from more

than 50 salt sheets in the Gulf of Mexico suggested that these had been emplaced intrusively like sills from near-surface diapirs into shallow, low-density, mostly clay-rich sediments (Nelson and Fairchild, 1989; Nelson, 1991) (Figure 23). These data showed reflectors onlapping a thin clastic carapace over the bulging salt sheet. The thickness of this carapace averaged only about 120 m. Sediments deeper than 300 m were estimated to be too strong to admit sills of salt. Nelson and Fairchild (1989) also deduced that after emplacement, many sheets were further inflated by salt (Fletcher et al., 1995).

Shapes of Passive Diapirs and Submarine Salt Glaciers

Vendeville and Jackson (1990, 1991, 1992a) proposed that the cross-sectional shape of a passive (downbuilt) diapir was the result of interplay between sedimentation and upwelling salt. Specifically, the dip of the salt–sediment contact was proportional to the ratio between the local aggradation rate and the net rise (or lateral spreading) rate of diapiric salt (the net rise rate being the gross rise rate minus any dissolution or erosion of the salt). For passive diapirs not cut by faults, this simple relation could account for the variation in time and space of salt structures from conical to steep-sided to overhanging to tongues to sheets. In effect, allochthonous sheets would be an extreme end-member of the above process, spreading rapidly at the surface during periods of extremely slow local aggradation (Talbot, 1995).

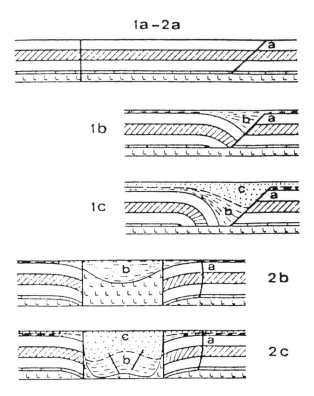

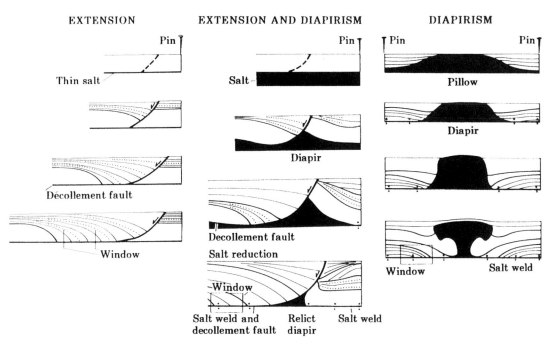

Figure 21—The concept of salt welding. (Top) Extensional breakup of overburden into separate rafts (1a–c) and the filling and inversion of a depocenter over a subsiding salt wall (2b–c) (from Burollet, 1975). (Bottom) Similar structural discordances produced by listric décollement fault (left), combined salt weld and décollement fault (center), and nonextensional salt weld (right) (from Jackson and Cramez, 1989).

Truncation of strata below salt tongues supported the view that many allochthonous sheets were extrusive. In the Gulf of Mexico, extrusion would necessarily have been underwater. Submarine salt glaciers were not a new idea. Trusheim (1960) speculated that greatly overhung diapirs were the remains of salt glaciers that extruded under water. He also described how solution breccias were intercalated in the top of the salt like a moraine. Talbot and Jarvis (1984) reported a marine veneer on the Kuh-e-Namak salt extrusion, which clearly implied submarine extrusion. The mechanics of submarine salt glaciers is investigated in detail by Fletcher et al. (1995), now the ruling hypothesis for emplacement of salt sheets.

Convergent and Divergent Gravity Gliding

Gravity gliding operates downslope, roughly normal to the shelf break. Parallel gliding on a homoclinal shelf

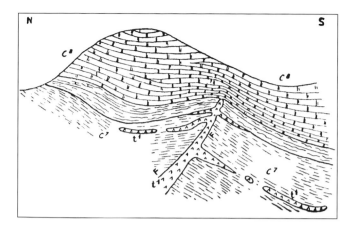

Figure 22—Ehrmann (1922) interpreted broad lateral flanges of Triassic salt (t[1], chevron pattern) as intrusive salt sills in the Gulf of Bejaïa (Bougie), Kabylie, Algeria.

Reactive Diapirism and Regional Extension

Beginning with Seidl (1926), several geologists have speculated that regional extension could initiate diapirism, but mechanical explanations were lacking until illustrated by physical modeling (Jackson and Vendeville, 1990; Vendeville and Jackson, 1992a). Diapiric walls rose beneath grabens created by regional extension at rates entirely determined by the extension rate (Figure 24). Extensional faulting thinned and weakened the overburden locally, creating tectonically induced differential loading. By this means, *reactive diapirs*—as they were called—could rise beneath overburdens of any thickness, density, or lithology. Once diapirs were near enough to the surface, they would typically transmute into active diapirs and forcefully intrude their thin roofs and reach the surface. Once there, diapirs would typically evolve to the passive rise stage (Nelson, 1989). Thereafter, the local aggradation rate, extension rate, or contraction rate would determine the structural style and mode of growth (Fletcher et al., 1995; Letouzey et al., 1995; Nilsen et al., 1995; Talbot, 1995).

These experiments also showed that large amounts of regional extension could be hidden remarkably well within a cross section containing salt structures. A passive diapiric wall (which reaches the surface) typically accommodates most or all of the regional extension simply by widening, whereas the adjoining blocks of overburden remain largely undeformed because of their greater strength. In contrast, for a reactive diapiric wall (which does not reach the surface), deep extension is accommodated partly by widening of the diapir; shallow extension is accommodated by faulting of its roof.

produces mainly two-dimensional structures: an updip domain of extension separated from a downdip domain of contraction by an undeformed domain of translation (Crans et al., 1980; Letouzey et al., 1995). Cobbold and Szatmari (1991) focused on the more common type of coastline, which is irregular. They found that gliding tends to be divergent off coastal salients but convergent off coastal reentrants. Divergent gliding produces strike-parallel extension, whereas convergent gliding produces strike-parallel contraction, neither of which can be restored by simple section balancing. Similar complexities also apply to linked zones of extension and contraction in gravity-spreading systems.

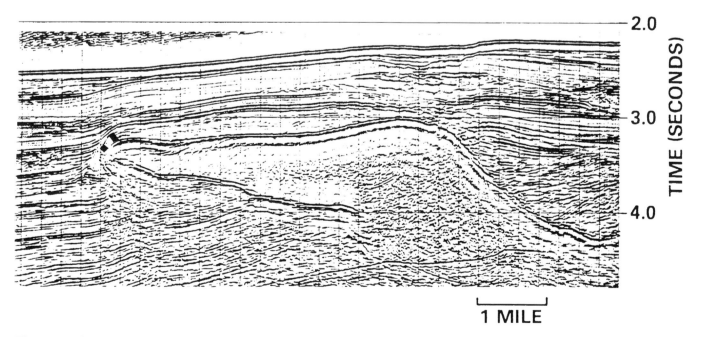

Figure 23—Onlap of frontal reflectors above a thin roof averaging only 100 m thick (black bars on left) was the principal evidence of shallow emplacement of allochthonous salt tongues (after Nelson, 1991).

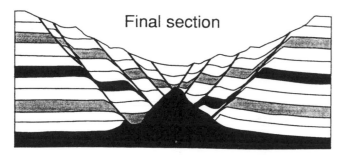

Final section

Restored section

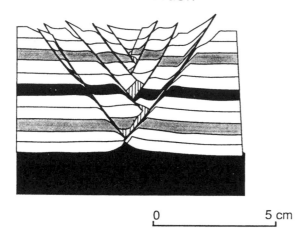

0 — 5 cm

Figure 24—Rise of a model reactive diapir due to tectonic differential loading induced by a graben formed by regional extension. Both the source layer and the overburden were initially tabular, as shown in the lower section. The restoration did not unstrain the fault blocks, demonstrating the local distortion within them; gaps are shown hachured (from Vendeville and Jackson, 1992a).

Diapiric Subsidence, Turtles, and Mock Turtles

The experiments cited in the preceding section showed that regional extension may cause a diapiric wall to widen between diverging blocks of overburden (Figure 25). This widening increases the demand on salt supply even though the diapir crest may not be rising. As the salt supply rate diminishes, the source layer thins due to extension and depositional loading, leading to redistribution of salt. Eventually, as the salt supply rate becomes too low, the roof of the widening diapir begins to sag, potentially forming a graben flanked by residual horns of salt (Figure 25E).

Turtle structures can also form. Trusheim (1960) associated turtle structure anticlines with the subsidence of pillow flanks as the pillow crests became diapiric and broke through their roofs (Figure 12); extension was not required to account for these structures. The experiments of Vendeville and Jackson (1992b) showed that turtle structure anticlines could also be formed by regional extension

as diapiric flanks subsided (Figure 25C–E). Burollet's (1975) concept of the crest of a diapir subsiding into two relict salt diapirs was also validated by these experiments and by reconstruction (Schultz-Ela, 1992). The synclinal depocenter separating these diapiric relics grounded on the basement then inverted into an anticline. This anticline was termed a *mock turtle structure* to distinguish it from a *turtle structure*, from which it differed by forming above a diapir (rather than between diapirs) and by missing stratigraphic section at its base (Figure 25E).

POSTSCRIPT

The Hedberg conference that led to this book was convened to disseminate, evaluate, and consolidate the many new ideas in salt tectonics that were circulating in the early 1990s. The conference was fortuitously held on the eve of a spectacular renaissance of industry interest in salt tectonics caused by a burgeoning subsalt play in the Gulf of Mexico. Amid the dust raised by this exploration boom and by the new, only partially tested ideas of the last few years, it is a reviewer's exacting task to distinguish wheat from chaff long before these become sifted naturally. Added to the difficulty of recognizing the durability and quality of a new concept is the inherent bias of an active researcher compared with a detached observer.

At present, the fluid era and the brittle era are glaringly contrasted in black-and-white opposites: fluid versus brittle, density versus strength, gravity forces versus lateral forces, and so on. This counterpoint is typical of any early period, when a new (brittle) paradigm is pushed to its limits to explore its implications. Useful pieces from the previous (fluid) paradigm may lie discarded, like Stille's orogenic engine, which rusted for decades and was then refurbished with the recognition of inversion-related salt tectonics.

It is premature for the brittle era to be viewed in full perspective, but some brief reflections on shades of gray in two aspects of salt tectonics soften strong contrasts and should check any overweening confidence that we fully understand how salt tectonics works.

In theory, a supercomputer groaning under the weight of appropriate algorithms and incorporating a realistically large number of variables could simulate a full range of salt tectonics by forward modeling. Although such a simulation is valid and could be repeatedly calibrated against natural structures, it could never actually verify the hypotheses built into the algorithms (Oreskes et al., 1994). Moreover, such a model cannot be built by even the most brilliant minds because we lack full data to be entered. Most lacking may be data on how overburden deforms. On geologic time scales, salt is a Newtonian viscous or power law fluid, whereas shale seems to deform by pervasive brittle shear. Yet both diapiric rock types can yield similar structural styles—apparently because they share a common overburden. The style of tectonics seems to be controlled by the overburden rather than by the composition of the diapiric rock.

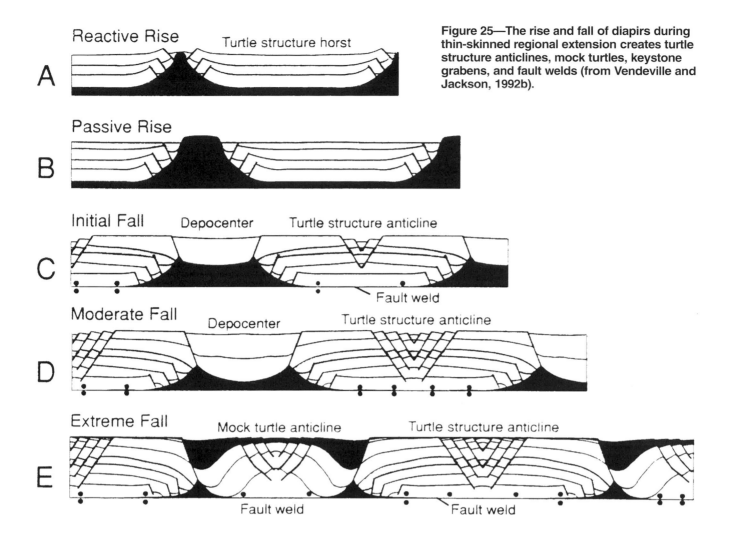

Figure 25—The rise and fall of diapirs during thin-skinned regional extension creates turtle structure anticlines, mock turtles, keystone grabens, and fault welds (from Vendeville and Jackson, 1992b).

What do we know of this overburden? Current modeling of salt tectonics is dominated by an isotropic overburden deforming as a brittle but previously broken rock, governed by Byerlee's law (e.g., Byerlee, 1978). This prefractured state is popular among modelers because its behavior is well understood and is robustly uniform for a large range of rocks (excluding shale). Moreover, Byerlee's law is a conservative approach: less deviatoric stress is required to deform rock already fractured in a wide range of orientations than to generate new fractures. However, overburdens accumulate in complex variety. What if the overburden itself contains halite, as in the Great Kavir or offshore Yemen? What are the different roles of disseminated halite or layered halite in weakening the overburden? What proportion of halite, and other evaporites, is necessary before we should treat the overburden as fluid? And what of shale? Overpressuring is known to promote fracturing in shale, but what if shale is prefractured? What is the role of diagenesis and compaction (downward, upward, and lateral)? Without these data, our knowledge of salt tectonics is broad but skeletal.

Then there is the third dimension. At the close of the fluid era, the long-ignored three-dimensionality of fluids overturning by spoke circulation was finally examined in detail. We should not have to wait that long for the brittle equivalent. Restrained by the two-dimensional nature of much seismic data, of section restoration, and of numerical modeling, our knowledge of the full three-dimensional world of salt tectonics lags far behind the two-dimensional representation. Technology is allowing increased exploration of the third dimension using seismic data volumes and three-dimensional processing, visualization, and structural restoration. As we better comprehend three-dimensional interactions, we should improve our interpretation of currently ambiguous processes such as salt withdrawal versus extension.

The history of salt tectonics reveals a familiar pattern of science. Once-dominant paradigms are supplanted by new ones. Next it will be the turn of the brittle paradigm to be dismantled and its durable parts recycled. Visionary pioneers are often excoriated by their peers but are eventually venerated for the same hypothesis. Looking back over the last century to the first dim perceptions of salt tectonics, we see a scientific landscape strewn not only with outmoded ideas but also with the conceptual seeds of the next revolution in salt tectonics.

Acknowledgments   *I am grateful to the following colleagues who kindly supplied copies of antique papers in Spanish, French, and German: Bernhard Knipping (Clausthal), Cecilio Quesada (Madrid), Antonio Ribeiro (Lisbon), and Maura Sans (Barcelona). I also thank BP Exploration for inviting me to participate in their salt tectonic field trip to Algeria, which included the salt dome shown in Figure 1; Chris Talbot for providing an invaluable and creative perspective on the whole paper but especially the postscript; Carlos Cramez for supplying the paper by Oreskes et al.; David Stephens for photographing all the illustrations; Kirt Kempter for digitally stitching Figure 1; Hongxing Ge for redrawing Figure 11; Bruno Vendeville and Giovanni Guglielmo for translating certain words; Amanda Masterson and Tucker Hentz for stylistic improvements; and Dan Schultz-Ela, Tom Nelson, and especially Sig Snelson for refereeing and polishing the paper. Research was funded by the Texas Advanced Research Program and by the Applied Geodynamics Laboratory consortium comprising Agip S.p.A., Amoco Production Company, Anadarko Petroleum Corporation, Arco Oil and Gas Company and Vastar Resources, Inc., BP Exploration Inc., Chevron, Conoco Inc. and Du Pont Corporation, Exxon Production Research Company, The Louisiana Land and Exploration Company, Marathon Oil Company, Mobil Research and Development Corporation, Petroleo Brasileiro S.A., Phillips Petroleum Company, Société Nationale Elf Aquitaine Production, Statoil, Texaco Inc., and Total Minatome Corporation. This paper is published by permission of the Director, Bureau of Economic Geology.*

REFERENCES CITED

Amery, G. B., 1969, Structure of Sigsbee Scarp, Gulf of Mexico: AAPG Bulletin, v. 53, p. 2480–2482.

Amery, G. B., 1978, Structure of continental slope, northern Gulf of Mexico, *in* A. H. Bouma, G. T. Moore, and J. M. Coleman, eds., Framework, facies, and oil-trapping characteristics of the upper continental margin Tulsa, Oklahoma, AAPG Studies in Geology 7, p. 141–153.

Arrhenius, S., 1913, Zur Physik der Salzlagerstätten: Meddelanden Vetensskapsakademiens Nobelinstitut, v. 2, no. 20, p. 1–25.

Bailey, E. B., 1931, Salt plugs: Geological Magazine, v. 68, p. 335–336.

Balk, R., 1936, Structure elements of domes: AAPG Bulletin, v. 20, p. 51–67.

Balk, R., 1949, Structure of Grand Saline salt dome, Van Zandt County, Texas: AAPG Bulletin, v. 33, p. 1791–1829.

Balk, R., 1953, Salt structure of Jefferson Island salt dome, Iberia and Vermilion parishes, Louisiana: AAPG Bulletin, v. 37, p. 2455–2474.

Bally, A. W., 1981, Thoughts on the tectonics of folded belts, *in* K. R. McClay and N. J. Price, eds., Thrust and nappe tectonics: Geological Society of London Special Publication 9, p. 13–32.

Bally, A. W., P. L. Gordy, and G. A. Stewart, 1966, Structure, seismic data, and orogenic evolution of southern Canadian Rocky Mountains: Canadian Petroleum Geology Bulletin, v. 14, p. 337–381.

Barton, D. C., 1933, Mechanics of formation of salt domes with special reference to Gulf Coast salt domes of Texas and Louisiana: AAPG Bulletin, v. 17, p. 1025–1083.

Bishop, R. S., 1978, Mechanism for emplacement of piercement diapirs: AAPG Bulletin, v. 62, p. 1561–1583.

Blickwede, J. J., and T. A. Queffelec, 1988, Perdido foldbelt: a new deep-water frontier in western Gulf of Mexico (abs.): AAPG Bulletin, v. 72, p. 163.

Braunstein, J., and G. D. O'Brien, 1968, Indexed bibliography of diapirism and diapirs, *in* J. Braunstein and G. D. O'Brien, eds., Diapirism and diapirs: AAPG Memoir 8, p. 358–414.

Brunstrom, R. G. W., and P. J. Walmsley, 1969, Permian Evaporites in North Sea Basin: AAPG Bulletin, v. 53, p. 870–883.

Bryant, W. R., J. W. Antoine, M. Ewing, and B. R. Jones, 1968, Structure of the Mexican continental shelf and slope, Gulf of Mexico: AAPG Bulletin, v. 52, p. 1204–1228.

Burollet, P. F., 1975, Tectonique en radeaux en Angola: Société Géologique de France Bulletin, v. 17, no. 4, p. 503–504.

Byerlee, J. D., 1978, Friction of rocks: Pure and Applied Geophysics, v. 116, p. 615–626.

Carter, N. L., and F. D. Hansen, 1983, Creep of rocksalt: Tectonophysics, v. 92, p. 275–333.

Choffat, P., 1882, Note préliminaire sur les vallées tiphonique et les éruptions d'ophite et de teschenite en Portugal: Société Géologique de France Bulletin, s. 3, v. 10, p. 267–288.

Cloos, E., 1968, Experimental analysis of Gulf Coast fracture patterns: AAPG Bulletin, v. 52, p. 420–444.

Cobbold, P. R., and P. Szatmari, 1991, Radial gravitational gliding on passive margins: Tectonophysics, v. 188, p. 249–289.

Cobbold, P., E. Rossello, and B. Vendeville, 1989, Some experiments on interacting sedimentation and deformation above salt horizons: Société Géologique de France Bulletin, v. 8, no. 3, p. 453–460.

Cobbold, P., 1995, Seismic and experimental evidence for thin-skinned horizontal shortening by convergent radial gliding on evaporites, deep water Santos Basin, Brazil, *in* M. P. A. Jackson, D. G. Roberts, and S. Snelson, eds., Salt tectonics: a global perspective: AAPG Memoir 65, this volume.

Correa Perez, I., and J. Gutierrez y Acosta, 1983, Interpretacion gravimetrica y magnetometrica del occidente de la Cuenca Salina del Istmo: Revista del Instituto Mexicano del Petroleo, v. 15, no. 4, p. 5–25.

Coward, M., and S. Stewart, 1995, Salt-influenced structures in the Mesozoic–Tertiary cover of the southern North Sea, U.K., *in* M. P. A. Jackson, D. G. Roberts, and S. Snelson, eds., Salt tectonics: a global perspective: AAPG Memoir 65, this volume.

Crans, W., G. Mandl, and J. Harembourne, 1980, On the theory of growth faulting: a geomechanical delta model based on gravity sliding: Journal of Petroleum Geology, v. 2, p. 265–307.

Daly, B. J., 1967, Numerical study of two fluid Rayleigh-Taylor instability: The Physics of Fluids, v. 10, p. 297–307.

De Böckh, H., G. M. Lees, and F. D. S. Richardson, 1929, Contribution to the stratigraphy and tectonics of the Iranian ranges, *in* J. W. Gregory, ed., The structure of Asia: London, Methuen, p. 58–176.

de Cizancourt, H., 1934, Plissements disharmoniques et diapirisme. Sur la tectonique des terrains saliferes: Société Géologique de France Bulletin, v. 5, p. 181–200.

DeGolyer, E., 1925, Origin of North American salt domes: AAPG Bulletin, v. 9, p. 831–874.

Diegel, F. A., and R. W. Cook, 1990, Palinspastic reconstruction of salt-withdrawal growth-fault systems, northern Gulf of Mexico (abs.): GSA Abstracts with Programs, v. 72, no. 7, p. A48.

Diegel, F. A., J. F. Karlo, D. C. Schuster, R. C. Shoup, and P. R. Tauvers, 1995, Cenozoic structural evolution and tectonostratigraphic framework of the Northern Gulf Coast continental margin, *in* M. P. A. Jackson, D. G. Roberts, and S. Snelson, eds., Salt tectonics: a global perspective: AAPG Memoir 65, this volume.

Dixon, J. M., 1975, Finite strain and progressive deformation in models of diapiric structures: Tectonophysics, v. 28, p. 89–124.

Douglas, R. J. W., 1950, Callum Creek, Langford Creek, and Gap map areas, Alberta: Geological Survey of Canada Memoir 255, 124 p.

Duval, B., C. Cramez, and M. P. A. Jackson, 1992, Raft tectonics in the Kwanza basin, Angola: Marine and Petroleum Geology, v. 9, no. 4, p. 389–404.

Ehrmann, F., 1922, De la situation du Trias et son rôle tectonique dans la Kabylie des Babourgs: Société Géologique de France Bulletin, s. 4, v. 22, p. 36–47.

Escher, B. G., and P. H. Kuenen, 1929, Experiments in connection with salt domes: Leidsche Geologiese Meddelanden, v. 3, no. 3 (2), p. 151–182.

Ewing, M., and J. Antoine, 1966, New seismic data concerning sediments and diapiric structures in Sigsbee Deep and upper continental slope, Gulf of Mexico: AAPG Bulletin, v. 50, p. 479–504.

Fenneman, N. M., 1906, Oil fields of the Texas–Louisiana Gulf Coastal Plain: USGS Bulletin 282, 146 p.

Fletcher, R. C., 1972, Application of a mathematical model to the emplacement of mantled gneiss domes: American Journal of Science, v. 272, p. 197–216.

Fletcher, R. C., M. R. Hudec, and I. A. Watson, 1995, Salt glacier and composite sediment–salt glacier models for the emplacement and early burial of allochthonous salt sheets, *in* M. P. A. Jackson, D. G. Roberts, and S. Snelson, eds., Salt tectonics: a global perspective: AAPG Memoir 65, this volume.

Garrison, L. E., and R. G. Martin, 1973, Geologic structures in the Gulf of Mexico Basin: USGS Professional Paper 773, 85 p.

Gibbs, A. D., 1983, Balanced cross-section construction from seismic sections in areas of extensional tectonics: Journal of Structural Geology, v. 5, no. 2, p. 153–160.

Gussow, W. C., 1968, Salt diapirism: importance of temperature, and energy source of emplacement, *in* J. Braunstein and G. D. O'Brien, eds., Diapirism and diapirs: AAPG Memoir 8, p. 16–52.

Halbouty, M., 1967, Salt domes—Gulf region, United States and Mexico: Houston, Texas, Gulf Publishing Company, 425 p.

Handin, J., and R. V. Hager, 1958, Experimental deformation of sedimentary rocks under confining pressure: tests at high temperature: AAPG Bulletin, v. 42, p. 2892–2934.

Harris, G. D., 1907, Notes on the geology of the Winnfield Sheet: Baton Rouge, Geological Survey of Louisiana, Report of 1907, Bulletin 5.

Harris, G. D., and A. C. Veatch, 1899, A preliminary report on the geology of Louisiana: Geological Survey of Louisiana Report, Baton Rouge, Louisiana, p. 9–138.

Harrison, J. C., 1995, Tectonics and kinematics of a foreland folded belt influenced by salt, Arctic Canada, *in* M. P. A. Jackson, D. G. Roberts, and S. Snelson, eds., Salt tectonics: a global perspective: AAPG Memoir 65, this volume.

Harrison, J. V., 1930, The geology of some salt-plugs in Laristan, southern Persia: Geological Society of London Quarterly Journal, v. 86, p. 463–522.

Harrison, J. V., 1931, Salt domes in Persia: Institution of Petroleum Technologists Journal, v. 17, p. 300–320.

Harrison, T. S., 1927, Colorado–Utah salt domes: AAPG Bulletin, v. 11, no. 2, p. 111–133.

Hayes, C. W., and W. Kennedy, 1903, Oil fields of the Texas-Louisiana Gulf coastal plain: USGS Bulletin 212.

Hildgard, E. W., 1872, On the geology of lower Louisiana and the salt deposits of Petite Anse Island: Smithsonian Contributions, separate no. 248, p. 32–34.

Hill, R. T., 1902, The Beaumont oil field with notes on the other oil fields of the Texas region: Journal of the Franklin Institute, v. 154, p. 273–274.

Hossack, J. R., and D. B. McGuinness, 1990, Balanced sections and the development of fault and salt structures in the Gulf of Mexico (GOM) (abs.): GSA Abstracts with Programs, v. 22, no. 7, p. A48.

Hossack, J. R., 1995, Geometric rules of section balancing for salt structures, *in* M. P. A. Jackson, D. G. Roberts, and S. Snelson, eds., Salt tectonics: a global perspective: AAPG Memoir 65, this volume.

Hubbert, M. K., 1937, Theory of scale models as applied to the study of geologic structures: GSA Bulletin, v. 48, p. 1459–1520.

Hubbert, M. K., 1951, Mechanical basis for certain familiar geologic structures: GSA Bulletin, v. 62, p. 355–372.

Hughes, D. J., 1968, Salt tectonics as related to several Smackover fields along the northeast rim of the Gulf of Mexico Basin: Gulf Coast Association of Geological Societies Transactions, v. 18, p. 320–330.

Humphris, C. C., Jr., 1978, Salt movement on continental slope, northern Gulf of Mexico, *in* A. H. Bouma, G. T. Moore, and J. M. Coleman, eds., Framework, facies, and oil-trapping characteristics of the upper continental margin: AAPG Studies in Geology 7, p. 69–86.

Jackson, M. P. A., and R. R. Cornelius, 1985, Tertiary salt diapirs exposed at different structural levels in the Great Kavir (Dasht-i Kavir) south of Semnan, north-central Iran: a remote-sensing study of their internal structure and shape: The University of Texas at Austin, Bureau of Economic Geology Open File Report OF-WTWI-1985-22, 108 p.

Jackson, M. P. A., and C. Cramez, 1989, Seismic recognition of salt welds in salt tectonics regimes: GCS-SEPM Tenth Annual Research Conference, Program and Abstracts, Houston, Texas, p. 66–89.

Jackson, M. P. A., and C. J. Talbot, 1989a, Anatomy of mushroom-shaped diapirs: Journal of Structural Geology, v. 11, p. 211–230.

Jackson, M. P. A., and C. J. Talbot, 1989b, Salt canopies: SEPM Gulf Coast Section, 10th Annual Research Conference, Program and Extended Abstracts, Houston, Texas, p. 72–78.

Jackson, M. P. A., and B. C. Vendeville, 1990, The rise and fall of diapirs during thin-skinned extension (abs.): AAPG Bulletin, v. 74, p. 683.

Jackson, M. P. A., R. R. Cornelius, C. H. Craig, and C. J. Talbot, 1987, The Great Kavir salt canopy: a major new class of salt structure (abs.): GSA Abstracts with Programs, v. 19, no. 7, p. 714.

Jackson, M. P. A., C. J. Talbot, and R. R. Cornelius, 1988, Centrifuge modeling of the effects of aggradation and progradation on syndepositional salt structures: The

University of Texas at Austin, Bureau of Economic Geology Report of Investigations No. 173, 93 p.

Jackson, M. P. A., R. R. Cornelius, C. H. Craig, A. Gansser, J. Stöcklin, and C. J. Talbot, 1990, Salt diapirs of the Great Kavir, Central Iran: GSA Memoir 177, 139 p.

Johnston, J., and L. H. Adams, 1913, On the effect of high pressures on the physical and chemical behavior of solids: American Journal of Science, s. 4, v. 35, p. 205–253.

Kennedy, W., 1892, A section from Terrell, Kaufman County, to Sabine Pass on the Gulf of Mexico: Geological Survey of Texas Third Annual Report, Austin, Texas, p. 41–125.

Kent, P. E., 1958, Recent studies of south Persian salt plugs: AAPG Bulletin, v. 42, p. 2951–2972.

Kent, P. E., 1966, Temperature conditions of salt dome intrusions: Nature, v. 211, p. 1387.

Kleinhanns, K., 1914, Die Abhängigkeit der Plastizität des Steinsalzes von umgebenden Medium: Zeitschrift für Physik, v. 15, p. 362–363.

Kupfer, D. H., 1976, Shear zones inside Gulf Coast stocks help delineate spines of movement: AAPG Bulletin, v. 60, p. 1434–1447.

Lachmann, R., 1910, Über autoplaste (nichttektonische) Formelemente im Bau der Salzgesteine Norddeutschlands: Deutschen Geologischen Gesellschaft Monatsberict, v. 62, p. 113–116.

Lees, G. M., 1927, Salzgletscher in Persien: Mitteilungen Geologische Gesellschaft Wien, v. 20, p. 29–34.

Lees, G. M., 1938, The geology of the oilfield belt of Iran and Iraq, *in* Science of petroleum, Vol. 1: London, Oxford University Press, p. 140–148.

Lerche, O., 1893, A preliminary report upon the hills of Louisiana, north of Vicksburg, Shreveport, and Pacific Railroad, Louisiana State Experiment Stations: Geology and Agriculture, no. 1, p. 27.

Letouzey, J., B. Colletta, R. Vially, and J. C. Chermette, 1995, Evolution of salt-related structures in compressional settings, *in* M. P. A. Jackson, D. G. Roberts, and S. Snelson, eds., Salt tectonics: a global perspective: AAPG Memoir 65, this volume.

Link, T. A., 1930, Experiments relating to salt-dome structures: AAPG Bulletin, v. 14, p. 483–508.

Lockett, S. H., 1871, Report of the Topographical Survey of Louisiana, Louisiana State University: Report of Superintendent for 1870, New Orleans, p. 16–26.

Lohest, M., 1921, A propos des plis diapirs rappel de quelques principes de tectonique: Société Géologique de Belgique Annales, v. 44, p. B94–B107.

Lotze, F., 1934, Über "Autochthone Klippen" mit Beispilen aus den Westlichen Pyrenäen: Nachrichten Gesellschaft Wissenschaften Göttingen, Mathematik-Physik, v. 1, p. 1–10.

Lotze, F., 1957, Steinsalz und Kalisalze, v. 1: Berlin, Gebrüder Borntraeger, 466 p.

McGill, G. E., and A. W. Stromquist, 1979, The grabens of Canyonlands National Park, Utah: geometry, mechanics, and kinematics: Journal of Geophysical Research, v. 84, no. B9, p. 4547–4563.

Mrazec, L., 1907, Despre cute cu simbure de strapungere [On folds with piercing cores]: Society of Stiite Bulletin, Romania, v. 16, p. 6–8.

Nelson, T. H., 1989, Style of salt diapirs as a function of the stage of evolution and the nature of the encasing sediments: GCS-SEPM Tenth Annual Research Conference, Program and Abstracts, Houston, Texas, p. 109–110.

Nelson, T. H., 1991, Salt tectonics and listric-normal faulting, *in* A. Salvador, ed., The Gulf of Mexico basin: GSA, Boulder, Colorado, v. J, p. 73–89.

Nelson, T. H., and L. H. Fairchild, 1989, Emplacement and evolution of salt sills in northern Gulf of Mexico (abs.): AAPG Bulletin, v. 73, no. 3, p. 395.

Nettleton, L. L., 1934, Fluid mechanics of salt domes: AAPG Bulletin, v. 18, p. 1175–1204.

Nettleton, L. L., 1943, Recent experimental and geophysical evidence of mechanics of salt-dome formation: AAPG Bulletin, v. 27, no. 1, p. 51–63.

Nettleton, L. L., 1955, History of concepts of Gulf Coast salt-dome formation: AAPG Bulletin, v. 39, p. 2373–2383.

Nettleton, L. L., and T. A. Elkins, 1947, Geologic models made from granular materials: American Geophysical Union Transactions, v. 28, p. 451–466.

Nilsen, K. T, B. C. Vendeville, and J.-T. Johansen, 1995, Influence of regional tectonics on halokinesis in the Nordkapp Basin, Barents Sea, *in* M. P. A. Jackson, D. G. Roberts, and S. Snelson, eds., Salt tectonics: a global perspective: AAPG Memoir 65, this volume.

O'Brien, C. A. E., 1957, Salt diapirism in South Persia: Geologie en Mijnbouw, v. 19, p. 357–376.

Oreskes, N., K. Shrader-Frechette, and K. Belitz, 1994, Verification, validation, and confirmation of numerical models in the earth sciences: Science, v. 263, p. 641–646.

Parker, T. J., and A. N. McDowell, 1955, Model studies of salt-dome tectonics: AAPG Bulletin, v. 39, p. 2384–2470.

Peel, F. J., C. J. Travis, and J. R. Hossack, 1995, Genetic structural provinces and salt tectonics of the Cenozoic offshore U.S. Gulf of Mexico: a preliminary analysis, *in* M. P. A. Jackson, D. G. Roberts, and S. Snelson, eds., Salt tectonics: a global perspective: AAPG Memoir 65, this volume.

Posepny, F., 1871, Studien aus dem Salinargebiete Siebenbürgens: Kaiserlich-Königlichen Geologischen Reichsanstalt Jahrbuch, v. 21, p. 123–186.

Quarles, M., Jr., 1953, Salt ridge hypothesis on the origin of Texas Gulf Coast type of faulting: AAPG Bulletin, v. 37, p. 489–508.

Ramberg, H., 1967, Gravity, deformation and the Earth's crust as studied by centrifuge models: London, Academic Press, 214 p.

Ramberg, H., 1981, Gravity, deformation and the Earth's crust in theory, experiments and geological application: London, Academic Press, 452 p.

Rettger, R. E., 1935, Experiments on soft-rock deformation: AAPG Bulletin, v. 19, p. 271–292.

Reyer, E., 1888, Theoretische geologie: Stuttgart, E. Schweizerbart'sche Verlagshandlung.

Richter-Bernburg, G., and W. Schott, 1959, The structural development of northwest German salt domes and their importance for oil accumulation, *in* Fifth World Petroleum Congress, p. 1–13.

Rios, J. M., 1948, Diapirismo: Instituto Geológico y Minero de España Boletín, v. 60, 390 p.

Rosenkrans, R. R., and D. J. Marr, 1967, Modern seismic exploration of the Gulf Coast Smackover trend: Geophysics, v. 32, p. 184–206.

Rowan, M. G., 1995, Structural styles and evolution of allochthonous salt, central Louisiana outer shelf and upper slope, *in* M. P. A. Jackson, D. G. Roberts, and S. Snelson, eds., Salt tectonics: a global perspective: AAPG Memoir 65, this volume.

Sans, M., and J. Vergés, 1995, Fold development related to contractional salt tectonics, southeastern Pyrenean thrust front, Spain, *in* M. P. A. Jackson, D. G. Roberts, and S. Snelson, eds., Salt tectonics: a global perspective: AAPG Memoir 65, this volume.

Schmeling, H., 1987, On the relation between initial conditions and late stages of Rayleigh-Taylor instabilities: Tectonophysics, v. 133, p. 16–31.

Schmeling, H., 1988, Numerical models of Rayleigh-Taylor instabilities superimposed upon convection: Geological Institutions of the University of Uppsala Bulletin , N. S., v. 14, p. 95–109.

Schultz-Ela, D. D., 1992, Restoration of cross sections to constrain deformation processes of extensional terranes: Marine and Petroleum Geology, v. 9, no. 4, p. 372–388.

Schuster, D. C., 1995, Deformation of allochthonous salt and evolution of related salt/structural systems, eastern Louisiana Gulf Coast, *in* M. P. A. Jackson, D. G. Roberts, and S. Snelson, eds., Salt tectonics: a global perspective: AAPG Memoir 65, this volume.

Seidl, E., 1926, Salztektonik und Zerrung: Deutschen Geologischen Gesellschaft Zeitschrift, v. 78.

Seni, S. J., and M. P. A. Jackson, 1983a, Evolution of salt structures, East Texas diapir province, part 1: sedimentary record of halokinesis: AAPG Bulletin, v. 67, p. 1219–1244.

Seni, S. J., and M. P. A. Jackson, 1983b, Evolution of salt structures, East Texas diapir province, part 2: patterns and rates of halokinesis: AAPG Bulletin, v. 67, p. 1245–1274.

Spiers, C. J., J. L. Urai, G. S. Lister, J. N. Boland, and H. J. Zwart, 1986, The influence of fluid-rock interaction on the rheology of salt rock: Nuclear Science and Technology, EUR 10399 EN, 131 p.

Stier, K., 1915, Strukturbild des Benther Salzgebirges: Jahresbericht des niedersächsischer geologischer Verein, Hannover, v. 8, p. 1–15.

Stille, H., 1910, Faltung des Deutschen Bodens und des Salzgebirges: Kali Zeitschrift, v. 5, no. 16, p. 17.

Stille, H., 1917, Injektiv faltung und damit zusammenhängende Erscheinungen: Geologische Rundschau, v. 8, p. 89–142.

Stille, H., 1925, The upthrust of the salt masses of Germany: AAPG Bulletin, v. 9, p. 417–441.

Sumner, H. S., B. A. Robison, W. K. Dirks, and J. C. Holliday, 1990, Morphology and evolution of salt/mini-basin systems: lower shelf and upper slope, central offshore Louisiana (abs.): GSA Abstracts with Programs, v. 72, no. 7, p. A48.

Talbot, C. J., 1978, Halokinesis and thermal convection: Nature, v. 273, p. 739–741.

Talbot, C. J., 1979, Fold trains in a glacier of salt in southern Iran: Journal of Structural Geology, v. 1, p. 5–18.

Talbot, C. J., 1981, Sliding and other deformation mechanisms in a glacier of salt ,S. Iran, *in* K. R. McClay and N. J. Price, eds., Thrust and nappe tectonics: Geological Society of London Special Publication 9, p. 173–183.

Talbot, C. J., 1992, *Quo Vadis* tectonophysics: with a pinch of salt!: Journal of Geodynamics, v. 16, p. 1–20.

Talbot, C. J., 1995, Molding of salt diapirs by stiff overburden, *in* M. P. A. Jackson, D. G. Roberts, and S. Snelson, eds., Salt tectonics: a global perspective: AAPG Memoir 65, this volume.

Talbot, C. J., and M. P. A. Jackson, 1987, Internal kinematics of salt diapirs: AAPG Bulletin, v. 71, p. 1068–1093.

Talbot, C. J., and R. J. Jarvis, 1984, Age, budget and dynamics of an active salt extrusion in Iran: Journal of Structural Geology, v. 6, p. 521–533.

Talbot, C. J., and E. A. Rogers, 1980, Seasonal movements in a salt glacier in Iran: Science, v. 208, p. 395–397.

Talbot, C. J., C. P. Tully, and P. J. E. Woods, 1982, The structural geology of Boulby (Potash) Mine, Cleveland, United Kingdom: Tectonophysics, v. 85, p. 167–204.

Talbot, C. J., P. Rönnlund, H. Schmeling, H. Koyi, and M. P. A. Jackson, 1991, Diapiric spoke patterns: Tectonophysics, v. 188, p. 187–201.

Thomassy, R., 1863, Supplément à la géologie de la Louisiana: Ile. Petite-Anse: Société Géologique de France Bulletin, v. 2, no. 20, p. 542–544.

Torrey, P. D., and C. E. Fralich, 1926, An experimental study of the origin of salt domes: Journal of Geology, v. 34, no. 3, p. 224–234.

Trusheim, F., 1957, Über Halokinese und ihre Bedeutung für die strukturelle Entwicklung Norddeutschlands: Deutschen Geologischen Gesellschaft Zeitschrift, v. 109, p. 111–151.

Trusheim, F., 1960, Mechanism of salt migration in northern Germany: AAPG Bulletin, v. 44, p. 1519–1540.

Urai, J. L., 1983, Water-assisted dynamic recrystallization and weakening in polycrystalline bischofite: Tectonophysics, v. 96, p. 125–157.

Urai, J. L., 1985, Water-enhanced dynamic recrystallization and solution transfer in experimentally deformed carnallite: Tectonophysics, v. 120, p. 285–317.

Urai, J. L., C. J. Spiers, H. J. Zwart, and G. S. Lister, 1986, Weakening of rock salt by water during long-term creep: Nature, v. 324, p. 554–557.

Van der Gracht, W. A. I. M. v. W., 1917, The saline domes of north-western Europe: Southwestern Association of Petroleum Geology, v. 1, p. 85–92.

Veatch, A. C., 1899, The Five Islands, Louisiana State Experiment Stations: Geology and Agriculture, p. 259–260.

Veatch, A. C., 1902, The salines of North Louisiana: Geological Survey of Louisiana, Report of 1902, Baton Rouge, p. 47–100.

Vendeville, B., 1988, Scale-models of basement-induced extension: Comptes Rendus de l'Académie des Sciences de Paris, s. II, v. 307, p. 1013–1019.

Vendeville, B. C., 1989, Scaled experiments on the interaction between salt flow and overburden faulting during syndepositional extension: SEPM Gulf Coast Section, 10th Annual Research Conference, Program and Extended Abstracts, Houston, Texas, p. 131–135.

Vendeville, B.C., and P. R. Cobbold, 1987, Synsedimentary gravitational sliding and listric normal growth faults: insights from scaled physical models: Comptes Rendus de l'Académie des Sciences de Paris, s. II, v. 305, p. 1313–1319.

Vendeville, B.C., and P. R. Cobbold, 1988, How normal faulting and sedimentation interact to produce listric fault profiles and stratigraphic wedges: Journal of Structural Geology, v. 10, p. 649–659.

Vendeville, B. C., and M. P. A. Jackson, 1990, Physical modeling of the growth of extensional and contractional salt tongues on continental slopes (abs.): AAPG Bulletin, v. 74, p. 784.

Vendeville, B. C., and M. P. A. Jackson, 1991, Deposition, extension, and the shape of downbuilding diapirs: AAPG Bulletin, v. 75, no. 3, p. 687–688.

Vendeville, B. C., and M. P. A. Jackson, 1992a, The rise of diapirs during thin-skinned extension: Marine and Petroleum Geology, v. 9, p. 331–353.

Vendeville, B. C., and M. P. A. Jackson, 1992b, The fall of diapirs during thin-skinned extension: Marine and Petroleum Geology, v. 9, p. 354–371.

Vendeville, B., P. R. Cobbold, P. Davy, J. P. Brun, and P. Choukroune, 1987, Physical models of extensional tectonics at various scales, *in* M. P. Coward, J. F. Dewey, and P. L. Hancock, eds., Continental extensional tectonics: Geological Society of London, Special Publication No. 28, p. 95–107.

Ville, L., 1856, Notice géologique sur les salines des Zahrez et les gites de sel gemme de Rang el Melah et d'Ain Hadjera (Algerie): Annales des Mines, v. 15, p. 351–410.

Wade, A., 1931, Intrusive salt bodies in coastal Asir, south western Arabia: Institute of Petroleum Technologists Journal, v. 17, p. 321–330 and 357–361.

Wenkert, D. D., 1979, The flow of salt glaciers: Geophysical Research Letters, v. 6, p. 523–526.

Withjack, M. O., and C. Scheiner, 1982, Fault patterns associated with domes—an experimental and analytical study: AAPG Bulletin, v. 66, p. 302–316.

Worrall, D. M., and S. Snelson, 1989, Evolution of the northern Gulf of Mexico, with emphasis on Cenozoic growth faulting and the role of salt, *in* A. W. Bally and A. R. Palmer, eds., The geology of North America—an overview: GSA, Boulder, Colorado, v. A, p. 97–138.

Yovanovitch, B., 1922, La géologie du pétrole au Maroc: Société Géologique de France Bulletin, v. 22, p. 234–245.

Hossack, J., 1995, Geometric rules of section balancing for salt structures, *in* M. P. A.
Jackson, D. G. Roberts, and S. Snelson, eds., Salt tectonics: a global perspective:
AAPG Memoir 65, p. 29–40.

Chapter 2

Geometric Rules of Section Balancing for Salt Structures

Jake Hossack

BP Exploration
Uxbridge, Middlesex
U.K.

Abstract

Restored sections provide not only a measure of the viability of structural interpretations but also have the ability to recreate the geometry of the structures through geologic time. Geologists have known for a long time that section balancing is more difficult in salt structures because of the ability of the salt to flow in and out of the plane of section and also to dissolve and thereby violate constant volume considerations. However, the surrounding sediments generally deform by brittle-plastic processes and are less able to flow out of the plane of a properly chosen section. The pragmatic approach is to restore sections by assuming constant-area conditions for the sediment structures alone and to leave the salt area as gaps that may change in area through time. Most restorations of salt structures suggest that throughout long periods of geologic time, salt remains at or close to the depositional surface and that volume reductions of up to 50% are possible in nature.

Salt structures usually involve regional displacements of the salt and its surrounding sediments so that extension in one place has to be balanced by basement extension or cover contraction in another. A key aid to the recognition of contraction and extension is the regional elevation of reference horizons. Generally, salt withdrawal and extensional faulting drop reference beds below regional elevation, whereas salt pillowing , salt sheet formation, and contraction will raise beds above regional elevation. In the Gulf of Mexico, the updip extensional growth faulting and salt withdrawal are balanced by the formation of downdip allochthonous salt sheets and fold and thrust belts, so that the total linear strain across the sediment cover is zero. The extension and contraction are linked by a series of salt and fault welds that lie at several structural levels.

INTRODUCTION

Trusheim (1960) used palinspastic sketches to illustrate the evolution of North German diapirs and to help the reader follow the complicated evolution of diapirs and salt-withdrawal basins through geologic time. With the increased availability of computers, section balancing of salt tectonics has become more rigorous (Moretti and Larrere, 1989; Rowan and Kligfield, 1989; Worrall and Snelson, 1989; Schultz-Ela, 1992). Generally, the shape changes undergone by a salt body are more extreme than those undergone by hanging walls during extensional and contractional deformation, so that the geologist has a more difficult task in recreating the evolution of the salt structures. Salt is also prone to disappear through time by flow and dissolution or to change cross-sectional area by flow in and out of the plane of section (Jenyon, 1987) . Hence, section restoration is the only way in which salt volumes can be recreated in the past from present-day data sets.

Many salt bodies and their surrounding withdrawal basins have complicated three-dimensional geometries implying that salt generally flows through the plane of any geologic section. Hence, the usual assumptions of two-dimensional area preservation (Dahlstrom, 1969; Hossack, 1979) do not apply to salt. This has probably caused many geologists to conclude that salt restorations are not worthwhile or reliable and that not much can be learned from palinspastic salt restorations.

I have a more optimistic view. Recent work carried out by the Applied Geodynamics Laboratory in Austin, Texas, has suggested that, although salt is geologically a viscous fluid that can flow easily through a structure, the surrounding sediments that constrain the salt will behave as a brittle-plastic material during deformation (Weijermars et al., 1993; Jackson and Vendeville, 1994). Most of the laboratory models described from the Applied Geodynamics Laboratory (Vendeville and Jackson, 1992a,b; Jackson and Vendeville, 1994) initiate the salt structures by general two-dimensional brittle deforma-

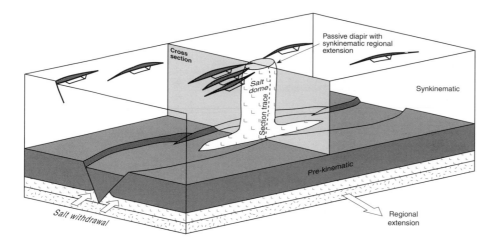

Figure 1—Three-dimensional model of a salt diapir, based on Applied Geodynamics Laboratory experiments, with brittle-plastic sediments above and surrounding viscous salt after reactive, active, and passive phases of diapirism. The initiating linear graben in the prekinematic overburden localized extensional faults in the synkinematic sediments. Plane of cross section is drawn in the direction of regional extension.

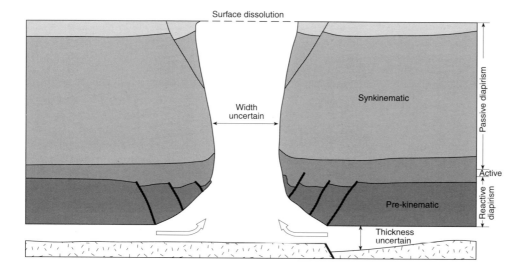

Figure 2—A schematic present-day cross section through the model of Figure 1 with the salt area left blank. The blank areas represent the greatest uncertainty in the section and in the subsequent restoration process.

tion in extension and less commonly in contraction. Although salt-related structures commonly vary along strike, the model sediments do not strain in this direction to any great extent. Thus, it is generally possible to balance the sediments in the models and, by analogy, the sediments in nature at two-dimensional, constant-area conditions (following the rules of Dahlstrom, 1969) in the main transport direction. This may be a contentious view. M. Rowan (personal communication, 1994) believes it is impossible to find a regional transport direction in an area as complicated as the Gulf of Mexico, whereas Peel et al. (1995) have noted that the salt structures of the northern Gulf of Mexico can be interpreted as an amalgamation of three nonsynchronous, two-dimensional, gravity-spreading systems. These systems are oriented parallel to the continental slope and have extension at their updip ends, contraction at their toes, and strike-slip transfer zones at their margins. Peel et al. (1995) further note that there is a fractal distribution of these gravity systems on all scales and that they are basically two-dimensional deformation systems. Hence, I remain optimistic about two-dimensional salt balancing.

Vendeville and Jackson (1992a,b) and Jackson and Vendeville (1994) have documented that model salt diapirs evolve through several distinct stages of development. Stage one is reactive diapirism, which initiates linear salt walls below the floors of linear grabens. Stage two corresponds to a brief phase of active upbuilding, where the salt attempts to lift up and can pierce its overburden. Once at the model surface, the salt is able to keep up with sedimentation, as the diapir grows, by passive downbuilding during the third stage. As long as there is sufficient unimpeded supply of salt from the mother layer, the diapir top remains at the surface as its total height increases with sedimentation. There is a general tendency for the diapir to contract upward toward a smaller, more circular cross section as it passes into the passive phase. If natural diapirs are similar to the model diapirs, a salt dome should have a geometry similar to that shown in Figures 1 and 2. The viscous salt will be totally encased in brittle-plastic sediment (Weijermars et al., 1993), which deforms regionally in a two-dimensional manner. There will be a regional strike and a regional transport direction normal to this direction. Salt can withdraw into the diapir from

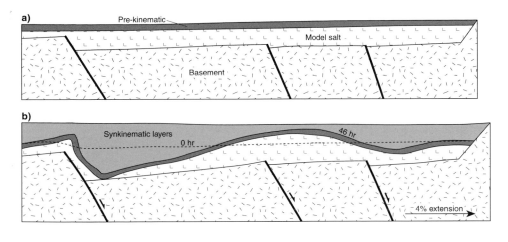

Figure 3—Sequential development of a physical model (Weston et al., 1993) during extension, showing (a) the initial model and (b) the model after 46 hours. A 4% basement-induced extension has created a withdrawal syncline on the left and a salt pillow on the right. The dashed line in (b) shows the initial or regional elevation of the top of the prekinematic section. In the withdrawal syncline, the reference bed after 46 hours was lowered below regional elevation, whereas in the contractional pillow, the same bed was raised above regional elevation.

both the regional strike and dip direction. A cross section drawn in the regional transport direction for the sediment deformation (Figure 2) can be restored in a constant-area, two-dimensional fashion with areas of uncertainty in the width of the diapir and the thickness of the mother salt layer. A pragmatic approach is to restore the section, leaving the salt area as a blank space, and to restore the sediment blocks around the salt, which will show the shape and potential volume of the salt gap back through time.

SECTION VALIDATION

Traditionally, section balancing has been used to test the viability and admissibility of a geologic interpretation (Dahlstrom, 1969; Elliott, 1983). Interpretations that cannot be restored successfully are deemed to be in error and should be redrawn until they can be balanced. Generally, a single geologic data set can be restored in more than one way, so that several solutions should be created. Errors can also be introduced by incorrect stratigraphic correlations, inadequate depth conversions, and inaccurate decompaction routines, but these errors will not be considered in this review. The use of computer packages such as GEOSEC (Rowan, 1994) and LOCACE (Moretti and Larrere, 1989) makes the development of multiple solutions easier to achieve and forces an interpreter to be more rigorous in interpretation. There is still no agreed method in the literature for restoration mechanisms for sediments associated with salt structures. Two methods are generally available in restoration programs: *flexural slip* and *inclined simple shear*. The former is used in contractional thrust belts so that constant bed lengths are retained in constant-area deformations. Extensional fault restorations usually use inclined shear restorations that do not retain constant bed lengths. I have used both shear and flexural slip methods in my salt restorations, the former where the style is dominated by extension and salt withdrawal and the latter where contraction is dominant.

One of the key concepts used in section balancing is the idea of *regional elevation*. This is the level to which a

key bed will return in the undeformed state and is an essential component of area-balance calculations (Hossack, 1979). Estimating regional elevation is often a repetitive trial-and-error process of projecting the level of key beds into the section from an area where there has been no deformation. Traditionally, the dip and elevation of reference beds are projected in from the foreland in thrust belt restorations or from the footwall into the hanging wall in extensional restorations. Regional elevation is a powerful tool to decide the style of deformation because reference beds are raised above regional elevation during contraction and lowered below regional elevation during extension. During inversion (Cooper and Williams, 1989), the regional concept is even more powerful because reference beds beneath the null point of a fault remain below regional elevation after deformation, whereas beds above the point rise above regional. Hence, the null point may be used in some circumstances to delineate regional elevation.

A description of regional elevation also plays a key role in salt balancing. Figure 3 describes a laboratory model (Weston et al., 1993) in which a rigid basement made up of tilted domino blocks, overlain by model salt and a higher sedimentary overburden, was extended. The extension initiated salt withdrawal and pillowing. The deformation can be defined by the initial and final positions of a prekinematic marker bed. The position of the bed in the undeformed state is shown in Figure 3 as line 0, which is the regional elevation of the bed. Surprisingly, in spite of the overall extension of the model, the salt withdrawal and differential loading in one place create a salt pillow that rises above regional elevation in another. This suggests a linked model (Figure 4) that can be internally balanced even where there is no basement extension. The differential loading and withdrawal create an accommodation space that becomes filled with sediment. Under perfect two-dimensional constant-area conditions, the excess area of the accommodation space below regional elevation (labeled A in Figure 3) (Hossack, 1979) will equal the area of excess section of the salt pillow (A') or a reference sedimentary horizon above regional elevation

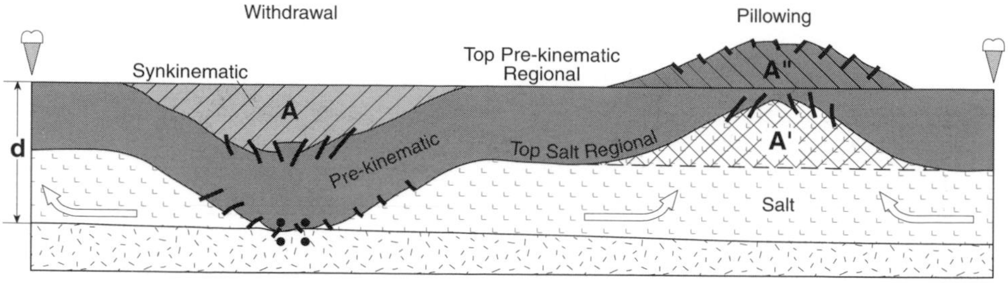

Figure 4—A schematic section showing linked salt withdrawal and pillowing. The section is internally balanced by area and line length. The area of excess section below regional elevation (A) is equal to the areas of excess section above regional elevation (A' or A'').

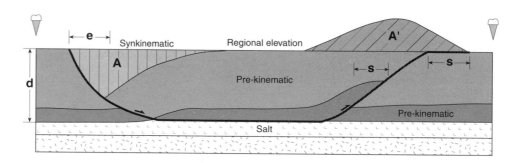

Figure 5—A schematic section showing linked extensional faulting and thrusting where the extension (e) is balanced by the shortening (s). The salt layer acts as a detachment linking the extension and contraction. The deformation can be driven either by gravitational gliding or spreading.

elsewhere (A''). The model is independent of the driving force for the deformation, so that the sediment loading can create its own accommodation space or the sediment merely fills a preexisting space. In the general case of three-dimensional strain, the three areas of excess section need not be equal. Under these conditions, perfect line length balancing becomes impossible.

Another self-balancing model is suggested in Figure 5, where extension is linked to contraction by a detachment on the top of a salt layer. The extension drops a reference bed in the prekinematic section below regional elevation to create sediment accommodation space (labeled A in Figure 5), which is balanced by the area of contractional excess section in the fold (A'). The linked system can be driven either by downslope gravity gliding or gravity spreading of a wedge. In general, with salt flow in and out of the section, the areas will not balance and line length may not be preserved. However, the brittle-plastic deformation in the encasing sediments may constrain them to retain approximate constant bed lengths. A comparison of Figures 4 and 5 suggests that extension and salt withdrawal are equivalent because they both lower reference horizons below regional elevation; conversely, salt pillowing and folding are equivalent because they raise beds above regional.

More complicated models can be created that link extension and withdrawal with pillowing and folding that can be internally balanced (Figure 6). A further complication that can be added to these conceptual balancing models is the creation of an allochthonous salt sheet (Worrall and Snelson, 1989; West, 1989; Duval et al., 1992). Consider that updip extension has created a downdip contractional salt pillow that balances the extension (A = A_1 in Figure 7). Imagine that in some unspecified way, the salt in the core of the pillow (A_1) is siphoned off and transferred to a higher structural level to create the alloch-

thonous salt sheet (A_3). During the transfer, the pillow deflates and the prekinematic beds return to regional elevation onto a salt weld (Jackson and Cramez, 1989) so that areas A_1 and A_2 disappear. A conceptual stem, which may also be a salt weld, connects the deflated pillow with the salt sheet. In the perfect two-dimensional case, the extensional excess section (A in Figure 7) equals the excess area of the original pillow (A_1), which in turn equals the area of salt in the sheet (A_3). Allochthonous salt sheets are equivalent to contraction and pillowing as they raise rock—in this case salt—above regional elevation. The model does not depend on the mode of transfer of the salt to a higher level. It could be intruded as a salt sill (Nelson and Fairchild, 1989) or extruded across the ground surface as a salt glacier (McGuinness and Hossack, 1993a,b; Fletcher et al., 1995).

All these conceptual models are combined in Figure 8 into a self-balancing system that links extension and salt withdrawal to downdip folding and thrusting and allochthonous salt canopies. Various parts of the system are linked by fault welds and salt welds (Hossack and McGuinness, 1990) and detachments. The higher salt canopies in turn can form new detachment systems so that the weld detachments can occur at several different levels in the section. This model explains the enigma that faced Worrall and Snelson (1989), who were unable to balance one of their regional cross sections in the Gulf of Mexico because 40 km of updip extension on listric growth faults could only be balanced by 5 km of shortening in a downdip fold and thrust belt. The balancing problem was compounded by an age difference between the growth faults and the thrusts. However, a large salt canopy, close to the sea floor, lies in their section between the extension faults and the thrusts, so that the missing extension can conceptually be taken up by the formation of the allochthonous salt sheet.

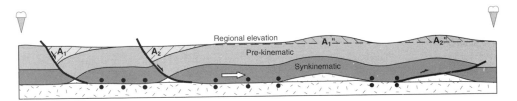

Figure 6—A schematic section that links extension, salt withdrawal, pillowing, and folding. In a perfect constant-area deformation, $A_1 + A_2 = A_1'' + A_2''$.

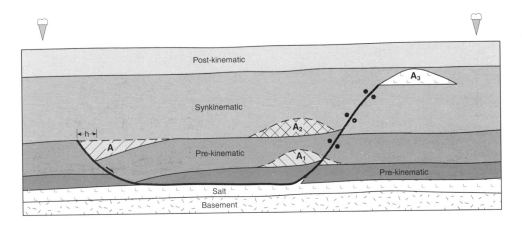

Figure 7—Formation of an allochthonous salt sheet. Updip extension with excess area (A) is balanced by downdip contractional salt pillowing (A_1 or $A_{2)}$. The allochthonous salt sheet is created by removing the salt from the core of the pillow (A_1) to the sheet higher in the section (A_3) and allowing the pillow (A_1) to deflate.

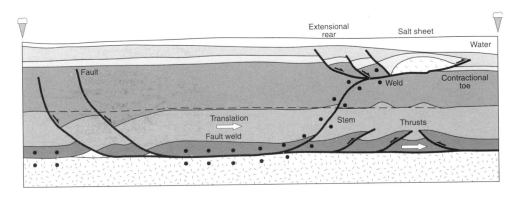

Figure 8—A totally linked system, based on sections from the Gulf of Mexico (Peel et al., 1995), which balances extension and salt withdrawal by folding, thrusting, and allochthonous salt sheets. Linkages among the different parts of the tectonic system are provided by salt welds, fault welds, and detachments.

UNDERSTANDING THE SECTION

As well as providing confidence in the admissibility and viability of a section, section balancing has a more important role to play in the study of salt structures—helping the interpreter to understand the development of the salt body and its relationships to the surrounding sediments through time. From a snapshot in the present, it is possible to reconstruct many millions of years of geologic history.

My first example is a sandbox model from the Applied Geodynamics Laboratory. Experiment 146 helped to confirm a geologic process that explains the creation of allochthonous salt sheets by glacial extrusion of salt across the sea floor (Fletcher et al., 1993; Hudec et al., 1993; McGuinness and Hossack, 1993a,b) (Figure 9). Previously, the allochthonous salt sheets in the Gulf of Mexico were believed to be intrusive sills (Nelson and Fairchild, 1989). Internally at BP, D. McGuinness and I developed a glacier model based on several backstripped sections from the Gulf of Mexico that seemed to show salt sheets spreading out through time over the sea floor at

the sediment–water interface. Independent model 146 was an exciting confirmation that glacial spreading of salt was mechanically reasonable. The model created a sub-horizontal salt sheet that climbs periodically up through ramps in a sedimentary section and is overlain by synkinematic sediment in salt-withdrawal basins. We did not have access to the complete experimental results, and so to understand how such a structure could have developed, McGuinness backstripped the section on LOCACE. M. Jackson (personal communication, 1993) confirmed that McGuinness' restoration, which was largely unconstrained, accurately recreated the structural development of the model. Jackson provided constraints from a full data set of the experiment, which have been incorporated into a second LOCACE restoration using flexural slip algorithms (Figure 9) to better constrain the geometries.

The restoration confirms that the salt body began as a salt wedge at the left end of the section, which was overlain by a thin prekinematic sedimentary roof. As a result of sedimentary progradation from the left, the salt wedge folded the prekinematic section into a large box fold that failed by local extensional tectonic excision on its frontal

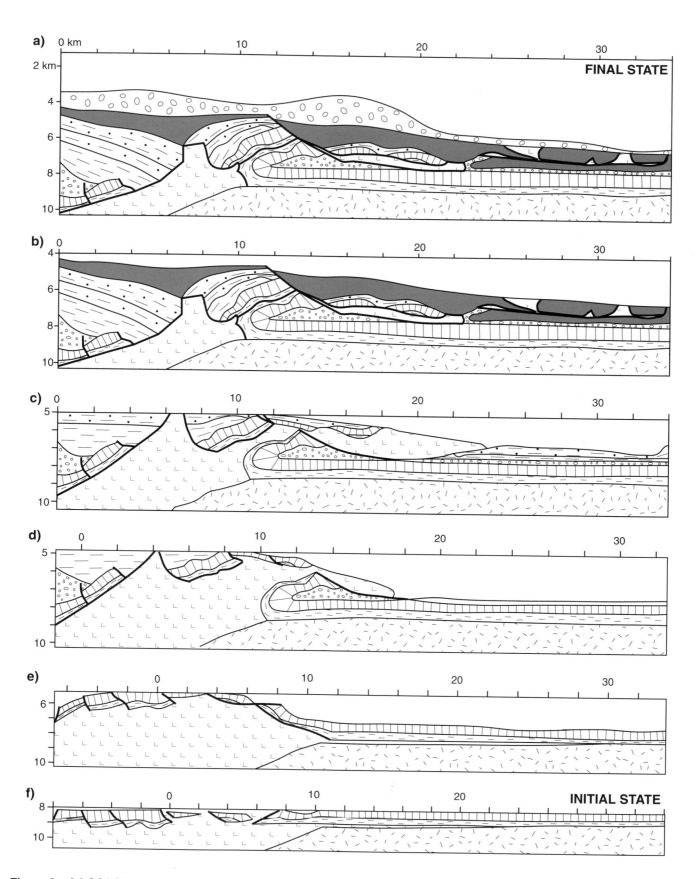

Figure 9—A LOCACE restoration of Applied Geodynamics Laboratory experiment 146, which created an allochthonous salt sheet by glacier-like spreading of the model salt across the surface of the model.

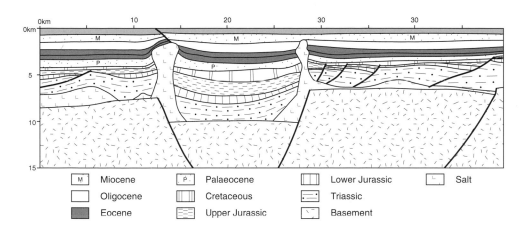

Figure 10—Present-day depth section through the Machar and Medan diapirs in the Eastern Trough of the Central Graben, North Sea.

limb, exposing salt at the model surface. Forward in time, the salt spread as a glacier over the surface of the model, and the frontal limb of the fold was overturned by a flap rotation into a recumbent syncline that was overridden by the salt glacier. As the salt glacier moved forward, it carried with it a stretched roof of the original prekinematic section as a series of separate rafts. Eventually the rafts overrode their equivalent stratigraphic section in the recumbent footwall syncline beneath the salt to produce a triple repetition of the same stratigraphy above and below the salt. Continuing sediment progradation from the left built sediment wedges out on top of the salt glacier, which continued to spread and climb section to the right to form a successive salt canopy at the right end of the section. The canopies became segmented by continuing sediment deposition and salt withdrawal, so that remnant parts of the canopy are connected by salt welds (Jackson and Cramez, 1989; West, 1989). The initial and improved versions of this restoration confirm the importance of backstripping in recreating the geologic history of a section correctly.

The second example (Figure 10) is a regional depth section through the Eastern Trough of the Central North Sea Graben (Foster and Rattey, 1993). The section is through the Machar and Medan salt diapirs, which are interpreted to sit above the basement faults that form the margins to a graben in the Eastern Trough. The diapirs root down into the Permian Zechstein mother salt layer, which existed both within the graben trough and the marginal highs. Both diapirs have steep-sided contacts that penetrate up through Mesozoic and Tertiary stratigraphy, proving that the diapirs have had a long history of evolution. The end of diapirism is marked by the age of the oldest sediments that cover the diapir crest. The two diapirs have become dormant at different geologic times: the Machar diapir on the left reaches up to near the top of the Paleocene, whereas the Medan diapir on the right has penetrated into a younger section as high as the middle Miocene. The Paleocene–middle Miocene section of the Machar diapir, however, is folded and stratigraphically thinned above the diapir crest, suggesting a more gradual decay in the diapir history compared to the Medan.

Figure 11 shows a LOCACE restoration of the section without decompaction back to the Triassic. Regionally, salt diapirism was initiated in the Early Triassic (Smith et al., 1993), and this restoration suggests that 5 km of cover extension in the Triassic was present locally to initiate the early reactive diapirism that formed the salt walls that subsequently developed into the diapirs. By the end of the Triassic, the diapirs were fully developed and had reached the passive phase of development. They continued to keep pace with regional sedimentation by remaining close to the earth's surface and growing upward by downbuilding as the sediments accumulated around them. Passive diapirism continued throughout the Jurassic, Cretaceous, and early Paleocene, during which time both crests of the diapirs were continually at the surface through 260 m.y. of geologic time. The Machar diapir began to be covered by sediment in the Paleocene as the grounding of the flanking blocks cut off its salt supply, whereas the Medan continued to grow. During this time, the Machar still showed some activity, as the Paleocene–middle Miocene sediments above the diapir crest were arched above regional elevation in a rejuvenated phase of attempted active piercement.

By the late Miocene, both diapirs had become completely inactive and were covered by upper Tertiary sediments. The timing of the basement faulting is uncertain. Assuming that there was a Permian and Early Triassic phase of rifting in the Central Graben (Smith et al., 1993) followed by a Late Jurassic rift event, the restorations show slip on the basement faults both in the Triassic and Jurassic. However, to minimize the vertical thickness of salt in the section, the cover has been pinned onto basement at the right end of the section (Figure 11). Hence, the restorations provide an estimate of the minimum salt thicknesses. The earliest time at which the overburden is judged to have welded onto basement is at the end of the Miocene because stratigraphic thickness changes related to salt withdrawal in the Eastern Trough from the Triassic to late Miocene. If sediment decompaction had been applied during the backstripping, the sediment columns and diapirs would have been respectively thicker and higher in the center of the section. However, the timing relationships to the welding should not change

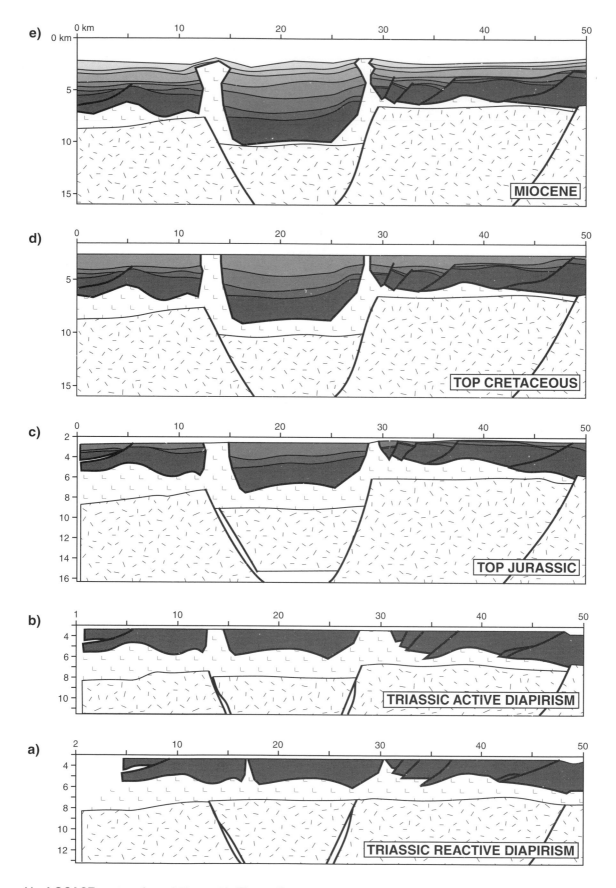

Figure 11—LOCACE restoration of Figure 10. The sediment layers were not decompacted during the backstripping.

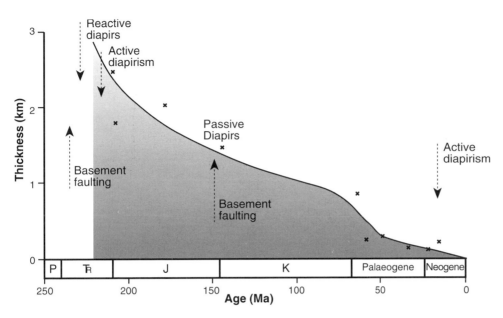

Figure 12—Change in apparent source layer thickness in the deepest part of the basin through time as measured in the LOCACE restorations shown in Figure 11a–e.

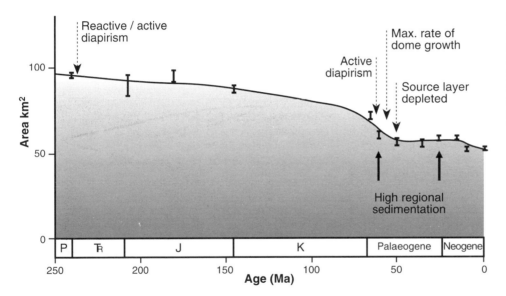

Figure 13—Change in the total area of the salt in the restored sections of Figure 11 through time. Timing of the various stages of diapir development is indicated. The area, which is a minimum estimate, has reduced by 50% since the Permian.

because the geometric effects of the ending of salt-withdrawal thickness changes should be unaffected by compaction unless there are some considerable lateral facies changes. Some simple measurements were taken from the nondecompacted stratigraphic thicknesses and areas of the salt and sediment in the thickest and deepest part of the trough through time, and these provide further background on the geologic history (Figures 12–15). With decompaction, the reported rates of sedimentation and diapir growth would be higher, but the relative rates and times of maximum growth would not change.

Figure 12 plots the change in the source layer thickness from the reactive diapir stage in the Triassic to the final death of the diapirs in the Neogene. The area of salt at each stage of the restoration depends on the geometric relationships between the cover and the basement extensions in the section, but these are assumed identical here.

The source layer thickness appears to have decayed since the Triassic on two successive exponential decay curves. The first curve shows a rapid decrease in layer thickness back to the early Paleogene, close to the time of the Machar reactivation, then a farther exponential decay throughout the Neogene.

Figure 13 plots the apparent minimum change in the salt area through geologic time. The restorations suggest a minimum area change of 100 km^2 in the Triassic that decays in a similar double-stepped manner to a present-day minimum of 50 km^2—an area loss of 50%. Again, the step in the graph coincides with the step in the salt thickness graph (Figure 12) and with the time of maximum dome growth.

Variations in the rates of growth of the Machar dome through time also show two phases in the history (Figure 14). Dome growth was initiated in the Late Triassic–Early

Figure 14—Maximum gross rate of increase in height or growth of the Machar dome through time without decompaction effects being estimated. Applying decompaction would increase the absolute rates but would not change the relative shape of the graph.

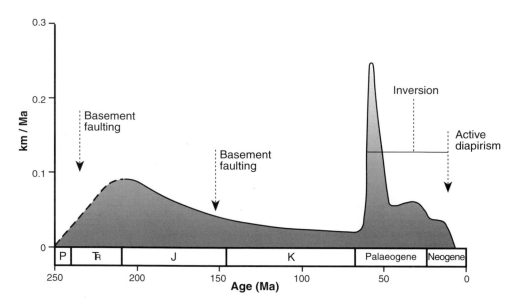

Jurassic with a rejuvenation in the Paleogene, which coincides with the regional Alpine inversion event. The earliest phase of dome growth is a result of initiation by the Triassic rift phase, whereas the regional Late Jurassic rift event did not appear to have had any effect on the growth of these diapirs. This might suggest that Jurassic rifting was not an important event in this area. The Tertiary rejuvenation did not correspond to a known rift event. However, the rejuvenation seems to coincide with several pulses of more rapid regional sedimentation (Figure 15). Sedimentation rates were higher in the Triassic and Jurassic during the regional rifting events. However, the Tertiary peaks in the sediment aggradation rate coincide with major uplift of the United Kingdom and the Atlantic margin and with the Alpine inversion in the North Sea (Ziegler, 1990). Two explanations are possible for the attempted reactivation of the Machar diapir, which was able to lift the overlying sediments above regional elevation. Either the increased sediment loading at the time of active aggradation increased the buoyancy forces in the salt source layer that still remained to cause the uplift, or the active compression of the Alpine inversion was able to squeeze the diapir and lift its lid (Nilsen et al., 1995).

CONCLUSIONS

Palinspastic restorations of geologic sections through salt structures can test the validity of geologic interpretations. However, more importantly, the restorations provide a kinematic view of the development of the structures through time. As for all other types of palinspastic restorations, care has to be taken in choosing the orientation of the section so that the line is parallel to the displacement vectors of the faults in the surrounding sediment carapace. Hence, constant-area conditions will apply to the sediments during the restoration. Salt can easily flow in a viscous manner in and out of the plane of

section. Pragmatic experience suggests that the salt is best left as a gap in the section between the surrounding sediment fault blocks. These spaces represent the area of greatest uncertainty in the restoration. During the active and passive phases of salt diapirism, which generally correspond to the largest part of the development history of salt diapirs and allochthonous salt sheets, salt is nearly always at or very close to the earth's surface. Hence, this is the time when salt dissolution is most active. Generally, the apparent area of salt in the restored sections nearly always increases backward through time, which corresponds to true loss of salt volume forward in time. The Eastern Trough example described here has apparently lost 50% of its salt area between diapir initiation in the Triassic and the death of the diapirs in the Miocene. Throughout this time, the diapirs were either continuously exposed at the surface or were covered by a thin condensed sediment section that periodically slumped from the top toward the sides of the diapir to expose the salt to the elements (R. Anderton, personal communication, 1993). Salt dissolution thus adjusted to the salt supply rate so that salt was able to stay at the same level as the sediment depositional surface.

Salt diapirs and allochthonous salt sheets can be formed in both extensional and contractional environments, and in some basins, both tectonic styles are combined in linked self-balancing systems. Salt withdrawal and extension are geometrically equivalent because they both lower reference horizons in the overburden below regional elevation. Salt pillowing, folding, and allochthonous salt sheets are equivalent to contraction because they raise reference horizons above regional. In a self-balancing system, such as the Gulf of Mexico, the updip salt withdrawal and extensional growth faults can be conceptually balanced downdip by folding, thrusting, and salt canopy formation. Where compensating canopies or fold and thrust belts are absent, the extension faults initiating the diapirism must be balanced by equivalent extension in the basement.

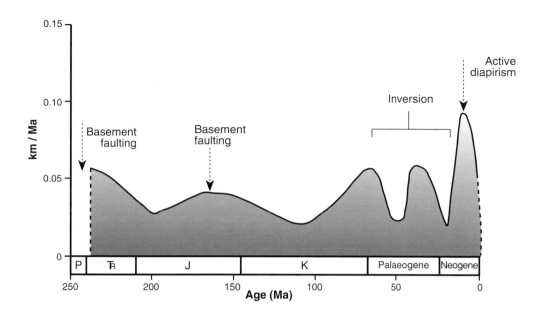

Figure 15—Estimate of regional aggradation rates in the center of the basin without decompaction effects being estimated. Applying decompaction would increase the absolute rates but would not change the relative shape of the graph.

Acknowledgments—The ideas in this paper matured over several years during discussions with many BP colleagues. I would particularly like to thank Dorie McGuinness, Frank Peel, Chris Travis, and Roger Anderton for their help and criticism over the years. Martin Jackson and the Applied Geodynamics Laboratory group provided an external sense check on many occasions, and I would like to thank Martin, Mark Rowan, and an unknown reviewer for pointing out many logical, stylistic, and drafting errors in my original manuscript, which helped to improve the final product.

REFERENCES CITED

Cooper, M. A., and G. D. Williams, eds., 1989, Inversion tectonics: Geologic Society of London, Special Publication, no. 44, 376 p.

Dahlstrom, C. D. A., 1969, Balanced cross sections: Canadian Journal of Earth Sciences, v. 6, p. 743–754.

Duval, B., C. Cramez, and M. P. A. Jackson, 1992, Raft tectonics in the Kwanza basin, Angola: Marine and Petroleum Geology, v. 9, p. 389–404.

Elliott, D., 1983, The construction of balanced cross sections: Journal of Structural Geology, v. 5, p. 101.

Fletcher, R. C., M. R. Hudec, and I. A. Watson, 1993, Salt glacier model for the emplacement of an allochthonous salt sheet (abs.): AAPG, International Hedberg Research Conference Abstracts, Bath, U.K., Sept. 13–17, p. 50–52.

Fletcher, R. C., M. R. Hudec, and I. A. Watson, 1995, Salt glacier and composite sediment–salt glacier models for the emplacement and early burial of allochthonous salt sheets, *in* M. P. A. Jackson, D. G. Roberts, and S. Snelson, eds., Salt tectonics: a global perspective: AAPG Memoir 65, this volume.

Foster, P. T., and P. R. Rattey, 1993, The evolution of a fractured chalk reservoir: Machar oilfield, U.K. North Sea, *in* J. R. Parker, ed., Petroleum geology of northwest Europe: Proceedings of the Fourth Conference, Geologic Society of London, p. 1445–1452.

Hossack, J. R., 1979, The use of cross sections in the calculation of orogenic contraction: Journal Geologic Society of London, v. 136, p. 705–711.

Hossack, J. R., and D. B. McGuinness, 1990, Balanced sections and the development of fault and salt structures in the Gulf of Mexico: GSA, Abstracts with Programs, v. 22, no. 7, p. A48.

Hudec, M. R., R. C. Fletcher, and I. A. Watson, 1993, The composite salt glacier: extension of the salt glacier model to post-burial conditions (abs.): AAPG, International Hedberg Research Conference Abstracts, Bath, U.K., Sept. 13–17. p. 91–92.

Jackson, M. P. A., and C. Cramez, 1989, Seismic recognition of salt welds in salt tectonics regimes: SEPM Gulf Coast Section, 10th Annual Research Conference, Houston, Program and Extended Abstracts, p. 66–71.

Jackson, M. P. A., and B. Vendeville, 1994, Regional extension as a geologic trigger for diapirism: GSA Bulletin, v. 106, p. 57–73.

Jenyon, M. K., 1987, Salt tectonics: New York, Elsevier, 191 p.

McGuinness, D. B., and J. R. Hossack, 1993a, The development of allochthonous salt sheets as controlled by the rates of extension, sedimentation and salt supply (abs.): AAPG, International Hedberg Research Conference Abstracts, Bath, U.K., Sept. 13–17. p. 116–118.

McGuinness, D. B., and J. R. Hossack, 1993b, The development of allochthonous salt sheets as controlled by the rates of extension, sedimentation, and salt supply (abs.): SEPM Gulf Coast Section, 14th Annual Research Conference, Program and Extended Abstracts, Houston, p. 127–139.

Moretti, I., and M. Larrere, 1989, LOCACE, computer-aided construction of balanced geologic cross sections: Geobyte, v. 4, no. 5, p. 16–24.

Nilsen, K. T., B. C. Vendeville, and J.-T. Johansen, 1995, Influence of regional tectonics on halokinesis in the Nordkapp Basin, Barents Sea, *in* M. P. A. Jackson, D. G. Roberts, and S. Snelson, eds., Salt tectonics: a global perspective: AAPG Memoir 65, this volume.

Nelson, T. H., and L. H. Fairchild, 1989, Emplacement and evolution of salt sills in northern Gulf of Mexico (abs.): AAPG Bulletin, v. 73, p. 395.

Peel, F. J., C. J. Travis, and J. R. Hossack, 1995, Genetic structural provinces and salt tectonics of the Cenozoic offshore U.S. Gulf of Mexico: a preliminary analysis, *in* M. P. A. Jackson, D. G. Roberts, and S. Snelson, eds., Salt tectonics: a global perspective: AAPG Memoir 65, this volume.

Rowan, M. G., 1994, A systematic technique for the sequential restoration of salt structures: Tectonophysics, v. 228, p. 331–348.

Rowan, M. G., and R. Kligfield, 1989, Cross section restoration and balancing as an aid to seismic interpretation in extensional terranes: AAPG Bulletin, v. 73, p. 955–966.

Schultz-Ela, D. D., 1992, Restoration of cross sections to constrain deformation processes of extensional terranes: Marine and Petroleum Geology, v. 9, p. 372–388.

Smith, R. I., N. Hodgson, and M. Fulton, 1993, Salt control on Triassic reservoir distribution, UKCS Central North Sea, *in* J. R. Parker, ed., Petroleum geology of northwest Europe: Proceedings of the Fourth Conference, Geologic Society of London, p. 547–557.

Trusheim, F., 1960, Mechanics of salt migration in northern Germany: AAPG Bulletin, v. 44, p. 1519–1540.

Vendeville, B. C., and M. P. A. Jackson, 1992a, The rise of diapirs during thin-skinned extension: Marine and Petroleum Geology, v. 9, p. 331–353.

Vendeville, B. C., and M. P. A. Jackson, 1992b, The fall of diapirs during thin-skinned extension: Marine and Petroleum Geology, v. 9, p. 354–371.

Weijermars, R., M. P. A. Jackson, and B. Vendeville, 1993, Rheological and tectonic modeling of salt provinces: Tectonophysics, v. 217, p. 143–174.

West, D. B., 1989, Model for salt deformation on deep margin of Central Gulf of Mexico Basin: AAPG Bulletin, v. 73, p. 1472–1482.

Weston, P. J., I. Davidson, and M. W. Insley, 1993, Physical modelling of North Sea salt diapirism, *in* J. R. Parker, ed., Petroleum geology of northwest Europe: Proceedings of the Fourth Conference, Geologic Society of London. p. 559–567.

Worrall, D. M., and S. Snelson, 1989, Evolution of the northern Gulf of Mexico, with emphasis on Cenozoic growth faulting and the role of salt, *in* A. W. Bally and A. R. Palmer, eds., The geology of North America—an overview, vol. A: GSA, Boulder, Colorado, p. 97–138.

Ziegler, P. A., 1990, Geologic atlas of western and central Europe (second edition): The Hague, Shell International Petroleum.

Letouzey, J., B. Colletta, R. Vially, and J. C. Chermette, 1995, Evolution of salt-
related structures in compressional settings, *in* M. P. A. Jackson, D. G. Roberts,
and S. Snelson, eds., Salt tectonics: a global perspective: AAPG Memoir 65,
p. 41–60.

Chapter 3

Evolution of Salt-Related Structures in Compressional Settings

J. Letouzey

B. Colletta

R. Vially

Institut Français du Pétrole
Rueil-Malmaison, France

J. C. Chermette

Total
Paris La Défense, France

Abstract

Sandbox experiments analyzed by computerized X-ray tomography provide relevant models of salt-related contractional structures and improve understanding of the relative importance of the many parameters influencing structural style. In front of thin-skinned fold and thrust belts, the salt layers provide décollement surfaces, which allow the horizontal strain to propagate far toward the edge of the foreland. As shortening increases, older structures forming in front of the system can be overtaken by out-of-sequence faulting and folding. The very low friction coefficient of salt layers induces a symmetric stress system. This promotes pop-up structures rather than asymmetric thrust faults. Salt extrusions are related to former salt ridges or salt walls squeezed by compression and dragged along thrust planes or to local low-pressure zones along crestal tear faults during folding. The salt that spreads out from the fault is rapidly dissolved. The resultant surface collapse structures are progressively filled by a mixture of Recent sediments and reprecipitated evaporites. Salt pinch-outs, either depositional or structural in origin, are a major controlling factor of the deformation geometry in fold and thrust belts. They trigger, either locally or regionally, contractional structures, including folds and thrusts, in rapidly prograding passive margins deforming by gravity gliding. In this structural context, salt pinch-outs also thicken due to differential loading and gravity spreading. The structural complexity in inverted grabens or in basement-involved orogenic belts where salt is present is the outcome of many factors. The salt thickness, the preexisting extensional structures, the synsalt and postsalt rifting, and the related distribution of older salt structures and sediments all localize folds and thrusts during later contraction. The relative orientation of the former extensional structures to the younger shortening structures largely controls the style of inversion (fault reactivation versus forced folding and short-cuts). Salt is the main detachment level between the folded cover rocks and the underlying faulted basement. However, secondary detachments, which are common in the overburden, add further complexities—triangle zones in the cores of anticlines and fish-tailed periclinal terminations.

INTRODUCTION

At shallow depths and low temperatures, rock salt and other evaporites are much weaker than other rocks. They flow at high strain rates when submitted to shear stress (Carter et al., 1993; Vendeville et al., 1995). Salt may flow by halokinesis even when no regional stresses are applied, either compressional or extensional. Three main models have been proposed for the driving mechanism of halokinesis (Jackson and Talbot, 1991):

1. buoyancy, relying on the density differential between salt and its overburden;
2. differential loading, where topography plays a major part; and
3. thermal convection (Talbot, 1978; Jackson and Galloway, 1984).

These mechanisms are still active when a regional stress regime is superimposed. We examine in this paper how the presence of interbedded salt or other evaporites affects the deformation style during regional contraction. The results of physical modeling are compared with field and subsurface examples.

Evaporitic basins that were later shortened are common; some examples of this type are shown in Figure 1. We shall investigate three distinct structural environments (Figure 2):

1. thin-skinned deformation found in fold and thrust belts;
2. shortening at the toe of prograding and gravity-gliding systems; and
3. intracratonic inverted basins, with or without components of wrenching.

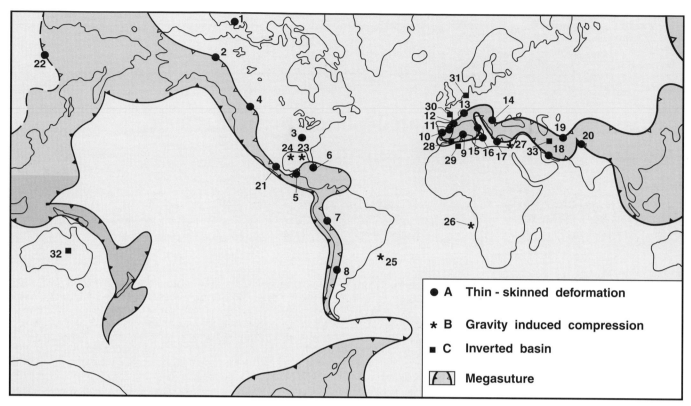

Figure 1—Location map of examples of compressive deformation involving a layer of evaporites, usually including salt. Key to numbered symbols (A, B, and C correspond to three parts of Figure 2): (A) Black circles (●), thin-skinned deformation in fold and thrust belts (modified from Davis and Engelder, 1985): 1, Parry Islands (Canada, salt age, Ordovician); 2, Franklins (Canada, Cambrian); 3, Appalachians (U.S.A., Silurian); 4, Wyoming (U.S.A., Upper Jurassic); 5, Guatemala (Middle Jurassic or Cretaceous?); 6, Cuba (Middle Jurassic); 7, Santiago (Peru, Permian); 8, Neuquen (Argentina, Jurassic–Cretaceous); 9, Rif-Tellian (Algeria, Morocco, and Tunisia, Triassic); 10, Betics (Spain, Triassic); 11, Iberics (Spain, Triassic); 12, Pyrénées (France and Spain, Triassic); 13, southern Alps, Jura (France, Triassic); 14, Carpathians (Romania, lower Miocene); 15, Apennines (Italy, Triassic); 16, Messinian (eastern Mediterranean, upper Miocene; 17, Mediterranean Ridge (eastern Mediterranean, upper Miocene); 18, Zagros (Iran, Infracambrian); 19, Tadjiks (C.E.I., Jurassic); 20, Salt Range (Pakistan, Infracambrian); 21, Sierra Madre Oriental (Mexico, Jurassic); 22, Verkhoyansks (C.E.I., Cretaceous). (B) Stars (*), shortening related to gravity gliding at the toe of massively prograding systems: 23, Mississippi Fan; 24, Perdido (Gulf of Mexico, Middle Jurassic); 25, Campos, Santos (Brazil, Lower Cretaceous); 26, West Africa (Angola, Gabon, Lower Cretaceous); 27, Nile delta, (Mediterranean, upper Miocene). (C) Black squares (■), intracratonic inverted basins: 28, Atlas (Algeria, Morocco, Tunisia, Triassic); 29, Triassic basin (Algeria, Triassic); 30, Aquitaine Basin (France, Triassic); 31, Southern North Sea (Permian); 32, Amadeus (Australia, Precambrian); 33, central Iran (Eocene–Miocene).

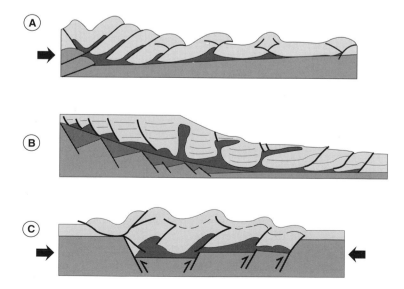

Figure 2—(A) Thin-skinned deformation in front of fold and thrust belt. Driving mechanism is the shortening in the inner foldbelt (A-subduction). Basement can be involved in the inner part of the fold and thrust belt, as shown here. (B) Gravity gliding-related shortening at the toe of a continental slope. The driving mechanism is both gravity spreading or gliding associated with the progressive tilting of the margin and gravity spreading of the salt layer under differential loading by prograding deposits. The basement is not involved in the deformation. (C) Intracratonic inverted basin. The driving mechanism is regional contraction orthogonal or oblique to the preexisting graben. The basement, salt, and overburden are all shortened together.

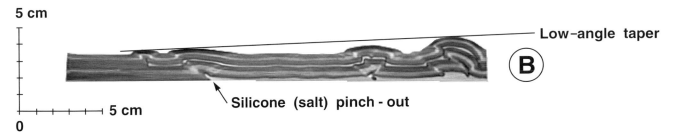

5 cm

0

Figure 3—Vertical sections (computerized tomographic scan images) of two experimental models showing the influence of basal friction on thrust imbrication and taper geometry. Amount of shortening is about the same in both experiments. (A) Purely brittle model composed of sand and Pyrex powder. The frontal pop-up structure is slightly tilted toward the foreland. Imbricate thrusts are stacked to create a high-angle taper. (B) Brittle and viscous model composed of sand, Pyrex powder, and silicone putty (SGM 36, Dow Corning). Viscous deformation at the base of the model induces a symmetric pop-up and a low-angle taper.

PHYSICAL EXPERIMENTS

Physical modeling has been used for a long time to simulate salt-induced structures. Institut Français du Pétrole (IFP) and its industrial partners Total and Elf-Aquitaine have systematically investigated the development of faults and related structures in shortened evaporitic basins using scaled experiments visualized by X-ray tomography.

To simulate geologic structures, we performed a series of sandbox experiments in a normal gravity field. Two kinds of analog materials were used: dry, granular materials to simulate brittle rocks and a viscous Newtonian material to simulate ductile rocks. Brittle analogs consist of dry sand and Pyrex powder with a grain size of 100 μm. Sand and Pyrex have Coulomb behavior, negligible cohesion, and internal friction angles of 31° and 41°, respectively. Their slight density contrast (1450 kg/m^3 for the sand versus about 1200 kg/m^3 for the Pyrex powder) is large enough to visualize distinct layers on the X-ray images (Figure 3). The ductile analog consists of silicone putty (PDMS, variety SGM 36 of Dow Corning) with Newtonian behavior at low strain rate, a viscosity of 10^{-4} Pa s, and a density of 970 kg/m^3. Computerized X-ray tomography applied to analog sandbox experiments allows the kinematic evolution and the three-dimensional geometry to be analyzed without interrupting or destroying the deforming models (Colletta et al., 1991). In granular materials, faults correspond to dilatant zones, which attenuate X-rays less and are thus clearly visible on computerized images. Apart from models simulating gravitational collapse (see below), all models were deformed by moving the vertical walls or the bottom of the box at a velocity of about 1 cm/hr. Distinct boxes and

boundary conditions were used to simulate the distinct tectonic regimes discussed in this paper. When boundary conditions are perfectly repeated, the resulting structures obtained in the deformed models are also repeated, showing that structures are not random but are strongly controlled by rheology and boundary conditions (Colletta et al., 1991).

THIN-SKINNED FOLD AND THRUST BELTS

In several mountain fronts (Figure 1), where the basement is not involved in the deformation, evaporite layers are preferred detachment levels (Davis and Engelder, 1985). Experiments (Figure 3) show that the deformation style in thin-skinned foldbelts critically depends on the resistance to sliding or, more specifically, on the friction coefficient along the detachment level (Mandl and Shippam, 1981; Mulugeta, 1988; Colletta et al., 1991; Huiqi et al., 1992; Calassou et al., 1993; Sassi et al., 1993). This dependence shows up both in the propagation of the deformation and in its style of folding.

Propagation of Deformation

Experiments demonstrate that the lower the basal friction coefficient, the farther the thrusts propagate outward, as predicted by the model of Davis et al. (1983). Low-friction detachments also result in lower taper angles and thinner cross sections (Figure 3). All these observations manifest the easy propagation along the

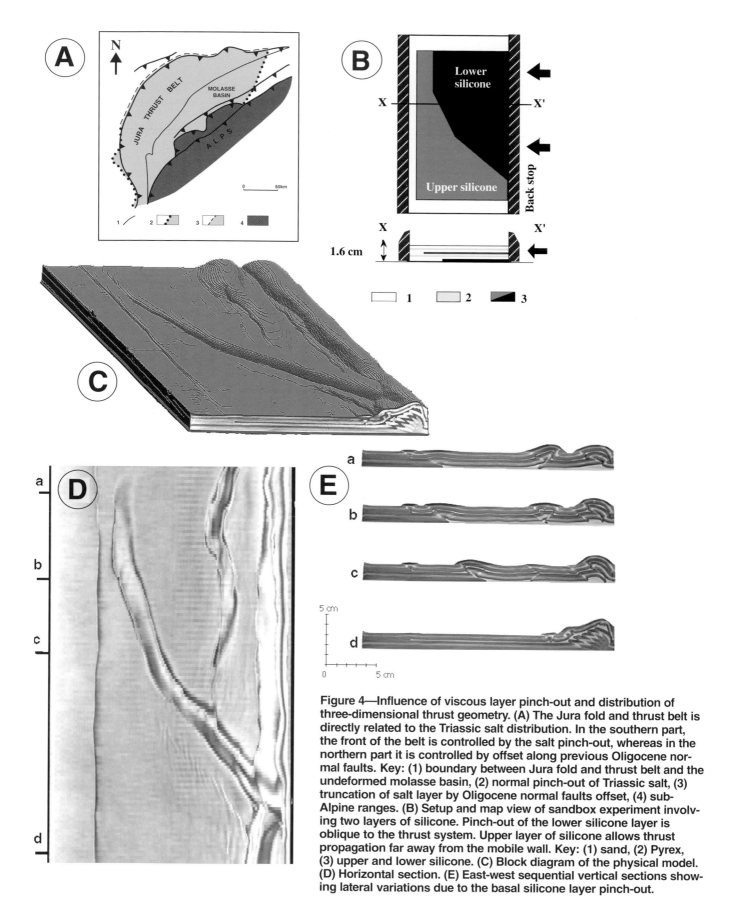

Figure 4—Influence of viscous layer pinch-out and distribution of three-dimensional thrust geometry. (A) The Jura fold and thrust belt is directly related to the Triassic salt distribution. In the southern part, the front of the belt is controlled by the salt pinch-out, whereas in the northern part it is controlled by offset along previous Oligocene normal faults. Key: (1) boundary between Jura fold and thrust belt and the undeformed molasse basin, (2) normal pinch-out of Triassic salt, (3) truncation of salt layer by Oligocene normal faults offset, (4) subAlpine ranges. (B) Setup and map view of sandbox experiment involving two layers of silicone. Pinch-out of the lower silicone layer is oblique to the thrust system. Upper layer of silicone allows thrust propagation far away from the mobile wall. Key: (1) sand, (2) Pyrex, (3) upper and lower silicone. (C) Block diagram of the physical model. (D) Horizontal section. (E) East-west sequential vertical sections showing lateral variations due to the basal silicone layer pinch-out.

low-resistance salt layer. Thus, folding and thrusting take place over a wide belt; shortening is spread over a large distance rather than concentrated (Figure 4).

A foldbelt with a low-angle taper projects far outward, as exemplified by the Jura fold and thrust belt. Located in front of the main Alps, the Jura deformation is largely controlled by the shape and thickness of an underlying Triassic salt basin (Figure 4A) (Philippe, 1994). The deformation front corresponds to the salt edge, which is a depositional pinch-out to the south but is of tectonic origin to the north, where the salt layer is truncated by Oligocene normal faults on the border of the Bresse graben.

Style of Folds

Our experiments also confirm the Hafner model (Hafner, 1951), which states that the bisector of the two shear planes associated with horizontal compression must tilt forward as basal friction increases, thus preventing backthrusting from propagating on a large scale (Figure 3). Where basal friction is low, this bisector remains nearly vertical and thrusts have no preferred vergence. Thus, both backthrusts and forethrusts develop freely, leading to overall symmetry. If secondary detachments (other evaporites or shale) overlie the main salt, blind thrusts, triangle zones, and pop-up structures are likely to form. These characteristics are clearly illustrated in the physical models (Figures 4B–E, 5) and are widely documented by field examples, which we discuss next.

Field Examples

Parry Islands Foldbelt (Canadian Arctic)

This is a well-studied example of a foldbelt detached over a salt layer (Harrison and Bally, 1988; Harrison, 1995). In the main detachment, provided by Ordovician halite-bearing evaporites originally up to 800 m thick, significant diapirism is absent. Evaporites tended to migrate toward the anticlinal cores as a consequence of long-wavelength buckling. Evaporites structurally thicken to up to 2200 m in the core of the anticlines and thin to less than 60 m below the synclines. Anticlines are long and narrow, having steep, symmetric limbs, whereas synclines are broad and flat-bottomed. Seismic data show that the anticlines are complex at depth, a secondary décollement within overlying Eifelian mudstones promoting pop-up and "fishtail" structures.

Jura (France and Switzerland)

The previously mentioned Jura mountains broadly fit the same picture over a Triassic evaporitic complex (Muschelkalk and Keuper, from the classic German Triassic succession) that includes anhydrite and salt (Laubscher, 1977). Box fold anticlines and broad flat-bottomed synclines are characteristic. Several shale layers in the sedimentary sequence above the salt provide secondary detachments and disharmony that explain the steeply dipping limbs of the folds (Philippe, 1994).

Zagros and Fars (Iran)

These folds are again box shaped and symmetric. The main detachment is in the deep Hormuz Salt of late Precambrian age (Farhoudi, 1978). Miocene evaporites provide secondary décollements, where pairs of shallow thrusts root with opposite vergence toward the crest of the anticline, as shown in the models of Figures 4E and 5.

Unlike the two previous areas, piercing diapirs are common in Iran because of a probably thick original salt and deep burial, which lowered viscosity by increasing the temperature. The mechanism that pushed the salt upward through thick overlying sediments is still unclear. However, diapiric evolution after salt flow into the anticline cores and thrusts, or salt extrusions into tear faults, which have been observed in the Atlas ranges of Algeria (Vially et al., 1994) (see Figure 17), may be the cause here.

Salt Injection

In all the experiments of thrust propagation over a silicone layer, the silicone was dragged along the thrust planes but stayed in normal stratigraphic contact with the overlying thrust sheet. The same observations are often made in natural settings (e.g., the Salt Ranges of Pakistan; Baker et al., 1988). However, in the Carpathians of Romania, structures detached over Miocene salt, which is known from subsurface data to be injected within younger sediments along thrust planes (Paraschiv and Olteanu, 1970). Similar features are observed in the sub-Alpine ranges of southern France, where Triassic gypsum outcrops in anomalous contacts with both the hanging walls and footwalls of thrusts. It is not known whether this is true injection related to compression or the deformed expression of previous diapirism.

SHORTENING ASSOCIATED WITH GRAVITY GLIDING OF PROGRADING SYSTEMS

Rapidly prograding systems undergoing upslope extension induced by gravity gliding are fairly common in passive margins. Shortening zones located at their toes are associated with either undercompacted shales related to rapid progradation (e.g., Niger delta, Tarakan delta offshore eastern Borneo, and offshore northeastern Brazil) or salt (Figures 1, 2). The best-documented cases of the latter come from the Gulf of Mexico (Perdido and Mississippi Fan foldbelts), but examples are also known from the South Atlantic margins (Campos and Santos basins, offshore Brazil; Cobbold et al., 1995; Mohriak et al., 1995; and West Africa, north of the Walvis Ridge) and from the eastern Mediterranean (Nile deep sea fan) (see also Heaton et al., 1995, on Yemen). Contraction is driven by gravity gliding associated with the progressive tilting of the margin and by gravity spreading of the salt layer under the differential loading related to its progradation.

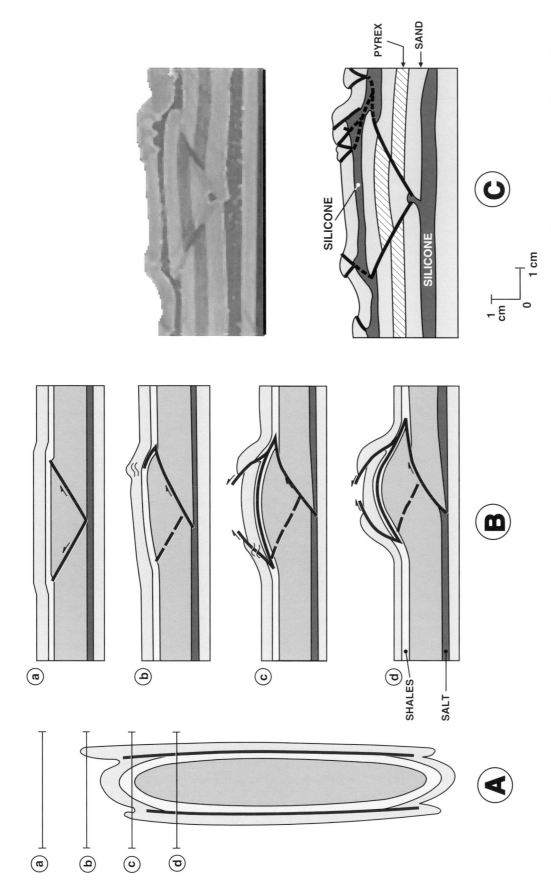

Figure 5—Detachment along the salt layer and along secondary detachments. (A) Schematic map and (B) four sections illustrating the fold style in the Atlas mountains of Algeria. Anticlines are commonly tight and symmetric with small thrusts verging toward the anticlinal hinge. After sufficient erosion, the core of the anticline shows complex thrust imbrications. Notice the absence of fault-propagation folding and thrust emergence in the synclines flanking the anticline. Shale layers in the overburden (Upper Jurassic in western Atlas and Upper Cretaceous in eastern Atlas) induced secondary décollements during folding and enabled imbrications, underthrusting, and pop-up structures. The emergence of small backthrusts creates fish-tail structures in the periclinal termination of the anticlines (Vially et al., 1994). (C) Physical modeling of a pop-up structure rooted on the main décollement. Its size is proportional to the thickness of the uplifted overburden. The two boundary faults curve upward into the upper décollement. Further shortening is accommodated by minor thrusts that develop pop-up structures.

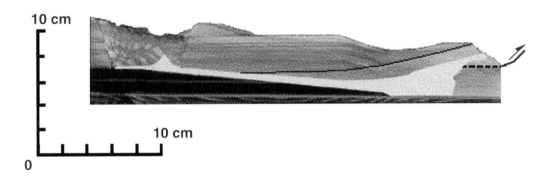

10 cm

10 cm

0

Figure 6—Vertical sections (computerized tomographic scan images) of a physical model simulating gravitational deformation of a brittle over-burden above a viscous layer.

In all these areas, salt was deposited during the postrift stage, so that thermal subsidence results in a basinward tilt after salt deposition. This tilt is only a few degrees (1°–3°) at the décollement in these areas and is not, by itself, the driving mechanism. Instead, deformation is driven by the general subsidence and progradation, which maintains a surface slope.

Experimental Results

Model simulations of gravity gliding have been published by Cobbold et al. (1989). Their experiments, where silicone putty was used to simulate salt and dry sand was used to simulate sediments, were designed to model a continental margin. The rigid base of the model was made of an upper (platform) and a lower (basin) flat linked by a sloping plane tilted 4° toward the base. Sediments were not added during deformation. The experiment resulted in extension near the upper hinge line and folds and thrusts in a local compressive regime near the base of the slope. The central domain glided downslope without being deformed.

Only a few gravity-gliding experiments simulating both extensive and contractional domains have been completed at IFP because the size of the experimental box is strongly limited by the capacity of the scanner. The experiment shown in Figure 6 featured aggradation and progradation while deformation was taking place. In the final deformed state, extension was limited to a narrow domain near the updip silicone pinch-out. The central part glided basinward without deforming (as in Cobbold et al., 1989).

However, such undeformed zones that have merely glided downslope are almost unknown from the seismic record. Extension is generally widespread down to the abyssal plain. More work is necessary to bridge the gap between experiments and observation.

Gravity spreading is illustrated in the Figure 6 experiment by the thinning of the silicone layer below the prograding infill and thickening in the basinal area. Gravity sliding of the sediment pile also contributed to this thickening. The downslope silicone putty edge was marked by a flexure in the overlying sand and a scarp at the surface. A silicone tongue formed above a thrust fault in the youngest sand layers.

Field Examples

Offshore Brazil

In the Santos and Campos basins of offshore Brazil, gravitationally induced salt flow and gravity gliding over a salt layer of Aptian age dominated postrift structural evolution (Cobbold et al., 1995; Mohriak et al., 1995). Thin-skinned extension is widely distributed over the Campos shelf and slope. However, in the northwestern Santos Basin, they are located in a narrow strip along the upslope salt pinch-out and along a large growth fault at the toe. In between, the Santos Basin shows some evidence of the thick Mesozoic and Cenozoic cover of the salt being slightly translated downward as a large, basically undeformed slumping sheet. The Aptian salt layer is currently very thick in the deeper part of both basins and below the abyssal plain of the São Paulo Plateau (Mohriak et al., 1995). There, shortening is demonstrated by the presence of thrust faults and growth folds (Cobbold et al., 1995). The basinward limit of the salt is evidenced by a large salt tongue. This limit corresponds to a prominent scarp on the sea bottom.

West African Margin

On the other South Atlantic margin, in the Congo and Kwanza basins, the thin-skinned décollement started over the same Aptian salt as soon as the first overlying sediments were deposited (Duval et al., 1992). Present extension rates can locally reach values as high as 800%. This extension is locally accommodated on the platform itself by reverse faulting, where the translation is blocked by basement paleohighs (e.g., at Cabo Ledo; Duval et al., 1992), and is regionally accommodated by salt ridges and thrust faults downslope. The basinward limit of the salt at the toe of the Angolan margin is marked by a scarp on the sea bottom. Data from unpublished oceanographic seismic profiles (CEPM) suggest the presence of salt sheets and shortening features here.

Eastern Mediterranean

In the eastern Mediterranean, the gravity sliding and extensional structures to the north of the Nile delta face the contractional structures of the Mediterranean Ridge (Sage and Letouzey, 1990). Shortening appears to be partially accommodated by broad folds cored by salt ridges. South of Eratosthenes seamount, the basinward limit of the salt is locally marked by a sea floor scarp that is possibly related to a large underlying salt stock. This feature may be interpreted as the result of piling up of the salt where movement is blocked by the relief of the seamount.

Perdido and Mississippi Fan

The Perdido and Mississippi Fan foldbelts of the Gulf of Mexico are the best documented examples of downdip contractional domains related to salt pinch-out (Jackson and Galloway, 1984; Worrall and Snelson, 1989; Wu et al., 1990a,b; Ewing and Lopez, 1991; Galloway et al., 1991; Weimer and Buffler, 1992; Wu, 1993; Diegel, 1995; Peel et al., 1995; Schuster, 1995).

According to Wu (1993), the Perdido foldbelt was formed in deep water as a response to updip loading by rapidly prograding Eocene and Oligocene depocenters when movements were blocked along the basinward edge of the Middle Jurassic salt. The Mississippi Fan foldbelt was formed in the same way during middle Miocene time when the depocenter shifted eastward. In both belts, thrusting was coeval with major extension, growth faulting, and massive salt diapirism updip on the shelf and slope. The advancing wedge of sediments progressively squeezed seaward an increasing mass of underlying salt, which accumulated at the toe of the slope, whereas contractional structures formed at the salt pinch-out. When the salt mass evolved to an allochthonous nappe within the younger sediments, thrust faults at the front ceased being active.

One question remains open: what was the role of shortening in the rise of the salt and formation of the allochthonous salt sheets? Was the shortening associated with the downdip glide of the sediments able to contract and squeeze the former diapiric salt stocks and overthrust the salt within unconsolidated sediments? Or was rise of the salt primarily due to extension and differential loading?

Role of Downdip Salt Pinch-Outs

The previous examples show that the basinward salt pinch-out plays a most important part in forming these frontal contractional structures by increasing the friction coefficient at the base of the sedimentary pile and thus preventing further translation. On the contrary, basins such as the Western Mediterranean, where the salt does not pinch out but rather extends across the basin, do not show any evidence of thrust-faulted zones (Figure 7). Updip extension over the Messinian salt is accommodated at the toe of the Rhone delta by buckling and folding above salt ridges (Biju-Duval et al., 1979).

INVERTED GRABENS

Graben inversion is common in intracratonic basins. Under regional shortening that is orthogonal or oblique to preexisting structures, inversion typically squeezes the hanging wall blocks upward. The basement, salt, and overlying sediments are all shortened (Figures 1, 2). The examples presented here have roughly the same tectonic history: the salt was deposited within the prerift to synrift series and was further rifted prior to being buried under thick postrift sediments and later inverted. Examples include the southern North Sea, western Alps, and Saharan Atlas.

However, graben reactivation in the presence of salt is not restricted to intracratonic grabens. In orogenic belts along former passive margins, grabens belonging to the rift stage of the margin can also be inverted. The example of the western Alps has been described by several authors (Lemoine et al., 1986; Gillcrist et al., 1987; De Graciansky

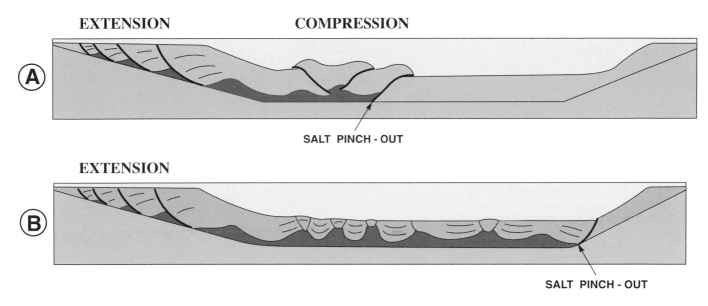

Figure 7—Basinward salt pinch-out controls the development of contractional structures, especially in gravity-driven, pro-grading systems. (A) Increasing frictional resistance at the base of the sedimentary pile prevents further basinward translation of the sediments. Folds and thrusts form along the salt pinch-out. (B) Conversely, in basins where the salt extends across the whole basin, shortening is distributed by buckle folds having collapsed crests without thrust zones.

et al., 1989). These orogenic belts most likely initially evolved as described below. The older structures were then exaggerated and transformed by further shortening.

Field Examples

Southern North Sea

Seismic data across the southern North Sea grabens suggest that the amount of shortening and inversion strongly depends on the orientation of the former graben (Voigt, 1962; Ziegler, 1975, 1989, 1990). The Broad Fourteens Basin is a northwest-striking Permian–Jurassic graben inverted in the Late Cretaceous–early Tertiary (Remmelts, 1995). The graben extensional faults were reactivated by north-south to NNE-SSW compression (Letouzey, 1986), forming en echelon folds and thrusts along the inverted graben boundaries. Those thrusts are rooted at the salt décollement (Van Wijhe, 1987a,b; Roelofsen and De Boer, 1991).

Where the salt layer is absent (to the south and west of this basin), inversion of northwest-striking grabens resulted in forced folds and flower type structures in the cover located above deep basement faults (Badley et al., 1989; Dronkers and Morozek, 1991). The salt ridges and growth faults formed during early extension control the location of the subsequent low-angle thrusts. Even during contraction, diapirism locally induces crestal faults above salt walls or stocks and tilting of the basin causes growth faults (Hayward and Graham, 1989; Roelofsen and De Boer, 1991; Huyghe, 1992). Uplift and erosion locally exceed 2500 m, for an estimated shortening of 10–12% (Huyghe, 1992).

The nearby Sole Pit High is also an inverted and folded graben, superposed on the same northwest-trending

Permian–Jurassic direction (Glennie and Boegner, 1981; Van Hoorn, 1987). Uplift is about 1500 m at the core of the fold. The main difference with the previous basin is that the salt layer is thick and extends beyond the graben.

Deformation within the cover took place mostly near the southern border of the Jurassic basin along the deep basement fault. Rocks flexed without apparent faulting at seismic scale. The large amount of uplift and folding shows that, below the graben infill, the basement rocks were shortened even though the basement faults retained normal offset. North of the Sole Pit High, diapirs were remobilized during inversion. Regionally, uplift and inversion were much weaker in grabens that strike at a low angle to the regional compression, such as the Dutch Central, Horn, and Gluckstadt grabens (Heybroek, 1975; Ziegler, 1990), than in the Sole Pit High or Broad Fourteens Basin, which strike at high angles to the direction of compression.

Atlas Foldbelt

The Atlas foldbelt, which extends from the Hauts Atlas in Morocco to the Tellian Atlas in Tunisia, was initially a thick intracratonic graben in which up to 8 km of marine Mesozoic sediments were deposited above thick Upper Triassic salt. Rifting and block faulting were already active during salt deposition and controlled the evaporite thickness and facies (Figure 8) (Vially et al., 1994). Seismic data indicate that salt flow and diapirism were limited during the Mesozoic and were partially induced by underlying basement faulting. In contrast, on the Saharan platform, the salt remains nondiapiric in the "Triassic basin" in spite of a thickness that reaches 1500 m in some areas and later contraction (Boudjema, 1987).

During the Tertiary, the African and European plates

Figure 8—(Top) Northern Algeria tectonic map showing extension of the Triassic–Lower Jurassic salt (after Vially et al., 1994). Key to map: A–A', B–B', and C–C' show locations of sections below; (D) location of Figure 14A; (E) location of Figure 17B; (F) location of Figure 18B; (G) location of Figure 18A; (H) Triassic and Jurassic grabens inverted during the middle Cretaceous (Boudjema, 1987). (Bottom) A–A', B–B', and C–C' are schematic regional sections (locations shown on map).

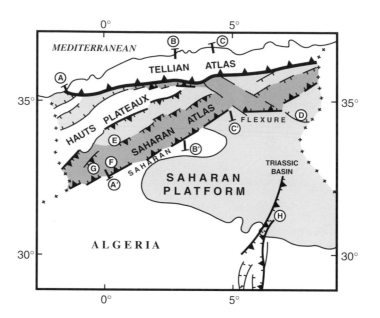

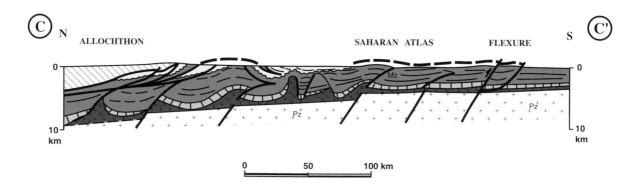

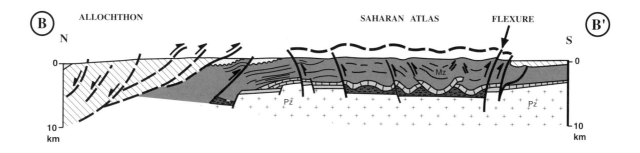

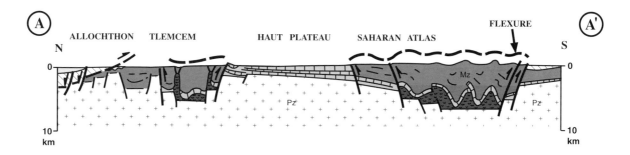

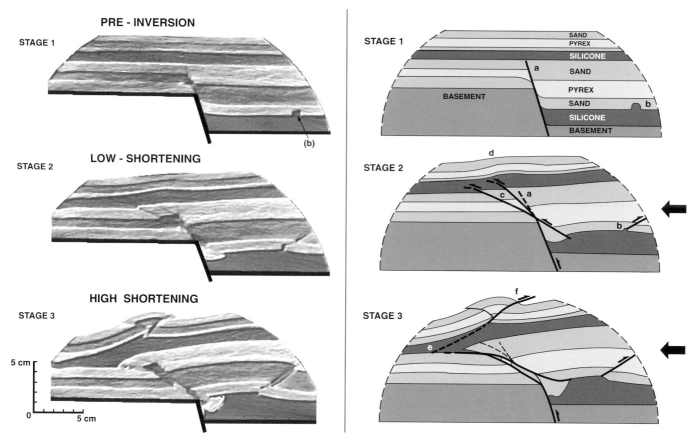

Figure 9—(Left) Computerized tomographic scan images and (right) interpreted drawings of vertical sections through a physical model simulating a normal fault at three stages of inversion. The upper viscous layer induces pop-up structures and backthrusting in the uppermost brittle unit. Stage 1 is preinversion. Stage 2 shows inversion of basement fault and shortening of the hanging wall, which induce low-angle reverse faults (c). These create a short-cut fault across the upper part of the footwall block and the lower part of the hanging-wall block. The original normal fault (a) is not reactivated. Squeezing of the small diapir (b) nucleates a backthrust. In stage 3, increased shortening induces a pop-up structure (e) and backthrusting in the uppermost brittle unit.

converged and progressively inverted the former extensional structures. The main contraction took place during the Neogene; Triassic salt acted as the main décollement over the whole area. In the north, the Tellian domain is a fold and thrust belt along the former Tethyan margin. South of the Hauts Plateaux, the Atlas domain is a typical folded and uplifted inverted graben (Vially et al., 1994). In eastern Algeria and Tunisia, these two domains merge and result from both basement inversion and thin-skinned contraction above the salt (Figure 8).

Modeling

Modeling with suitable detail the structural complexity of the southern North Sea or the Atlas Mountains is still difficult. However, one can use simple sandbox experiments to explore the relative influence of each parameter on the structural style. Publications on inversion (Koopman et al., 1987; McClay, 1989; Buchanan and McClay, 1991) almost never deal with either oblique or salt-related inversion.

Several experiments were carried out to investigate how displacement along a steeply dipping preexisting fault influences the overlying structures. For simplification, in most of these experiments, basement deformation was simulated by rigid body translation along the faults (their dip was fixed at 70° in all the experiments presented in this paper). Subsidence of the block associated with sand sedimentation first simulated extension and normal faulting. Either reverse dip-slip or oblique-slip along these simulated basement faults was then applied to the hanging wall blocks; shortening was induced in the cover by means of a vertical backstop.

Influence of Dip and Obliquity of Preexisting Normal Faults

Several experiments to analyze the reactivation of faults quantitatively by compression were carried out at IFP prior to our study of inverted structures. Precut faults with variable dips and orientations were introduced in sand models. Frictional resistance was significantly reduced along such precut discontinuities to about 50% of

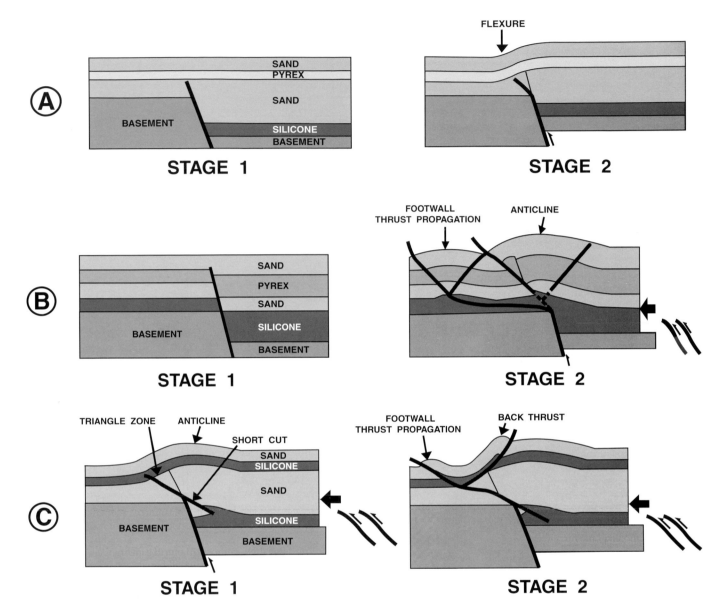

Figure 10—Schematic evolution of inverted normal faults in three distinct physical models. (A) Simple inversion of basement normal fault induces a new low-angle reverse fault in the brittle overburden and flexure. The new reverse fault branches off the older normal fault, which is partially reactivated. (B) Inversion of the basement normal fault and shortening of the hanging wall block occurs with a continuous silicone detachment above the basement. The older normal fault is not reactivated but is passively transported and rotated by new thrusts and pop-up structures. (C) Inversion of basement normal fault and shortening of the hanging wall block having two silicone detachments produce a new low-angle reverse fault (short cut). The older normal fault is not reactivated and is replaced by a short-cut fault. After shortening and inversion increase, a pop-up structure, forethrust, and backthrust can develop above the uppermost silicone detachment.

its value in homogeneous sand. Precuts also have the same mechanical properties as faults created in sand by shear dilatancy. These experiments show only faults with orientations that are similar to those that would be naturally created by superposed compression and would likely be reactivated. Precut faults in sand with dips higher than 60° and lower than 10° and faults with an obliquity lower than 30° to the compression direction were never reactivated (Sassi et al., 1993). The Mohr-Coulomb analysis of fault reactivation experiments also shows that the

angular domain of reactivation varies with the stress ratio and increases when the friction on the plane of weakness is lowered (e.g., by increasing fluid pressure). Interactions between faults were also important in fault reactivation (Sassi et al., 1993).

In most of the inversion experiments, the Mohr-Coulomb behavior of the material prevented steeply dipping normal faults within the sedimentary cover from being reactivated by contraction. These steep faults were generally passively transported over new reverse faults

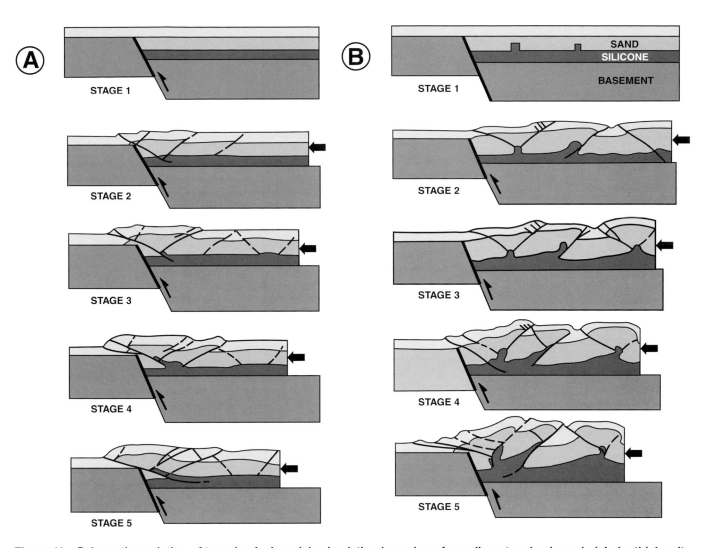

Figure 11—Schematic evolution of two physical models simulating inversion of a sedimentary basin underlain by thick salt. (A) With a tabular silicone layer, the basement fault controls the contractional structures, which propagate backward (to the right). The first low-angle reverse fault develops as a short-cut structure. (B) The location and development of contractional structures are directly related to the squeezing of preexisting silicone diapirs.

(e.g., see the faults bounding the grabens in Figures 9 and 10 and the collapse graben above the silicone ridge in Figure 16). Exceptions to this rule are usually observed in the following special settings:

- oblique shortening (less than 30° from the original fault direction in experiments),
- tilting of the preexisting normal fault plane during deformation, and
- forced slip along the fault when large hanging wall uplifts are formed.

Orthogonal Inversion

The previous observations explain why, in natural examples where the horizontal compressive stress directions are nearly orthogonal to the extensional structures and where these are bounded by high-angle normal faults, much of the inversion is accommodated by forced folding rather than by slip along the former faults

(Letouzey, 1990). New low-angle reverse faults rooted in the décollement and originating at the crest of the upthrown basement block passively transport the former high-angle normal fault planes. Short-cut geometries are thus initiated (Figures 9, 10, 11, 12). Seismic profiles over the contact zone between the Saharan Atlas and the Hauts Plateaux commonly show short-cut structures. Here the direction of the former normal faults is orthogonal to compression (Vially et al., 1994).

Oblique Inversion

If the horizontal compression is oblique to the former normal fault, a strike-slip component of motion is induced that allows former high-angle normal faults to be reactivated (Letouzey, 1990; Letouzey et al., 1990). In addition to conventional inversion features such as forced folds and short cuts over basement faults, strike slip will form en echelon synthetic faults in the hanging walls of preexisting basement faults (Figure 13). In Algeria, the

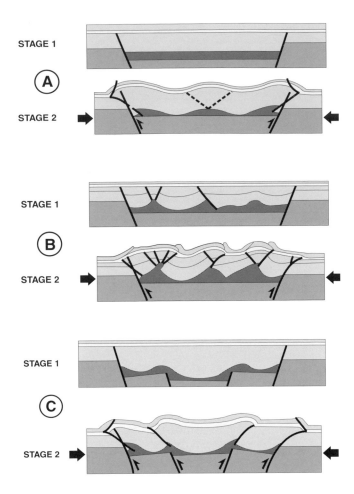

Figure 12—Schematic evolution of an inverted basement graben containing salt. (A) Folds and short-cut faults develop in homogeneous strata above the inverted basement faults. Fold hinges in the graben are perpendicular to the regional compression. (B) Thrusts and folds are localized by preexisting salt structures. (C) Thrusts and folds are localized by reactivation of basement faults.

Saharan flexure shows such en echelon faulting where the regional shortening was oblique to the deeper normal faults (Figure 14) (Vially et al., 1994).

In the Atlas Mountains of Algeria, oblique tectonic inversion explains the location and orientation of the major structures, especially on the basin edges such as the Saharan flexure and its associated folds. But inside the Atlas domain itself, seismic sections commonly show salt pillows concordant to bedding and anticlines having a constant thickness of Mesozoic section above the Triassic salt décollement. Folds of this kind, which are not caused by basement faulting or diapirism, trend orthogonally to the regional Neogene compression (Figure 15) (Vially et al., 1994).

Inversion of Diapirs and Salt-Induced Structures

Physical models suggest that precontractional salt tectonics and diapirism influence the location of thrusts and folds. After salt deposition, renewed extension on the underlying basement faults causes salt to flow toward the footwall crests, especially to the salt ridges above footwall blocks (Vendeville, 1987; Vendeville and Jackson, 1992a,b; Nalpas and Brun, 1993). During contraction and inversion, folds and thrusts are initiated where thickness variations are present in the overburden (Figures 10B, 12B, 16A). Such phenomena are illustrated by the thrust faults and structures fringing the Broad Fourteens Basin (Hayward and Graham, 1989; Huyghe, 1992). Thinning of the overburden over salt ridges or salt walls localizes thrusts and folds during contraction even when they are not related to basement normal faults (Figures 11B, 16A). This localization only occurs in the experiments when the preexisting diapirs are favorably oriented to the contraction. Moreover, isolated symmetric salt domes do not seem to influence thrust locations (compare with Nilsen et al., 1995). Normal fault offsets on the top of the salt can also localize thrusts (Figure 16B).

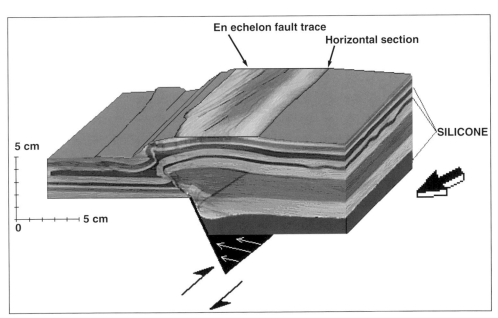

Figure 13—Block diagram built from computerized tomographic scan images of a physical model of oblique inversion along a basement fault. White arrows show slip directions. Note the en echelon faults in the arched uppermost brittle layer, which contains two secondary detachments.

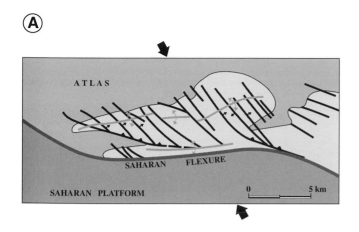

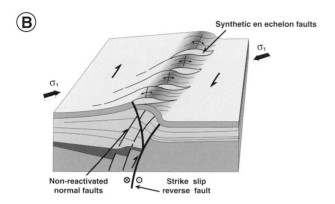

Figure 14—Example of structures related to oblique inversion. (A) Map view of the structures along the Saharan flexure between Biskra (Algeria) and the Tunisian border (see Figure 8 for location). The northwest-southeast Neogene compression (large arrows) was oblique to the Mesozoic normal faults. A wrenching component during inversion caused flexure and folding. On the scale of the structure, the hinges of the anticlines are crosscut by oblique-striking faults with oblique slip. Seismic profiles elsewhere along the Saharan flexure indicate that the Cretaceous normal faults were deformed but not reactivated. (B) Schematic block diagram of the oblique inversion structures shown above (compare to Figure 13).

The experiment shown in Figure 11B suggests that salt ridges or salt walls can be squeezed during contraction. Salt is then dragged along the thrust planes and soles of the thrust sheets. This process, combined with erosion and salt extrusion, initiates collapse structures along thrust faults in the Atlas Mountains (Figure 17). Here, some salt extrusions are also related to crestal tear faults or to wrench faults (Figure 18).

Secondary Disharmony Levels

As for any foldbelt detached on salt, the structural complexity in inverted grabens is influenced by sec-

ondary detachments. For example, multiple detachments explain structural differences between the eastern and western domains of the Saharan Atlas in Algeria (Vially et al., 1994).

An upper décollement in the experiments (Figures 5, 9) promotes backthrusting and pop-up structures, which allow flexures to develop above blind basement faults (e.g., Saharan flexure; Vially et al., 1994). Secondary detachments also influence fold geometries within the inverted graben, such as in the Hauts Atlas domain. These folds are several kilometers long and pop-up shaped in map view. Smaller backthrusts emerge near the crest and induce fish-tailed periclinal terminations (like those in Figure 5).

CONCLUSIONS

Salt layers directly influence the tectonic style of contractional structures. Analysis of field and subsurface data and physical models elucidates contractional salt tectonics involving inversion, oblique slip, and thin-skinned shortening.

In contractional regimes, the geometry of thrust belts is controlled by the weak salt décollement. Because of the very low strength of the salt layers, deformation in the overlying more competent beds can be transmitted far away from zones of crustal shortening, inducing wide fold and thrust belts having a very low angle taper. Symmetric pop-up structures develop above nearly frictionless salt strata. Thrust sequence is partially controlled by the salt layer. Older structures in front of the system are overtaken by out-of-sequence faulting and folding as shortening increases. Contractional salt tectonics may also be driven by gravity at the foot of prograding continental slopes. Differential loading and gravity spreading cause local salt thickening. Salt tongues associated with thrust faults can develop.

During graben inversion, weak salt layers induce complex structures such as pop-up and fish-tail structures. The distribution of older salt structures and synrift sediments influences the location and geometry of thrusts and folds during later shortening. Contrasted rheology in the sedimentary pile generally increases the complexity of contractional structures, inducing major disharmony between the suprasalt and subsalt sediments.

Acknowledgments *This study is part of a research program on the modeling of complex structures jointly supported by IFP, Total, and Elf Aquitaine. Jean-Marie Mengus and Pascal Balé were very helpful in conducting the experiments, and Patrick Le Foll ably drafted the diagrams. Corinne Ledard and Valérie Lacoste are thanked for typing the manuscript, which was edited by Martin Jackson.*

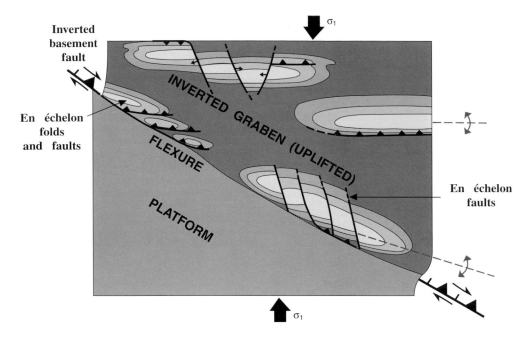

Figure 15—Summary sketch of the structures formed by oblique inversion, exemplified by the Saharan flexure separating the Atlas domain, northeastern Algeria, and Saharan platform. The Atlas domain is a typical folded and uplifted inverted graben above thick salt. In addition to conventional inversion structures such as forced folds over the basement faults (Saharan flexure), a strike-slip component forms en echelon folds in the footwalls of basement faults. Fold axial traces are slightly oblique to the flexure and are crosscut by synthetic crestal tear faults. Within the Atlas domain, folds that are unrelated to basement fault inversion or diapirism trend perpendicular to the regional Tertiary compression.

REFERENCES CITED

Badley, M. E., J. D. Price, and L. C. Backshall, 1989, Inversion reactivated faults and related structures: seismic examples from the southern North Sea, *in* M. A. Cooper and G. D. Williams, eds., Inversion tectonics: Geological Society of London Special Publication, v. 44, p. 201–219.

Baker, D. M., R. J. Lillie, R. S. Yeats, G. D. Johnson, M. Yousuf, and A. S. H. Zamin, 1988, Development of Himalayan frontal thrust zone: Salt Range, Pakistan: Geology, v. 16, p. 3–7.

Biju-Duval, B., J. Letouzey, and L. Montadert, 1979, Variety of margins and deep basins in the Mediterranean, *in* J. S. Watkins, L. Montadert, and P. N. Dickerson, eds., Geologic and geophysical investigations of continental margins: AAPG Memoir 29, p. 293–317.

Boudjema, A., 1987, Evolution structurale du bassin pétrolier "Triasique" du Sahara Nord Oriental (Algérie): Thèse, Universitie de Paris XI, Orsay, 290 p.

Buchanan, P. G., and K. R. McClay, 1991, Sandbox experiments of inverted listric and planar fault systems: Tectonophysics, v. 188, p. 97–115.

Calassou, S., C. Larroque, and J. Malavieille, 1993, Transfer zones of deformation in thrust wedges: an experimental study: Tectonophysics, v. 221, p. 325–344.

Carter, N. L., S. T. Horseman, J. E. Russell, and J. Handin, 1993, Rheology of rocksalt: Journal of Structural Geology, v. 15, p. 1257–1271.

Cobbold, P., 1995, seismic and experimental evidence for thin-skinned horizontal shortening by convergent radial gliding on evaporites, deep water Santos Basin, Brazil, *in* M. P. A. Jackson, D. G. Roberts, and S. Snelson, eds., Salt tectonics: a global perspective: AAPG Memoir 65, this volume.

Cobbold, P. R., E. Rossello, and B. Vendeville, 1989, Some experiments on interacting sedimentation and deformation above salt horizons: Bulletin de la Société Géologique de France, v. 3, p. 453–460.

Colletta, B., J. Letouzey, R. Pinedo, J. F. Ballard, and P. Bale, 1991, Computerized X-ray tomography analysis of sandbox models: examples of thin-skinned thrust systems: Geology, v. 19, p. 1063–1067.

Davis, D. M., and T. Engelder, 1985, The role of salt in fold and thrust belts: Tectonophysics, v. 119, p. 67–88.

Davis, D. M., J. Suppe, and F. A. Dahlen, 1983, Mechanics of fold and thrust belts and accretionary wedges: Journal of Geophysical Research, v. 88, p. 1153–1172.

De Graciansky, P. C., G. Dardeau, M. Lemoine, and P. Tricart, 1989, The inverted margin of the French Alps and foreland basin inversion, *in* M. A. Cooper and G. D. Williams, eds., Inversion tectonics: Geological Society of London Special Publication, v. 44, p. 87–104.

Diegel, F. A., J. F. Karlo, D. C. Schuster, R. C. Shoup, and P. R. Tauvers, 1995, Cenozoic structural evolution and tectonostratigraphic framework of the Northern Gulf Coast continental margin, *in* M. P. A. Jackson, D. G. Roberts, and S. Snelson, eds., Salt tectonics: a global perspective: AAPG Memoir 65, this volume.

Dronkers, A. J., and F. J. Morozek, 1991, Inverted basins of the Netherlands: First Break, v. 9, p. 409–425.

Duval, B., C. Cramez, and M. P. A. Jackson, 1992, Raft tectonics in the Kwanza Basin, Angola: Marine and Petroleum Geology, v. 9, p. 389–404.

Ewing, T. E., and R. F. Lopez, 1991, Principal structural features, Gulf of Mexico Basin, *in* A. Salvador, ed., The Gulf of Mexico Basin: the geology of North America: GSA, v. J, p. 31–52.

Farhoudi, G., 1978, A comparison of Zagros geology to island arcs: Journal of Geology, v. 86, p. 323–334.

Galloway, W. E., D. G. Bebout, W. L. Fisher, J. B. Dunlap, Jr., R. Cabrera-Castro, J. E. Lugo-Rivera, and T. M. Scotte, 1991, Cenozoic, *in* A. Salvador, ed., The Gulf of Mexico Basin: the geology of North America: GSA, v. J, p. 245–324.

Gillcrist, R., M. Coward, and J. L. Mugnier, 1987, Structural inversion and its controls: examples from the Alpine foreland and the French Alps: Geodinamica Acta, v. 1, p. 5–34.

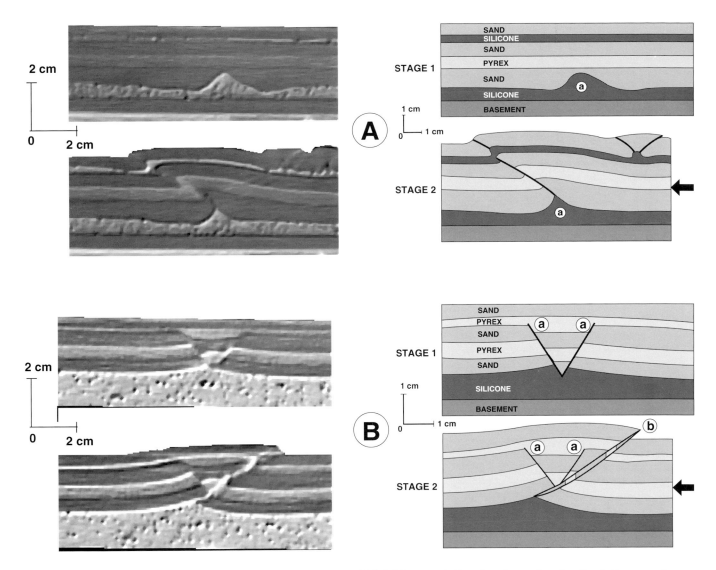

Figure 16—(Left) Computerized tomographic scan images and (right) interpretative drawings showing the influence of pre-existing structures within the overburden of physical models. (A) Thinning of the overburden over salt ridges or salt walls, or (B) steps at the top of the salt due to normal faulting, localize low-angle thrusts and pop-up structures during compression, where the preexisting structures are perpendicular or oblique to the shortening direction. Nevertheless, steeply dipping normal faults within the overburden (e.g., crestal graben above salt ridge) and isolated salt domes were not reactivated but instead were passively transported above low-angle thrusts.

Glennie, K. W., and P. L. E. Boegner, 1981, Sole Pit inversion tectonics, *in* L. V. Illing and G. D. Hobson, eds., Petroleum geology of the continental shelf of north-west Europe: London, Heyden and Son, p. 110–120.

Hafner, W., 1951, Stress distributions and faulting: GSA Bulletin, v. 62, p. 373–398.

Harrison, J. C., 1995, Tectonics and kinematics of a foreland folded belt influenced by salt, Arctic Canada, *in* M. P. A. Jackson, D. G. Roberts, and S. Snelson, eds., Salt tectonics: a global perspective: AAPG Memoir 65, this volume.

Harrison, J. C., and A. W. Bally, 1988, Cross-sections of the Parry islands fold belt on Melville Island, Canadian Arctic islands: implications for the timing and kinematic history of some thin-skinned décollement systems: Canadian Petroleum Geology Bulletin, v. 36, p. 311–332.

Hayward, A. B., and R. H. Graham, 1989, Some geometrical characteristics of inversion, *in* M. A. Cooper and G. D.

Williams, eds., Inversion tectonics: Geological Society of London Special Publication, v. 44, p. 17–39.

Heaton, R. C., M. P. A. Jackson, M. Bamahmoud, and A. S. O. Nani, 1995, Superposed Neogene extension, contraction, and salt canopy emplacement in the Yemeni Red Sea, *in* M. P. A. Jackson, D. G. Roberts, and S. Snelson, eds., Salt tectonics: a global perspective: AAPG Memoir 65, this volume.

Heybroek, P., 1975, On the structure of the Dutch part of the Central North Sea Graben, *in* A. W. Woodland, ed., Petroleum geology of the continental shelf of north-west Europe: London, Applied Science, p. 339–349.

Huiqi, L., K. R. McClay, and D. Powell, 1992, Physical models of thrust wedges, *in* K. R. McClay, ed., Thrust tectonics: London, Chapman & Hall, p. 71–81.

Huyghe, P., 1992, Enregistrement sédimentaire des déformations intraplaques: l'exemple de l'inversion structurale

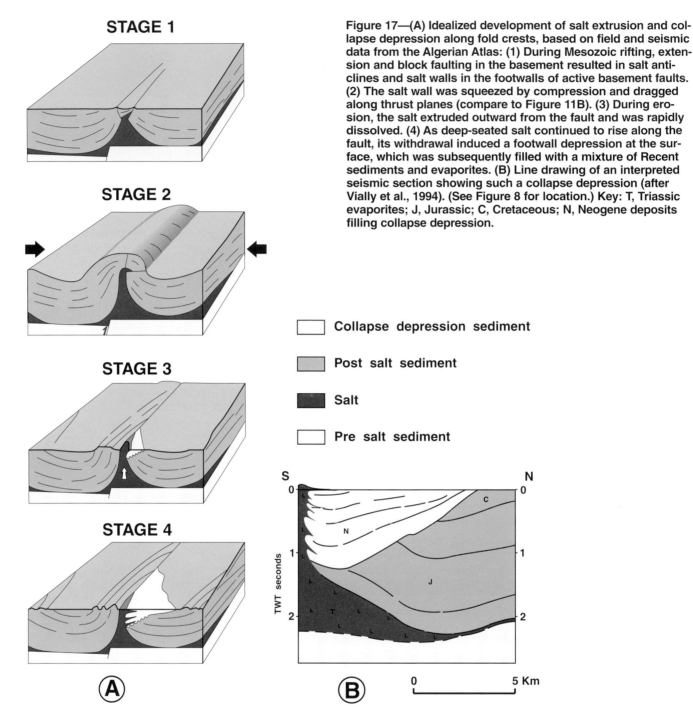

STAGE 1

STAGE 2

STAGE 3

STAGE 4

Ⓐ

Ⓑ

Figure 17—(A) Idealized development of salt extrusion and collapse depression along fold crests, based on field and seismic data from the Algerian Atlas: (1) During Mesozoic rifting, extension and block faulting in the basement resulted in salt anticlines and salt walls in the footwalls of active basement faults. (2) The salt wall was squeezed by compression and dragged along thrust planes (compare to Figure 11B). (3) During erosion, the salt extruded outward from the fault and was rapidly dissolved. (4) As deep-seated salt continued to rise along the fault, its withdrawal induced a footwall depression at the surface, which was subsequently filled with a mixture of Recent sediments and evaporites. (B) Line drawing of an interpreted seismic section showing such a collapse depression (after Vially et al., 1994). (See Figure 8 for location.) Key: T, Triassic evaporites; J, Jurassic; C, Cretaceous; N, Neogene deposits filling collapse depression.

Collapse depression sediment

Post salt sediment

Salt

Pre salt sediment

d'un bassin de la Mer du Nord: Thèse, University of Grenoble, 258 p.

Jackson, M. P. A., and W. E. Galloway, 1984, Structural and depositional styles of Gulf Coast Tertiary continental margins: application to hydrocarbon exploration: AAPG Continuing Education Course Notes, v. 25, 226 p.

Jackson, M. P. A., and C. J. Talbot, 1991, A glossary of salt tectonics: The University of Texas at Austin, Bureau of Ecoomic Geology Geological Circular 91-4, 44 p.

Koopman, A., A. Speksnijder, and W. T. Horsfield, 1987, Sandbox model studies of inversion tectonics: Tectonophysics, v. 137, p. 379–388.

Laubscher, H. P., 1977, Fold development in the Jura: Tectonophysics, v. 37, p. 337–362.

Lemoine, M., T. Bas, A. Arnaud-Vanneau, H. Arnaud, T. Dumont, M. Gidon, M. Bourbon, P. C. De Graciansky, J. L. Rudkiewicz, J. Megard-Gailli, and P. Tricard, 1986, The continental margin of the Mesozoic Tethys in the Western Alps: Marine and Petroleum Geology, v. 3, p. 179–199.

Letouzey, J., 1986, Cenozoic paleo-stress pattern in the Alpine Foreland and structural interpretation in a platform basin: Tectonophysics, v. 132, p. 215–231.

Letouzey, J., 1990, Fault reactivation, inversion and fold-thrust belt, *in* J. Letouzey, ed., Petroleum and tectonics in mobile belts: Paris, Technip, p. 101–128.

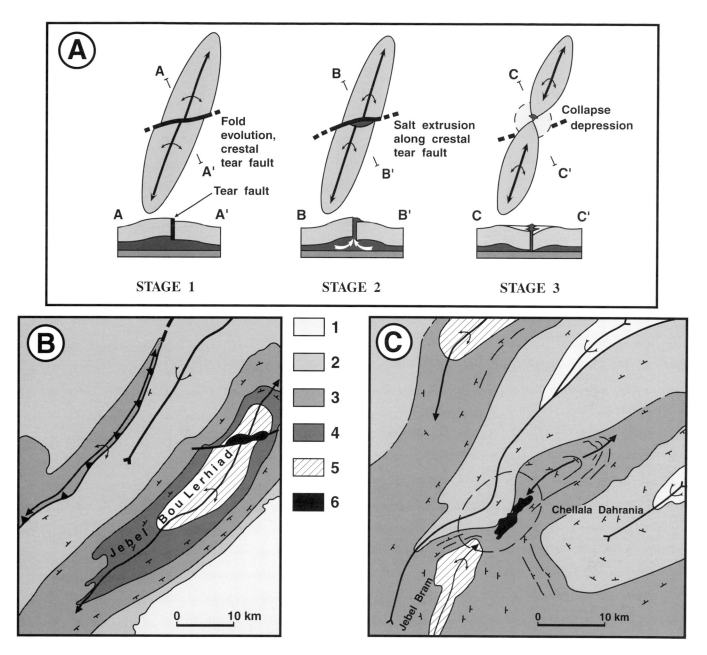

Figure 18—Salt extrusions from a crestal tear fault during contraction. (A) Stage 1 shows flow of salt contributing to growth of the fold by passively flowing into its core, creating a salt-cored anticline. Stage 2 is during folding when the salt flowed rapidly into low-pressure, pull-apart zones and extruded outward from an emergent diapir on the fault. In stage 3, the salt glacier was rapidly dissolved. Deep-seated salt withdrawal around the diapir initiated circular depressions at the surface, which became filled by Recent sediments. (B) Map of Jebel Bou Lerhiad (Algerian Atlas; see Figure 8 for location) showing an example of stage 2. Key: (1) Tertiary, (2) Cretaceous, (3) Upper Jurassic, (4) Middle Jurassic, (5) Lower Jurassic, (6) Triassic. (C) Map of Chellala Dahrania (Algerian Atlas; see Figure 8 for location) showing an example of stage 3. Key: (1) Upper Cretaceous, (2) Aptian–Albian, (3) Neocomian, (4) Jurassic, (5) Triassic.

Letouzey, J., Ph. Werner, and A. Marty, 1990, Fault reactivation and structural inversion: backarc and intraplate compressive deformations; example of the eastern Sunda shelf (Indonesia): Tectonophysics, v. 183, p. 341–362.

Mandl, G., and G. K. Shippam, 1981, Mechanical model of thrust sheet gliding and imbrication, *in* K. R. McClay and N. J. Price, eds., Thrust and nappe tectonics: Geological Society of London Special Publication, v. 9, p. 79–97.

McClay, K. R., 1989, Analogue models of inversion tectonics, *in* M. A. Cooper and G. D. Williams, eds., Inversion tectonics: Geological Society of London, Special Publication, No. 44, p. 41–59.

Mohriak, W. U., J. M. Macedo, R. T. Castellani, H. D. Rangel, A. Z. N. Barros, M. A. L. Latgé, A. M. P. Mizusaki, P. Szatmari, L. S. Demercian, J. G. Rizzo, and J. R. Aires, 1995, Salt tectonics and structural styles in the deep water province of the Cabo Frio region, Rio De Janeiro, Brazil, *in* M. P. A. Jackson, D. G. Roberts, and S. Snelson, eds., Salt tectonics: a global perspective: AAPG Memoir 65, this volume.

Mulugeta, G., 1988, Modelling the geometry of Coulomb thrust wedges: Journal of Structural Geology, v. 10, p. 847–859.

Nalpas, T., and J. P. Brun, 1993, Salt flow and diapirism related to extension at crustal scale: Tectonophysics, v. 228, p. 349–362.

Nilsen, K. T, B. C. Vendeville, and J.-T. Johansen, 1995, Influence of regional tectonics on halokinesis in the Nordkapp Basin, Barents Sea, *in* M. P. A. Jackson, D. G. Roberts, and S. Snelson, eds., Salt tectonics: a global perspective: AAPG Memoir 65, this volume.

Paraschiv, D., and Gh. Olteanu, 1970, Oil fields in Miocene–Pliocene zone of eastern Carpathians (district of Ploiesti), *in* M. T. Nalmouty, ed., Geology of giant petroleum fields: AAPG Memoir 4, p. 399–427.

Peel, F. J., C. J. Travis, and J. R. Hossack, 1995, Genetic structural provinces and salt tectonics of the Cenozoic offshore U.S. Gulf of Mexico: a preliminary analysis, *in* M. P. A. Jackson, D. G. Roberts, and S. Snelson, eds., Salt tectonics: a global perspective: AAPG Memoir 65, this volume.

Philippe, Y., 1994, Transfer zone in the southern Jura thrust belt (eastern France): geometry, development and comparison with analogue modelling experiments, *in* A. Mascle, ed., Exploration and petroleum geology of France: EAPG Memoir 4, Berlin, Springer-Verlag, p. 327–346.

Remmelts, G., 1995, Fault-related salt tectonics in the southern North Sea, The Netherlands, *in* M. P. A. Jackson, D. G. Roberts, and S. Snelson, eds., Salt tectonics: a global perspective: AAPG Memoir 65, this volume.

Roelofsen, J. W., and W. D. De Boer, 1991, Geology of the Lower Cretaceous Q/1 oil fields, Broad Fourteens basin, The Netherlands, *in* A. M. Spencer, ed., Generation, accumulation and production of Europe's hydrocarbons: Oxford, Oxford University Press, p. 203–216.

Sage, L., and J. Letouzey, 1990, Convergence of the African and Eurasian plates in the eastern Mediterranean, *in* J. Letouzey, ed., Petroleum and tectonics in mobile belts: Paris, Technip, p. 49–68.

Sassi, W., B. Colletta, P. Balé, and T. Paquereau, 1993, Modelling of structural complexity in sedimentary basins: the role of preexisting faults in thrust tectonics: Tectonophysics, v. 226, p. 97–112.

Schuster, D. C., 1995, Deformation of allochthonous salt and evolution of related salt/structural systems, eastern Louisiana Gulf Coast, *in* M. P. A. Jackson, D. G. Roberts, and S. Snelson, eds., Salt tectonics: a global perspective: AAPG Memoir 65, this volume.

Talbot, C. J., 1978, Halokinesis and thermal convection: Nature, v. 273, p. 739–741.

Van Hoorn, B., 1987, Structural evolution, timing, and tectonic style of the Sole Pit inversion: Tectonophysics, v. 137, p. 239–284.

Van Wijhe, D. H., 1987a, Structural evolution of inverted basins in the Dutch offshore: Tectonophysics, v. 137, p. 171–219.

Van Wijhe, D. H., 1987b, The structural evolution of the Broad Fourteens basin, *in* J. Brooks and K. Glennie, eds., Petroleum geology of north-west Europe: London, Graham and Trotman, p. 315–323.

Vendeville, B. C., 1987, Champs de failles et tectonique en extension: modélisation expérimentale: Thèse de 3ème cycle, Rennes, v. 15, 392 p.

Vendeville, B. C., Hongxing Ge, and M. P. A. Jackson, 1995, Scale models of salt tectonics during basement-involved extension: Petroleum Geoscience, v. 1, no. 2, p. 179–183.

Vendeville, B. C., and M. P. A. Jackson, 1992a, The rise of diapirs during thin-skinned extension: Marine and Petroleum Geology, v. 9, p. 331–353.

Vendeville, B. C., and M. P. A. Jackson, 1992b, The fall of diapirs during thin-skinned extension: Marine and Petroleum Geology, v. 9, p. 354–371.

Vially, R., J. Letouzey, F. Bénard, N. Haddadi, G. Desforges, H. Askri, and A. Boudjema, 1994, Basin inversion along the worth African Margin, the Saharan Atlas (Algeria), *in* F. Roure , ed., Peritethyan platforms: Paris, Technip, p. 79–118.

Voigt, W. A., 1962, Über Randtröge vor Schollenrändern und ihre Bedeutung im Gebiet der Mitteleuropäischen senke und angrenzender Gebiete: Zeitschrift der Deutschen Geologischen, v. 114, p. 378–418.

Weimer, P., and R. T. Buffler, 1992, Structural geology and evolution of the Mississippi Fan foldbelt, deep Gulf of Mexico: AAPG Bulletin, v. 76, p. 225–251.

Worrall, D. M., and S. Snelson, 1989, Evolution of the northern Gulf of Mexico, with emphasis on Cenozoic growth faulting and the role of salt, *in* A. W. Bally and A. R. Palmer, eds., An overview: the geology of North America: GSA, v. A, p. 97–138.

Wu, S., P. R. Vail, and C. Cramez, 1990a, Allochthonous salt, structure, and stratigraphy of the northeastern Gulf of Mexico, part I: stratigraphy: Marine and Petroleum Geology, v. 7, p. 318–333.

Wu, S., A. W. Bally, and C. Cramez, 1990b, Allochthonous salt, structure, and stratigraphy of the northeastern Gulf of Mexico, part II: structure: Marine and Petroleum Geology, v. 7, p. 334–370.

Wu, S., 1993, Salt and slope tectonics offshore Louisiana: Ph.D. dissertation, Rice University, Houston, Texas, 251 p.

Ziegler, P. A., 1975, Geologic evolution of North Sea and its tectonic framework: AAPG Bulletin, v. 59, p. 1073–1097.

Ziegler, P. A., 1989, Geodynamic model for Alpine intra-plate compressional deformation in western and central Europe, *in* M. A. Cooper and G. D. Williams, eds., Inversion tectonics: Geology Society of London Special Publication, v. 44, p. 63–85.

Ziegler, P. A., 1990, Geologic atlas of western and central Europe: The Hague, Shell International Petroleum, Maatstschappij B.V., 239 p.

Talbot, C. J., 1995, Molding of salt diapirs by stiff overburden, *in* M. P. A. Jackson,
D. G. Roberts, and S. Snelson, eds., Salt tectonics: a global perspective: AAPG
Memoir 65, p. 61–75.

Chapter 4

Molding of Salt Diapirs by Stiff Overburden

C. J. Talbot

Hans Ramberg Tectonic Laboratory
Institute of Earth Sciences
Uppsala University
Uppsala, Sweden

Abstract

Although active diapirs must deform the overburdens they pierce, the shape of passive (downbuilt or syn-depositional) diapirs is formed or molded by their overburdens. Molding of salt diapirs is simplified here to profiles of diapirs entirely downbuilt in effectively rigid overburden. The dips of salt–sediment contacts are shaped by the interaction of two processes: local net accumulation of overburden (A = deposition minus compaction) at rate $\dot{A}$ and the net increase in relief of salt structures (R = salt rise minus dissolution) at rate $\dot{R}$. Steady kinematic molding ratios, $\dot{R}$ / $\dot{A}$, forward model realistic dips of molded salt contacts, α, at particular depths using $\dot{R}$ / $\dot{A}$ or $\dot{A}$ / $\dot{R}$ = tan $\alpha/2$. Rising or falling ratios of incremental molding forward model complete diapir profiles. Conversely, molding histories can be read by backstripping profiles of downbuilt diapirs.

Salt diapirs are downbuilt in a field of downbuilding (100 > $\dot{R}$ / $\dot{A}$ > 0.01), that is bounded by burial and extrusion. Within this range, aggradation faster than salt can rise ($\dot{R}$ / $\dot{A}$ < 1) molds tapering (narrowing-upward) top contacts of salt. Accumulation of overburden slower than salt rises ($\dot{R}$ / $\dot{A}$ > 1) molds flaring (widening-upward) salt contacts. Below this range (where $\dot{R}$ / $\dot{A}$ < ~0.01), the top contact of the salt is eclipsed (temporarily buried to depths from which it can still upbuild) or even occluded (buried below its critical roof thickness and thus unable to rise again autonomously). Occluded salt is either dissolved at depth or rises in reactivated diapirs after exhumation or faulting of overburden that is not rigid. Where $\dot{R}$ / $\dot{A}$ > ~100, salt emerges like a fountain and extrudes sheets of allochthonous salt. Extruded salt is recycled back into the ocean by dissolution at the surface or after burial and reactivation in another cycle.

INTRODUCTION

Diapirs (from the Greek word *diapeirein*, "to pierce") are ductile intrusions, whatever their dynamics. Salt is ductile when strains of 0.1% or more take longer than years, whereas the overburden may fault as well as flow simultaneously. Stiff overburdens can contain, shape, or form passive and reactive diapirs or be deformed by active diapirs. The forming of passive salt diapirs by stiff sediments accumulating around them will be referred to here as *molding*.

In the 60 years since Nettleton (1934) began his physical modeling, the shapes and patterns of diapirs that actively upbuild through fluid overburdens have been reasonably well understood (Jackson and Talbot, 1994). Physical models of the single gravity overturn of viscous overburdens overlying layers of salt (e.g., Jackson et al., 1990) are geometrically, kinematically, and dynamically similar to the spoke patterns of multiple overturns of viscous fluids due to vigorous thermal convection (Talbot et al., 1991).

In an hourglass, loose grains of dry sand flow like fluids. However, beds of sand exceeding a critical thickness (Nettleton and Elkins, 1947; Vendeville and Jackson, 1992c) become strong enough to be impenetrable to diapirs unless they deform by faulting (Jackson and Vendeville, 1994). Thus, it has become increasingly common to model sedimentary basins using overburdens of loose particulate sands (Cobbold et al., 1989; Vendeville and Jackson, 1992a,b,c), collections of jostling fault blocks (Davison et al., 1993), or plastics that flow at nonlinear rates (Talbot, 1992; Podladchikov et al., 1993).

This work explores how salt diapirs are molded by syndiapiric deposition of stiff overburden around them. Barton (1933) introduced the concept of downbuilding to account for salt structures that pierce clastic sediments along the Gulf Coast of Texas and Louisiana. Instead of a

$\dot{f} \sim \dot{d}$

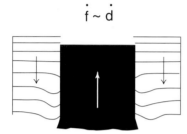

Overburden
A = accumulation
$\dot{A}$ = net rate of local accumulation
$\dot{A} = \dot{s} - \dot{c}$ (rate of sedimentation
                              minus compaction)

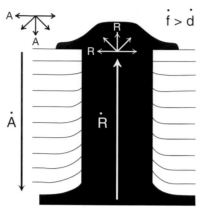

Rising salt structures
R = salt rise
$\dot{R}$ = net rise rate
$\dot{R} = \dot{f} - \dot{d}$ (flow rate minus dissolution

Figure 1—In Barton's (1933) concept of passive piercement, salt diapirs downbuild with crests that remain emergent as their overburden accumulates around them. In contrast to active diapirs that deform overburden already in place, the shapes of passive emergent diapirs are molded by stiff overburden. The net local accumulation, A, of effectively rigid clastic sediments occurs at rate $\dot{A}$ (rate of aggradation minus compaction). The relief of salt diapirs, R, downbuilt by stiff overburden occurs at net rate $\dot{R}$ (flow rate of salt minus dissolution). It is assumed that both salt and sediment spread sideways as fast as they rise or accumulate, respectively.

buried salt layer actively upthrusting toward the surface through an overburden already in place, Barton (1933) argued that cores of initially tabular salt layers remained passive but emergent near the sea floor while the top of the surrounding salt was buried ever deeper by clastic sediments (see Jackson, 1995, his figure 9).

For downbuilt structures to form, some of the unconsolidated and water-saturated overburden must sink differentially before it becomes denser than salt. In modern terminology (Jackson and Talbot, 1994), minibasins, deepening and consolidating to become depopods or depotroughs, drape intervening passive diapirs that remain emergent near the sediment surface. The crests of downbuilding salt structures remain near the sediment surface, while the base of the salt sinks (Figure 1). However, Barton (1933) himself emphasized that few salt structures are likely to form by one mechanism alone; passive downbuilding and active upbuilding commonly go hand in hand (Jackson et al., 1988).

According to some numerical modeling (e.g., Podladchikov et al., 1993), overburdens that strain more than two orders of magnitude more slowly than salt can be considered as effectively rigid. Fluids fill rigid containers whatever their shapes; this means that unconstrained salt flows sideways as fast as it rises. The shape of the containing overburden that molds passive salt diapirs depends on the rate at which the overburden accumulates and the rate at which the salt rises and spreads against the sediments.

SIMPLIFICATIONS OF SALT MOLDING

The shapes of molded salt structures depend on the cumulative interplay of several complex processes with rates that are poorly known (McGuinness and Hossack, 1993; Vendeville and Jackson, 1993). This discussion of molding of downbuilding salt structures by stiff overburden thus starts by using several gross simplifications, some of which will be relaxed in later sections.

The shape of downbuilt diapirs are described here by α, the dip of the salt–sediment contact molded between particular intervals in space and time. The dip of the diapiric contact is forward and backward modeled in time using a kinematic ratio $\dot{R}$ / $\dot{A}$, referred to here as the *molding ratio*. Later, attention will be focused on the change of the incremental molding ratio over time.

The local net accumulation of overburden thickness, A, occurs at rate $\dot{A}$; this is the rate of aggradation of overburden minus its vertical compaction. Sediment aggradation and compaction (and any salt dissolution) are assumed to be laterally uniform beneath a flat top surface. Considering all overburden as effectively rigid neglects complications due to variations in overburden strength.

The increase in structural relief of the diapir, R, occurs at a net rate, $\dot{R}$, which is the rate of vertical rise of salt diapirs minus any salt dissolution. Salt structures can rise $(+ \dot{R})$ or fall $(- \dot{R})$ because of horizontal extension or shortening of their overburden (e.g., Vendeville and

Jackson, 1992a,b; Jackson et al., 1994a) due to faulting caused by lateral forces in overburden or basement (McGuinness and Hossack, 1993). The effect of lateral forces on the salt and both its overburden and basement is important in many areas but are neglected here (but see Jackson et al., 1994b). This neglect of lateral forces and the assumptions of rigid basement and overburden are probably the most limiting assumptions made here.

Flow lines in the salt can converge or diverge, such that $\dot{R} / \dot{A}$ varies around salt diapirs of different geometries (Talbot, 1993). To avoid such complications, only profiles of upright, symmetric salt stocks or walls downbuilt with steady $\dot{R} / \dot{A}$ are considered here initially.

The net rate $\dot{R}$ depends on the effective viscosity of the salt; the thickness, pressure, and pressure gradient in the salt source; and the shape, dimensions, and boundary conditions of the salt body (Weijermars et al., 1993). Crystalline salt flows so much faster than consolidated clastic rocks that salt flow can be expected to dominate most deformation involving salt. However, although salt flow has an upper limit (meters per day?) (Talbot and Rogers, 1980), the molding ratio involves clastic sediments as well as clastic rocks; loose sediments can move even faster in air or water than crystalline salt can flow. The pressure that drives $\dot{R}$ depends mainly on the mass of overburden loading the source, the thickness of the source layer, and any lateral tectonic forces. In effect, $\dot{R}$ usually depends on tectonic strains, depletion of the source layer, and the aggradation rate averaged over a long time. In contrast, the local rate at which an overburden accumulates can change much faster. In essence, because the viscosities of air and water are lower than the viscosity of crystalline salt, $\dot{R}$ is likely to change much more slowly than $\dot{A}$.

FORWARD MODELS OF SALT MOLDING

Constant Dip at Steady Molding Ratio

We assume that shapes of passive diapirs are controlled by competing vertical rates of sediment accumulation $\dot{A}$ and salt rise $\dot{R}$. An additional assumption is that laterally unconfined salt can spread as fast sideways as it can rise and that sediments can be carried sideways as fast as they can accumulate downward. As long as the vertical rates result in horizontal motions, constant $\dot{A}$ and $\dot{R}$ will lead to a salt–sediment contact with constant dip, α. With all the simplifying assumptions, the dip of the salt–sediment contact, α, is related to the vertical rates by $\tan \alpha/2 = \dot{R} / \dot{A}$. The remainder of this work is merely an exploration of the implications of this simple relationship.

Curves of $\dot{R} / \dot{A}$ or $\dot{A} / \dot{R} = \tan \alpha/2$ on Figure 2 show how the dip of molded salt contacts relate to the logarithm of a steady molding ratio, $\dot{R} / \dot{A}$. Because it is impractical to measure dips to accuracies closer than $\pm 0.5°$, the molding ratio can only be sensibly inferred in the four orders of magnitude between 100 and 0.01. The field of salt molding is divided into three regimes and five subregimes on Figure 2. A regime of molded cylindrical contacts ($\alpha = 90° \pm 10°$) separates a flaring regime, in which the lower salt contact widens upward (top half, Figure 2), from a tapering regime, in which the upper salt contact narrows upward (bottom half, Figure 2). The flaring regime is subdivided into subregimes of flaring and extrusion. The tapering regime is subdivided into subregimes of eclipse (buried but active) and occlusion (buried and inactive but retaining the potential to rise buoyantly).

Cylindrical Regime

In Barton's original concept of downbuilding (1933), clastic aggradation and compaction kept pace with salt rise minus salt dissolution so that the structures in the top of the salt remained near the surface of clastic sedimentation. Barton assumed that molded salt contacts remained subvertical ($\alpha \approx 90° \pm 10°$) throughout downbuilding. Here, subvertical salt contacts of a downbuilt diapir are interpreted to mean steady $\dot{R} / \dot{A} \approx 1$. Narrow flanges or waists represent small or temporary variations in the molding ratio $\dot{R} / \dot{A}$.

Tapering Regime

In the bottom half of Figure 2, $\dot{R} / \dot{A} < 1$ and top salt contacts are molded such that they converge upward or taper. Molded salt contacts that are overstepped by overburden are referred to as being in the tapering regime. A loose correlation between $\dot{A}$ and grain size (Galloway and Williams, 1991) suggests that tapering top salt contacts are likely to be molded by coarser grained clastic sediments.

Significant episodes when $\dot{A}$ is only slightly higher than $\dot{R}$, that is, when $0.1 < \dot{R} / \dot{A} < 1$, lead to molding of conical salt contacts dipping between $80°$ and ~$11.5°$. Deep tapering diapirs are likely to be obscured in routine seismic reflection surveys. However, conical shoulders or crests of diapirs molded to these tapers are known in many basins (e.g., Jackson and Seni, 1984).

Eclipse by Roof Thinner than Critical Thickness

If the molding ratio is significantly less than unity for significant times ($\dot{R} / \dot{A} < 0.01$), then aggradation buries

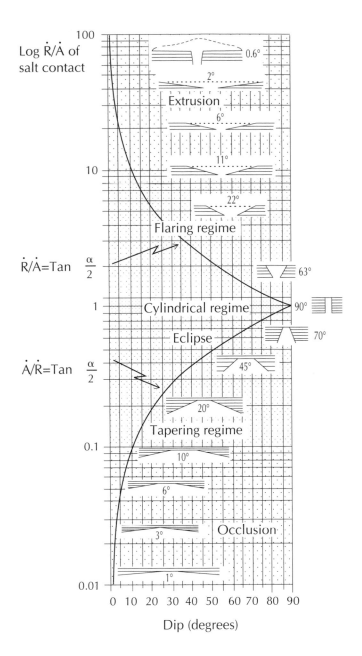

Figure 2—Plot showing how the contact dip (α) of passive salt diapirs can relate to a steady-state molding ratio, $\dot{R}/\dot{A}$. Salt molding can be divided into several regimes with different shading: $\dot{R}/\dot{A} > 1$ (top half), where molded salt contacts flare outward upward (and $\dot{R}/\dot{A} = \tan \alpha/2$), and $\dot{R}/\dot{A} < 1$ (bottom half), where molded top salt contacts taper inward upward (and $\dot{A}/\dot{R} = \tan \alpha/2$). Flaring and tapering regimes are divided by a regime of cylindrical contacts where $\alpha = 90°$ ($\pm 10°$) because the molding ratio $\dot{R}/\dot{A} \approx 1$. Schematic cross sections show the resulting form of the molded diapirs.

salt faster than it can develop or maintain surface relief. Many emergent diapirs have subhorizontal crests truncated by dissolution. Salt contacts dipping <0.5° are likely to be capped by insoluble residues left after salt dissolution, whether or not they are also buried by clastic overburden. There is a critical roof thickness (indicated to be ~20% of the entire thickness of the overburden flanking the diapir; Schultz-Ela et al., 1993; Jackson et al., 1994b) below which the diapir can pierce actively and above which it cannot. Salt diapirs that are buried to depths less than their critical roof thickness (Nettleton and Elkins, 1947; Vendeville and Jackson, 1992c) are referred to here as being in the regime of eclipse (Figure 2). Eclipsed diapirs may appear static and dead but could still grow by active upbuilding and pierce their overburden roofs (Schultz-Ela et al., 1993).

Occlusion by Roof Thicker than Critical Thickness

Burying diapirs deeper than their critical roof thickness not only eclipses passive diapirs but also occludes them until they are incapable of piercing actively. However, until occluded salt bodies are dissolved at depth, they maintain their potential to pierce actively if their denser overburden is ever thinned by erosion or deformation.

Significant phases when $\dot{R}/\dot{A} < 1$ can lead to autochthonous salt sequences being buried fast enough to suppress initiation of salt relief. An extreme case was the eclipse and occlusion of Hormuz salt by platform carbonates for about 300 m.y. (Middle Cambrian–Jurassic) around the Persian Gulf (Kent, 1958; Jackson and Vendeville, 1994). Because the buoyancy of rooted salt structures increases with their relief, it takes deeper burial to eclipse or occlude tall diapirs (i.e., $\dot{R}/\dot{A} < 1$ for longer times) than to eclipse or occlude short diapirs.

Flaring Regime

In the top half of Figure 2, $\dot{R}/\dot{A} > 1$ and rising salt spreads over aggrading sediments so that the dips of molded salt contacts flare outward as one goes upward. Molded salt contacts that climb over effectively rigid surroundings are referred to here as being in the flaring salt contact regime (Figure 2). Most flaring salt contacts are the bottom contacts of allochthonous salt bodies. The shape of the diapiric crest during significant episodes of $\dot{R}/\dot{A} > 1$ is not specified here but is generally either subhorizontal or parabolic (Figure 2, dashed contacts) (Talbot, 1993).

Extrusion

When molding ratios are in the range $100 > \dot{R}/\dot{A} > 10$, the dip, α, of the bottom contact flares between 0.6° and

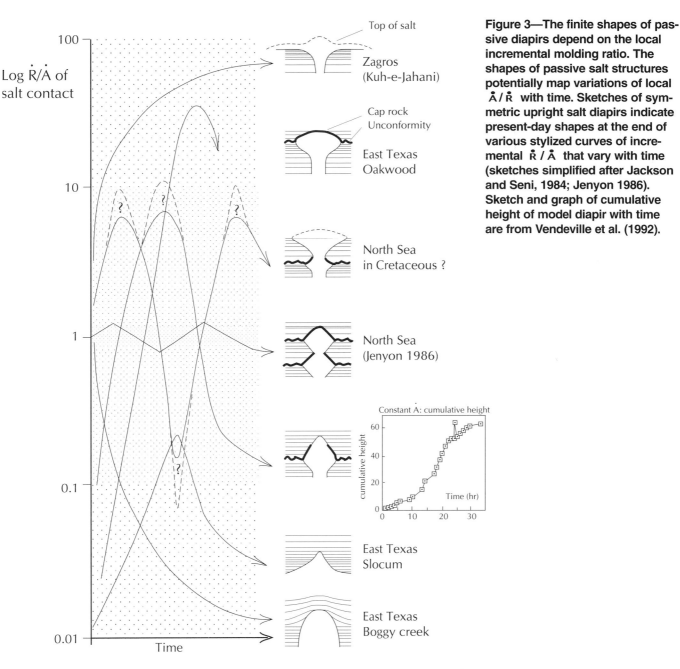

Figure 3—The finite shapes of passive diapirs depend on the local incremental molding ratio. The shapes of passive salt structures potentially map variations of local $\dot{A}$/$\dot{R}$ with time. Sketches of symmetric upright salt diapirs indicate present-day shapes at the end of various stylized curves of incremental $\dot{R}$/$\dot{A}$ that vary with time (sketches simplified after Jackson and Seni, 1984; Jenyon 1986). Sketch and graph of cumulative height of model diapir with time are from Vendeville et al. (1992).

about 11.5°. In the flaring regime, salt is likely to puddle and spread up and over the bottom salt contact, as in the viscous salt bergs (Talbot, 1993) in the Great Kavir of central Iran, where $\dot{A}$ is negative (Jackson et al., 1990). In the top half of the flaring regime (Figure 2) is a subregime of salt extrusion ($\dot{R}$ / $\dot{A}$ > ~10) where salt emerges as a fountain below its piezometric level (Talbot, 1993) and above its level of neutral buoyancy (called the level of isostatic equilibrium by Barton, 1933). The salt fountain spreads out over bottom salt contacts with α < ~0.6°. Salt extrusion can spread over substrates that dip away from the orifice ($-\alpha$), as in the Zagros Mountains of southern Iran, where $\dot{A}$ is essentially zero (Talbot and Jarvis, 1984), or in the Gulf of Mexico, where $\dot{A}$ is positive (Wu et al., 1990).

Finite Diapir Shapes by Incremental Molding Ratio

The range of potential profiles of salt contacts molded by steady rise and aggradation (Figure 2) broadens considerably when it is acknowledged that the molding ratio $\dot{R}$ / $\dot{A}$ can vary with time. In practice, all the many factors molding salt contacts are likely to change continuously, such that the sketches in Figure 2 can be put together in any combination to mold diapirs of any shape. The history of the finite molding ratio depends on the local history of the incremental molding ratio and whether it is positive and rising or negative and falling. Sketches to the right of stylized curves in Figure 3 show the results of some likely salt molding histories.

Although $\dot{R}$ and $\dot{A}$ already account for salt dissolution and compaction, dashed curves on Figure 3 indicate their likely overshoot effects during changes in sign of the incremental molding ratio with time. Upward overshoots (dashed) in the flaring regime indicate episodes when the salt dissolution rate, $\dot{D}$, exceeded the rate of flow of emergent salt, $\dot{F}$. The ratio $\dot{F}/\dot{D}$ controls the shape of the top salt boundary considered elsewhere (Talbot, 1993). Salt extrudes where $\dot{F} > \dot{D}$ and is dissolved as fast as it rises when $\dot{D} \approx \dot{F}$ or faster when $\dot{F} < \dot{D}$.

Insoluble residues that cap salt diapirs can be inferred as significant episodes when $\dot{D} \approx \dot{F}$ and choked breccia pipes as significant phases when $\dot{D} > \dot{F}$. Subsurface salt dissolution leads to salt welds (Jackson and Cramez, 1989) that can be primary and subhorizontal at depth or secondary in the form of breccia pipes beneath surface craters (Ala, 1974). The record of salt extrusion (when $\dot{F} \gg \dot{D}$) may be cryptic after salt has dissolved following dispersal of former cap and roof rocks over disconformities or unconformities reaching far beyond their vents onshore (Kent, 1958) or offshore (Talbot, 1993).

Downward overshoots (dashed, Figure 3) on curves in the regime of tapering top salt contacts indicate accumulation of overburden roofs during eclipse or occlusion. Counteraction of eclipse by active upbuilding is recorded by upward bulging (with or without faulting) of layers where the roof is not rigid. Curves plotting the two lowermost sketches in Figure 3 lead to the occlusion of diapirs below critical roof thicknesses, a situation ripe for future reactivation (considered later).

SPECIAL CASES OF MOLDING RATIO

Constant Aggradation Rate

Diapirs in the cylindrical regime are inferred to record Barton's constant $\dot{R}/\dot{A}$ but not necessarily constant aggradation rate, $\dot{A}$. Experiments with molded diapirs downbuilt in a few days (Vendeville et al., 1992) or a month (C. J. Talbot, unpublished data) emphasized that even a constant aggradation rate does not necessarily mean either a constant $\dot{R}$ or a constant molding ratio. Even when the aggradation rate is held constant, $\dot{R}$ changes with time through the three phases shown in the cumulative growth curve in Figure 3. The first phase (when $\dot{R}/\dot{A} < 1$) is marked by slow but accelerating $\dot{R}$ while the source material flows between planar floor and roof (Vendeville et al., 1992; Jackson et al., 1994a). The second phase (when $\dot{R}/\dot{A} > 1$) is marked by salt contacts that flare as $\dot{R}$ and the incremental molding ratio increase as salt rises in a diapir gaining relief and therefore buoyancy. The third phase ($\dot{R}/\dot{A} < 1$) is marked by a tapering salt crest when $\dot{R}$ decelerates as the source is exhausted; continued aggradation leads to diapir eclipse and occlusion.

Figure 4 considers in more detail the special case of diapirs molded by steady aggradation rate, $\dot{A}$, allowing for typical variations in $\dot{R}$ over the growth history of the diapir. Initial deposition and burial of salt (Figure 4a) can lead to downbuilding of a passive salt pillow, active upbuilding of a salt mullion, or reactive growth of a salt roller, depending on whether the local force field is due to gravity (directly or indirectly) through thin-skinned lateral shortening or lateral extension (Vendeville and Jackson, 1992a,b,c). As the relief of the salt structure increases, the molding ratio rises further (Figure 4b), tapering top salt contacts steepen, and the peripheral sink migrates inward to reach the salt contact where it becomes vertical in the cylindrical regime (Figure 4c). If the molding ratio continues to rise through the flaring regime, salt structures may bulge above the aggrading surface and become self-sustaining as clastic sediments bypass rather than overstep the salt (Figure 4c). For diapirs to grow in the flaring regime, any insoluble residues or superposed overburden must be conveyed off the spreading salt before they consolidate to an impenetrable lid. Both subaerial (Talbot and Jarvis, 1984) and submarine (Wu et al., 1990) salt domes rising like fountains above their level of neutral buoyancy (Figure 4d) have been shown to shed both insoluble cap rocks and superposed marine sediments. When the source is isolated or depleted, the fountainheads collapse to puddles (Figure 4e) and no longer convey all their cap rocks to the snout (Talbot, 1993). They are then likely to be buried as dissolution overtakes the flow rate of emergent salt, $\dot{F}$. Whether the salt sequence is eventually converted to its dispersed or *in situ* insoluble record after one (as in Figure 4) or more phases of $\dot{R}/\dot{A} > 1$ depends on the local and regional geology.

Islands of cap rock rising above nonaggrading portions of the U.S. Gulf Coast exemplify molding ratios passing through the cylindrical regime (Figure 4f) as their molding ratios fall through the regime of cylindrical contacts ($\dot{R}/\dot{A} \approx 1$). If the source layer is diminished, exhausted, or pinched off, or if the diapir is eclipsed, the molding ratio falls through the tapering regime. Cap rocks begin to clog the vent rather than being carried beyond. Eclipse may be followed by occlusion, such as in the East Texas Basin (Figures 4i, j) (Seni and Jackson, 1983). Alternatively, eclipse may lead directly to active or reactive piercement, such as in the Gulf of Mexico (Wu et al., 1990; Rowan, 1995), the North Sea (Davison et al., 1993), and the Nordkapp Basin (Koyi et al., 1993, 1995). Active or reactive diapirs piercing formerly occluding roofs distort or fault nonrigid overburden (Figure 4h) and no longer have passively molded contacts.

Molding Ratio Increases Basinward

The forward models in Figures 3 and 4 display the upright symmetry expected of individual diapirs having

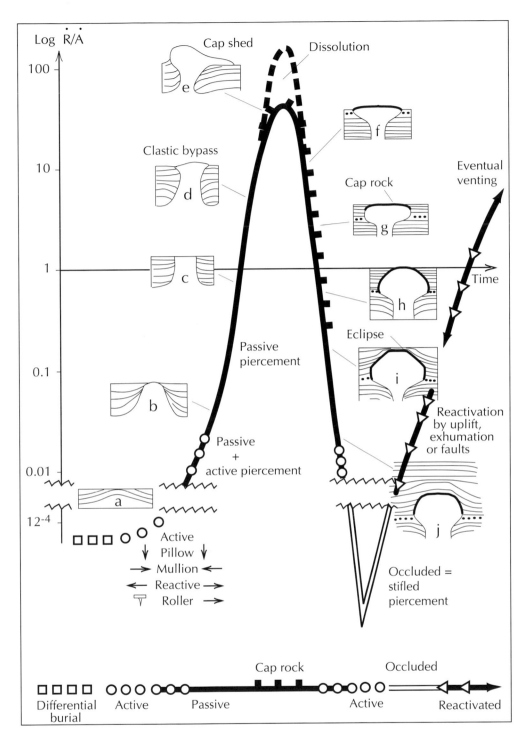

Figure 4—Diagrammatic history of a salt diapir molded by constant aggradation rate $\dot{\mathbf{A}}$. The sketches (a–j) portray some of the stages indicated by the molding ratio curve. (See the text for more details.)

molding ratios that vary in time rather than space. Figure 5 illustrates the result of relaxing one of the original simplifications and allowing the molding ratio to vary in space as well as time. In Figure 5, $\dot{\mathbf{R}} / \dot{\mathbf{A}}$ increases toward an opening ocean basin as aggradation rate diminishes seaward. Upright symmetric stocks become increasingly asymmetric slopeward and pass through salt sheets to giant salt nappes (Diegel et al., 1995; Peel et al., 1995; Rowan, 1995; Schuster, 1995). Even with constant $\dot{\mathbf{R}}$, the sign of the downslope molding ratio changes across

Figure 5. Subtle flanges, waists, and overhangs molded by alternations of taper and flare under the shelf are exaggerated on the slope by increasing molding ratios, such that they form salt flats and ramps entirely in the flaring regime (Figure 5). This amplification of structural asymmetry offshore is further emphasized by salt dissolution being lower in seawater than in rainwater or fresh groundwater onshore (Talbot, 1993). Asymmetry is exaggerated even further by thin-skinned gravity spreading of the whole slope (Talbot, 1992).

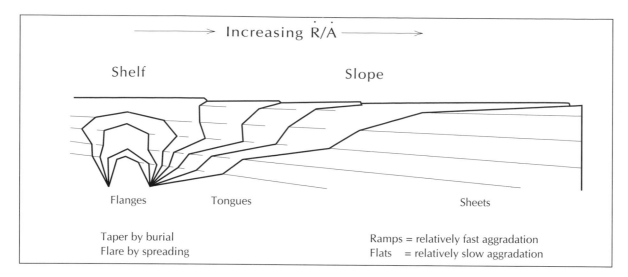

Figure 5—Molded diapirs become increasingly asymmetric to the right as the aggradation rate decreases and the molding ratio increases basinward. Alternations of high and low molding ratios result in alternations of flare and taper that mold subtle flanges around symmetric salt stocks beneath the shelf. Flanged diapirs can be eclipsed or occluded. Tapers beneath the shelf can overturn to ramps beneath the slope that become increasingly gentle basinward so that steep ramps alternate with gentle flats along the bases of salt nappes.

BACKWARD MODELING

Forward models of molding ratios that vary in both space and time (Figure 5) hint at the potential of reading molding histories from the dip of salt contacts molded during particular intervals of space in time. This potential is illustrated here first by inferring histories of molding ratio from profiles of individual diapirs and then by comparing the molding histories of suites of diapirs of Louann salt.

Individual Diapir in East Texas Basin

There are sufficient data in Jackson and Seni (1984) to infer the histories of molding ratios for 15 diapirs in the East Texas basin. True-scale profiles in this atlas were converted to graphs of dip α (the angle between the salt contact and congruent bedding) with time. Figure 6 shows the results of reading the dips of molded salt contacts as time maps of $\dot{R} / \dot{A}$ for a single representative example from the East Texas basin: Steen Dome (see also Jackson et al., 1994a, for other examples of this approach). The dashed curve charting the history of the logarithmic molding ratio of Steen Dome begins abruptly in the Early Cretaceous (120 Ma) because the stem and roots of this diapir were too deep to be resolved by the data available. The oldest visible molded salt contacts still flare as a result of a 110-m.y. episode of salt rise (Figure 6). The molding ratio paused as it fell through the cylindrical regime near 100 Ma—a pause that might have been buffered by salt dissolution and signaled by a disconformity in the surrounding rocks.

A dip in the curve at 90 Ma may indicate an episode of eclipse by comparatively rapid aggradation followed by upbuilding that distorted the formerly eclipsing sequence. Salt dissolution alone may have kept an 80-Ma peak in molding ratio within the tapering regime. Emergent diapirs still accumulating cap rocks above horizontal dissolution surfaces in the East Texas Basin are presumably still rooted. Their salt has been dissolved as fast as it has risen throughout a ~70 Ma hiatus in aggradation. Quiescence of emergent Louann (Jurassic) salt for so long suggests that the closed interior East Texas Basin reached stability in the Late Cretaceous (Seni and Jackson, 1983).

Individual Diapirs in Gulf of Mexico

A similar Mesozoic history to that in the East Texas Basin may eventually be read in equivalent rocks at depth offshore. However, the clastic sediments that bypassed the East Texas Basin for the last 70 Ma instead surrounded downbuilding and spreading salt diapirs and sheets in the adjacent Gulf of Mexico.

Figure 7 illustrates the results of inferring molding histories for most of the Tertiary from the highly flared bottom contact of the Sigsbee salt nappe illustrated by Worrall and Snelson (1989). This salt is usually considered to be driven by the superincumbent continental margin that is advancing by both sedimentary progradation and gravity spreading. The bottom contact of the Sigsbee nappe ramped as steeply as 11°–14° in the Eocene, suggesting that the salt then slowed at ~52 Ma to spread only four to five times faster downslope than the local clastic sediments aggraded. However, even then, the lowest

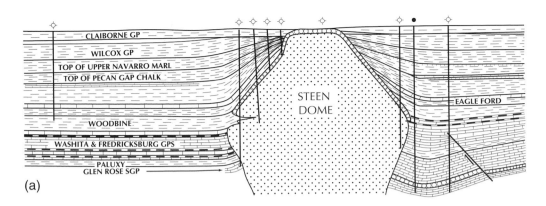

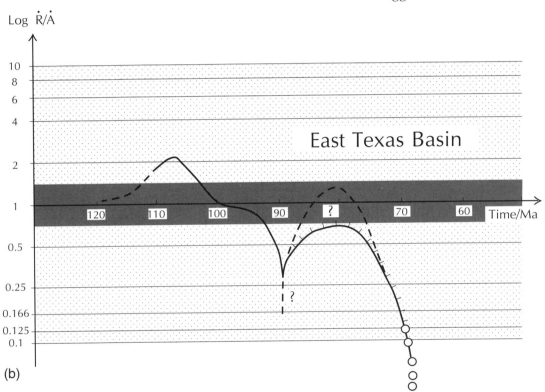

Figure 6—(a) The original dips of salt contacts in a true scale profile of a representative salt stock in the East Texas basin (Jackson and Seni, 1984). **(b)** The dips graphed as a molding ratio history. See key at bottom of Figure 4 for explanation of line patterns.

molding ratio for the nappe remained in the flaring regime. Since ~56 Ma, the molding ratio of the most distal salt nappe rarely fell below $\dot{R} / \dot{A} \approx 10$ (Figure 7), whereas smaller nearer shore salt sheets illustrated by Wu et al. (1990) successively rose into the flaring salt contact regime (Figure 7).

Figure 8 explores whether aggradation rates near the Sigsbee salt nappe can be inferred from the flare of the bottom salt contact by assuming a constant $\dot{R}$. The Paleogene accumulation rate plotted against time (right, Figure 8) was calculated using the solid $\dot{R} / \dot{A}$ curve in Figure 7, assuming that $\dot{R} = 100$ m/m.y. and that compaction can be neglected at this scale. The other graphs in Figure 8 plot deposition rate against time measured directly from borehole logs of the clastic sediments hundreds of kilometers away on the shelf (Galloway and

Williams, 1991). Figure 8 shows that clastic aggradation rates indirectly read from a small-scale profile of the Sigsbee salt nappe on the slope do not correlate particularly well with more direct measurements on the shelf. However, there is sufficient correlation to wonder what the differences imply in terms of local dynamics (e.g., shifts in regional depocenters).

Suites of Diapirs

Figure 9 combines the backward models of molding histories inferred for each of four of the age groups of salt stocks distinguished by Seni and Jackson (1983) in the East Texas Basin (Figure 7) together with the Sigsbee salt nappe and smaller salt diapirs offshore Gulf of Mexico (Figure 8). Curves of molding history are distinctive for

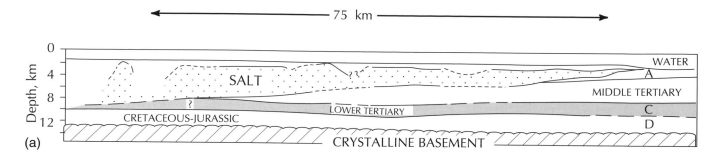

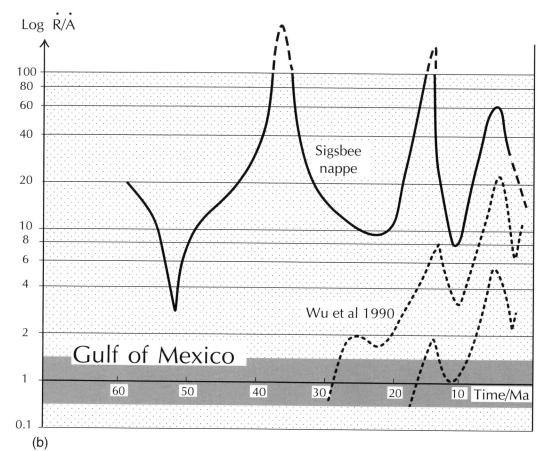

Figure 7—(a) Cross section of the Sigsbee salt nappe (from Worrall and Snelson, 1989). (b) The solid curve backward models the history of molding ratio for most of the Tertiary from the flared bottom of the salt nappe. The dashed curves plot the molding history from two smaller salt sheets nearer shore (shown in Wu et al., 1990, his figure 24).

each individual diapir, but most share some qualitative features. Unfortunately, the histories of molding ratios for onshore and offshore structures shown in Figure 9 do not overlap in time. However, molding ratios for diapirs in each basin tend to peak together whether they are flaring or tapering. Each of the two suites of salt diapirs shown in Figure 9 have incremental molding ratios that tend to vary harmonically, even when their finite molding ratios are in different regimes. Figure 9 also suggests that dissolution by fresh groundwater in temperate climates may buffer salt rise so that both rising and falling molding histories tend to pause in the regime of cylindrical salt contacts ($\mathring{R}/\mathring{A} \approx 1$), causing a ledge on the curves.

Peaks in paths of molding history in the flaring contact regime can be inferred to represent episodes of salt extrusion when distal allochthonous salt spread by flaring over adjoining sediments. On the shore or shelf, molding ratios that peak in the tapering top salt regime are represented only by caprock-mantled salt hips, shoulders, or crests draped by bedding dragged upward by active piercement following eclipse. Offshore, these same peaks are likely to be represented by salt flats. Troughs in molding ratio represent salt ramps in the larger salt nappes and, in the smaller salt sheets offshore, episodes of eclipse ended by upbuilding (Koyi, 1991) or reactivation (Wu et al., 1990). Equivalent troughs appear to correlate better offshore than onshore, suggesting that eclipse and occlusion are more local in the closed basin onshore than in the open basin offshore where salt sheet spreading may record pulses of downdip contraction associated with updip extension of complete provinces (Peel et al., 1993, 1995; Diegel, 1995).

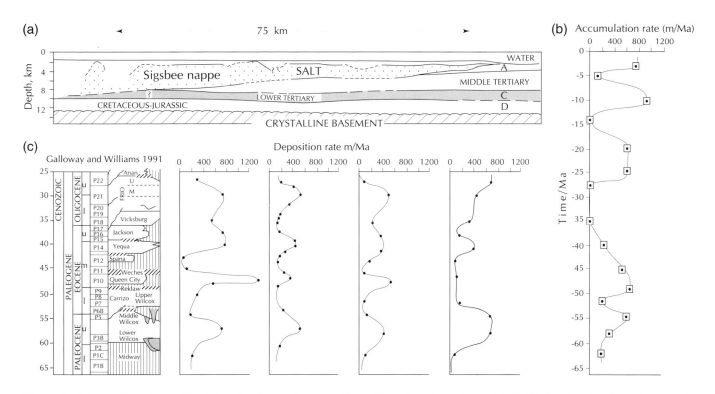

Figure 8—(a) Small-scale profile of the Sigsbee salt nappe (from Worrall and Snelson, 1989). (b) Graph showing history of aggradation rates inferred from the flare of the bottom salt contact in the salt nappe, assuming a constant Ȧ = 100 m/m.y. (c) Graphs of deposition rates (also in m/m.y.) measured directly from clastic sediments logged in boreholes on the continental shelf hundreds of kilometers away by Galloway and Williams (1991).

DISCUSSION

Harmonic molding histories such as those in Figure 9 point to basinwide influences on any combination of Å, Ṙ, or Ṙ / Å. In general, aggradation rates are likely to rise during progradation and fall during regression. Shared and distinctive features of curves in Figure 8 offer another route to checking and refining local details of sequence stratigraphy. Histories recorded by a few tens of meters of cap rocks have already been shown to correlate with magnetic stripes hundreds of kilometers wide in the ocean floor (Kyle et al., 1987). Similarly, the shapes of individual salt diapirs distill local nuances in the stratigraphy of the whole basin.

Some of the regional effects that can be discerned through local details of molding histories are doubtless due to tectonics. Thus, Mesozoic faulting in the intracontinental East Texas Basin was obviously minor compared to the post-Mesozoic collapse of the slope into the opening intercontinental Gulf of Mexico. Prograding deltas massage salt substrates through pores in their overburden. The shapes of molded salt diapirs record every sigh in salt pressure whether or not there have been lateral forces in addition to the aggraded load. Molded salt contacts not only sensitively monitor the interplay between aggradation and salt escape but can even amplify some components. Downbuilt diapirs may be passive, but the dips of their molded salt contacts are active pressure gauges. Salt diapirs rise or fall with lateral shortening or extension (Jackson and Vendeville, 1990; Vendeville and Jackson, 1992a,b). They also flare or taper with sediment progradation or regression in closed basins, or exude with flare upward over salt ramps and flats or downslope over the sea floor in open basins.

Diapirs are molded by the interplay of several complex processes, each operating at rates that change continuously. Not only are *these* rates poorly known, but also the rates at which they change and the times taken for their associated lags and feedbacks. In a collection of 16 papers on salt tectonic modeling (Cobbold, 1993), only one paper (Podladchikov et al., 1993) modeled erosion of overburden and dissolution of salt. Even that paper simplified the erosion of clastic sediments and the dissolution of salt by combining them into a single "redistribution" process that they showed can accelerate active diapiric upbuilding by about an order of magnitude.

Two other studies emphasized that we cannot rely on forward models alone. Thus, Vendeville et al. (1992) noted that experimental diapirs rise too fast when scaled to nature. Weijermars et al. (1993) advised taking into account the effective viscosity of salt only when it is extrusive; in nonemergent modes, they assumed that the rate of salt flow can be simplified to infinitely fast. Such comments remind us of a fundamental rule in all stratigraphic

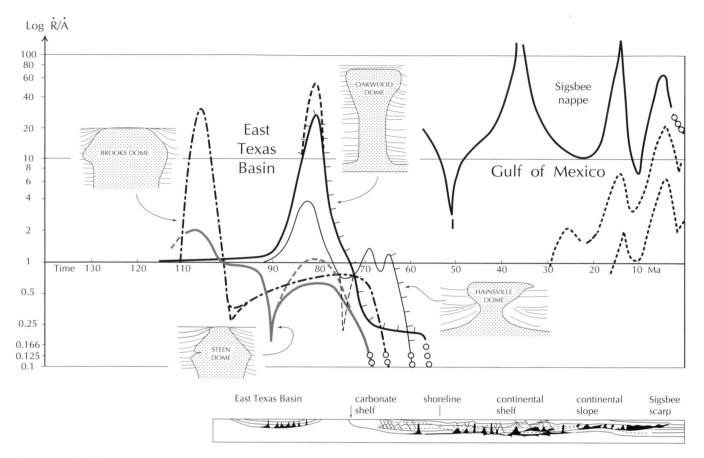

Figure 9—Molding histories of diapirs representative of four different age groups in the closed interior East Texas basin compared with salt sheets and nappes offshore in the spreading slope of the Gulf of Mexico. Each of the symmetric upright stocks in the closed onshore basin were probably buried as their sources were exhausted locally. Tertiary sediments that bypassed the interior basin accumulated around highly asymmetric passive diapirs of the same Louann salt under the slope spreading into the open gulf by sedimentary and tectonic progradation (data from Seni and Jackson, 1983; Worrall and Snelson, 1989; Wu et al., 1990).

studies: the only rates that can be properly constrained are those that fit within the window used to view them (Sadler and Strauss, 1990). Only the fastest rates are visible in the short times available for laboratory models (Figure 10), and we need more constraints on natural rates. The salt growth and deformation rates plotted against the duration of observation in Figure 10 are nowhere near exhaustive, but they indicate considerable overlap between growth rates of salt structures and aggradation rates of fluvial sediments and illustrate the relevance of the simplified approach to salt molding taken here.

Regimes of salt molding shown in Figure 2 are not symmetric above and below the cylindrical regime. Even occlusion of salt is likely to be only temporary over eons. That is why we know of no surviving salt older than ~800 Ma, although pseudomorphs after halite demonstrate that much older salt sequences must have been lost to dissolution (with or without extrusion) before their overlying basin sequences reached greenschist facies (Buick and Dunlop, 1990). Whether they are autochthonous salt beds

of low relief or allochthonous salt diapirs of high relief, however deeply they are occluded, all buried salt bodies will inevitably recycle back into the ocean. This is likely either by dissolution at depth or after dissipation near the surface, probably after rising as active or reactive diapirs. Similarly, although even vigorous salt extrusions in the flaring regime can be temporarily buried back into the tapering regime, salt will inevitably be removed from the geologic record during sustained periods of $\dot{R} / \dot{A} > 1$. The most likely records of such past episodes of salt extrusion are in the depocenters over areas of rapid salt withdrawal and in any spreads of insoluble residues along disconformities or unconformities in the overburden nearby (Kent, 1958).

Episodes of passive molding almost certainly alternate with active and reactive diapirism (with or without faulting). Retrospective distinction of these two processes requires the identification of (1) true dips using cut-off angles, (2) the location of the axis of any peripheral sink, and (3) any drag of overburden up or down against the

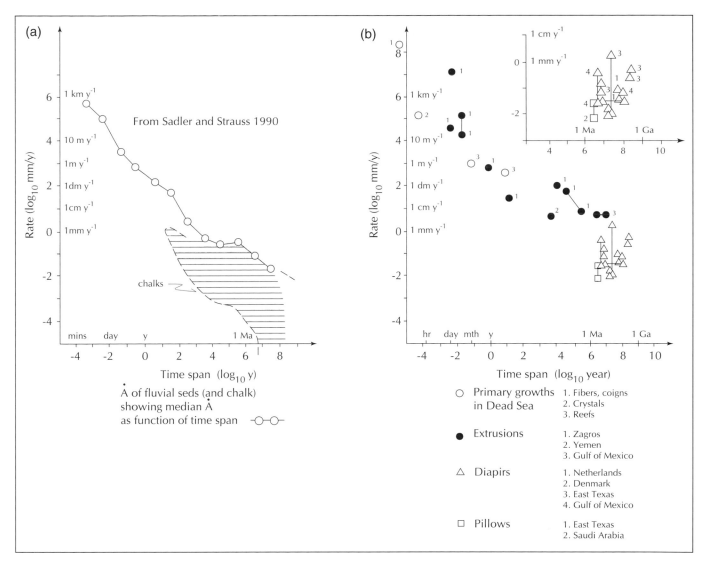

Figure 10—Comparison of rates of alluvial sediment accumulation and growth of natural salt structures plotted against the time window in which they are viewed. Both curves show that the smaller the time window, the faster the recorded rate. Thus, both sedimentation and salt structures grow episodically. (a) Rates of sedimentation of fluvial sediments (and chalk) correlate with the time frame in which they are viewed (from Sadler and Strauss, 1990). (b) Rates of growth of primary and secondary salt structures plotted against the time frame of measurement. Data were chosen to illustrate the effect of the time window and are not exhaustive. (Data for Saudi Arabia from Edgell, 1992; The Netherlands, Remmelts et al., 1993; extrusions in Zagros, Wenkert, 1979, Talbot and Rogers, 1980, Talbot and Jarvis, 1984, and author's unpublished 1994 measurements; Yemen, Davison et al., in press; Gulf of Mexico, Wu et al., 1990, and Talbot, 1993; halite crystal and reef growth in southern Dead Sea; author's unpublished observations.)

salt (Jackson and Seni, 1984). However, in gross terms, identifying passive growth may be relatively simple. Passively molded diapirs are likely to have irregular blunt profiles with primary angularities that match vertical facies changes in the adjacent overburden. By contrast, active diapirs are likely to have smooth, sinuous profiles strained into conformity with layering in adjoining overburdens. Active diapirism can smooth contacts molded to earlier angularity, but only faulting can sculpt abrupt angles into profiles smoothed by earlier active diapirism.

Acknowledgments   This work was triggered by a conversation with Martin Jackson and Bruno Vendeville in December 1989 about Martin's April 1988 notion that the angle of climb of the bottom contact of a salt sheet relates to $\dot{R}/\dot{A}$. Presented at the AAPG Hedberg conference in September 1993, the main version was written while on sabbatical at the Geophysical Laboratory of the Carnegie Institution, Washington, D.C., in early 1994. I gratefully acknowledge comments on various versions by Neil Irvine, Bruno Vendeville, and in particular, Martin Jackson and Peter Cobbold.

REFERENCES CITED

Ala, M. A., 1974, Salt diapirism in southern Iran: AAPG Bulletin v. 58, p. 1758–1770.

Barton, D. C., 1933, Mechanics and formation of salt domes with special reference to Gulf coast domes of Texas and Louisiana: AAPG Bulletin, v. 17, p. 1025–1083.

Buick, R., and J. S. R. Dunlop, 1990, Evaporitic sediments of Early Archean age from the Warrawoona Group, North Pole, Western Australia: Sedimentology, v. 37, p. 247–277.

Cobbold, P., ed., 1993, New insights into salt tectonics: Tectonophysics, v. 228, p. 1–448.

Cobbold, P., E. Rosello, and B. Vendeville, 1989, Some experiments on interacting sedimentation and deformation above salt horizons: Bulletin de la Société de Géologique de France, v. 5, p. 453–460.

Davison, I., M. Insey, M. Harper, P. Weston, D. Blundell, K. McClay, and A. Qualington, 1993, Physical modelling of overburden deformation around salt diapirs: Tectonophysics , v. 228, p. 255–274.

Davison, I., D. Bosence, I. Alsop, and M. H. Al-Aawah, in press, Deformation and sedimentation around active Miocene salt diapirs on the Tihama Plain, NW Yemen, *in* Salt tectonics: Geology Society of London Special Publication, No. 100.

Diegel, F. A., J. F. Karlo, D. C. Schuster, R. C. Shoup, and P. R. Tauvers, 1995, Cenozoic structural evolution and tectono-stratigraphic framework of the Northern Gulf Coast continental margin, *in* M. P. A. Jackson, D. G. Roberts, and S. Snelson, eds., Salt tectonics: a global perspective: AAPG Memoir 65, this volume.

Edgell, H. S., 1992, Basement tectonics of Saudi Arabia as related to oil field structures, *in* M. J. Rickard et al., eds., Basement tectonics: The Netherlands, Kluer Academic Publishers, p. 169–193.

Galloway, W. E., and T. A. Williams, 1991, Sediment accumulation rates in time and space: Paleogene genetic stratigraphic sequences of the northwestern Gulf of Mexico Basin: Geology, v. 19, p. 986–989.

Jackson, M. P. A., 1995, Retrospective salt tectonics, *in* M. P. A. Jackson, D. G. Roberts, and S. Snelson, eds., Salt tectonics: a global perspective: AAPG Memoir 65, this volume.

Jackson, M. P. A., and C. Cramez, 1989, Seismic recognition of salt welds in salt tectonics regimes, *in* Gulf of Mexico salt tectonics, associated processes and exploration potential: SEPM Gulf Coast Section, 10th Annual Research Conference, Program and Extended Abstracts, Houston, Texas, p. 66–71.

Jackson, M. P. A., and S. J. Seni, 1984, Atlas of salt domes in the East Texas basin: The University of Texas at Austin, Bureau of Economic Geology Report of Investigations No. 140, 102 p.

Jackson, M. P. A., and C. J. Talbot, 1986, External shapes, strain rates, and dynamics of salt structures: GSA Bulletin, v. 97, p. 305–328.

Jackson, M. P. A., and C. J. Talbot, 1994, Advances in salt tectonics, *in* P. L. Hancock, ed., Continental deformation: Oxford, U.K., Pergamon Press and International Union of Geological Sciences, p. 159–180.

Jackson, M. P. A., and B. C. Vendeville, 1990, The rise and fall of diapirs during thin-skinned extension: AAPG Bulletin, v. 74, no. 5, p. 683.

Jackson, M. P. A., and B. C. Vendeville, 1994, Regional extension as a geologic trigger for diapirism: GSA Bulletin, v. 106, p. 57–73.

Jackson, M. P. A., C. J. Talbot, and R. R. Cornelius, 1988, Centrifuge modeling of the effects of aggradation and progradation on syndepositional salt structures: The University of Texas at Austin, Bureau of Economic Geology Report of Investigations No. 173, 93 p.

Jackson, M. P. A., R. R. Cornelius, C. R. Craig, A. Gansser, J. Stöcklin, and C. J. Talbot, 1990, Salt diapirs of the Great Kavir, central Iran: GSA Memoir 177, 139 p.

Jackson, M. P. A., B. C. Vendeville, and D. D. Schultz-Ela, 1994a, Structural dynamics of salt systems: Annual Review Earth and Planetary Science, v. 22, p. 93–117.

Jackson, M. P. A., B. C. Vendeville, and D. D. Schultz-Ela, 1994b, Salt related structures in the Gulf of Mexico: a field guide for geophysicists: The Leading Edge, August 1994, p. 837–842.

Jenyon, M. K., 1986, Salt tectonics: London, Elsevier, Applied Science, 191 p.

Kent, P. E., 1958, Recent studies of South Persian salt plugs: AAPG Bulletin, v. 422, p. 2951–2972.

Koyi, H., 1991, Mushroom diapirs penetrating into highly viscous overburden: Geology, v. 19, p. 1229–1232.

Koyi, H., C. J. Talbot, and B. O. Tørudbakken, 1993, Salt diapirs in the southwest Nordkapp Basin: analogue modeling: Tectonophysics, v. 228, p. 167–188.

Koyi, H., C. J. Talbot, and B. O. Tørudbakken, 1995, Salt tectonics in the northeast Nordkapp basin, southwestern Barents Sea, *in* M. P. A. Jackson, D. G. Roberts, and S. Snelson, eds., Salt tectonics: a global perspective: AAPG Memoir 65, this volume.

Kyle, J. R., M. R. Ulrich, and W. Gose, 1987, Textural and paleomagnetic evidence for the mechanism and timing of anhydrite cap rock formation, Winfield salt dome, Louisiana, *in* I. Lerche and J. J. O´Brien, eds., Dynamical geology of salt and related structures: Orlando, Florida, Academic Press, p. 497–542.

McGuinness, D. B., and J. R. Hossack, 1993, The development of allochthonous salt sheets as controlled by the rates of extension, sedimentation, and salt supply, *in* Rates of geologic processes: GCS-SEPM 14th Annual Research conference ", December 1993, p. 127-149.

Nettleton, L. L., 1934, Fluid mechanics of salt domes: AAPG Bulletin, v. 81, p. 1175-1204.

Nettleton, L. L., and T. A. Elkins, 1947, Geological materials made from granular materials: EOS, Transactions American Geophysical Union, v. 28, p. 451–466.

Peel, F. J., C. J. Travis, J. R. Hossack, and D. B. McGuinness, 1993, Structural provinces in the cover sediments in the U.S. Gulf of Mexico basin; linked systems of extension, compression, and salt movement (abs.): AAPG Annual Convention Official Program, New Orleans, p. 164.

Peel, F. J., C. J. Travis, and J. R. Hossack, 1995, Genetic structural provinces and salt tectonics of the Cenozoic offshore U.S. Gulf of Mexico: a preliminary analysis, *in* M. P. A. Jackson, D. G. Roberts, and S. Snelson, eds., Salt tectonics: a global perspective: AAPG Memoir 65, this volume.

Podladchikov, Y., C. J. Talbot, and A. N. B. Poliakov, 1993, Numerical modelling of complex diapirs: Tectonophysics, v. 228, p. 167–188.

Remmelts, G., E. Muyzert, D. J. van Rees, M. C. Geluk, C. C. de Ruyter, and A. F. B. Wildenborg, 1993, Evaluation of salt bodies and their overburden in The Netherlands for the disposal for radioactive waste, b. Salt movement: Report to Rijks Geologische Dienst, 87 p.

Rowan, M. G., 1995, Structural styles and evolution of allochthonous salt, central Louisiana outer shelf and upper slope, *in* M. P. A. Jackson, D. G. Roberts, and S. Snelson, eds., Salt tectonics: a global perspective: AAPG Memoir 65, this volume.

Sadler, P. M., and D. J. Strauss, 1990, Estimation of completeness of stratigraphical sections using empirical data and theoretical models: Journal of the Geological Society of London, v. 147, p. 471–486.

Schultz-Ela, D. D., M. P. A. Jackson, and B. C. Vendeville, 1993, Mechanics of active salt diapirism: Tectonophysics, v. 228, p. 275–312.

Schuster, D. C., 1995, Deformation of allochthonous salt and evolution of related salt/structural systems, eastern Louisiana Gulf Coast, *in* M. P. A. Jackson, D. G. Roberts, and S. Snelson, eds., Salt tectonics: a global perspective: AAPG Memoir 65, this volume.

Seni, S. J., and M. P. A. Jackson, 1983, Evolution of salt structures, East Texas diapir province, part 2: Patterns and rates of halokinesis: AAPG Bulletin, v. 67, p. 1245–1274.

Talbot, C. J., 1992, Centrifuged models of Gulf of Mexico profiles: Marine and Petroleum Geology, v. 9, p. 412–432.

Talbot, C. J., 1993, Spreading of salt structures in the Gulf of Mexico: Tectonophysics, v. 228, p. 151–166.

Talbot, C. J., and R. J. Jarvis, 1984, Age, budget, and dynamics of an active salt extrusion in Iran: Journal of Structural Geology, v. 6, p. 521–533.

Talbot, C. J., and E. A. Rogers, 1980, Seasonal movements in a salt glacier in Iran: Science, Washington, v. 208, p. 395–397.

Talbot, C. J., H. Schmeling, P. Rönnlund, H. Koyi, and M. P. A. Jackson, 1991, Diapiric spoke patterns: Tectonophysics, v. 188, p. 187-201.

Vendeville, B. C., and M. P. A. Jackson, 1990, Physical modeling of the growth of extensional and contractional salt tongues on continental slopes (abs.): AAPG Bulletin, v. 74, p. 784.

Vendeville, B. C., and M. P. A. Jackson, 1991, Deposition, extension, and the shape of downbuilding salt diapirs (abs.): AAPG Bulletin, v. 75, p. 687–689.

Vendeville, B. C., and M. P. A. Jackson, 1992a, The rise of diapirs during thin-skinned extensions: Marine and Petroleum Geology, v. 9, p. 331-353.

Vendeville, B. C., and M. P. A. Jackson, 1992b, The fall of diapirs during thin-skinned extensions: Marine and Petroleum Geology, v. 9, p. 354-371.

Vendeville, B. C., and M. P. A. Jackson, 1992c, Critical roof thickness of active diapirs: EOS Transactions, American Geophysical Union, v. 73, no. 27, 572 p.

Vendeville, B. C., and M. P. A. Jackson, 1993, Rates of extension and deposition determine whether growth faults or salt diapirs form, *in* Rates of geologic processes: GCS-SEPM 14th Annual Research Conference, December 1993, p. 269–276.

Vendeville, B. C., M. P. A. Jackson, and R. Weijermars, 1992, Can diapirs rise that fast? (and why): GSA, Abstracts with Programs, v. 224, p. A145.

Weijermars, R., M. P. A. Jackson, and B. C. Vendeville, 1993, Rheological and tectonic modelling of salt provinces: Tectonophysics, v. 217, p. 143–174.

Wenkert, D. D., 1979, The flow of salt glaciers: Geophysical Research Letters, v. 6, p. 523–526.

Worrall, D. M., and S. Snelson, 1989, Evolution of the northern Gulf of Mexico, with emphasis on Cenozoic growth faulting and the role of salt, *in* A. W. Bally and A. R. Palmer, eds., The geology of North America—an overview: GSA, v. A, p. 97–138.

Wu, S., A. W. Bally, C. and Cramez, 1990, Allochthonous salt, structure, and stratigraphy of the northeastern Gulf of Mexico, part 11, structure: Marine and Petroleum Geology, v. 7, p. 334–370.

Fletcher, R. C., M. R. Hudec, and I. A. Watson, 1995, Salt glacier and composite
sediment–salt glacier models for the emplacement and early burial of
allochthonous salt sheets, *in* M. P. A. Jackson, D. G. Roberts, and S. Snelson,
eds., Salt tectonics: a global perspective: AAPG Memoir 65, p. 77–108.

Chapter 5

Salt Glacier and Composite Sediment–Salt Glacier Models for the Emplacement and Early Burial of Allochthonous Salt Sheets

Raymond C. Fletcher

Exxon Production Research Company
Houston, Texas
U.S.A.

Present address:
Department of Geosciences
New Mexico Institute of Mining and Technology
Socorro, New Mexico, U.S.A.

Michael R. Hudec

Ian A. Watson

Exxon Production Research Company
Houston, Texas
U.S.A.

Abstract

Allochthonous salt sheets in the northern Gulf of Mexico were emplaced as extrusive "salt glaciers" at the sediment–water interface. Massive dissolution was suppressed by a thin carapace of pelagic sediments. During emplacement, several hundred meters of bathymetric relief restricted rapid sedimentation to outside the glacial margins. The glaciers acted as sediment dams, influencing the transport and deposition of sediment from an upslope source. Because of contemporaneous sedimentation, the base of the glaciers climbed upward in all directions away from their feeder stocks, and successive sedimentary horizons were truncated against it. The local slope at the base of the sheets is equal to the local rate of sedimentation divided by the local rate of salt advance. Alternating episodes of slow and rapid sedimentation gave rise to a basal salt surface of alternating flats and ramps, which are preserved. Many salt sheets have nearly circular map patterns but are strongly asymmetric. Feeder stocks occur near upslope edges, and base-of-salt slopes are greater updip of the feeder. The asymmetry is due to more rapid sedimentation at the upslope edge and to slower advance induced by the smaller hydraulic head between the salt fountain and the upslope edge compared to the downslope edge.

Rapid emplacement of the Mickey salt sheet (Mitchell dome) from a preexisting salt stock took ~4 m.y., as ~1 km of sediment was deposited. A three-dimensional geomechanical model for the rapid salt emplacement yields the following relationship for the diapir's downdip radius versus time: $R(t) \approx Mt^q \approx B[(\rho - \rho_w)gK^3/\eta]^{1/8}t^q$, where M, q, B, and K are constants related to salt supply into the sheet, ρ and ρ_w are the densities of salt and water, g is the acceleration of gravity, η is salt viscosity, and t is a model time extrapolated back to zero sheet volume at $t = 0$. The advance history of the Mickey salt sheet is equally well fitted by two histories of salt supply, corresponding to values of $q = 1/2$ and $q = 1$ in the above expression. The model requires that the volume of the sheet grew as $V \approx Kt$ (for $q = 1/2$) or $V \approx Kt^{7/3}$ (for $q = 1$). Fits to the advance history can be used to determine the remaining constants. From the expression for M, salt viscosities $\eta \approx 8.3 \times 10^{18}$ ($q = 1/2$) and $\eta \approx 4.8 \times 10^{18}$ Pa s ($q = 1$) are obtained, consistent with experimental data on salt creep.

Once salt extrusion ceases, a large fraction of the glacier's topographic relief is lost, but the steep shoulder at the downslope edge is maintained. Sediment influx concentrated at the updip edge maintains a sloping surface, and a glacier-like flow continues within a composite salt–sediment glacier. If a minibasin forms near the updip edge, further downdip advance can be substantial. Velocities on the surface of a composite glacier indicate that overburden particles above the leading edge can move 1.5 times as fast as the sheet advances, resulting in a tractor tread model for near-toe kinematics. That the sedimentary carapace of the glacier moves faster than the sheet advances suggests that extension in the sedimentary veneer generally exceeds salt sheet advance. Burial of the toe results in cessation of advance, but updip minibasin deepening and downdip salt diapir growth continue as long as the surface remains sloped and the finite-strength sediment in and around the buried sheet does not establish a mechanically stable configuration. Relative buoyancy between salt and sediment influence late-stage development.

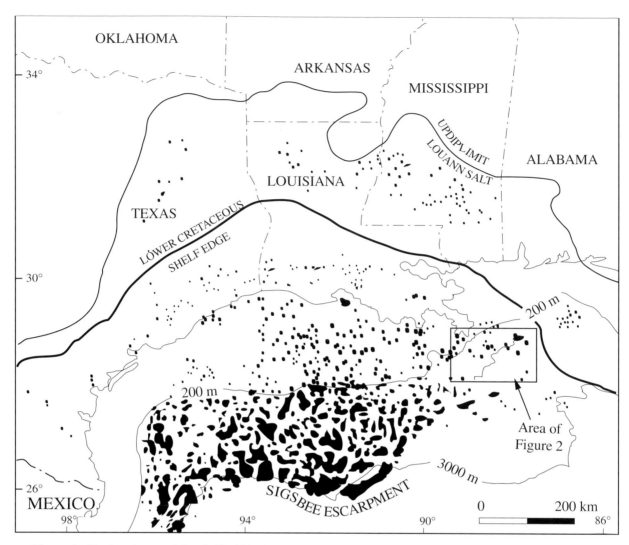

Figure 1—Regional map of the Gulf of Mexico showing the distribution of shallow salt (after Martin, 1978).

INTRODUCTION

Near-surface allochthonous salt sheets are striking components in the tectonic structure of the northern U.S. Gulf of Mexico (Figure 1). They occur notably along the Sigsbee escarpment associated with the Pliocene–Pleistocene depocenter. More deeply buried structures likely originated by the modification of salt sheets during burial, generally with the salt displaced into diapirs or sheets at higher stratigraphic levels (e.g., Seni, 1992). Their emplacement has been variously attributed to both intrusive and extrusive mechanisms. Nelson and Fairchild (1989a,b) proposed that salt sheets in the Gulf of Mexico were emplaced as sills intruded within 300 m of the sea floor. Their interpretation of the sedimentary record during emplacement implies rapid intrusion as thin sheets about 100 m thick, sometimes cutting upward in the section, followed by inflation to thicknesses in excess of 1 km. This mechanism is equivalent to the emplacement of igneous laccoliths by inflation of older tabular sills (Pollard and Johnson, 1973). Nelson and

Fairchild (1989a,b) proposed that intrusion preferentially takes place above a level of neutral density contrast, preferably into a layer of weak shale overlying stronger sediments. The sill mechanism circumvents the problem of massive dissolution that would seem likely if salt were extruded at the seafloor.

Improved understanding and seismic imaging of sub-salt structure and stratigraphy has led several recent authors to propose that salt sheets are initially emplaced at the sediment–water interface as glaciers (McGuinness and Hossack, 1993; Talbot, 1994; Fletcher et al., 1995; Hudec et al., 1995). In the present paper, we use detailed observations from the Mickey salt sheet (Mitchell dome) (Figure 2), along with observations from other Gulf of Mexico sheets, to infer that salt sheets are emplaced as extrusive glaciers at the sea bottom. The key observations supporting this model have only become available with good subsalt seismic imaging. We then use a quantitative geomechanical model to simulate the advance history and form of the Mickey sheet. The model yields an estimate for salt viscosity consistent with experimental data

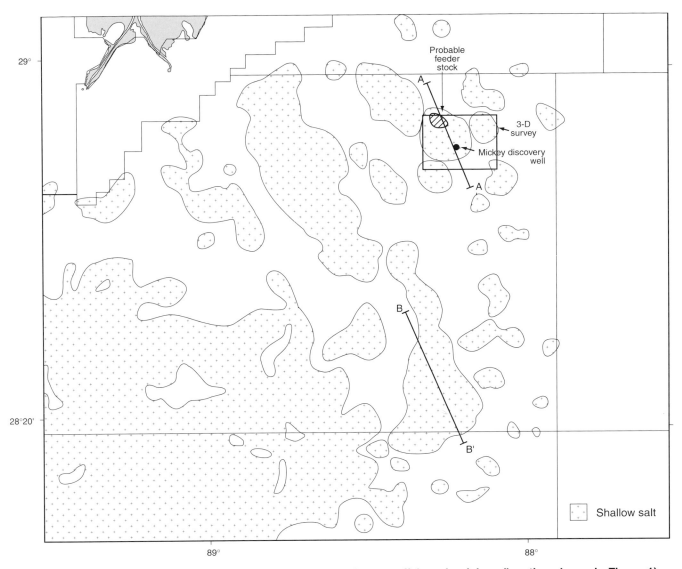

Figure 2—Map showing the distribution of salt sheets in the study area, offshore Louisiana (location shown in Figure 1). Section lines A–A' and B–B' show the locations of two-dimensional seismic profiles (Figures 3, 20). The small box shows the outline of a three-dimensional seismic survey over the Mickey sheet (Figures 8, 22, 23).

on salt creep. The results support the model and provide a detailed picture of active emplacement and subsequent topographic relaxation after cessation of salt flow from the source feeder. An extension of the glacier model applies to the evolution of a sheet during early burial and provides a mechanism for the formation of minibasins.

ARGUMENTS FOR AN EXTRUSIVE ORIGIN FOR ALLOCHTHONOUS SALT SHEETS

Interpreted seismic and depth profiles across the center of the Mickey salt sheet, Mississippi Canyon area, Gulf of Mexico (Figures 2–4), show many of the key relationships used to support an extrusive origin for salt sheets. The profile (Figures 3, 4) is oriented in approximately the

regional dip direction and passes close to the position of maximum salt thickness, inferred to be the pinched-off feeder stock. The base of the salt and the subsalt sediments are well imaged. Note the following features in Figures 3 and 4:

1. Sedimentary horizons are truncated against the base of the salt.
2. The mean slope of the base is 6.5° over the downdip part of the profile; a smaller scale stair step form is superposed on this.
3. Sediments beneath the salt are essentially undeformed.
4. The first sedimentary units above the salt onlap the top of the salt.
5. The edge of the sheet is blunt and about 400 m high.
6. Although the sheet is nearly circular in map view

Figure 3—Two-dimensional time-migrated seismic profile across the Mickey salt sheet. Line location shown in Figure 2. Courtesy of Geco-Prakla.

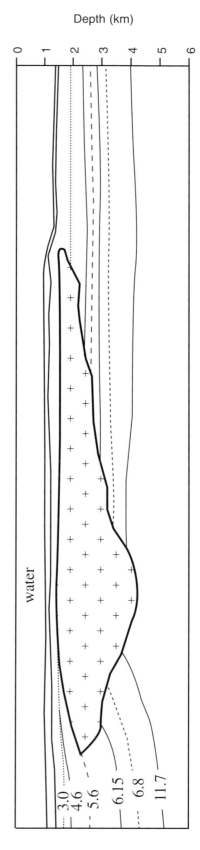

Figure 4—Schematic cross section at 1:1 scale across the Mickey salt sheet, constructed from a two-dimensional prestack depth-migrated seismic line acquired along the same line of section as Figure 3. Horizon labels are ages in millions of years.

A) Present day

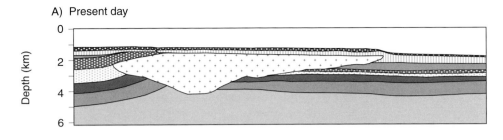

B) Hypothetical pre-intrusion configuration

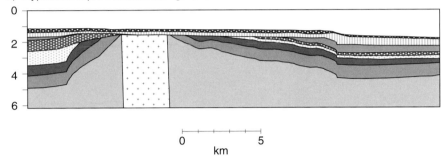

0   5
km

Figure 5—Two-dimensional restoration of the Mickey salt sheet (Figure 4), assuming that the sheet was emplaced by intrusion. Sediment deformation was restored by vertical shear against the present bathymetry. Compaction effects are not included. The restoration shows an unreasonable preinjection configuration (B), leading us to reject the intrusion hypothesis.

(Figure 2), the interpreted feeder stock is upslope from the center, so the sheet is strongly asymmetric.

7. Sediments drape and thin by roughly 30% over the toe of the sheet, apparently as a result of both depositional onlap and slumping.

We argue that observations (1), (3), and (4) are inconsistent with intrusion of the salt sheet. For an intrusion model to work, the roof and floor of the salt sheet should restore to a geologically reasonable configuration. Restoration of the Mickey salt sheet to a hypothetical preintrusion configuration (Figure 5) yields an implausible structure, leading us to reject an intrusive mechanism.

However, observations (1)–(6) can readily be explained as the result of the advance of a salt glacier from an extrusive source over the sea floor (Figures 6, 7). A dynamically supported fountain over the feeder forms a bathymetric high that restricts most sedimentation to outside the sheet margin and provides the relief necessary for lateral glacier flow. Given this model, we account for the above observations as follows:

1. Sedimentation occurs around the periphery of the glacier, so the glacier climbs upsection as it advances away from the feeder stock. Sedimentary horizons are truncated against the base; the age of the truncated horizon marks the time that the glacier advanced past that point (Figure 6).

2. If β is the slope in radians of the base of the sheet in the direction of advance, S the local rate of sedimentation, and dR/dt the local rate of advance, then we have

$$\beta = S/(dR/dt) \qquad (1)$$

A staircase form for the base of the salt is the prod-

uct of changes in the relative rates of sheet advance and sedimentation. If the rate of advance is dominated by the long-period variation in salt supply and is insensitive to local conditions at the edge of the glacier, the smaller scale stair steps in the base will be tied to variations in sedimentation rate. The glacier is assumed to advance roughly in the manner of a tank tread, rather than as a rigid salt mass sliding above a basal thrust. As a consequence, the constructional form of the base is preserved. A seismic dip map on the base of the Mickey salt sheet and one of its neighbors (Figure 8) shows that ramps in the salt base tend to form concentric to the feeder stock over at least half of the sheet circumference. In our model, the ramps are time synchronous, everywhere marking the subsalt truncation of a single interval. The trace of the ramp marks the edge of the salt sheet at the time of deposition of that interval.

3. If the tank-tread assumption is correct, substantial deformation of subsalt units does not occur. Parts of the sedimentary carapace that slump from the toe of the glacier could, however, be overrun by the advancing sheet. A layer of disrupted material beneath the sheet may thus exist. Disrupted zones have been interpreted immediately beneath salt sheets in many Gulf of Mexico subsalt wells (D. Moore, oral presentation to the Houston Geological Society, 1994).

4. During burial, the first strata deposited on top of the sheet are expected to onlap the existing salt bathymetry. At Mickey, such units appear on the seismic lines to onlap directly against salt (see Figures 6, 7), although it is more likely that the salt was directly covered by a thin carapace of pelagic

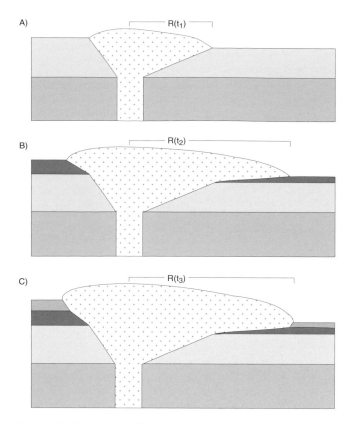

Figure 6—Schematic diagrams showing three stages in the advance of a salt glacier. **(A)** Salt glacier is at the surface at time $t = t_1$. **(B)** At a later time $t = t_2$, the glacier has advanced both updip and downdip. Sedimentation is contemporaneous with glacier advance, so the glacier climbs upsection as it advances. Sedimentation is more rapid updip, so the base of the salt is steeper in that direction. The glacier is advancing faster downdip because the downdip sediment surface is lower than the updip surface, providing greater driving force (head) for downdip flow. **(C)** At a later time $t = t_3$, sedimentation was more rapid than the glacial advance rate during this interval, so the base of salt is steeper, yielding the beginning of a stair-step form.

sediment only a few tens of meters thick. Such a carapace would limit dissolution at the glacier surface but would have negligible mechanical effect. Wells penetrating several Gulf of Mexico salt sheets have indeed cut highly condensed pelagic sections above salt that are equivalent in age to sediments truncated at the base (McGuinness and Hossack, 1993). As these authors point out, repeated section above and below salt definitively excludes an intrusive origin.

5. Later, we show that a steep escarpment, several hundred meters high, is maintained at the toe of an advancing salt glacier.

6. The geomechanical model indicates that an active salt glacier typically has several hundred meters of relief on the sea floor. It therefore acts as a dam to

sediment transport, tending to pond sediment updip. Note the greater thickness of sediment layers updip of the Mickey sheet (Figure 4). The tendency of withdrawal basins to form updip of a feeder diapir also produces greater sediment accommodation updip. Faster sedimentation rates updip of the sheet result in a steeper base of salt, according to Eq. (1), yielding the observed asymmetry. The slower rate of glacier advance in the updip direction is a result of the smaller head between the maximum height above the "salt fountain" (Talbot, 1994), above the source diapir, and the glacier edge.

7. The drape of sediments over the toe of salt sheets was a major argument initially used to interpret allochthonous salt sheets as intrusive (Nelson and Fairchild, 1989a,b). In the salt glacier model, this geometry is interpreted to be a combination of depositional drape over a preexisting bathymetric escarpment and minor late-stage thickening of the salt during early burial.

In summary, observations around little-deformed salt sheets suggest that they were emplaced extrusively as glaciers rather than intrusively as sills.

GEOMECHANICAL MODEL OF A SALT GLACIER

At present, the conditions that give rise to episodes of salt sheet emplacement on a regional scale are not well understood. A necessary condition is that salt stocks or diapirs reach the sea bottom. "Downbuilding" of a near-surface diapir could proceed without the emplacement of a salt sheet, but some additional factor favors the rapid extrusion of salt that leads to sheet emplacement. We take up the modeling of sheet emplacement at this point. The modeling seeks to answer the following questions. Can a model be devised that simultaneously satisfies the constraints provided by (1) the advance history in the downdip direction, (2) the sheet volume, and (3) the rheologic properties of salt, as determined from creep experiments? What new insights are provided by a quantitative dynamic model satisfying these constraints?

Model Formulation

Salt extruded from a source diapir forms a salt fountain, from which it spreads radially outward. We evaluate the effects of different regimens of supply, in which the rate of increase in the volume of the sheet Q decreases with time, is constant, or increases with time.

The salt is treated as a uniform viscous fluid. Salt viscosity is a function of temperature (Jackson and Talbot, 1991), but a temperature difference of 20°C changes salt viscosity by only a factor of about two, which, from the present results, is not significant. Maximum deviatoric

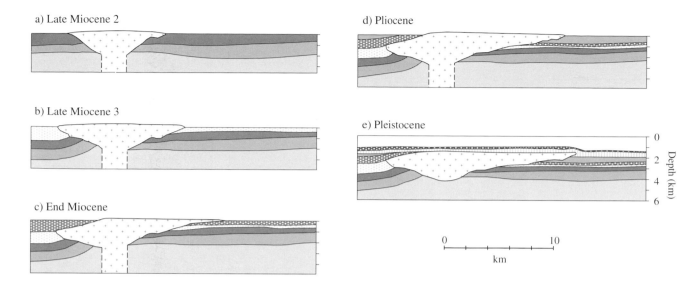

Figure 7—Two-dimensional restoration of the Mickey salt sheet (Figure 4), assuming that the sheet was emplaced as a glacier. Sediment deformation was restored by vertical simple shear, and the effects of sediment compaction were included. The shape of the surface of the salt at each stage in the restoration was computed from our three-dimensional geomechanical model (see Figures 16, 17).

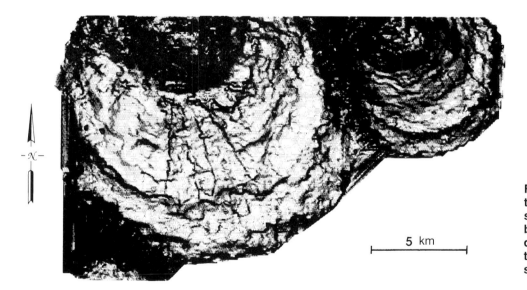

Figure 8—Seismic dip map on the base of salt for the Mickey salt sheet and one of its neighbors. Steeper dips are show as darker bands. Location of the three-dimensional survey is shown in Figure 2.

stress levels reached during emplacement, about 0.01–0.1 MPa, are small enough that the salt is well approximated as a viscous fluid.

It is assumed that no slip takes place between the salt and the sediment at the base of the sheet. The substrate is treated as rigid, so that later compaction, deformation due to emplacement, and faulting associated with salt withdrawal at depth can be ignored.

The basal surface onto which the salt spreads is progressively constructed by sedimentation outside its periphery (Figure 6). In principle, it would be possible to track the varying slope of the base of the sheet from Eq. (1) (as done by Talbot, 1995), but this would involve specifying the local rate of sedimentation and its variation around the periphery of the sheet and over time. In the most refined of our models, however, we treat the spread-

ing of a glacier into a "container basin" of preassigned form approximating the observed base of the Mickey sheet. Evidently, the container basin outside the current glacier edge is irrelevant to the model.

Mathematical Approximation

Lubrication theory (Huppert, 1982; Emerman and Turcotte, 1983) provides an adequate approximation of the dynamics of a spreading sheet. A flow treatable by lubrication theory is dominated by subhorizontal shear, whereas subhorizontal shortening or extension is much smaller. The theory applies to a sheet with a thickness that is small relative to its horizontal dimension, a condition only satisfied after the sheet has undergone some spreading. However, comparison with experiments (Huppert,

1982) shows that the theory does well in describing sheet spreading, even if, as must be the case, this ratio is not initially large.

The approximations used in lubrication theory also break down at the steep toe of the sheet, but, again, the forms of spreading sheets are in good agreement with those observed experimentally. The transition from upward flow in the feeder stock to outward radial flow is not accounted for. Indeed, the addition of material from a source diapir cannot be distinguished in our model from deposition at the surface, as in an ice glacier. Later we use this feature to formulate and analyze a model for a composite salt–sediment glacier, fed in whole or in part by sedimentation onto its upper surface.

The theory (see the Chapter Appendix) results in an evolution equation for sheet thickness, h, as a function of time:

$$\partial h/\partial t \approx [(\rho - \rho_w)g/3\eta]\{\partial/\partial x[h^3(\partial h/\partial x + \beta_x)] + \partial/\partial y[h^3(\partial h/\partial y + \beta_y)]\} + \Delta \qquad (2a)$$

where x and y are horizontal coordinates with origins at the center of the feeder stock and the vertex of the container basin. The horizontal coordinate x is in the regional downdip direction and the horizontal coordinate y in the regional strike direction. See Table 1 for the definition of other variables.

The quantity Δ is the net rate of addition of volume to a vertical column of the sheet, per unit area, through both its basal and upper surface. Thus, it may account for both the addition of material from the feeder stock, over a limited area near the coordinate origin, and any dissolution of salt at the upper surface. Because salt is added only near the coordinate origin and the details are unimportant, it is generally sufficient to specify only the rate of increase in sheet volume, $Q = Q(t)$. The function Δ is then used only to describe the distribution of the rate of surface dissolution over the sheet; Δ is negative for dissolution. For an axisymmetric sheet, (2a) reduces to the simpler form

$$\partial h/\partial t \approx [(\rho - \rho_w)g/3\eta](1/r)\partial/\partial r[rh^3(\partial h/\partial r + \beta)] + \Delta \quad (2b)$$

where β is now the basal slope of salt in the radial direction of downdip advance.

The stress and velocity throughout the sheet are also obtained in the course of deriving Eq. (2). These depend on the current form of the sheet. In the axisymmetric case, the shear stress is

$$\sigma_{rz} \approx (\rho - \rho_w)g(\partial h/\partial r + \beta)(z - h) \qquad (3a)$$

and the horizontal radial component of velocity is

$$v_r \approx [(\rho - \rho_w)g/(2\eta)](\partial h/\partial r + \beta)(z^2 - 2hz) \qquad (3b)$$

Both the stress differences, $\sigma_{rr} - \sigma_{zz}$ and $\sigma_{\theta\theta} - \sigma_{zz}$, and the vertical component of the velocity, v_z, are smaller, second-order quantities.

Axisymmetric Glacier Spreading onto a Planar Horizontal Surface: Model of Advance History and Sheet Form

The dynamics of a spreading salt glacier has two fundamental controls: (1) the tendency of the salt to spread laterally from a salt fountain at which a certain regimen of supply is maintained and (2) the lateral constraint imposed by contemporaneous sedimentation through the buildup of a "container basin."

A closed-form similarity solution of Eq. (2b) is available for an axisymmetric glacier spreading onto a planar, horizontal surface (Huppert, 1982). *Similarity* refers to a reduction in the partial differential equation (2b) in the two independent variables, r and t, to an ordinary differential equation in the single variable $\xi = r/R(t)$, where $R(t)$ is the radius of the sheet. This is accomplished by adopting specific functional forms for h and R, given below, and by imposing certain other restrictions such as on the regimen of salt supply.

The closed-form results clearly show the dependence of glacier dynamics on physical properties such as the viscosity and density and on the regimen of salt supply. The axisymmetric model provides a first cut at an analysis of the downdip advance history of the Mickey sheet. Two features of the Mickey sheet not directly accounted for by this model—the shallow conical form of its base and its updip–downdip asymmetry—are treated later in a three-dimensional numerical model.

The similarity solution requires a regimen of salt supply in which the sheet volume increases according to a relationship of the form

$$V(t) = K_\alpha t^\alpha \qquad (4a)$$

where K_α and α are constants. The rate of volume increase is

$$Q = Q(t) = dV/dt = \alpha K_\alpha t^{\alpha-1} \qquad (4b)$$

The rate of salt supply from the source diapir, Q_Σ, is only equal to Q if there is no dissolution; otherwise, it is greater.

The similarity solution is obtained by adopting the forms

$$R(t) = Mt^q \qquad (5a)$$

and

$$h(r, t) = Ct^p\Psi(\xi) \qquad (5b)$$

From the early part of the analysis (see Chapter Appendix), we find that

$$q = (3\alpha + 1)/8$$
$$p = (\alpha - 1)/4 \qquad (5c)$$

To incorporate the effect of dissolution, we must specify the form of the rate of dissolution per unit area, $\Delta(r, t)$. To satisfy the conditions for a similarity solution, $\Delta(r, t)$

Table 1—Definition of Variables Used in this Paper

Variable	Definition
A	constant in expression for shape function of axisymmetric sheet (Ψ)
α	exponent of time (t) in expression for sheet volume (V) or for cross-sectional area of salt ridge
B	constant in expression for the downdip radius of sheet
β, $\beta(r)$	basal slope of axisymmetric sheet
β_x, β_y	components of slope in x and y directions
β_0, β_1	slopes used in model of three-dimensional container basin
C	coefficient in similarity solution expression for sheet thickness
C^*	C in zero dissolution case
C_{sat}, C_w	concentrations of salt in saturated solution and in seawater
D_{xx}, D_{xz}, D_{zz}	components of rate of deformation in spreading salt ridge
D_{rr}, $D_{\theta\theta}$, D_{zz}, D_{rz}	nonzero components of rate of deformation tensor in axisymmetric sheet
δ	thickness of pelagic carapace
d_0	constant in dissolution rate
Δ	specific rate of dissolution on surface of salt sheet
D	diffusivity of salt in water
ϕ	sediment porosity
η	uniform viscosity of sheet
g	acceleration of gravity
h, $h(r,t)$, $h(x,y,t)$	vertical thickness of sheet, as function of position and time
K, K_α	constant coefficient in expression for sheet volume, without and with volume regimen designated
K', K_α'	nondimensional form of coefficients
L	half-width of spreading salt ridge
M	coefficient in similarity solution form for sheet radius
M'	nondimensional form of M in finite-difference solution
M^*	M in zero dissolution case
m	factor involved in modification of similarity solution in the case of dissolution
p	exponent of time (t) in similarity solution expression for sheet thickness (h)
q	exponent of time (t) in similarity solution expression for sheet radius (R)
Q, $Q(t)$	flux, or rate of salt supply from source diapir
σ	mean (normal) stress, $\sigma = (1/3)(\sigma_{xx} + \sigma_{yy} + \sigma_{zz})$
r, θ, z	radial, circumferential, and vertical coordinates for axisymmetric sheet
R, $R(t)$, $R(V)$	radius of axisymmetric sheet, with designation as function of time (t) or of sheet volume (V)
R'	nondimensional form of R used in finite-difference solution
ρ, ρ_w	densities of salt and water
S	rate of sedimentation
σ_{xx}, σ_{xy}, σ_{xz}, σ_{yy}, σ_{yz}, σ_{zz}	components of stress in Cartesian system x, y, z
σ_{rr}, σ_{rz}, σ_{zz}, $\sigma_{\theta\theta}$	nonzero components of stress in axisymmetric sheet
t	time from initiation of sheet emplacement
t^*	scale time in finite-difference solution
t'	nondimensional time, t/t^*, in finite-difference solution
t_f, t_f'	time at end of emplacement; $t_f' = t_f/t^*$
τ	tortuosity of diffusion pathway
u, v, w	components of velocity in sheet, referred to Cartesian coordinates x, y, z
v_r, v_z	components of velocity in axisymmetric sheet
V, $V(t)$	sheet volume, with designation as function of time
x, y, z	two horizontal and one vertical coordinates
x^*	length scale in finite-difference solution
ξ	reduced variable, $r/R(t)$, used in similarity solution for axisymmetric sheet
Ψ, $\Psi(\xi)$	shape function in similarity solution for axisymmetric sheet
Ψ'	derivative of Ψ with respect to ξ
z_0	height of base of salt above horizontal reference level

must have a special time dependence. For simplicity, we take Δ to be uniform over the surface of the sheet, but other r dependences could be used. Then,

$$\Delta(r, t) = -d_0 t^{(\alpha-5)/4} \qquad (6a)$$

where d_0 is a constant with SI units of m sec$^{(1-\alpha)/4}$.

The net rate of salt being dissolved at the current sheet radius is

$$Q_{\text{diss}} = \pi R^2 \Delta = \pi M^2 d_0 t^{\alpha-1} \qquad (6b)$$

Then, the total flux from the source diapir must equal this plus the amount contributing to the volume increase of the sheet, or

$$\begin{aligned} Q_\Sigma &= Q + Q_{\text{diss}} \\ &= \alpha K_\alpha [1 + (\pi M^2 d_0 / \alpha K_\alpha)] t^{\alpha-1} \\ &= Q(1 + f) \end{aligned} \qquad (6c)$$

The total flux is partitioned between dissolution loss and the increase in glacier volume by a constant ratio, f. In essence, it is the choice of this simple relationship that permits the existence of a similarity solution.

Carrying out the remaining analysis, we obtain

$$\begin{aligned} M &= m^{3/2} M^* \\ C &= m C^* \\ M^* &= 0.584[(\rho - \rho_w) g K_\alpha{}^3 / (q\eta)]^{1/8} \end{aligned}$$

and

$$C^* = 1.01\{\eta K_\alpha / [q^{1/3}(\rho - \rho_w) g]\}^{1/4} \qquad (7a)$$

The factor m is given by

$$m^4 = 1 - (8\alpha / [15(3\alpha + 1)]) f \qquad (7b)$$

This quantity, which accounts for the effect of dissolution, is independent of the choice of physical parameters and depends, through α, only on the regimen of the salt supply and the proportion of it dissolved, f. For zero dissolution, $m = 1$.

Finally, using an approximate solution for the function Ψ, the radial profile of sheet thickness is given by the following equation:

$$\begin{aligned} h(r,t) &\approx Ct^p\{(3q)^{1/3}(1 - \xi)^{1/3} + (d_0/5qC)(1 - \xi) \\ &\quad + [(4q - 1)/24q](3q)^{1/3}(1 - \xi)^{4/3}] \end{aligned} \qquad (8)$$

Modeling the Downdip Advance History of the Mickey Sheet

Let us apply the above result for $R(t)$ to fit the downdip advance history of the Mickey sheet (Table 2, Figure 9) assuming zero dissolution. The model then depends only on the parameters α, K_α, η, and $(\rho - \rho_w)g$. Because the last is known, the fit involves fixing the first three.

Because the Mickey sheet is not axisymmetric and its base is not horizontal, we must construct an approximating sheet with these properties. We form an axisymmetric

Table 2—Advance History of the Mickey Salt Sheet in the Downdip Direction

Horizon Age (Ma)	Time (m.y.)	Sheet Radius (km)	Height of Base (km)
11.7	0	1.10	0.09
9.0	2.7	1.33	0.17
8.3	3.4	2.78	0.78
7.7	4.0	2.88	0.84
6.8	4.9	3.08	0.99
6.15	5.55	4.35	1.17
5.6	6.1	6.63	1.56
4.6	7.1	8.95	1.78
3.0	8.7	11.43	2.20
0.0	11.7	11.93	2.45

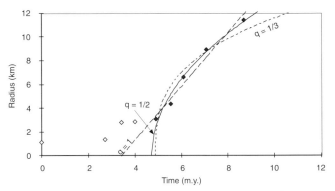

Figure 9—Data points showing the radius of the Mickey salt sheet in the downdip direction versus time. The five solid data points lie in the interval of rapid advance, about 4 m.y. in duration. Best-fit curves through these points are for the relationships $R(t) = Mt^q$, for values of the power q = 1/3, 1/2, and 1. Model time, t, in each case is measured forward from a fictitious origin at which the sheet had zero radius. The fictitious interval of active emplacement, during which the sheet was supplied with salt, is denoted as t_f. Values of the constants M, K_α, η, and t_f are given in Table 3.

sheet by swinging the downdip section of the natural sheet about a vertical axis through the inferred position of the center of the feeder stock. This sheet has a volume $V \approx 340$ km^3, about 1.6 times the actual volume of about 205 km^3.

The model time, t, is measured from the inception of spreading, at which time the axisymmetric model sheet has zero volume. If t_f is the time at which the total volume was emplaced, $V(t_f) = 340$ km^3. We use the axisymmetric model to treat only the episode of rapid spreading, during which the greater volume of the sheet was emplaced.

Because the Mickey sheet had finite volume and radius prior to the episode of rapid spreading, the model time of initiation, $t = 0$, is fictitious. However, if the regimen of supply to a sheet is changed and then maintained over a substantial period, any deviation from the form that the sheet would have if it had been spreading under the regimen of supply since an initiation at $t = 0$ will

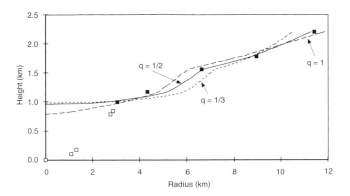

Figure 10—Shape of the base of the Mickey salt sheet as determined by the same data points used in Figure 9. Curves through these points are for the best-fit advance histories shown in Figure 9 for q = 1/3, 1/2, and 1. In each case, the modeled advance history was combined with the actual sedimentation history at the Mickey sheet to derive its basal shape.

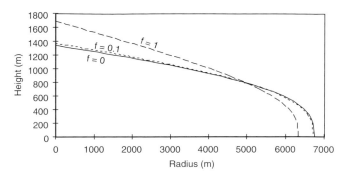

Figure 11—The effects of dissolution on the shape of the top of salt for an axisymmetric sheet spreading onto a flat surface. The curves shown are derived from Eq. (8) assuming α = 7/3 and t = 3 m.y., with other parameters obtained from Table 3. Shapes of the top of salt are shown for zero dissolution (f = 0), 9% dissolution (f = 0.1), and 50% dissolution (f = 1). A sheet with more dissolution requires a greater flow of salt up the feeder to yield the same net flux and thus has a higher fountain and a steeper overall slope.

rapidly disappear. That is, the initial condition of the sheet before the episode would be "forgotten." Such an adjustment to the new regimen should take no more time than that required to double the initial volume, a modest fraction of the approximate 4-m.y. period of rapid advance. Thus, the prior history of the salt mass can be ignored, and application of the model only requires that the relationship $V \propto t^\alpha$ apply over the period of rapid advance. This period is spanned by the five solid data points in Figure 9.

The approximation of spreading into a conical basin by spreading onto an horizontal surface should be adequate if the basin is sufficiently shallow that the rate of spreading is dominated by internal resistance to flow rather than by the geometric constraint of a container. We provisionally assume this to be so before describing a more realistic simulation.

Of the input variables needed for the model, the history of salt supply to the Mickey sheet during its rapid advance is the least well constrained. If the rate of volume increase were constant, $Q = K_1 = V(t_f)/t_f \approx 85$ km^3/m.y. for the model sheet, approximating t_f as 4 m.y. Here, in addition to a regimen of constant Q that results in radial advance $R \propto t^{1/2}$, we consider a regimen for which $R \propto t^{1/3}$ and $Q \propto t^{-2/3}$ dies off with time and a regimen for which $R \propto t$ and $Q \propto t^{4/3}$ increases with time.

A fit of Eq. (5a), with a specified value of q, to the five data points spanning the period of rapid advance (Figure 9) yields a value for the constant M and the model time required to emplace the sheet, t_f, at which the radius reaches 11 km. The fits for q = 1/3, 1/2, and 1 are shown in Figure 9, and the values of M and t_f are given in Table 3. The best fits are for q = 1/2 and q = 1, but the fit for q = 1/3 is also acceptable.

The parameter $K_\alpha = 340 t_f^{-\alpha}$ (km^3/m.y.$^\alpha$) is also available from the fit. Because we have assumed zero dissolution (m = 1), the only unknown in Eq. (7a) is the viscosity, η. Estimates for η in Table 3 differ by only a factor of three

and are comparable to, though slightly larger than, those obtained from the laboratory determined flow law for salt (Jackson and Talbot, 1991) for a temperature of ~20°–40°C, a grain size of ~1 mm to 1 cm, and a shear stress of ~0.01–0.1 MPa, estimated from the model using Eq. (3a).

The advance history, which tracks the lateral position of the front of the glacier through time, can be combined with the sedimentation history, which gives the elevation of the front of the sheet through time, to construct a shape for the base of salt. Shapes have been calculated for the base of the Mickey sheet using the known sedimentation history (Table 2) and the best-fit advance curves for q = 1/3, 1/2, and 1 (Figure 10). Given that the same sedimentation history was used for each of the models, it is not surprising that the models that provide the best fit to the advance history (q = 1/2 and q = 1) also provide the best fit to the shape of the salt base.

Overall, the advance history and basal shape of the Mickey sheet appear relatively insensitive to the assumed supply history. Excellent fits are possible for flux histories that are either constant (q = 1/2, α = 1) or increasing (q = 1, α = 7/3) with time. A constant flux seems most intuitive geologically, but other regimes could exist, especially if dissolution were taken into account.

Effect of Salt Dissolution

We now consider what effect dissolution of the adopted form in Eq. (6a) has on the dynamics of an axisymmetric spreading sheet. A rate of dissolution can be estimated by determining the steady-state diffusion of salt through a porous pelagic carapace of uniform thickness, δ. Using the standard equations for steady-state diffusion through a layer, we obtain

$$\Delta = -\phi\tau D(C_{\text{sat}} - C_{\text{w}})/(\delta\rho) \qquad (9a)$$

where τ is the tortuosity of diffusion pathways, ϕ is the

Table 3a—Parameters for the Best-Fit Curves to the Mickey Sheet Advance History, Assuming Axisymmetric Geometry (SI Units)

q	α	M (m/sec^q)	t_f (sec)	K_α (m^3/sec$^\alpha$)	η (10^{18} Pa s)
1/3	5/9	2.13×10^{-1}	1.38×10^{14}	4.73×10^3	11.3
1/2	1	1.00×10^{-3}	1.20×10^{14}	2.84×10^{-3}	6.73
1	7/3	7.13×10^{-11}	1.54×10^{14}	2.67×10^{-22}	4.31

Table 3b—Parameters for the Best-Fit Curves to the Mickey Sheet Advance History, Assuming Axisymmetric Geometry (Alternative Units)

q	α	M (km/m.y.^{q})	t_f (m.y.)	K_α (km^3/m.y.$^\alpha$)	η (10^{18} Pa s)
1/3	5/9	6.73	4.39	149	11.3
1/2	1	5.64	3.80	89.5	6.73
1	7/3	2.25	4.89	8.38	4.31

porosity of the sediment, $D \approx 10^{-9}$ m^2/sec is the ionic diffusivity in water, $C_{sat} \approx 300$ kg/m^3 is the saturation concentration, $C_w \approx 30$ kg/m^3 is the seawater concentration, and $\rho = 2150$ kg/m^3 is the salt density. Using maximal values $\tau \approx 1$ and $\phi \approx 0.5$,

$$\Delta \approx -[(6.3 \times 10^{-11})/\delta] \text{ m/sec} \qquad (9b)$$

with δ expressed in meters. Expressed in units that are more easily comprehended, $\Delta \approx (20/\delta)$ cm/year, with δ still expressed in meters.

If Eq. (9b) is to conform to that required for the similarity solution of Eq. (6a), the thickness of the pelagic carapace would have to vary as

$$\delta = c_0 t^{[(5-\alpha)/4]} \qquad (9c)$$

For the two preferred models, the dependences are $\delta \propto t$ for $\alpha = 1$ and $\delta \propto t^{2/3}$ for $\alpha = 7/3$. In either case, the thickness would grow at a rate about proportional to time. In more complicated models, a changing and nonuniform porosity structure of the carapace could play a role in addition to its thickness.

The meaning of the constants d_0 in Eq. (6a) and c_0 in Eq. (9c) is difficult to comprehend. Ignoring the dependence of M on f, we use the values of M and K_α from Table 3 in Eq. (6c) to yield

$$d_0/f = 3.89 \times 10^{-2} \text{m/sec}^{1/3} \qquad (\alpha = 7/3)$$
$$d_0/f = 8.95 \times 10^2 \text{m/sec}^{1/3} \qquad (\alpha = 1) \qquad (10a)$$

and, equating Eqs. (6a) and (9a) for Δ, and using Eq. (9c),

$$c_0 = (1/f)(1.61 \times 10^{-9} \text{m/sec}^{2/3}) \qquad (\alpha = 7/3)$$
$$c_0 = (1/f)(7.01 \times 10^{-14} \text{m/sec}^{2/3}) \qquad (\alpha = 1) \qquad (10b)$$

The relationships of Eqs. (10b) and (9c) can be used to determine the thickness of the pelagic carapace required to fit the advance history, assuming that some fraction, f,

of the salt has been dissolved. Given that the carapace of the Mickey sheet is below seismic resolution after 5 m.y. of emplacement, we can use this to make a minimum estimate of the fraction of salt dissolved. For example, for $\alpha = 7/3$ and $f = 0.1$, the carapace was 26 m thick after 2 m.y. and 48 m thick at the end of emplacement at 5 m.y. Note that the thickness varies as $1/f$. Because a 50-m-thick carapace ought to be visible on seismic profiles, this simple model for dissolution suggests that, for $\alpha = 7/3$, more than 10% of the extruded salt must have been dissolved. A similar calculation for $\alpha = 1$ indicates that a carapace of 50 m or less requires 20% dissolution over 5 m.y. Qualitatively this result is not surprising because the constant flux emplacement history ($\alpha = 1$) has more of the salt exposed to dissolution for a longer time than the increasing flux history.

The quantity m, which enters the expression for M in Eq. (7a), affects the calculated value of the salt viscosity. For $\alpha = 7/3$, Eq. (7b) yields $m^4 = 1 - 0.155f$. Even when half of the salt issuing from the source diapir is dissolved, or $f = 1$, $m = 0.92$, not greatly different from unity. Because $M \propto (m^{12}/\eta)^{1/8} \propto [(1 - 0.155f)^3/\eta]^{1/8}$, the value obtained from the advance data implies a smaller salt viscosity for a larger ratio f, but even for $f = 1$, the reduction is only by the modest factor of 0.6. If $\alpha = 1$, the effect of dissolution is even smaller ($m = 0.96$).

Conversely, for a given viscosity, dissolution will result in less advance for a salt sheet of a given volume, but again, the effect will be small. The effect is also seen in the form of the profile (Figure 11). A sheet being dissolved will rise more steeply toward the salt fountain. The resulting greater flow rate compensates for the salt loss by dissolution. The effect on the profile can be measured by the ratio of the second dissolution-related coefficient in Eq. (8) to the leading coefficient. For example, if $\alpha = 7/3$, then the second coefficient in Eq. (8) becomes $d_0/[5(3)^{1/3}C]$. Ignoring the insignificant effect of dissolution on C, through the quantity m, we have

$$C = 1.785 \times 10^{-2} \text{m/sec}^{1/3}$$

and using Eq. (10a),

$$d_0/[5(3)^{1/3}C] = 0.3f \qquad (11)$$

Thus, for $f = 0.1$, the height of the salt fountain is only about 3% greater than that for the zero dissolution case; for $f = 1$, it is a substantial but still modest 30% higher.

Topographic Relief of a Salt Sheet

The radial profile of sheet thickness in Eq. (8) can be used to estimate the bathymetric relief of a salt sheet. Because the axisymmetric model sheet spreads across a horizontal surface, the relief relative to the downdip edge of the sheet is the full sheet thickness, $h(r, t)$. For a sheet spreading into a shallow basin constructed by contemporaneous sedimentation, only a fraction of the thickness variation contributes to bathymetric relief. For the Mickey sheet, the axisymmetric profile is best viewed as an estimate of bathymetry only along the profile from the salt fountain to the downdip edge.

Because the approximation is crude, we use only the leading term in Eq. (8) to estimate the form of the salt top during active emplacement. We can then write

$$h(r, t) \approx h_0(t)(1 - r/R)^{1/3} \qquad (12a)$$

where from Eq. (8) we get

$$h_0(t) = 1.01\{\eta K_\alpha/[q^{1/3}(\rho - \rho_w)g]\}^{1/4}(3q)^{1/3}t^{[(2q-1)/3]} \qquad (12b)$$

In evaluating Eq. (12b), it is more convenient to use kilometers and million years as units, which requires converting the factor $\{\eta K_\alpha/[(\rho - \rho_w)g]\}$ to one in km m.y. Noting that η typically has a value close to 10^{18} Pa s, we can write $\eta = N \times 10^{18}$ Pa s. Using $\rho - \rho_w = 1150$ kg/m³ and $g = 9.8$ m/sec², the factor can be written as $2.808 \times 10^{-3} NK_\alpha$, where K_α is expressed in units of km³/m.y.$^\alpha$, and both K_α and N are taken from Table 3.

We carried out the computation for all three cases in Table 3. The results are, for $\alpha = 5/9$ ($q = 1/3$),

$$h_0(t) = 1.63t^{-1/9} \qquad (12c)$$

for $\alpha = 1$ ($q = 1/2$),

$$h_0(t) = 1.40 \qquad (12d)$$

and, for $\alpha = 7/3$ ($q = 1$),

$$h_0(t) = 0.822t^{1/3} \qquad (12e)$$

where h_0 is in kilometers and t is in million years.

Profiles have been computed at $t = 1$ m.y., $t = 3$ m.y., and $t = t_f$ for each of the three cases (Figure 12). Note that for constant flux ($\alpha = 1$), the maximum height is also constant. When the flux dies off with time ($\alpha = 5/9$), the maximum height also dies off weakly, and when the flux increases strongly with time ($\alpha = 7/3$), the height grows at a modest rate.

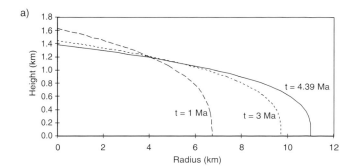

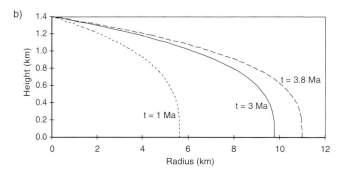

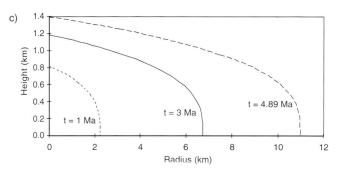

Figure 12—Profiles of the top of salt through time, calculated using Eqs. (12c–e) for an axisymmetric sheet spreading onto a flat surface. Profiles are shown at $t = 1$ m.y., 3 m.y., and t_f for (a) waning salt flux ($\alpha = 5/9$), (b) constant flux ($\alpha = 1$), and (c) increasing flux ($\alpha = 7/3$). Note that an escarpment with several hundred meters of relief exists at the downdip margin of the sheet in each case.

A second aspect of the profiles is their dependence on the scaled distance from the axis, $\xi = r/R$, given by the dimensionless quantity $(1 - \xi)^{1/3}$. Essentially, for zero dissolution, all sheet profiles can be plotted on the same master curve if the ordinate is ξ and the abscissa is h/h_0 (Figure 13). The form is that of a broad "pancake," with a gentle slope away from the axis and a steep shoulder in which the height drops off more and more rapidly toward the margin. The thickness is still $h/h_0 = 0.46$ at 90% of the sheet radius.

By these estimates, active salt sheets will have large topographic relief and the relief will be concentrated at their advancing edges. An important consequence is that active salt glaciers are expected to be effective dams to sediment transport, and submarine channels must divert around them. If salt sheets coalesce to form a continuous barrier, areas downdip of the sheets may be sand starved.

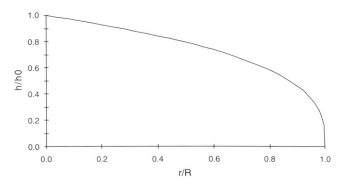

Figure 13—Master curve for the shape of all axisymmetric salt sheets spreading onto a horizontal surface, from Eq. (12a). This shape assumes no dissolution. Both the height and radius are normalized to their maximum values.

Fully Three-Dimensional Simulation of the Mickey Sheet

By means of a three-dimensional model, implemented numerically, we can account for the asymmetry of the natural sheet. All of the results developed and discussed for the simpler model of an axisymmetric sheet spreading onto a horizontal surface could be developed by numerical means, but the labor is much greater. As a consequence, we only consider the case of zero dissolution. Computations were carried out for both of our preferred regimens of supply $V \propto t^{7/3}$ and $V \propto t$. Results are described in detail for the $\alpha = 7/3$ case, with a brief summary of the results for $\alpha = 1$.

The form of the base of the Mickey sheet (Figure 14) can be approximated by tilting a symmetric conical basin with uniform slope angle $\beta_0 = 0.271$ (15.5°) about a horizontal axis normal to the direction of the regional slope by the angle $\beta_1 = 0.157$ (9°). The expression for the height of the basin above its vertex at the origin is

$$z_0(x, y) = \beta_0 r + \beta_1 x \qquad (13a)$$

where x is a horizontal coordinate in the downdip direction, y is the horizontal direction regional slope-parallel direction, and $r = (x^2 + y^2)^{1/2}$. The maximum and minimum slope angles, which occur in the profile plane along the x-axis, are $\beta_0 + \beta_1 = 0.428$ (24.5°) and $\beta_0 - \beta_1 = 0.113$ (6.5°). The components of the slope of the basal surface, used in the evolution equation (2a) are

$$\beta_x = \partial z_0 / \partial x = \beta_0(x/r) + \beta_1$$
$$\beta_y = \partial z_0 / \partial y = \beta_0(y/r) \qquad (13b)$$

In carrying out the numerical solution of Eq. (2a), the quantity $h(x, y, t)$ is computed on a square grid of discrete points $\{x, y\}$ by means of a standard finite-difference method. Salt is fed at a small number of surface elements centered on grid points about the apex of the basin by incrementing the values of h at these grid points at each time step. The amount of the increment and its variation with time are selected to give a regimen of the type in Eq. (4a).

Figure 14—Comparison of contours for the base of the Mickey salt sheet and the asymmetric model "container basin." Both are contoured at the same interval.

Smoothed contours of Mickey base salt at arbitrary interval in area of 3-D seismic survey

Contours of model container basin

A characteristic length, x^*, and time, t^*, are used to cast Eq. (2b) into dimensionless form. In so doing, it is convenient to set

$$(\rho - \rho_w)gx^*t^*/(3\eta) = 1 \qquad (14a)$$

The length x^* is picked as the grid spacing; the choice made, $x^* = 0.4$ km, provides an adequate representation of the final sheet form. The model results depend only on the parameter α and the dimensionless constant $K_\alpha' = K_\alpha[(t^{*\alpha})/(x^*)^3]$, which together specify the regimen of salt supply.

The model provides the full geometry of the spreading sheet as a function of time. Here, we are particularly interested in the downdip radius, which we continue to denote as R. The computation yields

$$R' = R/x^* = R'(t/t^*, \alpha, K') \qquad (14b)$$

Having chosen $\alpha = 7/3$, the model is run for a series of values of K'. For each value, a set of discrete values of R' is obtained.

A fit of the model to the observations from the natural sheet requires that the dimensionless volume of the model sheet equals $V' = 205/(x^*)^3 = 3200$ at the same time that its downdip radius equals $R' = 11/x^* = 27.5$. For a given α, only a single value of K' will yield this result.

Using $x^* = 0.4$ km and an estimate of salt viscosity of $\eta \approx 3 \times 10^{18}$ Pa s, Eq. (14a) yields $t^* \approx 0.06$ m.y., so that the sheet should reach its final form at $t_f' \approx 4$ m.y./$t^* \approx 67$. Because

$$V' = K'(t')^{7/3} \qquad (14c)$$

the desired value is roughly $K' \approx 3400/(67)^{7/3} = 0.17$.

To make the results applicable to salt sheets of varying volume, still supposing the same regimen of salt supply is applicable, we carried out numerical computations for a wide range of K' values (Figure 15). Interpolation is required to find the value of K' yielding the desired pair (R', V').

We found that the simple functional relationship for spreading of an axisymmetric sheet onto a horizontal surface as in Eq. (5a) still holds to a good approximation for the downdip radius, the only difference being in the constant multiplier. The relationship is $R' \approx M't'$ or, substituting for the volume from Eq. (14c),

$$R'(V', 7/3, K') \approx [M'/(K')^{3/7}](V')^{3/7}$$
$$\approx [c/(K')^{3/56}](V')^{3/7} \qquad (15a)$$

The results computed for the values $K' = 0.234$, 0.386, 1, and 3.86 yield values of the constant c that vary only between 0.82 and 0.79, with the larger value holding for the lowest value of K'. Because the estimated value of K' is closest to that value, we set $c = 0.82$; the pair $(R', V') = (27.5, 3200)$ corresponds to $K' = 0.24$.

Using $t_f = 4.89$ m.y. (Table 3), $V = 205$ km³, and the inferred regimen of supply, $V = Kt^{7/3}$, we obtain $K = 5.05$ km³/m.y.$^{7/3}$. Combining these with the estimate for M (Table 3), we obtain the revised estimate of $\eta \approx 4.8 \times 10^{18}$ Pa s. These values, along with the results of an equivalent analysis for $\alpha = 1$, are presented in Table 4.

Recasting the relationship of Eq. (15a) into one in dimensioned quantities and in time rather than volume, we obtain

$$R(t) = 0.715[(\rho - \rho_w)gK^3/\eta]^{1/8}t \qquad (\alpha = 7/3)$$
$$R(t) = 0.791[(\rho - \rho_w)gK^3/\eta]^{1/8}t^{1/2} \qquad (\alpha = 1) \qquad (15b)$$

The equivalent relationships for the axisymmetric sheet spreading onto an horizontal surface, combining Eqs. (5a) and (7a), differ only in the numerical coefficients, which are 0.584 ($\alpha = 7/3$) and 0.536 ($\alpha = 1$). These are roughly 70–80% the values in Eq. (15b). That is, although the three-dimensional sheets are flowing into a conical, constraining basin, the *downdip* rates of advance are greater than that for the sheets spreading onto a horizontal surface! This is because the form of the basin focuses the bulk of the salt flow in the downdip direction; this effect more than offsets the constraining effect of the basin.

The relationship in Eq. (15a) can only approximate the behavior of a sheet spreading into a basin because the exact behavior cannot correspond to a similarity solution. Consider the limiting form of a sheet as it continues to fill a basin. The salt already in the basin will tend to relax toward a form in which the upper surface of the sheet is perfectly flat. For the considered basin, this form, determined solely from its geometry, is

$$R' = 2.96(V')^{1/3} \qquad (15c)$$

As more and more fluid is added to the sheet, the incre-

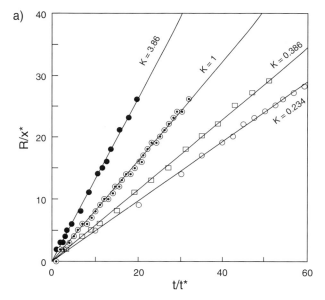

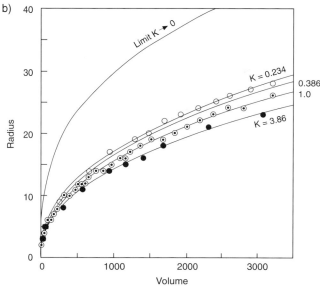

Figure 15—Downdip radial advance into the asymmetric conical model basin, as determined from the numerical results for the $\alpha = 7/3$ case. (a) Linear fits to $R/x^* = M'(t/t^*)$, for $K' = 0.234$, 0.386, 1, and 3.86. (b) Fits of the form $R/x^* = C[V/(x^*)^3]^{3/7}$ for the same values of K' and the geometric limit $R/x^* = 2.98[V/(x^*)^3]^{1/3}$.

ment of volume added per unit time relative to the volume already present decreases according to

$$(1/V)(dV/dt) = \alpha t^{-1} \qquad (15d)$$

Thus, the perturbing effect of the incoming fluid will constantly decrease, and the form of Eq. (15c) will be approached more and more closely. But initially, the relationship of Eq. (15a) will hold. Because similarity requires that one or the other relationship hold at all times, a similarity solution cannot exist.

While the downdip advance of the three-dimensional model for the Mickey sheet falls far from the geometric

Table 4a—Parameters for the Best-Fit Curves to the Mickey Sheet Advance History, Fully Three-Dimensional Model (SI Units)

q	α	M (m/sec^q)	t_f (sec)	K_α (m^3/sec$^\alpha$)	η (10^{18} Pa s)
1/2	1	1.00×10^{-3}	1.20×10^{14}	1.71×10^{-3}	8.3
1	7/3	7.13×10^{-11}	1.54×10^{14}	1.61×10^{-22}	4.8

Table 4b—Parameters for the Best-Fit Curves to the Mickey Sheet Advance History, Fully Three-Dimensional Model (Alternative Units)

q	α	M (km/m.y.^{q})	t_f (m.y.)	K_α (km^3/m.y.$^\alpha$)	η (10^{18} Pa s)
1/2	1	5.64	3.80	53.9	8.3
1	7/3	2.25	4.89	5.05	4.8

limit of Eq. (15c), the updip advance of the model falls closer to it because the constraint of the container (steeper in this direction) has a much greater influence.

The salt flux to a spreading glacier will follow a schedule that is more complicated than that considered here. Whereas the model rate of supply increases constantly with time and then is abruptly shut off, the rate of supply for a natural sheet will decrease continuously to zero.

Form of Sheet, Topographic Relaxation after Cessation of Salt Supply, and Comparison to Present Form of Mickey Sheet

A three-dimensional model with $K' = 0.24$ was run to obtain the form of the sheet at intermediate times $t = 2.5$ and 4 m.y., with the final time $t_f = 5.06$ m.y. (The choice of $t_f = 5.06$ m.y. rather than 4.89 m.y. was based on a preliminary fit to the Mickey geometry.) Salt supply was then shut off, and the relaxed forms at times $t_f + 0.25$ m.y. and $t_f + 1$ m.y. were determined.

Updip–downdip profiles of the model are shown in Figure 16a. During active salt supply, the relief between the salt fountain and the updip edge decreases from about 350 m during the earlier two stages to about 300 m at the last stage. The relief between the fountain and the downdip edge is somewhat larger during the later stages, ranging from about 1000 to 1200 m over the three stages shown. Recall that in the axisymmetric model, the relief at final emplacement was 1000 m but the increase in relief was more marked.

After only 0.25 m.y. of relaxation, the salt fountain (as represented by a height maximum at the position of the source diapir) has vanished, and the maximum height on the sheet is now at its updip edge. The steep shoulder of the sheet, about 700 m at the downdip edge but only about 150 m at the updip edge, still remains. In the second stage, 1 m.y. after cessation of salt supply, the sheet has a much gentler overall slope and the steep shoulder at the downdip edge is still present but only about two-thirds of its initial height. Most importantly, the salt at the upper edge of the sheet has flowed down from its high point on the container wall, and there is no more steep shoulder there (Figure 16b). Between these two stages is the point at which nonpelagic sediments start to be transported

onto the salt sheet and burial of the sheet begins.

The supply of salt to the Mickey sheet likely shut off about 3 Ma. This is signaled by the large increase in the basal slope in the downdip direction (see Figure 4), from 6.5° to 18°, an increase that we attribute to a decrease in advance rate over the last 0.5–1 km of advance rather than to a sudden increase in sedimentation rate. This aspect of the Mickey sheet is not included in the model. However, the amount of advance during 1 m.y. after cessation of salt supply is roughly equal to that attributed to the constant volume advance of the Mickey sheet. The relaxed model sheet has about the same maximum height as the Mickey sheet, although the overall slope of the upper surface of the model sheet is, curiously, less than that of Mickey. Part of the difference in the shapes of the sheets near the toe might be attributed to the absence of a steeper rim to the model basin along its downdip edge. Another difference is that the complex structure at the updip end of the Mickey sheet is not included in the model.

Contour maps of the form of the upper surface of the model sheet are shown for the end of active emplacement (Figure 17a) and after 1 m.y. of relaxation at constant volume (Figure 17b). The dip section diameters of the sheet in these two stages are about 15% larger than the strike section diameter. These figures illustrate the steep shoulder of the sheet, which is continuous but of variable height at the end of active emplacement and is discontinuous and present only over the lower half of the sheet after 1 m.y. of relaxation.

EVOLUTION OF SALT SHEET DURING EARLY BURIAL: COMPOSITE SALT–SEDIMENT GLACIER

Model of a Composite Salt–Sediment Glacier

Most salt sheets are buried today beneath hundreds to thousands of meters of sediment. Their burial and subsequent deformation have played major roles in the evolu-a)

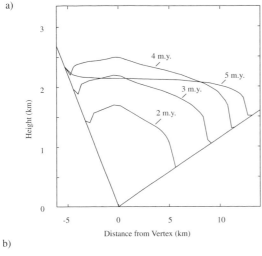

b)

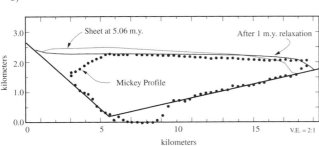

Figure 16—(a) Downdip profiles of model sheet at *t* = 2, 3, and 4 m.y. during active emplacement, and then after 1 m.y. of topographic relaxation at constant volume (5-m.y. curve). (b) Comparison of the model profiles at the end of active emplacement and after 1 m.y. of topographic relaxation, with the profile of the natural Mickey sheet.

a)

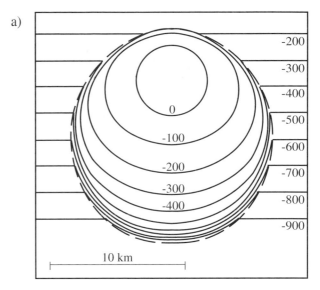

b)

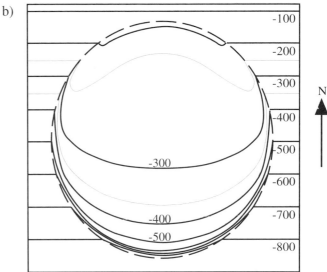

Figure 17—(a) Contours of the surface of the model sheet at the time of final emplacement, for α = 7/3, *t* = 5.06 m.y. (b) The same, but after 1 m.y. of topographic relaxation at constant salt volume. Contours are in meters.

tion of many structures and hydrocarbon accumulations in the Gulf of Mexico (e.g., Worrall and Snelson, 1989; Diegel and Cook, 1990; Sumner et al., 1990; Wu et al., 1990; Diegel et al., 1995; Peel et al., 1995; Schuster, 1995). We now extend the salt glacier model to apply to the evolution of salt sheets during early burial. During this stage, we call the evolving structure a *composite salt–sediment glacier.*

When the supply of salt up the feeder is pinched off, the salt fountain relaxes and sediment begins to prograde onto the sheet from its updip edge. Relief is still present at the downdip end, and the composite glacier continues to increase in volume by sedimentation on the top rather than by salt feed from the bottom. Thus, the sheet continues to advance, maintaining a steep shoulder at its downdip edge (Figure 18). As long as the sediment layer is only a modest fraction of the total thickness, the dynamics of this composite glacier can still be roughly approximated by that of a salt-only glacier.

For this later phase of evolution, the axisymmetric model is of little use. The locus of sediment supply is now the updip edge rather than a more central salt fountain, and the updip–downdip polarity of the glacier assumes

an even more central role. Accordingly, we shall instead consider a two-dimensional plane flow model in which the along-strike dimension is unboundedly large. (A sheet of finite "strike" dimension can be imagined to be bounded laterally by two vertical, frictionless, rigid walls any desired distance apart. The model is still the same two-dimensional one. A more realistic model would take into account the three-dimensional geometry of the glacier and the possibility that structures, especially minibasins, may have irregular geometries or be eccentrically located on the glacier.) Another motive for considering a two-dimensional model is that much thinking about such structures is two-dimensional, especially as it involves operations carried out in a single cross section.

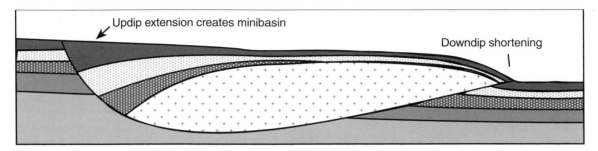

Figure 18—Schematic diagram of a composite salt–sediment glacier. Sediments on top of the glacier are carried along with the underlying salt, producing a pull-apart at the updip end and contractional structures at the downdip toe.

Advancing Composite Glacier Model

Consider a two-dimensional sheet in the form of a low ridge, spreading in two directions onto a horizontal surface from a central axis. If this were a salt glacier, we would think of salt supplied from an axial salt wall. Here, however, we imagine that sediment is being added to the surface of the sheet, chiefly near its axis. Because the two-way flow possesses a mirror plane at the axis, we can think of the axis as the updip stationary edge of a "half-sheet." This is the desired model, the addition of sediment occurring near the updip edge.

Ignoring the difference between the rheological behavior of salt and sediment, we treat the whole as a single viscous sheet. An analysis is carried out using lubrication theory, but we do not give the details (see Huppert, 1982). If the volume of the half-sheet per unit length in the strike direction is given by Eq. (4a), its width is found to be

$$L(t) = M^*t^{q^*} \tag{16a}$$

where

$$q^* = (1/5)(3\alpha + 1) \tag{16b}$$

The form of the quantity M^*, analogous to M, is not needed, because we are not concerned with fitting data.

The result of central interest is the horizontal component of velocity, evaluated at the glacier surface:

$$u(x, h; t) \approx (3/2)(dL/dt)$$
$$\{1 + (1/4)[(3\alpha - 4)/(3\alpha + 1)](1 - x/L)\} \tag{16c}$$

The horizontal component of the surface rate of deformation is

$$D_{xx}(x, h; t) \approx [\partial u/\partial x](x, h; t) \approx -(3/40)(3\alpha - 4)t^{-1} \tag{16d}$$

The rate of deformation is uniform over the sheet to this approximation. The vertical rate of deformation is $D_{zz} = -D_{xx}$. The rate of deformation decreases as $1/t$ from the initiation of spreading and depends otherwise only on α. At $t = 1$ m.y., the surface rate of deformation is $\approx 3 \times 10^{-14}$/sec, a magnitude comparable to that in mountain building. These results are for sediment added at the updip edge, but they are insensitive to other distributions concentrated over the upper part of the sheet.

If surface velocity is normalized to the rate at which the sheet as a whole is advancing, dL/dt, the distribution of velocities across the sheet is solely a function of α (Figure 19). For $\alpha < 4/3$, the updip end of the sheet moves more slowly than downdip parts, causing extension within the sheet ($D_{xx} > 0$). If $\alpha = 4/3$, all overburden particles on top of the sheet have the same velocity, so the surface layer is transported as an inextensible lamina ($D_{xx} = 0$). If $\alpha > 4/3$, updip parts move faster than downdip parts, causing contraction on the surface of the sheet ($D_{xx} < 0$). If it is assumed that significant thicknesses of sediment are deposited only on top of salt glaciers during periods of waning or of no salt supply from the feeder ($\alpha < 1$), the model correctly predicts that extension should be the dominant structural style above composite glaciers.

Whatever the regimen of supply (the value of α), the horizontal velocity of particles at the toe of the sheet is given by

$$u(L, h; t) \approx (3/2)(dL/dt) \tag{17a}$$

Accordingly, particles on top of the leading edge of the sheet are moving 1.5 times as fast as the sheet is advancing. Because a particle at the surface cannot move beyond the end of the sheet, this result would appear impossible. Recall, however, that lubrication theory does not provide a detailed picture of the flow near the toe of the sheet. The interpretation of this result can be explained, however, by a "tractor tread" model in which sediment cover is "consumed" by moving around the toe and overridden by the advancing glacier at a rate that is half the rate of advance.

This interpretation is confirmed by observation of the equivalent process at the downdip edges of subaerial salt glaciers and at the edges of advancing silicic lava domes and flows (Fink, 1983). In both, slumping at the steep front, rather than continuous flow, is a major contributor to transfer of material from the upper surface to the lower surface of the glacier. The layers of disrupted material below salt sheets in the Gulf of Mexico, interpreted from subsalt well data referred to earlier, are likely formed in this manner.

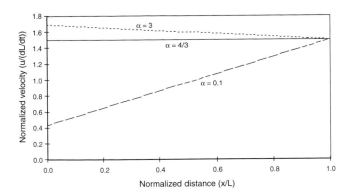

Figure 19—Particle velocity on the surface of a composite salt–sediment glacier, plotted as a function of position on the glacier. Particle velocity is normalized to the advance rate of the sheet, dL/dt. For α < 4/3, the updip part of the sheet moves more slowly than downdip parts, implying intrasheet extension. For α > 4/3, the reverse is true, implying shortening on the glacier surface. Note that in all cases the particle velocity at the front of the glacier is equal to 3/2 times the advance rate of the sheet.

The horizontal velocity at the updip edge of the sheet can be expressed as

$$u(0, h; t) \approx \{45\alpha/[8(3\alpha + 1)]\}(dL/dt) \qquad (17b)$$

We interpret this velocity as the rate of generation of new surface area at the updip edge of the glacier. Again, lubrication theory does not provide the detailed flow at the axis, and moreover, the result is derived from an approximation in the form of an expansion about the toe of the glacier, so that it need not provide a good approximation at the updip edge. However, several lines of reasoning suggest that Eq. (17b) has substantial validity:

1. If the sheet is merely spreading at constant volume, $\alpha = 0$ and $u(0, h; t) = 0$, and no new surface area is created. Here, an exact similarity solution, rather than an expansion about the toe of the sheet, is available; for this, the velocity at the origin vanishes.
2. On silicic domes and lava flows, new surface area is created above the feeder conduit. The mechanism is via the continuous downward propagation of a vertical tension fracture through the brittle surface layer, at a "crease structure" (Anderson and Fink, 1992). The surfaces of the crack are then pulled into a horizontal orientation to form the new surface area. Flow banding within the solidified part of the lava is then oriented vertically below the new surface.
3. On a composite salt–sediment glacier, the normal fault structure at a breakaway zone might be appropriately interpreted as a producer of new surface area because a surface particle on the hanging wall is moving at a finite horizontal velocity away from the footwall (Figure 18).

Implication of the Model for "Restoration" of the Sediment Structure

The picture of the surface kinematics provided in the composite glacier model is strikingly different from that of other kinematic models. For example, it is commonly held that, in some sense, shortening and extension balance in a structure produced by spreading under gravity (Price, 1971). This notion plays a central role in current thinking about the tectonics of the Gulf of Mexico sediment–salt wedge. Our model suggests that any balance between shortening and extension must take into account the difference between a plane flow and a radially diverging flow. The latter is important in any spreading wedge having large strike-parallel dimension relative to its dip-parallel dimension.

From the above model, extension of the updip end of the two-dimensional sheet need not be balanced by shortening at its downdip margin (16c). The reason for this is simply that the sheet is not constrained laterally, either when it spreads onto a horizontal surface or when it spreads into an upward-sloping conical basin. In the latter case, horizons are truncated against the base of the sheet, but the base of the model sheet is not a thrust fault.

The kinematic model of McGuinness and Hossack (1993) estimated the position of the truncation of a sedimentary horizon against the base of the sheet on the basis of data from above the salt sheet. McGuinness and Hossack (1993, their figure 7) suggested that updip extension is equal in magnitude to the advance of the toe of the glacier and is recorded by thrusting or folding in the overlying section (Huber, 1989; Seni, 1992). As a consequence, McGuinness and Hossack (1993) suggested that the advance history of a salt sheet can be reconstructed, and the position of subsalt cutoffs determined, by sequentially restoring the extension in updip normal faults.

Our model predicts a very different relationship between the magnitudes of glacier advance and above-salt extension. Particles immediately above the leading edge of the sheet move at 3/2 times the advance rate of the sheet (Eq. 17a), so over any given interval of time, sediments above salt extend 1.5 times as much as the sheet advances. Furthermore, for low values of α, the extension is distributed across the entire length of the sheet and is not necessarily concentrated at the updip end. Restoration of above-salt extension may still be used to restore the advance history of a sheet and determine the position of basal cutoffs because the rate of salt advance is 2/3 that of extension . However, great care should be taken to ensure that all of the above-salt extension is documented before attempting such a restoration.

We illustrate some of the utility of this by restoring the advance history of a real composite salt glacier (Figures 20, 21). By comparison, it is interesting to note that this same salt sheet was restored as an intrusion by Moretti et al. (1990). Our data on the advance history of this sheet are shown in Table 5. The advance history is less well known than at the Mickey sheet owing to the lesser quality of subsalt data and the lack of subsalt well information. However, the data are sufficient to constrain the

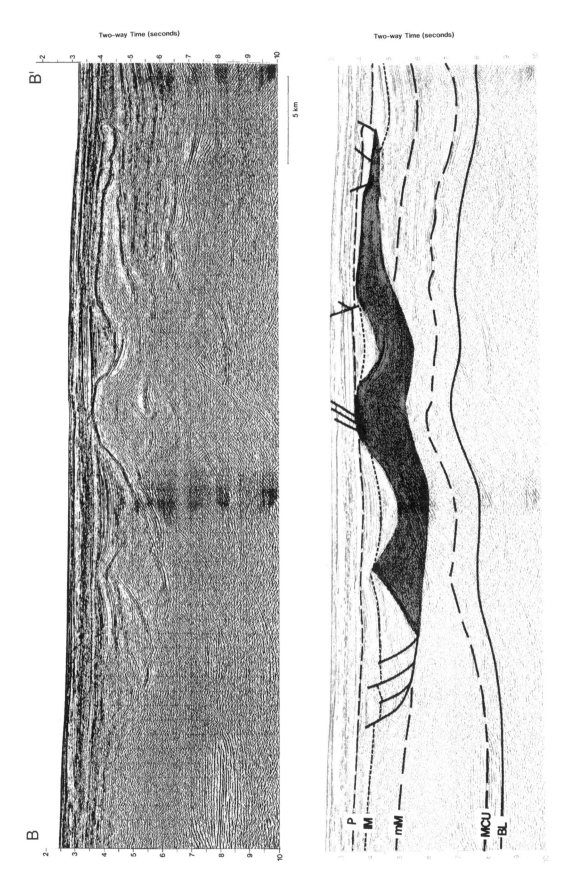

Figure 20—Uninterpreted (top) and interpreted (bottom) two-dimensional time-migrated seismic profile over a composite salt-sediment glacier, offshore Louisiana. Location is shown in Figure 2. Interpreted horizons: P = Pleistocene (1.2 Ma), lM = late Miocene (7.1 Ma), mM = middle Miocene (11.7 Ma), MCU = middle Cretaceous unconformity, BL = Base of Louann. Courtesy of Geco-Prakla.

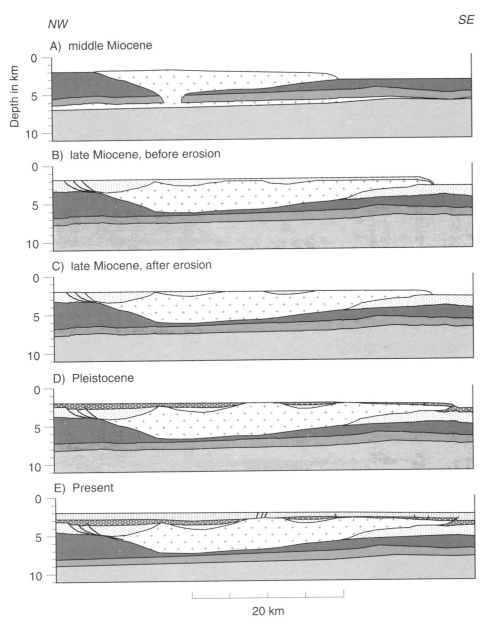

Figure 21—Restoration of the seismic line shown in Figure 20. Depth conversion was done using constant layer velocities. All blocks were restored by vertical simple shear. The effects of decompaction are not included. The positions of the three minibasins on top of the sheet in the restoration were calculated using Eq. (16c) to determine their velocities with respect to one another, assuming $\alpha = 0.039$ (see text for discussion).

advance history over at least part of the composite glacier phase (Pliocene–Pleistocene horizons). During this time, the downdip end of the sheet advanced 1120 m, while the updip end underwent roughly 220 m of visible extension (determined by adding the heaves on the faults). Our model predicts that the surface of the composite glacier underwent $1120 \times 3/2 = 1680$ m of extension during this interval. Most of this hypothetical extension is assumed to be cryptic because the sedimentary veneer of that age is either not present or below seismic resolution over much of the sheet.

Knowing $u(0) = 220/3.4$ m/m.y. and determining dL/dt from the advance history, we can solve Eq. (17b) for α, yielding a value of $\alpha = 0.039$. This suggests that advance took place under conditions of near-constant volume spreading. Knowing α, we can compute the velocity of all points on the surface of the glacier and use this information in building the restoration (Figure 21).

Several points from this restoration are worth emphasizing because they are in our experience applicable to restoration of composite glaciers in general:

1. The positions of the minibasins at each stage of the restoration were calculated by Eq. (16c). The minibasins are carried along with the glacier as it advances. They do not just sink into the salt.
2. A composite glacier can advance with its updip end buried and the downdip end exposed. If this occurs, it is impossible to estimate the total surface extension of the glacier (and thereby determine subsalt cutoff positions) because of the lack of any sediments to record the deformation over the "salt-only" parts of the sheet. We could not have restored the advance history of this sheet without independent evidence from the subsalt cutoffs.
3. The salt sheet experienced only 1–2% of its

Table 5—Advance History of a Composite Glacier in the Downdip Direction

Horizon Name	Horizon Age (Ma)	Sheet Radius (km)
middle Miocene	11.7	23.00
late Miocene	7.1	30.40
Pleistocene	1.2	31.52

advance after the toe was buried. Most composite glaciers stop shortly after burial of the toe, probably because burial of the toe tends to coincide with the loss of relief on the glacier surface. Composite glaciers can, however, be rejuvenated if the toe is reexposed by erosion or slumping of the overlying sediments.

Interpretation of Structures in Surface Sediments of the Mickey Sheet

To interpret the structures in the surface sediments of the Mickey sheet, we need to determine surface strain for a glacier experiencing radially diverging flow. We approximate these by considering the surface kinematics for an axisymmetric sheet. The radial, circumferential, and vertical rates of deformation at the surface of an axisymmetric sheet spreading onto an horizontal surface without dissolution are as follows (from the Chapter Appendix):

$$D_{rr} \approx -[3(3\alpha - 1)/16]t^{-1}$$
$$D_{\theta\theta} \approx [3(3\alpha + 1)/16]t^{-1}$$
$$D_{zz} \approx -(3/8)t^{-1} \qquad (20)$$

The circumferential component, $D_{\theta\theta}$, is always positive, corresponding to circumferential extension. The radial component is positive, corresponding to radial extension, for $\alpha < 1/3$, but is negative, corresponding to radial shortening, for larger values. Note, however, that radial shortening does not imply thrusting or folding here because the vertical component is always negative, corresponding to thinning of the surface layer. Radial shortening instead corresponds to a component of strike-slip deformation.

For a sheet spreading at constant volume, $\alpha = 0$, the case appropriate to the phase of late relaxation,

$$D_{rr} = D_{\theta\theta} \approx (3/16)t^{-1}$$
$$D_{zz} \approx -(3/8)t^{-1} \qquad (21a)$$

All surface elements undergo isotropic in-plane extension. Structures formed in a pristine surface layer should have no preferred orientation, such as normal faults with a map pattern similar to mud cracks on a horizontal mud surface. However, if sediments begin to be deposited over the sheet while $\alpha > 0$, the sediment layer is apt to have an array of faults already established. Stretching would most likely be taken up by reactivation of older faults.

For constant ($\alpha = 1$) and increasing ($\alpha = 7/3$) rates of salt supply, selected as most appropriate to the Mickey structure, we obtain the following:

$$\alpha = 7/3 \quad D_{rr} \approx -(9/8)t^{-1} \qquad \alpha = 1 \quad D_{rr} \approx -(3/8)t^{-1}$$
$$D_{\theta\theta} \approx (12/8)t^{-1} \qquad\qquad D_{\theta\theta} \approx (3/4)t^{-1}$$
$$D_{zz} \approx -(3/8)t^{-1} \qquad\qquad D_{zz} \approx -(3/8)t^{-1}$$
$$(21b)$$

In both cases, radial shortening and vertical thinning are balanced by circumferential extension. Consequently, faults in a thin sediment layer will be hybrid in nature, with radial strike-slip dominating over normal dip-slip by a factor of four to one.

The upper surfaces of the Mickey sheet and its neighbor (Figures 22, 23) offer some evidence that Eqs. (21a, b) provide a realistic description of processes active on top of a salt–sediment glacier. The most prominent features on top salt in both sheets are a series of radially trending escarpments, here interpreted as faults offsetting the top of salt. Circumferential benches are also present, although less prominent. Our tentative interpretation of the radial faults is that they originated as strike-slip faults in the pelagic carapace during the phase of active salt supply and were then reactivated as normal faults during near-constant volume relaxation after the feeder was pinched off. Preliminary investigation of the circumferential benches suggests that they might mark points of onlap of individual beds.

Development of Updip Minibasins and Downdip Salt Swells after Cessation of Sheet Advance

If an advancing composite salt–sediment glacier represents the second major phase in salt sheet evolution, the third phase is represented by structures formed when glacial advance ceases. As long as substantial relief is present between the updip edge of the glacier and a steep-shouldered glacier toe, the glacier will advance. Cessation of advance is thus reasonably tied to the burial of the downdip shoulder by sediment at a rate sufficiently large to overcome the tendency of the advancing glacier to climb upsection.

A "glacier" model remains appropriate as long as a gradient exists on the surface of a weak sediment layer overlying a continuous, thick layer of salt whose lateral dimension is still much greater than the thickness of the overlying sediment. Salt flows in a manner tending to reduce the surface slope, thinning below the updip part of the sediment cover and thickening beneath the downdip part.

We consider a model in which sediment deforms in conformity with salt. We treat the sediment as a viscous fluid with the same density and viscosity as the salt, the same approximation used in our model for an advancing

A

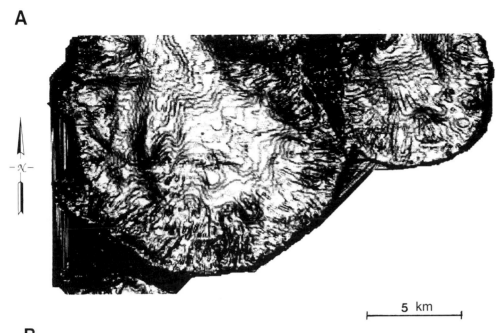

Figure 22—(A) Seismic dip and (B) azimuth maps from three-dimensional seismic data on the top of salt for the Mickey sheet and a neighboring salt sheet (area outlined in Figure 2). The dip map shows a series of monoclinal steps (dark bands) in the salt, tentatively interpreted to mark the position of onlap of individual beds. The azimuth map highlights a series of features trending radially outward from the position of the former feeder stock, here interpreted as faults offsetting the top of salt.

5 km

B

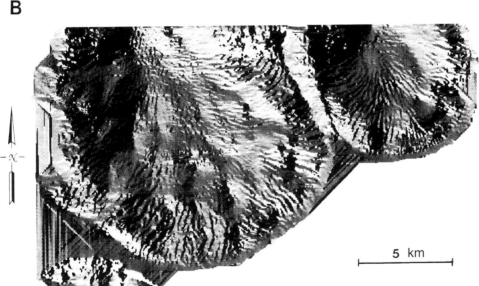

5 km

composite glacier. Like that model, the one considered here is two-dimensional. Sedimentation concentrated above the updip subsiding part of the sediment layer will produce a minibasin, and thickening of salt below the rising part will produce an incipient salt swell.

The initial salt sheet is modeled in a rectangular box extending an indefinitely large distance along strike (Figure 24). The upper surface of the sheet slopes gently to the right at a constant and uniform angle, θ. The sheet is bounded laterally by vertical, frictionless, and rigid walls. Although the type of lateral boundary is chosen for mathematical convenience, it can be identified with natural features. The updip wall, along which the motion of material is downward, may be thought of as a vertical "normal fault." The downdip wall, where material moves upward, may either be thought of as a vertical "thrust fault" or the edge of a salt swell.

The salt sheet is initially 1 km thick and 5 km wide. Only the aspect ratio 5:1, chosen rather arbitrarily, is binding, so one could think of the sheet as 10 km wide and 2 km thick, for example. The absence of any sediment layer prior to this phase of development is likewise an arbitrary choice. In the context of the present single-fluid model, one could simply "draw in" the form of any desired initial salt–sediment interface.

It was assumed for this model that the surface slope, θ, remains constant through time, and that there is no net aggradation of material. The volume of the composite structure thus remains constant. Sediment is added where the topography-induced flow has a downward component; sediment and salt are eroded where the flow has an upward component.

The flow in the model is therefore determined by how sedimentary processes induce and maintain topography

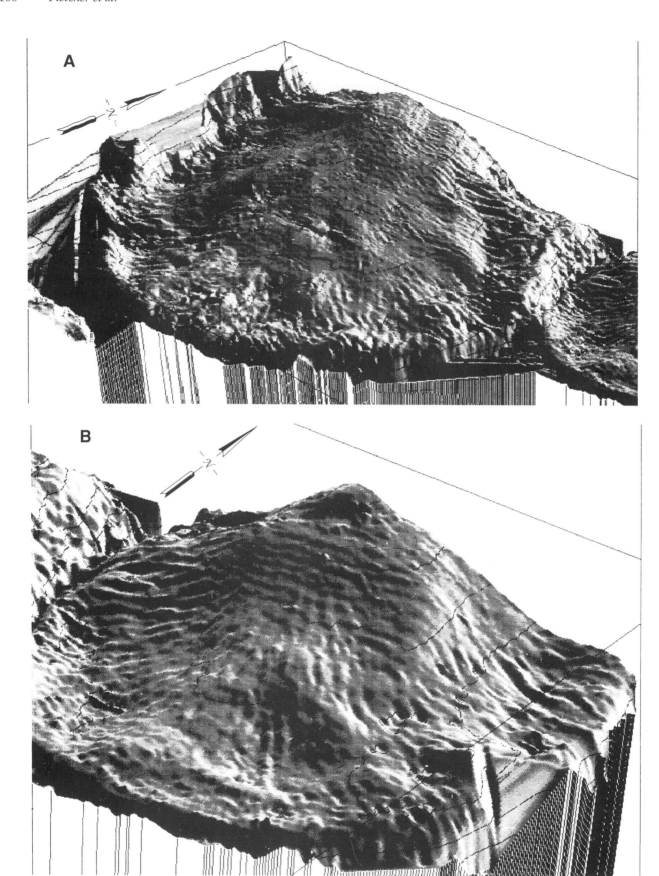

Figure 23—Perspective views of (A) the Mickey sheet and (B) a neighboring salt sheet, derived from three-dimensional seismic interpretation (area shown in Figure 2). View is from the southeast. Both the circumferential ridges and radial trends observed in Figure 22 are visible.

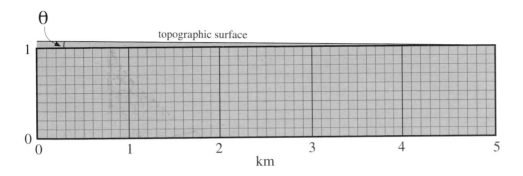

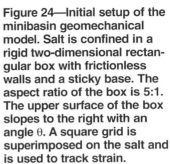

Figure 24—Initial setup of the minibasin geomechanical model. Salt is confined in a rigid two-dimensional rectangular box with frictionless walls and a sticky base. The aspect ratio of the box is 5:1. The upper surface of the box slopes to the right with an angle θ. A square grid is superimposed on the salt and is used to track strain.

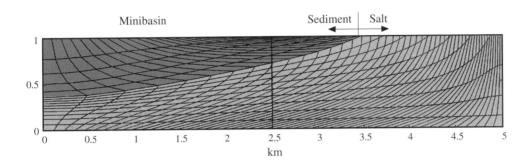

Figure 25—Results of the minibasin model after five dark layers of sediment have been added. Salt has subsided beneath the left end of the model, forming a minibasin. Salt has risen beneath the right end of the model to form a salt swell, the crest of which has been eroded off. The initially square grid in the salt has been deformed. Subhorizontal lines above the salt are sediment increments added at equal intervals, and steeper lines are sediment streamlines. The topographic slope is not shown.

at its surface. The equations governing this flow are not discussed here but are derived from a series solution satisfying the equations for plane flow of a viscous fluid in a rectangular box (Patton and Fletcher, 1995). Likewise the rates of flow and subsidence, as determined from model dimensions, surface slope, density, and salt viscosity, are not presented here. The principal aim is to show the glacier-like nature of the flow during this phase in salt sheet evolution.

The model structure for a single stage later is shown in Figure 25. Five layers of sediment have been added to the model at regular intervals to maintain a constant and uniform slope, θ. The surface slope is small and is not shown in Figure 25. The requirement of a constant slope has led to deposition in the updip minibasin and erosion above the downdip salt swell.

Grids in both the salt and sediment parts of the model have different meanings. The lines within the salt represent a deformed, originally square grid (Figure 24). Because the sediment is arbitrarily treated as having the same properties as salt, the strain pattern does not correspond to extrusion from beneath the minibasin. Indeed, the strain recorded in the sediments is quantitatively much greater than that expected in a natural minibasin.

In the sedimentary part of the model, the lines subparallel to the salt contact represent successive sediment horizons. To the right of 2.5 km, older sediment horizons are truncated against the youngest. The set of steeply plunging lines near the left wall are sediment streamlines.

Because we chose to consider the case of no net aggradation, the flow is steady and the streamlines are also particle paths. The forms of the streamlines are symmetric about the vertical surface at 2.5 km, so they can be visualized for much of the upper half of the model.

Near the lateral walls, the flow is dominantly vertical, but a transition to mainly horizontal flow takes place over a distance roughly equal to the total thickness. The fluid sticks to the basal surface of the model. The inverse relationship between streamline spacing and speed indicates a maximum, purely horizontal velocity at the surface at the half-way point. Because the transition length is on the order of the total thickness, the flow is dominantly horizontal and glacier-like over the middle 60% of the model.

Downdip Toe Structures and Late Stage "Inflation"

One prominent feature of the Mickey sheet that has yet to be wholly explained is the steep bathymetric shoulder at the downdip edge of the sheet (Figures 3, 4). This lies above the ~300 m of sediment deposited after extrusion of the sheet was complete. Such a feature, which is also present on all of the adjacent sheets, has been cited by Nelson and Fairchild (1989a, b) as a principal piece of evidence for "inflation" during the second phase of their intrusive emplacement mechanism.

Although the wholesale emplacement of a salt sheet by intrusion has been effectively eliminated by observations not available to Nelson and Fairchild (1989a, b), a process of late stage intrusive "inflation" remains a possible mechanism for the formation of a downdip bathymetric shoulder. Another mechanism is depositional drape of sediment onto a relict shoulder remaining from the last phase of glacier advance. The development of a salt swell complementary to a minibasin, as in the present model, leads to thickening of the salt and overlying sediments by horizontal shortening at the downdip end of the structure. In the absence of contact with a frictionless wall, the thickening would likewise result in a monoclinal fold.

End of Composite Salt–Sediment Glacier Phase

In nature, the end of the phase just modeled becomes more likely when (1) the minibasin gets close to the base of the salt sheet, (2) the sea bottom becomes nearly horizontal, or (3) the overlying sediment layer becomes much thicker than the salt layer. Sediments do have finite strength, so these factors tend to retard deformation driven solely by gravity. Once minibasins get sufficiently deep for density inversion to occur, buoyancy-driven salt tectonism may develop as a distinct mechanism of deformation.

SUMMARY AND CONCLUSIONS

Geologic relationships conclusively point to the emplacement of allochthonous salt sheets in the northern Gulf of Mexico as extrusive submarine salt glaciers. Critical relationships supporting this conclusion are as follows: (1) repetition of section above and below salt on some salt sheets (McGuinness and Hossack, 1993), (2) truncation of sedimentary units against the base of the salt, (3) apparent lack of structures related to injection around many sheets, and (4) onlap of units against the top of the salt. These factors render restoration of salt sheets as intrusive bodies implausible but are consistent with extrusive emplacement. A fully three-dimensional geomechanical model based on lubrication theory succeeds in reproducing the advance history of the Mickey salt sheet using reasonable estimates for salt viscosity.

The geomechanical model suggests that, during emplacement, a salt sheet forms a bathymetric high with several hundred meters of relief. Actively spreading salt sheets are thus barriers to sediment transport, either ponding sediments updip or diverting them around the margin. Most sedimentation is restricted to outside the periphery of the sheet, and only a thin pelagic carapace protects it from wholesale dissolution. A carapace of only a few tens of meters is calculated to be enough to significantly retard dissolution.

During sedimentation, the extruding salt glacier climbs upsection away from the feeder as progressively younger units are covered by the advancing sheet. The slope of the base of salt is thus a function of the relative rates of sedimentation and glacier advance. A flat basal slope corresponds to rapid advance and/or slow sedimentation, whereas a more steeply inclined base reflects slower advance and/or rapid sedimentation. The staircase shape of the base of many salt sheets is therefore interpreted as a constructional form, recording variations in the advance and sedimentation rates during emplacement of the glacier.

Burial of a salt sheet generally follows pinch-off of the feeder diapir, which allows the topography of the sheet to relax, especially at its updip end. Sediments can then prograde across the sheet, beginning the burial process. Even during burial, however, a sheet typically retains several hundred meters of relief at its downdip end, so the sheet continues to advance as a composite salt–sediment glacier. Sediments deposited on top of the salt in a composite glacier are deformed in conformity with the underlying salt. They are carried along with the glacier as it moves, commonly producing an extensional breakaway zone at the updip end. The extensional breakaway zone may be the initial cause for many of the fault-bounded minibasins located at the updip ends of salt sheets.

Near the toe of a composite glacier, the model indicates that overburden particles above the salt glacier move faster than the salt advances, causing them to be slumped off the toe and overridden like a tractor tread. This means for thin sedimentary carapaces that no necessary balance exists between the magnitudes of extension and contraction above advancing composite salt–sediment glaciers.

After cessation of sheet advance, salt may continue to move within the sheet as a result of sediment loading. Salt flows out from under minibasins to form salt swells beneath less deeply buried parts of the sheet. This may partially explain some of the late stage inflation observed beneath the toes of some shallowly buried sheets.

Acknowledgments   *The authors would like to thank the management of Exxon Production Research Company and Exxon Exploration Company for permission to publish these results. Geco-Prakla is also thanked for permission to reproduce the seismic data. We thank Steve Greenlee for his three-dimensional seismic interpretation of top and base salt at the Mickey sheet. Reviews by Mike McGroder, Sig Snelson, Tom Nelson, Jay Jackson, and several anonymous reviewers contributed greatly to the quality of the manuscript.*

REFERENCES CITED

Anderson, S. W., and J. H. Fink, 1992, Crease structures: indicators of emplacement rates and surface stress regimes of lava flows: GSA Bulletin, v. 104, p. 615–625.

Diegel, F. A., and R. W. Cook, 1990, Palinspastic reconstructions of salt-withdrawal growth-fault systems, northern Gulf of Mexico (abs.): GSA Abstracts with Programs, v. 22, p. A48.

Diegel, F. A., J. F. Karlo, D. C. Schuster, R. C. Shoup, and P. R. Tauvers, 1995, Cenozoic structural evolution and tectono-stratigraphic framework of the Northern Gulf Coast continental margin, *in* M. P. A. Jackson, D. G. Roberts, and S. Snelson, eds., Salt tectonics: a global perspective: AAPG Memoir 65, this volume.

Emerman, S. H., and D. L. Turcotte, 1983, A fluid model for the shape of accretionary wedges: Earth and Planetary Science Letters, v. 63, p. 379–384.

Fink, J. H., 1983, Structure and emplacement of a rhyolitic obsidian flow: Little Glass Mountain, northern California: GSA Bulletin, v. 94, p. 362–380.

Fletcher, R. C., M. R. Hudec, and I. A. Watson, 1995, Salt glacier model for the emplacement of an allochthonous salt sheet (abs): AAPG Annual Convention Official Program, p. 29A.

Huber, W. F., 1989, Ewing Bank thrust fault zone, Gulf of Mexico, and its relationship to salt sill emplacement, *in* Gulf of Mexico Salt Tectonics, Associated Processes and Exploration Potential: SEPM Gulf Coast Section, 10th Annual Research Conference, Program and Extended Abstracts, Houston, Texas, p. 60–65.

Hudec, M. R., R. A. Fletcher, and I. A. Watson, 1995, The composite salt glacier: extension of the salt glacier model to post-burial conditions (abs): AAPG Annual Convention Official Program, p. 45A.

Huppert, H. E., 1982, The propagation of two-dimensional and axisymmetric viscous gravity currents over a rigid horizontal surface: Journal of Fluid Mechanics, v. 121, p. 43–58.

Jackson, M. P. A., and C. J. Talbot, 1991, A glossary of salt tectonics: The University of Texas at Austin, Bureau of Economic Geology Geological Circular No. 91-4, 44 p.

McGuinness, D. B., and J. R. Hossack, 1993, The development of allochthonous salt sheets as controlled by rates of extension, sedimentation, and salt supply, *in* J. M. Armentrout, R. Bloch, H. C. Olson, and B. F. Perkins, eds., Rates of Geological Processes: SEPM Gulf Coast Section, 14th Annual Research Conference, Program with Papers, p. 127–139.

Martin, R. G., Jr., 1978, Northern and eastern Gulf of Mexico continental margin: stratigraphic and structural framework, *in* A. H. Bouma, G. T. Moore, and J. M. Coleman, eds., Framework, facies, and oil-trapping characteristics of the upper continental margin: AAPG Studies in Geology 7, p. 21–42.

Moretti, I., S. Wu, and A. W. Bally, 1990, Computerized balanced cross-section LOCACE to reconstruct an allochthonous salt sheet, offshore Louisiana: Marine and Petroleum Geology, v. 7, p. 371–377.

Nelson, T. H., and L. H. Fairchild, 1989a, Emplacement and evolution of salt sills in northern Gulf of Mexico (abs.): AAPG Bulletin, v. 73, p. 395.

Nelson, T. H., and L. H. Fairchild, 1989b, Emplacement and evolution of salt sills: Houston Geological Society Bulletin, v. 32, p. 6–7.

Patton, T. L., and R. C. Fletcher, 1995, Mathematical block-motion model for deformation of a layer above a buried fault of arbitrary dip and sense of slip: Journal of Structural Geology, v. 17, p. 1455–1472.

Peel, F. J., C. J. Travis, and J. R. Hossack, 1995, Genetic structural provinces and salt tectonics of the Cenozoic offshore U.S. Gulf of Mexico: a preliminary analysis, *in* M. P. A. Jackson, D. G. Roberts, and S. Snelson, eds., Salt tectonics: a global perspective: AAPG Memoir 65, this volume.

Pollard, D. D., and A. M. Johnson, 1973, Mechanics of growth of some laccolithic intrusions in the Henry Mountains, Utah II: Tectonophysics, v. 18, p. 311–354.

Price, R. A., 1971, Gravitational sliding and the foreland thrust and fold belt of the North American Cordillera: discussion: GSA Bulletin, v. 82, p. 1133–1138.

Schuster, D. C., 1995, Deformation of allochthonous salt and evolution of related salt–structural systems, eastern Louisiana Gulf Coast, *in* M. P. A. Jackson, D. G. Roberts, and S. Snelson, eds., Salt tectonics: a global perspective: AAPG Memoir 65, this volume.

Seni, S., 1992, Evolution of salt structures during burial of salt sheets on the slope, northern Gulf of Mexico: Marine and Petroleum Geology, v. 9, p. 452–468.

Sumner, H. S., B. A. Robison, W. K. Dirks, and J. C. Holliday, 1990, Morphology and evolution of salt/minibasin systems: lower shelf and upper slope, central offshore Louisiana (abs): GSA Abstracts with Programs, p. A48.

Talbot, C. J., 1994, Spreading of salt structures in the Gulf of Mexico: Tectonophysics, v. 228, p. 151–166.

Talbot, C. J., 1995, Molding of salt diapirs by stiff overburden, *in* M. P. A. Jackson, D. G. Roberts, and S. Snelson, eds., Salt tectonics: a global perspective: AAPG Memoir 65, this volume.

Worrall, D. M., and S. Snelson, 1989, Evolution of the northern Gulf of Mexico, with emphasis on Cenozoic growth faulting and the role of salt, *in* A. W. Bally and A. R. Palmer, eds., The Geology of North America; an overview: GSA Decade of North American Geology, v. A, p. 97–138.

Wu, S., A. W. Bally, and C. Cramez, 1990, Allochthonous salt, structure, and stratigraphy of the northeastern Gulf of Mexico, part II, structure: Marine and Petroleum Geology, v. 7, p. 334–370.

Chapter Appendix

ANALYSIS OF SPREADING SALT GLACIER

A *salt sheet* is a body with a horizontal dimension to thickness ratio that is relatively large throughout much of its evolution. For example, the sheet presented in this paper has a final diameter to thickness ratio of about 9. This suggests the application of lubrication theory analysis (Huppert, 1982; Emerman and Turcotte, 1983) as a means of obtaining a description of the dynamics of a spreading salt glacier. *Lubrication theory* provides a good approximation to a flow in which the magnitude of the vertical gradient of a quantity is much greater than that of its horizontal gradient, so that, for example, horizontal rate of shear ($\approx \partial u/\partial z$) dominates the horizontal rate of extension or shortening ($\partial u/\partial x$). In the final model, a sloping rather than horizontal basal surface is treated, so the analysis is given in detail.

The equations of equilibrium are approximated as

$$\partial \sigma_{zz}/\partial x + \partial \sigma_{xz}/\partial z \approx 0 \tag{A1a}$$
$$\partial \sigma_{zz}/\partial y + \partial \sigma_{yz}/\partial z \approx 0 \tag{A1b}$$
$$\partial \sigma_{zz}/\partial z \approx \rho g \tag{A1c}$$

where, because

$$\sigma_{xx} - \sigma_{zz} = 2\eta(\partial u/\partial x - \partial w/\partial z) = -2\eta(2\partial u/\partial x + \partial v/\partial y)$$
$$\sigma_{yy} - \sigma_{zz} = 2\eta(\partial v/\partial y - \partial w/\partial z) = -2\eta(\partial u/\partial x + 2\partial v/\partial y) \tag{A2}$$

the differences between the other normal stresses and σ_{zz} are ignored. Integration of Eq. (A1c) and application of the condition that the normal traction be continuous at the upper surface gives

$$\sigma_{zz}(x, z_0 + h) \approx -\rho_w g(D - h - z_0) \tag{A3a}$$

followed by the use of this result in the independent integration of Eqs. (A1a) and (A1b), with the conditions on vanishing of the shear tractions,

$$\sigma_{xz}(x, z_0 + h) \approx 0 \tag{A3b}$$
$$\sigma_{yz}(x, z_0 + h) \approx 0 \tag{A3c}$$

yields

$$\sigma_{zz} \approx \rho g(z - z_0 - h) - \rho_w g(D - h - z_0) \tag{A4a}$$
$$\sigma_{xz} \approx (\rho - \rho_w)g(\partial h/\partial x + \beta_x)(z - h - z_0) \tag{A4b}$$
$$\sigma_{yz} \approx (\rho - \rho_w)g(\partial h/\partial y + \beta_y)(z - h - z_0) \tag{A4c}$$

where

$$\beta_x = \partial z_0/\partial x$$
$$\beta_y = \partial z_0/\partial y \tag{A4d}$$

Approximating the rate of deformation components,

the constitutive relationships for an isotropic viscous fluid give

$$\partial u/\partial z \approx \sigma_{xz}/\eta$$
$$\partial v/\partial z \approx \sigma_{yz}/\eta \tag{A5}$$

Substitution from Eq. (A4) into (A5) and integration, using the no-slip conditions,

$$u(x, y, z_0) = 0$$
$$v(x, y, z_0) = 0 \tag{A6}$$

yields

$$u \approx [(\rho - \rho_w)g/\eta](\partial h/\partial x + \beta_x)$$
$$[(1/2)(z^2 - z_0^2) - (h + z_0)(z - z_0)] \tag{A7a}$$
$$v \approx [(\rho - \rho_w)g/\eta](\partial h/\partial y + \beta_y)$$
$$[(1/2)(z^2 - z_0^2) - (h + z_0)(z - z_0)] \tag{A7b}$$

The evolution equation for the sheet thickness, $h(x, y, t)$, is based on a global conservation condition for the volume of fluid contained in a column with a base of area $dA = dxdxy$ centered at the position (x, y),

$$\partial h/\partial t = -(\partial J_x/\partial x + \partial J_y/\partial y) + \Delta(x,y) \tag{A8}$$

where J_x and J_y are the components of a horizontal vector describing the net flux of fluid volume per unit area oriented normal to the vector. $\Delta(x, y)$ is a source term describing the net addition or removal of fluid volume to the column per unit area, in this case, by upward flow through its base, by "sedimentation" at its upper surface ($\Delta > 0$), or by dissolution or "erosion" at its surface ($\Delta < 0$). However,

$$J_x = \int_{z_0}^{z_0+h} u(x, y, z, t)dz$$
$$= (\rho - \rho_w)g/3\eta]h^3(\partial h/\partial x + \beta_x) \tag{A9}$$

with an equivalent expression for J_y. Substitution in Eq. (A8) yields the desired evolution equation:

$$\partial h/\partial t \approx [(\rho - \rho_w)g/3\eta]$$
$$\{\partial/\partial x[h^3(\partial h/\partial x + \beta_x)] + \partial/\partial y[h^3(\partial h/\partial y + \beta_y)]\} + \Delta \tag{A10}$$

Taking into account a diverging outward flow for the axisymmetric case in writing the equation equivalent to Eq. (A8), the equivalent of Eq. (A10) is

$$\partial h/\partial t \approx [(\rho - \rho_w)g/3\eta](1/r)\partial/\partial r[rh^3(\partial h/\partial r + \beta)] + \Delta \tag{A11}$$

Similarity Solution with Surface Dissolution

The present development follows Huppert (1982) but adds the effect of surface dissolution. Under sufficiently regular conditions, the evolution equation for the axisymmetric case admits of a *similarity solution* of the form

$$h(r, t) = Ct^p \Psi(r/R) \qquad \text{(A12a)}$$

where the radius of the sheet is

$$R(t) = Mt^q \qquad \text{(A12b)}$$

Provided that the quantities $\beta(r, t)$ and $\Delta(r, t)$ have suitable forms, substitution of Eq. (A12) into Eq. (A11) permits this partial differential equation in two variables, r and t, to be reduced to a single ordinary differential equation in the similarity variable $\xi = r/R$.

The form of the solution in Eq. (A12) is adopted to achieve this mathematical simplification, and further restrictive conditions on the form of the axisymmetric basal surface and of specific rate of addition or removal of fluid, $\Delta(r,t)$, must be imposed as well. The tacit assumption made is that such a solution can be useful to us in understanding the dynamics of a spreading salt sheet, even though it may not possess all the desired features of the natural structure.

Performing the substitution, we have

$$Ct^{p-1}(p\Psi - \xi q \Psi')$$
$$= [(\rho - \rho_w)g/3\eta](C^4/M^2)t^{4p-2q}\{(1/\xi)[\xi\Psi^3(\Psi' + \phi)]' + \Delta \qquad \text{(A13a)}$$

where

$$\phi = \beta(M/C)t^{(q-p)} \qquad \text{(A13b)}$$

To meet the conditions for a similarity solution, the powers of t on both sides of the equation must be equal, and the functions remaining must be functions of ξ alone, or

$$3p = 2q - 1 \qquad \text{(A14a)}$$
$$\Delta = -D^*(\xi)t^{(p-1)} \qquad \text{(A14b)}$$
$$\text{and} \quad \beta = B^*(\xi)t^{[-(q+1)/3]} \qquad \text{(A14c)}$$

Because the sheet advances, q is a positive number. If we use a basic assumption of the present model that the substrate not deform, the basal slope of the sheet at any position, r, cannot depend on t, or $\partial\beta/\partial t = 0$. Application of this condition to Eq. (A14c) results in

$$B^*(x) = B^*_0\xi^{[-(q+1)/3q]} \qquad \text{(A15a)}$$

and, substituting into Eq. (A14c),

$$\beta = \beta(r) = [B^*_0/M^{1/q}]r^{-[(q+1)/3q]} \qquad \text{(A15b)}$$

That is, the slope of the basal surface must decrease outward according to the indicated power of r and must tend to an unboundedly large value as $r \to 0$. The latter requirement does not create any difficulty because we can associate the large slope with the edge of the feeder diapir. However, according to Eq. (1), the outward decrease in slope must be attributed to some combination of a decrease in the rate of sedimentation or increase in the rate of advance. Without going into detail, we find that these are unattractive alternatives for the present example and thus choose to consider only the case of spreading onto a horizontal surface, $\beta = 0$. We retain a nonzero Δ with the required dependence of Eq. (A14b).

Without loss of generality, we can set

$$[(\rho - \rho_w)g/3\eta](C^3/M^2) = 1 \qquad \text{(A16a)}$$

because the remaining function can absorb an arbitrary constant. Applying the other conditions, Eq. (A13a) becomes

$$p\Psi - \xi q\Psi' = (1/\xi)(\xi\Psi^3\Psi')' - D^*/C \qquad \text{(A16b)}$$

For Eq. (A16b) to be tractable, we must deal with the complex form of the function $D^*(\xi)$ required to describe the flow into the sheet from a local region near the origin, corresponding to the source diapir and, in the general case, dissolution distributed smoothly over the upper surface of the sheet. To do this, we use the following artifice. The function D^* is reserved solely for the description of surface dissolution, and we also specify that the volume of the sheet increase according to the regimen

$$V(t) = Kt^\alpha \qquad \text{(A16c)}$$

The latter condition amounts to supposing that material is added to the sheet within a small region centered at the origin, $r = 0$, and that the details are unimportant.

But using Eq. (A12),

$$V(t) = \int_0^R h(r, t)(2\pi r dr)$$
$$= Ct^p(Mt^q)^2 \int_0^1 \xi\Psi(\xi)d\xi \qquad \text{(A17)}$$

and from Eq. (A16c),

$$p + 2q = \alpha \qquad \text{(A17b)}$$
$$2\pi CM^2 \int_0^1 \xi\Psi(\xi)d\xi = K \qquad \text{(A17c)}$$

Note also that the net rate of dissolution over the surface of the sheet using Eq. (A14) is

$$dV_{\text{diss}}/dt = \int_0^R D^*(\xi)t^{p-1}(2\pi r dr)$$

$$= 2\pi M^2 t^{[2q+(p-1)]} \int_0^1 \xi D^*(\xi)\delta\xi \qquad \text{(A18)}$$

From Eqs. (A14a) and (A17b),

$$p = (\alpha - 1)/4$$
$$q = (3\alpha + 1)/8 \qquad \text{(A19)}$$

Then the power of t in Eq. (A18) is $2q + p - 1 = \alpha - 1$. The net rate of dissolution is thus a constant times the rate of increase in sheet volume, or

$$dV/dt = f\,dV_{diss}/dt \qquad \text{(A20a)}$$

or, integrating,

$$V = fV_{diss} \qquad \text{(A20b)}$$

Accordingly, the net flux of salt from the source diapir is

$$Q_\Sigma = Q + Q_{diss} = (1+f)Q \qquad \text{(A20c)}$$

It is essentially this regularity in behavior that permits the existence of a simple, similarity solution.

To complete the solution, it is necessary to assign a functional form to $D^*(\xi)$ and then to solve Eq. (A16b) for the function $\Psi(\xi)$. A further condition on Ψ is given by the requirement

$$h(R, t) = 0 \qquad \text{(A21a)}$$

or, from Eq. (A12a),

$$\Psi(1) = 0 \qquad \text{(A21b)}$$

Here, we assume a uniform specific rate of dissolution, so that

$$D^*(\xi) = d_0 \qquad \text{(A22)}$$

which is a constant.

In the case $\alpha = 0$ and $d_0 = 0$, corresponding to a sheet of constant volume that is not undergoing dissolution, Eq. (A16b) has an exact solution. Substituting from Eq. (A19), $p = -1/4$ and $q = 1/8$, Eq. (A16b) becomes

$$-(1/8)(2\xi\Psi + \xi^2\Psi') = -(1/8)(\xi^2\Psi)' = (\xi\Psi^3\Psi')' \qquad \text{(A23a)}$$

Integrating once and using Eq. (A21b), we have

$$-(1/8)\xi = \Psi^2\Psi' \qquad \text{(A23b)}$$

Integrating again and using Eq. (A21b) gives

$$\Psi^3 = (3/16)(1 - \xi^2)$$

or $\quad \Psi = (3/16)^{1/3}(1 - \xi^2)^{1/3} \qquad \text{(A23c)}$

Other approximate solutions are obtained by expansion in powers of $1 - \xi$. For example, substituting

$$\Psi(\xi) \approx A(1 - \xi)^r$$

which satisfies the condition of Eq. (A21b) into Eq. (A16b) yields, to lowest order,

$$\Psi(\xi) \approx (3q)^{1/3}(1 - \xi)^{1/3} \qquad \text{(A25a)}$$

Note that the effect of dissolution does not appear to lowest order. For $\alpha = 0$, and $q = 1/8$, Eq. (A25a) is indeed the lowest order term in the expansion of Eq. (A23c) in powers of $1 - \xi$.

Retaining higher order terms,

$$\Psi(\xi) \approx (3q)^{1/3}(1 - \xi)^{1/3} + (d_0/5qC)(1 - \xi) + [(4q - 1)/(24q)](3q)^{1/3}(1 - \xi)^{4/3} \qquad \text{(A25b)}$$

Again, for $\alpha = 0$ and $d_0 = 0$, Eq. (A25b) yields the first two terms in the expansion of Eq. (A23c).

Using Eq. (A25b) to perform the integration in Eq. (A17c), we obtain

$$2\pi CM^2\{(9/28)(3q)^{1/3} + (d_0/30qC) + (3/70)[(4q - 1)/8q](3q)^{1/3}\} = K \qquad \text{(A26a)}$$

Because, for values of α appropriate to the present application, the third term in braces is a small correction to the first term, we discard it. From Eqs. (A17) and (A18),

$$\pi M^2 d_0 = \alpha f K \qquad \text{(A26b)}$$

Using this to substitute for D_0^* in Eq. (A26a) and Eq. (A19) for q, we have

$$\pi(mC^*)(m^{3/2}M^*)^2\{(9/14)(3q)\}^{1/3} = Km^4 \qquad \text{(A26c)}$$

where

$$m^4 = 1 - \{(8\alpha)/[15(3\alpha + 1)]\}f \qquad \text{(A26d)}$$

and C^* and M^* are the values of C and M when $m = 1$.

Surface Rates of Deformation for Axisymmetric Spreading onto a Horizontal Surface

The radial component of velocity in the axisymmetric case for a horizontal basal surface is

$$v_r \approx [(\rho - \rho_w)g/2\eta]\partial h/\partial r(z^2 - 2zh) \qquad \text{(A27a)}$$

The normal components of the rate of deformation tensor are

$$D_{rr} = \partial v_r/\partial r$$

$$D_{\theta\theta} = v_r/r \qquad \text{(A27b)}$$

and, by incompressibility,

$$D_{zz} = -(D_{rr} + D_{\theta\theta}) \qquad \text{(A27c)}$$

Substituting Eq. (A27a) into Eq. (A27b) and then evaluating the resulting quantities at the surface $z = h$, we have

$$D_{rr}(r, h) \approx -[(\rho - \rho_w)g/2\eta](\partial/\partial r)(h^2\partial h/\partial r) \qquad \text{(A28a)}$$
$$D_{\theta\theta}(r, h) \approx -[(\rho - \rho_w)g/2\eta](1/r)(h^2\partial h/\partial r) \qquad \text{(A28b)}$$

Using Eqs. (A12), (A16a), and (A19), we can reduce these expressions to

$$D_{rr}(r, h) \approx -(3/2)t^{-1}(\Psi^2\Psi')' \qquad \text{(A29a)}$$
$$D_{\theta\theta}(r, h) \approx -(3/2)t^{-1}(1/\xi)(\Psi^2\Psi') \qquad \text{(A29b)}$$

Using the approximation Eq. (A25b), we obtain

$$\Psi^2\Psi' \approx -(1/8)(3\alpha + 1) - (9/14)f\alpha(1 - \xi)^{2/3}$$
$$- (1/8)(3\alpha - 1)(1 - \xi) \qquad \text{(A29c)}$$

From these results, the expressions in Eq. (20) in the text are readily computed.

Diegel, F. A., J. F. Karlo, D. C. Schuster, R. C. Shoup, and P. R. Tauvers, 1995, Cenozoic structural evolution and tectono-stratigraphic framework of the northern Gulf coast continental margin, *in* M. P. A. Jackson, D. G. Roberts, and S. Snelson, eds., Salt tectonics: a global perspective: AAPG Memoir 65, p. 109–151.

Chapter 6

Cenozoic Structural Evolution and Tectono-Stratigraphic Framework of the Northern Gulf Coast Continental Margin

F. A. Diegel

E & P Technology Company
Shell Exploration and Production Company
Houston, Texas, U.S.A.

J. F. Karlo

Pecten International Company
Houston, Texas, U.S.A.

D. C. Schuster

Consultant
Olmstead Falls, Ohio, U.S.A.

R. C. Shoup

Shell Offshore Inc.
New Orleans, Louisiana, U.S.A.

P. R. Tauvers

Shell Offshore Inc.
New Orleans, Louisiana, U.S.A.

Abstract

The Cenozoic structural evolution of the northern Gulf of Mexico Basin is controlled by progradation over deforming, largely allochthonous salt structures derived from an underlying autochthonous Jurassic salt. The wide variety of structural styles is due to a combination of (1) original distribution of Jurassic and Mesozoic salt structures, (2) different slope depositional environments during the Cenozoic, and (3) varying degrees of salt withdrawal from allochthonous salt sheets. Tectono-stratigraphic provinces describe regions of contrasting structural styles and ages. Provinces include (1) a contractional foldbelt province, (2) a tabular salt–minibasin province, (3) a Pliocene–Pleistocene detachment province, (4) a salt dome–minibasin province, (5) an Oligocene–Miocene detachment province, (6) a lower Oligocene Vicksburg detachment province, (7) an upper Eocene detachment province, and (8) the Wilcox growth fault province of Paleocene–Eocene age.

Within several tectono-stratigraphic provinces, shale-based detachment systems, dominated by lateral extension, and allochthonous salt-based detachment systems, dominated by subsidence, can be distinguished by geometry, palinspastic reconstructions, and subsidence analysis. Many shale-based detachments are linked downdip to deeper salt-based detachments. Large extensions above detachments are typically balanced by salt withdrawal.

Salt-withdrawal minibasins with flanking salt bodies occur as both isolated structural systems and components of salt-based detachment systems. During progradation, progressive salt withdrawal from tabular salt bodies on the slope formed salt-bounded minibasins which, on the shelf, evolved into minibasins bounded by arcuate growth faults and remnant salt bodies. Associated secondary salt bodies above allochthonous salt evolved from pillows, ridges, and massifs to leaning domes and steep-sided stocks.

Allochthonous salt tongues spread from inclined salt bodies that appear as feeder faults when collapsed. Coalesced salt tongues from multiple feeders formed canopies, which provided subsidence potential for further cycles of salt withdrawal. The Sigsbee escarpment is the bathymetric expression of salt flows that have overridden the abyssal plain tens of kilometers since the Paleogene. The distribution and palinspastic reconstruction of Oligocene–Miocene salt-based detachments and minibasins suggest that a Paleogene salt canopy, covering large areas of the present onshore and shelf, may have extended as far as the Sigsbee salt mass.

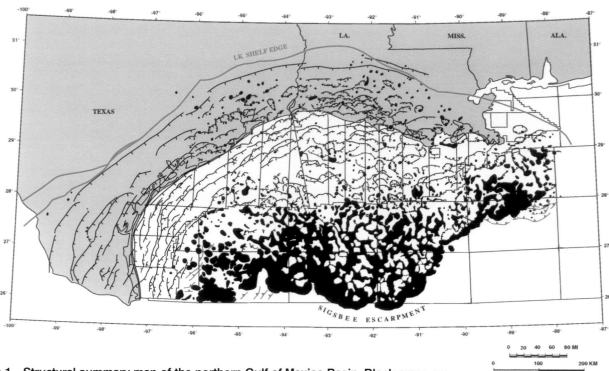

Figure 1—Structural summary map of the northern Gulf of Mexico Basin. Black areas are shallow salt bodies. Tick marks are on the downthrown side of major growth faults: black = seaward dipping; red = landward dipping (counter-regional); blue = thrust faults.

INTRODUCTION

New concepts, seismic data, and hydrocarbon exploration in deeper water led to a revolution in the understanding of Cenozoic tectonics of the northern Gulf of Mexico continental margin in the 1980s. In particular, recognition of allochthonous salt bodies combined with quantitative palinspastic reconstruction changed the prevailing view of the northern Gulf of Mexico Basin from a passive margin with vertical rooted salt stocks and massifs with intervening steep growth faults, to a complex mosaic of diachronous detachment fault systems and variously deformed allochthonous salt sheets.

The modern history of Gulf Coast structural studies as expressed in published literature began with the recognition of the Sigsbee escarpment as a salt overthrust at the toe of the slope (Amery, 1969). The profound significance of this observation was first considered by Humphris (1978), who proposed large-scale basinward flow of salt and subsequent withdrawal by downbuilding of slope sediments deposited on top of the moving salt mass. In the same volume (Bouma et al., 1978), which represents a turning point, Martin (1978) reviewed the stratigraphic and structural framework of the Gulf Coast with the contemporary understanding of margin progradation over autochthonous Louann salt with attendant rooted vertical stocks and steep growth faults apparently related to flow of deeply buried shale and salt masses. An allochthonous salt canopy in Iran (Jackson and Cornelius,

1985) was recognized at a time when the petroleum industry was interpreting allochthonous salt wings and sheets on seismic reflection profiles of the outer shelf and slope, offshore Louisiana.

In 1989, many of these interpretations and concepts of Gulf Coast salt tectonics were presented, including several contributions from industry (GCS SEPM 10th Annual Conference in 1989). In the same year, Worrall and Snelson (1989) used quantitative palinspastic reconstructions to show how Humphris' (1978) model for basinward salt flow applied to large-scale growth fault systems of the Texas shelf, and Jackson and Cramez (1989) discussed the recognition of salt welds on allochthonous and autochthonous salt. Sumner et al. (1990) described the three-dimensional structure of minibasins developed by Pliocene–Pleistocene evacuation of allochthonous salt derived from counter-regional (northward-dipping) feeders on the Louisiana outer shelf. Diegel and Cook (1990) and Diegel and Schuster (1990) used structural reconstructions incorporating subsidence analysis to constrain the geometry and thickness of evacuated allochthonous salt in the onshore and shelf of Louisiana even in areas where shallow salt bodies are no longer present. Studies of Gulf Coast salt tectonics since then have focused in areas of relatively recent hydrocarbon exploration efforts in the outer shelf and upper slope (e.g., Huber, 1989; West, 1989; Wu et al., 1990; Seni, 1992; Rowan et al., 1994). In this chapter, we show that the new concepts are equally applicable to coastal and inner shelf areas.

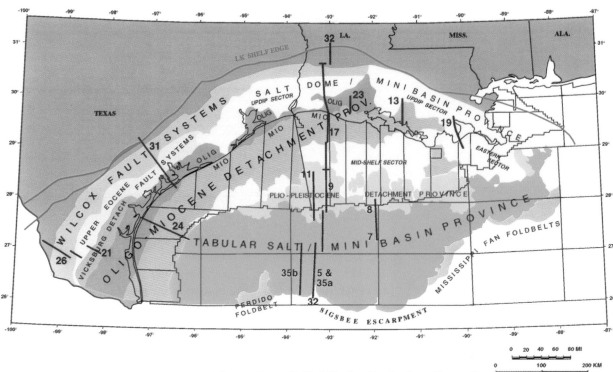

Figure 2—Tectono-stratigraphic provinces of the northern Gulf of Mexico Basin. Locations of profiles described in the text are indexed by figure numbers (in red).

Typically, reviews of Gulf Coast salt structures (e.g., Martin, 1978; Worrall and Snelson, 1989) begin with a description of the low-relief structures at the updip basin margin, proceed to the high-relief salt stocks of coastal Louisiana, and then describe the more complex leaning stocks and allochthonous salt wings and sheets of the outer shelf and slope. That approach is logical based on the evolutionary deformation sequences described by Trusheim (1960) and Seni and Jackson (1983) for progradation across autochthonous salt, which thickens into the axis of relatively simple cratonic basins. However, the scale and complexity of the Gulf Coast continental margin are more clearly understood by proceeding in the opposite direction, from the abyssal plain to the coastal areas. This inverse approach has the advantage of using shallow and well-imaged structures on the modern slope as analogs for the early history of structures that are now more fully developed and deeply buried beneath neritic and continental sections. This approach also allows us to revisit the less recently studied but still actively explored areas of the inner shelf and onshore Gulf Coast and to provide a consistent and comprehensive tectono-stratigraphic framework of the Gulf Coast from a modern perspective.

In this chapter, tectono-stratigraphic provinces are defined and described. We analyze the evolution and origin of these provinces with the aid of selected palinspastic reconstructions of two-dimensional cross sections, deep-structure maps, and subsidence analysis. The similarity of structures in the more basinward provinces to reconstructed structures farther landward provides addi-

tional analogs for determining the early history of the older structures. The relationships between adjacent provinces are discussed when appropriate. Finally, alternative palinspastic reconstructions of a transect from the Cretaceous margin to the modern abyssal plain in western Louisiana are presented to address the overall Cenozoic structural evolution of the basin. The Chapter Appendix describes the reconstruction and subsidence analysis methodologies and the inferred salt budget and magnitude of salt dissolution during the Cenozoic.

TECTONO-STRATIGRAPHIC PROVINCES

Overview

Construction of a regional framework of Gulf Coast structure entails several difficulties. These include the great three-dimensional complexity of the structures, the large variability along strike as well as landward and basinward, and the uncertainty of deep structures in many areas that are not well imaged by contemporary seismic methods. Only the first-order structures are adequately reflected on a structural summary map (Figure 1), which does little to reflect the deep structure and genetic relationships.

A tectono-stratigraphic province map (Figures 2, 3) illustrates eight distinct regions defined by contiguous

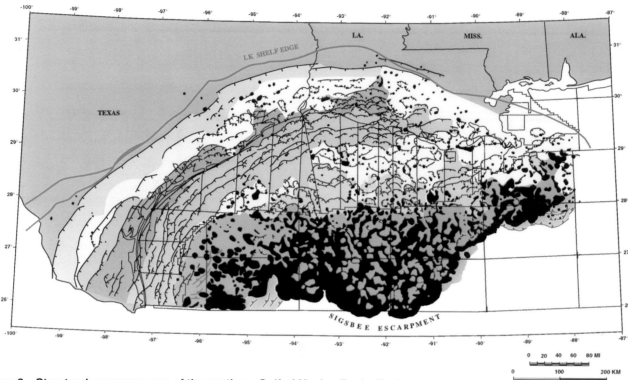

Figure 3—Structural summary map of the northern Gulf of Mexico Basin. Tectono-stratigraphic provinces are color coded as in Figure 2 and faults as in Figure 1.

areas of similar structural style. The eight nongenetic provinces discussed here are (1) a contractional foldbelt province at the toe of slope, (2) a tabular salt–minibasin province on the slope, (3) a Pliocene–Pleistocene detachment province on the outer shelf, (4) a salt dome–minibasin province, (5) an Oligocene–Miocene detachment province onshore and on the shelf, (6) an Oligocene Vicksburg detachment province onshore Texas, (7) an upper Eocene detachment province, and (8) the Wilcox growth fault province of Paleocene–Eocene age.

The province map is necessarily a poor representation of the structural complexity at multiple levels, and particularly on the slope, many reasonable subdivisions are possible to reflect some of the significant changes in structural style and degree of structural development in this active structural environment. Also, the nature and origin of middle slope contractional belts on the Texas slope (Figure 1) are not discussed in this chapter. The primary subdivision of the shelf and onshore areas is that between provinces dominated by listric growth faults soling on subhorizontal detachments and the large salt dome–minibasin province. The salt dome–minibasin province can be further subdivided geographically into updip, eastern, and mid-shelf sectors. The individual detachment provinces are distinguished by age of expanded section, but we also interpret a fundamental genetic distinction between those detachments that are salt welds (Pliocene–Pleistocene and Oligocene–Miocene detachment provinces) and those that are purely sliding surfaces

not directly related to salt withdrawal. In addition, many of the structures in the tabular salt–minibasin province of the slope represent earlier stages of structural evolution than the more structurally evolved provinces on the shelf and onshore.

The toe-of-slope contractional foldbelt provinces include the Perdido foldbelt of Oligocene age in Texas (Blickwede and Queffelec, 1988; Weimer and Buffler, 1992) and the Mississippi Fan foldbelt of Miocene–Pliocene age in eastern Louisiana (Weimer and Buffler, 1992; Wu et al., 1990). These salt-floored fold and thrust systems apparently formed at the basinward margin of autochthonous salt. Note that these systems are of different ages and separated geographically by a wide zone lacking known contractional deformation. (See the above references for more information on these provinces.)

The *tabular salt–minibasin province* is characterized by extensive salt sheets with intervening deep-water sediment-filled minibasins. Most of these minibasins form bathymetric lows today. The *Pliocene–Pleistocene detachment province* includes areas of evacuated allochthonous salt along detachments for listric growth faults as well as remnant allochthonous or "secondary" salt domes and wings in the area of the Pliocene–Pleistocene shelf margin depocenters. The *salt dome–minibasin province* is characterized by salt stocks and intervening shelf minibasins bounded by large-displacement, arcuate, and dominantly counter-regional growth faults. This province is highly diachronous: structures and related depocenters range

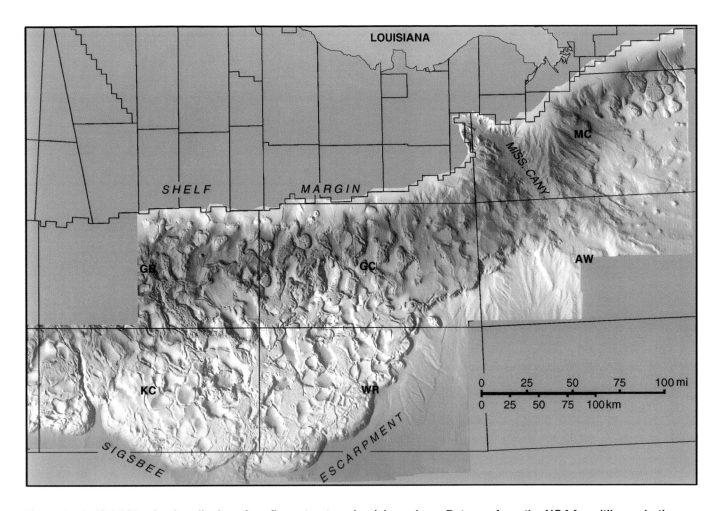

Figure 4—Artificial illumination display of seafloor structure, Louisiana slope. Data are from the NOAA multibeam bathymetric survey. Vertical exaggeration is 10 times; illumination is from the west-southwest. Colors from gray, brown, light blue, to dark blue indicate shallow to deep water. OCS protraction areas are outlined in black: KC = Keathley Canyon, GB = Garden Banks, GC = Green Canyon, WR = Walker Ridge, MC = Mississippi Canyon, and AW = Atwater.

in age from Eocene to Pleistocene updip to the modern shelf margin. The *Oligocene–Miocene detachment province* is large and complex but is characterized by listric down-to-the-basin growth faults that sole in the Paleogene section. The *Oligocene Vicksburg detachment system* in onshore Texas contains sand-prone Vicksburg deltaic sediments greatly expanded by a listric down-to-the-basin fault system that soles in Eocene Jackson shales. The *upper Eocene detachment province* includes several listric detachment-based fault systems expanding the upper part of the Eocene section. In the Paleocene–Eocene *Wilcox growth fault province* in southern Texas are found the oldest major growth fault systems downdip of the middle Cretaceous margin. Like the upper Eocene fault systems, the geometries of these systems are variable along strike, with listric faults either soling directly on the autochthonous Louann salt or in the upper Cretaceous section before stepping down to the Louann level. A more complete description and analysis of the individual provinces follow.

Tabular Salt–Minibasin Province

The tabular salt–minibasin province covers most of the continental slope along the northern Gulf of Mexico margin, stretching from Mexico to eastern Louisiana between the shelf margin and the Sigsbee escarpment at the toe of the slope. Although much variability is present in this large region, allochthonous salt tongues or "tabular" salt with intervening sediment-filled minibasins represent its dominant structural style (Figure 3). We use the term *tabular salt* to refer to laterally extensive salt bodies with flat tops. The term *salt sheet* refers to allochthonous salt with a subhorizontal top and base, and *salt wing* means a less extensive allochthonous salt body with a demonstrable base.

The bathymetry of the modern Louisiana slope reflects the profound influence of salt tectonics and sedimentation in the deep-water environment (Figure 4). First-order features include the prominent Sigsbee escarpment, the expression of a large salt body overriding the abyssal

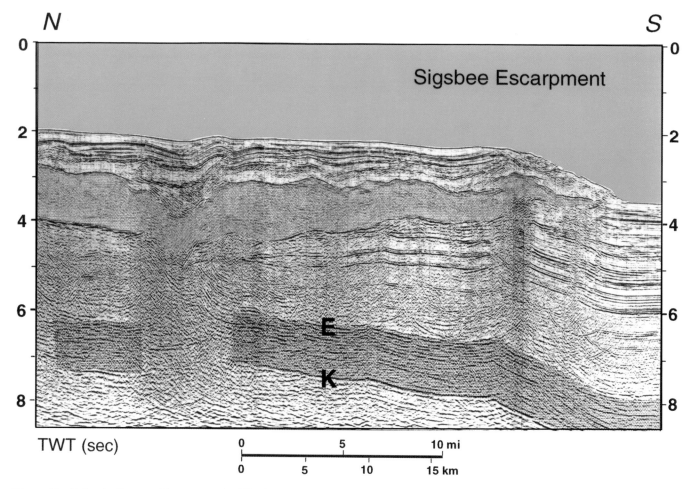

Figure 5—Seismic time profile across the Sigsbee escarpment, western Louisiana slope. Eocene(?) (E) and middle Cretaceous (K) reflectors are pulled up beneath the relatively fast allochthonous salt layer (green). There is no evidence of a foldbelt here. See Figure 2 for location.

plain, the Mississippi Canyon, and the upper part of the Mississippi Fan.

The western Keathley Canyon area and the southern part of the eastern Keathley Canyon and Walker Ridge are underlain by a contiguous canopy of coalesced allochthonous salt (Figure 4). The western part is covered by a thin sedimentary cover forming nascent polygonal minibasins above allochthonous salt and separated by crestal grabens on salt ridges. The southern Keathley Canyon and Walker Ridge area, just landward of the escarpment, is underlain by tabular salt near the seafloor (Figure 5). The dominant features of the central part of the Louisiana slope are the deep and currently sediment-starved minibasins surrounded by interconnected shallow salt bodies.

Isolated salt bodies and interconnected minibasins surrounded by arcuate growth fault systems also occur in eastern Green Canyon and western Atwater areas (Figure 4), where recent sedimentation has reduced the bathymetric relief relative to areas farther west. Areas northeast of Mississippi Canyon are dominated by erosion, large slides, and isolated allochthonous salt bodies forming

bathymetric highs with distinct convex outlines. Slope minibasins expressed as bathymetric lows commonly contain sediments greater than 6 km thick either symmetrically or asymmetrically ponded in basins tilted southward (Figure 6).

In general, there is a gradual transition from isolated minibasins surrounded by contiguous salt in the lower slope to isolated salt bodies surrounded by interconnected fault-bounded minibasins near the shelf margin. This transition reflects progressive deformation during progradation of the margin across allochthonous salt. A seismic profile in the middle slope shows an early stage of sedimentation above allochthonous salt (Figure 7). The perched basin in Figure 7 is beginning to subside into the salt, whereas faults with seafloor expression indicate a contemporaneous sliding downslope. Normal faults occur at the northern end, and reverse faults occur at the southern end.

A profile (Figure 8) just to the north of Figure 7 shows the result of progradation of the shelf margin across the northern end of another allochthonous salt body. Shelf strata, expanded on listric growth faults, forced evacua-

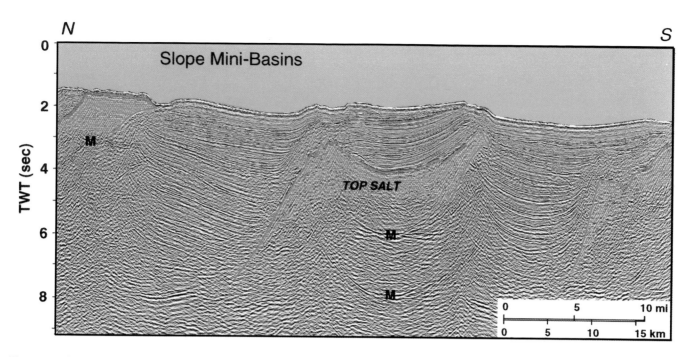

Figure 6—Slope minibasins, offshore Louisiana. Two deep basins are separated by a shallow basin perched above allochthonous salt. Events marked "M" are multiple reflections. See Figure 2 for location.

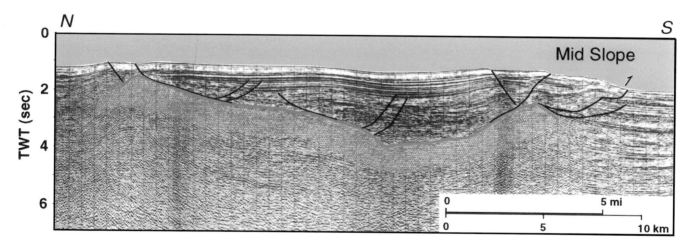

Figure 7—Seismic profile from the middle slope of Louisiana. Down-to the-south normal faults at the north end are linked to thrust structures at the south end. Sliding perched basin is beginning to subside into shallow salt (green). See Figure 2 for location.

tion of the northern end of the salt body to form a weld connected to a south-leaning salt massif. Strata within the basin thin rapidly onto the massif and show evidence of erosion. Normal faults occur at the south end of the basin where reverse faults similar to those in Figure 7 may have been present. The updip salt weld represents the downdip portion of the Pliocene–Pleistocene detachment province.

Pliocene–Pleistocene Detachment Province

Sumner et al. (1990, p. 48) divided the Pliocene–Pleistocene detachment province into separate regions of "organized" and "disorganized" roho systems. The organized systems occur in the western and eastern parts of the area and are underlain by extensive salt welds, or *rohos*. The disorganized systems occur in the central area where a combination of residual salt wings, evacuation surfaces, and windows between salt bodies forms a more complex structure.

We restrict the term *roho* to the characteristic discontinuous, high-amplitude seismic reflections caused by remnant salt along welds (Jackson and Cramez, 1989), also called *salt-evacuation surfaces* or *salt-withdrawal surfaces*. We also refer to the Pliocene–Pleistocene salt-withdrawal structures of the outer shelf as *roho systems*, after Sumner

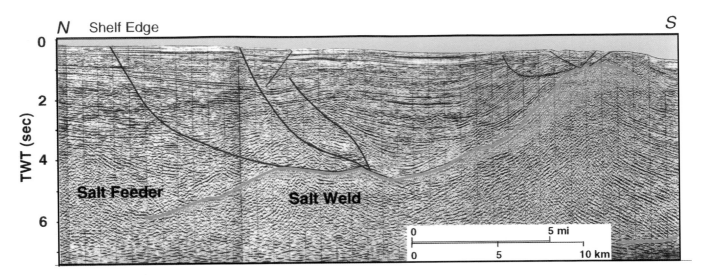

Figure 8—Seismic profile from the shelf edge of Louisiana. Shelf margin sedimentation and associated listric growth faults collapsed the north end of a salt body to produce a weld (green) just south of a counter-regional salt feeder. The southern end of the salt body remains near the seafloor with shelf margin sediments onlapping and thinning onto the southern flank.

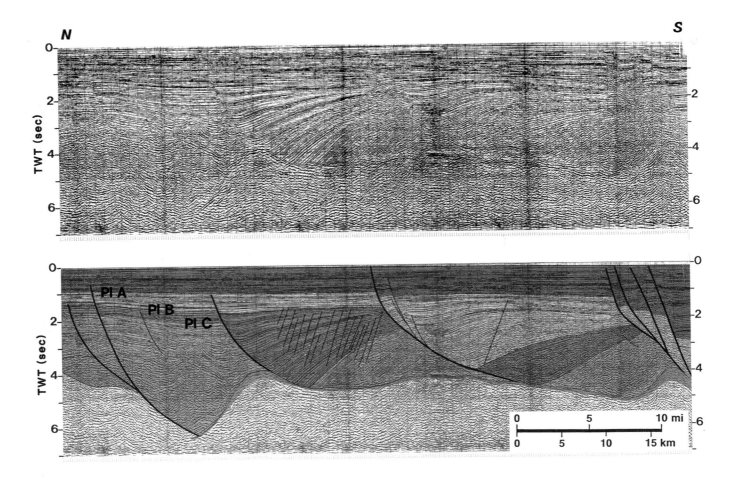

Figure 9—Uninterpreted (top) and interpreted (bottom) seismic profile across an organized roho system, western Louisiana outer shelf, showing roho reflections along the detachment for Pliocene–Pleistocene listric growth faults. A north-dipping counter-regional salt feeder is interpreted at the north end of the subhorizontal salt weld (green). Pl A, B, C = three successive Pliocene–Pleistocene levels. See Figure 2 for location.

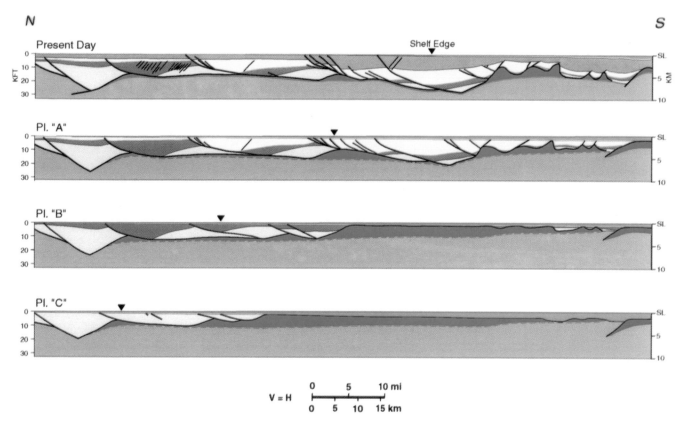

Figure 10—Reconstruction of depth-converted seismic profile based in part on Figure 9. Pl. A, B, C = three successive Pliocene–Pleistocene levels. Large extensions are balanced by reduction of a shallow allochthonous salt body (green). Inverted triangles represent the approximate position of the paleoshelf margin based on paleontologic interpretation of depositional environments. The base of salt through time is based only on the structural relief above salt. Reconstruction by the PREP method (see Chapter Appendix).

et al. (1990), but choose the more general genetic term *salt-withdrawal fault system* or *salt-based detachment system* for similar structures without roho reflections.

Organized roho areas of the outer shelf show large amounts of extension by listric down-to-the-basin growth faults that expand Pliocene–Pleistocene sediments above the salt welds (Figure 9). Although some contractional structures exist locally, they do not balance the cumulative extensions. Palinspastic reconstruction suggests that extension is balanced by withdrawal of tabular salt originally present near the seafloor (Figure 10). (See Chapter Appendix for more details on reconstruction methods.) Another reconstructed example from the Pliocene–Pleistocene detachment province illustrates the complete collapse of a shelf-margin minibasin onto a salt weld (Figures 11, 12). The former north-dipping thinning wedge of sediment is now collapsed to form a complexly faulted turtle structure above the salt weld. The evolution of the toe of the former salt mass includes possible initial thrusting, followed by inversion into a counter-regional extensional fault during deformation of the salt massif into salt domes along the counter-regional faults.

In the last two examples (Figures 10, 12), we reconstructed the top of salt by undoing growth fault motions and flattening to a reasonable seafloor constrained by the position of the paleoshelf margins. The base of salt in the reconstructions is only inferred from the minimum structural relief above the salt through time. Diegel and Cook (1990) and Rowan (1994) presented strategies for incorporating subsidence analysis to independently constrain salt thicknesses through geologic time. This technique (see Chapter Appendix), combined with the structural reconstructions, is a powerful tool in deducing the early history of more deeply buried structural systems of the inner shelf and onshore areas. The large salt-withdrawal component of subsidence, estimated by backstripping in southern Louisiana, provides a solution to the long-standing problem of how space was created for thick Cenozoic shallow water deposits late in the history of a passive margin initiated in Jurassic time (see Chapter Appendix).

Oligocene–Miocene Detachment Province

The Oligocene–Miocene detachment province covers most of the modern slope and parts of coastal onshore Texas and Louisiana (Figures 2, 3). This is a region of large-displacement, dominantly down-to-the-basin listric growth faults that sole on a regional detachment above the Paleogene section. The updip limit of the detachment is irregular and crosses the more linear trends of the Oligocene and Miocene depocenters (e.g., Winker, 1982).

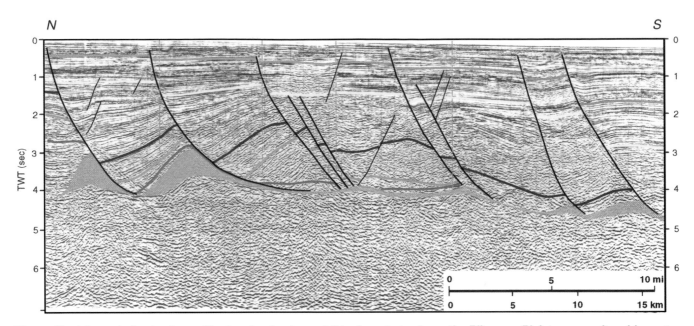

Figure 11—Interpreted seismic profile showing basinward-thinning strata above the Pliocene–Pleistocene salt weld, western Louisiana outer shelf. See Figure 2 for location.

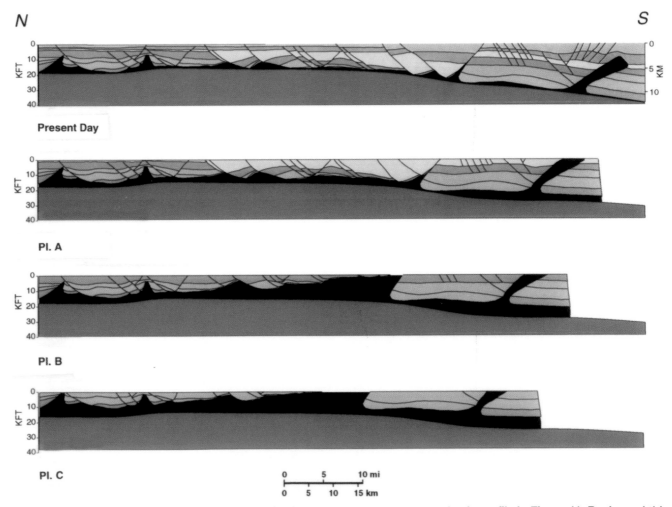

Figure 12—Reconstruction of depth-converted seismic profile based in part on seismic profile in Figure 11. Basinward-thinning strata above the weld restore to an onlapping configuration on the south flank of a presently evacuated salt body. Reconstruction by the MESH method (see Chapter Appendix). Pl. A, B, C = three sucessive Pliocene–Pleistocene levels.

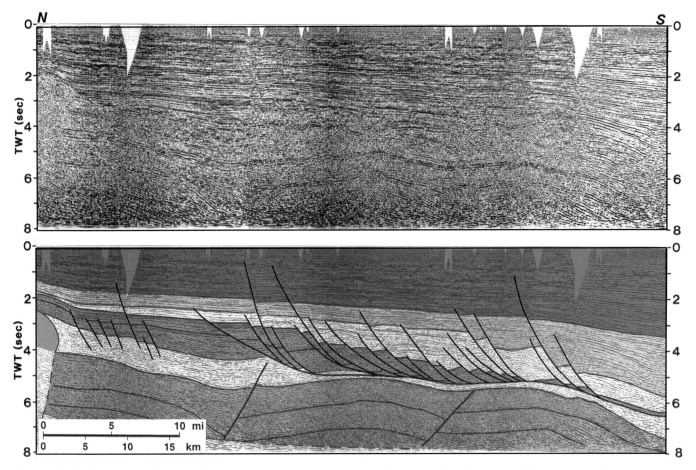

Figure 13—Uninterpreted (top) and interpreted (bottom) seismic profile across the Oligocene–Miocene detachment, onshore southern Louisiana. Listric normal faults with greatly expanded Miocene strata (upper colorless interval) sole above through-going Eocene and older strata. Large counter-regional faults beneath the detachment are interpreted to be feeders for allochthonous salt since evacuated from the detachment. See Figure 2 for location.

Another characteristic of this province is the great thickness of deltaic sediments above the detachment, usually exceeding 5 km.

Depth conversion of an interpreted seismic profile (Figures 13, 14) from onshore southern Louisiana illustrates the magnitude of the subsidence problem in this province. Wells have penetrated Miocene neritic sediments as deep as 6 km below sea level. This remarkable stacking of deltaic sandstone reservoirs helps make southern Louisiana one of the world's great petroleum provinces. Thermal and isostatic subsidence alone cannot account for more than 6 km of shallow-water sediment deposited in the late Tertiary on a passive margin where rifting occurred in the Jurassic, but subsidence can be balanced isostatically with salt withdrawal (see Chapter Appendix). This technique for estimating salt withdrawal also accounts for rotation and translation of fault blocks; in this and other examples from the Oligocene–Miocene detachment, extensional faulting does not account for the large subsidence anomalies.

Estimated salt thicknesses using the method in the Chapter Appendix are used in the alternative reconstructions (Figures 15, 16) of the southern Louisiana cross section (Figure 15). Although this technique estimates the amount of salt withdrawal, it does not locate the level of the evacuation surface. End-member models show that either salt was withdrawn from the autochthonous Louann level (Figure 15) or the detachment for listric growth faults represents a salt weld that formerly contained a thick, allochthonous salt body (Figure 16).

We prefer the allochthonous model for several reasons. (1) Salt penetrations occur along the Oligocene–Miocene detachment in western Louisiana. (2) The geometries resemble those above the shallower Pliocene–Pleistocene detachment systems, and we know of no example of a listric detachment formed in response to a rolling fold of salt beneath several kilometers of deep-water sediments. (3) The thinning wedge above the detachment suggests a collapsed onlap similar to examples from the outer shelf described previously. (4) Sub-detachment counter-regional faults provide a means for extrusion of the Louann salt to the level of the Oligocene seafloor. Finally, (5) it seems

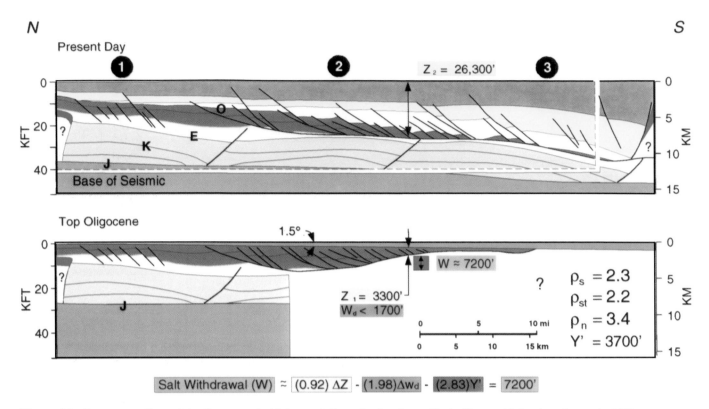

Figure 14—Reconstruction of depth-converted interpretation of seismic profile in Figure 13 (top) to the end of Oligocene time (bottom). Backstripping of expanded Miocene deltaic sediments leaves a basinward-thinning wedge of Oligocene sediments (purple) above the detachment. Assuming no salt withdrawal since the end of Oligocene at position 1, backstripping indicates 1.1 km of excess subsidence in this area, certainly an overestimate. The change in sediment overburden thickness from the end of Oligocene to the present is 7 km. Change in water depth (Δw_d) is less than 500 m. Isostatic balance estimates salt withdrawal of greater than 2 km of salt. If change in water depth is less or excess thermal subsidence is less, then more salt withdrawal is required for isostatic balance. (See discussion on subsidence and salt withdrawal in Chapter Appendix). Reconstruction is by the PREP method.

mechanically unlikely that a thick salt layer would remain undeformed beneath kilometers of sediments that thicken into counter-regional growth faults of pre-Oligocene age.

The detachment system discussed in the previous example extends across a large part of coastal Louisiana and Texas and comprises the Oligocene–Miocene detachment province. Also, similar deep penetrations of Miocene deltaic sediments are well known across the entire area. Geometric analogs to known salt-based detachment systems of the Pliocene–Pleistocene depocenters, as well as palinspastic reconstructions and subsidence analysis, imply that large areas of the shelf may have been underlain by allochthonous salt sheets that were evacuated by progradation of the Miocene deltaic margin. This scenario probably applies even in parts of the Oligocene–Miocene detachment province in western Louisiana and Texas, where few shallow salt domes occur. We interpret this entire province to be a salt-based detachment with salt emplacement at an allochthonous level in the Paleogene and subsequent salt evacuation during progradation of the late Oligocene–late Miocene shelf margin.

A regional seismic profile from western Louisiana illustrates the scale of the Oligocene–Miocene detachment system in an area where the detachment is relatively shallow and well imaged (Figure 17). Sub-detachment strata, seismically correlated to Eocene and Cretaceous rocks penetrated updip, are well imaged. A marked discordance occurs along the detachment: sub-detachment strata extend across the entire profile with relatively even thickness, whereas deltaic units above the detachment are greatly expanded but thin rapidly basinward to be replaced by successively younger strata. This pattern of expansion and thinning reflects the progressive evacuation of allochthonous salt during progradation of the shelf. Two wells are shown on this profile where reported salt penetrations occur at the level of detachment, basinward of sub-detachment counter-regional structures that may have acted as feeders for allochthonous salt. The map distribution of these feeders (Figure 18) suggests multiple sources for a probably extensive Paleogene salt canopy that is now reduced to a weld.

The irregular landward edge of the Oligocene–Miocene detachment system in southeastern Texas and Louisiana corresponds to the landward limit of the con-

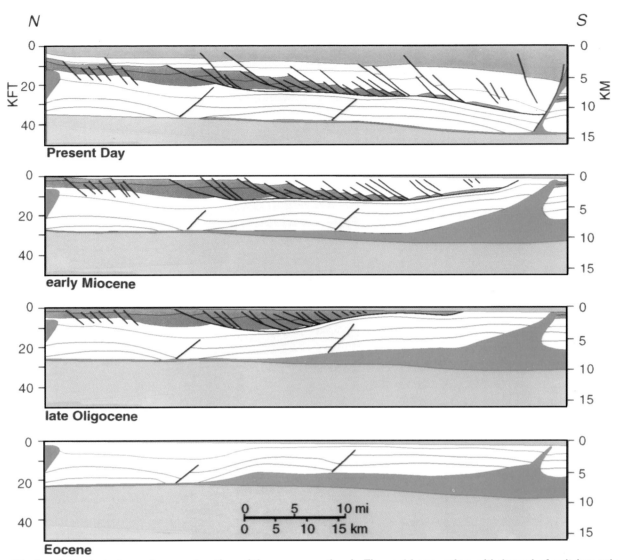

Figure 15—Isostatically balanced reconstruction of the cross section in Figure 14 assuming withdrawal of salt (green) from the autochthonous salt level. This interpretation is rejected in favor of an allochthonous salt model (Figure 16). Reconstruction above detachment is by the PREP method.

tinuous salt canopy on the Paleogene slope. The age of earliest evacuation of the canopy also varies with its updip extent. Landward reentrants in the canopy edge were the earliest evacuated areas, and basinward promontories were evacuated later. The earliest evacuation corresponds to early Frio deltaic deposition in the early Oligocene of central Louisiana. But just to the east, the detachment reaches only as far landward as the present coast within the early Miocene depocenters.

Salt Dome–Minibasin Province

The salt dome–minibasin province (Figures 2, 3) is divided geographically into updip, eastern, and midshelf sectors. All of the sectors share the same structural style that defines this nongenetic province—salt stocks and intervening shelf minibasins bounded by large-displacement, arcuate, and dominantly counter-

regional growth faults. Unlike the mid-shelf sector, the updip and eastern sectors are composed of isolated structural systems surrounded by areas of relatively simple structure.

Updip and Eastern Sectors

The landward edge of the Oligocene–Miocene detachment is interpreted as the updip limit of a continuous Paleogene salt canopy, but isolated allochthonous salt bodies occur in the updip and eastern sectors of the salt dome–minibasin province. Dominantly down-to-the-basin listric growth faults of the detachment province formed in areas of extensively coalesced allochthonous salt, but isolated minibasins rimmed by arcuate faults and flanking salt domes formed during evacuation of isolated allochthonous salt bodies of the updip and eastern sectors of the salt dome–minibasin province.

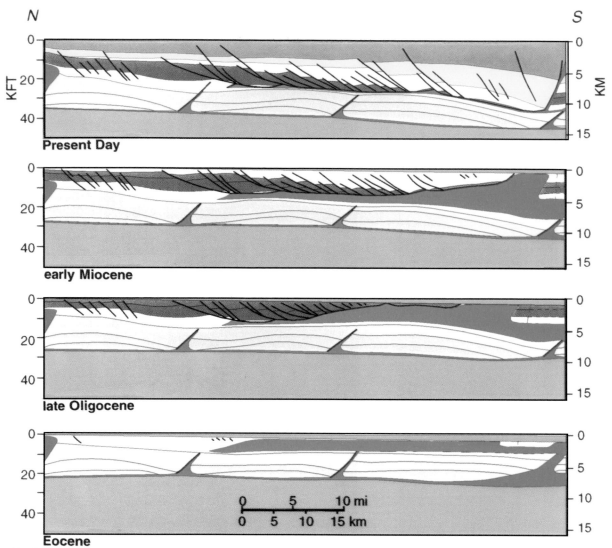

N   S

KFT / KM

Present Day

early Miocene

late Oligocene

Eocene

Figure 16—Isostatically balanced reconstruction of the cross section in Figure 14 assuming a two-stage evolution: (1) extrusion of salt (green) into a canopy near the seafloor by Eocene–Oligocene time (bottom section) and (2) evacuation of allochthonous salt by prograding Miocene depocenters to produce a salt weld (middle two sections). In this model, counter-regional faults below the detachment are interpreted as collapsed salt bodies that acted as feeders during the salt emplacement. Reconstruction above detachment is by the PREP method.

Palinspastic reconstruction of a cross section through coastal southeastern Louisiana illustrates the structural evolution of the eastern sector of the province (Figure 19). The present-day cross section shows a minibasin bounded on the south by a large displacement counter-regional fault and bounded on the north by a smaller displacement down-to-the-south growth fault. Both of these faults sole within the Paleogene section, well above the Jurassic salt horizon. South-leaning salt domes occur along the counter-regional fault east and west out of the plane of section (Schuster, 1995). The soling horizon connects the shallow counter-regional fault to a deeper counter-regional fault to form a stepped-counter-regional system (Schuster, 1993, 1995).

The large apparent extension above the soling horizon is much greater than the extension in the Mesozoic and

Paleogene section. The section is balanced by including an isolated salt body at the soling horizon. This structure evolved in two distinct phases: (1) extrusion of an allochthonous salt body near the seafloor in Paleogene time followed by (2) evacuation of that salt body to form a minibasin floored by a salt weld and bounded by salt-withdrawal faults and leaning salt domes along the counter-regional fault (Figure 19).

Two distinct structural styles—salt-based detachments and stepped counter-regional fault systems—formed during shelf margin progradation in southern Louisiana. Where the allochthonous salt coalesced to form a continuous canopy, salt-based detachment systems developed. Conversely, where the salt bodies were isolated, salt-floored minibasins and marginal salt domes formed. The modern Louisiana slope is a direct analog for the

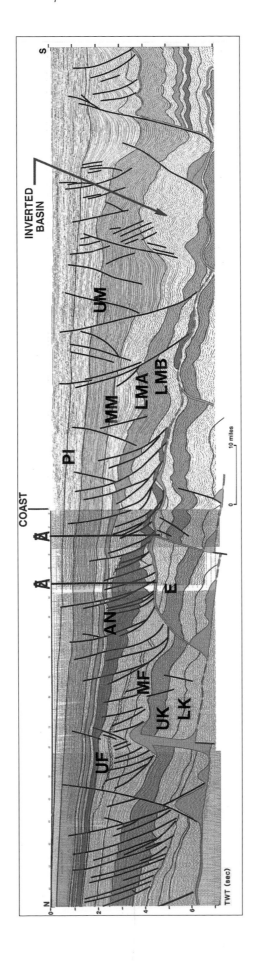

Figure 17—Interpreted seismic profile across the Oligocene–Miocene detachment province and mid-shelf salt dome–minibasin province. Strata above the salt detachment (green) form a series of expanded wedges that thin basinward above more isopachous subdetachment Eocene and older strata, which are deformed by counter-regional faults. Well symbols indicate wells that penetrate salt at the level of detachment. Pl = lower Pleistocene, UM = upper Miocene, MM = middle Miocene, LMA and LMB = lower Miocene, UF = Oligocene upper Frio, MF = Oligocene middle Frio, E = Eocene, UK = Upper Cretaceous, and LK = Lower Cretaceous. See Figure 2 for location. This vertically exaggerated profile is shown at true scale in Figure 35 (folded insert).

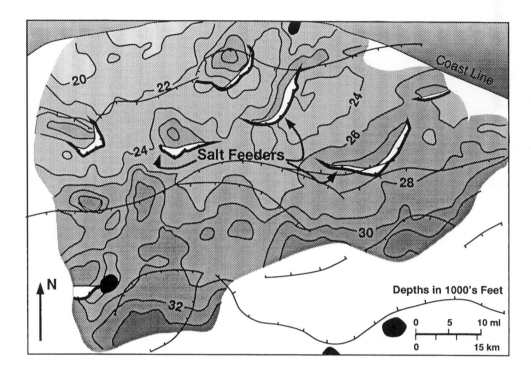

Figure 18—Structure contour map on Eocene subdetachment horizon illustrating subdetachment minibasins and counter-regional faults interpreted as feeders for a now evacuated allochthonous Paleogene salt canopy.

Paleogene slope before deformation of allochthonous salt. The modern bathymetry (Figure 4) shows the outlines of isolated allochthonous salt in the easternmost Louisiana slope and a more continuous canopy to the west.

Mid-Shelf Sector

The structural style of the mid-shelf sector of the salt dome–minibasin province is similar to the updip and eastern sectors of the province. The mid-shelf minibasins generally contain younger deltaic sediments, and the deep structure is obscured by deep burial. Unlike the more isolated fault systems of the updip and eastern sectors, the counter-regional faults of the mid-shelf sector form a linked network across much of the shelf. Also, although characterized by a different structural style than the Oligocene–Miocene detachment province, this sector is probably genetically related to it.

Salt-based detachment systems terminate basinward either in minibasins bounded by counter-regional faults or in thrust complexes related to the forward edge of a salt sheet (e.g., Sumner, 1990; Schuster, 1995). In the former case, salt domes occur around the edges of the minibasins, most commonly along the counter-regional faults. The reconstruction of the previous onshore example (Figure 16) shows the evolution of the basinward margin of a salt massif into a minibasin bounded by a counter-regional fault and associated salt dome.

This evolutionary scenario is also evident when comparing typical dip cross sections in sequence from the lower slope to onshore (Figure 20). The lower slope example (Figure 20a) shows extensive allochthonous salt near the seafloor; the upper slope example (Figure 20b) shows the initiation of subsidence where basinward slid-

ing is accomplished by a linked slip system of down-to-the-basin normal faults at the landward end of the salt body and basinward-directed thrusts at the basinward end. The shelf margin example (Figure 20c) shows complete collapse of the landward part of the salt body to form a weld beneath listric normal faults and onlap onto a south-leaning asymmetric salt massif similar to the second stage in previous reconstructions of both the Pliocene–Pleistocene and Oligocene–Miocene salt-based detachments (Figures 12, 16). The outer shelf example (Figure 20d) shows complete evacuation of an allochthonous salt body by formation of a counter-regional fault at the southern end. The inner shelf example (Figure 20e) is geometrically similar to the outer shelf example except that it is more deeply buried. It is also similar to the reconstructed onshore example (Figure 16).

The salt dome–minibasin style of structure occurs in isolation within areas of discrete allochthonous salt bodies, as in the updip and eastern sectors of the province, but it also occurs as the basinward part of many salt-based detachment systems. It is likely that the mid-shelf sector of the salt dome–minibasin province bears this relation to the adjacent Oligocene–Miocene detachment. If the Oligocene–Miocene and mid-shelf provinces are related this way, then the large minibasins in the mid-shelf area may also be floored by allochthonous salt at the Paleogene level rather than being rooted directly to the Jurassic Louann salt horizon. The interpreted regional seismic profile in Figure 17 shows the relationship between the Oligocene–Miocene salt-based detachment and the mid-shelf sector of the salt dome–minibasin province. This profile was chosen to avoid salt domes, but the large counter-regional faults at the southern end of the section are linked to salt domes out of the plane of the section (Figure 3).

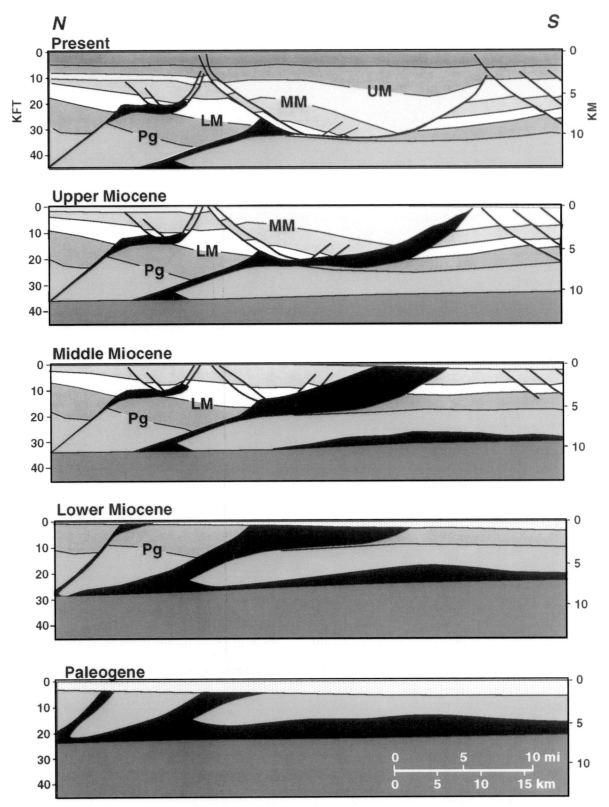

Figure 19—Reconstruction of a stepped counter-regional system formed by evacuation of an isolated allochthonous salt body (Schuster, 1995). UM = upper Miocene, MM = middle Miocene, LM = lower Miocene, Pg = Paleogene. See Figure 2 for location.

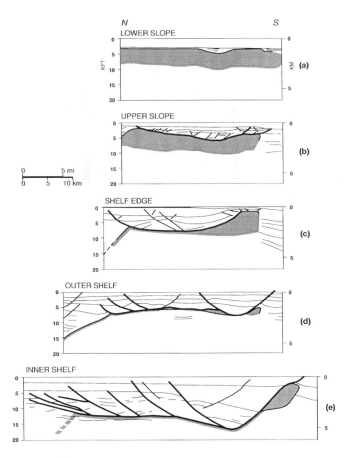

Figure 20—Evolution of salt withdrawal fault systems above allochthonous salt (gray) as shown by a series of true-scale depth sections from the present-day Louisiana Gulf Coast. The present-day structures from (a) deep water to (e) onshore represent evolutionary stages as the continental margin prograded across allochthonous salt. These deep-water examples are analogs for the early history of fully developed fault systems onshore and on the inner shelf.

Oligocene Vicksburg Detachment System

Palinspastic Analysis and Comparison with Salt-Based Detachments

Not all detachments in the northern Gulf of Mexico Basin are salt-withdrawal fault systems. A large shale-based detachment system is recognized onshore in southern Texas in the lower Oligocene Vicksburg productive trend (e.g., Honea, 1956; Combes, 1993) (Figures 2, 21). The well-imaged detachment surface is about 700 m below the top of the Eocene Jackson shale, which is often penetrated along the detachment. Although this fault system shares a superficial similarity to the salt-withdrawal detachment systems previously discussed, it is geometrically distinct. The superficial similarities include the presence of expanded deltaic sediments above listric normal faults that sole into a subhorizontal detachment surface.

The profound differences are apparent in reconstructed depth cross sections (compare Figures 22, 23). In this shale-based detachment system, the expanded sequences are younger landward in contrast to salt-based examples (Figures 10, 12, 16, 23), where expanded sequences prograde basinward. The base of the reconstructed sediments remains sub-horizontal in the shale-based example, unlike the characteristic basinward onlap configuration in reconstructed salt-based detachment systems. Growth faults above salt-based detachments generally become younger basinward, but reconstructions of the Vicksburg detachment indicate periodic landward backstepping of the active growth fault. Extension increases with age above the Vicksburg detachment's conveyor belt. In contrast, in salt-withdrawal systems such the Oligocene–Miocene detachment system, a wave of extension moves basinward with the prograding depocenter such that all the faulted strata, regardless of age, are extended about the same amount, but at different times. Above salt-based detachments a zone of extension in the upper slope and outer shelf progrades along with the margin. Older growth faults are stranded on the shelf rather than continuously translated along the detachment by cumulative extension recurring at the head of the fault system, as in the Vicksburg fault system.

Unlike the salt-withdrawal fault systems, the shale-based Vicksburg detachment is an example of extreme extension. The oldest units in the Vicksburg example (Figure 22) were translated horizontally more than 16 km, with all the extension accumulated across a fault zone 2.4 km in restored horizontal width (over 600% extension). In contrast, the oldest sediments in the salt-withdrawal example (Figure 23) show about 3.2 km of horizontal translation distributed over a zone of faulting 16 km wide in the reconstructed state (about 20% extension). Salt withdrawal during extension resulted in about 2.1 km of vertical motion, or about 70% of the horizontal extension in the Louisiana example. About 1.2 km of vertical motion during extension occurred in the Vicksburg example, or only about 7% of the horizontal movement.

Numerous reconstructions, including those presented here, indicate that salt-withdrawal and shale-based detachment systems can be distinguished using palinspastic reconstruction independent of confirming evidence such as salt penetrations. Unambiguous reconstructions are, however, dependent on the availability of reliable biostratigraphic control. Reconstructions are only diagnostic back to the age of the deepest reliable stratigraphic correlation across the fault system. In the absence of deep well control with reliable biostratigraphic markers, interpretations of fault system evolution are as speculative as the correlations. Large changes in speculative correlation across growth faults result in radically different reconstructed geometries. Correlations based solely on seismic character across large growth faults are often misleading or completely useless. In the absence of deep biostratigraphic control, apparently conservative correlations (i.e., minimized fault displacements) tend to make reconstructed salt-withdrawal systems appear to be shale-based slide systems.

W   E

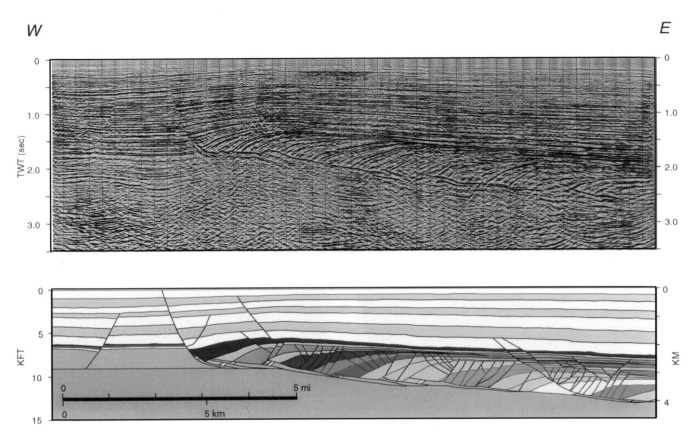

Figure 21—Uninterpreted seismic profile (top) and depth-converted interpretation (bottom) of the Vicksburg detachment system, onshore southern Texas. The strong reflector at the base of the rotated section is the interface between faster Vicksburg Formation rocks and slower Eocene shales (dull green). Offsets of this reflector are interpreted to be fault slices of Eocene shales on the hanging wall of the detachment. See Figure 2 for location.

Relationship of Vicksburg Detachment System to Oligocene–Miocene Detachment Province

Because of the limited dip extent of the cross section, the previous Vicksburg reconstruction (Figure 22) does not address downdip compensation of extension. The relationship to the next youngest extensional fault system of Oligocene Frio age, however, is similar to the relationship of a perched Miocene detachment system to the Oligocene–Miocene master detachment on the Texas shelf (Figure 24). The perched detachment overlies a deeper detachment that extends basinward beneath younger extensional fault systems.

Reconstruction of the perched detachment (Figure 25) shows extreme extension and a lower Miocene geometry similar to the Vicksburg example, with no indication of allochthonous salt at the perched level. The restored onlapping wedge geometry of the subperched detachment section (Figure 25, Oligocene) suggests that salt withdrawal occurred at this deeper but still allochthonous level. This model is consistent with the previously presented interpretation that the Oligocene–Miocene detachment represents an extensive, time-transgressive salt weld.

The relationship of the Vicksburg detachment to the Oligocene–Miocene detachment beneath expanded Frio

sediments may be similar. We know of one cored salt penetration at the level of detachment for Frio growth faults in onshore southern Texas that is hundreds of kilometers distant from known shallow salt domes along strike. The extreme extension in these sections is probably taken up by a reduction in the length of salt near the seafloor. Although the timing of the Perdido folds is appropriate for some of the updip extensional fault systems, the magnitude and duration of contraction are insufficient for balancing the updip extensional fault systems (Worrall and Snelson, 1989).

Wilcox Fault Province of Southern Texas

Description

The oldest Tertiary growth fault system in the northern Gulf of Mexico Basin is the Paleocene–Eocene Wilcox fault system (Figure 2). Although this system varies greatly along strike, its base is relatively shallow and well imaged in southern Texas (Figure 26). The deep structure of the southern Texas Wilcox fault system is unlike those previously discussed. The most prominent feature of the trend is the great expansion (more than tenfold) of Wilcox deltaic strata confined to narrow depotroughs. These

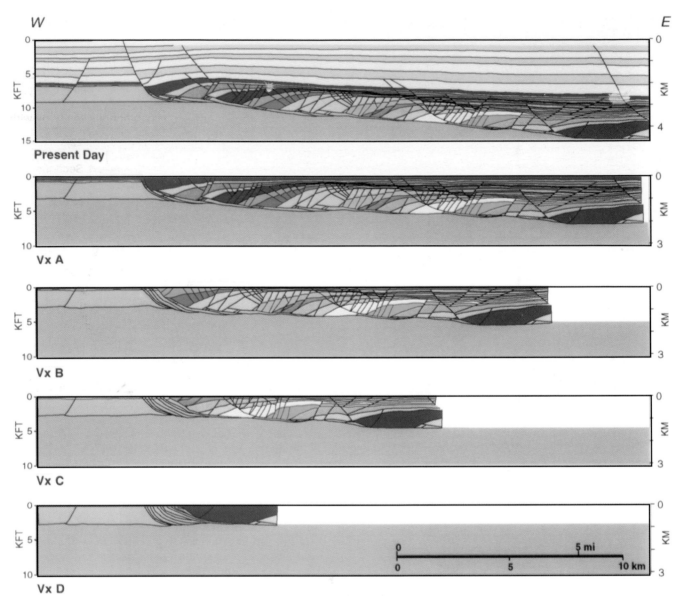

W   E

Present Day

Vx A

Vx B

Vx C

Vx D

Figure 22—Reconstruction of cross section in Figure 21. The restored base of the Vicksburg section remained subhorizontal through time. Slices of Eocene shales (dull green) were stripped off the footwall and carried eastward along the detachment. This sliding system does not directly involve salt withdrawal. Reconstruction is by the MESH method. Vx A, B, C indicate four successive Vicksburg levels.

depotroughs are also characterized by the apparent absence of Cretaceous strata, which are well imaged outside the troughs (Figure 26).

The landward edge of the troughs is the locus of the complex Wilcox growth fault system, which expands the upper Wilcox section by about a factor of ten. The complex imbricate fan of down-to-the-basin growth faults merges downward into major fault planes that sole at the Jurassic Louann salt level, apparently directly overlain by Paleogene strata. The basinward edge of the Eocene-filled depotroughs is bounded by counter-regional faults that extend to the Louann salt level and have Cretaceous strata on their footwalls.

Palinspastic Analysis and Alternative Interpretations

The reconstruction of part of this profile (Figure 26) shows the creation of space for the Wilcox depotrough by collapse of an autochthonous Mesozoic salt massif (Figure 27). The width of these massifs at the end of the Cretaceous is not constrained by the reconstruction, which shows a maximum Tertiary extension model with minimum width of the Cretaceous salt massifs. The opposite end-member, pinning the basinward Mesozoic block at the eastern end of the section, is also geometrically admissible, resulting in wide salt massifs and no net extension in the Tertiary. In either case, this reconstruction

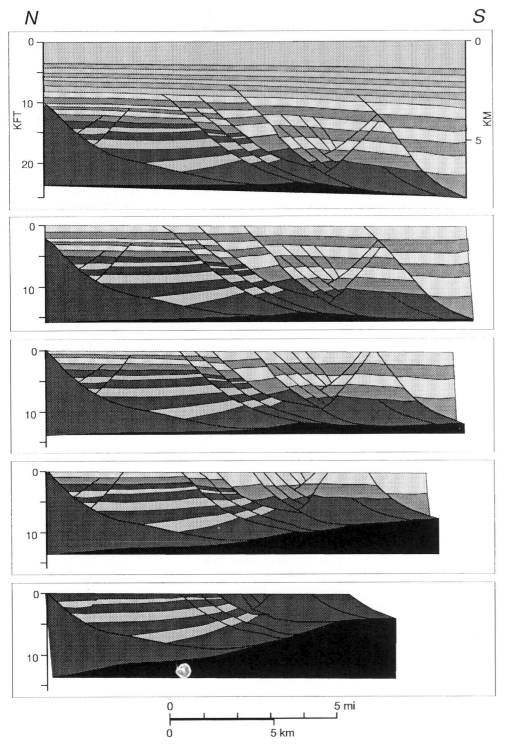

Figure 23—Reconstruction above a salt-based detachment in the onshore southern Louisiana Frio trend. Contrast the characteristic reconstructed geometry of a basinward thinning wedge onlapping a now evacuated allochthonous salt body with a shale-based detachment system (Figure 22). Reconstruction is by the MESH method. Sections from bottom to top represent successive stages of evolution from Oligocene to present day.

does not address the formation of the salt walls in Cretaceous time. The two possibilities are (1) thinning of the Lower Cretaceous cover by postdepositional extension in the Late Cretaceous, or (2) syndepositional growth throughout the Cretaceous without extreme extension. The first mechanism has been proposed for the evolution of similar salt–depotrough structures in the Kwanza Basin (Verrier and Castello-Branco, 1972; Duval et al., 1992; Lundin, 1992; Vendeville and Jackson, 1992b).

Arguments in favor of the extensional model (not shown) include documentation of the mechanism by physical modeling (Vendeville and Jackson, 1992a,b) and the generally isopachous nature of the Lower Cretaceous strata. The extensional hypothesis, however, requires basinward sliding of at least 40 km, and no contractional structures of the appropriate age and magnitude are known to exist. Extreme extension could be compensated by large contraction of salt width in the downdip salt

NW   SE

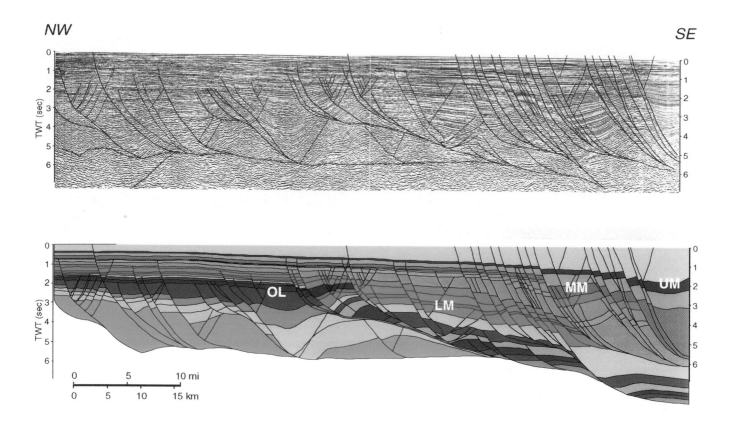

Figure 24—Interpreted seismic profile (top) and detailed drawing (bottom) across the Oligocene–Miocene detachment province on the south-central Texas shelf. The middle Miocene perched detachment (center) is interpreted as an analog for the Vicksburg shale-based detachment; it is connected downdip to the deeper Oligocene–Miocene detachment interpreted as an extensive salt weld. UM = upper Miocene, MM = middle Miocene, LM = lower Miocene, OL = Oligocene. See Figure 2 for location.

basin or by a hidden thrustbelt beneath the salt on the poorly known Texas slope. Although there are extensional structures in the Lower Cretaceous section, the irregular shape of the collapse edge (Figures 28, 29) suggests that extension alone may not account for the origin of the salt walls.

Details of the geometry of Wilcox growth faults are controlled by the salients and reentrants in the collapse edge of the Cretaceous strata onto the Louann salt horizon. At the northeastern end of the map area, the large Wilcox depotrough abruptly terminates but is replaced northward by a separate trough that is offset to the west. At the southern end of the map area, the eastern margin of the trough is not mapped, but the western edge has an abrupt offset that overlies the position of a basement wrench fault system (Figure 30). These steep basement faults offset the base of Louann salt and are possibly coeval with Louann deposition. Additional displacement on these faults formed a northwest-trending anticline during Paleocene deformation of the Sierra Madre and Coahuila foldbelts in northeastern Mexico.

The irregular edges of the troughs do not match well, suggesting either that the Lower Cretaceous deep-water equivalent deposits onlapped existing salt walls or that complex internal deformation has greatly altered the shape of these edges. The blunt terminations, in particular, are difficult to restore without intervening salt bodies or large tear faults. Northwest-trending offsets may represent different initial positions of Cretaceous salt walls rather than large tear faults. The different initial positions could be caused by original salt thickness changes across Jurassic wrench faults. On the downdropped side of these faults, thicker salt farther landward might result in formation of a salt wall farther updip on that side of the fault.

Whatever their origin, collapse of large autochthonous salt walls created space for Wilcox depotroughs and related growth fault systems. The southern Texas deep structure, possibly basement controlled, is distinctly different from the isolated counter-regional withdrawal basins beneath the Oligocene–Miocene detachment offshore western Louisiana (Figure 18). Although large salt walls existed on the Cretaceous slope in southern Texas, isolated pillows or diapirs, which later became feeders for allochthonous salt, existed in southern Louisiana. Similar salt walls may have existed in the Louisiana Wilcox trend as well, and sub-detachment withdrawal basins may occur beneath the Texas shelf.

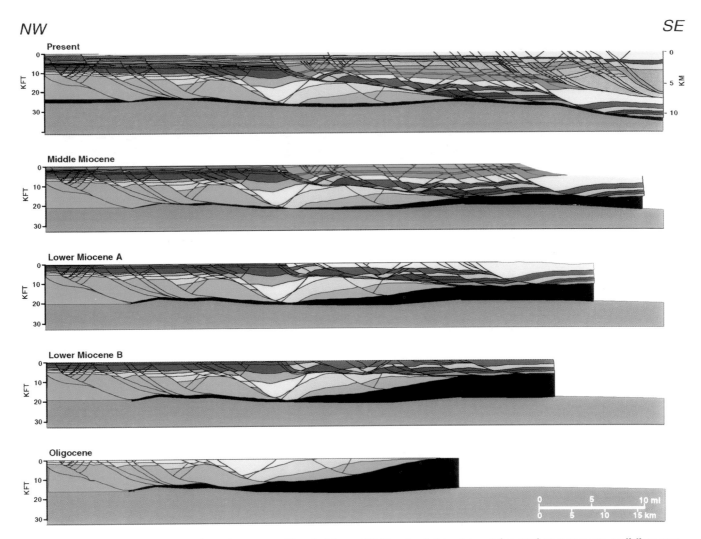

Figure 25—True-scale reconstruction of cross section in Figure 24. Perched detachment (center) restores as a sliding surface above a basinward-thinning wedge that collapsed onto the developing salt weld above Eocene and older strata. Salt is shown in black. Reconstruction is by the MESH method. See Figure 2 for location.

Relationship to Oligocene–Miocene Detachment Province

A true-scale regional reconstruction (Figure 31) across the onshore part of the central Texas Gulf Coast shows the nature of the transition from the Wilcox depotroughs to the Oligocene–Miocene detachment system. In this part of Texas, the imbricate fan of the Wilcox fault system has widened to form a perched detachment above Upper Cretaceous strata, but it still roots into a broad depotrough with most, if not all, of the Mesozoic section absent above the autochthonous Louann salt horizon. This depotrough is overlapped by a younger Eocene perched detachment that may terminate in a poorly known depotrough beneath thick Eocene shales. The seaward end of that trough is interpreted to be the feeder system for allochthonous salt subsequently evacuated by progradation of the Oligocene Frio shelf margin. Again, the width of the salt walls is unconstrained by the reconstructions of the late Eocene and Late Cretaceous.

PROVINCE RELATIONSHIPS ALONG WESTERN LOUISIANA TRANSECT

The limitations of subregional reconstructions are apparent from the Texas examples just presented, in which salt-bounded blocks of sediment are not laterally constrained by reconstruction and the relative magnitudes of extension and salt reduction are not determined. Inclusion of salt withdrawal in cross section reconstruction produces an extra degree of freedom compared to typical thrust belt reconstructions. Although the backstripping approach is a powerful way to reconstruct syndepositional structures, the results are completely dependent on the stratigraphic correlations. There are no geometric rules for deducing the deep structure of salt-withdrawal fault systems with nonrigid footwalls. Accurate reconstruction of these systems is dependent on seismic geometries and stratigraphic correlations. The requirement to choose a composite profile that minimizes

W   E

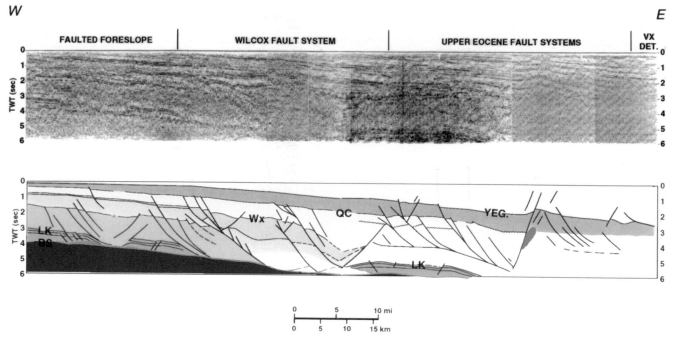

Figure 26—Uninterpreted regional seismic profile (top) and interpreted drawing (bottom) across the Wilcox and upper Eocene fault systems, onshore southern Texas. YEG = Eocene Yegua, QC = Eocene Queen City, Wx = Eocene Wilcox, LK = Lower Cretaceous, BS = base of Jurassic Louann salt. See Figure 2 for location.

out-of-plane three-dimensional effects presents an additional burden. Regional reconstructions that cross the entire basin provide additional constraints as well as an opportunity to illustrate models of overall structural evolution.

Diegel and Schuster (1990) presented two such regional reconstructions. One is in the eastern Gulf through the isolated systems of the eastern part of the salt dome–minibasin province (Figure 19 is extracted from that reconstruction; see also Schuster, 1995, this volume). The other one is in western Louisiana (Figure 32) and is discussed here and is reconstructed in Figures 33, 34, and 35 (folded insert). The western Louisiana transect is in the center of the basin and crosses (1) the Wilcox fault system, (2) an upper Eocene fault system, (3) the onshore salt dome–minibasin province, (4) the Oligocene–Miocene detachment system and related mid-shelf salt dome–minibasin province, (5) a Pliocene–Pleistocene organized roho system, and (6) the tabular salt–minibasin province of the slope.

The reconstructed western Louisiana cross section has the advantages of being in the complex central part of the basin and being relatively well imaged at deep levels. Still, it is important to separate well-constrained parts of this section from speculative parts without reliable seismic geometries and correlations. To separate interpretation from speculation, we include two types of cross sections: (1) an interpreted seismic profile and depth-converted frame cross section showing only reliable correlations and seismic geometries (Figure 35 [enclosure]) and (2) speculative, alternative cross sections completed to the pre-Louann basement (Figures 33–35).

Description

The Cretaceous carbonate margin and Louann salt horizon are imaged at the northern end of the profile. The profile crosses the Wilcox fault system, which terminates in a laterally extensive depotrough. This depotrough is bounded on the south by counter-regional faults and associated south-leaning salt domes of the updip sector of the salt dome–minibasin province. A small upper Eocene fault system overlies this trough. This laterally discontinuous detachment system is a relatively superficial structure within the depotrough and is therefore not subdivided from the updip sector of the salt dome–minibasin province in Figures 2 and 3. The landward edge of the Oligocene–Miocene detachment is overlain by expanded middle Oligocene Frio deltaic strata. Successive younger late Oligocene–early Miocene depocenters occur basinward, and a middle Miocene depocenter is located in the mid-shelf sector of the salt dome–minibasin province.

High-amplitude continuous seismic reflectors correlated to Cretaceous and Eocene chalks persist beneath the Oligocene–Miocene detachment and are deformed into counter-regional fault-bounded minibasins (see Figure 18). The Oligocene–Miocene detachment surface is not imaged beyond the salt dome–minibasin province. Southward, the roho-based Pliocene–Pleistocene fault systems continue to the salt massifs at the shelf edge. Well control indicates that this shallow detachment overlies middle Miocene strata. Presently, the roho reflection along the detachment represents the effective base of reliable seismic geometries in this area. The slope portion of

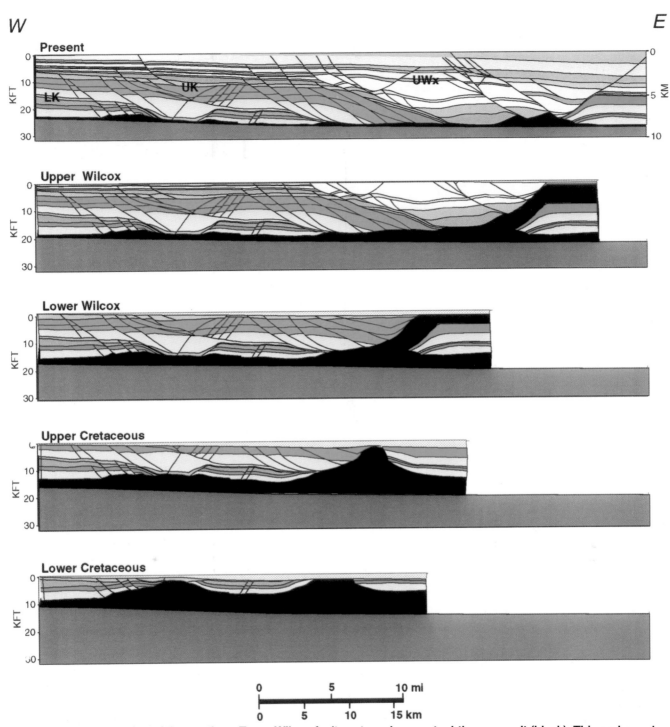

W   E

Figure 27—Reconstruction of the southern Texas Wilcox fault system above autochthonous salt (black). This end-member model assumes large extension during Wilcox deposition, but a continuum of models, trading the width of Upper Cretaceous salt walls for Tertiary extension, is geometrically possible. LK = Lower Cretaceous, UK = Upper Cretaceous, UWx = Upper Eocene Wilcox. Sea water is stippled. Reconstruction is by the MESH method.

the profile crosses a tabular salt body without an imaged base, as well as two deep minibasins updip of the salt mass that extends southward to the Sigsbee escarpment. Flat-lying abyssal plain strata are clearly imaged for 70 km under the Sigsbee salt mass. There is no fold and thrust belt at the toe of the slope here. Small structures within the depth-converted subsalt reflections may be small contractional structures or artifacts of the approximate depth conversion assuming vertical ray paths beneath thick salt.

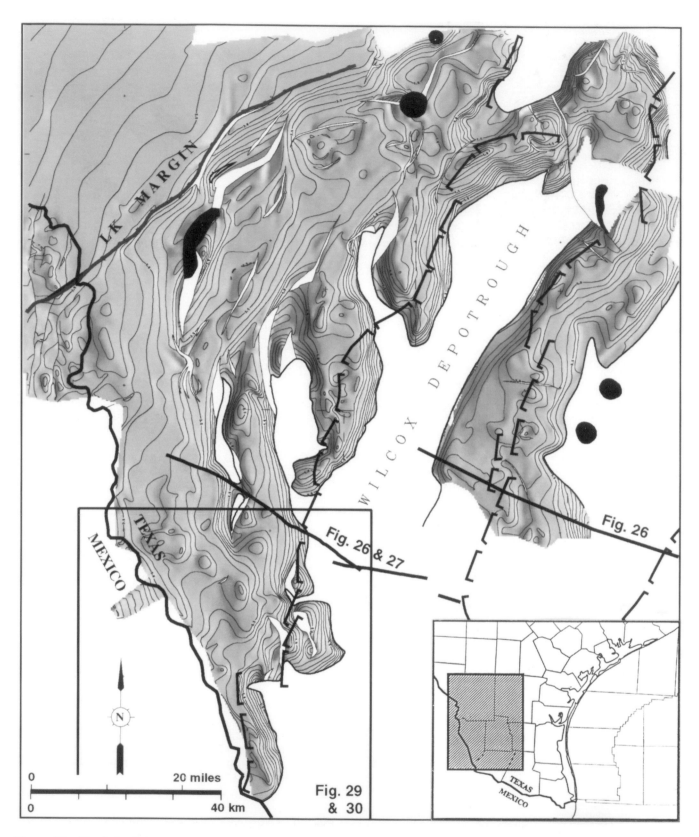

Figure 28—Shaded relief map of the structure on a Lower Cretaceous horizon, southern Texas. Lower Cretaceous strata are absent from prominent depotroughs for Wilcox deposition. Bracket symbols indicate generalized traces of Tertiary growth faults. LK = Lower Cretaceous. See inset map for location.

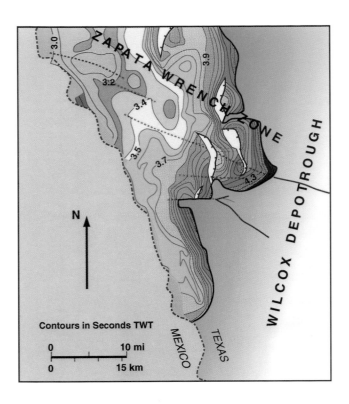

Figure 29—Seismic time–structure map of Lower Cretaceous horizon near the Mexican border showing off-set of collapse edge (outlined in Figure 30) along the trend of basement faults (purple) that offset the base of the Louann. Northwest-trending anticline represents reactivation of the fault system during Laramide compression related to the Coahuila foldbelt in northeastern Mexico.

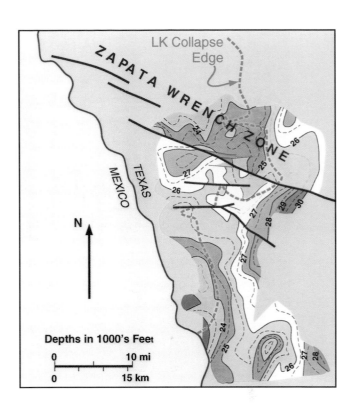

Figure 30—Pre-Louann basement structure in the same area as Figure 29 showing basement faults (purple).

Regional Palinspastic Analysis: Alternative Models

Two alternative speculative sections based on the frame section were restored to investigate end-member scenarios for the evolution of the north-central Gulf of Mexico Basin (Figures 32, 33, 35 [enclosure]). Model I (Figure 33) extrapolates the base of the Sigsbee salt mass directly to the Louann level, as suggested by Worrall and Snelson (1989). In this model, the base of the mid-slope tabular salt body is also rooted to the autochthonous Louann level. Likewise, speculative feeder systems for the Pliocene–Pleistocene roho systems are shown rooted directly to the autochthonous salt. The updip, better con-strained part of the cross section is the same in both mod-els. Model II (Figures 34, 35 [enclosure]), also consistent with the frame section, differs from model I in extending the Paleogene weld of coastal Louisiana beneath the outer shelf and slope to connect to the base of the Sigsbee salt mass. In model II, the middle slope tabular salt is shown as relatively thin, but this minor difference is inde-pendent of the main difference between models I and II. In model II, the feeders for the Pliocene–Pleistocene welds are rooted to a deeper allochthonous salt weld at the Paleogene level.

Additional seismic observations support the continua-tion of the base of the Sigsbee salt above a Paleogene hori-zon as in model II. On the frame section, the base of the Sigsbee salt body cuts down to the level of Eocene(?) abyssal plain strata before being obscured by a deep basin, but seismic profiles just west of the cross section show that the base of the Sigsbee salt body extends an additional 30 km northward, subparallel to and above flat-lying Eocene(?) and older strata (Figure 36). Although the details of the speculative parts of both sections are conjectural, several aspects of the reconstructions (described below) lead us to favor model II. Both models restore to a similar structural style at the end of Cretaceous time: low-relief asymmetric salt bodies devel-oped under a pelagic cover. At the northern end of the section, a thicker Cretaceous section updip of a salt mas-sif reflects basinward flow of autochthonous salt into the massif. Initiation of these salt structures is not addressed by the reconstructions. In model I, the Sigsbee salt over-thrust was initiated by the end of the Cretaceous; it began later in model II at about the same time as other extru-sions farther landward.

In both models, most of the salt structures evolved into leaning stocks by the end of Eocene Wilcox deposition. In model I, the Sigsbee salt mass was up to 3.7 km thick with more than 60 km of overthrust. The northernmost salt body on the section remained constrained by shelf mar-gin deposition into a possibly overhung stock near the paleoshelf margin. The entire profile to the south was in the bathyal environment at this time. Although the

(Text continues on p. 142.)

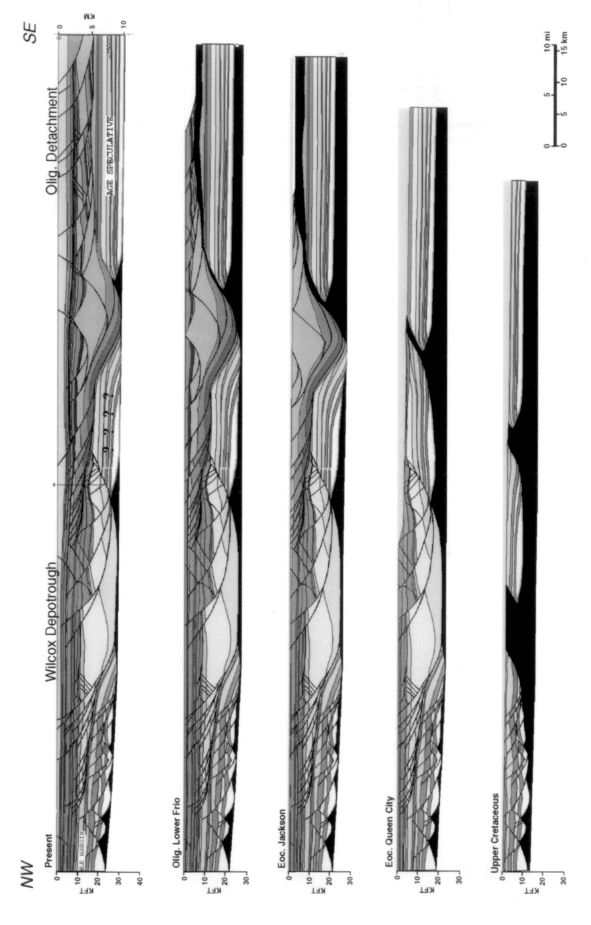

Figure 31—Reconstruction of a regional cross section across the Wilcox, upper Eocene, and Frio fault systems, onshore central Texas. The Wilcox fault system is perched above autochthonous salt (black). The Frio-age growth faults are interpreted to sole into an evacuated Paleogene canopy now forming the Oligocene–Miocene salt-based detachment that continues offshore (Figure 24). LK = Lower Cretaceous. See Figure 2 for location.

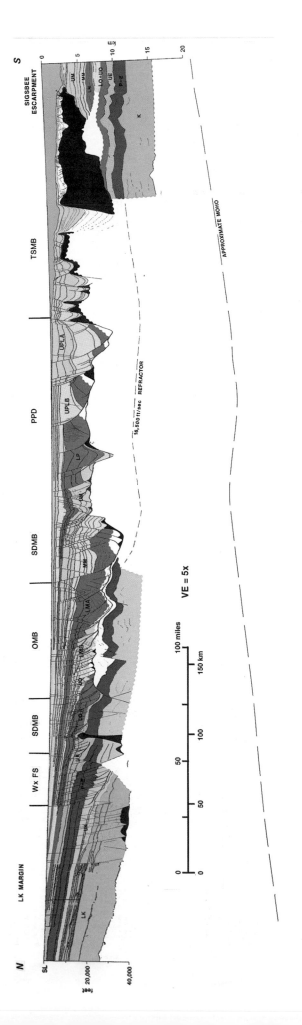

Figure 32. Five times vertically exaggerated frame section (showing only reliable correlations and observed seismic geometries) across western Louisiana, from the Lower Cretaceous margin to the abyssal plain. A true-scale section is shown in Figure 35 (folded insert). Salt is shown in black. WxFS = Wilcox fault system, SDMB = salt dome–minibasin province, OMB = Oligocene–Miocene detachment province, PPD = Pliocene–Pleistocene detachment province, TSMB = tabular salt–minibasin province, LK = Lower Cretaceous, UK = Upper Cretaceous, P-E = Paleocene–Eocene, UE = upper Eocene, LO = lower Oligocene, UO = upper Oligocene, LM, LMB, LMA = lower Miocene, MM = middle Miocene, UM = upper Miocene. See Figure 2 for location.

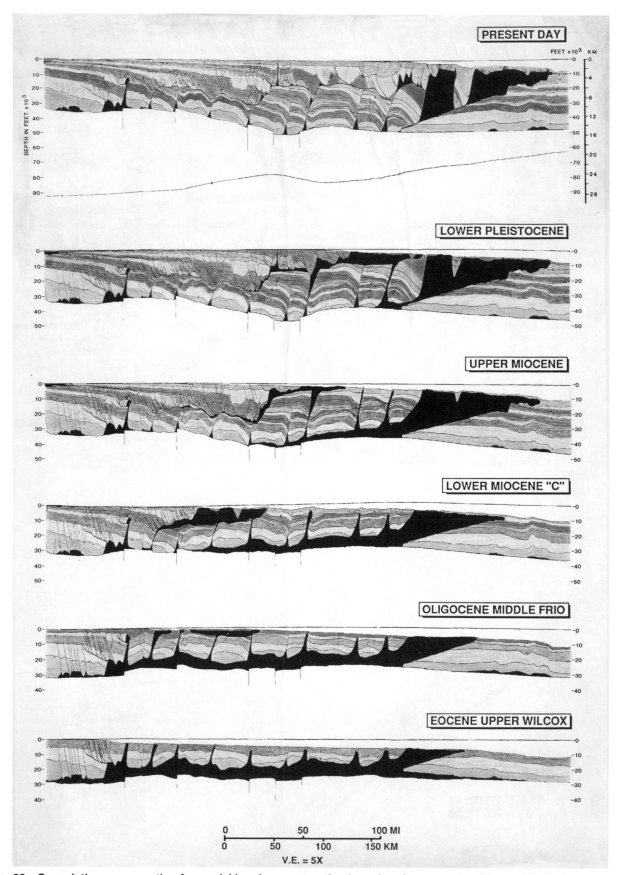

Figure 33—Speculative cross section for model I and reconstruction based on frame section (Figure 32). Reduced and squeezed from true-scale original; only selected stages are shown. Salt is shown in black. Base of Sigsbee salt mass is connected directly to the autochthonous Louann salt layer. This model is rejected in favor of model II (Figures 34, 35). Reconstruction is by the PREP method.

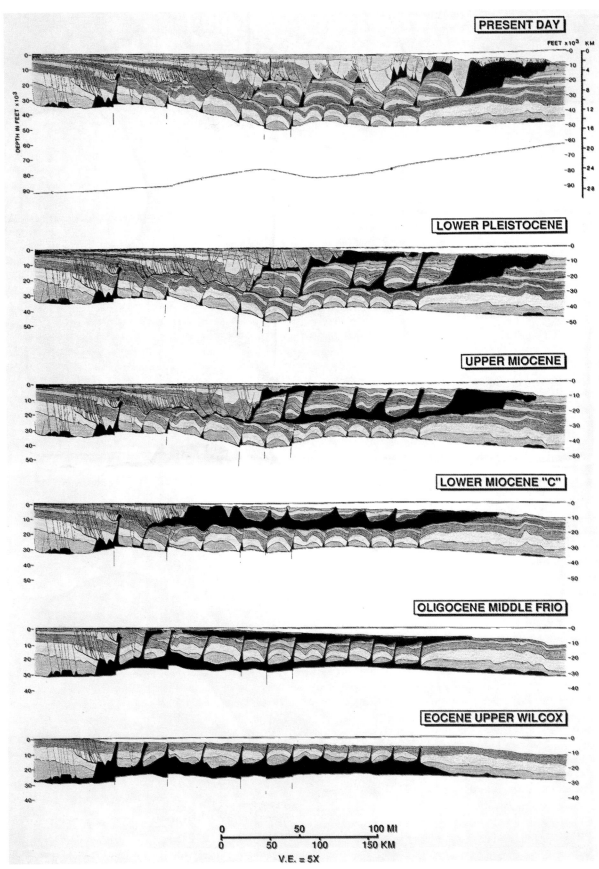

Figure 34—Speculative cross section for model II and reconstruction based on frame section (Figure 32). Reduced and squeezed from true-scale original; only selected stages are shown. Salt is shown in black. This model, which we prefer, connects the base of the Sigsbee allochthonous salt to the Paleogene canopy level. Reconstruction is by the PREP method.

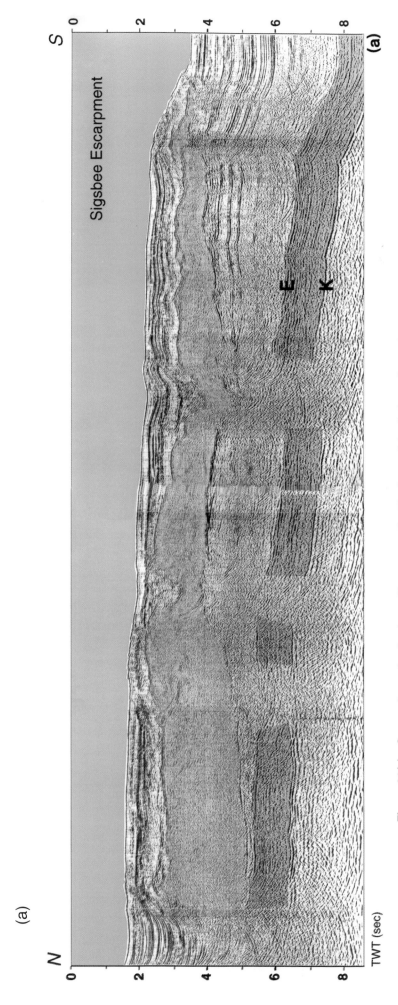

Figure 36(a)—Comparison of seismic profiles across the Sigsbee salt body (green), southwestern Louisiana slope. Profile used for the frame section and reconstructions (Figures 32, 33, 34, 35) shows the base of the Sigsbee salt sheet reaching as deep as a reflector tentatively correlated to the Eocene (top of brown interval).

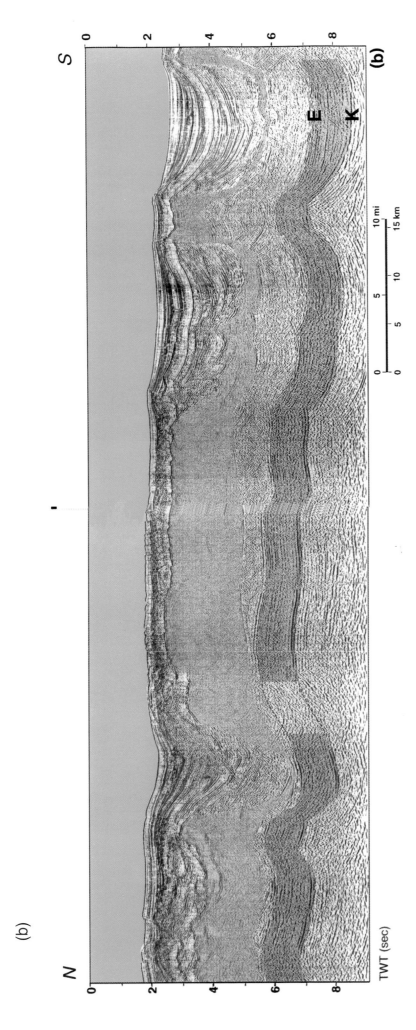

Figure 36(b)—Profile farther northwest shows the base of salt (green) continuing above through-going Eocene(?) and older abyssal plain strata (brown interval) for an additional 30 km northward. Apparent structure on the base of salt and subsalt strata is due to the velocity contrast of salt and sediment. E = Eocene, K = Cretaceous. See Figure 2 for locations.

142   *Diegel*

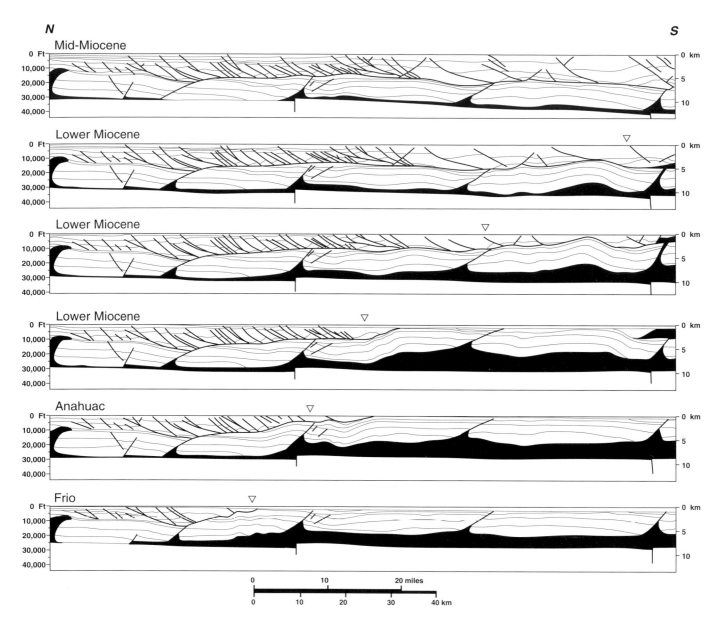

Figure 37—Alternative rejected reconstruction of onshore and inner shelf portions of frame section (Figure 32) assuming withdrawal of salt from the autochthonous level rather than a Paleogene salt canopy. Reconstruction is by the PREP method.

present-day section in model II appears more complicated than that in model II by inclusion of an additional level of allochthonous salt, model II is simpler in the restored upper Eocene section. In model II, the extrusion of allochthonous salt began over the entire bathyal part of the section, with the precursor to the Sigsbee salt mass forming as the most basinward of these flows. In model I, extrusion occurred only from the slope feeders that were updip of the imaged Paleogene detachment. Other salt bodies (excluding the Sigsbee salt) remained constrained into leaning stocks, even though they were in the same environment.

By middle Frio time, the salt canopy in model II was complete, but some salt was still at the autochthonous

level. The middle Frio marked the end of about 25 m.y. of relatively low rates of sedimentation in the Louisiana Gulf Coast that followed Eocene Wilcox deposition. This interval, almost as long as all the remaining Oligocene, Miocene, and Pliocene combined, is probably represented by less than 600 m of sediment beneath the western Louisiana shelf. The exact timing of the extrusion of salt during this condensed interval is unconstrained. Initiation and coalescence of salt flows did not necessarily occur precisely at the same time throughout the basin.

In model I, the Paleogene salt extrusion occurred only on the upper half of the slope. Salt remained at the autochthonous level on the lower slope. Although there is no direct evidence for a Paleogene canopy on the lower slope, as in model II, it is likely that sedimentation rates were even lower in this more distal position. Thus, any existing stocks were less constrained by sedimentation and more likely to flow into allochthonous sheets near the seafloor.

In late Oligocene time, middle Frio deposition represented the renewal of clastic progradation that continues to the Recent. In both models, Frio deltaic sedimentation began to prograde across the completely coalesced salt canopy, and the first major salt-based detachment faulting began. In model II, several minibasins formed over the canopy on the slope. The age of initiation and the geometry of these postulated minibasins are unconstrained, and the interpretation shown in model II is only one of several possible scenarios. The ages of the basins could be synchronous, younging to the south, or more irregular, depending on deep-water sediment dispersal patterns. In model I, the lower slope remained a relatively sediment-starved region with continued downbuilding of sediment between old salt stocks.

The structures initiated in Oligocene Frio time continued through Anahuac, early–middle Miocene in both models. The Paleogene salt evacuation surface was created by deltaic progradation and related listric growth faulting that progressively collapsed the salt canopy. The Sigsbee salt body continued to grow and override abyssal sediments. In model I, autochthonous rooted diapirs continued downbuilding, and in model II, minibasins continued to deepen on the slope. In the early Miocene of both models, a large minibasin in the modern mid-shelf region inverted to become a faulted turtle structure above allochthonous salt. In model I, the basinward end of the canopy formed a Sigsbee-like salt overthrust that climbed section and overrode Frio–middle Miocene slope sediments.

Salt withdrawal from the autochthonous level is a possible alternative to the Paleogene canopy indicated in both models I and II for present-day coastal Louisiana and the inner shelf. The implications of autochthonous solution include collapse of onlap onto a large rolling fold within the 4-km-thick subdetachment stratigraphy (Figure 37). There are several arguments against this autochthonous salt model. (1) It is inconsistent with allochthonous salt penetrations onshore and (2) inconsistent with mapped subdetachment salt-collapse structures (Figure 18). (3) It is unlikely that thick salt would remain undeformed beneath 4 km of Cretaceous–Eocene sediments until Oligocene time. (4) Although geometrically admissible, it is unlikely that a shale-based gravity slide would develop over a detachment surface dipping steeply landward, and (5) it is also unlikely that the 4-km-thick subdetachment section could be deformed by a rolling fold mechanism requiring folding and unfolding. Finally, (6) the salt-based detachment model is preferred because analogs in the Pliocene–Pleistocene trend are well known whereas no example of a rolling fold and backward-sloping detachment is known to us.

A possible objection to the regionally extensive Paleogene canopy proposed in model II is that many areas, particularly the western Louisiana inner shelf, are devoid of salt domes or other remnant shallow salt. However, salt-withdrawal fault systems are not always associated with remnant shallow salt, and large areas of salt evacuation may be difficult to recognize. Both models I and II imply efficient salt evacuation by lateral flow and/or dissolution. Although there is abundant remnant shallow salt in the central Louisiana Pliocene–Pleistocene detachment province offshore, considerably less is present in the same province offshore western Louisiana, where the salt-based detachment is also documented by drilling. In the East and West Cameron outer shelf in this province, there is a region of about 80 km dip extent and 65 km strike extent underlain by a salt weld without shallow secondary salt domes present (Figure 3). In another example in southern Texas, a salt interval was cored at the level of detachment for Frio growth faults in southern Texas, where the nearest salt dome is 80 km away in an older fault system and the nearest known shallow salt in the same age fault system is over 300 km away. These results challenge the dogma that the distribution of salt domes in a basin reflects the distribution of original salt deposition.

By late Miocene time, thin salt flows formed in the upper slope of both models I and II. In model I, this salt had sources at both the autochthonous and allochthonous Paleogene levels. In model II, this salt was fed entirely from allochthonous salt at the allochthonous Paleogene level. By Pliocene time in both models, additional shallow salt flows formed farther downdip. These flows coalesced on this line of section to form an organized roho system, but farther east toward the depocenter, flows were constrained by higher sedimentation rates and remained isolated to form a disorganized roho system. In model II, the flows rooted to the evacuating Paleogene canopy. As the Miocene flows continued to inflate, the updip parts were deformed and evacuated by Pliocene sediments at the prograding shelf edge. From Pleistocene to Recent time, allochthonous salt extruded in the Miocene–Pliocene was largely evacuated into domes out of the plane of section and basinward by continued progradation of the shelf margin depocenters. Both models imply significant loss of salt from the plane of the section, partly due to accumulation in salt domes out of the plane, but probably largely due to dissolution (see Chapter Appendix).

The four main differences in tectono-stratigraphic evolution highlighted by the alternative reconstructions of models I and II are given in Table 1.

In summary, model II is favored mainly for two reasons. First, seismic observation of the base of salt subparallel to Paleogene horizons on the mid-slope is consistent with model II. Second, although the present-day structure is simpler in model I, the restored Oligocene structure is simpler in model II, and there is no apparent reason to restrict salt extrusion to only the upper part of the Paleogene slope.

Table 1—Summary of Alternative Reconstructions in Model I (Figure 33) and Preferred Model II (Figures 34, 35)

Event or Feature	Model I	Model II
Sigsbee salt overthrust initiated	Before Paleogene	During Paleogene
Allochthonous salt extrusion	On upper part of Paleogene slope only	On entire Paleogene slope
Miocene slope sedimentation	Around down-building, leaning stocks	In slope minibasins on allochthonous salt
Feeders for allochthonous salt on outer shelf	Rooted directly to Louann salt level	Rooted to allochthonous Paleogene canopy

DISCUSSION AND CONCLUSIONS

Our current understanding of the structural evolution of the northern Gulf of Mexico Basin is based on improved seismic imaging, deep structural mapping, palinspastic analysis using biostratigraphic correlations, and an analog approach that uses developing structures on the modern slope to understand the early history of older structures on the shelf and onshore. There are still large areas of the basin where the deep structure is obscure and reliable correlations are impossible. In these areas, interpretations and reconstructions are necessarily speculative. Improved imaging and analysis have changed our understanding of Gulf Coast evolution, and it is reasonable to assume that these advances will continue.

The use of modern and Pliocene–Pleistocene analogs for the early history of older structures, although striking in many cases, may be limited by dramatic changes in sedimentation rates and styles of deep-water sediment dispersal through time. Palinspastic reconstruction results are limited by lack of adequate seismic imaging and stratigraphic correlations in many areas. Reconstructions typically provide viable alternative evolutionary scenarios but not unique solutions. Salt-withdrawal estimates based on backstripping are limited by uncertainties in paleowater depths and residual tectonic subsidence (an error in water depth produces twice the error in salt withdrawal, and an error in tectonic subsidence produces 2.8 times the error; see Chapter Appendix and Figure A-1). Other sources of error include uncertainties in densities, velocities, and decompaction histories, as well as unaccounted flexural effects and complexities in the thermal history of the margin due to rapid sedimentation and complex salt structures. Subsidence analysis for salt withdrawal is useful for finding first-order phenomena such as distinquishing salt-based from shale-based detachment systems, but it is unlikely to be a sensitive indicator of paleobathymetry or sea level changes.

Large-scale salt withdrawal provides a solution to the long-standing problem of production of accommodation space for extremely thick deltaic sections in the Cenozoic. Our observations and analyses argue for large-scale evacuation of a Paleogene salt canopy that extended across most of the margin, from the present onshore to the present middle slope, from southern Texas to central Louisiana. We interpret salt-withdrawal features updip and to the east of this canopy as more isolated structures rooted to the autochthonous level or as

isolated allochthonous salt bodies not coalesced into a canopy. The location of this transition is not well known to us in many areas. Younger allochthonous salt structures in the outer shelf and upper slope from southern Texas to central Louisiana are tentatively interpreted to be rooted to this older allochthonous level rather than the autochthonous level. This scenario is probably misleading in its simplicity. The complexity, variety, and three-dimensional nature of structures in the region present many additional problems.

The tectono-stratigraphic provinces described here are nongenetic, but we have presented interpretations of their origins and interrelationships. Although subject to refinement and realignment of boundaries, the provinces may remain useful first-order divisions even as new data become available and new concepts are developed. The variation in structural style from the salt dome–minibasin province to the salt-based detachment provinces is interpreted in two ways. Salt domes and related counter-regional fault-bounded minibasins occur either as (1) a downdip component of the fully evolved salt-based detachment system (mid-shelf sector) or as (2) evacuated allochthonous salt bodies that never coalesced into an extensive sheet (updip and eastern sectors).

The variation in structural style from the tabular salt–minibasin province to the salt dome–minibasin province is probably a difference in the extent of salt withdrawal, with basins on the slope surrounded by tabular salt evolving into fault-bounded basins flanked by residual salt domes, mainly on the shelf. The presence of contractional structures at the toe of allochthonous salt bodies may also be due to the extent of salt withdrawal, with early contractional systems developed on the slope inverting to become large-displacement counter-regional faults on the shelf. Variation within the Pliocene–Pleistocene roho systems, from organized to disorganized areas, remains unexplained. The comparison of sub-detachment structure beneath the Oligocene–Miocene detachment (Figure 18) and the deep structure beneath the southern Texas Wilcox fault system (Figure 28) highlights the influence of Mesozoic salt structures, and possibly pre-Louann structures, in determining the style and geometry of Tertiary growth fault systems.

Worrall and Snelson (1989) noted a difference in structural style between the Louisiana and Texas parts of the Oligocene–Miocene detachment province, with more linear faults typical of Texas and more arcuate fault patterns in Louisiana. This difference is quantitative rather than qualitative. Similar structures occur on both sides of the state line, but perched detachments are more common and extensive in Texas, whereas regional fault trends are more arcuate in Louisiana. Even within the "linear" fault trends of Texas and western Louisiana, detailed mapping usually shows linear fault systems to be composed of complexly nested arcuate faults. Worrall and Snelson (1989) attributed these differences to the dominance on the Texas shelf of Tertiary strandplain and barrier island depositional environments as contrasted to the alluvial and deltaic environments more typical of offshore Louisiana.

It is unlikely that geologically rapid shifts of depositional environment on the shelf would radically change the geometry of an active fault system that spans dozens of sequences and began in the slope environment. We suggest a slight modification of this concept. Perhaps the style of deposition initially deforming allochthonous salt on the slope is the most important factor determining the ultimate structural style, even though changes in depositional style on the slope are likely to be related to different depositional styles on the shelf (e.g., line sources or point sources for deep-water deposition related to differing shelf environments). The origin of perched detachments may also be linked to the presence of shales likely for detachment, such as the Jackson and Anahuac shales of Texas.

The extent and origin of the toe-of-slope foldbelts are also not well known. Foldbelts may be entirely absent in some areas or perhaps merely obscured by allochthonous salt overriding the basinward depositional limit of autochthonous salt. The timing of the known foldbelts does not appear to correlate in a simple way to the timing of updip extension, which persisted throughout the Cenozoic (compare Peel et al., 1995). Perhaps changes in slope or reduction of deep salt, which compensates for most of the extension, are important.

A final question raised by this discussion is the uniqueness of northern Gulf Coast allochthonous salt structures. Extensive allochthonous salt, including salt canopies, is reported from few salt basins (e.g., Great Kavir, Jackson and Cornelius, 1985; Jackson et al., 1990; Isthmian salt basin, southern Mexico, Correa Perez and Gutierez y Acosta, 1983). Are Gulf Coast style allochthonous salt structures more common, but unrecognized, or are the scale and complexity of salt-related structures in the Gulf Coast unique?

Acknowledgments *The work presented in this chapter relied on the interpretations and ideas of a large number of Shell Oil Company staff over many years. In particular, the groundbreaking work in the 1960s by C. C. Roripaugh, J. M. Beall, and others led to an early understanding of allochthonous salt structures when the results of seismic surveys were more ambiguous than those from current techniques. The term roho was derived in mock comparison to the moho. Based on seismic refraction experiments in this province during the late 1960s, Roripaugh and others at Shell recognized that the high-amplitude discontinuous reflectors were residual salt on evacuation surfaces. Also at Shell, in the 1970s, D. M. Worrall and S. Snelson pioneered computer-aided reconstruction techniques for analysis of growth fault and salt structures. A 5-year research project in the 1980s on the Cenozoic tectono-stratigraphic evolution of the northern Gulf of Mexico Basin provided a broad understanding of the complex structural framework across the basin from Florida to Mexico and the Cretaceous margin to the abyssal plain. The project, including both research and exploration staff, was planned by S. Snelson, who supervised the team at Shell Development Company (R. M. Coughlin, A. D. Scardina, F. A. Diegel, and D. C. Schuster). C. L. Conrad, C. C. Roripaugh, and S. C. Reeve at various times supervised the Shell Offshore team (C. F. Lobo, J. F. Karlo, and R. C. Shoup). R. N. Nicholas, M. L. Long, and other management staff at Shell provided critical support for Shell's long-term investment in this project and related research. J. C. Holliday and H. S. Sumner added an important regional study of the Louisiana outer shelf at Shell Offshore Inc.; S. A. Goetsch, C. J. Ando, and P. R. Tauvers undertook additional research projects on regional structural evolution of the Gulf Coast at Shell Development Co. Other Shell staff making critical onshore contributions from Shell Western E&P Inc. included E. J. Laflure and M. R. Lentini. The special interest and paleontologic support provided by E. B. Picou contributed greatly to these projects. This work also depended on the acquisition and processing of seismic data, seismic interpretations, maps, stratigraphic correlations, paleontologic analyses, and ideas of many other Shell staff involved in Gulf Coast exploration. R. W. Cook and C. E. Harvie developed Shell's proprietary workstation reconstruction program used to restore many of the cross sections presented here. E. E. White provided expert graphics support for all Shell Development studies cited here and for this chapter. Critical reviews by F. J. Peel, S. Snelson, M. P. A Jackson, and an anonymous reviewer have improved the quality of the work substantially.*

REFERENCES CITED

Amery, G. B., 1969, Structure of the Sigsbee scarp, Gulf of Mexico: AAPG Bulletin, v. 53, p. 2480–2482.

Balk, R., 1949, Structure of Grand Saline salt dome, Van Zandt County, Texas: AAPG Bulletin, v. 33, p. 1791–1829.

Balk, R., 1953, Salt structure of Jefferson Island salt dome, Iberia and Vermilion parishes, Louisiana: AAPG Bulletin, v. 37, p. 2455–2474.

Barton, D. C., C. H. Ritz, and M. Hickey, 1933, Gulf Coast geosyncline: AAPG Bulletin, v. 17, p. 1446–1458.

Bennett, S. S., and J. S. Hanor, 1987, Dynamics of subsurface salt dissolution at the Welsh dome, Louisiana Gulf Coast, *in* I. Lerche and J. J. O'Brien, eds., Dynamical geology of salt and related structures: Orlando, Academic Press, p. 653–678.

Blickwede, J. J., and T. A. Queffelec, 1988, Perdido foldbelt: a new deep water frontier in western Gulf of Mexico (abs.), AAPG Bulletin, v. 72, p. 163.

Bouma, A. H., G. T. Moore, and J. M. Coleman, eds., 1978, Framework, facies, and oil-trapping characteristics of the upper continental margin: AAPG Studies in Geology #7, 326 p.

Burk, C. A., M. Ewing, J. L. Worzel, A. O. Beall, Jr., W. A. Berggren, D. Bukry, A. G. Fischer, and E. A. Pessagno, Jr., 1969, Deep-sea drilling into the Challenger Knoll, central Gulf of Mexico: AAPG Bulletin, v. 53, p. 1338–1347.

Combes, J. M., 1993, The Vicksburg Formation of Texas: depositional systems distribution, sequence stratigraphy, and petroleum geology: AAPG Bulletin, v. 77, p. 1942–1970.

Correa Perez, I. and J. Gutierrez y Acosta, 1983, Interpretacion gravimetrica y magnetometrica del occidente de la Cuenca Salina del Istmo: Revista del Instituto Mexicano del Petroleo, v. 15, p. 5–25.

Diegel, F. A., and R. W. Cook, 1990, Palinspastic reconstruction of salt withdrawal growth fault systems, northern Gulf of Mexico (abs.): GSA Annual Meeting, Programs with Abstracts, Dallas, Texas, p. 48.

Diegel, F. A., and D. C. Schuster, 1990, Regional cross sections and palinspastic reconstructions, northern Gulf of Mexico (abs.): GSA Annual Meeting, Programs with Abstracts, Dallas, Texas, p. 66.

Duval, B., C. Cramez, and M. P. A. Jackson, 1992, Raft tectonics in the Kwanza basin, Angola: Marine and Petroleum Geology, v. 9, p. 389–404.

Fletcher, R. C., 1995, Salt glacier and composite sediment–salt glacier models for the emplacement and early burial of allochthonous salt sheets, in M. P. A. Jackson, D. G. Roberts, and S. Snelson, eds., Salt tectonics: a global perspective: AAPG Memoir 65, this volume.

Goldman, M. I., 1933, Origin of the anhydrite cap rock of American salt domes: USGS Professional Paper No. 175, p. 83–114.

Halbouty, M. T., 1979, Salt domes, Gulf region, United States and Mexico (2nd ed.): Houston, Texas, Gulf Publishing Company, 425 p.

Hanor, J. S., 1983, Fifty years of thought on the origin and evolution of subsurface brines, in S. J. Boardsman, ed., Revolution in the Earth Sciences: Dubuque, Iowa, Kendall/Hunt, p. 99–110.

Honea, J. W., 1956, Sam Fordyce–Vanderbilt fault system of southwest Texas: Gulf Coast Association of Geological Societies Transactions, v. 6, p. 51–54.

Hoy, R. B., R. M. Foose, and J. B. O'Neill, Jr., 1962, Structure of Winnfield salt dome, Winn Parish, Louisiana: AAPG Bulletin, v. 46, p. 1444–1459.

Huber, W. F., 1989, Ewing Bank thrust fault zone, Gulf of Mexico, and its relationship to salt sill emplacement: : SEPM Gulf Coast Section, 10th Annual Research Conference, Program and Extended Abstracts, Houston, Texas, p. 60–65.

Humphris, C. C., Jr., 1978, Salt movement on continental slope, northern Gulf of Mexico, in A. H. Bouma, G. T. Moore, and J. M. Coleman, eds., Framework, facies and oil-trapping characteristics of the upper continental margin: AAPG Studies in Geology #7, p. 69–86.

Jackson, M. P. A., and R. R. Cornelius, 1985, Tertiary salt diapirs exposed at different structural levels in the Great Kavir (Dasht–I Kavir) south of Semen, north-central Iran: a remote sensing study of their internal structure and shape: The University of Texas at Austin, Bureau of Economic Geology Open File Report OF–WTWI, p. 108.

Jackson, M. P. A., and C. Cramez, 1989, Seismic recognition of salt welds in salt tectonics regimes (abs.): SEPM Gulf Coast Section, 10th Annual Research Conference, Program and Extended Abstracts, Houston, Texas, p. 66–71.

Jackson, M. P. A., R. R. Cornelius, C. H. Craig, A. Gansser, J. Stocklin, and C. J. Talbot, 1990, Geology and dynamics of a remarkable salt diapir province in the Great Kavir, central Iran: GSA Memoir 177, 139 p.

Kupfer, D. H., 1962, Structure of Morton Salt Company mine, Weeks Island salt dome, Louisiana: AAPG Bulletin, v. 46, p. 1460–1467.

Lehner, P., 1969, Salt tectonics and Pleistocene stratigraphy on continental slope of northern Gulf of Mexico: AAPG Bulletin, v. 53, p. 2431–2479.

Le Pichon, X., and J. C. Sibuet, 1981, Passive margins: a model of formation: Journal of Geophysical Research, v. 86, p. 3708–3720.

Lundin, E. R., 1992, Thin-skinned extensional tectonics on a salt detachment, northern Kwanza basin, Angola: Marine and Petroleum Geology, v. 9, p. 405–411.

Manheim, F. T., and J. L. Bischoff, 1968, Composition and origin of interstitial brines in Shell Oil Company drill holes on the northern continental slope of the Gulf of Mexico (abs.), GSA Annual Meeting, Programs with Abstracts, Mexico City, Mexico, p. 189.

Martin, R. G., 1978, Northern and eastern Gulf of Mexico continental margin: stratigraphic and structural framework, in A. H. Bouma, G. T. Moore, and J. M. Coleman, eds., Framework, facies, and oil–trapping characteristics of the upper continental margin: AAPG Studies in Geology #7, p. 21–42.

McKenzie, D., 1978, Some remarks on the development of sedimentary basins: Earth and Planetary Sciences Letters, v. 40, p. 25–32.

Parsons, B., and J. G. Sclater, 1977, An analysis of the variation of ocean floor bathymetry and heat flow with age: Journal of Geophysical Research, v. 82, p. 803–827.

Peel, F. J., C. J. Travis, and J. R. Hossack, 1995, Genetic structural provinces and salt tectonics of the Cenozoic offshore U.S. Gulf of Mexico: a preliminary analysis, in M. P. A. Jackson, D. G. Roberts, and S. Snelson, eds., Salt tectonics: a global perspective: AAPG Memoir 65, this volume.

Rowan, M. G., 1994, A systematic technique for the sequential restoration of salt structures: Tectonophysics, v. 228, p. 331–348.

Rowan, M. G., B. C. McBride, and P. Weimer, 1994, Salt geometry and Pliocene–Pleistocene evolution of Ewing Bank and northern Green Canyon, offshore Louisiana (abs.): AAPG Annual Convention, Program with Abstracts, Denver, Colorado, p. 247.

Sawyer, D. S., 1985, Total tectonic subsidence: a parameter for distinguishing crust type at the U.S. Atlantic continental margin: Journal of Geophysical Research, v. 90, p. 7751–7769.

Schuster, D. C., 1993, Deformation of allochthonous salt and evolution of related structural systems, eastern Louisiana Gulf Coast (abs.): AAPG Annual Convention, Program with Abstracts, New Orleans, Louisiana, p. 179.

Schuster, D. C., 1995, Deformation of allochthonous salt and evolution of related salt–structural systems, eastern Louisiana Gulf Coast, in M. P. A. Jackson, D. G. Roberts, and S. Snelson, eds., Salt tectonics: a global perspective: AAPG Memoir 65, this volume.

Seni, S. J., 1992, Evolution of salt structures during burial of salt sheets on the slope, northern Gulf of Mexico: Marine and Petroleum Geology, v. 9, p. 452–468.

Seni, S. J., and M. P. A. Jackson, 1983, Evolution of salt structures, east Texas diapir province, part 2: patterns and rates of halokinesis: AAPG Bulletin, v. 67, p. 1245–1274.

Steckler, M. S., and A. B. Watts, 1978, Subsidence of the Atlantic-type continental margin off New York: Earth and Planetary Sciences Letters, v. 41, p. 1–13.

Sumner, H. S., B. A. Robison, W. K. Dirks, and J. C. Holliday, 1990, Morphology and evolution of salt/minibasin systems: lower shelf and upper slope, central offshore Louisiana (abs.): GSA Annual Meeting, Programs with Abstracts, Dallas, Texas, p. 48.

Trabant, P. K., and B. J. Presley, 1978, Orca Basin, anoxic depression on the continental slope, northwest Gulf of Mexico, *in* A. H. Bouma, G. T. Moore, and J. M. Coleman, eds., Framework, facies and oil-trapping characteristics of the upper continental margin: AAPG Studies in Geology #7, p. 303–312.

Trusheim, F., 1960, Mechanism of salt migration in northern Germany: AAPG Bulletin, v. 44, p. 1519–1540.

Vendeville, B. C., and M. P. A. Jackson, 1992a, The rise of diapirs during thin-skinned extension: Marine and Petroleum Geology, v. 9, p. 331–353.

Vendeville, B. C., and M. P. A. Jackson, 1992b, The fall of diapirs during thin-skinned extension: Marine and Petroleum Geology, v. 9, p. 354–371.

Verrier, G., and F. Castello-Branco, 1972, Le Fosse Tertiaire et le Gisement de Quenguela-Nord (Bassin du Cuanza): Revue de L'Institut Francais du Petrole, v. 27, p. 51–72.

Weimer, P., and R. T. Buffler, 1992, Structural geology and evolution of the Mississippi Fan foldbelt, deep Gulf of Mexico: AAPG Bulletin, v. 76., p. 225–251.

West, D. B., 1989, Model for salt deformation of central Gulf of Mexico basin: AAPG Bulletin, v. 73, p. 1472–1482.

Williams, C. A., 1975, Seafloor spreading in the Bay of Biscay and its relationship to the North Atlantic: Earth and Planetary Sciences Letters, v. 24, p. 440–456.

Winker, C. D., 1982, Cenozoic shelf margins, northwestern Gulf of Mexico basin: Gulf Coast Association of Geological Societies Transactions, v. 32, p. 427–448.

Worrall, D. M., and S. Snelson, 1989, Evolution of the northern Gulf of Mexico, with emphasis on Cenozoic growth faulting and the role of salt, *in* A. W. Bally and A. R. Palmer, eds., The geology of North America: an overview: GSA Decade of North American Geology, v. A, p. 97–138.

Wu, S., A., A. W. Bally, and C. Cramez, 1990, Allochthonous salt, structure, and stratigraphy of the northeastern Gulf of Mexico, part II: structure: Marine and Petroleum Geology, v. 7, p. 334–370.

Chapter Appendix

Reconstruction Techniques

Two reconstruction techniques were used to restore syndepositional faulting in this study: the proprietary PREP computer program described by Worrall and Snelson (1989) and a proprietary finite-element program called MESH (Diegel and Cook, 1990). This finite-element technique preserves area, accounts for decompaction, and minimizes the shape change within fault blocks. The horizontal component of unfaulted bed length is preserved, and footwalls are not assumed to be rigid. The method used for each reconstruction is noted in the captions. Paleobathymetric slopes are assumed to be constant through time with respect to the position of the prograding shelf margin, with a maximum slope of 1.5°. Because of the probability of three-dimensional flow and dissolution, salt area is not preserved. Instead, salt thicknesses through time are estimated from a one-dimensional isostatic calculation described below and by Diegel and Cook (1990).

Subsidence and Salt Withdrawal

A fundamental problem of Gulf Coast geology is to explain the great thickness of shallow water Tertiary strata on a Mesozoic passive margin. Barton et al. (1933, p. 1457) clearly recognized the problem early in the history of Gulf Coast exploration:

> The Gulf Coast geosyncline arouses isostatic meditation. . . . Isostatically, the Gulf Coast geosyncline must be, and for a long time must have been, negatively out of equilibrium. Subsidence continued, however, and presumably must have increased the lack of isostatic equilibrium, as the progressive depression of the basement has

increased the negative gravity anomaly. The movement, therefore, has been the reverse of what would be expected from the theory of isostasy.

Barton et al. (1933) were correct in noting that isostatic loading could not account for Gulf Coast subsidence and suggested that basement subsidence was caused by yielding of the crust beneath the sediment load to form a "geosyncline." They concluded their argument (p. 1458) with a note of uncertainty about this interpretation, however:

> The surface in the Gulf Coast seems to have remained nearly at sea-level. . . . The subsidence, therefore, seems more probably to be the effect of the sedimentation and to have tended to compensate it. But the subsidence can not be the effect of a movement toward isostatic equilibrium under the effect of the extra load of the sediments. . . . [The subsidence] seems more easily explainable not as an effect of the sedimentation but of some dynamic cause. . . . But the close equivalence of subsidence and sedimentation is not so easily explained by such a dynamic cause.

Thermal and isostatic subsidence alone cannot account for over 6 km of shallow water sediment deposited in the upper Tertiary on a passive margin where rifting occurred in the Jurassic. The rate of thermal subsidence of a passive margin decreases exponentially with time and is also proportional to crustal attenuation, where oceanic crust represents a maximum thermal subsidence case (Parsons and Sclater, 1977; McKenzie, 1978; Le Pichon and Sibuet, 1981; Sawyer, 1985). Therefore, by comparison with the North Atlantic (Williams, 1975; Sawyer, 1985) and with theoretical subsidence histories (McKenzie, 1978; Le Pichon and Sibuet, 1981), a reasonable maximum for the excess subsidence (as defined in Figure A-1) since rifting is about 2.3 km (equal to ~3.2 km

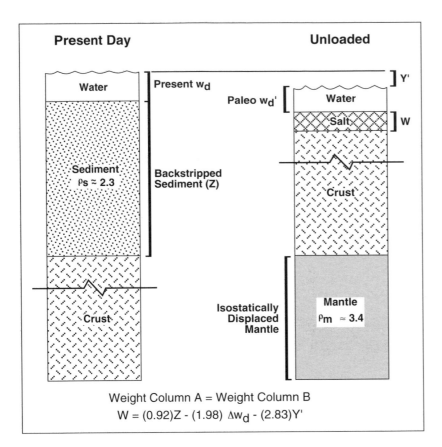

Present Day   **Unloaded**

Water

Present w_d

Paleo w_d'

Water

] Y'

Salt

] W

Sediment
$\rho_s \approx 2.3$

Backstripped
Sediment (Z)

Crust

Crust

Isostatically
Displaced
Mantle

Mantle
$\rho_m \approx 3.4$

Weight Column A = Weight Column B
$$W = (0.92)Z - (1.98)\,\Delta w_d - (2.83)Y'$$

Figure A-1—One-dimensional isostatic balance including salt withdrawal and paleobathymetry. Excess subsidence is defined by Y'.

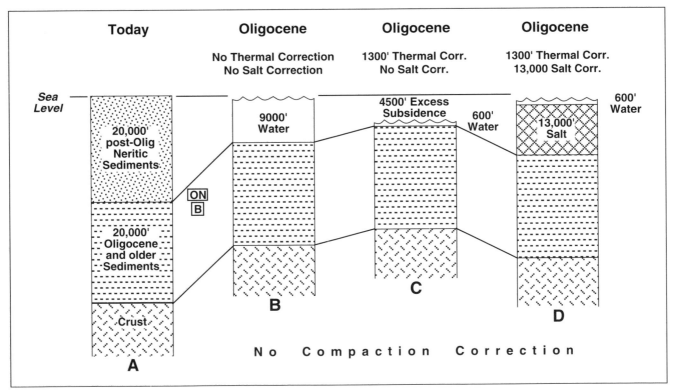

Today   **Oligocene**   **Oligocene**   **Oligocene**

No Thermal Correction
No Salt Correction

1300' Thermal Corr.
No Salt Corr.

1300' Thermal Corr.
13,000 Salt Corr.

Sea
Level

20,000'
post-Olig
Neritic
Sediments

9000'
Water

4500' Excess
Subsidence

600'
Water

600'
Water

13,000'
Salt

ON
B

20,000'
Oligocene
and older
Sediments

Crust

A   **B**   **C**   **D**

No   Compaction   Correction

Figure A-2—Magnitude of the subsidence problem in southern Louisiana. Backstripping of 6 km of sediments (column A) requires about 2.7 km of change in water depth (column B). Using a reasonable estimate of 200 m change in water depth still leaves 1.4 km of unaccounted excess subsidence (column C) that could be balanced by 400 m of excess thermal subsidence and 4 km of salt withdrawal (column D).

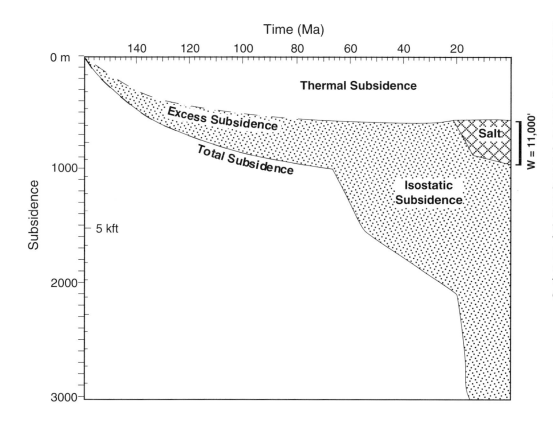

Figure A-3—Graph of subsidence versus age using restored decompacted thicknesses above basement at the south end of Figure 14 (onshore southern Louisiana). Dashed lines are extrapolated to Jurassic time based on depth to the Louann salt level. The backstripped excess subsidence curve is interpreted to represent thermal subsidence, very nearly flat by 50 Ma. The large anomaly in Miocene time (last 20 m.y.) is interpreted to be the result of salt withdrawal.

of conventional tectonic subsidence, including a hypothetical water column). Significant thermal subsidence continued for about 150 m.y. after rifting, but the bulk of this subsidence occurred in the first 100 m.y., with an exponential decline. Even a linear distribution of the total excess subsidence over 150 m.y. suggests that 400 m is a conservative estimate for the maximum excess thermal subsidence since the end of the Oligocene on the Gulf Coast margin.

For the post-Oligocene subsidence to be the result of isostatic loading, a top Oligocene water depth of about 2700 m would be necessary (Figure A-2), but we know from interpretation of depositional environments and faunal picks that this area was on the continental shelf at that time. Even given crude approximations of densities and ignoring decompaction, the magnitude of this discrepancy is impressive. Additional space created by a component of thermal subsidence is also insufficient to account for this subsidence. Using salt withdrawal as an unknown and using an estimated paleowater depth, a simple one-dimensional Airy isostatic model estimates the magnitude of salt withdrawal (Figure A-1). A plot of the total subsidence and backstripped subsidence as a function of age highlights the profound subsidence anomaly in Miocene time (Figure A-3). As noted by Barton et al. (1933), this subsidence anomaly migrates basinward with the prograding depocenters (Figure A-4). Salt withdrawal is the "dynamic mechanism" sought by Barton. Thick Louann salt, deposited in the Mesozoic, in effect stored the early subsidence of the basin for reuse by the prograding Cenozoic clastic margin that displaced the weak salt.

The backstripping technique (Steckler and Watts, 1978) can also be applied to deformed cross sections. Rather than using thicknesses from a single well, we reconstructed the fault motions above the detachment and measured changes in overburden thickness accounting for lateral translation and rotation of fault blocks. Estimated salt thickness was then added to the reconstruction (see Figures 15, 16). Although this technique estimates the amount of salt withdrawal, it does not locate the level of the evacuation surface. Two possible models are that the salt was withdrawn from the autochthonous Louann level (Figure 15) or that the detachment for listric growth faults represents a salt weld that formerly contained a thick, allochthonous salt body (Figure 16).

We prefer the allochthonous model for several reasons. (1) Salt penetrations occur along the Oligocene–Miocene detachment in western Louisiana. (2) The geometries resemble those above the shallower Pliocene–Pleistocene detachment systems, and we know of no example of a listric detachment formed in response to a rolling fold of salt beneath several kilometers of deep-water sediments. (3) The thinning wedge above the detachment suggests a collapsed onlap similar to the previous described examples from the outer shelf. (4) Subdetachment counter-regional faults provide a means for extrusion of the Louann salt to the level of the Oligocene seafloor. Finally, (5) it seems mechanically unlikely that a thick salt layer would remain undeformed beneath thousands of meters of sediments that thicken into counter-regional growth faults of pre-Oligocene age.

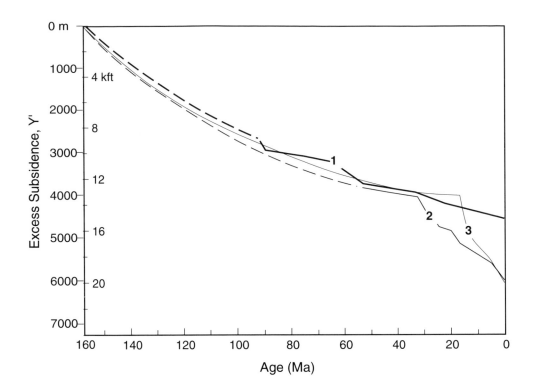

Figure A-4—Graph of back-stripped excess subsidence versus age for three positions indicated in Figure 14. Dashed lines are extrapolated to Jurassic time based on depth to the Louann salt level. No subsidence anomaly is apparent at position 1, but successive positions southward (2 and 3) show a large anomaly of excess subsidence following the prograding depocenters. This anomaly is interpreted as a wave of salt-withdrawal subsidence.

Salt Budget and Dissolution

Thick accumulations of shallow-water sediments and presumed lack of significant tectonic subsidence indicate that a large amount of salt-withdrawal subsidence has occurred in the Gulf Coast basin. Certainly a large component of lateral flow of salt is reflected in the Sigsbee salt mass. Although the Sigsbee salt body along the reconstructed profile may not be representative regionally, its area above the Paleogene level is about 320 km², which corresponds to an average salt thickness of 1100 m over the 290-km length of the section from the head of the Oligocene–Miocene detachment to the updip end of the Sigsbee salt mass. This area of salt accounts for less than half of the original 2.4-km salt thickness estimated from subsidence analysis using the method discussed previously. Shallow salt bodies out of the plane of the section probably account for a small part of the remainder, with dissolution completing the salt balance.

Although downbuilding relative to sediments is necessary to explain the height of Gulf Coast salt stocks, evidence of structural truncation and caprock indicates that there has also been considerable upward flow compensated by dissolution. Accumulation of anhydrite residue in caprock implies that thousands of feet of salt have been removed from the crests of onshore salt domes (Goldman, 1933). This interpretation is also supported by the truncation of vertical foliation observed in shallow salt mines (Balk, 1949, 1953; Hoy et al., 1962; Kupfer, 1962). Bennett and Hanor (1987) attributed increased formation water salinity in the vicinity of Welsh salt dome, onshore southern Louisiana, to active dissolution. They estimate that a minimum of 6 km³ of salt was dissolved

into the present formation waters. Although salt was penetrated at 2050 m depth at Welsh, no caprock was reported. Seni and Jackson (1983), on the basis of withdrawal basin volume, estimated that almost half of the mobilized salt of the East Texas salt basin was dissolved (380 km³ of a total volume of 800 km³). Caprock, although common in East Texas, is not thick enough to account for all of this volume loss. Seni and Jackson (1983) inferred that the loss occurred by erosion and dissolution at the seafloor rather than solely by circulating groundwater.

Evidence of salt dissolution is not limited to the onshore area. Average Gulf Coast formation waters are more than four times more saline than seawater, and many authors believe this is because of salt dissolution (for review, see Hanor, 1983). If halite is exposed to seawater, either directly by uplift and erosion, sea level drop, or extrusion or indirectly by contact with flowing pore water, it will dissolve. Although caprock is only reported from 5 of 77 cored offshore salt domes (Halbouty, 1979), one of the Eureka cores in the upper slope encountered 36 m (117 ft) of anhydrite caprock above a salt massif (Lehner, 1969). This amount of caprock would require a minimum of about 2300 ft (700 m) of salt dissolution (assuming an average 5% anhydrite content of Louann salt). Manheim and Bischoff (1968) reported salinity gradients approaching saturation in formation waters encountered by Eureka core holes near salt bodies in the upper slope and interpreted these gradients as indicators of active slope salt dissolution. At least one slope minibasin is known to have a stable brine pool (Trabant and Presley, 1978). This brine occurrence was discovered when pore waters in research cores were found to be eight times more saline than seawater.

Salt dissolution is undoubtedly occurring in other parts of the slope even though the seafloor structure does not always allow stable brine pools to form. Burk et al. (1969) reported caprock in core recovered from the Challenger Knoll in the Sigsbee abyssal plain. Apparently, circulating meteoric water is not necessary for salt dissolution to occur. Caprock on the shelf may be periodically exposed at the seafloor by extrusion or sea level fluctuations and removed by erosion. The allochthonous Sigsbee salt mass overrode the abyssal plain sediments with its upper surface at or near the seafloor throughout the entire Cenozoic era. Similarly, an extensive Paleogene salt canopy extruded near the sea floor would have provided the opportunity for large amounts of dissolution in the past. Without an impermeable pelagic mud drape, we might expect cumulative dissolution to be more extensive than implied by the reconstructions presented here (compare Fletcher et al., 1995, in this volume).

Peel, F. J., C. J. Travis, and J. R. Hossack, 1995, Genetic structural provinces and salt
tectonics of the Cenozoic offshore U.S. Gulf of Mexico: a preliminary analysis,
in M. P. A. Jackson, D. G. Roberts, and S. Snelson, eds., Salt tectonics: a global perspective: AAPG Memoir 65, p. 153–175.

Genetic Structural Provinces and Salt Tectonics of the Cenozoic Offshore U.S. Gulf of Mexico: A Preliminary Analysis

F. J. Peel

C. J. Travis

BP Exploration Inc.
Houston, Texas, U.S.A.

J. R. Hossack

BP Exploration Inc.
Stockley Park, Middlesex, U.K.

Abstract

Structures in the Cenozoic section of the U.S. Gulf of Mexico margin are thin-skinned, gravity-driven, and powered by the deposition of sediment on the shelf and upper slope. Deformation driven by sedimentation takes the form of salt displacement (including diapirism, salt withdrawal, and salt canopy formation), plus seaward gravity spreading and sliding. Lateral flow of salt gives rise to the emplacement of large-scale salt canopies of different ages. Lateral tectonic movement of both sediment and salt results in linked systems on a wide range of scales. We identify four structural provinces that contain distinct groups of structural elements believed to be genetically related: (1) far-eastern Gulf, in which no major Cenozoic deformation is seen; (2) eastern Gulf, defined mainly by a middle–late Miocene linked system of extension and contraction; (3) central Gulf, in which Oligocene updip extension was absorbed within a preexisting giant salt canopy; and (4) western Gulf, defined by several Paleogene–middle Miocene linked systems of extension and contraction. The ages and extents of each linked system match the major foci of sediment input to the shelf.

INTRODUCTION

Geologic understanding of the Gulf of Mexico Basin has evolved slowly, driven by incremental advances in data quality and availability. In particular, an understanding of the Cenozoic stratigraphic and structural framework of the northern Gulf of Mexico has developed in stages, advancing from the basin margins toward its center and following the progressive shift of oil and gas exploration activity offshore into deeper U.S. waters.

Over the last decade, a number of key papers have been published that describe the structural geometries, salt tectonics, and stratigraphy of parts of the northern Gulf of Mexico. It is becoming clear that there are considerable changes in the age and style of the dominant geologic processes along the width of the northern Gulf of Mexico Basin margin. For example, Worrall and Snelson (1989) published two regional balanced cross sections demonstrating different structural styles in the Louisiana and Texas portions of the Gulf of Mexico. In this volume, Diegel et al. (1995) include several detailed regional sections that show further along-strike variations. Liro (1992) demonstrated along-strike variation in structural style along the downdip limit of allochthonous salts.

Recent publications focusing on smaller areas have shown that a consistent structural style can be seen across subregions. For example, Weimer and Buffler (1992) and Wu et al. (1990a,b) have established a framework for the structure and evolution of salt canopies and contractional folds in part of the northeastern Gulf of Mexico. The age and style of salt emplacement and contraction are consistent across this area but completely different from that seen, for example, around the Perdido foldbelt in the northwestern Gulf (Worrall and Snelson, 1989).

The aim of this chapter is to describe further the variations in structural style and age across the northern Gulf of Mexico margin and to begin to establish large-scale structural linkages. Major regional-scale structural processes on the margin include the formation of giant allochthonous salt bodies, updip extension in the form of growth faulting, and downdip contraction in the form of lower slope folding, thrusting, and shortening within salt canopies. We illustrate these features with a suite of six regional-scale retrodeformable structural cross sections and three palinspastic restorations. The interpretations are based on seismic, well, and gravity data. In addition, we define areas in which the ages of salt canopy emplacement and subsequent spreading are broadly similar.

Finally, we define four major genetic structural provinces, each of which consists of complexes of structural elements of similar age arranged systematically. The locations of the lateral boundaries of some of these provinces are somewhat speculative; they are commonly masked by shallow salt or overprinted by local effects such as salt withdrawal basin development. Some of the boundaries may be diffuse zones rather than discrete edges. We present the structural provinces as a starting point for further discussion. We hope that with further work and the ever-increasing quality and availability of seismic data, the location and nature of the provinces and their boundaries may become better defined.

The observations and interpretations presented here are based on work carried out at BP Exploration during 1991–1993 in support of a regional analysis of the hydrocarbon habitat and prospectivity of the northern Gulf of Mexico. Seismic data were interpreted at a 2-mile spacing along the whole margin. The regional lines included here were based on paper seismic sections up to 50 ft long. Clearly, it is impractical to include representative supporting data on this scale, and line fragments are insufficient to show the whole structure. Therefore, this chapter presents the working ideas of the BP regional group, which are impossible to document fully in a work of this size.

GEOLOGIC FRAMEWORK

The focus of this paper is the structure of the Cenozoic section of the northern Gulf of Mexico. However, the structure and stratigraphy of the underlying section exert a significant influence on the development of the Cenozoic, and some key points need to be addressed here.

Basement and Synrift Section

Although various models exist for the opening of the Gulf of Mexico (e.g., Pindell and Dewey, 1982; Pindell, 1985; Salvador, 1987), there is consensus that the main period of rifting occurred in the Middle Jurassic, followed by oceanic spreading in the middle of the basin continuing possibly into the Late Jurassic. The Middle Jurassic rifting of the continental margin created an extensive set of deep grabens separated by elevated horst blocks (Hossack and Matson, 1993). At least in the easternmost Gulf, the supply of synrift sediments was not sufficient to bury this topography, and some basement highs probably remained as significant paleobathymetric features into the Cenozoic (see Figure 4). Remnant highs were an important influence on the subsequent stratigraphic development, for example by controlling the position of Cretaceous platform margins (McFarlan and Menes, 1991; Sohl et al., 1991) and on later structural styles through their influence on autochthonous salt distribution. There is some evidence of Late Jurassic basement faulting on the periphery of the basin (Thomas, 1988), but

in the offshore area there is no documented evidence of subsequent active basement faulting.

Age and Distribution of the Autochthonous Salt

The principal autochthonous salt in the northern Gulf of Mexico is the Upper Jurassic Louann Salt (e.g., Salvador, 1991a). This salt does not appear to have been deposited as a uniform blanket (see Figure 4); Buffler (1989) and Salvador (1991a) have shown that it is absent over most of the oceanic crust and varies considerably in thickness elsewhere. The interpretation of Salvador (1991a) is that the salt was synrift and its distribution was controlled by the geometry of active rifting. An alternative possibility, suggested by our reconstructions, is that the salt was deposited early in the postrift stage and that its distribution was controlled by rift-generated topography that had not yet been buried. There are insufficient data to determine which model is correct, but either way, the salt was likely to have been thin or absent over the crests of horst blocks and thickest over the grabens. In turn, any lateral variations in original salt thickness may have influenced the structural style and internal variability of the overlying stratigraphy.

Structural and Stratigraphic Style of the Post–Louann Salt Section

Considerable along-strike variation occurs in the structural style of the post-rift Mesozoic and Cenozoic section, and no single line can represent the entire northern Gulf of Mexico margin. However, a number of features are common to the whole study area, illustrated by a 5:1 vertically exaggerated geologic cross section through the north-central Gulf margin (Figure 1). These common factors are (1) a high volume of Cenozoic clastic sediment, which has caused the shelf margin to prograde several hundred kilometers; (2) deep subsidence of the basement under the depocenter, mainly as a result of sediment loading; (3) abundant salt, mobilized up into the postrift section as diapirs and salt canopies; (4) sediments not underlain by salt that have a continuous layer cake stratigraphy devoid of major internal structures; and (5) sediments overlying salt (autochthonous or allochthonous) that vary considerably in structure and lateral stratigraphic thickness.

Many major structural elements vary laterally across the northern Gulf margin. The most significant lateral variations are (1) the thickness of the Cenozoic sediment sequence and the age of maximum sediment input (e.g., Galloway et al., 1991; Pulham et al., 1994); (2) the amount of salt present in the section; (3) the extent of major salt canopies and the age of their emplacement; (4) the extent to which the salt canopies have been consumed by salt-withdrawal basins; and (5) the extent and age of movement of large-scale linked systems of extension and contraction.

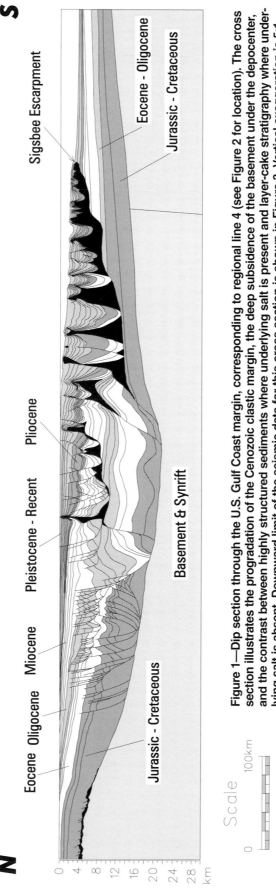

Figure 1—Dip section through the U.S. Gulf Coast margin, corresponding to regional line 4 (see Figure 2 for location). The cross section illustrates the progradation of the Cenozoic clastic margin, the deep subsidence of the basement under the depocenter, and the contrast between highly structured sediments where underlying salt is present and layer-cake stratigraphy where underlying salt is absent. Downward limit of the seismic data for this cross section is shown in Figure 3. Vertical exaggeration is 5:1.

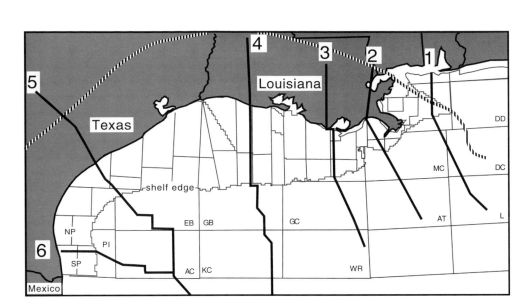

Figure 2—Location of regional seismic lines 1 to 6. Onshore area is shaded. Offshore divisions correspond to U.S. protraction areas: AC = Alaminos Canyon, AT = Atwater Valley, DC = DeSoto Canyon, DD = Destin Dome, EB = East Breaks, GB = Garden Banks, GC = Green Canyon, KC = Keathley Canyon, L = Lloyd, MC = Mississippi Canyon, NP = North Padre, PI = Port Isabel, and WR = Walker Ridge. Hatched line denotes Lower Cretaceous shelf edge and Florida Escarpment offshore.

The Upper Jurassic–Recent interval, where it overlies salt, is strongly affected by sedimentation-driven gravity tectonics. The resulting structures take two forms:

1. Salt displacement is driven by vertical sediment movement, generally dominated by salt withdrawal and diapirism on the paleoshelf and upper slope and by canopy emplacement plus salt withdrawal on the middle to lower slope.
2. Seaward gravity sliding or spreading occurs in which updip extensional growth faulting, typically in the shelf and upper slope, is linked to an equivalent amount of downdip contraction, typically in the middle to lower paleoslope. As postulated by Hossack (1995) and illustrated by examples later in this chapter, this contraction may take a variety of forms, such as shortening of a salt canopy or development of fold and thrust belts. This process gives rise to what are herein termed *linked systems* on a wide range of scales.

The similarities and differences in structural style and timing in the Cenozoic northern Gulf of Mexico resulting from these tectonic processes and the variables just described are the focus of this chapter.

REGIONAL GEOLOGIC CROSS SECTIONS

The regional model described here is based primarily on the interpretation of a grid of seismic data, typically spaced 2×2 miles or less, across the shelf and slope of offshore Texas and Louisiana. In addition, we interpreted six regional seismic lines, extending where possible from the updip limit of the basin to the abyssal plain (Figure 2). Each regional section was depth converted and the geologic model extended down to the top of basement (Figure 3). Below the limit of seismic and well data, inter-

pretation of the deep structure was based on gravity and magnetic data, geometric constraints provided by section balancing and restoration (Peel, 1993), and examination of adjacent areas having better data. A dotted line on each section indicates the downward limit of seismic control. By means of the computer restoration program LOCACE (Moretti and Larrere, 1989), the depth sections were then palinspastically restored to the paleostructure at a number of key time periods.

LOCACE carries out constant-area restorations in two dimensions using either vertical or inclined shear or flexural slip algorithms. This is generally a reasonable assumption for palinspastic clastic sediment restorations, but there are inherent dangers in assuming constant area for salt through geologic time (Hossack, 1995). In LOCACE restorations, the restoration process unstrains and rearranges the sedimentary blocks. The shape of salt bodies at past times is defined by the spaces between the restored blocks. Within reason, the volume of the salt is allowed to vary through time to accommodate the effects of salt dissolution and out-of-plane movement. Another major uncertainty in restorations where salt is involved is the recreation of paleotopography or paleobathymetry. The interpreter has to reconstruct this for each stage of the restoration using evidence for paleowater depth provided by lithology, seismic facies, and biostratigraphy. The restoration process usually provides additional constraints on the original interpretations that subsequently may have to be modified through iterative repetitions of the process. The six structural cross sections presented here (Figures 3a–f) were originally constructed and restored at 1:1 scale; they are presented here with a vertical exaggeration of 5:1 ratio unless otherwise specified.

Line 1: Present-Day Structure

Line 1 (Figure 3a) is a dip line across the shelf and upper slope in Viosca Knoll and northeastern Mississippi Canyon protraction areas (the locations of the regional lines and offshore protraction areas are given in Figure 2). It then passes into deep water through the Lloyd protraction area, where it runs parallel to the Florida escarpment (hatched line in Figure 2). Here, the Cenozoic section is relatively thin and mostly undeformed, except in areas adjacent to salt stocks and minor allochthonous salt sheets. As a result, the post-Louann Mesozoic section and the base autochthonous salt surface are clearly imaged on seismic data over much of the area. Where it overlies autochthonous salt, the post-Louann Mesozoic section is highly variable in thickness and contains abundant syn-depositional structures.

North of the present-day shelf edge (onshore and in Main Pass and the Mississippi–Alabama shelf) are a number of small-scale growth faults, mostly of Late Jurassic age. These are interpreted to be the product of southward gravity sliding, detaching on the salt, within the late Mesozoic carbonate platform. The structural style here is similar to that described for the adjacent Destin Dome protraction area by MacRae and Watkins (1993). Near the edge of the platform, the gravity sliding system on the shelf is interpreted to terminate in a set of thrusts. The location of the late Mesozoic carbonate platform edge appears to correspond to the edge of a major basement horst block defined by seismic, magnetic, and gravity data (Hossack and Matson, 1993). Farther south, a broad basement high is the westward continuation of the Florida Middle Ground arch. Between the platform edge and the crest of this high, the Mesozoic is affected by large-scale salt withdrawal basins, turtle structures, and diapirs.

Between the crest of the basement high and the southward pinch-out of the Louann Salt, the Mesozoic section is thinner and highly deformed. Based partially on the observation that the top-Jurassic datum is elevated above its regional level in this area, a zone of poorly imaged structures in the Mesozoic is speculatively interpreted as Cretaceous thrust faults, possibly related to out-of-plane southwest-directed gravity sliding at the foot of the Florida escarpment.

South of this zone, the section becomes abruptly much simpler, having a continuous layer cake stratigraphy and no obvious internal structures. Because this change corresponds to the southernmost evidence of salt diapirism and withdrawal, it is interpreted to coincide with the southeastern pinch-out of autochthonous salt.

Little evidence exists in this section of significant deformation during the Cenozoic, other than adjacent to salt stocks. In particular, there is no evidence of major horizontal movement of Cenozoic age. Local salt diapirs and small allochthonous sheets exist, but there are no regional-scale salt canopies. The structures seen on line 1 are typical of a wide area of the northeastern Gulf of Mexico, which corresponds to the far-eastern province (see Figure 12).

The observations made on this line may provide a clue to the structures that are likely to occur farther west, where the Mesozoic is too deeply buried to be clearly imaged. Wherever Mesozoic strata overlie autochthonous salt, they are probably deformed and laterally variable in thickness. Given the lateral variability in structural style seen in the east, it is difficult to predict the precise nature of that structure in the offshore areas farther west.

Line 1: Palinspastic Restorations

Line 1 is restored to its hypothetical paleostructure in middle Cretaceous, Late Jurassic, and Middle Jurassic time (Figure 4). The basic restoration to middle Cretaceous time principally involved reconstruction using vertical simple shear to a paleodepositional surface, reversing the later effects of differential salt withdrawal. In the restoration, we assumed that the surface north of the middle Cretaceous shelf edge was horizontal and near sea level. The surface over the basement high at the southern end of the reconstruction was assumed to be overlain by a minimum original thickness of salt. The original inclination of the middle Cretaceous paleoslope between the base of the carbonate platform and the crest of the basement high is uncertain, but was assumed to have had a uniform gradient. It could have been slightly concave upward, having less salt on the autochthonous level, or convex upward, having more salt. Whatever the restoration chosen, the main style of deformation between middle Cretaceous and the present day has been evacuation of salt from the autochthonous level into salt stocks.

Restoration to Late Jurassic time (Figure 4) again involved dominantly vertical shear, reversing the effects of earlier differential salt withdrawal. The uncertainty in defining the inclination and depth of the paleodepositional surface is more severe here. We opted for a conservative restoration involving a concave upward surface for the paleoslope, thereby minimizing the thickness of salt required in the reconstruction. We believe this to be the minimum reasonable thickness of salt; any less would require an unreasonably steep and probably unstable northward slope in the southern half of the line. We see evidence of gravity sliding elsewhere occurring on gentler slopes, but none here.

Restoration to Middle Jurassic time (at the end of Louann Salt deposition) mainly involved block rotation of the basement to remove the effects of thermal subsidence and isostatic flexure. We assumed that the northern and southern limits of thick Louann Salt occurred at the same topographic elevation.

The main points illustrated by the restorations are as follows:

1. Most deformation was related to differential salt withdrawal from the autochthonous salt level.
2. Although there is evidence for out-of-plane lateral movement of the cover in the form of gravity sliding or spreading during the Mesozoic, no evidence exists of significant lateral movement during the Cenozoic.

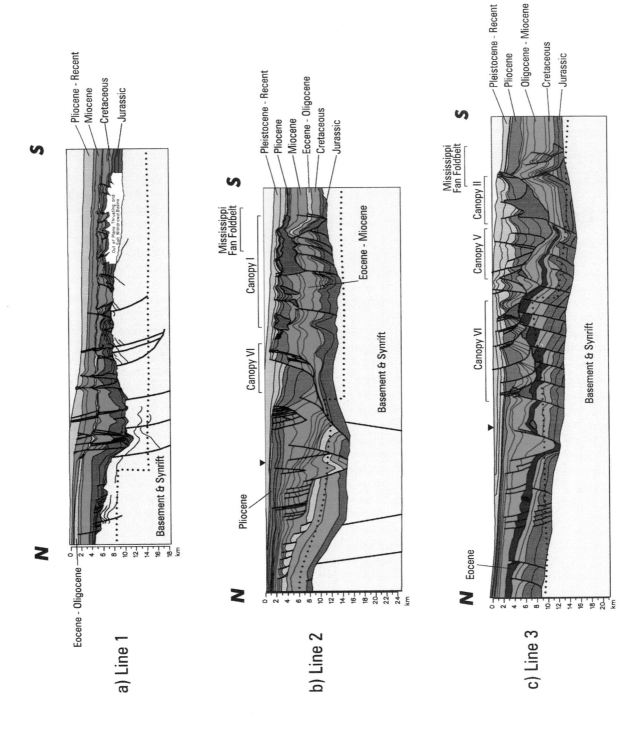

Scale

0   100km

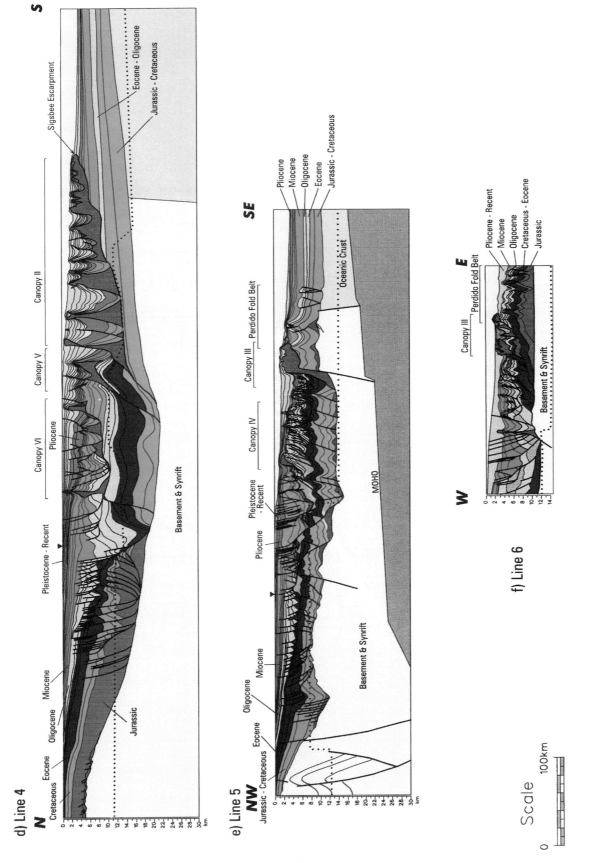

d) Line 4

e) Line 5

f) Line 6

Scale

0   100km

Figure 3—(a–c) Regional seismic lines 1, 2, and 3 (on facing page) and (d–f) lines 4, 5, and 6 (on this page), shown with 5:1 vertical exaggeration (see Figure 2 for locations). Dotted lines show lower limit of seismic data. Inverted black triangles show location of shoreline. Mottled green in line 5 (e) denotes possible highly deformed or diapiric shale. See text for discussion of individual lines.

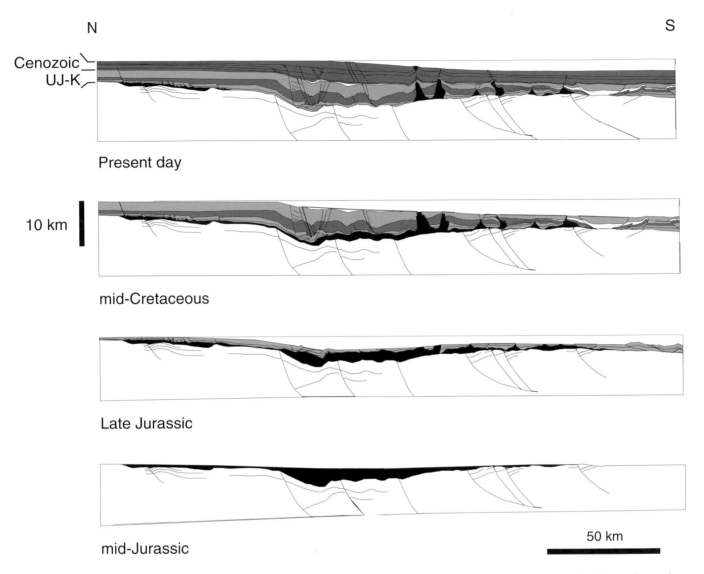

N   S

Cenozoic
UJ-K

Present day

10 km

mid-Cretaceous

Late Jurassic

50 km

mid-Jurassic

Figure 4—Restoration of regional line 1, displayed with 2:1 vertical exaggeration. Key: black = salt; UJ-K = Upper Jurassic (post-Louann Salt) and Cretaceous.

3. The restoration suggests that a considerable thickness of Louann Salt was originally present in the graben and thinned over the basement highs. Most of the postulated original salt has been lost from the section, probably by near-surface dissolution during the Mesozoic rather than out-of-plane movement because there are no salt bodies along strike large enough to have absorbed this volume of salt.
4. Salt-withdrawal basins developed above the autochthonous salt during the Late Jurassic–Cretaceous were the major control on depositional thickness variations on the slope during that period.
5. The original distribution and thickness of the Louann Salt were strongly controlled by basement topography produced during the earlier rifting episode.

Line 2: Present-Day Structure

Line 2 (Figure 3b) runs from onshore southeastern Louisiana to the upper slope in western Mississippi Canyon and into central Atwater Valley protraction areas (see Figure 2 for location). From north to south, the structural elements of the line are as follows:

1. In the updip part of the line, several thin-skinned extensional growth faults of Paleogene–early Miocene age sole out within the Paleogene.
2. The present-day shelf is dominated by a huge counter-regional salt-withdrawal basin filled mostly with middle Miocene–Pliocene strata. The basin subsided into an allochthonous salt canopy of Paleogene age (canopy VI, Figure 3b). The salt has been totally evacuated, and the base of the former canopy is now seen as an extensive salt weld. In

the line of the section, the basinward edge of the salt-withdrawal basin is an inclined salt weld. Along strike, this edge contains inclined salt bodies. Detailed descriptions of this area with seismic examples are given in this volume by Schuster (1995) and Diegel et al. (1995).

3. The Cenozoic of the upper slope is dominated by salt stocks and small isolated salt sheets. The structure of the Upper Jurassic–Cretaceous interval appears to be dominated by salt withdrawal, similar in scale to that in line 1 (Figure 3a). There does not appear to be any major extensional faulting of middle–late Miocene age here.

4. The middle slope contains a large salt canopy (canopy I). The intersection of horizons with the bottom surface of the salt indicates that this canopy formed in middle–late Miocene time (see later discussion).

5. Beneath and downdip from this canopy is a fold and thrust belt of middle–late Miocene age, the Mississippi Fan foldbelt (Weimer and Buffler, 1989, 1992), shortened by about 10 km. The foldbelt developed at the same time as the main sediment input to the shelf occurred; the coincidence in their ages suggests a causal link between the two processes.

Line 3: Present-Day Structure

Line 3 (Figure 3c) runs onshore through central Louisiana to the upper slope in Green Canyon and into Walker Ridge protraction areas (Figure 2). From north to south, the structural elements of the section are as follows:

1. Onshore there is a set of thin-skinned extensional growth faults of Paleogene age.

2. The present-day shelf is dominated by salt-withdrawal basins of various styles, with major listric south-dipping growth faults rooting into a salt detachment. Most of the basin fill is of middle–late Miocene age. The basin subsided into an allochthonous salt canopy of Paleogene age (canopy VI). The salt has been incompletely evacuated, and the base of the former canopy is now an extensive salt weld linking isolated salt bodies. We have drawn a number of domino faulting blocks beneath this weld. These blocks are speculative because only their tops can be seen on seismic data; other interpretations are possible.

3. The Cenozoic of the upper slope is dominated by salt stocks and small isolated salt sheets. The Mesozoic structure is uncertain because it is poorly imaged on seismic data.

4. The middle slope contains a large salt canopy (canopy II, Figure 3c). The intersection of horizons with the bottom surface of the salt indicates that this canopy expanded in middle–late Miocene time, but it may have initiated earlier than this.

5. Beneath and downdip from this canopy is a fold

and thrust belt of middle–late Miocene age, which is a continuation of the Mississippi Fan foldbelt. We interpret about 10 km of shortening in this section. Again, the foldbelt developed at a time of major sediment input to the shelf.

Although there are significant differences in appearance and in some elements of structural style between lines 2 and 3 (Figure 3b,c), most of the major structural elements occur in both of them (such as the major foldbelt in the lower slope and the extensive salt canopies on the shelf and upper slope). Far more similarity exists between these lines than between lines 1 and 2.

Line 4: Present-Day Structure

Line 4 (Figure 3d) runs onshore through western Louisiana to the continental slope in Garden Banks and Keathley Canyon protraction areas and then into the continental rise in Mexican waters (Figure 2). This section can be compared to a nearby section in Diegel et al. (1995). From north to south, the structural elements of the section are as follows:

1. Onshore, a system of thin-skinned extensional growth faults of Late Cretaceous–early Eocene age is interpreted to have detached on autochthonous salt.

2. Above and south of this deep extensional system, a shallower system of thin-skinned extensional growth faults, mainly of Oligocene age, detaches on a horizon near the top of the Eocene. The system terminates south of the present-day shoreline, where the basal detachment descends sharply beneath the base of the seismic data.

3. Between the present-day coast and shelf edge, the section is dominated by salt-withdrawal basins of various styles, with many down-to-the-south listric growth faults rooting into salt-based detachments to the north and a counter-regional withdrawal style to the south (Diegel et al., 1995; Schuster, 1995). The basins contain sediments of Oligocene–Pleistocene age. The basins subsided into an allochthonous salt canopy, or set of canopies (canopy complex VI, Figure 3d), which probably formed in the Paleogene. This line of section was selected for its relative structural simplicity; the salt geometries are more complex along strike. Horizons interpreted to be of early Miocene age subcrop onto base salt near the southern limit of the canopy, indicating that canopy spreading continued into the early Miocene. The salt has been incompletely evacuated, and the base of the former canopy is now an extensive salt weld linking isolated salt bodies.

4. The present-day continental slope is dominated by a single vast salt canopy (canopy II) more than 200 km long. On the upper slope, several counter-regional salt-withdrawal basins subside into this

canopy, including the Auger subbasin (McGee et al., 1994). Tentative seismic correlation from the shelf into these basins suggests that the oldest sediments within them are probably of middle Miocene age and that most of the sedimentary infill is of Pliocene–Pleistocene age. The base of the salt canopy can be identified in a few locations around this area on seismic data, but the subsalt interpretation here is not constrained by seismic data.

5. On the middle slope, the character of the salt-withdrawal basins changes; most have subsided vertically and symmetrically, unlike those landward. However, in relatively recent time (Pleistocene?), several withdrawal basins appear to have collided so that the margin of one basin is thrust over the adjacent basin. Again, the base of canopy II can be identified in a few locations around this area on seismic data, constraining the depth of the base of the canopy, but the subsalt interpretation here is not constrained by seismic data.

6. In the lower slope, the withdrawal basins are separated by salt walls and there is no evidence of withdrawal basin collision. The base of the salt is clearly imaged over the southern 70 km of the canopy, and correlated seismic horizons have been mapped up to their intersection with the base of salt surface. After depth conversion of the seismic data, these horizons are undeformed and continuous with the continental rise stratigraphy. The undeformed layer cake nature of the continental rise section, without diapirs or evidence of salt withdrawal, suggests that it is not underlain by autochthonous salt. There is no equivalent of the Mississippi Fan foldbelt here.

One of the critical elements in constructing this section was defining the geometry of the base of salt surface and the subsalt stratigraphy under the giant slope canopy (canopy II). As noted earlier, the base of this canopy is directly constrained by seismic data in the south and by isolated areas of seismic imaging in the north, defining a fairly smooth surface with steep sections (ramps) and subhorizontal sections (flats). The continental rise stratigraphy, defined by seismic data south of the canopy and under its southern 70 km, forms a simple, undeformed wedge, thickening northward. We extrapolated this wedge northward under the canopy up to its intersection with the base salt surface. This predicts that, under the deepest part of the canopy, the base of the salt sits on Upper Cretaceous strata. However, given the uncertainty involved and the probable thin lower Paleogene section in this area, the base of the canopy could be anywhere from middle Cretaceous to lower Oligocene (compare Diegel et al., 1995).

Line 4: Palinspastic Restorations

Line 4 is restored in Figure 5 to its paleostructure at intervals from Late Jurassic to present day. A difficulty faced when restoring cross sections that involve salt is the possibility that an unknown amount of salt may have been lost from the section through time due to out-of-plane movement or dissolution. The issue of out-of-plane movement can by addressed by determining whether adjacent areas show evidence of having absorbed excess salt. As discussed in the restoration of line 1, changing the shape of the paleoseafloor in the restoration adds more or less salt to the restored section. Also, on long regional lines, significant vertical movement and tilting of the basement through time are possible due to thermal subsidence and isostatic loading. This affects the volume of salt incorporated into the restored section. One approach to defining the basement movement used by Rowan (1994) is to estimate the isostatic and flexural effects mathematically. The approach used to restore the basement in this section was to model the basement as two cantilever blocks and use the stratigraphy above them to constrain the amount of rotation (Peel, 1993). The dip of the northern basement cantilever is constrained by rotating the paleotop surface of the carbonate platform or clastic shelf back to horizontal. The dip of the southern cantilever is constrained by rotating the paleocontinental rise sediments back to the dip of the modern continental rise. If these rotations are correct, the two basement blocks should match at their junction.

The restoration to Late Jurassic (Figure 5e) was obtained by stripping off all post-Louann overburden. The restoration suggests that the salt was originally concentrated within a large basin several hundred kilometers north of its present concentration.

The restoration for the end of Cretaceous time (Figure 5d) shows the Mesozoic carbonate platform building out over the salt basin and displacing it southward. Our estimate of the base of the canopy level (described earlier) indicates that the main salt canopy could have begun forming as early as this time (as shown here).

The restoration for the end of Oligocene time (Figure 5c) shows the clastic shelf margin building southward. We are confident that the main slope canopy was in place by then, in addition to one or more salt canopies in the upper slope. In this restoration, a large amount of Oligocene extension on the shelf was transferred into *en bloc* southward translation of the upper slope; this movement was absorbed by shortening and thickening of the main slope canopy (canopy II).

Salt-withdrawal basins above canopy II had probably begun forming by middle Miocene time. As the regional focus of deposition moved into this area in the Pliocene–Pleistocene (Galloway et al., 1991), the growth of the basins accelerated. The restoration for the end of the Pliocene (Figure 5b) shows that the clastic shelf margin had by that time advanced across the more northern complex of canopies and that rapid deposition had begun in salt-withdrawal basins on top of the main slope canopy. At this time, the salt-withdrawal basins were still separated from one another by salt. The canopy had effectively reached its present extent.

Comparison with the present-day section (Figure 5a)

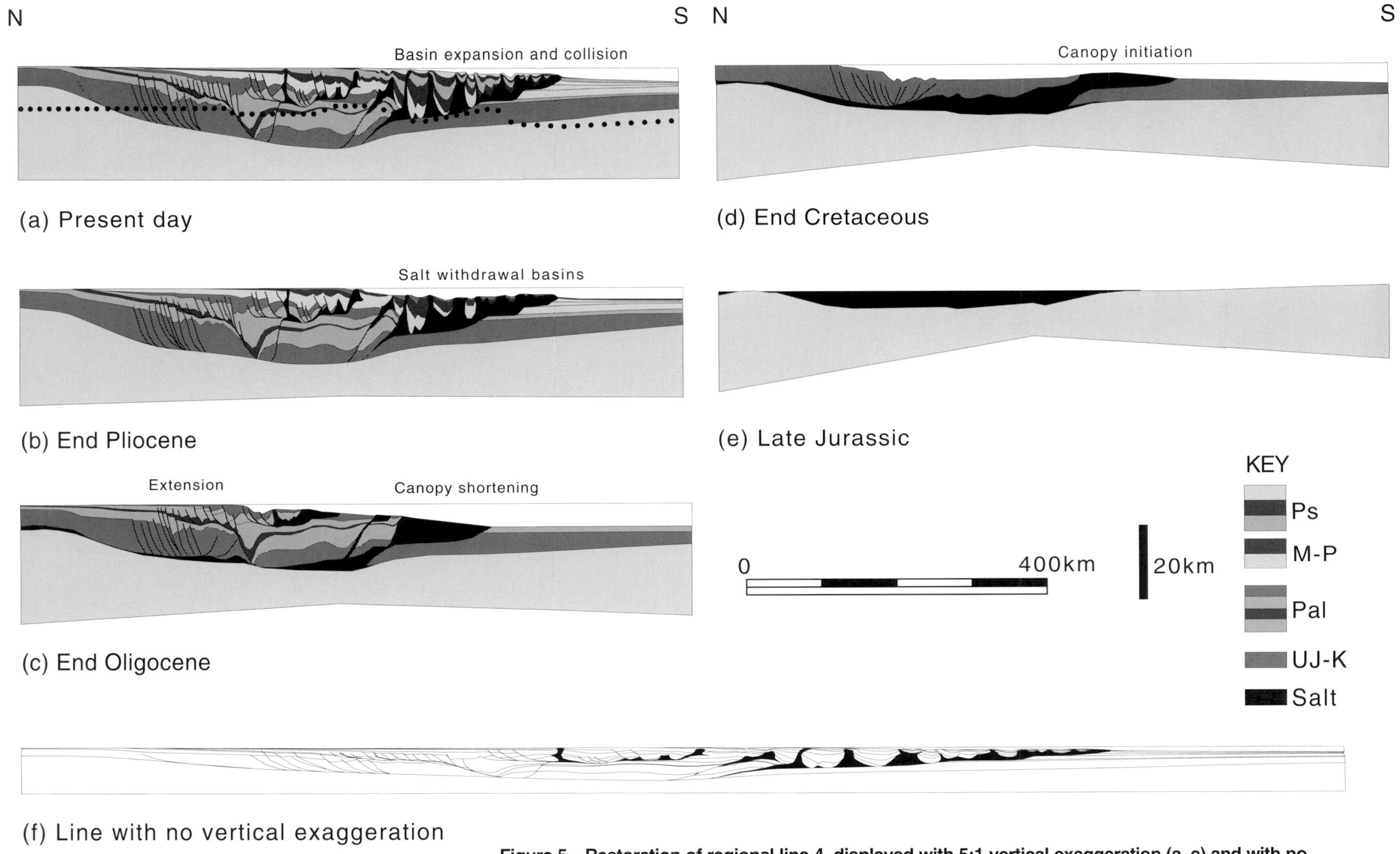

S

N

S N

Canopy initiation

(d) End Cretaceous

Basin expansion and collision

(a) Present day

Salt withdrawal basins

(b) End Pliocene

Canopy shortening

Extension

(c) End Oligocene

(e) Late Jurassic

KEY

Ps

M-P

Pal

UJ-K

Salt

20km

0   400km

(f) Line with no vertical exaggeration

Figure 5—Restoration of regional line 4, displayed with 5:1 vertical exaggeration (a–e) and with no vertical exaggeration (f). Key: black = salt; UJ-K = Upper Jurassic (post-Louann Salt) and Cretaceous; Pal = Paleogene; M-P = Miocene–Pliocene; Ps = Pleistocene. Dotted line in part (a) shows lower limit of seismic data.

shows that widening of the withdrawal basins, particularly in the upper slope, caused a progressively intensifying space problem. The margins of some basins have been thrust over the edge of their neighboring withdrawal basins.

Line 5: Present-Day Structure

Line 5 (Figure 3e), originally interpreted and constructed by R. G. Matson, runs from the Llano uplift in onshore Texas to the slope in East Breaks and Alaminos Canyon protraction areas, then into the continental rise in Mexican waters (Figure 2). From northwest to southeast, the structural elements of this section are as follows:

1. In the northwest, a system of thin-skinned extensional growth faults of Late Cretaceous–early Eocene age detaches on top of the basement (within autochthonous salt?).

2. Above and south of this system, a shallower system of thin-skinned extensional growth faults detaches on a horizon within the lower Eocene. The total amount of extension of middle Eocene–Oligocene age is at least 50 km.

3. The present-day shelf and upper slope contain several other elements. In addition to extensional faults detaching at the lower Eocene level, large Miocene faults are interpreted as cutting down to the autochthonous salt level. On the present-day slope, a discordance in the lower Eocene appears to be a salt weld, with salt bodies rising from it. We do not have the data to determine how far updip this possible early Eocene canopy originally extended. The amount of extension of early–middle Miocene age is about 10 km. Within this area, part of the section is seismically irresolvable and may correspond to highly deformed shale (mottled green on Figure 3e).

4. On the middle slope, three different levels of strata have very different structural styles. The Upper Jurassic–Cretaceous and possibly lower Eocene interval appears relatively simple, containing broad, low folds. Above a lower Eocene(?) décollement, the upper Eocene–Oligocene section is intensely deformed and has been interpreted here as a strongly shortened fold and thrust belt detaching on a welded former canopy (canopy IV). Several salt bodies rising from the décollement level identify it as a salt weld. The age of this deformation is not precisely constrained; there appears to be a strong late Oligocene component, but the shortening may have lasted longer. The folded section is onlapped by an essentially undeformed Pliocene–Pleistocene section.

5. On the lower slope is a laterally extensive, but in this section, narrow salt canopy (canopy III).

6. Southwest of the salt canopy is the Perdido foldbelt (Martin, 1980), which is a zone of large northeast-trending anticlines. We interpret the age of the

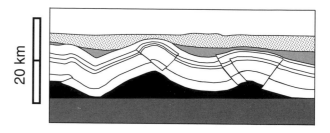

(a) Salt-cored buckle fold model

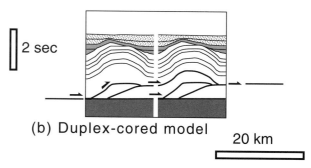

(b) Duplex-cored model

20 km

Figure 6—Alternative models for the structural style of the Perdido foldbelt. (a) Part of regional line 6 displayed without vertical exaggeration showing salt-cored buckle folds. (b) A section (in seismic time) adapted from the model of Mount et al. (1990). The folds are interpreted here to be cored by a thrust duplex. We have added a second such fold to show that this model requires a cumulative transfer of thrust displacement from a lower to a higher décollement surface.

Perdido folding as dominantly latest Oligocene–middle Miocene, based on the estimated age of the oldest horizons seen to onlap the folds.

Published models for the development of the Perdido folds include (1) salt-cored buckle folds with a relatively minor amount of thrusting of the limbs (e.g., Worrall and Snelson, 1989), as shown in Figure 6a; and (2) fault bend folds cored by a duplex involving a larger amount of thrusting (Mount et al., 1990), as shown in Figure 6b. The duplex model requires that, under each anticline, ~7 km of contractional displacement is transferred from the basal décollement to a conjectured higher décollement. At the downdip limit of the foldbelt, the total displacement on this higher level would be 20–30 km, passing southeastward into the abyssal plain. This contractional displacement would then terminate in a separate belt of folds or imbricate thrusts; no such contraction is seen in the abyssal plain. Therefore, we favor the interpretation of these structures as salt-cored buckle folds, involving about 5 km of horizontal contraction.

Line 6: Present-Day Structure

Line 6 (Figure 3f) begins at the southern Texas coastline, crosses the present shelf in North Padre Island protraction area, continues across the slope in Port Isabel and

Alaminos Canyon protraction areas, and into the continental rise in Mexican waters (Figure 2). Unlike line 5, this line does not extend onshore to the updip limit of the basin. The structure of the onshore continuation, as described by Diegel et al. (1995), is dominated by thin-skinned extension and raft tectonics similar to those described in the Kwanza Basin of Angola (Lundin, 1992), detaching on autochthonous salt and a higher detachment system within the Paleogene. A seismic line (Figure 7) located near line 6 illustrates many of the features characteristic of this line. From west to east, the structural elements of the section are as follows:

1. A set of down-to-the-basin growth faults of Oligocene age represents only the eastern end of a much larger system (Diegel et al., 1995).
2. Large growth faults of late Oligocene–middle Miocene age, probably detaching on autochthonous salt, involve about 20 km of stratal extension.
3. The upper slope contains a previously undocumented thin-skinned fold and thrust belt of late Oligocene–middle Miocene age, which we call the Port Isabel foldbelt after the protraction area in which it is best developed.
4. A complex of salt canopies and salt sheets (canopy III) occurs, of which a younger, higher level set was emplaced between late Eocene and middle Miocene time. These appear to be derived from an older allochthonous canopy (possibly Eocene age).
5. Below and downdip from the salt canopy is the Perdido foldbelt, which we interpret to be mainly of latest Oligocene–middle Miocene age, similar to the Port Isabel foldbelt.

Line 6: Palinspastic Restorations

Figure 8 displays a set of sequential palinspastic restorations of line 6 without vertical exaggeration. Figures 8c, d, e show three reconstructions of the section during the late Oligocene. Figure 8e is a reconstruction for a middle Frio Formation surface (upper Oligocene). The exact age is uncertain because there is no well tie to the horizon palinspastically restored in the central part of the section. The geometry around the large salt body to the west at this time suggests that its development up to this stage had occurred by downbuilding (Barton, 1933), by which the top of the salt had remained near the seabed while sediments subsided dominantly by vertical salt withdrawal around it. Over time, the amount of salt remaining to withdraw from the source layer level progressively decreased, while the copious influx of clastic sediments continued. As a result, the accommodation space for these sediments was increasingly generated in Figures 8d, c, b by down-to-the-basin growth faulting rather than by vertical salt withdrawal. The large salt body collapsed by lateral extension (sections d, e). This evolution of this salt body is a good example of sequential rise and fall of diapirs during thin-skinned extension described by Vendeville and Jackson (1992a,b).

The growth faults sole out on décollements at several stratigraphic levels, including autochthonous salt, intra-Eocene, and intra-Oligocene, as seen in Figure 8b. Extension in the growth faults is linked to downdip contraction at a range of scales, from the 10-km scale of system x–x' to the 100-km scale of system y–y' in Figure 8b. At least 10 km of displacement was transferred along the autochthonous salt level between the shelf growth faults and the Perdido foldbelt. This important observation indicates that very large scale, linked slip systems exist in the northern Gulf of Mexico margin.

In the final stage of the evolution of this section, from late Miocene to Pleistocene, the principal locus of sediment input had switched away from this area toward the east, and the area became structurally less active. The compressional structures in the slope were inactive, and they were passively buried during this stage, as shown in Figure 8a.

SALT CANOPIES

It has long been recognized that the northern Gulf of Mexico margin contains a vast volume of allochthonous salt (e.g., Martin, 1980). This mainly takes the form of complexes of diapirs and small salt sheets (typically less than 20 km in downdip extent) as well as several vast (much greater than 20 km extent), laterally extensive, sub-horizontal allochthonous salt bodies (Simmons, 1992). In general, this chapter follows the terminology of Jackson and Talbot (1991), who suggested that allochthonous sub-horizontal salt bodies should be called *sheets* if they have a single stem and *canopies* if they have multiple stems. In some of the giant salt bodies, multiple stems can be identified and the bodies are therefore true canopies. However, in most cases, the stems are poorly imaged or invisible on available seismic data, often making it difficult to distinguish between sheets and canopies. However, because several stems have been identified in far smaller canopies (e.g., Schuster, 1995), we consider it highly improbable that the great salt structures could be fed by a single feeder, so they are most likely to be true canopies in the sense of Jackson and Talbot (1991). We have chosen here to refer to all large, laterally extensive allochthonous salt bodies as canopies, even if multiple stems have not been documented.

Recent evidence of seismic and well data (Nelson, 1991; McGuinness and Hossack, 1993; Fletcher et al., 1995), physical modeling (Jackson et al., 1994), and onshore salt glaciers (Talbot and Jarvis, 1984) indicates that canopies most commonly form by glacier-like extrusion along or close to the paleoseabed rather than by sill-like intrusion at depth (as proposed by O'Brien and Lerche, 1988; Nelson and Fairchild, 1989). An important implication of the glacier model is that the past extent of salt sheets may be defined by the termination of dated horizons against the lower surface of the salt body. Hence, by mapping such cutoffs, the history of initiation and spreading of the major salt canopies can be reconstructed.

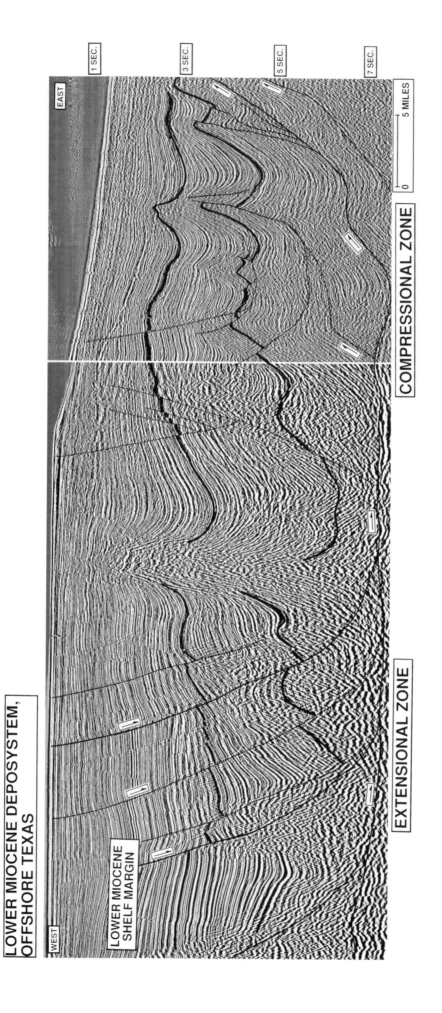

LOWER MIOCENE DEPOSYSTEM, OFFSHORE TEXAS

WEST

LOWER MIOCENE SHELF MARGIN

EXTENSIONAL ZONE

COMPRESSIONAL ZONE

EAST

1 SEC.

3 SEC.

5 SEC.

7 SEC.

0   5 MILES

Figure 7—Seismic section through the northwestern U.S. Gulf of Mexico shelf and slope. This line shows an intermediate-scale linked system of extension and compression on the order of 100 km wide. (Data courtesy of TGS-Calibre Geophysical Company and GECO-PRAKLA.)

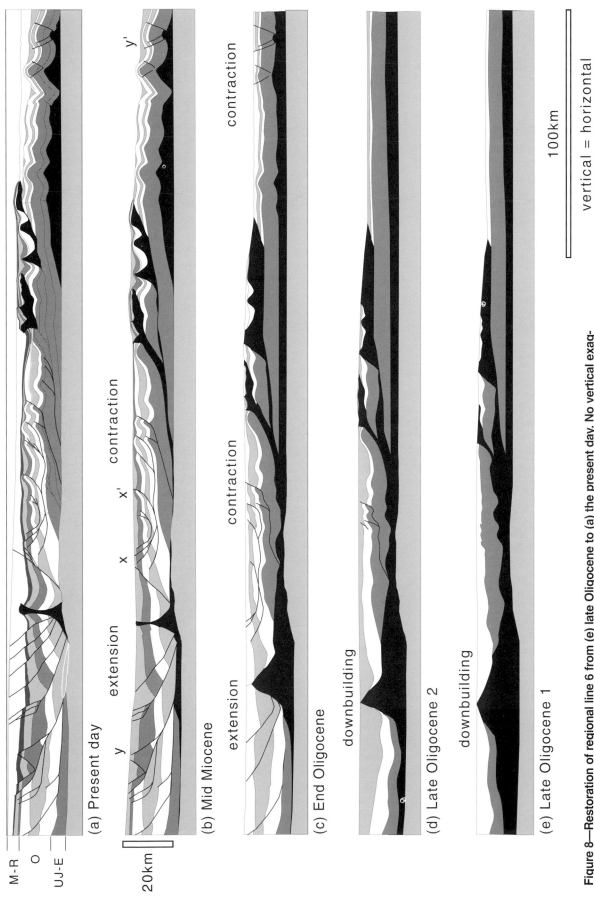

Figure 8—Restoration of regional line 6 from (e) late Oligocene to (a) the present day. No vertical exaggeration. Key: black = salt; UJ-E = post–Louann Upper Jurassic, Cretaceous, and Eocene; O = Oligocene; M-R = Miocene–Recent. The segments x–x' and y–y' (in part b) are referred to in the text.

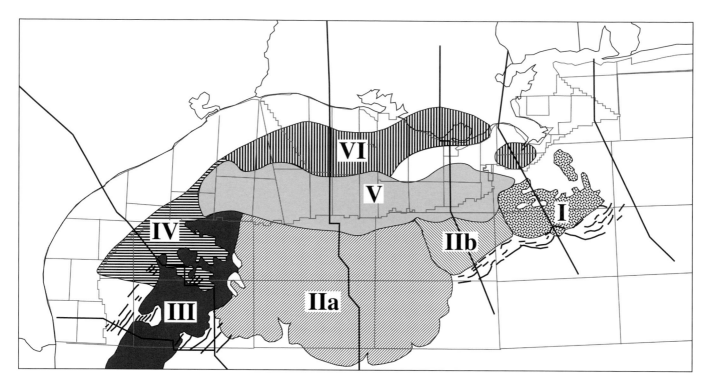

Figure 9—Major allochthonous salt canopies and genetically related canopy complexes. Numbers I to VI correspond to canopies I to VI defined in the text. The areas outlined are large single canopies and sets of genetically related sheets; not shown here are many smaller individual diapirs and small allochthonous salt bodies. Additional large canopies and canopy complexes may exist farther north. The age of formation and the geographic limits of the canopies are relatively well defined for I and II but more poorly constrained farther west and north. Heavy lines south of canopies I and IIb and both east and west of canopy III indicate fold axial traces in Mississippi Fan, Port Isabel, and Perdido foldbelts, respectively.

This approach has been used to restore salt sheets to past geometries in two-dimensional sections (e.g., figure 21 of Wu et al., 1990b; figure 3 of Moretti et al., 1990; figure 14 of Seni, 1992). McGuinness and Hossack (1993) used the approach to restore salt body geometries in map form. Ramps and flats on the base salt surface indicate relatively slow and relatively rapid periods of canopy spreading, respectively (Jackson and Talbot, 1991).

Interpreters at BP have set up a regional correlation of seismic horizons across the shelf and slope of the northern Gulf of Mexico (e.g., Pulham et al., 1994), based on seismic and well data and also on published data (including Shaub et al., 1984; Weimer, 1990; Wu et al., 1990a; Weimer and Buffler, 1992). Where possible, these seismic horizons have been interpreted or extrapolated beneath the salt up to their cutoff against the basal salt surface. The salt distorts the seismic image of the underlying strata because of velocity contrasts and raypath bending. Thus, to confirm the geometric viability of the subsalt interpretation, a number of lines were depth converted. Using these data, key horizon cutoffs on the base salt were mapped, and from this, the emplacement and spreading histories of major salt bodies were inferred. Because the quality of subsalt imaging and the precision of dating of the seismic horizons vary considerably, some salt bodies were reconstructed more confidently than others.

Despite the intrinsic uncertainties, we were able to identify that the northern Gulf margin contains distinct giant allochthonous salt canopies and canopy complexes with different histories of initiation, spreading, and consumption by salt-withdrawal basins. The existence of separate canopies has previously been recognized; here we have tried to define their lateral extent and age of formation.

Large salt canopies and canopy complexes with a common emplacement history are shown in Figure 9. Some of these areas contain a single connected body of salt, in other words, a single canopy (e.g., canopy II). Other areas contain complexes of separate canopies that have not coalesced (e.g., canopy complex V). In other areas, we have not been able to determine whether the salt is a single canopy or a complex of separate canopies because of their depth, structural complexity, or poor seismic imaging.

Because of their lack of regional extent, minor salt masses in the Mississippi Canyon protraction area in the northeastern Gulf of Mexico have not been shown in Figure 9. For the location of these smaller masses, refer to Martin (1980), Wu et al. (1990a,b), Diegel et al. (1995), or Schuster (1995).

The canopies and canopy complexes shown in Figure 9 represent the original maximum extent of the salt. In some cases, the salt is largely intact, such as in canopy I and the downdip part of canopies IIa and IIb. In other areas, the salt in the original canopy has been largely displaced by salt-withdrawal basins and today is mainly represented as a residual salt weld (canopies IV and V).

Characteristics of Canopies and Canopy Complexes

Canopy I

Canopy I (Figure 9) lies on the borders of the Mississippi Canyon and Atwater protraction areas. On line 2 (Figure 3), it is the large salt body that partly overlies the Mississippi Fan foldbelt. The canopy is largely intact, having relatively few withdrawal basins above it. As illustrated by Wu et al. (1990b, his foldout 5a), the base of the salt is commonly well imaged on seismic data in this area, and some seismic horizons can be interpreted up to their termination on the base of the salt. The canopy initiated in the middle Miocene and continued spreading into the late Miocene.

Canopy II

Canopy II is a vast contiguous volume of salt lying west of canopy I that occupies the upper and middle slope in the central part of the margin. (We are uncertain whether the area labeled IIb in Figure 9 is genetically part of canopy IIa or is a separate unit.) On the upper slope, the salt of canopy II is largely displaced by salt-withdrawal basins that are now in contact with the base of the canopy. On the lower slope, the salt is more intact and salt-withdrawal basins are smaller and still floored by salt.

In contrast to the middle–late Miocene spreading of canopy I, canopy II initiated originally in the present-day upper slope in the early Paleogene or possibly latest Cretaceous. Our correlation of the ages of the subsalt horizons indicates that spreading occurred during two episodes, one in the late Oligocene–early Miocene and another in the middle–late Miocene. A final minor phase of spreading occurred in the late Pliocene.

Canopy III

The age of formation and spreading of canopy III is more poorly constrained, but we suggest that it formed in the Paleogene and continued to spread in the early Miocene.

Canopy IV

Canopy IV is interpreted to be a paleo-canopy, from which the majority of the salt has been removed by salt withdrawal. Today it is an extensive salt weld from which diapirs rise (regional line 5, Figure 3a). Because there are no well ties to the subweld stratigraphy, the age of emplacement of canopy IV is not constrained. We conjecture that it formed in the early Eocene.

Canopies V and VI

Northwest of canopy VI, there is evidence of abundant allochthonous salt. This study did not include regional mapping of this vast area, so we have not been able to determine how many distinct canopies actually exist in this area. In our reconnaissance, we have tentatively identified canopy complex V, which may have formed between the early–middle Miocene and the early Pliocene, and updip, canopy complex VI, which possibly formed in the Eocene–Oligocene.

Driving Mechanisms of Salt Canopy Spreading

The driving mechanisms of canopy spreading are not fully understood. There is evidence that in small allochthonous bodies, the spreading rate is closely related to the rate of salt supply up a stem into the salt body, modulated by the rate of sedimentation (McGuinness and Hossack, 1993; Fletcher et al., 1995). These may also be the controlling factors for the initial spreading of larger bodies. However, large, older canopies (e.g., canopy II) show a multiphase history of spreading, in which spreading episodes occurred long after the initial salt emplacement. Here, the spreading is probably driven by redistribution of salt within the canopy rather than by injection of additional salt into it. As previously mentioned, the spreading history of canopy II indicates that a major influx of sediment on top of the canopy during the Pliocene–Pleistocene rapidly increased the effective volume of the canopy but did not result in any significant spreading.

LINKED SYSTEMS

There appears to have been no extension or contraction of the subsalt basement of the northern Gulf of Mexico Basin during the Cenozoic. Because the margin is bounded both updip and downdip by regions without Cenozoic deformation, pin lines for regional section restoration can be placed through both ends of the section. Therefore, every increment of Cenozoic extension seen in one part of the section must be linked to an equal amount of contraction elsewhere in the section. Deformation may be balanced locally or transmitted far downslope and is accomplished by a variety of mechanisms (Hossack, 1995). For example, a gravity slide block is usually bounded by updip extensional faulting, downdip folding and thrusting, and lateral edges consisting of strike-slip fault zones (e.g., Figure 10). Wu et al. (1990b) showed an interpretation of extensional faulting balanced by downdip contraction of a salt body and emplacement of the excess salt as an allochthonous sheet.

A discrete linked system of structural elements that form a closed loop in map pattern and all move at the same time is referred to here as a *cell*. Such a cell can be very small (e.g., an individual shelf margin failure detaching at a structurally shallow level) or very large (e.g., a major shelf growth fault linked to lower slope thrusting by a detachment along autochthonous salt). The linked system developed on top of a small allochthonous salt sheet shown in Figure 10 is a good example of a small-

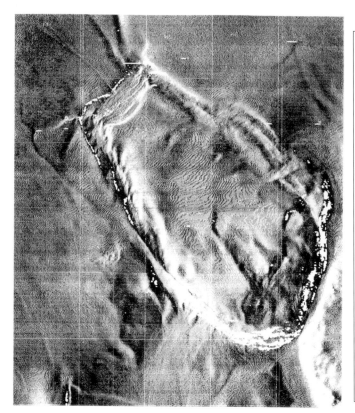

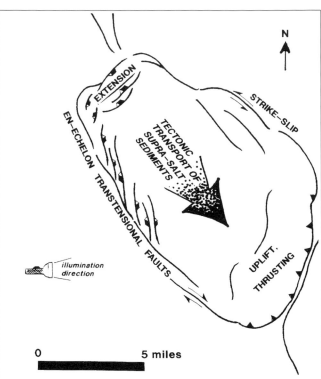

Figure 10—Three-dimensional seismic image (left) and an interpretative drawing (right) of a near-seabed surface. The seismic image is shaded according to dip azimuth to produce an artificial illumination effect, showing a small-scale linked system overlying an allochthonous salt sheet. In cross section, the system is an evolving salt-withdrawal basin of the salt-based detachment style of Diegel (1995).

scale structural cell. Large cells may contain smaller cells within them, such as the small slide carried on the back of a larger slide shown in Figure 11. This example is itself entirely contained within a large 60-mile-long cell (Figures 7, 8), contained within a still larger cell (several hundred miles in dimension) linking the Perdido foldbelt to growth faults on the shelf (Figures 8, 12).

Based on our analysis of regional cross sections, the northern Gulf of Mexico margin appears to contain a large number of structural cells, large and small, of varying ages. Our challenge is to group these cells in a geologically meaningful way.

CENOZOIC STRUCTURAL PROVINCES

Interpretation of a closely spaced grid of regional seismic lines across the continental slope shows that large areas share a set of common major structural elements and cells. For example, the Mesozoic growth faulting and downdip contraction on regional line 1 (Figure 3a) are similar to features on dip lines in the Florida escarpment to the east in the Desoto Canyon area (MacRae and Watkins, 1993). Regional line 2 (Figure 3b) contains essentially the same type of structures as regional line 3 (Figure 3c) (summarized in the section entitled Eastern Province). Between these areas of similarity, the structural style changes across a narrow zone.

Areas of similar structural character also show a common tectonostratigraphic history and in general have a similar history of canopy emplacement. For example, in the westernmost Gulf (characterized by lines 5 and 6 in Figures 3e, f), major deformation of Eocene–early Miocene age correlates with the time of major sediment input to the area. By late Miocene time, the locus of major sediment input to the Gulf had shifted far to the east, to the area characterized by lines 2 and 3, along with an eastward shift of the locus of major structural activity in the form of extensional faulting, salt sheet emplacement, folding, and thrusting (Figure 3).

This coincidence of major sediment input with major structuring, along with an apparent lack of basement involved deformation, leads us to conclude that Cenozoic structural activity in the northern Gulf of Mexico is entirely gravity driven, induced by sediment loading and topographic gradients, and that it is thin-skinned, detaching on or above the Jurassic Louann Salt. During the Cenozoic, clastic sediment deposition renewed the shelf–slope topographic gradients and supplied sediment on top of the salt; thus, the location of sediment depocenters was the dominant control on the location of structural activity.

Based on this conclusion, it would seem appropriate to group all contemporaneous structural cells by region and by association with major sediment input sites to the basin. This goal is complicated, however, because struc-

DIP SECTION

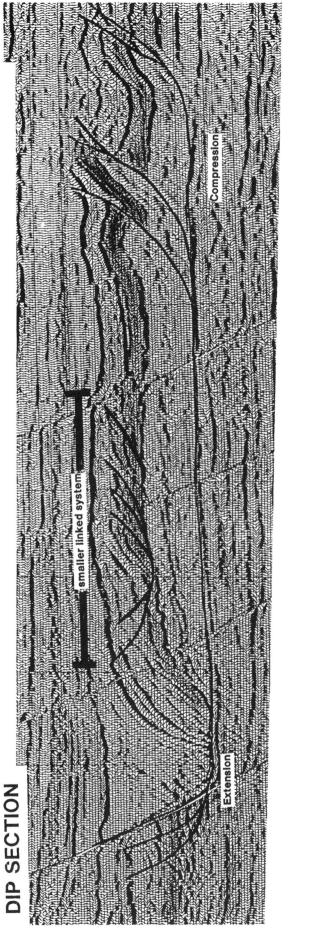

Compression

smaller linked system

Extension

STRIKE SECTION

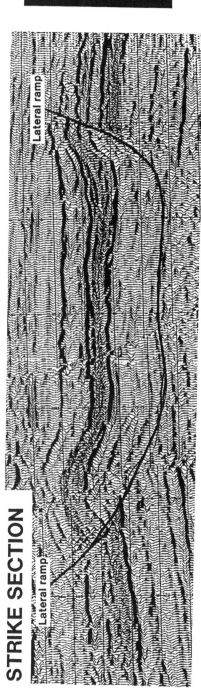

Lateral ramp

Lateral ramp

0.5 s

Figure 11—Dip and strike seismic sections showing a small-scale example of a linked system of extension and compression on the order of 10 km long. A smaller linked system is carried on the back of the larger system. This example as a whole is carried within the larger system shown in Figure 7. (Data courtesy of TGS-Calibre Geophysical Company and GECO-PRAKLA.)

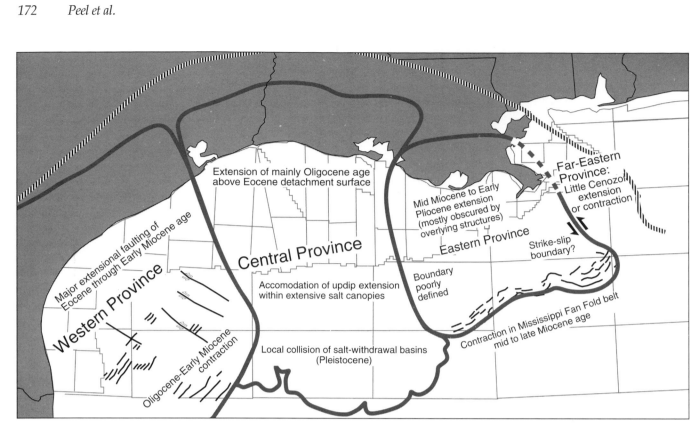

Figure 12—Division of the northern Gulf of Mexico margin into Cenozoic structural provinces. Some of the major characteristics of each province are labeled. Hatched line denotes Lower Cretaceous shelf edge and Florida Escarpment offshore.

tural cells of different age and genesis may overlap in space, such that boundaries between groups of related cells are diffuse or obscured. In addition, seismic data quality may be insufficient to clearly image major lateral boundaries. Although we are unable at this time to precisely define and map all groups of genetically and temporally related structural cells, we can clearly recognize large regions of similar structural character that also share a common tectonostratigraphic history and general history of canopy emplacement. These areas are referred to here as *structural provinces*.

We currently recognize four major structural provinces in the Cenozoic of the northern Gulf of Mexico Basin (Figure 12). The provinces are the same as those of Apps et al. (1994). The locations of the province boundaries are tentative; they are commonly masked by shallow salt or overprinted by local structures such as salt-withdrawal basins. Some boundaries may be diffuse zones rather than discrete edges. The structural provinces may also incorporate groups of cells of different ages and origin (e.g., the central province includes both Oligocene and Pliocene–Pleistocene linked systems that are not directly related), but the overall structure and tectono-stratigraphic history of each province are internally consistent and distinct from adjacent provinces. We present these provinces as a starting point for further discussion and investigation. We hope that in the future, after further work and due to the ever-increasing quality and availability of seismic data, the location and the nature of the provinces and their boundaries may become better defined.

Far-Eastern Province

The far-eastern province is characterized by the absence of large-scale Cenozoic horizontal movement and salt canopies (see regional line 1 in Figure 3a). The lateral boundaries of this province are defined by major changes in structural style and stratigraphy. The eastern boundary is the Florida escarpment. We speculate that the western boundary may be a thin-skinned strike-slip transfer zone, defining the eastern limit of the Mississippi Fan foldbelt (Weimer and Buffler, 1989). At the western boundary of the province is a continuous dip-elongate trend of shallow salt bodies, which may define the transfer zone.

Eastern Province

The structure of the eastern province is illustrated by regional lines 2 and 3 (Figure 3b,c). This province is characterized by a combination of the following features: (1) a major linked system of extension and contraction, principally of middle–late Miocene age, connecting extension probably located under the present-day shelf to contraction in the Mississippi Fan foldbelt; (2) a large salt canopy (canopy I) on the present-day middle slope, formed in middle–late Miocene; and (3) a largely evacuated salt canopy (canopy VI) located under the present-day shelf, emplaced in the Paleogene (see Schuster, 1995).

The western edge of the eastern province is poorly defined and is tentatively projected northward from the western visible limit of the Mississippi Fan foldbelt. Future work may help to define the location and nature of the western boundary.

The Cenozoic section here is thicker than in the far-eastern province and is especially thick in the salt-with-drawal basins under the present-day shelf. The timing of the major sediment supply to the shelf (middle Miocene–Pliocene) corresponded to the timing of movement of the large-scale linked system in this province, indicating that lateral movement was gravity-driven and powered by sediment supply to the shelf and upper slope.

Central Province

The structure of the central province is illustrated by regional line 4 (Figure 3d). This province is characterized by the following features: (1) a vast salt sheet (canopy II) dominating the whole slope; (2) canopies V and VI under the present-day shelf, which are now extensively consumed by salt-withdrawal basins; (3) thin-skinned extension, mainly of Oligocene age, under the shelf and onshore area detaching near the top of the Eocene; (4) a separate thin-skinned extension system in the onshore area of Late Cretaceous–early Eocene age, detaching on a lower level, probably autochthonous salt; and (5) absence of a lower slope contractional belt.

A palinspastic restoration of regional line 4 suggests an explanation for the absence of a lower slope contractional belt in the central province to match the observed Oligocene updip extension (Figure 5). When extension was occurring, the southern part of the section consisted of a giant unbroken salt layer having little or no sediment cover. The extending sediments were able to move without significant resistance into the back of this salt body, accommodated by shortening and thickening of the salt. This is a good example of Hossack's (1995) contention that extension in salt terranes must be balanced by equivalent downdip contraction, but that this contraction may take a number of forms that do not necessarily involve shortening of cover sediments.

Western Province

The structure of the western province is illustrated by regional lines 5 and 6 (Figures 3e, f). This province is characterized by the following features: (1) a major linked system of Eocene–middle Miocene extension and contraction, connecting updip and onshore extension with contraction in foldbelts on the middle slope (e.g., Port Isabel foldbelt) and on the lower slope (Perdido foldbelt); (2) a large salt canopy (canopy III) on the middle to lower slope, emplaced in the late Eocene–middle Miocene; and (3) an older canopy (canopy IV), probably formed in the Eocene, that originally covered much of Corpus Christi, western East Breaks, and western Alaminos Canyon protraction areas. This canopy is now almost entirely displaced by sediment and is expressed as a regional salt weld.

CONCLUSIONS

1. Structures within the Cenozoic of the northern Gulf of Mexico margin are thin-skinned, gravity-driven, and powered by the deposition of sediment on the shelf and upper slope.
2. Major regional-scale structural processes on the basin margin include the formation of giant allochthonous salt bodies, updip extension by growth faulting, and downdip contraction by folding and thrusting or canopy shortening.
3. By assuming that salt sheets are emplaced and spread near-surface like glaciers, salt sheet growth can be mapped through time. The results of this analysis indicate that major salt canopies and canopy complexes developed and spread at different times between Late Cretaceous–early Paleogene and late Miocene in different areas of the Gulf.
4. Structurally linked systems on a wide range of scales were formed by lateral tectonic movement of sediment and salt. Major structural activity at any one time was controlled by the locus of major sediment input to the basin at that time.
5. We have tentatively recognized four major genetic structural provinces having speculative and partly diffuse lateral boundaries, each of which consists of complexes of structural elements of similar age arranged in a systematic fashion.

Acknowledgments   We are grateful to BP Exploration Inc. for permission to publish this paper and to TGS-Calibre for permission to include the seismic examples. Numerous colleagues and ex-colleagues have contributed information, interpretation, and ideas toward this study. In particular, we thank D. B. McGuinness for the original version of regional line 6, R. G. Matson for constructing line 5, and other members of the BP Houston Regional Group (G. Apps, C. Blankenship, K. Boyd, D. Brumfield, N. Piggot, and B. Yilmaz) for ideas and constructive discussion. We are also grateful to R. Fitzpatrick and D. Rule for their help in preparing Figures 1 and 3. Finally, we were considerably aided by the incisive and helpful comments of our reviewers (R. Buffler, F. Diegel, P. Tauvers, M. P. A. Jackson, and S. Snelson).

REFERENCES CITED

Apps, G. M., F. J. Peel, C. J. Travis, and C. A. Yeilding, 1994, Structural controls on Tertiary deep water deposition in the northern Gulf of Mexico, *in* P. Weimer, A. H. Bouma, and B. F. Perkins, eds., Submarine fans and turbidite systems: sequence stratigraphy, reservoir architecture, and production characteristics: SEPM Gulf Coast Section, 15th Annual Research Conference, Proceedings, Houston, Texas, p. 1–7.

Barton, D. C., 1933, Mechanics of formation of salt domes with special reference to Gulf Coast salt domes of Texas and Louisiana: AAPG Bulletin, v. 17, p. 1025–1083.

Buffler, R. T., 1989, Distribution of crust, distribution of salt, and the early evolution of the Gulf of Mexico basin: SEPM Gulf Coast Section, 10th Annual Research Conference, Program and Extended Abstracts, Houston, Texas, p. 25–27.

Diegel, F. A., J. F. Karlo, D. C. Schuster, R. C. Shoup, and P. R. Tauvers, 1995, Cenozoic structural evolution and tectonostratigraphic framework of the Northern Gulf Coast continental margin, *in* M. P. A. Jackson, D. G. Roberts, and S. Snelson, eds., Salt tectonics: a global perspective: AAPG Memoir 65, this volume.

Fletcher, R. C., 1995, Salt glacier and composite sediment–salt glacier models for the emplacement and early burial of allochthonous salt sheets, *in* M. P. A. Jackson, D. G. Roberts, and S. Snelson, eds., Salt tectonics: a global perspective: AAPG Memoir 65, this volume.

Galloway, W. E., D. G. Bebout, W. L. Fisher, J. B. Dunlap, Jr., R. Cabrera-Castro, J. E. Lugo-Rivera, and T. M. Scott, 1991, Cenozoic, *in* A. Salvador, ed., The geology of North America; the Gulf of Mexico basin: GSA Decade of North American Geology, v. J, p. 245–324.

Hossack, J. R., 1995, Geometric rules of section balancing for salt structures, *in* M. P. A. Jackson, D. G. Roberts, and S. Snelson, eds., Salt tectonics: a global perspective: AAPG Memoir 65, this volume.

Hossack, J. R., and R. Matson, 1993, Basement geometry of the Gulf of Mexico as deduced from gravity and magnetic anomalies and its influence on the deposition and creation of structures in the Louann Salt (abs.): AAPG, International Hedberg Research Conference Abstracts, Sept. 13–17, Bath, U.K., p. 88.

Jackson, M. P. A., and C. J. Talbot, 1991, A glossary of salt tectonics: : The University of Texas at Austin, Bureau of Economic Geology Geological Circular 91-4, 44 p.

Jackson, M. P. A., B. C. Vendeville, and D. D. Schultz-Ela, 1994, Structural dynamics of salt systems: Annual Review Earth Planetary Science, v. 22, p. 93–117.

Liro, L. M., 1992, Distribution of shallow salt structures, lower slope of the northern Gulf of Mexico, USA: Marine and Petroleum Geology, v. 9, p. 433–451.

Lundin, E. R., 1992, Thin-skinned extensional tectonics on a salt detachment, northern Kwanza basin, Angola: Marine and Petroleum Geology, v. 9, p. 405–411.

MacRae, G., and J. S. Watkins, 1993, Basin architecture, salt tectonics, and Upper Jurassic structural styles, Desoto Canyon salt basin, northeastern Gulf of Mexico: AAPG Bulletin, v. 77, p. 1809–1824.

Martin, R. G., 1980, Distribution of salt structures in the Gulf of Mexico: USGS Miscellaneous Field Studies Map MF-1213, 2 sheets, scale 1:250,000.

McGee, D. T., P. W. Bilinski, P. S. Gary, D. S. Pfeiffer, and J. L. Scheimann, 1994, Geologic models and reservoir geometries of Auger field, deepwater Gulf of Mexico (abs.): SEPM Gulf Coast Section, 15th Annual Research Conference, Proceedings, Houston, Texas, p. 245–256.

McFarlan, E., Jr., and L. S. Menes, 1991, Lower Cretaceous, *in* A. Salvador, ed., The geology of North America; the Gulf of Mexico basin: GSA Decade of North American Geology, v. J, p. 181–204.

McGuinness, D. B., and J. R. Hossack, 1993, The development of allochthonous salt sheets as controlled by the rates of extension, sedimentation, and salt supply: SEPM Gulf Coast Section, 14th Annual Research Conference, Proceedings, Houston, Texas, p. 127–139.

Moretti, I., and M. Larrere, 1989, LOCACE, computer-aided construction of balanced geological cross section: Geobyte, v. 4, no. 5, p. 16–24.

Moretti, I., S. Wu, and A. W. Bally, 1990, Computerized balanced cross section LOCACE to reconstruct an allochthonous salt sheet, offshore Louisiana: Marine and Petroleum Geology, v. 7, p. 371–377.

Mount, V. S., J. Suppe, and S. C. Hook, 1990, A forward modeling strategy for balancing cross sections: AAPG Bulletin, v. 74, no. 5, p. 521–531.

Nelson, T. H., 1991, Salt tectonics and listric-normal faulting, *in* A. Salvador, ed., The geology of North America; the Gulf of Mexico basin: GSA Decade of North American Geology, v. J, p. 73–89.

Nelson, T. H., and L. H. Fairchild, 1989, Emplacement and evolution of salt sills in the northern Gulf of Mexico (abs.): AAPG Bulletin, v. 73, p. 393.

O'Brien, J., and I. Lerche, 1988, Impact of heat flux anomalies around salt diapirs and salt sheets in the Gulf Coast on hydrocarbon maturity; models and observations: Gulf Coast Association of Geological Societies Transactions, v. 38, p. 231–243.

Peel, F. J., 1993, Structure, basin evolution, and salt tectonics of the central U.S. Gulf of Mexico derived from a regional cross section (abs.): AAPG, International Hedberg Research Conference Abstracts, Sept. 13–17, Bath, U.K., p. 119.

Pindell, J. L., 1985, Alleghenian reconstruction and subsequent evolution of the Gulf of Mexico, Bahamas and proto-Caribbean: Tectonics, v. 4, no. 1, p. 1–39.

Pindell, J. L., and J. F. Dewey, 1982, Permo-Triassic reconstruction of western Pangea and the evolution of the Gulf of Mexico/Caribbean region: Tectonics, v. 1, p. 179–212.

Pulham, A. J., G. M. Apps, C. A. Yeilding, K. P. Boyd, and J. M. Casey, 1994, A stratigraphic framework for the Miocene to Recent, northern Gulf of Mexico (abs.): AAPG Annual Convention, Denver, Colorado, Program with Abstracts, p. 239.

Rowan, M. G., 1994, A systematic technique for the sequential restoration of salt structures: Tectonophysics, v. 228, p. 331–348.

Salvador, A., 1987, Late Triassic–Jurassic paleogeography and the origin of the Gulf of Mexico: AAPG Bulletin, v. 71, p. 419–451.

Salvador, A., 1991a, Triassic–Jurassic, *in* A. Salvador, ed., The geology of North America; the Gulf of Mexico basin: GSA Decade of North American Geology, v. J, p. 131–180.

Salvador, A., ed., 1991b, The geology of North America; the Gulf of Mexico basin: GSA Decade of North American Geology, v. J, 568 p.

Schuster, D. C., 1995, Deformation of allochthonous salt and evolution of related salt–structural systems, eastern Louisiana Gulf Coast, *in* M. P. A. Jackson, D. G. Roberts, and S. Snelson, eds., Salt tectonics: a global perspective: AAPG Memoir 65, this volume.

Seni, S. J., 1992. Evolution of salt structures during burial of salt sheets on the slope, northern Gulf of Mexico: Marine and Petroleum Geology, v. 9, p. 452–468.

Shaub, F. J., R. T. Buffler, and J. G. Parsons, 1984, Seismic stratigraphic framework of deep, central Gulf of Mexico: AAPG Bulletin, v. 68, v. 11, p. 1790–1802.

Simmons, G. D., 1992, The regional distribution of salt in the northwestern Gulf of Mexico: styles of emplacement and implications for early tectonic history: Ph.D. dissertation, Texas A&M University, College Station, Texas, 183 p.

Sohl, N. F., E. Martinez, P. Salmeron-Urena, and F. Soto-Jaramillo, 1991, Upper Cretaceous, *in* A. Salvador, ed., The geology of North America; the Gulf of Mexico basin: GSA Decade of North American Geology, v. J, p. 205–244.

Talbot, C. J., and R. J. Jarvis, 1984, Age, budget and dynamics of an active salt extrusion in Iran: Journal of Structural Geology, v. 6, p. 521–533.

Thomas, W. A., 1988, Early Mesozoic faults of the northern Gulf coastal plain in the context of opening of the Atlantic Ocean, *in* W. Manspeizer, ed., Triassic–Jurassic rifting, continental breakup, and the origin of the Atlantic Ocean and passive margins, Developments in Geotectonics 22, part A: New York, Elsevier, p. 463–476.

Vendeville, B. C., and M. P. A. Jackson, 1992a, The rise of diapirs during thin-skinned extension: Marine and Petroleum Geology, v. 9, p. 331–353.

Vendeville, B. C., and M. P. A. Jackson, 1992b, The fall of diapirs during thin-skinned extension: Marine and Petroleum Geology, v. 9, p. 354–371.

Weimer, P., 1990, Sequence stratigraphy, facies geometries, and depositional history of the Mississippi Fan, Gulf of Mexico: AAPG Bulletin, v. 74, no. 4, p. 425–453.

Weimer, P., and R. T. Buffler, 1989, Structural geology of the Mississippi Fan fold belt, deep Gulf of Mexico: SEPM Gulf Coast Section, 10th Annual Research Conference, Program and Extended Abstracts, Houston, Texas, p. 148–150.

Weimer, P., and R. T. Buffler, 1992, Structural geology and evolution of the Mississippi Fan fold belt, deep Gulf of Mexico: AAPG Bulletin, v. 76, no. 2, p. 225–251.

Worrall, D., and S. Snelson, 1989, Evolution of the northern Gulf of Mexico, with emphasis on Cenozoic growth faulting and the role of salt, *in* A. W. Bally and A. R. Palmer, eds., The geology of North America; an overview: GSA Decade of North American Geology, v. A, p. 97–138.

Wu, S., A. W. Bally, and C. Cramez, 1990a, Allochthonous salt, structure, and stratigraphy of the northeastern Gulf of Mexico; part I, stratigraphy: Marine and Petroleum Geology, v. 7, p. 318–333.

Wu, S., A. W. Bally, and C. Cramez, 1990b, Allochthonous salt, structure, and stratigraphy of the northeastern Gulf of Mexico; part II, structure: Marine and Petroleum Geology, v. 7, p. 334–370.

Schuster, D. C., 1995, Deformation of allochthonous salt and evolution of related
salt–structural systems, Eastern Louisiana Gulf Coast, *in* M. P. A. Jackson,
D. G. Roberts, and S. Snelson, eds., Salt tectonics: a global perspective: AAPG
Memoir 65, p. 177–198.

Chapter 8

Deformation of Allochthonous Salt and Evolution of Related Salt–Structural Systems, Eastern Louisiana Gulf Coast

D. C. Schuster

Shell Offshore Inc.
New Orleans, Louisiana
U.S.A.

Present address:
Olmsted Falls, Ohio
U.S.A.

Abstract

Salt tectonics in the northern Gulf of Mexico involves both vertical diapirism and lateral silling or flow of salt into wings and tablets (sheets). Combinations of these two modes of salt deformation, concurrent with sediment loading and salt evacuation, have produced complex structures in the coastal and offshore region of southeastern Louisiana, a prolific oil and gas province. Many large growth faults and salt domes in the study area root into intra-Tertiary salt welds that were formerly occupied by allochthonous salt tablets. Two end-member structural systems involving evacuation of former tabular salt are recognized: *roho systems* and *stepped counter-regional systems*. Both end-member systems share a similar multi-staged evolution, including (1) initial formation of a south-leaning salt dome or wall sourced from the Jurassic salt level; (2) progressive development into a semitabular allochthonous salt body; and (3) subsequent loading, evacuation, and displacement of the tabular salt into secondary domes. In both systems, it is not uncommon to find salt displaced as much as 16–24 km south of its autochthonous source, connected by a horizontal salt weld to an updip, deflated counter-regional feeder.

Although both end-member structural systems may originate before loading of allochthonous salt having grossly similar geometry, their final structural configurations after loading and salt withdrawal are distinctly different. Roho systems are characterized by large-displacement, listric, south-dipping growth faults that sole into intra-Tertiary salt welds marked by high-amplitude reflections continuous with residual salt masses. Salt from the former salt tablets has been loaded and squeezed laterally and downdip. Stepped counter-regional systems, in contrast, comprise large salt domes and adjacent large-displacement, north-dipping growth faults that sole into intra-Tertiary salt welds before stepping down again farther north. Within the large salt-withdrawal basins north of the counter-regional faults are south-dipping strata that terminate onto subhorizontal salt welds.

Recognition of these more complex, deep-seated salt geometries should be factored into an analysis of hydrocarbon charge, migration, and trapping in light of the strong correlation between oil and salt–structural systems in the Gulf Coast.

INTRODUCTION

Most salt masses in the offshore northern Gulf of Mexico are allochthonous tabular bodies or floored salt domes that have been displaced both vertically and laterally from an autochthonous Jurassic stratigraphic level. Allochthonous salt in the region was first recognized along the Sigsbee Escarpment at the base of the continental slope (Amery, 1969; Lehner, 1969), where abyssal strata could be traced about 10 km beneath allochthonous salt. Subsequent studies have interpreted allochthonous salt sheets to be very widespread and often partially or wholly evacuated by overlying sediments (e.g., Humphris, 1979; Martin, 1984; Brooks, 1989; West, 1989; Worrall and Snelson, 1989; Wu et al., 1989, 1990; Diegel and Cook, 1990; Sumner et al., 1990; Diegel et al., 1993; Peel et al., 1993; Schuster, 1993).

This paper describes the loading and deformation of several salt–structural systems in the southeastern Louisiana Gulf Coast (Figure 1) that formed primarily by deformation of former tabular salt bodies. I concentrate on the kinematic relationships of deformed salt and related structures as inferred from interpreted 1:1 cross sections and reconstructions based primarily on seismic reflection data. The proprietary seismic data base for this study consisted of regional two-dimensional time-migrated profiles and included detailed interpretation and mapping of offshore regional seismic grids.

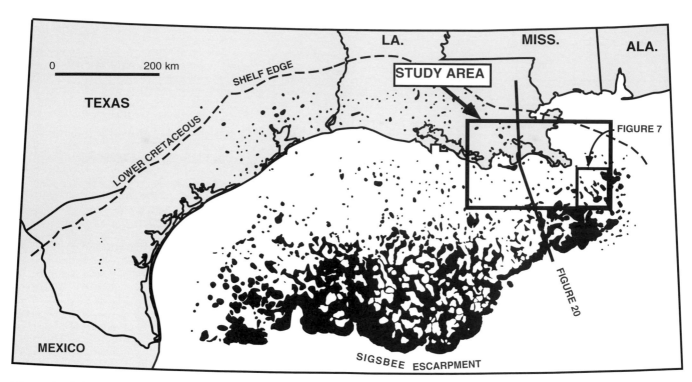

Figure 1—Salt map of the northern Gulf of Mexico showing location of the study area in southeastern Louisiana, onshore and offshore. Shallow salt bodies are in black. Also shown are the locations of two figures in this paper. (After Diegel et al., 1993.)

EMPLACEMENT OF ALLOCHTHONOUS SALT BODIES

Although shallow tabular salt bodies are widely recognized under the Louisiana continental slope, their mode of emplacement is not completely understood due in part to our inability to resolve subsalt structure under most salt tablets. Because of limited analog data from other salt basins, our current understanding of the emplacement mechanisms of tabular allochthonous salt is derived mostly from analysis of Gulf Coast data (e.g., Humphris, 1979; Jackson and Talbot, 1986; Cao et al., 1989; Worrall and Snelson, 1989; Peel et al., 1993; Jackson et al., 1994) and experimental and theoretical studies. An early important study by Trusheim (1960) of salt deformation in the European Zechstein documented the evolution of salt from autochthonous pillows to nearly detached, shallow diapirs with minor overhangs—but no allochthonous salt sheets. An excellent analysis of the subaerial salt diapirs in Iran and their evolution into an allochthonous salt canopy was presented by Jackson et al. (1990).

Experimental work (e.g., Ramberg, 1972; Jackson et al., 1985; Jackson and Vendeville, 1993) indicates that low-viscosity, low-density material (= salt) overlain by higher density material (= sedimentary cover) can invert to form structures similar to allochthonous salt tablets. The possible initiation and driving mechanisms for salt deformation, such as buoyancy, differential loading, thermal convection, and regional extension, have been discussed by

many authors (e.g., Trusheim, 1960; Woodbury et al., 1980; Jackson and Talbot, 1986; Kehle, 1988; Worrall and Snelson, 1989; Vendeville and Jackson, 1992a,b; Jackson and Vendeville, 1994).

Many allochthonous salt wings and tablets along the Louisiana Gulf Coast are presently, or were formerly, connected to deeper salt along north-dipping feeders at their northern ends (Larberg, 1983; Seni and Jackson, 1989; Diegel and Schuster, 1990; Sumner et al., 1990; Schuster, 1993; Walters, 1993). One example of a counter-regional feeder is shown in Figure 2, a seismic dip section across the updip end of an allochthonous salt tablet under the southeastern Louisiana upper slope. The shallow salt tablet angles downward at its northern end and eventually pinches out into a north-dipping, counter-regional "fault" (i.e., a deflated salt feeder, also termed a *secondary salt weld* by Jackson and Cramez, 1989), which extends down to presumed Jurassic basement. Feeders to other allochthonous salt bodies may be more completely evacuated of salt so that only the shallow, subhorizontal salt tablets remain. Similar north-dipping feeder geometries account for many leaning teardrop salt domes in the study area.

Figure 3 is a strike section illustrating the width of the feeder for the salt tablet (Figure 2). A salt-withdrawal syncline of Miocene–Pleistocene strata is situated above a narrow band of high-amplitude reflections, a zone once occupied by a thicker salt body. The width of the deflated feeder here is about half that of the related salt tablet to the south. The deflated feeder in Figure 3 dips away from the viewer, and the stratigraphy below the feeder belongs

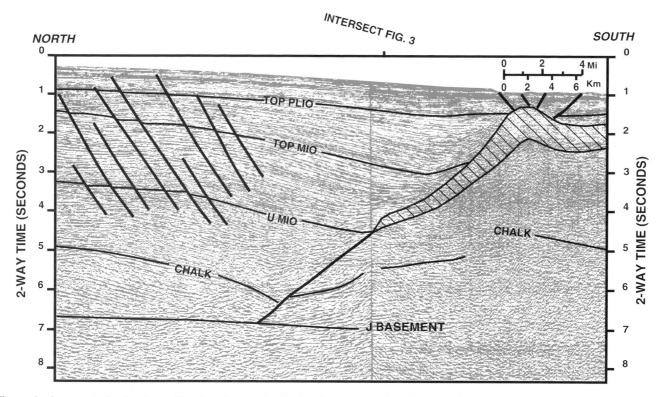

Figure 2—Interpreted seismic profile showing north-dipping (counter-regional) feeder for a shallow allochthonous salt tablet under the upper slope off southeastern Louisiana. Salt has been evacuated from the lower part of the feeder. Top Pliocene, top Miocene, and upper Miocene stratigraphic horizons are marked. Chalk = possible carbonate horizon; J basement = "basement" overlain by Jurassic autochthonous salt layer (salt now evacuated). Diagonal pattern = salt. See Figure 7 for location.

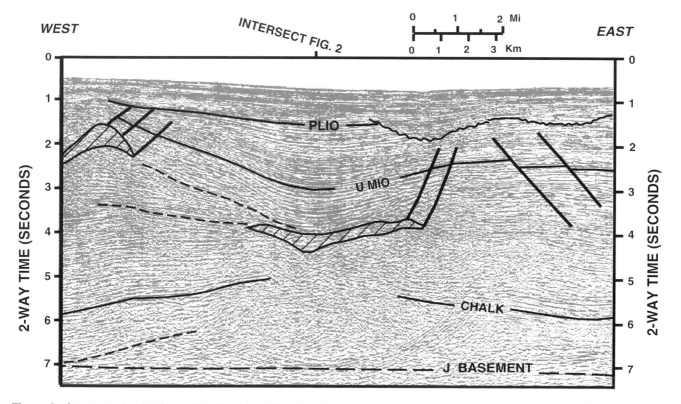

Figure 3—Interpreted seismic profile showing strike view through the counter-regional feeder for an allochthonous salt tablet. Section intersects Figure 2. Stratigraphic horizons are as in Figure 2. Diagonal pattern = salt. See Figure 7 for location.

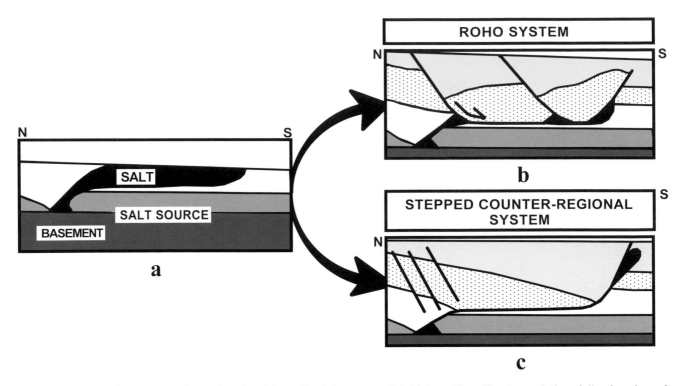

Figure 4—Schematic cross sections showing (a) an allochthonous salt tablet and its ultimate evolution, following depositional loading and salt evacuation, into either of two end-member structural systems: (b) a roho system, dominated by major down-to-the-basin growth faults soling into evacuated salt, or (c) a stepped counter-regional system, consisting of a large down-to-the-north growth fault.

to the "footwall block" under the tabular salt body. The older section forms an apparent turtle structure (Trusheim, 1960; Jackson and Talbot, 1991), indicating early autochthonous salt withdrawal—probably into the salt body that eventually fed the allochthonous salt tablet.

Although this salt body (Figure 2) and many others in the northern Gulf Coast have counter-regional feeders that extend to probable Jurassic basement, many salt feeders in the shelf and coastal region of Louisiana sole or root within the Tertiary section. These intermediate levels are salt evacuation phenomena termed *tertiary salt welds* by Jackson and Cramez (1989), in which strata formerly on top of salt masses have progressively subsided onto strata formerly below salt. These geometries and the evolution of these structural systems are the focus of this paper.

SALT–STRUCTURAL SYSTEMS FORMED BY EVACUATION OF ALLOCHTHONOUS SALT SHEETS

Introduction

Although it has been recognized for some time that evacuated and unevacuated intra-Tertiary salt sheets underlie much of the continental slope of the northern Gulf of Mexico, there has been increasing recognition that large areas of evacuated allochthonous salt underlie the

present-day shelf and coastal plain as well (e.g., Wu et al., 1989; Diegel and Cook, 1990; Sumner et al, 1990; Diegel et al., 1993; Schuster, 1993). In the study area (Figure 1), two end-member systems are recognized to have developed during progradational loading and evacuation of allochthonous salt (Figure 4). These end-members are termed *roho systems* and *stepped counter-regional systems* (Schuster, 1993). Although both systems originally had similar gross allochthonous salt architecture before loading, their final structures contrast markedly. [The term *roho* was coined about 1970 at Shell Oil Co. from "C. C. *Ro*ripaugh's (a Shell geophysicist) Mo*ho*" to describe a high-velocity zone characterized by a thin band of high-amplitude discontinuous reflectors that form the base of most of the faults and allochthonous salt features in an area.]

Roho systems (Figure 4b) are characterized by major, listric, down-to-the-basin growth faults that sole into intra-Tertiary salt evacuation surfaces (tertiary salt welds). In contrast, stepped counter-regional systems (Figure 4c) are characterized by major, listric, down-to-the-north growth "faults" that sole into salt evacuation surfaces. At the updip end of both systems, the subhorizontal salt evacuation surfaces step downward, typically forming deflated salt feeders (secondary salt welds).

The present study area of coastal southeastern Louisiana (Figure 1) is structurally less complex than coastal south-central or southwestern Louisiana (Diegel and Cook, 1990; Sumner et al., 1990; Diegel et al., 1993).

The more isolated structural systems described here may aid in the analysis of more complex structures that have characteristics of both end-members.

Example of a Roho System

An excellent example of a roho system is found 20 km (12 mi) southeast of the present Mississippi birdsfoot delta (Figure 1). This isolated, northwest-southeast trending roho system is 16–24 km (10–15 mi) wide and 72 km (45 mi) long. Parts of this structural system have been described previously as a product of salt evacuation (Wu et al., 1989, 1990). The system is floored seismically by high-amplitude, discontinuous reflections connecting residual salt masses and ranges in depth from 3 km (10,000 ft) updip to 4.5 km (15,000 ft) downdip.

The structural geometry of this roho system is shown in a seismic dip profile (Figure 5). The salt evacuation surface dips gently south from the head of the system basinward to the north flank of an allochthonous salt tablet. A separate allochthonous salt body farther downdip connects to the roho system to the east. Two major listric growth faults that sole into the updip end of the salt evacuation surface were active mainly in Pliocene–Pleistocene time. Well control is limited to the northern end of the system, and age correlations downdip are speculative. Much of the allochthonous salt in this system may have been derived from a counter-regional feeder under the northern end (shown in Figure 5). However, apparent offsets of the "chalk" horizon suggest two more salt feeders farther basinward. This implies that the intra-Tertiary allochthonous salt may originally not have been strictly continuous and that original gaps could have been smeared with salt during basinward translation of overlying strata.

A prominent erosional surface truncates north-dipping Miocene–Pliocene strata at the downdip end of the roho system (Figure 5). This unconformity may have formed during Pliocene submarine erosion of toe sediments elevated by overthrusting. Basinward gliding of these sediments along evacuated salt coincided with updip extensional growth faulting.

A seismic strike profile (Figure 6) across the roho system illustrates how the salt has been evacuated from the center and forced laterally into secondary salt bodies, often as edge ridges. The concave-up geometry of this strike section may be typical for both roho systems and many stepped counter-regional systems. This interpreted section shows tremendous expansion of Pliocene strata, although there has also been basinward translation of the expanded strata (out-of-plane movement). Consequently, the east-dipping fault at the western edge of the system probably has a considerable strike-slip component. The stratigraphy outside the roho system is relatively flat and undisturbed, except for deep turtlelike structures.

A geologic map of the roho system (Figure 7) shows the following: (1) the lateral extent of the salt evacuation surface, (2) residual allochthonous salt thicknesses, (3) major growth faults above the surface and probable strike-slip faults along the flanking margins, (4) counter-regional salt feeders, (5) toe structures along the basinward margins, (6) regions of tilted and eroded strata, and (7) the relationship of the roho system to the adjacent allochthonous salt masses.

The major down-to-the-basin growth faults on evacuated salt occur only at the updip end, whereas most of the salt is now basinward of the growth faults. Several wells have penetrated salt but did not reach subsalt strata. One well penetrated 140 m (450 ft) of salt in a region where no salt was seismically resolved. The most extensive regions of evacuated salt (generally less than 300 m, or 1000 ft, thick) occur in the updip part and along a central trough. Updip, residual salt occurs in east-west trending salt rollers in the footwalls of the two major down-to-the-basin growth faults (see Figure 5). Downdip, the salt masses trend more north-south, reflecting the effects of lateral salt migration into edge-parallel ridges. The salt occupying the southernmost part of the system may not have migrated very far. Toe thrusts and contractional folds along various downdip margins indicate basinward translation of strata. Broad zones of angular unconformities in the central and southern parts of the system (Figures 5, 7) may denote uplift and erosion, possibly due to basinward gliding and tilting.

Palinspastic Analysis of a Roho System

The roho system just described can be palinspastically restored in a variety of ways, depending on different age correlations, deep speculative interpretation, restoration algorithms used, and other factors. One possible scenario for the evolution of this roho system involves the formation of three allochthonous salt tablets having subsequent sediment loading, salt evacuation, and basinward translation of strata. This scenario is shown in Figure 8 and is based on a palinspastic reconstruction of a depth conversion of the seismic profile in Figure 5. The depth section was restored partially with the aid of proprietary computer programs. The restorations generally preserve line lengths, incorporate decompaction, and preserve paleo-bathymetric slopes relative to the prograding shelf margin. Salt area is not preserved. The reconstruction exercise deals with deformation and evacuation of allochthonous salt and the formation of structures on top of allochthonous salt (Miocene time to present). It does not address such issues as autochthonous salt thickness, early salt history, or rigorous isostatic compensation.

Prior to the Miocene, autochthonous salt diapirs had already started to develop shallow salt wings. Expansion of Miocene strata above the shallowest part of the counter-regional feeder at the north end of the roho system (Figure 5) suggests salt evacuation from a leaning diapir and basinward silling, or flow, into a growing allochthonous salt tablet. In middle–late Miocene time (Figure 8f), thin tabular salt, which was initially covered by a veneer of slope sediments, may have begun to inflate as more salt was evacuated from depth up through the feeders. The margins of the salt expanded laterally and climbed upsection until late Miocene time (Figure 6). By

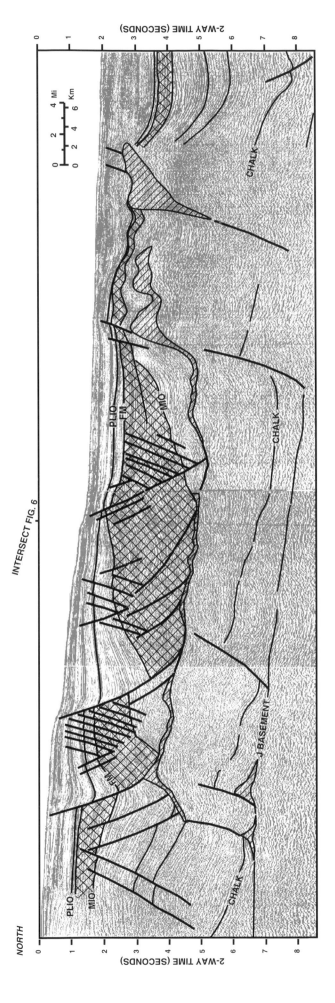

Figure 5—Interpreted seismic dip profile along a roho system. Large, listric growth faults sole into the evacuated salt surface (tertiary salt weld), along which are residual salt lenses. A counter-regional feeder under the head (left end) of the system contributed at least some of the salt, but other possible feeders may occur under the salt weld. FM = local field marker; chalk = possible carbonate horizon; J basement = "basement" overlain by remnant autochthonous Jurassic salt; diagonal pattern = salt; cross pattern = lower Pliocene stratigraphic interval. See Figure 7 for location.

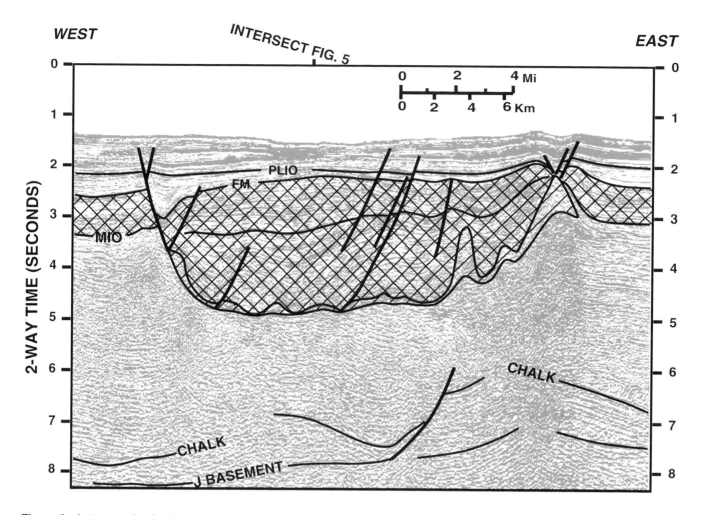

Figure 6—Interpreted seismic strike profile across a roho system showing the upward concavity of the evacuated salt surface. Allochthonous salt has been evacuated from the central region and squeezed out to the margins. Section intersects Figure 5. Stratigraphic horizons are as in Figure 5. Diagonal pattern = salt; cross pattern = lower Pliocene interal. See Figure 7 for location.

this time, prior to the main loading-evacuation phase, the tabular salt was about 1.5 km (5000 ft) thick and nearly as laterally extensive as the present roho system.

Loading of tabular salt (not the shallow part of the feeder) is interpreted as beginning in late Miocene time (Figure 8e). Major growth faulting initiated in latest Miocene, and updip extension was accompanied by salt evacuation, basinward translation, and thrusting of strata over salt at the toe of the system. Loading and growth-fault expansion continued into the Pleistocene, but significant gliding and downdip overthrusting ended in the Pliocene (Figure 8c). By this time, the toe overthrust was mostly eroded and buried. The reconstruction panel also shows a migrating front of basinward-thickening salt being pushed downdip as salt is removed from the section (laterally out of the plane, and by dissolution at or near the seafloor). In this scenario, nearly 24 km (15 mi) of extension observed updip is compensated downdip by both overthrusting (with erosion) and salt evacuation. A third shortening mechanism, folding of the overriding strata, occurs in front of toe thrusts (Figure 7).

Stepped Counter-Regional Systems: Bourbon Example

One example of a stepped counter-regional system is situated at the outer shelf south of the Mississippi delta (Figure 9) and involves the Bourbon (MC-311) and WD-133 salt domes. A seismic dip section through the Bourbon dome (Figure 10) shows the tripartite nature of a stepped counter-regional system. It consists of (1) a southward-leaning salt dome at the downdip end (on trend with large counter-regional growth faults), whose north-dipping flank soles into a (2) subhorizontal detachment-like horizon (salt weld), which (3) steps down (to Jurassic basement?) as the deep counter-regional segment of the system. The Bourbon dome is a secondary salt dome that is rooted in the subhorizontal salt weld rather than extending directly down to Jurassic basement. Gently folded strata below the soling horizon are probably turtle structures developed during early autochthonous salt deformation.

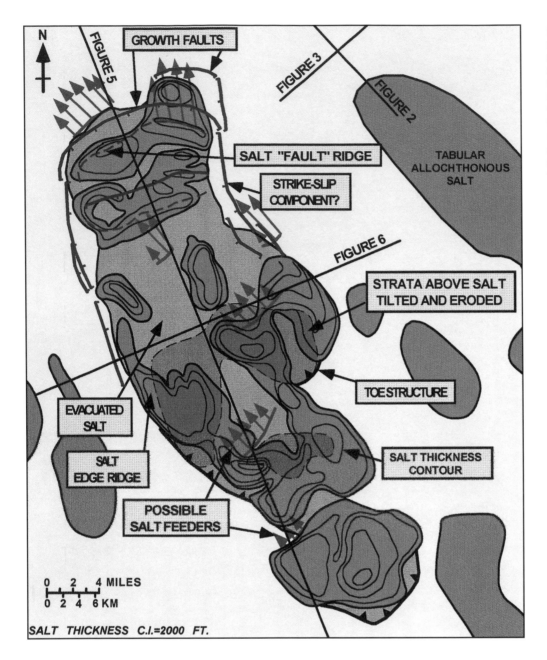

Figure 7—Salt–structure map of a roho system showing allochthonous salt (green), inferred salt feeders (red), evacuated salt (tan), major growth faults (blue), and contractional toe structures. Diagonal pattern = strata above salt tilted and eroded. See Figure 1 for location.

The structural geometry in an adjacent seismic profile (Figure 11) is similar to that shown in Figure 10, except that a shallow counter-regional fault replaces the overhung salt dome at the downdip end of the system. The 6.0–6.7 km (20,000–22,000 ft) deep soling horizon, nearly 16 km (10 mi) in length along dip, is a prominent discontinuity marked by south-dipping Miocene deltaic strata apparently downlapping onto older subhorizontal basin plain strata. Anomalous diffractions and out-of-plane reflections above the interpreted horizon in Figure 11 are related in part to the lateral curvature of the fault. Expansion of strata into the shallow counter-regional fault and onto the soling horizon results from evacuation of allochthonous salt throughout the late Cenozoic. An extensional fault system of similar age has formed in the hinge zone above the northern end of the soling horizon.

These keystone-style faults are extensional compensation structures formed in response to bending during rotation of the strata to the south and their subsidence onto the salt evacuation horizon.

A three-dimensional picture of this salt–structural system is needed to understand its development. A composite salt–structure map (Figure 12) displays salt, evacuated salt, and major faults. The Bourbon system is slightly elongate (east-northeast) and about 1000 km^2 (400 mi^2) in extent. The arcuate downdip margin of the system consists of large, shallow counter-regional growth faults on trend with the two shallow salt domes. The faults and northern salt flanks sole into a subhorizontal evacuated salt surface at 6–7 km (20,000–24,000 ft) depths. This surface dips more steeply to the north at two locations along its northern edge, possibly forming two deep deflated salt

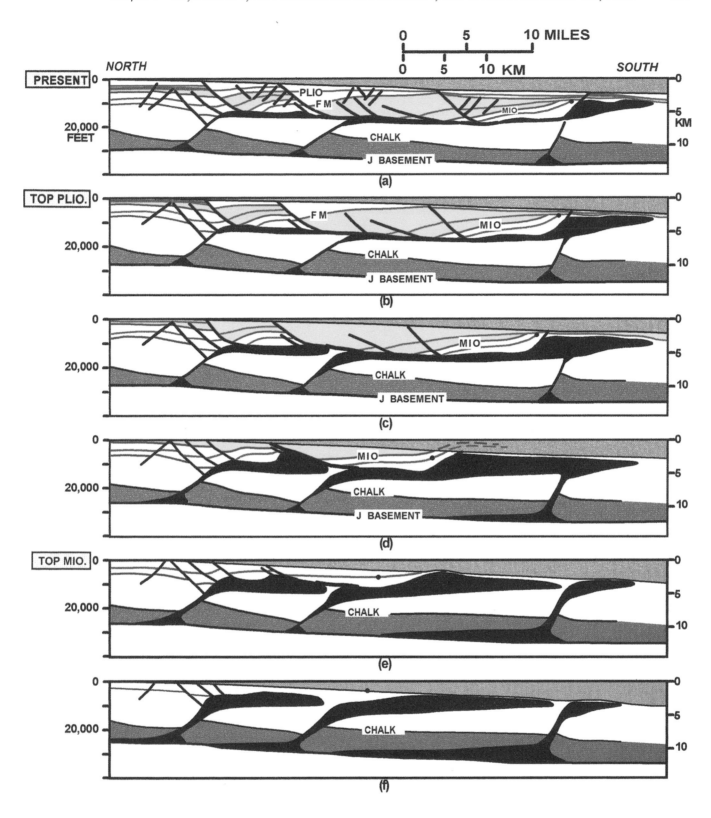

Figure 8—(a) True scale (1:1) depth section (converted from Figure 5) along a roho system. **(b–f)** Palinspastic reconstructions back to Miocene time showing one possible evolution. Stratigraphic horizons are as in Figure 5. Salt is shown in black.

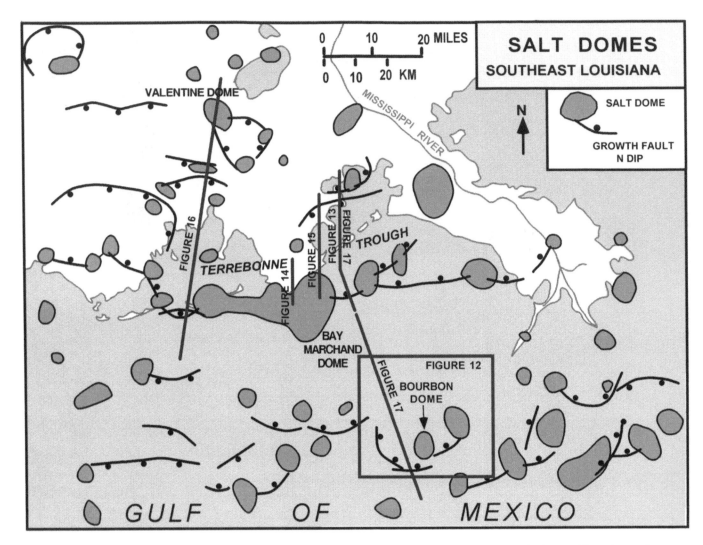

Figure 9—Map of onshore and offshore southeastern Louisiana showing salt domes, growth faults, and locations of seismic profiles and Figure 12.

feeders. The Bourbon dome is overhung southward, whereas the WD-133 dome farther east is a truncated cone whose base coincides with the evacuated salt horizon. Both domes, however, are interpreted as allochthonous, pinching out into an evacuated intra-Tertiary salt horizon.

Although shallow counter-regional faults strike into the south flanks of the domes, the exact relationship is not clear. West of Bourbon dome, several fault splays gradually curve around to the north (where there may be some possible strike-slip movement), forming the southwestern margin of the Bourbon system. In contrast, the counter-regional fault east of the dome is a single surface. The counter-regional faults are sediment-on-sediment slip surfaces along their shallower levels. However, at their deeper levels, where they begin to flatten, they are subsidence structures whose hanging walls and footwalls were once separated by salt (i.e., little or no sediment-on-

sediment slip ever took place). The position along the soling fault where this change occurs (the basinward limit of original salt) can be estimated through palinspastic reconstruction.

The deep, north-dipping feeders to the stepped counter-regional system were mapped to at least 9 km (30,000 ft) depth (Figure 12). Because the deep chalk event is offset by the deflated feeder (Figure 10), it is tentatively concluded that the deep feeder extends directly to the Jurassic salt source.

Stepped Counter-Regional Systems: Bay Marchand–Terrebonne Example

Another example of a stepped counter-regional system is located along the coast of southeastern Louisiana (Figure 9) and coincides with the Terrebonne trough (Limes and Stipe, 1959), an east-trending basin extending

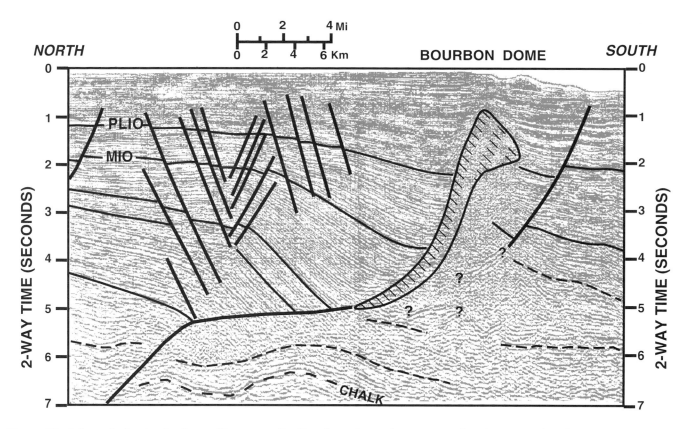

Figure 10—Interpreted seismic dip profile across the Bourbon stepped counter-regional system showing geometry of the Bourbon salt dome (diagonal pattern). Counter-regional fault south of dome may sole into the evacuated salt horizon. Chalk = possible carbonate horizon. See Figure 12 for location.

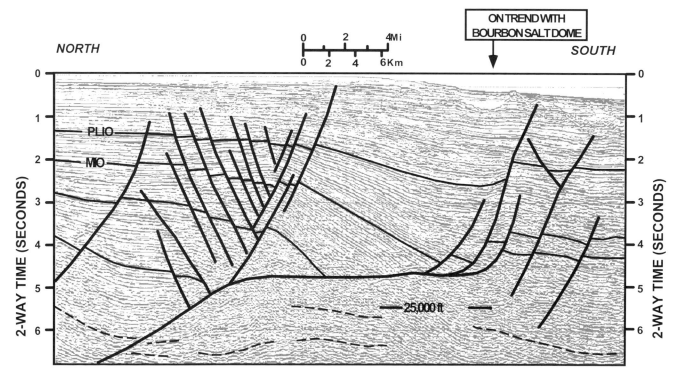

Figure 11—Interpreted seismic dip profile across the Bourbon stepped counter-regional system near Figure 10. No significant salt is apparent on the section. See Figure 12 for location.

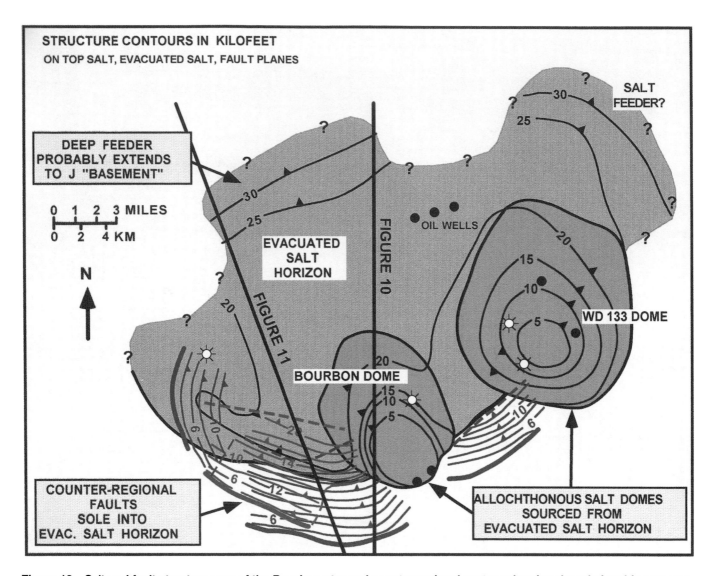

Figure 12—Salt and fault structure map of the Bourbon stepped counter-regional system showing the relationships among two secondary salt domes (green), major counter-regional faults (blue and red), and an evacuated salt horizon (tan). See Figure 9 for location.

more than 160 km (100 mi) to the west of the Mississippi Delta. The Terrebonne trough is a late Miocene salt withdrawal basin bounded on the south by large salt domes connected by major counter-regional growth faults. The basin-filling strata thicken southward, either into a north-dipping fault or onto the northern flank of a dome. The salt domes are the sites of large oil and gas fields such as Bay Marchand, Caillou Island, Grand Isle 16, and West Delta 30. Until the past decade, seismic reflection data across much of this trend were of poor quality because of acquisition problems in this swampy region.

The structure of this stepped counter-regional system is illustrated in Figure 13, a seismic dip profile extending southward across the Terrebonne trough to a buried counter-regional fault between the Bay Marchand and GI 16 salt domes offshore (see Figure 9). South-dipping upper Miocene strata expand into the north-dipping fault

(splays) at the south end of the system. The deeper strata terminate onto a subhorizontal intra-Tertiary horizon into which the counter-regional fault soles. Although subhorizontal bedding reflection segments are visible below the interpreted structural discontinuity, the surface itself is not well imaged in Figure 13. Seismic reflections from this soling horizon are denoted (with arrows) in Figure 14, an adjacent dip profile that extends over the northern flank of the Bay Marchand salt dome.

A deep, north-dipping seismic event in Figure 13 is apparently a deflated feeder to the Bay Marchand–Terrebonne system. South-dipping bedding reflections (early Miocene and older) terminate onto this secondary salt weld, which formerly connected shallow allochthonous salt to the Jurassic source horizon. This rare seismic expression of a deep feeder segment provides important documentation for the stepped counter-regional model.

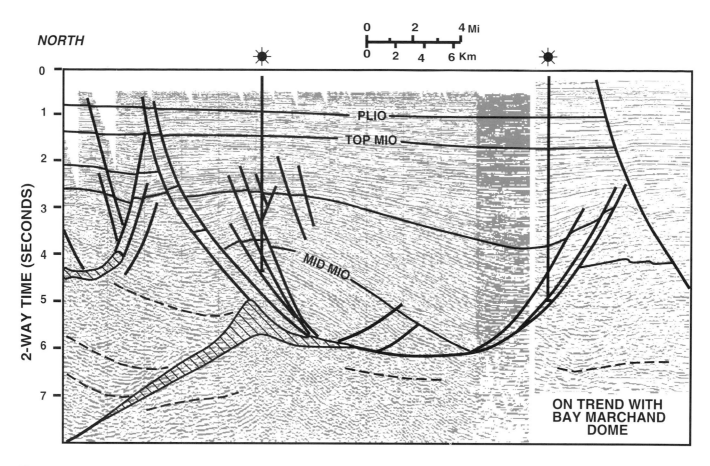

Figure 13—Interpreted seismic dip profile across the Bay Marchand–Terrebonne stepped counter-regional system east of Bay Marchand dome. Diagonal pattern = salt. See Figure 9 for location.

Large, south-dipping growth faults north of the rollover anticline in Figure 13 also sole into the intra-Tertiary horizon. This dual nature of opposing growth faults soling into a deep discontinuity implies evacuation of a buried allochthonous salt body. The evacuation surface is probably resting on strata of earliest Miocene age or older. Similar opposing faults in southeastern Louisiana were called *collapse fault systems* by Seglund (1974), who interpreted them as caused by the evacuation of autochthonous salt.

Figure 15 shows another interpreted seismic profile across the Bay Marchand stepped counter-regional system, extending through the Terrebonne Trough and northern flank of the Bay Marchand salt dome. Many of the structural elements in Figure 13 are again present here, except that, in lieu of a downdip counter-regional fault, the sediments thicken toward and onlap the salt dome. Although not imaged, the base of the Bay Marchand dome is inferred to merge northward into an intra-Tertiary evacuated salt horizon. Based on this interpretation, the salt in the large dome was derived from the autochthonous Jurassic source at least 32 km (20 mi) to the north.

Other salt domes on trend with Bay Marchand (see Figure 9) may also have been fed from evacuated alloch-thonous salt. A regional seismic dip profile 48 km (30 mi) to the west across the structural system west of Caillou Island Dome (Figure 16) suggests a similar stepped geometry. Because of limited available data, it could not be determined if the stepped counter-regional system along the Terrebonne trough is continuous or interrupted.

Palinspastic Analysis of Stepped Counter-Regional Systems

Squeezed versions of original 1:1 reconstruction sections across the Bay Marchand and Bourbon systems show an evolutionary scenario for the two stepped counter-regional systems (Figure 17). The dip section across coastal and offshore southeastern Louisiana, which avoids salt domes (see Figure 9), is reconstructed back to Eocene time to evaluate the structural evolution of allochthonous salt. Each salt system originated as a leaning stock or salt wall, then developed into a shallow allochthonous salt tongue, and finally was loaded and almost completely evacuated. The interpretation shows that in Miocene time, a significant part of each evacuated salt layer was occupied by a southward-thickening, sub-horizontal wedge of salt whose upper surface was close

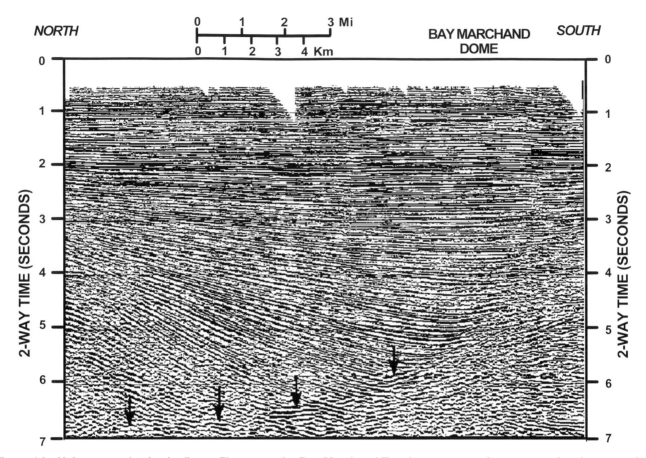

Figure 14—Uninterpreted seismic dip profile across the Bay Marchand-Terrebonne stepped counter-regional system showing reflections (arrows) from a deep soling horizon (top of salt or salt evacuation surface). North flank of the Bay Marchand dome is at south end of section. See Figure 9 for location.

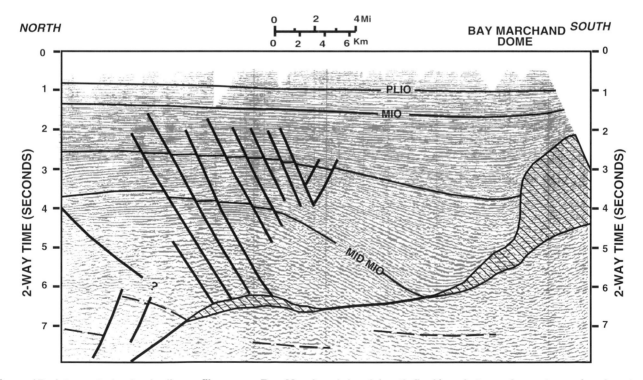

Figure 15—Interpreted seismic dip profile across Bay Marchand dome (north flank) and stepped counter-regional system. Base of secondary salt dome is speculative. Diagonal pattern = salt. See Figure 9 for location.

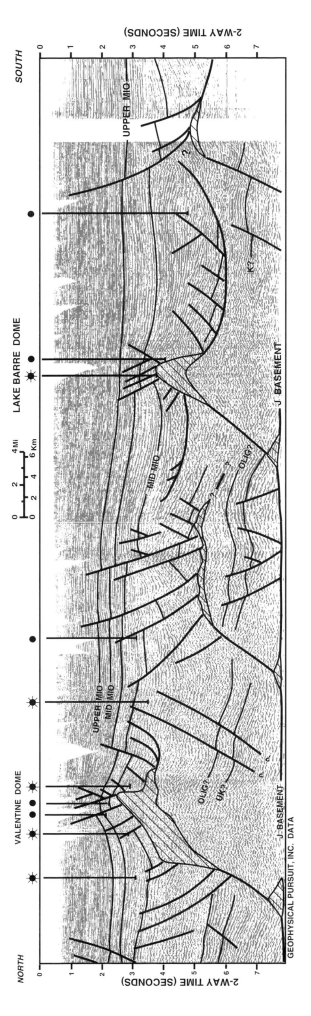

Figure 16—Regional seismic dip profile across southeastern Louisiana and coastal zone, showing the leaning Valentine dome, the eastern part of a roho system, and the western part of the Bay Marchand–Terrebonne stepped counter-regional system. J basement = "basement" overlain by remnant autochthonous Jurassic salt; diagonal pattern = salt. See Figures 9 and 19 for location.

to or at the sea bottom. Eventually, original unconformable onlaps against salt were rotated to form the present-day apparent downlaps and pseudo-faults on the evacuated salt horizons. It is this unique geometry that characterizes counter-regional growth "faults." Once most of the salt was evacuated, subsidence waned and shelf deposition prograded across each counter-regional fault.

The reconstruction of 1:1 sections involved several steps. First, stratigraphic line length and thickness (no decompaction) was preserved by hand restoration of more detailed scaled sections. Second, the salt area was allowed to "move" freely in and out of the section, which was otherwise selected to minimize out-of-plane movement. Third, paleobathymetric slopes were preserved relative to the shelf margin (determined from well paleo-data). The restored base of Jurassic salt was estimated from one-dimensional isostatic calculations (Diegel and Cook, 1990; Diegel et al., 1993) made from the perimeter of the stepped systems where allochthonous salt is absent. The base of allochthonous salt was then determined by also assuming that strata underneath the stepped system are pinned to the footwall of the shallow counter-regional fault and move with it as a single body.

The reconstructions indicate that the formation of stepped counter-regional systems did not involve significant displacement of hanging wall strata along the pseudo-fault surface. Instead, each system formed by evacuation of an unconformably onlapped salt body and by subsidence of the overlying sediments onto strata that were formerly underneath salt (shown schematically in Figure 18). When the sediment wedge above evacuated salt is palinspastically reconstructed, the resulting horizontal gap along the sea bottom and shallow void are filled with salt. Line length d (a bedding plane) in Figure 18b is not pulled back up the fault to connect with footwall line length e but instead is restored to horizontal without translation. This leaves a gap along the sea bottom that is occupied by very shallow salt. Any thin sediment cover above the salt may be present today as shale "sheath" along the evacuated salt horizon. Alternatively, strata above a stepped counter-regional system cannot be pulled palinspastically basinward to connect with corresponding strata downdip of the system because the existing horizontal displacement across the deep counter-regional feeder is tiny relative to the apparent displacement of strata across the shallow counter-regional "fault" (Figure 17).

The resulting restorations depicting considerable expanses of salt at or near the sea bottom may be difficult to envision because little evidence for such circumstances exists today. However, most of the tabular salt masses under the present lower slope were formed prior to the more recent glacial lowstands and have been subsequently buried. Although few offshore salt bodies in the northern Gulf of Mexico presently have much salt exposed at the sea bottom, some have fairly shallow crests (e.g., Bourbon dome at ~1000-ft depth) and may have been at least partially exposed during Pleistocene lowstands.

The volume of salt in the Bourbon stepped counter-regional system today (Figure 12) is considerably less than that estimated for the Miocene–Pliocene, based on regional mapping and analysis. Considerable salt has been lost, presumably by dissolution at or near the sea bottom. This is consistent with similar findings for reconstructions of the North Louisiana Salt Basin (Kupfer et al., 1976), the East Texas Basin (Seni and Jackson, 1984), and western Louisiana onshore and offshore (Diegel et al., 1993). Large volumes of salt can be dissolved by migrating pore fluids within the sediments surrounding domes, especially if driven by deep geopressures (Hanor, 1986). Jenyon (1984) cited evidence for salt dissolution and collapse of salt domes due to fluid migration along crestal faults. A considerable volume of salt can be dissolved at the sea bottom if extrusion and dissolution rates are comparable to those for subaerial salt glaciers (Jackson and Talbot, 1986). Further analysis is needed to assess the potential rates of dissolution during sea level lowstands when the salt could be exposed to sea water or air.

Regional Analysis

Analysis of both roho and stepped counter-regional systems demonstrates that evacuation of allochthonous salt is important in the formation of salt structures in offshore southeastern Louisiana. Figure 19 is a regional salt–structure map showing various salt evacuation systems along coastal and offshore southeastern Louisiana. The map portrays salt domes and allochthonous salt evacuation surfaces (tertiary salt welds), as well as deep-seated, deflated(?) salt feeders (secondary salt welds). Many regional and counter-regional growth faults, commonly connecting the salt domes, sole into evacuated salt horizons. Stepped counter-regional systems dominate the eastern part of the area in Figure 19, whereas roho-style systems are common in the west. Not all of the counter-regional systems are stepped; many deflated salt dome feeders (secondary salt welds) extend without interruption directly to presumed Jurassic basement.

A regional seismic dip profile northwest of Bay Marchand dome in southeastern Louisiana (Figure 16; located in Figure 19) displays some of the variety of salt–structural systems. The northernmost and simplest system is the Valentine Dome, whose (deflated?) feeder appears to extend directly to Jurassic basement. Relatively shallow counter-regional growth faults soling into the overhung southern edge of the dome may be related to a possible evacuated salt horizon further eastward, as speculated in Figure 19. Downdip from the Valentine Dome is a small roho-style system without an apparent toe structure. To the west, this evacuated allochthonous salt horizon may expand significantly in size (Figure 19). Immediately downdip of this roho system is a major stepped counter-regional system which underlies the Terrebonne trough and extends 32 km (20 mi) eastward to Bay Marchand Dome. The southernmost segment of this stepped counter-regional system is truncated by a regional growth fault at the head of a major roho system partially visible on Figure 16 and more

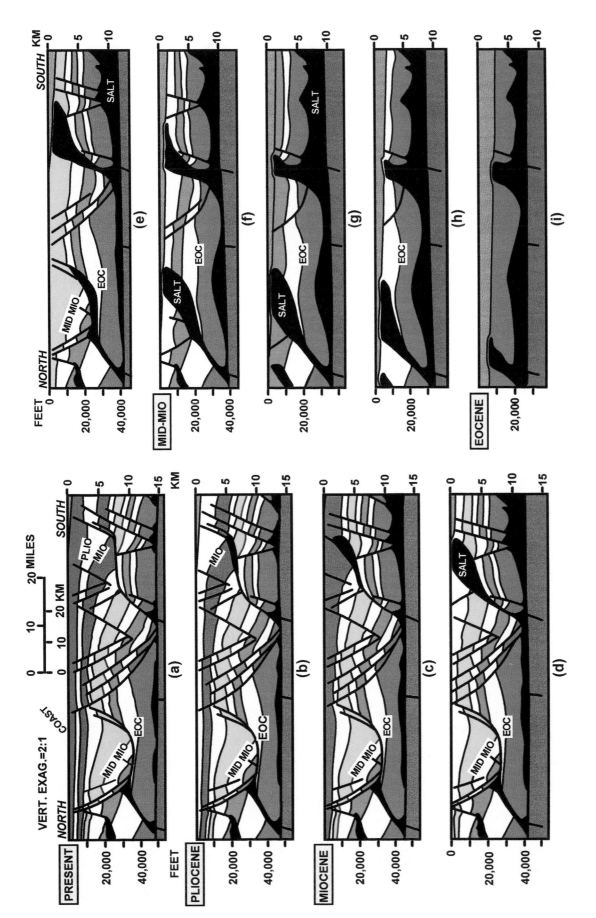

Figure 17—(a) Reconstruction section across the Bay Marchand–Terrebonne and Bourbon stepped counter-regional systems. (b–i) Sections showing the evolution of former allochthonous salt bodies (black) from Eocene to Pliocene time. Vertical exaggeration is 2:1. See Figures 9 and 19 for location.

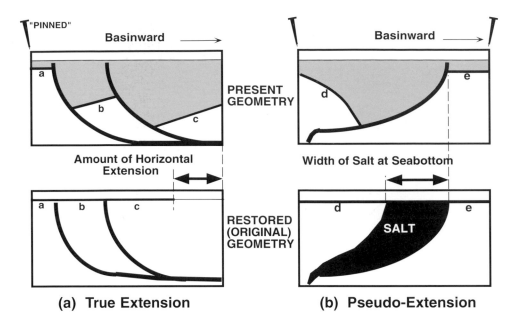

Figure 18—Schematic line length restorations showing (a) true extension versus (b) pseudo-extension due to salt evacuation and sediment subsidence. After reconstruction, line lengths *d* and *e* do not close together in the reconstruction because salt existed at or very near the sea bottom at time of deposition.

extensively shown in map view in Figure 19. If these allochthonous salt structures were closer together or only slightly more complex, their recognition as separate salt–structural systems would become very difficult.

The Bay Marchand-Terrebonne and Bourbon stepped counter-regional systems, as depicted in Figure 17, are placed in regional perspective in Figure 20, a cross section extending from southeastern Louisiana to the deep Gulf Basin (Diegel and Schuster, 1990). Updip of these allochthonous salt structures are salt domes with counter-regional feeders extending smoothly to Jurassic basement. Downdip are shallow allochthonous salt masses in various stages of deformation.

Differences Between Roho and Stepped Counter-Regional Systems

Both roho and stepped counter-regional systems have distinct histories, although both formed above allochthonous salt sheets formerly connected to presumed Jurassic basement. In roho systems, updip growth faults and toe structures downdip typically indicate basinward extension and gliding of strata above salt. When restoring down-to-the-south faults in a roho system, the line lengths are pulled back and true extension (see Figures 8, 18a) can be measured. In contrast, in stepped counter-regional systems, the rotational growth in the "hanging wall" section has developed without actual extension. The subhorizontal surface that now appears to be a fault is basically a subsided surface that originally was an unconformity of sediment onlapping a salt mass (see Figures 17, 18b). Interpretation of this subsided onlap unconformity as a detachment fault would require northward updip extension and gliding of the overlying strata. However, there are no known toe structures updip to have accommodated such extension; furthermore, exten-

sion updip would violate the general rule of gravity tectonics. Alternatively, palinspastically moving the footwall block basinward to produce the apparent extension generates a line length (volume) gap at the deep feeder, a situation unobserved in the data.

What conditions cause one mode of deformation to dominate over another? Before addressing this problem, it should be emphasized that these two systems are treated as idealized end-members. Allochthonous salt evacuation phenomena west of the study area possess attributes of both structural systems. One solution may relate to downdip length of the original tabular salt mass (F. A. Diegel, personal communication, 1989). A short tablet may not have the length necessary to yield a gravity gradient differential sufficient to allow basinward gliding and hence formation of large, roho-style listric growth faults at the head (updip end) of the evacuating tablet. The difference in deformation mode between roho and stepped counter-regional systems may also depend on variations in the (1) original thickness of the allochthonous salt body, (2) nature of initial salt deformation, (3) volume and rate of sedimentary influx, and (4) location and geometry of sedimentation.

ECONOMIC IMPLICATIONS

The interpretation of widespread allochthonous salt evacuation surfaces (tertiary salt welds) and related allochthonous salt domes in the study area of the southeast Louisiana Gulf Coast represents a fundamental change in the current perception of the deep structure of this economically important region. Improved and deeper seismic data, coupled with palinspastic analysis, should lead to substantial modifications of interpretations of deep salt–structural geometries here. Further, such analysis might better constrain our understanding of migration

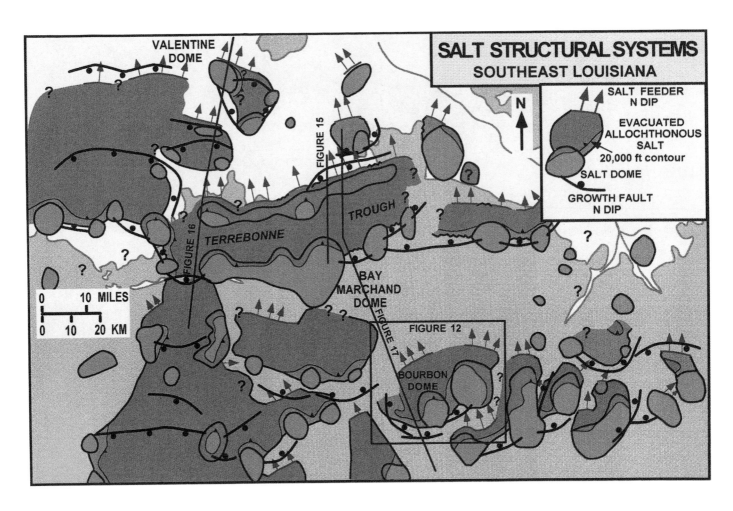

Figure 19—Salt–structure map of various salt domes and allochthonous salt evacuation systems, onshore and offshore southeastern Louisiana. A 20,000-ft (6-km) structural contour extends over salt (light and dark green) and evacuated salt (light and dark tan). Deep salt feeders (red arrows) probably extend to the Jurassic autochthonous salt.

pathways, source rock distribution, and hydrocarbon charge and mix. Subhorizontal salt evacuation surfaces can provide previously unrecognized migration barriers and, correspondingly, new potential traps.

A reinterpretation of many presumably deep-seated salt domes to perched, overhung, teardrop-shaped salt bodies holds particular exploration significance. Inferred traps against supposedly steeply overhung salt flanks may actually be far removed from salt. Recognition of salt masses as overhung wings or floored salt domes opens up new play opportunities under the southern flanks, many of which may be within economic range of the drill bit. Data wipe-out zones below shallow salt should not always be interpreted as deep salt. Better constraints on the evolutionary scenarios of these complex salt–structural systems may also benefit reservoir prediction. For example, if we can better determine the relationship between sedimentary loading history and structural growth, then recognition of diagnostic structures should yield clues on lithologic (i.e., reservoir) distribution.

CONCLUSIONS

1. Most salt bodies in the study area of southeastern Louisiana Gulf Coast are leaning domes, wings, or tablets connected to counter-regional (north-dipping) feeders, now usually subsided (secondary salt welds), which extend to deeper salt horizons.
2. Many salt domes, including some classic domes, root into intra-Tertiary evacuated salt horizons (tertiary salt welds) and do not extend directly to Jurassic basement. These domes are commonly associated with large counter-regional growth faults that sole into the same evacuated salt horizons.
3. Two types of structural systems related to evacuation of allochthonous tabular salt are present in southeastern Louisiana onshore and offshore. These two idealized end-members are roho systems and stepped counter-regional systems, both of which possess a multi-staged evolution. A roho system is characterized by listric down-to-the-

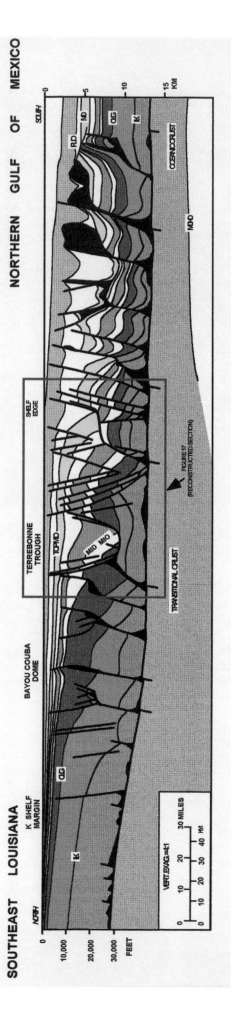

Figure 20—Regional dip section across the northern Gulf of Mexico, extending through southeastern Louisiana onshore to the base of the continental slope. Section shows the regional context of the Bay Marchand–Terrebonne and Bourbon stepped counter-regional systems. Salt shown in black. Faults below the autochthonous salt horizon (Jurassic) are speculative. Original 1:1 section has been squeezed to 4:1 vertical exaggeration, substantially steepening true dips. See Figure 1 for location.

south growth faults, whereas a stepped counter-regional system is characterized by a large north-dipping growth fault at the basinward end of the evacuated salt layer.

4. This improved understanding of the deep structure and salt geometry in the region may (1) provide a better structural framework for charge and migration and reservoir studies, (2) yield new insights into the reasons for hydrocarbon distribution and mix, and (3) aid paleomaturity studies through a better understanding of the structural evolution of complex salt systems.

Acknowledgments   This study was part of a regional project to gain a better understanding of the tectonostratigraphic framework of the northern Gulf of Mexico margin, undertaken by both research and operations staff from Shell Development Co., Shell Offshore, Inc., and Shell Western Exploration and Production, Inc. S. Snelson played a key role in the planning, coordination, and supervision of this long-term study. I thank Shell Oil Co. for permission to publish and many people for their technical contributions, ideas, and related research at Shell, including A. D. Scardina, J. F. Karlo, R. C. Shoup, F. A. Diegel, R. M. Coughlin, S. Snelson, D. M. Worrall, K. P. Guilbeau, W. R. Trojan, S. C. Reeve, E. B. Picou, H. S. Sumner, and J. C. Holliday. Many concepts involving evacuated allochthonous salt were initially proposed by C. C. Roripaugh, J. M. Beall, and others at Shell in the late 1960s. The two-dimensional seismic reflection profiles were provided by and published with permission from Geco-Prakla, Inc., Geophysical Pursuit, Inc., TGS-Calibre Geophysical Co., and Shell Offshore, Inc. E. E. White and J. L. Tassin assisted in the preparation of illustrations in this paper. C. S. Cameron, M. G. Rowan, and S. Snelson provided valuable comments and suggestions in their reviews, which greatly improved the original manuscript.

REFERENCES CITED

Amery, G. B., 1969, Structure of Sigsbee Scarp, Gulf of Mexico: AAPG Bulletin, v. 53, p. 2480–2482.

Brooks, R. O., 1989, Horizontal component of Gulf of Mexico salt tectonics, *in* Gulf of Mexico salt tectonics, associated processes and exploration potential: SEPM Gulf Coast Section, 10th Annual Research Conference, Program and Extended Abstracts, Houston, Texas, p. 22–24.

Cao, S., I. Lerche, and J. J. O'Brien, 1989, Moving salt sheets and the deformation and faulting of sedimentary formations: SEPM Gulf Coast Section, 10th Annual Research Conference, Program and Extended Abstracts, Houston, Texas, p. 28–38.

Diegel, F. A., and R. W. Cook, 1990, Palinspastic reconstruction of salt-withdrawal growth fault systems, northern Gulf of Mexico (abs.): GSA, Abstracts with Programs, Dallas, Texas, p. A48.

Diegel, F. A., and D. C. Schuster, 1990, Regional cross sections and palinspastic reconstructions, northern Gulf of Mexico (abs.): GSA, Abstracts with Programs, Dallas, Texas, p. A66.

Diegel, F. A., J. F. Karlo, D. C. Schuster, R. C. Shoup, and P. R. Tauvers, 1993, Cenozoic structural evolution and tectono-stratigraphic framework of the northern Gulf Coast continental margin (abs.): AAPG Annual Convention Official Program, New Orleans, Louisiana, p. 91.

Hanor, J. S., 1986, Evidence for large-scale vertical overturn of pore fluids in the Louisiana Gulf Coast (abs.): GSA, Abstracts with Programs, San Antonio, Texas, v. 18, p. 627.

Humphris, C. C., Jr., 1979, Salt movement on continental slope, northern Gulf of Mexico: AAPG Bulletin, v. 63, p. 782–798.

Jackson, M. P. A., and C. Cramez, 1989, Seismic recognition of salt welds in salt tectonic regimes, *in* Gulf of Mexico salt tectonics, associated processes and exploration potential: SEPM Gulf Coast Section, 10th Annual Research Conference, Program and Extended Abstracts, Houston, Texas, p. 66-71.

Jackson, M. P. A., and C. J. Talbot, 1986, External shapes, strain rates, and dynamics of salt structures: GSA Bulletin, v. 97, p. 305–323.

Jackson, M. P. A., and C. J. Talbot, 1991, A glossary of salt tectonics: The University of Texas at Austin, Bureau of Economic Geology, Geological Circular 91-4, 44 p.

Jackson, M. P. A., and B. C. Vendeville, 1993, Extreme over-thrusting and extension above allochthonous salt sheets emplaced during experimental progradation (abs.): AAPG Annual Convention Official Program, New Orleans, Louisiana, p. 122–123.

Jackson, M. P. A., and B. C. Vendeville, 1994, Regional extension as a geologic trigger for diapirism: GSA Bulletin, v. 106, p. 57–73.

Jackson, M. P. A., C. J. Talbot, and R. R. Cornelius, 1985, Centrifuge modeling of the effects of variable sedimentary loading on the external geometry, kinematics, and dynamics of syndepositional salt structures: The University of Texas at Austin, Bureau of Economic Geology Open File Report OF-WTWI-1985-21, p. 145.

Jackson, M. P. A., R. R. Cornelius, C. H. Craig, A. Gansser, J. Stöcklin, and C. J. Talbot, 1990, Salt diapirs of the Great Kavir, central Iran: GSA Memoir 177, 139 p.

Jackson, M. P. A., B. C. Vendeville, and D. D. Schultz-Ela, 1994, Structural dynamics of salt systems: Annual Reviews of Earth and Planetary Sciences, v. 22, p. 93–117.

Jenyon, M. K., 1984, Seismic response to collapse structures in the southern North Sea: Marine and Petroleum Geology, v. 1, p. 27–36.

Kehle, R. D., 1988, The origin of salt structures, *in* B. C. Schreiber, ed., Evaporites and hydrocarbons: New York, Columbia University Press, p. 345–404.

Kupfer, D. H., C. T. Crowe, and J. M. Hessenbruch, 1976, North Louisiana Basin and salt movements (halokinetics): Gulf Coast Association Geological Society Transactions, v. 26, p. 94–110.

Larberg, G. M., 1983, Contra-regional faulting: salt with-drawal compensation, offshore Louisiana, Gulf of Mexico, *in* A. W. Bally, ed., Seismic expression of structural styles: AAPG Studies in Geology 15, v. 2.3.2, p. 42–44.

Lehner, P., 1969, Salt tectonics and Pleistocene stratigraphy on continental slope of northern Gulf of Mexico: AAPG Bulletin, v. 53, p. 2431–2479.

Limes, C. L., and J. C. Stipe, 1959, Occurrence of Miocene oil in south Louisiana: Gulf Coast Association Geological Society Transactions, v. 9, p. 77–90.

Martin, R. G., 1984, Diapiric trends in the deep water Gulf basin, *in* Characteristics of Gulf Basin deep-water sediments and their exploration potential: SEPM Gulf Coast

Section, 5th Annual Research Conference, Program and Abstracts, Austin, Texas, p. 60–62.

Peel, F. J., C. J. Travis, J. R. Hossack, and D. B. McGuinness, 1993, Structural provinces in the cover sediments of the U.S. Gulf of Mexico basin: linked systems of extension, compression and salt movement (abs.): AAPG Annual Convention Official Program, New Orleans, Louisiana, p. 164.

Ramberg, H., 1972, Experimental and theoretical study of salt dome evolution: Unesco, Geology of Saline Deposits, Proceedings Hanover Symposium 1968, p. 247–251.

Schuster, D. C., 1993, Deformation of allochthonous salt and evolution of related structural systems, eastern Louisiana Gulf Coast (abs.): AAPG Annual Convention Official Program, New Orleans, Louisiana, p. 179.

Seglund, J. A., 1974, Collapse-fault systems of Louisiana Gulf Coast: AAPG Bulletin, v. 58, p. 2389–2398.

Seni, S. J., and M. P. A. Jackson, 1984, Sedimentary record of Cretaceous and Tertiary salt movement, East Texas basin: Times, rates, and volumes of salt flow and their implications to nuclear waste isolation and petroleum exploration: The University of Texas at Austin, Bureau of Economic Geology Report of Investigations 139, 89 p.

Seni, S. J., and M. P. A. Jackson, 1989, Counter-regional growth faults and salt sheet emplacement, northern Gulf of Mexico, *in* Gulf of Mexico salt tectonics, associated processes and exploration potential: SEPM Gulf Coast Section, 10th Annual Research Conference, Program and Extended Abstracts, Houston, Texas, p. 116–121.

Sumner, H. S., B. A. Robison, W. K. Dirks, and J. C. Holiday, 1990, Morphology and evolution of salt/mini-basin systems: lower shelf and upper slope, central offshore Louisiana (abs.): GSA, Abstracts with Programs, Dallas, Texas, p. A48.

Trusheim, F., 1960, Mechanism of salt migration in northern Germany: AAPG Bulletin, v. 44, p. 1519–1540.

Vendeville, B. C., and M. P. A. Jackson, 1992a, The rise of diapirs during thin-skinned extension: Marine and Petroleum Geology, v. 9, p. 331–353.

Vendeville, B. C., and M. P. A. Jackson, 1992b, The fall of diapirs during thin-skinned extension: Marine and Petroleum Geology, v. 9, p. 354–371.

Walters, R. D., 1993, Reconstruction of allochthonous salt emplacement from 3-D seismic reflection data, northern Gulf of Mexico: AAPG Bulletin, v. 77, p. 813–841.

West, D. B., 1989, Model for salt deformation on deep margin of central Gulf of Mexico basin: AAPG Bulletin, v. 73, p. 1472–1482.

Woodbury, H. O., I. B. Murray, Jr., and R. E. Osborne, 1980, Diapirs and their relation to hydrocarbon accumulation, *in* A. D. Miall, ed., Facts and principles of world petroleum occurrence: Canadian Society of Petroleum Geologists Memoir 6, p. 119–142.

Worrall, D. M., and S. Snelson, 1989, Evolution of the northern Gulf of Mexico, with emphasis on Cenozoic growth faulting and the role of salt, *in* A. W. Bally and A. R. Palmer, eds., The Geology of North America; an overview: GSA Decade of North American Geology, v. A, p. 97–138.

Wu, S., C. Cramez, A. W. Bally, and P. R. Vail, 1989, Evolution of allochthonous salt in the Mississippi Canyon area, *in* Gulf of Mexico salt tectonics, associated processes and exploration potential: SEPM Gulf Coast Section, 10th Annual Research Conference, Program and Extended Abstracts, Houston, Texas, p. 161–165.

Wu, S., A. W. Bally, and C. Cramez, 1990, Allochthonous salt, structure and stratigraphy of the northeastern Gulf of Mexico, part II: structure: Marine and Petroleum Geology, v. 7, p. 334–371.

Rowan, M. G., 1995, Structural styles and evolution of allochthonous salt, central Louisiana outer shelf and upper slope, *in* M. P. A. Jackson, D. G. Roberts, and S. Snelson, eds., Salt tectonics: a global perspective: AAPG Memoir 65, p. 199–228.

Chapter 9

Structural Styles and Evolution of Allochthonous Salt, Central Louisiana Outer Shelf and Upper Slope

Mark G. Rowan

*Department of Geological Sciences
   and Energy & Minerals Applied Research Center
University of Colorado
Boulder, Colorado, U.S.A.*

Abstract

Seismic interpretation and section restoration are combined with recent models of salt deformation to describe the geometry and evolution of allochthonous salt from the central Louisiana outer shelf and upper slope. Scattered salt bodies are connected by a complex system of diachronous salt welds or remnant salt having two end-member geometries: (1) regionally extensive, subhorizontal sheets bounded by north-dipping (counter-regional) feeders and characterized by common listric growth faults that may accommodate significant extension; and (2) elliptical depressions bounded by dipping salt welds and arcuate growth faults that accommodate little extension.

Most salt bodies in the study area were emplaced at or near the sea floor and grew by downbuilding (passive diapirism). Reactive and active diapirs are rare. The former are confined to the updip margins of shallow salt sheets, and the latter may occur basinward of major salt-withdrawal minibasins. Many salt bodies along the downdip margins of sheets have been modified by contraction.

Two end-member evolutionary models account for the range of observed structural styles. In "counter-regional" systems, which are more typical of the shelf, salt rises through south-leaning feeder stocks and flows both downdip and along strike to form allochthonous sheets. In "salt stock canopy" systems, which are more typical of the upper slope, bulb-shaped salt stocks expand outward and form salt canopies. Subsequent gravitational collapse and sedimentary loading form bowl-shaped minibasins, from which salt is displaced into allochthonous tongues and remnant salt bodies.

INTRODUCTION

Our understanding of salt tectonics has increased significantly in recent years through the contributions of many researchers in academia and industry. The advances result from improved seismic data, more realistic physical modeling, and the sequential restoration of cross sections. The AAPG Hedberg Research Conference on Salt Tectonics, held in Bath, U.K., during September 1993, highlighted many of these advances and provided the stimulus for further work. In this chapter, I evaluate recent models of salt deformation and apply them to allochthonous salt from offshore central Louisiana.

Humphris (1978) first suggested that large allochthonous salt nappes are a common feature of the Gulf of Mexico. This thesis has been confirmed by improved seismic imaging, which clearly demonstrates that tabular salt sheets exist at all scales (e.g., Worrall and Snelson,

1989; Wu et al., 1990; Nelson, 1991). In turn, a better understanding of salt geometries has led to numerous models of salt evolution that attempt to explain the different levels of allochthonous salt, the varying geometries of salt bodies, the large growth faults, and the contractional structures observed in the Gulf (e.g., Worrall and Snelson, 1989; West, 1989; Diegel and Cook, 1990; Hossack and McGuinness, 1990; Wu et al., 1990; Seni, 1992, 1994; McGuinness and Hossack, 1993; Diegel et al., 1993, 1995; Schuster, 1993, 1995; Fletcher et al., 1995; Peel et al., 1995). More recently, new seismic processing techniques such as prestack depth migration (e.g., Ratcliff, 1993) provide clearer pictures of base of salt and subsalt geometries, but these have yet to be incorporated into published models of salt sheet evolution.

Most models for the emplacement of allochthonous salt envision salt moving up and basinward from deeper sources along north-dipping (counter-regional) feeder systems (e.g., Worrall and Snelson, 1989; West, 1989; Diegel and Cook, 1990; Wu et al., 1990; Walters, 1993; Schuster, 1993, 1995; Diegel et al., 1993, 1995; Peel et al., 1995). Examples of *counter-regional systems* given by Schuster (1995) show distinct levels of allochthonous salt, each having only minor relief along the weld (or equivalent base of salt) where the sheet may cut up-section at its flanking margins and toe. The only significant relief in the weld/base of salt system occurs where two levels are connected by north-dipping collapsed feeders. Counter-regional systems may include abundant listric growth faults that sole into the welds or salt sheets and that can accommodate significant extension.

Another model for allochthonous salt emplacement envisions salt stocks that gradually spread outward in all directions to form overhanging flanks (McGuinness and Hossack, 1993; Hodgkins and O'Brien, 1994; House and Pritchett, 1994; Rowan et al., 1994b). The overhung salt stocks eventually collapse and become sedimentary mini-basins. The evacuated salt moves up and laterally to form sheets, tongues, and other secondary salt bodies. Overhung stocks may merge to form salt stock canopies (Jackson and Talbot, 1989, 1991) either before collapse or during loading and associated salt flow. These *salt stock canopy systems*, in contrast to the counter-regional systems, are characterized by the following: (1) significant relief on the base of salt, where elliptical lows are postulated to be underlain by feeder stocks; (2) salt welds that dip in all directions; and (3) relatively rare growth faults having only minor amounts of extension (Rowan et al., 1994b).

The initiation, growth, and collapse of individual salt bodies, whether allochthonous or autochthonous, have been investigated by Vendeville and Jackson (1992a,b). By modeling overburden as a brittle rather than ductile material, they were able to generate structures that appear to be realistic analogs of many natural features. They recognized three types, or phases, of salt evolution: *reactive, active,* and *passive diapirism.* The latter two stages of salt body growth were originally defined by Nelson (1991), and passive diapirism is equivalent to downbuilding (Barton, 1933), in which salt bodies increase in relative height as adjacent basins thicken by sinking into an underlying source salt layer. Whereas reactive and active diapirs initiate and grow beneath a significant thickness of overburden, passive diapirs grow at or close to the sea floor beneath only a thin veneer of draped sediments.

Recent studies suggest that salt sheets and tongues also form at the sea floor as extrusions rather than as salt sills. McGuinness and Hossack (1993) argued that salt sheets advance basinward by extruding as salt glaciers over the sea floor. Jackson and Talbot (1991), Talbot (1993), and Fletcher et al. (1995) suggest that salt bodies connected to an adequate deep source form active, topographically elevated salt "fountains" at the sea floor. These salt masses may spread out laterally during times of reduced sedimentation or increased salt flow (Vendeville and Jackson, 1991; Talbot, 1995), forming bodies that range from symmetric salt stocks having pronounced subhorizontal overhangs to asymmetric salt tongues or "glaciers"

(Talbot, 1993; Fletcher et al., 1995). Once the deep salt source is depleted or cut off by local salt welds, or the salt fountain is drowned by rapid sedimentation, the topographic high collapses by means of gravity spreading or downslope gravity gliding, thereby initiating basin formation (Fletcher et al., 1995).

In this chapter, various models of salt tectonics are integrated with the results of seismic interpretation and section restoration to describe and explain the structural styles beneath the outer shelf and upper slope of offshore central Louisiana (Figure 1). I first address salient aspects of the observed structural geometries: (1) the mapped salt distribution and the geometry of salt welds; (2) the presence and growth of reactive, active, and passive diapirs; (3) the distribution and significance of extensional growth faults; and (4) the occurrence of contractional structures. Next, the observed geometries are evaluated in the context of two end-member scenarios for the evolution of allochthonous salt: the counter-regional model of Schuster (1995) and the salt stock canopy model of Rowan et al. (1994b). Finally, I briefly discuss the implications for facies development, hydrocarbon migration and entrapment, and subsalt exploration.

The data consist of a grid of two-dimensional seismic profiles of various vintages from the 1980s; line spacings are usually 1.6 km (1 mi) between north-south lines and 3–5 km (2–3 mi) between east-west lines. In addition, wireline logs from over 170 wells and biostratigraphic data from over 120 wells were used in the seismic interpretation (see Weimer et al., 1994).

STRUCTURAL STYLES

Salt System Geometry

The study area is dominated by numerous circular to irregularly shaped salt bodies connected at several different levels by either thin tabular salt or salt welds (Figure 1). This composite salt system has considerable structural relief. The tops of some salt bodies are shallower than 1.0 sec TWT (two-way traveltime), and some welds are deeper than 7.0 sec TWT. Different levels can be separated by overhangs (Figures 1, 2) or connected by dipping salt welds (Figures 1, 3). Dipping welds are not limited to counter-regional (northward-dipping) orientations; instead, they occur in all orientations and define elliptical lows in the deep salt welds (see Rowan et al., 1994b). These elliptical depressions characterize most of the upper slope in the study area (Figure 1) and contrast with regionally extensive, subhorizontal salt sheets and welds typical of both the shelf region to the north (Schuster, 1995) and the Sigsbee Nappe system to the south (e.g., West, 1989; Worrall and Snelson, 1989; Wu et al., 1990; Simmons, 1992; Seni, 1994; Diegel et al., 1995).

A seismic profile through one elliptical depression is illustrated in Figure 4. The deep weld connects a shallow, tabular salt sheet on the north to an allochthonous salt body on the south and underlies a synclinal minibasin.

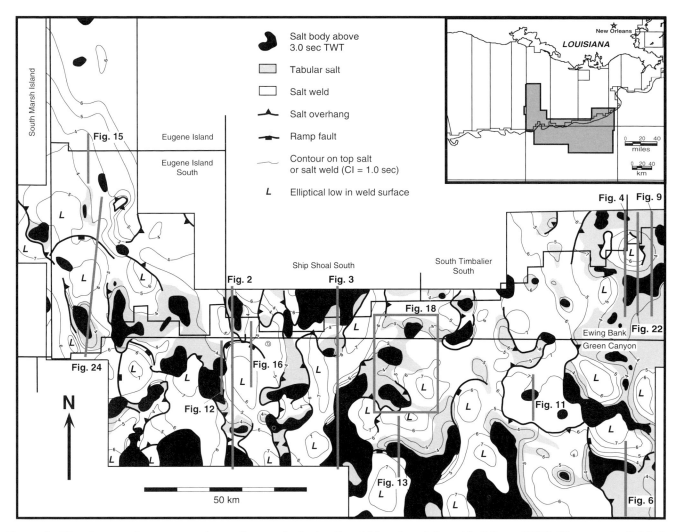

Figure 1—Structure contour map on top salt or equivalent salt weld (contour interval is 1.0 sec TWT). The area is divided into salt bodies (shallower than 3.0 sec), approximately tabular salt, and salt welds (salt generally <100 msec in thickness). The map also shows areas where shallow salt overhangs deeper levels, elliptical lows in the deep salt weld, and ramp faults that arc around the lows. The locations of illustrated seismic profiles and detailed maps are indicated, and the location of the study area and the shelf edge are shown on the inset map.

Where it dips southward, the salt weld truncates approximately horizontal reflections under the weld. Because this geometry is analogous to *salt ramps* at the base of salt tongues (Jackson and Talbot, 1991; McGuinness and Hossack, 1993), dipping welds that cut up-section in any direction are also called *ramps* here.

The ramp in Figure 4 curves around to both the east and west, forming the northern half of a circular low in the weld (Figure 1). Subweld truncations are imaged on seismic data around more than 180° of the arc (see Figures 9, 22). Although similar relationships are observed beneath other ramps in the study area (see Figures 16, 24), reflections beneath most dipping salt weld surfaces are usually unclear. Nevertheless, based on the elliptical map patterns of the lows in the deep welds and occasional hints of relatively undeformed subweld reflections, it is speculated that most of these lows have geometries similar to those shown in Figure 4.

Salt Body Geometries

Salt bodies in the study area display a wide variety of structural styles, as evidenced by their varying geometries, topographic expression at the sea floor, and relationships with surrounding sediments. Based on the results of experimental modeling, Vendeville and Jackson (1992a) proposed that salt bodies be classified into three types: (1) reactive diapirs, which initiate and grow beneath grabens formed by extension of the overburden; (2) active diapirs, which rise by shouldering aside and piercing a tectonically thinned and weakened overburden; and (3) passive diapirs, which originate at the sea floor and grow in height by downbuilding. Furthermore, they suggested that most diapirism is triggered by regional extension (Jackson and Vendeville, 1994) and that their three styles often represent successive stages in the evolutionary history of a single salt body (Vendeville

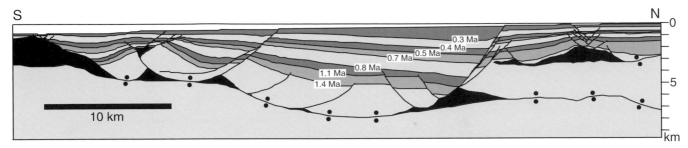

Figure 2—True scale (1:1) depth section of an interpreted seismic profile (see Figures 1, 5, 27 for location). Water is white, salt is black, and salt welds are shown by pairs of dots. From north to south, the section shows the following: a shallow salt sheet that overhangs a deeper weld; a ramp fault above a salt roller where the deep weld dips to the south; a minibasin containing north-dipping strata; a north-dipping weld branching into another growth fault; a contractional anticline cored by a pinched-off salt body that has accommodated some thrusting; and a shallow, tabular salt body. Stratigraphic interpretation by A. Navarro (1994); data courtesy of Halliburton.

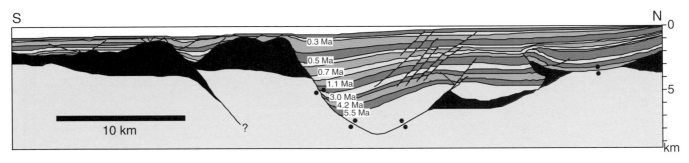

Figure 3—True scale (1:1) depth section of an interpreted seismic profile (see Figures 1, 5, 27 for location). Water is white, salt is black, and salt welds are shown by pairs of dots. From north to south, the section shows the following: a shallow salt sheet/weld; a thrust fault coring an eroded contractional anticline where the weld ramps down to the south; a mini-basin containing a south-dipping monocline whose hinge zone is cut by normal faults; a north-dipping (counterregional) weld connected to a shallow, tabular salt body; and a second tabular sheet bounded by a north-dipping feeder. Stratigraphic interpretation by Z. Acosta (1994); data courtesy of Halliburton.

and Jackson, 1992a). If the overburden is thin and uneven, however, the first two stages may be bypassed, so that salt bodies are initiated and evolve passively (Vendeville and Jackson, 1992a). This scenario is supported by many palinspastic restorations in the Gulf of Mexico (e.g., Worrall and Snelson, 1989; Diegel and Cook, 1990; Rowan, 1993; Rowan et al., 1994a; Diegel et al., 1995).

In this section, seismic interpretation and palinspastic restoration are used to document the presence and growth of reactive, active, and passive diapirs in the study area. Interpreted seismic profiles were converted to depth using vertical ray paths and depth-dependent velocity functions derived from nearby well data and then restored using the software package GEOSEC and the method of Rowan (1993). Salt bodies chosen are either still active or have only recently stopped growing, and restorations were constructed back to the oldest horizon that could be correlated with confidence (based on well penetrations). Thus, the initiation and early histories of the salt bodies are generally not shown.

Reactive Diapirs

Reactive diapirs appear to be rare in the study area (Figure 5). The best example (Figure 6) is along the southern margin of an elliptical depression (Figure 1) and at the

northwestern corner of a salt sheet that forms part of the Sigsbee nappe just to the southeast (as mapped by Wu, 1993; Seni, 1994). This salt body has distinct characteristics that differentiate it from most other salt bodies in the study area and are compatible with a reactive origin (see Vendeville and Jackson, 1992a). It is (1) elongate in an east-west direction, (2) triangular in profile view (Figure 6), (3) overlain by an extensional graben, (4) flanked by normal faults that are progressively older down the sides of the salt body, and (5) located along the updip edge of a salt sheet.

One possible restoration of the salt body demonstrates a reactive style of growth (Figure 7). In this interpretation, 5.2 km of basinward sliding of the overburden in the last 1.1 Ma was accommodated by extension at the updip edge of the allochthonous sheet, resulting in 1.7 km of vertical growth of the reactive salt body. Whether this postulated magnitude of extension and rafting of the overburden is balanced downdip by some combination of thrusting and folding (with possible erosion), salt extrusion at the Sigsbee Escarpment, and salt dissolution has not been determined.

Because of the uncertainties inherent in section restoration, other evolutionary scenarios may be geometrically admissible. For example, the restorations to 0.5 and 1.1

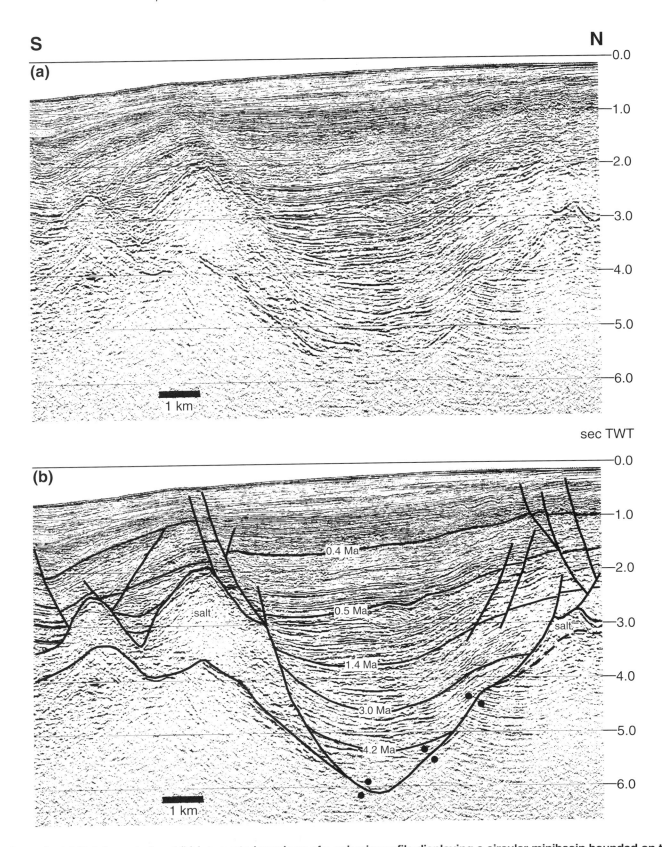

Figure 4—(a) Uninterpreted and (b) interpreted versions of a seismic profile displaying a circular minibasin bounded on the north by a south-dipping salt weld and on the south by a salt sheet and overlying north-dipping growth fault (see Figures 1, 5, 27 for location). The south-dipping weld (ramp) separates south-dipping strata in the minibasin from nearly horizontal subweld reflections (see Figures 9, 22). Salt welds are indicated by pairs of dots. Stratigraphic interpretation by F. Budhijanto (1995); data courtesy of Halliburton.

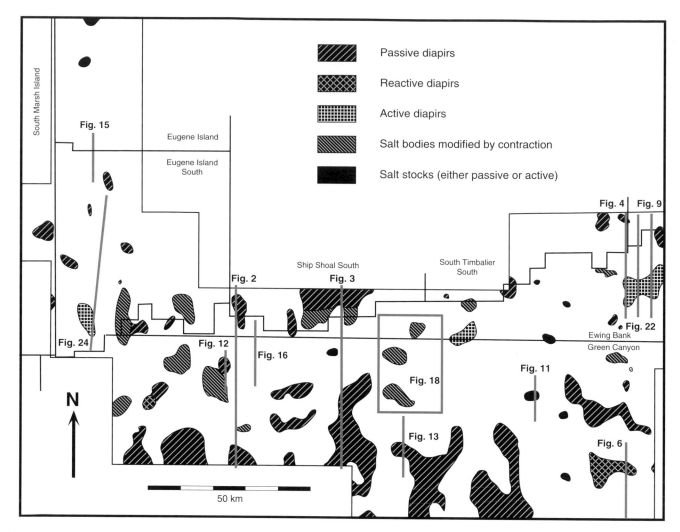

Figure 5—Map showing distribution of different types of salt bodies. Salt bodies shallower than 3.0 sec TWT (Figure 1) are classified as passive, reactive, or active diapirs; salt bodies modified by contraction; or salt stocks. The locations of illustrated seismic profiles and detailed maps are indicated.

Ma (Figures 7c, d) could be constructed using different assumptions to show salt at the sea floor and only 1.6 km of extension in the last 1.1 Ma (Figure 8). The corresponding evolution would suggest that a passive diapir having a steep northern flank became buried as extension rates increased after 0.5 Ma and that it subsequently grew by reactive means. This interpretation is considered improbable for several reasons. (1) It is inconsistent with the results of physical modeling (Vendeville and Jackson, 1992a), and (2) it is incompatible with the observed thickening toward the salt of sequences older than 0.5 Ma. (3) It requires that an originally vertical salt flank be modified by complex deformation into a gently inclined flank (compare Figures 8a, 7b), and (4) it shows that the salt body decreases in height between 0.5 and 0.4 Ma as the surrounding sediments are buried and compacted. Finally, (5) no mechanism (such as major changes in sedimentation rate) can be identified that would have triggered such a change in structural style and salt body growth.

Active Diapirs

The occurrence of active diapirism, involving uplift or piercement and pushing aside of overlying strata, is difficult to determine. Characteristic features of active diapirism should include a combination of the following features: elevation of the roof above regional datum, arching of adjacent beds on all sides of a salt body, crestal normal faults, flanking reverse faults, and a lack of thinning or facies changes in strata that have been intruded if these predate intrusion (Schultz-Ela et al., 1993). However, interpretation of such features can be ambiguous. First, strong deformation of the adjacent beds may be hidden by salt overhangs or, if visible, may be caused by contractional modification of a passive diapir (see below and Nilsen et al., 1995). Second, reverse faults may be reactivated later as normal faults, and third, some passive diapirs can also have negligible effects on sediment thickness or facies development because associated sea floor relief may be small and localized (see Figure 12).

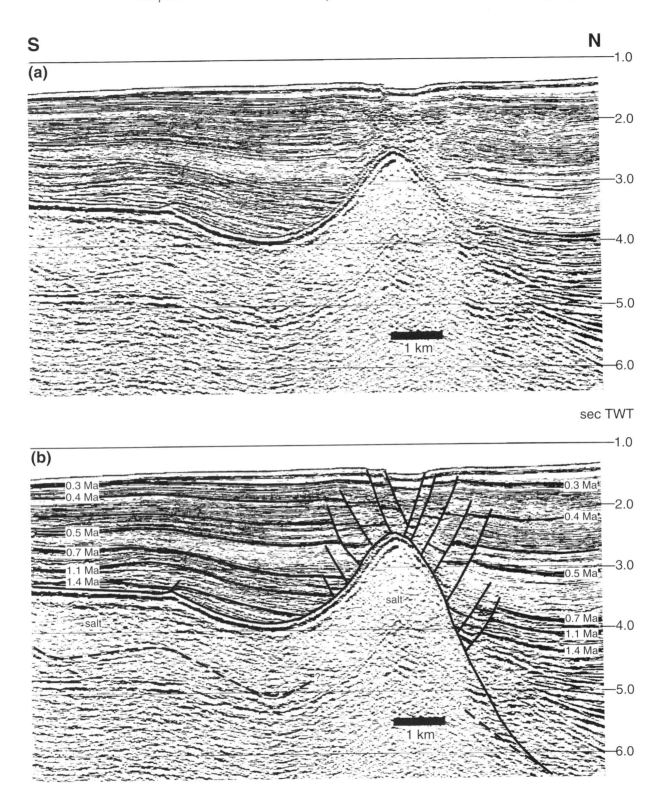

Figure 6—(a) Uninterpreted and (b) interpreted versions of a seismic profile displaying a reactive diapir at the updip margin of a tabular salt sheet (see Figures 1, 5, 27 for location). The salt body is overlain by a crestal graben and faults that get progressively older down the salt flanks. Stratigraphic interpretation by F. Budhijanto (1995); data courtesy of Halliburton.

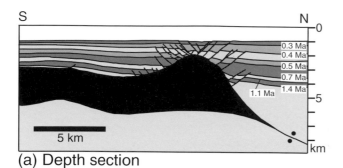

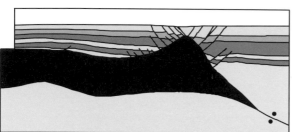

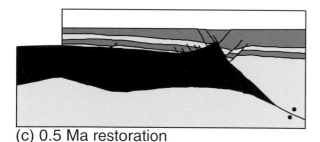

Figure 7—(a) True scale (1:1) depth section and (b–d) sequential restorations of the seismic profile in Figure 6. In the past 1.1 Ma, basinward sliding of strata above the salt sheet (shown by the change in length of the suprasalt section) has been accommodated by extension at the updip edge of the sheet and the resultant initiation and growth of a reactive diapir. Salt weld is indicated by pairs of dots. Depth conversion and restorations were carried out using GEOSEC, assuming that horizon cutoffs on either side of the salt body were originally in contact with each other.

There are several instances in the study area where salt bodies have pushed aside and raised overburden above its regional elevation (Figure 5), but no clear examples of true piercement occur. The salt body in Figure 9 is bound-

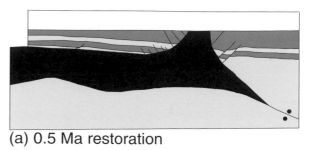

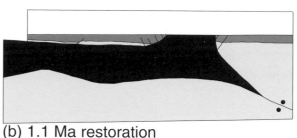

Figure 8—Alternative restorations of the seismic profile in Figure 6 to (a) 0.5 Ma and (b) 1.1 Ma. These were made by assuming that cutoffs of these horizons were originally separated by salt at the sea floor (compare with Figures 7c, d). Scale and symbols same as in Figure 7. This interpretation is considered unlikely.

ed on the north by a growth fault, and restoration shows that, above the salt body, the footwall of this fault has been pushed up and rotated during the last 0.5 Ma as salt was withdrawn from beneath the adjacent basin (Figure 10). Although growth on the fault is associated with subsidence of the hanging wall into the underlying salt, most of the extension was actually accommodated by the basinward displacement of the footwall above the salt sheet (7.0 km of movement in the last 1.1 Ma). The history of this movement is in sharp contrast to that interpreted for the sheet restored in Figure 7, where downdip gliding of the overburden created extensional faulting and reactive diapirism. In the example of Figure 10, active diapirism appears to have driven much of the downdip movement.

Possible candidates for true piercement diapirs in the study area are a series of narrow stocks located in a north-northwest trending deep trough in eastern Ewing Bank and Green Canyon (Figures 5). These are anomalous and unlike any other salt bodies in the region in that they are isolated, roughly cylindrical, and deeply rooted to depths between 5.5 and 7.5 sec TWT (Figure 11). There is some apparent deformation and very little thickness change in the surrounding and overlying sediments. The stocks appear to be located over deep salt rollers (Seni, 1992), suggesting that they may have originated as reactive diapirs, but whether their subsequent growth was by active or passive means cannot be determined from the observed geometries. Restoration cannot resolve the issue inasmuch as both types of history could be reconstructed depending on the methods used.

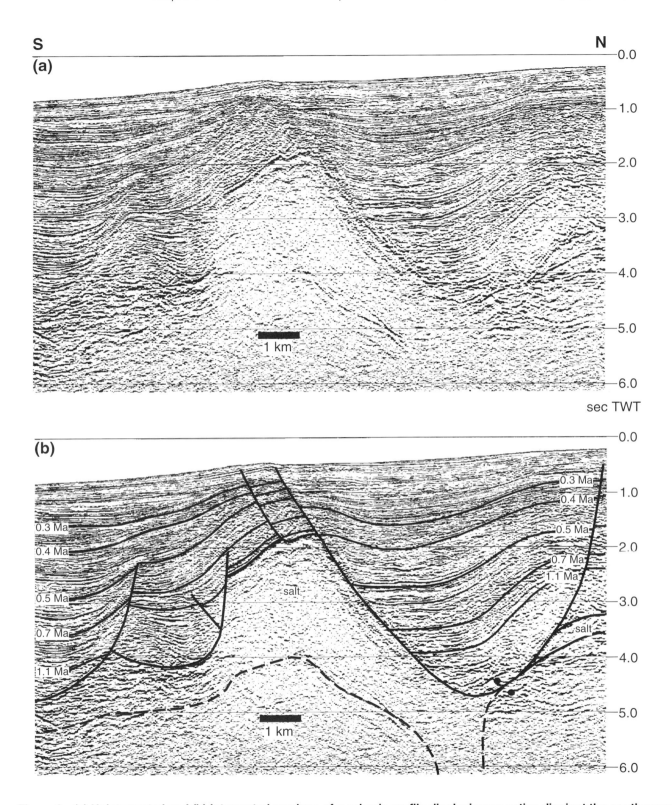

Figure 9—(a) Uninterpreted and (b) interpreted versions of a seismic profile displaying an active diapir at the southern edge of a circular minibasin (see Figures 1, 5, 27 for location; see also Figures 4, 22 across the same feature). The southern flank of the salt body is uplifted and rotated in the footwall of a north-dipping growth fault. Hints of the deep stem to this salt body are seen on this and adjacent profiles. Salt weld is indicated by pairs of dots. Stratigraphic interpretation by F. Budhijanto (1995); data courtesy of Halliburton.

S   N
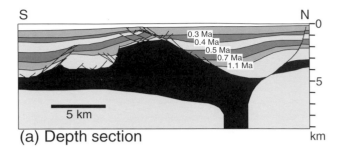

(a) Depth section

(b) 0.3 Ma restoration

(c) 0.5 Ma restoration

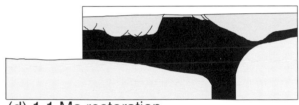

(d) 1.1 Ma restoration

Figure 10—(a) True scale 1:1 depth section and (b–d) sequential restorations of the seismic profile in Figure 9. During the last 1.1 Ma, depositional loading on the northern flank of the salt body resulted in uplift and rotation of the southern flank (active diapirism), widening of the salt body, and basinward translation of the footwall of the overlying growth fault (shown by the change in length of the suprasalt section). Note that the ramp fault at the north edge of the minibasin appears to have formed after the underlying salt weld. Depth conversion and restorations carried out using GEOSEC.

Passive Diapirs

Passive diapirs (those that grow by downbuilding) are common in the study area, in contrast to reactive and active diapirs (Figure 5). They vary considerably in both plan view and profile view geometries and in their bathymetric expression. The example in Figure 12 is symmetric in profile and has apparently overhanging flanks. The minor relief at the sea floor and only slight upturn of adjacent sediments above 1.5 sec TWT (the observed upturn is mostly a velocity effect) indicate a passive evolution in the last 0.5 Ma. More significant upturn of deeper strata, especially along the northern flank, suggests a possibly different early history.

Passive salt bodies are more commonly asymmetric, generally having one overhanging edge underlying a prominent scarp at the sea floor (e.g., Figure 13).

(below)

Figure 11—(a) Uninterpreted and (b) interpreted versions of a seismic profile displaying a cylindrical salt stock (see Figures 1, 5, 27 for location). At least some active diapirism is indicated by the folding and uplift of the 0.4 Ma horizon. Salt weld is indicated by pairs of dots. Stratigraphic interpretation by P. Varnai; data courtesy of Halliburton.

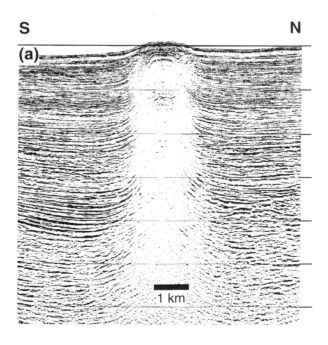

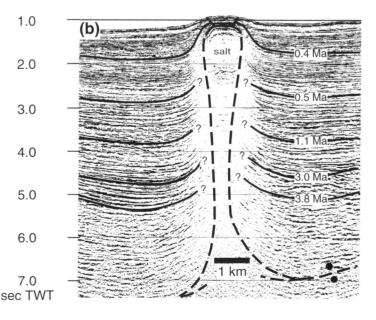

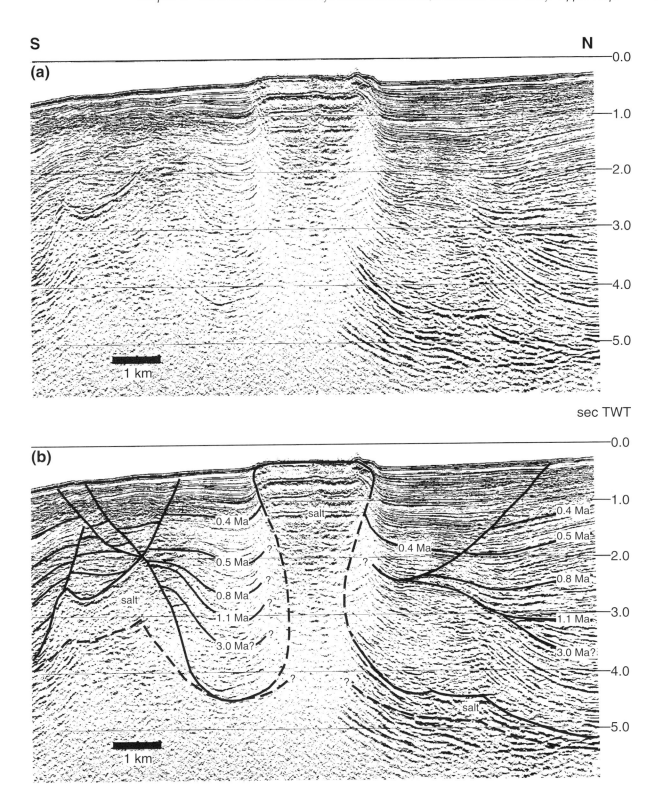

Figure 12—(a) Uninterpreted and (b) interpreted versions of a seismic profile displaying a symmetric passive diapir (see Figures 1, 5, 27 for location). The relative lack of uplift and folding of shallow strata (except for drag along the salt contact) indicates passive growth during the most recent evolutionary stages. More folding and erosional truncation at depth on the north flank suggest a possible earlier phase of contractional modification. The growth fault on the north flank is highly oblique to the section. Stratigraphic interpretation by A. Navarro (1994); data courtesy of Halliburton.

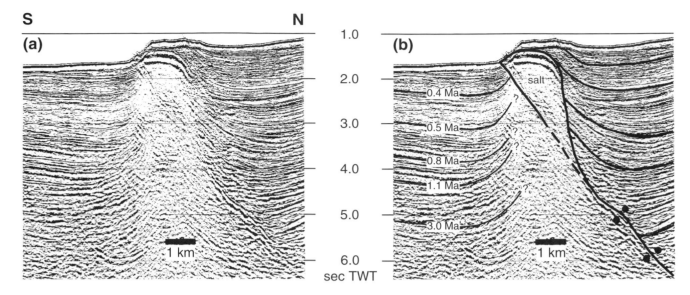

Figure 13—(a) Uninterpreted and (b) interpreted versions of a seismic profile displaying an asymmetric, passive diapir (see Figures 1, 5, 27 for location). The overhanging edge and the sea floor scarp are typical of the salt bodies in the southern part of the study area, although most bodies are wider and have a greater salt volume. Folding and uplift of strata on the north flank suggest a possible component of either active diapirism or contractional modification. Salt weld is indicated by pairs of dots. Stratigraphic interpretation by Z. Acosta (1994); data courtesy of Halliburton.

Restoration shows that the salt body remained at the sea floor and gradually moved basinward for at least the last 3.0 Ma (Figure 14). The geometry of the northward-dipping salt weld and the overlying sediments requires that the overburden move back to the north during restoration, that is, that the section shortened forward in time. This may be misleading, however, because of possible out-of-the-plane movement; the profile crosses a salt body that extends to the southeast (Figure 1), and both salt and overburden may have a transport component in this direction. Either way, the salt body grew primarily by passive diapirism. Other examples of restored asymmetric passive diapirs in the northern Gulf of Mexico are given by Worrall and Snelson (1989) and Schuster (1995).

The difference between passive salt bodies that grow vertically and those that grow asymmetrically to form overhangs may be related to sedimentation rates and patterns. Most overhangs are located on the downdip flanks of salt bodies, where sedimentation rates might have been relatively low. The downdip (southeastern) flank of the symmetric salt body in Figure 12, however, is on the margin of a major fault-bounded depocenter (Figure 1), where high sedimentation rates may have inhibited lateral flow of salt. Other cylindrical stocks are found almost exclusively in a deep north-trending depotrough in the eastern part of the study area (Figure 1).

Passive diapirs are generally located on shallow salt sheets or along the downdip and lateral flanks of elliptical minibasins (Figures 1, 5). Restorations throughout the study area show that most of the salt bodies evolved in a passive, downbuilding manner, at least since the oldest restored stages (Rowan et al., 1994a,b). Of 28 restored salt bodies, 1 formed by reactive diapirism, 2 can be classified as active diapirs, 17 grew passively, and 8 are passive

diapirs that were subsequently modified by contraction (see later). Growth of passive diapirs can be similar to the example in Figure 14 in which a pod of salt moves up and laterally as adjacent sediments impinge upon it. Growth can also occur as salt is evacuated from either tabular bodies or beneath major minibasins and gradually flows into flanking salt bodies that continue to grow at the sea floor (e.g., Worrall and Snelson, 1989; Seni, 1992; Rowan, 1993). Growth probably ceases once the deep source is depleted or closed off by welds or when sedimentation rates increase sufficiently to bury the salt body.

Discussion

Interpreted geometries and restorations both indicate an apparent scarcity of reactive and active diapirism in the study area. However, restorations generally illustrate only the most recent stages of salt body growth because neither seismic nor well data are typically adequate to constrain details of initiation and early growth. Thus, it is difficult to know whether the many examples of passive diapirs could have originally gone through reactive and active phases. Where initial overburden above a salt sheet was thin, any faults and sequence thickness increases characteristic of reactive diapirism may be below seismic resolution.

There are, however, several possible reasons why examples of reactive diapirism rarely appear in this area of widespread allochthonous salt. First, experiments reported by Vendeville and Jackson (1993) suggest that reactive diapirs grow only if deposition is slow relative to extension. When deposition is rapid relative to extension, growth faulting occurs without any associated growth of reactive diapirs. The high rates of deposition in offshore Louisiana may be enough in most cases to counter the

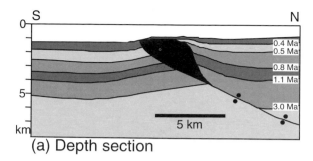

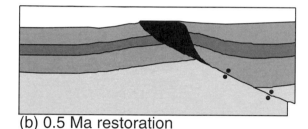

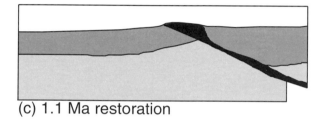

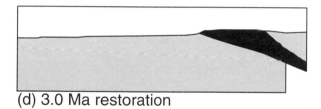

Figure 14—(a) True scale 1:1 depth section and (b–d) sequential restorations of the seismic profile in Figure 13. The restorations show that the salt body gradually moved up and laterally at the sea floor (passive diapirism) during sedimentary loading and 1.5 km of contraction over the last 3.0 Ma. The change in salt area (and possibly the apparent shortening) is probably due to out-of-plane flow (see text). Salt weld is indicated by pairs of dots. Depth conversion and restorations were carried out using GEOSEC.

effects of overburden extension above shallow salt sheets, such that reactive diapirism is suppressed.

A more likely explanation for the apparent scarcity of reactive diapirism is that the boundary conditions of the Vendeville and Jackson (1992a) experiments may be inappropriate for the analysis of secondary salt bodies that have developed above allochthonous salt sheets. In their models, a uniform layer of "salt" is first covered by "sediments" before extension is artificially induced and salt mobilization begins. Given these starting conditions, with a thick and brittle overburden, diapirism can initiate only

by reactive mechanisms. Allochthonous salt sheets, however, are probably emplaced at the sea floor (McGuinness and Hossack, 1993; Talbot, 1993; Fletcher et al., 1995), and those in the study area initially lacked much, if any, sedimentary overburden. As these sheets were buried by sedimentation, salt was redistributed. Some salt may have flowed to the front of the sheet as it advanced basinward at the sea floor, and some may have moved into secondary salt bodies that grew as a result of downbuilding between depositional lobes. Thus, most salt bodies in the study area can be explained as having developed passively, without significant extension. Locally, however, reactive diapirism may have occurred due to extension above a buried salt sheet (Figures 6, 7).

Growth Faults

Normal faults are also ubiquitous here, ranging from small-displacement, nearly planar faults to large listric growth faults. Small-displacement faults are commonly induced by bending of strata. They may occur in a variety of settings: within the hinge zones of large counter-regional salt withdrawal monoclines (Figure 3), along the crests of contractional anticlines (see Figures 17, 19), and in the hanging walls of larger growth faults (see Figure 16). In addition, small planar faults are common above tabular salt (Figures 9, 10) and are characteristic of reactive diapirs (Figure 6).

Larger growth faults have locations and orientations controlled by the morphology of the composite salt weld system. Three settings can be distinguished. First, systems of growth faults occur within and along the margins of tabular salt sheets or their equivalent salt welds. These are the *roho faults* of Sumner et al. (1991), Schuster (1995), and Diegel et al. (1995) and are characterized by listric geometries, moderate to strong rotation of beds, and salt rollers in the footwalls (Figure 15). In map view, the strike of these growth faults mimics sheet geometries, in which internal faults strike roughly perpendicular to the downslope extension direction and marginal faults commonly parallel the lateral edges of the sheet. Roho faults can accommodate significant extension and are rarely associated with reactive diapirs.

Large growth faults are also associated with the few active diapirs identified here. Uplifted and tilted strata on downdip flanks of such salt bodies form the footwalls to north-dipping growth faults that merge with the updip flanks of the salt (Figure 9). Unlike the counter-regional faults of Schuster (1995), these north-dipping faults can accommodate significant extension. The kinematics of this extension are anomalous: instead of the hanging wall subsiding and sliding basinward, as is the case for south-dipping growth faults, the footwall is elevated and rotated by active diapirism and is progressively displaced away from the hanging wall as the salt body widens (Figure 10).

The third type of relatively large-displacement growth faults occurs above the ramps that define the margins of elliptical depressions in the salt weld surface (Figure 16; see also Figures 2, 24). These faults, here called *ramp faults,*

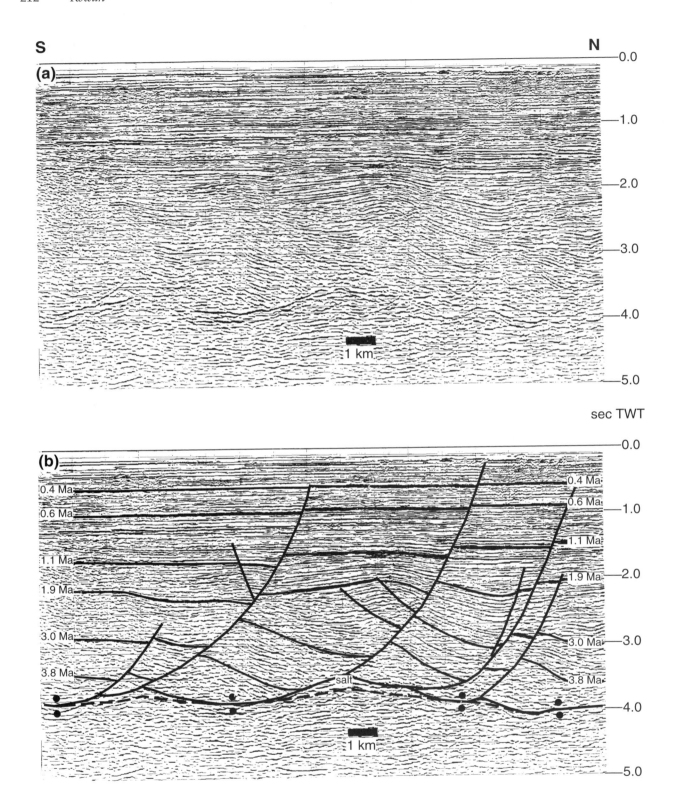

Figure 15—(a) Uninterpreted and **(b)** interpreted versions of a seismic profile displaying a roho-style (Sumner et al., 1991; Diegel et al., 1995; Schuster, 1995) fault system (see Figures 1, 5, 27 for location). The profile is oblique to the fault strikes but still shows the characteristic geometry. A series of down-to-the-basin listric growth faults sole into a level salt weld (indicated by pairs of dots). Hanging walls show prominent rollover; footwalls appear to have small salt rollers. Stratigraphic interpretation by P. Flemings and P. Weimer; data courtesy of Halliburton.

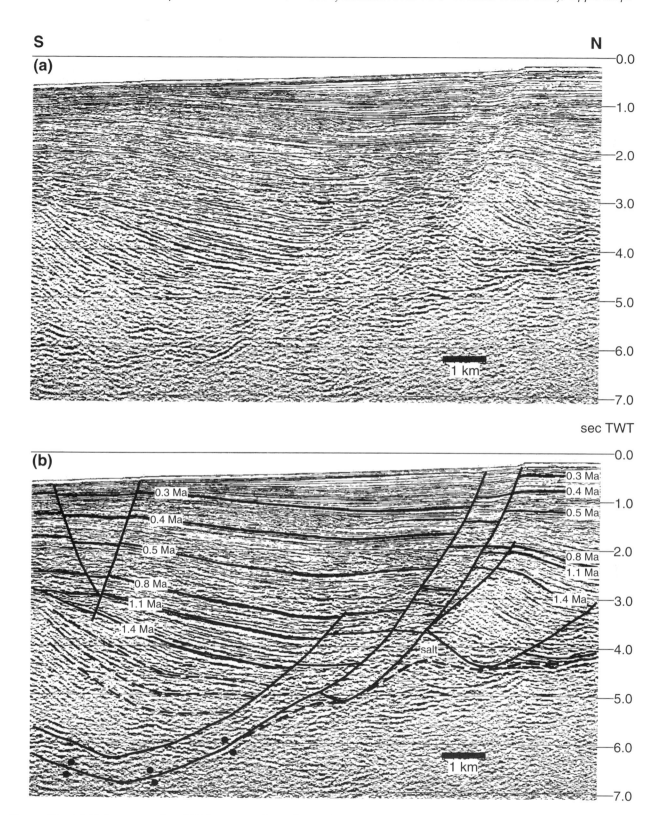

Figure 16—(a) Uninterpreted and (b) interpreted versions of a seismic profile displaying ramp faults. These are located over a south-dipping ramp in the salt weld (indicated by pairs of dots) that is underlain by subhorizontal reflections (see Figures 1, 5, 27 for location). Sequences younger than 1.4 Ma thicken into the fault system; older horizons (dashed lines) diverge away from the faults and into the basin center. Stratigraphic interpretation by A. Navarro (1994); data courtesy of Halliburton.

are strongly arcuate in plan view. The faults are found on the updip or lateral edges of six of the elliptical lows within the study area (Figure 1), thereby defining the landward margins of some minibasins. The absence of such faults around the perimeters of other elliptical lows apparently results typically from the encroachment of allochthonous salt sheets from the north. For example, the northeastern margin of the depression shown in Figures 4 and 9 has been partially overridden by a shallow sheet (see Figures 1, 22), and in Figure 17, contractional structures over the ramp may have inhibited growth faulting. Elsewhere, sedimentation patterns may dictate whether ramp faults form; they may be present only around major depocenters.

Contractional Structures

Contractional structures are being recognized with increasing frequency in allochthonous salt systems of the Gulf of Mexico (e.g., Huber, 1989; Hossack and McGuinness, 1990; Cook and D'Onfro, 1991; Sumner, 1991; Rowan et al., 1994a; Peel et al., 1995; Schuster, 1995). Thrust faults and associated contractional folds represent one of the ways in which extension above a salt sheet can be accommodated and balanced. In the study area, they usually form at the downdip edges of shallow sheets (Figure 1), either where the sheet terminates and overhangs a deeper level or where it is connected to a deeper level by a south-dipping ramp (Figure 3). More rarely, contractional structures are present south of some elliptical depressions (Figure 2).

The two anticlines illustrated in Figure 17 are interpreted as contractional in origin because of several features: (1) the distinctive asymmetric fold geometries, containing hanging walls raised above regional datum; (2) faint hints of low-angle faults that terminate at the synclinal hinge zones; (3) the location of these folds at the front of a salt sheet where the salt weld ramps down to the south (Figure 1); and (4) angular unconformities. These fold geometries and locations are similar to those of the thrusted fold documented by Huber (1989) in the study area.

The two apparent contractional structures in Figure 17 terminate laterally at salt bodies (Figure 18). These salt bodies are unusual and unlike most others here in that overlying and flanking sequences are strongly folded (Figure 19). The deformation is not just a local drag adjacent to the salt flanks but is expressed up to 3 km away. Restoration shows that the synclinal beds are only partially a result of subsidence into underlying salt. The beds have been rotated and uplifted above regional datum, especially along the southern flanks of the salt bodies (Figure 20). The restorations also show that the section has shortened through time, suggesting that the uplift and folding are due to contractional modification of preexisting salt bodies. This interpretation is supported by the proximity of these salt bodies to interpreted contractional folds and faults (Figure 18) and by the presence of the salt bodies near the front of a salt sheet. Other salt

bodies of similar geometry are usually found in similar locations (Figure 5). Conversely, salt bodies located within or along the updip and lateral margins of tabular salt, or along ridges between elliptical lows, are nowhere associated with this type of broad folding and uplift.

The shortening of the section through time (Figure 20) is a result of conserving bed length during restoration. If vertical or inclined simple shear had been used instead (see Rowan and Kligfield, 1989), the length of the section would remain constant. Thus, another possible interpretation is that these are true active diapirs and that the deformation was caused by buoyancy-driven piercement. Although a contractional origin is favored for the reasons stated earlier, the data and restorations are inconclusive. One way to distinguish these would be to use three-dimensional restoration to evaluate the strain compatibility between the folding around the salt bodies and over the adjacent thrust faults.

One of the implications of possible contractional modification of salt bodies is that vertical flow of salt can continue after depletion of the source layer (Nilsen et al., 1995). Salt is displaced upward as the salt body is squeezed and the overburden is folded. When carried to an extreme, contraction can close the stems of salt bodies so that adjacent sides are in contact, with the updip side often thrusted over the basinward side. Such structures, though rare, have been recognized in the study area (Figure 2) and in the Kwanza Basin of Angola (Duval et al., 1993).

EVOLUTION OF ALLOCHTHONOUS SALT

The geometries of salt bodies and growth faults described in preceding sections are compatible with two end-member models for the emplacement and evacuation of allochthonous salt. The first is the counter-regional model, in which allochthonous sheets are fed by north-dipping counter-regional salt stocks or walls (Schuster, 1995). The second is the salt stock canopy model of Rowan et al. (1994b), in which salt stocks feed overhangs that spread laterally and ultimately form allochthonous salt canopies. In the following sections, I will more fully describe these two end-members and argue that a combination of both models appears to be necessary to explain the observed geometries in the study area.

Counterregional Model

In the counter-regional model of salt sheet emplacement (Schuster, 1993, 1995), salt initially rises from the autochthonous layer, through south-leaning salt bodies, to flow basinward as tabular sheets (Figures 21a, b). Evacuation of salt from the feeders produces the counter-regional fault systems found along the northern margins of allochthonous salt sheets or their equivalent salt welds. Subsequent evacuation of the tabular salt sheets can occur

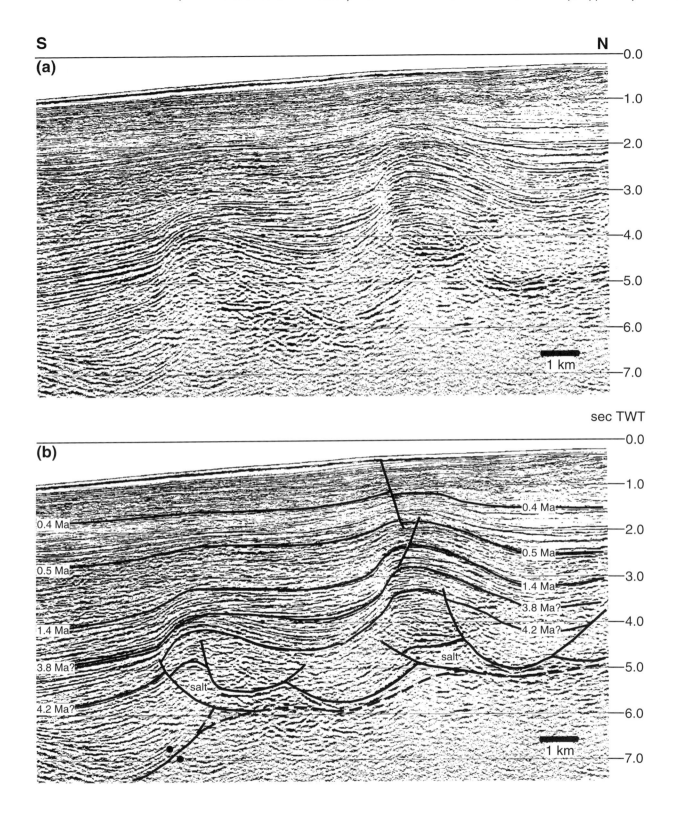

Figure 17—(a) Uninterpreted and (b) interpreted versions of a seismic profile displaying thrust faults and associated contractional anticlines (see Figure 18 for location). These structures are located over a south-dipping ramp in the salt weld (indicated by pairs of dots) that is underlain by subhorizontal reflections. The asymmetric fold geometries, uplift of the hanging walls above regional, erosional truncation, location at the southern edge of a shallow salt sheet, and faint hints of thrust faults all suggest a contractional origin. Stratigraphic interpretation by Z. Acosta (1994); data courtesy of Halliburton.

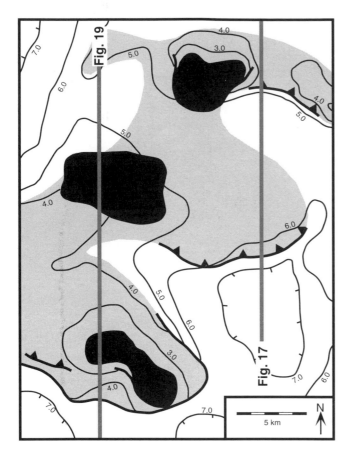

Figure 18—Detailed map of an area of contractional structures (see Figures 1, 5, 27 for location). The map shows structure contours (contour interval 1.0 sec TWT) on the top of salt or the equivalent salt weld. Salt above 3.0 sec is black, approximately tabular salt is gray, and salt welds (salt <100 msec thick) are white. Thick lines are salt overhangs; those with barbs indicate contractional thrusts and folds. Note how contractional structures terminate laterally toward salt bodies. The locations of Figures 17 and 19 are indicated.

in a variety of ways that are limited by two end-members. In *stepped counter-regional systems,* salt within the tabular sheet is gradually displaced into south-leaning salt bodies at the downdip margins of the sheets (Figure 21c). Evacuation of these salt bodies creates younger, shallower counter-regional systems overlain by prominent south-dipping rollovers of supra-salt strata (Figure 21d). In contrast, *roho systems* are marked by south-dipping listric growth faults that sole into the salt layer or weld (Figure 21e). Displaced salt flows laterally and downdip into salt bodies and may be extruded or dissolved at the sea floor.

Examination of the examples presented by Schuster (1995) shows that two key elements characterize counter-regional systems: (1) allochthonous salt is connected to deeper levels only along feeders that dip approximately to the north (Figure 21), and (2) counter-regional systems are dominated by areally extensive, subhorizontal tabular salt sheets or welds, often with remnant salt bodies con-

centrated along the lateral or downdip margins. There is little structural relief on the base salt or the equivalent weld other than that provided by the north-dipping feeders (Figure 21), in contrast to the geometries characteristic of the salt stock canopy model discussed next.

Salt Stock Canopy Model

The counter-regional model (Schuster, 1995) nicely accounts for the geometries of allochthonous salt observed in large parts of the Louisiana shelf and coastal regions. However, many of the features characteristic of the study area require another mode of emplacement. These features are (1) the elliptical depressions in the deep salt weld bounded by ramps that dip in all directions, (2) the presence of salt bodies over saddles between the depressions, (3) the relative lack of areally extensive level sheets or welds, and (4) the scarcity of large-displacement growth faults. Another model is needed to explain the origin and evolution of salt systems characterized by elliptical lows in deep salt welds.

Restorations of two profiles show different aspects of the evolution of elliptical depressions. A seismic line across one circular low in the northeastern corner of the study area is shown in Figure 22 (see also Figures 4, 9, 10). Restorations illustrate the gradual loading and evacuation of a bulb-shaped salt stock to form a circular minibasin (Figure 23); its stem is speculative. Salt withdrawal was asymmetric, occurring initially along the northern margin of the stock. Between 3.8 and 3.4 Ma, a shallow salt sheet advancing from the northeast overrode this margin. By 1.9 Ma, the asymmetry of the minibasin switched, possibly because of weld formation along its northern flank. Subsequent withdrawal of salt from the southern part of the remnant stock created a south-dipping growth wedge of sediments. The evacuated salt flowed into a marginal salt body that grew, at least in part, by active diapirism (see Figure 10).

The second restored profile (Figure 24), in the northwestern part of the study area (Figure 1), crosses another elliptical low with a salt weld similar to that of the previous example. During the last 3.8 Ma, extension along growth faults bounding the northern margin of the minibasin has been accommodated downdip both by flow of salt into a flanking body that widened and moved basinward and by folding and uplift of the overburden (Figure 24). Although the early history cannot be reconstructed due to a lack of deep well control, the elliptical shape of the depression (Figure 1), based on analogy with the previous example (Figure 23), suggests an origin as a broad, bulb-shaped stock.

Detailed analysis of the restoration (Figure 24) reveals systematic patterns in the timing and rates of extension and contraction. Listric, roho-style growth faults north of the minibasin were most active prior to 1.9 Ma, creating prominent rollovers and growth wedges (the southernmost of these can be seen at the northern end of Figure 24). The ramp fault at the northern margin of the elliptical low, in contrast, was most active between 1.9 and 0.4 Ma

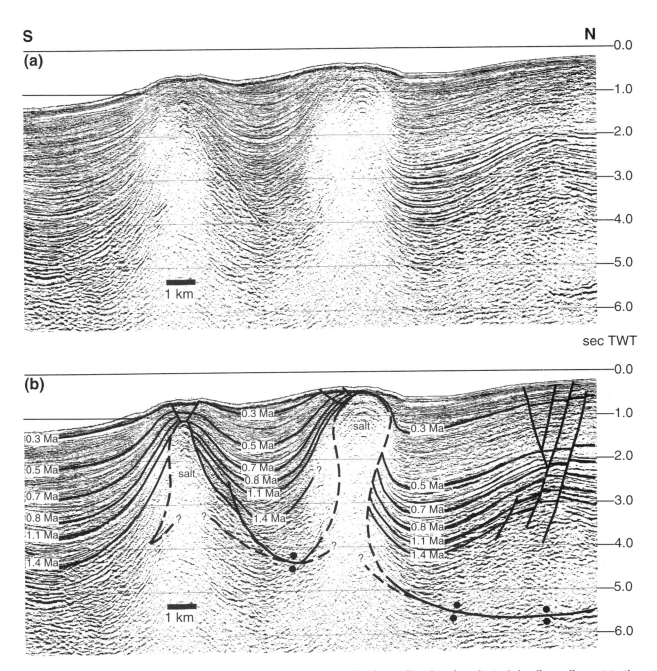

Figure 19—(a) Uninterpreted and (b) interpreted versions of a seismic profile showing the salt bodies adjacent to thrust terminations (see Figure 18 for location). The associated sediments are strongly uplifted and folded, which is thought to have resulted from contraction at the southern edge of a shallow salt sheet (see restoration in Figure 20). Salt welds are indicated by pairs of dots. Stratigraphic interpretation by Z. Acosta (1994); data courtesy of Halliburton.

(Figure 25); associated footwall splays and a prominent antithetic fault formed after 1.4 Ma. Contractional folding over the southern salt body was most active between 1.9 and 0.8 Ma (Figure 24), suggesting that it was a response to extension along the ramp fault.

The younger age of the ramp fault relative to the roho-style faults to the north could simply be due to the progradational basinward shift of depocenters. Another possibility, however, is related to the timing of salt evacuation and weld formation. Roho-style faults typically become less active once the underlying tabular salt is evacuated (Figure 24) (F. A. Diegel, personal communication, 1994). In contrast, extension rates on ramp faults appear to accelerate (Figure 25) roughly coeval with the timing of weld formation along the ramp (Figures 10, 24). Admittedly, an earlier history of more rapid extension on the ramp fault cannot be precluded because of the lack of deeper well data and accurate correlations. However, the geometries of strata older than 3.8 Ma (Figure 24a), if carried through the restorations, would show subhorizontal

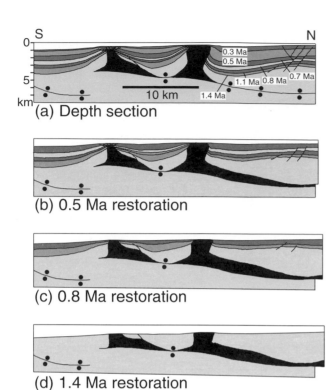

(a) Depth section

(b) 0.5 Ma restoration

(c) 0.8 Ma restoration

(d) 1.4 Ma restoration

Figure 20—(a) True-scale 1:1 depth section and (b–d) sequential restorations of the seismic profile in Figure 19. The restorations suggest that deformation around the salt bodies is due to a combination of salt withdrawal and contractional folding. Uplift and rotation are especially strong on the southern flanks of both salt bodies. Salt welds are indicated by pairs of dots. Depth conversion and restorations were carried out using GEOSEC.

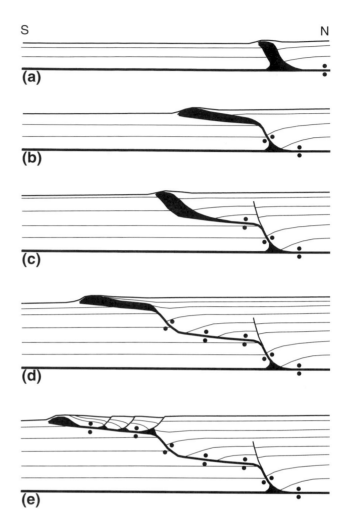

Figure 21—Schematic evolution of counterregional allochthonous salt systems, combining both the stepped counterregional (middle level) and roho (upper level) end-member models of Schuster (1993, 1995). Counterregional systems are characterized by north-dipping feeders and subhorizontal salt sheets (see text). Salt is black, welds are indicated by pairs of dots, and bedding geometries are shown by thin lines. No ages are inferred, and the section is not to scale.

dips and a lack of growth wedges at 3.8 Ma, consistent with an interpretation of little, if any, early extension on the ramp fault. The restorations show that rollover of the older strata was caused by the entire suprasalt section moving basinward over the ramp on the salt weld because of extension on faults located farther north (Figure 24c–e). The general pattern in elliptical lows is that deep strata show subhorizontal to ramp-parallel dips rather than rollover into the ramp (Figures 4, 9, 22). The implication is that early deformation was dominated by vertical subsidence into the salt rather than by growth faulting. For example, the ramp fault in Figure 16 switched from simple subsidence to hanging wall rollover just prior to 1.4 Ma (although it is not known whether this corresponds to weld formation).

Evacuation of salt from bulb-shaped stocks creates minibasins floored by elliptical lows in salt welds (Figure 23). The dips of the resulting weld ramps are thus interpreted as relics of the original overhanging interfaces between salt bodies and surrounding sediments and not as products of later deformation. This interpretation is compatible with observed subsalt reflections that are generally horizontal (Figures 4, 9, 16, 22).

A conceptual model for the evolution of a bulb-shaped stock into a bowl-shaped minibasin is shown in Figure 26 (see also Rowan et al., 1994b). The original salt stock (Figure 26a) grows by downbuilding, gradually expanding laterally due to relatively slow sedimentation rates or fast salt flow rates (Figure 26b). Models (Jackson and Talbot, 1991; Talbot, 1993, 1995; Fletcher et al., 1995) suggest that as long as there is an adequate supply of salt flowing up the feeder, the salt body forms an active salt fountain that is topographically elevated, with only a thin veneer of sediments draping the top (Figure 26a, b). Once the deep source is depleted or the feeder is shut off, the topographic high collapses gravitationally through a combination of gravity spreading and gravity gliding, thereby initiating basin formation and driving further extrusion of the toe over the sea floor (Figure 26c)

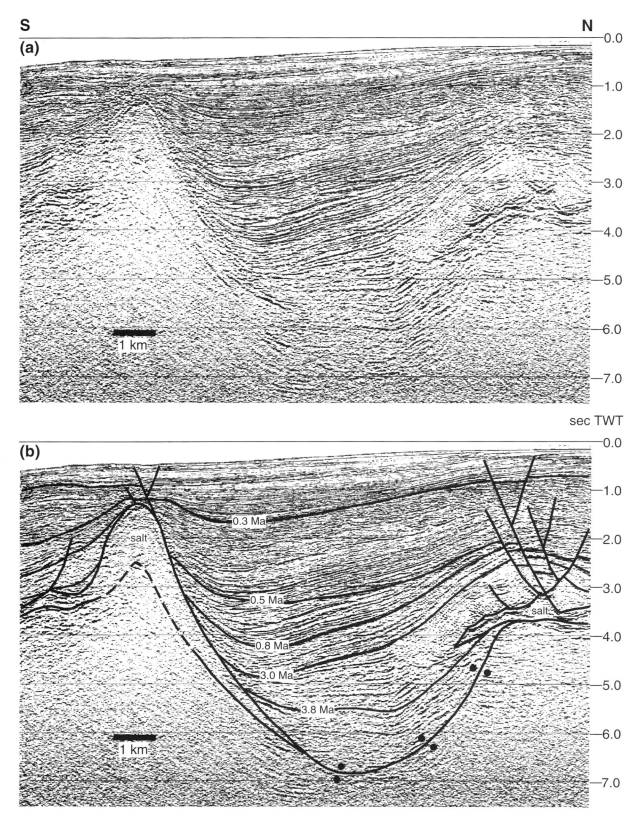

Figure 22—(a) Uninterpreted and (b) interpreted versions of a seismic profile displaying a bowl-shaped minibasin (see Figures 1, 5, 27 for location; see also Figures 4, 9 across the same feature). The minibasin is bounded on the south by a salt body and a thin salt sheet and on the north by a south-dipping ramp with underlying horizontal reflections. A shallow salt sheet has partially overridden the northern margin of the minibasin. Salt weld is indicated by pairs of dots. Stratigraphic interpretation by F. Budhijanto (1995); data courtesy of Halliburton.

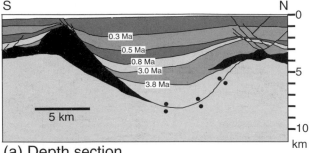

(a) Depth section

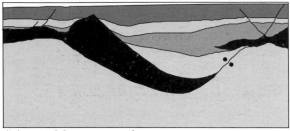

(b) 0.5 Ma restoration

(c) 3.0 Ma restoration

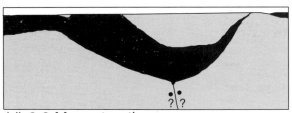

(d) 3.8 Ma restoration

(e) L. Miocene (?) restoration

Figure 23—(a) True-scale 1:1 depth section and (b–e) sequential restorations of the seismic profile in Figure 22. The restorations show the evolution of a bulb-shaped salt stock into a bowl-shaped minibasin after the deep salt source was depleted or cut off. Sedimentary loading and subsidence of the minibasin were asymmetric, initially occurring over the northern margin and then over the southern margin as salt was evacuated into the adjacent salt body and shallow sheet. Salt welds are indicated by pairs of dots. Depth conversion and restorations were carried out using GEOSEC.

(Fletcher et al., 1995). Displaced salt flows laterally and basinward into flanking diapirs that form topographic highs, so that the basin initially grows by salt withdrawal and aggradation (Figure 26d). Once the salt is evacuated, the basin can no longer grow vertically, and accelerated extension on ramp faults may then provide accommodation space for continued sedimentation as the basin grows laterally (Figure 26e). Any extension is balanced by contractional folds and thrusts, salt extrusion and possible dissolution, or the lateral squeezing and contractional modification of existing salt bodies. Because of the structural relief on the base of the minibasin, however, extension and associated contraction are limited, and the basin rapidly fills up and is bypassed as a local depocenter (Figure 26f).

Although reactive diapirism may trigger initial salt body and basin growth as the salt fountain collapses, differential loading quickly appears to become the dominant process. Jackson and Vendeville (1994) argued that sediments must be carbonates or be funneled into the same area for sedimentary loading to initiate and drive early salt body growth. Because uncompacted siliciclastic overburden is initially less dense than salt, the early overburden forms a topographic high while the surrounding salt is lower. Thus, they propose that the effects of any early differential loading are countered by deposition over the flanking salt. However, this argument is based on the assumption that salt is in a state of equilibrium. Salt bodies adjacent to growing depocenters are not static; dynamic flow of salt may keep flanking salt bodies elevated above the proto-basins irrespective of the density contrasts. Furthermore, where there are differences in sediment thickness or water depth above salt, the density contrast between sediments and sea water will dominate that between the sediments and salt, driving salt mobilization (Jackson and Talbot, 1986; F. A. Diegel, personal communication, 1994).

Separate bulb-shaped stocks that are close enough to one another will link up during lateral spreading to form salt stock canopies (hence the name of the model). The mapped geometry of deep salt welds throughout much of the study area is compatible with a canopy origin. The various elliptical lows represent the initially isolated bulb-shaped stocks, and the intervening saddles represent locations where adjacent salt bodies merged. The depth to the saddles, which generally varies between 4.0 and 6.0 sec TWT (Figure 1), depends on a combination of the spacing of the original salt bodies and the rates of lateral spreading. The salt stock canopy model is consistent with a composite salt body mapped by McGuinness and Hossack (1993, their figure 18), in which the base salt has three elliptical lows separated by saddles. When the salt is eventually evacuated from this canopy, the resultant weld will have a geometry similar to those mapped throughout the study area (Figure 1).

In the salt stock canopy model, salt displaced from bulb-shaped stocks eventually ends up in salt bodies located either over the saddles between adjacent lows or along the frontal or lateral edges of salt tongues that spread out from the canopy (Rowan et al., 1994a). An

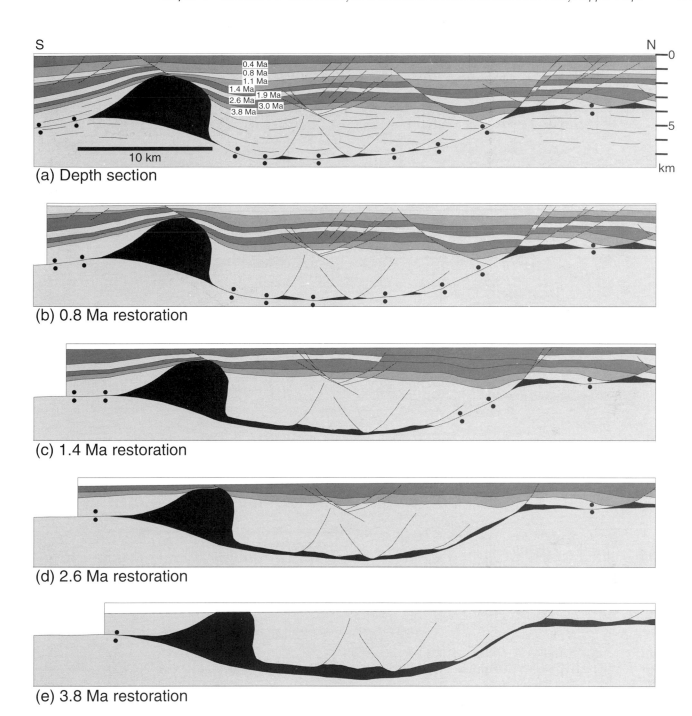

S   N

0.4 Ma
0.8 Ma
1.1 Ma
1.4 Ma 1.9 Ma
2.6 Ma 3.0 Ma
3.8 Ma

10 km

(a) Depth section

(b) 0.8 Ma restoration

(c) 1.4 Ma restoration

(d) 2.6 Ma restoration

(e) 3.8 Ma restoration

Figure 24—(a) True-scale 1:1 depth section and (b–e) sequential restorations of a composite section constructed from parallel and adjacent north-south seismic profiles (see Figures 1, 5, 27 for location). From north to south, the depth section shows a subhorizontal salt weld overlain by roho-style faulting, a ramp fault over the northern margin of an elliptical minibasin, and a large allochthonous salt body at the southern edge of the minibasin. The oldest restorations show the last phases of salt withdrawal from beneath the minibasin. Because of the two-dimensional nature of section restoration, extension on the roho and ramp faults is shown to be accommodated downdip by impingement of the minibasin into the salt body, which in turn grows actively by the uplifting, rotation, and downdip sliding of strata above its southern flank. However, some extension is possibly accommodated out of the plane of the section. Salt welds are indicated by pairs of dots. Data courtesy of Halliburton and TGS; depth conversion and restorations carried out using GEOSEC.

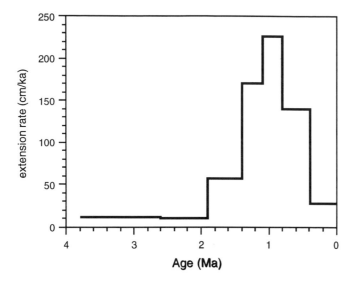

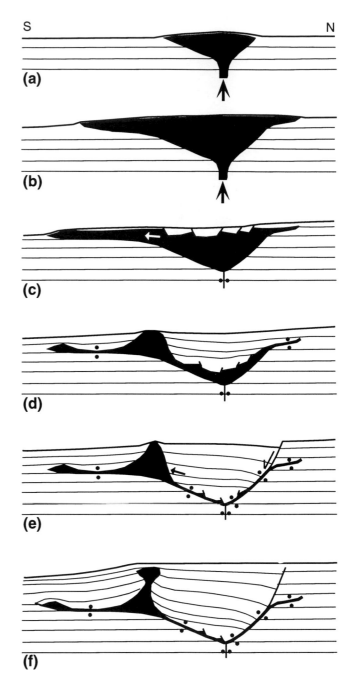

Figure 25—Plot of extension rate versus time for the ramp fault system of Figure 24. Extension was measured from the restorations by summing the heaves on the ramp fault and its associated footwall splays and hanging wall antithetic fault. The acceleration in extension rate appears to be roughly coeval to salt weld formation.

important ramification is that most salt bodies in the study area are secondary allochthonous features lacking deep roots. The saddles, or highs in the base of salt surface, are believed to represent the sutures in the original salt canopy. The old feeders are typically not seen but are speculated to lie beneath the deep lows of the salt weld, that is, below the deepest minibasins (see Figure 9). Lacking deep data, it is impossible to determine whether the inferred feeder stocks were passive products of downbuilding or developed as reactive diapirs during regional extension.

The salt stock canopy model is similar in many respects to the salt sheet deformation model published recently by House and Pritchett (1994) and Hodgkins and O'Brien (1994). Using modern seismic processing techniques such as prestack depth migration, they illustrate salt geometries from the outer shelf west of the study area (W. M. House, personal communication, 1994) that appear to represent various stages of the canopy model. For example, they show the initial bulb-shaped salt stock, partial collapse and basin formation, and complete evacuation represented by a dipping salt weld overlying horizontal subsalt reflections. Although they envision an evolution similar to that depicted in Figure 26, two major aspects of their model differ from the salt stock canopy model. First, they propose that salt body subsidence may be due to regional extension on a deeper salt level as well as to possible gravity spreading. Second, salt sheets in their model are thought to originate as sills intruded into surrounding sediments rather than as salt glaciers extruded at the surface.

Figure 26—Schematic evolution of (a) a bulb-shaped salt stock into (f) a bowl-shaped minibasin and downdip salt tongue. Merger of salt from two or more bulb-shaped stocks during either initial growth or subsequent loading creates allochthonous salt stock canopies that evolve into salt weld systems having significant structural relief. Salt is black (flow from a deep source is shown by vertical arrows), welds are indicated by pairs of dots, and bedding geometries are shown by thin lines. No ages are inferred, and the section is not to scale.

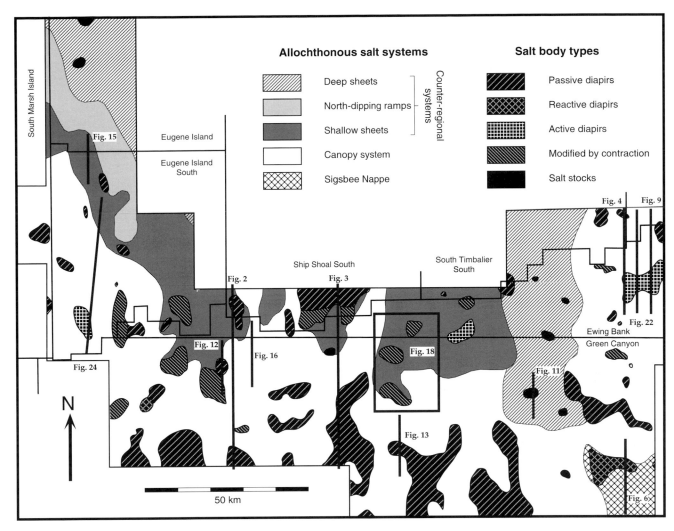

Figure 27—Map showing distribution of allochthonous salt systems and different types of salt bodies (from Figure 5). The boundaries between regions dominated by counterregional, salt stock canopy, and Sigsbee systems are approximate (see text). The locations of illustrated seismic profiles and detailed maps are indicated.

Discussion

Salt geometries and structural features in the study area suggest that both the counter-regional and salt stock canopy models are required to explain the full range of structural styles. Neither end-member by itself can account for the observed variation in salt weld morphology, the geometry of remnant salt bodies, and the abundance and shape of growth faults. The study area has been divided into regions in which one or the other model is dominant (Figure 27). The counter-regional end-member has been subdivided into deep salt sheets, major northward-dipping ramps, and shallow sheets. The boundaries on the map are approximate and subjective, in that the different systems merge in some areas.

Interpreted regional profiles show that the salt sheet at the southeastern corner of the map is the northwestern corner of a shallow sheet that extends southeastward to the Sigsbee Escarpment as the Sigsbee nappe (West, 1989;

Wu, 1993; Seni, 1994). Its relationship to the elliptical low just to its north (Figures 1, 6) suggests that this sheet formed, at least in part, from salt evacuated from a bulb-shaped salt stock. The size and extent of the canopy mapped by Seni (1994) suggest that it is probably fed by other stocks located to the east of the study area. Some might be bulb-shaped stocks, but others might be counter-regional feeders. For this reason, the Sigsbee system is differentiated from the counter-regional and salt stock canopy systems (Figure 27).

Comparison of Figure 27 with the salt structure contour map (Figure 1) shows that counter-regional systems are characterized by areas of relatively level tabular salt or welds, whereas salt stock canopy systems are dominated by deep salt welds having significant structural relief. The salt or weld morphology, in turn, determines the patterns of growth faulting. Counterregional systems contain abundant roho-style faults that may accommodate significant extension, whereas salt stock canopy systems typi-

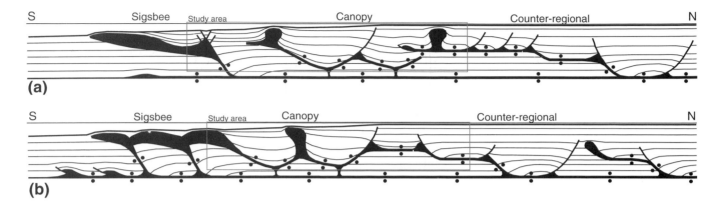

Figure 28—Speculative schematic cross sections extending from the Louisiana coastline to the abyssal Sigsbee Escarpment showing typical geometries and relationships of counterregional, salt stock canopy, and Sigsbee salt systems. The systems are shown as (a) separate and (b) connected; neither example necessarily represents geometries observed along any one seismic profile. The outline indicates the data available in this study. Geometries on the north are adapted from Schuster (1995) and those on the south from Simmons (1992). The sections are not to scale, either horizontally or vertically.

cally display only minor extension along arcuate ramp faults. There is also an apparent relationship between the structural styles of salt bodies and the different types of allochthonous salt systems in the study area (Figure 27). (1) Reactive diapirs are found at the updip margins of shallow sheets (part of the Sigsbee Nappe), and (2) active diapirs occur within the salt stock canopy system (just basinward of major elliptical lows). (3) Most circular salt stocks are rooted in deep sheets. (4) Salt bodies modified by contraction are common along the southern margins of shallow sheets, although some also occur downdip from canopy minibasins. (5) Passive diapirs are found in all environments. Similar relationships are seen in seismic profiles west of the study area in Garden Banks and East Breaks (Fiduk, 1994).

Within the study area is a rough north-south division of allochthonous salt systems. Counterregional systems, having both deep and shallow levels of salt sheets, characterize the north. To the south, the region is dominated by salt stock canopy systems. In the southeastern corner is the northern edge of the Sigsbee nappe. This pattern is consistent with data and interpretations published by other workers. To the north, much of the central Louisiana shelf appears to contain counter-regional systems (Schuster, 1993; 1995); to the south, salt of the lower slope makes up the allochthonous sheets of the Sigsbee nappe system (e.g., West, 1989; Worrall and Snelson, 1989; Wu et al., 1990; Simmons, 1992; Seni, 1994). To the west, remnant salt stock canopies have been identified that are along trend with some of the elliptical lows in the study area (Hodgkins and O'Brien, 1994; House and Pritchett, 1994).

The apparent north-south division of allochthonous salt systems is illustrated by two speculative and schematic cross sections extending approximately from the Louisiana shoreline to the Sigsbee Escarpment (Figure 28). Neither is intended to display the relationships actually existing on a single real profile; rather they combine and illustrate some of the geometries and styles

of allochthonous salt systems reported here and elsewhere. Furthermore, the sections represent only the central Louisiana offshore. In one section (Figure 28a), the different systems are separate. To the north is a counterregional system that steps up twice to a shallow roho system; the contractional toe of the shallow sheet has overridden a canopy system having a ramp fault and a salt body modified by contraction. To the south, the distinct Sigsbee nappe consists of a single sheet that has advanced southward past the original depositional limit of the Louann Salt. In the second section (Figure 28b), the various elements have merged into one composite allochthonous salt system. The southern of two stepped counter-regional systems feeds a shallow salt sheet; this has merged with the northern edge of a canopy system containing a remnant salt body over the saddle between two lows. A salt sheet sourced from the canopy has merged with counter-regional systems to form an amalgamated Sigsbee nappe, and the Sigsbee is overriding the Mississippi Fan foldbelt at the depositional edge of the autochthonous salt.

There is considerable uncertainty at depth. Counterregional feeders for the Sigsbee nappe in Figure 27 are shown to root in the autochthonous salt (based on observations south of the study area by Simmons, 1992), but they could also have originated from a deep allochthonous sheet (Diegel and Cook, 1990; Diegel et al., 1995). Similarly, the salt stock canopy systems in the study area may be sourced from a deep allochthonous salt sheet or directly from the Louann Salt. Because of the depth of the elliptical lows (>7.0 sec TWT) (see Hodgkins and O'Brien, 1994), the salt stocks are shown rooted in the autochthonous layer.

One of the implications of this composite model is that salt geometries observed in the lower slope, upper slope, and shelf do not necessarily represent immature, intermediate, and mature stages, respectively, in the evolution of allochthonous salt. For example, the salt ridges and massifs of northern Green Canyon are unlikely to evolve into salt geometries typical of mature counter-regional systems because the massifs are forming from a salt stock canopy having a different initial configuration of salt. Restorations of counter-regional systems from the inner shelf (Schuster, 1995) do not show earlier stages that resemble those seen today in much of the study area.

The regional salt map of Diegel et al. (1995, their figure 1) supports the division of the study area into counter-regional and salt stock canopy systems. Although these authors showed that most of the Louisiana shelf is characterized by linear regional and counter-regional growth faults, southern Eugene Island displays a distinct pattern of highly arcuate faults. I have classified this part of the shelf as a salt stock canopy system (Figure 27), dominated by elliptical minibasins and arcuate ramp faults (Rowan et al., 1994b). Thus, the salt and fault geometries of southern Eugene Island, rather than those typical of most of the shelf, represent the mature evolutionary stage of the salt systems presently found in northern Green Canyon.

ECONOMIC IMPLICATIONS

The structural styles and evolution of allochthonous salt have important ramifications for hydrocarbon exploration. For example, the various mechanisms of salt body growth have different effects on deformation of adjacent sediments and therefore on trap geometries. Passive diapirism has relatively little effect on bedding geometries, while reactive diapirism creates potential fault traps. Active diapirism and contractional modification create anticlinal folds and steep dips adjacent to the sealing salt. If the mechanism of salt body growth has changed with time, trap geometries probably vary with depth.

Salt body growth also influences syntectonic sedimentation. Different styles of diapirism have substantial effects on both facies development and thickness changes, primarily because of their variable bathymetric expression. Reactive diapirs create overlying fault-bounded grabens (Figure 6). Passive diapirs may create little sea floor relief (Figure 12), but are more often characterized by a steep scarp on one flank and a gentle slope on the other (Figure 13). Salt bodies modified by contraction create significant highs that extend away from the salt bodies themselves (Figure 19). Salt bodies that stop growing and become inactive, either because of the depletion of their source or the termination of tectonism, gradually become buried and have no topographic expression.

Thus, the bathymetry at any time is controlled by the distribution, size, and type of salt bodies that are currently growing. In turn, this relief plays a critical role in facies development by controlling the location of depocenters, the geometry of the sediment transport system, and the presence of scarps or slopes that may fail and generate slumps. It also helps to determine the distribution of sand within a minibasin. For example, although sands will rapidly pinch out on the flanks of a salt body that is rising because of regional contraction, there may be little change in thickness adjacent to a passive diapir and sands may thicken over a reactive diapir. Preliminary results in the study area indicate a strong correlation between facies development and salt tectonics (Weimer et al., 1994). For example, the presence of slumps is associated in both time and space with major phases of contractional modification of salt bodies in north-central Green Canyon and central Ewing Bank (Acosta, 1994).

The evolutionary models for the growth of counter-regional and salt stock canopy systems also have important ramifications for subsalt exploration. First, structural closure of salt welds, either within canopy systems or at the junction between canopy and counter-regional systems (Figure 28), provides subsalt highs that may be attractive exploration targets. The general trap and seal geometry is favorable, and hydrocarbon migration pathways may be focused into such areas. Second, the geometry and evolution of a given salt system can be used to predict the presence of sands beneath shallow salt. Salt glaciers that spread out from a source salt body during a period of slow sedimentation or fast salt movement will flow into topographic lows in the sea floor, between adjacent salt highs. This is exactly where sands may have been preferentially deposited during preceding low-stands. This will not always be true for every shallow salt sheet because facies depend on interactions between salt and the sediment transport system. Nevertheless, careful analysis of the salt system geometry and reconstruction of its evolutionary history can be powerful tools in the successful exploration for subsalt hydrocarbons.

CONCLUSIONS

1. Two end-member models of allochthonous salt systems are necessary to explain the range of structures in the central Louisiana outer shelf and upper slope: the counter-regional model of Schuster (1993, 1995) and the salt stock canopy model of Rowan et al. (1994b).
2. Counterregional systems have regionally extensive tabular salt sheets or salt welds that are connected to deeper levels at their updip margins by north-dipping collapsed feeders. The bases of the sheets may climb gradually both laterally and downdip but otherwise have little structural relief.
3. Salt stock canopy systems form by the lateral spreading and eventual amalgamation of isolated bulb-shaped salt stocks. Gravitational subsidence of elevated salt highs leads to basin initiation; subsequent sedimentary loading drives salt evacuation and creates minibasins floored by elliptical bowl-shaped salt welds.
4. In both systems, salt flow and growth occur at or near the sea floor. Because salt does not have to move through any significant thickness of over-

burden to form salt bodies, downbuilding (passive diapirism) is the dominant process of salt body growth.

5. Reactive diapirism appears to be rare, confined to the landward edges of shallow salt sheets having extended overburden. This does not preclude possible reactive initiation of salt body growth that cannot be resolved on seismic data.

6. Active diapirism is probably rare and, where documented, is typically characterized by the uplift and rotation of one flank of a salt body rather than by true piercement. Isolated cylindrical salt stocks may have grown by either active or passive means or by some combination of the two.

7. Some salt bodies are surrounded by strongly folded strata that have been uplifted above their regional depositional level. They are usually at the downdip margins of shallow salt sheets and are commonly associated with adjacent thrust faults. Thus, the folding is interpreted to be related to contractional modification of the salt bodies.

8. Counterregional systems are characterized by common down-to-the-basin listric growth faults that sole into the salt sheet or weld and may accommodate significant extension. In contrast, extension is rare in salt stock canopy systems because of the structural relief of the salt weld. The main faults are arcuate in plan view and located over the updip and lateral margins of the elliptical lows.

9. There is a general north-south division of the two systems within the study area: counter-regional systems dominate the outer shelf, whereas canopy systems are more typical of the upper slope (although the boundary does not follow the shelf margin).

10. Different styles of salt body growth create variable relief at the sea floor and thus have an important influence on facies development and trap geometry.

11. Subsalt hydrocarbon potential may be enhanced below saddles in salt stock canopy systems and where counter-regional and canopy systems have merged because of favorable trap geometries, the probable existence of channel sands, and possible focusing of hydrocarbon migration pathways.

Acknowledgments   I would like to thank many people for stimulating discussions on salt tectonics over the last several years, including Barry McBride, Paul Weimer, Fred Diegel, Mike Hudec, Jay Jackson, Ian Watson, Bruno Vendeville, Frank Peel, Dorie McGuinness, Carl Fiduk, Bob Ratliff, Peter Flemings, and Clint Moore. Barry McBride, Bruce Trudgill, William House, Martin Jackson, and especially Fred Diegel and Sig Snelson provided helpful reviews of the manuscript. I am indebted to Halliburton, TGS, and Exxon for seismic data; Marathon, Pennzoil, PaleoData, and Micro-Strat for well and biostratigraphic data; and CogniSeis, GeoQuest, and Landmark for software. Well correlationships and seismic stratigraphic interpretations were conducted by Peter Varnai, Fadjar Budhijanto, Zurilma Acosta, Rafael Martinez, Alonso Navarro, Peter Flemings, and Paul Weimer. This work was carried out under the auspices of the University of Colorado Gulf of Mexico Industrial Consortium (funded by Amoco, Anadarko, BHP, BP, Chevron, Conoco, CNG, Exxon, Marathon, Mobil, Pennzoil, Petrobras, Phillips, Shell, Texaco, Total, Union Pacific, and Unocal) and the Global Basins Research Network (funded by the DOE). Finally, I would like to thank Martin Jackson, Sig Snelson, and Dave Roberts for organizing and running such a superb conference.

REFERENCES CITED

Acosta, Z., 1994, Sequence stratigraphy of Plio-Pleistocene sediments in north-central Green Canyon and western Ewing Bank, northern Gulf of Mexico: Masters Thesis, University of Colorado, Boulder, 187 p.

Barton, D. C., 1933, Mechanics of formation of salt domes, with special reference to Gulf Coast salt domes of Texas and Louisiana: AAPG Bulletin, v. 17, p. 1025–1083.

Budhijanto, F. M., 1995, Sequence stratigraphy of Plio-Pleistocene sediments of the northeastern Green Canyon and eastern Ewing Bank area, northern Gulf of Mexico: Masters Thesis, University of Colorado, Boulder, 154 p.

Cook, D., and P. D'Onfro, 1991, Jolliet Field thrust fault structure and stratigraphy, Green Canyon Block 184, offshore Louisiana: Gulf Coast Association of Geological Societies, Transactions, v. 41, p. 100–121.

Diegel, F. A., and R. W. Cook, 1990, Palinspastic reconstructions of salt-withdrawal growth-fault systems, northern Gulf of Mexico (abs): GSA Abstracts with Programs, v. 22, p. A48.

Diegel, F. A., J. F. Karlo, D. C. Schuster, R. C. Shoup, and P. R. Tauvers, 1993, Cenozoic structural evolution and tectonostratigraphic framework of the northern Gulf Coast continental margin (abs.): AAPG Annual Convention, Program with Abstracts, New Orleans, Louisiana, p. 91.

Diegel, F. A., J. F. Karlo, D. C. Schuster, R. C. Shoup, and P. R. Tauvers, 1995, Cenozoic structural evolution and tectonostratigraphic framework of the Northern Gulf Coast continental margin, in M. P. A. Jackson, D. G. Roberts, and S. Snelson, eds., Salt tectonics: a global perspective: AAPG Memoir 65, this volume.

Duval, B., C. Cramez, D. D. Schultz-Ela, and M. P. A. Jackson, 1993, Extension, reactive diapirism, salt welding, and contraction at Cegonha, Kwanza basin, Angola (abs.): AAPG, International Hedberg Research Conference Abstracts, Sept. 13–17, Bath, U.K., p. 41–43.

Fiduk, J. C., 1994, Plio-Pleistocene evolution of the upper continental slope, Garden Banks and East Breaks areas, northwest Gulf of Mexico: Ph.D. dissertation, The University of Texas at Austin, 311 p.

Fletcher, R. C., 1995, Salt glacier and composite sediment–salt glacier models for the emplacement and early burial of allochthonous salt sheets, in M. P. A. Jackson, D. G. Roberts, and S. Snelson, eds., Salt tectonics: a global perspective: AAPG Memoir 65, this volume.

Hodgkins, M. A., and M. J. O'Brien, 1994, Salt sill deformation and its implications for subsalt exploration: The Leading Edge, v. 13, p. 849–851.

Hossack, J. R., and D. B. McGuinness, 1990, Balanced sections and the development of fault and salt structures in the Gulf of Mexico (GOM) (abs): GSA Abstracts with Programs, v. 22, p. A48.

House, W. M., and J. A. Pritchett, 1994, Salt deformation modeling through the use of enhanced seismic imaging techniques: The Leading Edge, v. 13, p. 844–848.

Huber, W. F., 1989, Ewing Bank thrust fault zone, Gulf of Mexico, and its relation to salt sill emplacement: SEPM Gulf Coast Section, 10th Annual Research Conference, Program and Extended Abstracts, Houston, Texas, p. 60–65.

Humphris, C. C., Jr., 1978, Salt movement on continental slope, northern Gulf of Mexico, *in* A. H. Bouma, G. T. Moore, and J. M. Coleman, eds., Framework, facies, and oil-trapping characteristics of the upper continental margin: AAPG Studies in Geology 7, p. 69–86.

Jackson, M. P. A., and C. J. Talbot, 1986, External shapes, strain rates, and dynamics of salt structures: GSA Bulletin, v. 97, p. 305–323.

Jackson, M. P. A., and C. J. Talbot, 1989, Salt canopies: Gulf Coast Section, SEPM, Tenth Annual Research Conference, Extended Abstracts, p. 72–78.

Jackson, M. P. A., and C. J. Talbot, 1991, A glossary of salt tectonics: The University of Texas at Austin, Bureau of Economic Geology Geological Circular No. 91-4, 44 p.

Jackson, M. P. A., and B. C. Vendeville, 1994, Regional extension as a geologic trigger for diapirism: GSA Bulletin, v. 106, p. 57–73.

McGuinness, D. B., and J. R. Hossack, 1993, The development of allochthonous salt sheets as controlled by the rates of extension, sedimentation, and salt supply, *in* J. M. Armentrout, R. Bloch, H. C. Olson, and B. F. Perkins, eds., Rates of geological processes: SEPM Gulf Coast Section, 14th Annual Research Conference, Program with Papers, Houston, Texas, p. 127–139.

Navarro, A. F., 1994, Sequence stratigraphy of Pleistocene sediments of northwestern Green Canyon area, northern Gulf of Mexico: Masters Thesis, University of Colorado, Boulder, 108 p.

Nelson, T. H., 1991, Salt tectonics and listric-normal faulting, *in* A. Salvador, ed., The Gulf of Mexico basin: GSA Decade of North American Geology, v. J, p. 73–89.

Nilsen, K. T, B. C. Vendeville, and J.-T. Johansen, 1995, Influence of regional tectonics on halokinesis in the Nordkapp Basin, Barents Sea, *in* M. P. A. Jackson, D. G. Roberts, and S. Snelson, eds., Salt tectonics: a global perspective: AAPG Memoir 65, this volume.

Peel, F. J., C. J. Travis, and J. R. Hossack, 1995, Genetic structural provinces and salt tectonics of the Cenozoic offshore U.S. Gulf of Mexico: a preliminary analysis, *in* M. P. A. Jackson, D. G. Roberts, and S. Snelson, eds., Salt tectonics: a global perspective: AAPG Memoir 65, this volume.

Ratcliff, D. W., 1993, New technologies improve seismic images of salt bodies: Oil & Gas Journal, v. 91, no. 39, p. 41–49.

Rowan, M. G., 1993, A systematic technique for the sequential restoration of salt structures: Tectonophysics, v. 228, p. 331–348.

Rowan, M. G., and R. Kligfield, 1989, Cross section restoration and balancing as aid to seismic interpretation in

extensional terranes: AAPG Bulletin, v. 73, p. 955–966.

Rowan, M. G., B. C. McBride, and P. Weimer, 1994a, Salt geometry and Plio-Pleistocene evolution of Ewing Bank and northern Green Canyon, offshore Louisiana (abs.): AAPG Annual Conference, Program with Abstracts, Denver, Colorado, p. 247.

Rowan, M. G., P. Weimer, and P. B. Flemings, 1994b, Three-dimensional geometry and evolution of a composite, multi-level salt system, western Eugene Island, offshore Louisiana: Gulf Coast Association of Geological Societies, Transactions, v. 44, p. 641–648.

Schultz-Ela, D. D., M. P. A. Jackson, and B, C. Vendeville, 1993, Mechanics of active salt diapirism: Tectonophysics, v. 228, p. 275–312.

Schuster, D. C., 1993, Deformation of allochthonous salt and evolution of related structural systems, eastern Louisiana Gulf Coast (abs.): AAPG Annual Convention, Program with Abstracts, New Orleans, Louisiana, p. 179.

Schuster, D. C., 1995, Deformation of allochthonous salt and evolution of related salt–structural systems, eastern Louisiana Gulf Coast, *in* M. P. A. Jackson, D. G. Roberts, and S. Snelson, eds., Salt tectonics: a global perspective: AAPG Memoir 65, this volume.

Seni, S. J., 1992, Evolution of salt structures during burial of salt sheets on the slope, northern Gulf of Mexico: Marine and Petroleum Geology, v. 9, p. 452–468.

Seni, S. J., 1994, Salt tectonics on the continental slope, northeast Green Canyon area, northern Gulf of Mexico: The University of Texas at Austin, Bureau of Economic Geology Report of Investigations No. 212, 102 p.

Simmons, G. R., 1992, The regional distribution of salt in the northwestern Gulf of Mexico: styles of emplacement and implications for early tectonic history: Ph.D. dissertation, Texas A&M University, College Station, Texas, 180 p.

Sumner, H. S., B. A. Robison, W. K. Dirks, and J. C. Holliday, 1991, Structural style of salt/minibasin systems: lower shelf and upper slope, central offshore Louisiana: Gulf Coast Association of Geological Societies, Transactions, v. 41, p. 582.

Talbot, C. J., 1993, Spreading of salt structures in the Gulf of Mexico: Tectonophysics, v. 228, p. 151–166.

Talbot, C. J., 1995, Molding of salt diapirs by stiff overburden, *in* M. P. A. Jackson, D. G. Roberts, and S. Snelson, eds., Salt tectonics: a global perspective: AAPG Memoir 65, this volume.

Vendeville, B. C., and M. P. A. Jackson, 1991, Deposition, extension, and the shape of downbuilding diapirs (abs.): AAPG Bulletin, v. 75, p. 687–688.

Vendeville, B. C., and M. P. A. Jackson, 1992a, The rise of diapirs during thin-skinned extension: Marine and Petroleum Geology, v. 9, p. 331–353.

Vendeville, B. C., and M. P. A. Jackson, 1992b, The fall of diapirs during thin-skinned extension: Marine and Petroleum Geology, v. 9, p. 354–371.

Vendeville, B. C., and M. P. A. Jackson, 1993, Rates of extension and deposition determine whether growth faults or salt diapirs form, *in* J. M. Armentrout, R. Bloch, H. C. Olson, and B. F. Perkins, eds., Rates of geological processes: SEPM Gulf Coast Section, 14th Annual Research Conference, Program with Papers, Houston, Texas, p. 263–268.

Walters, R. D., 1993, Reconstruction of allochthonous salt emplacement from 3-D seismic reflection data, northern Gulf of Mexico: AAPG Bulletin, v. 77, p. 813–841.

Weimer, P., P. Varnai, A. Navarro, Z. Acosta, F. Budhijanto, R. Martinez, M. Rowan, B. McBride, and T. Villamil, 1994, Sequence stratigraphy of Neogene turbidite systems, Green Canyon and Ewing Bank, northern Gulf of Mexico: preliminary results: SEPM Gulf Coast Section, 15th Annual Research Conference, Program with Papers, Houston, Texas, p. 383–399.

West, D. B., 1989, Model for salt deformation of central Gulf of Mexico basin: AAPG Bulletin, v. 73, p. 1472–1482.

Worrall, D. M., and S. Snelson, 1989, Evolution of the northern Gulf of Mexico, with emphasis on Cenozoic growth faulting and the role of salt, *in* A. W. Bally and A. R. Palmer, eds., The geology of North America: an overview: GSA Decade of North American Geology, v. A, p. 97–138.

Wu, S. A., 1993, Salt and slope tectonics, offshore Louisiana: Ph.D. dissertation, Rice University, Houston, Texas, 251 p.

Wu, S. A., A. W. Bally, and C. Cramez, 1990, Allochthonous salt, structure and stratigraphy of the northeastern Gulf of Mexico, part II: structure: Marine and Petroleum Geology, v. 7, p. 334–370.

Coward, M., and S. Stewart, 1995, Salt-influenced structures in the Mesozoic–Tertiary cover of the southern North Sea, U.K., in Jackson, M. P. A., D. G. Roberts, and S. Snelson, eds., Salt tectonics: a global perspective: AAPG Memoir 65, p. 229–250.

Chapter 10

Salt-Influenced Structures in the Mesozoic– Tertiary Cover of the Southern North Sea, U.K.

Mike Coward

Geology Department
Imperial College
London, U.K.

Simon Stewart

Amerada Hess Ltd.
London, U.K.

Abstract

A structural model encompassing the southern North Sea Basin west of the Central Graben has been developed that combines gravity gliding of the postsalt cover with basement tectonics. The basin differs from many salt basins in that it forms a closed system. Section construction and balancing through the cover of the North Sea need to take into account thin-skinned and thick-skinned extensions and contractions. The North Sea salt formed in Permian time in two large oval basins separated by the Mid North Sea High. The shape of these basins reflects variable patterns of thermal subsidence. Subsequent salt tectonics was governed by local graben structures and by regional uplift and subsidence.

Rifting initiated during the Triassic and allowed reactive and locally passive diapirs to develop in the postsalt cover. In the southern North Sea, the Dowsing graben system in the cover is offset from the Dowsing fault zone below the salt. This offset in extensional structures probably relates to the salt thickness and to the position of the surface hinge line that controlled the onset of gravity gliding in the postsalt section. Gravity gliding of the cover into the Triassic–Jurassic Sole Pit trough and away from zones of rift flank uplift was associated with Late Jurassic–Early Cretaceous extension in the Central North Sea; gliding caused asymmetric compressional pillows to develop downslope. Gravity spreading of the cover during the Late Cretaceous–early Tertiary was associated with tilting during thermal subsidence of the southern North Sea Basin, enhanced by pulses of tectonic inversion in the southern North Sea basement. The resultant glide tectonics formed new small grabens upslope and compressional pillows downslope. Where the compressional pillows were eroded sufficiently or faulted later, the salt broke through the thinned cover to produce new active and then passive diapirs, which drained the pillows to produce new rim synclines.

INTRODUCTION

Models of salt tectonics in the North Sea have been dominated by concepts of *halokinesis*, in which the salt and its cover are considered to be viscous media with no yield strength. The first comprehensive study of salt tectonics in the southern North Sea Basin was published by Christian (1969), who mapped both salt structures and underlying basement faults and concluded that no spatial relationship existed. This conclusion has also been reached in more recent studies in the Norwegian-Danish Basin (Hospers et al., 1988) and in the Central North Sea (Hodgson et al., 1992). Christian (1969) proposed that basin tilt may have been a significant control, noting the conformity between salt structure trend and shape of the basin. Several papers by Jenyon (1985, 1988a,b) and Jenyon and Cresswell (1987) described the evolution of various salt structures within the basin. Further papers have emphasized decoupling between the presalt and postsalt sections and provided examples of structural development in the postsalt sediments (Walker and Cooper, 1987; Cooke-Yarborough, 1991; Pritchard, 1991; Arthur, 1993; Oudmayer and de Jager, 1993).

Common to each of these studies is a focus on specific salt structures or salt tectonic processes within a small part of the basin. The present paper attempts to place the examples of salt tectonics within a coherent structural model covering the Mesozoic and Tertiary evolution of all of the southern North Sea Basin west of the Central Graben. Regional mapping using several nonexclusive seismic surveys (Figure 1) and released wells suggests that major structures in the postsalt sequence can be described by basin-wide, gravity-driven structural models that incorporate Upper Permian salt as the key decoupling horizon.

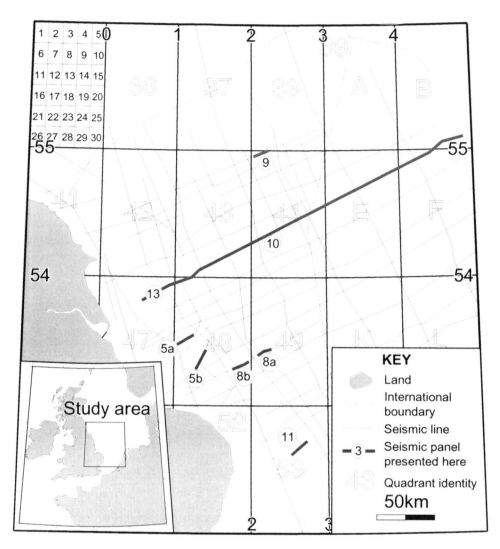

Figure 1—Map of southern North Sea study area (see inset for location) with Universal Transverse Mercator projection showing international boundaries and regional seismic lines used in this study. Latitude and longitude lines (labeled in degrees) constitute quadrant boundaries. Quadrants are labeled with numbers in the U.K. sector and letters in the Dutch sector. U.K. sector quadrants are subdivided into license blocks, as shown in Q35 (upper left); such blocks are referenced in the format 35/1 and so on.

KEY

Land

International boundary

Seismic line

— 3 — Seismic panel presented here

Quadrant identity

50km

BASEMENT STRUCTURE

The basement of the southern North Sea comprises upper Precambrian intrusions and volcanic rocks overlain by Devonian and Carboniferous sedimentary rocks. Deep seismic data shows the basement fabric to be generally weak, but there are zones of locally strong reflectors that may represent late Precambrian or Caledonian tectonic events (Blundell et al., 1991). The basement fabric trends approximately northwest to north-northwest. The upper Paleozoic rocks were deposited in a series of north-northwest striking grabens, possibly reworking earlier basement fabrics inverted during the Late Carboniferous (Coward, 1993).

Reverse faulting associated with Late Carboniferous basin inversion is recorded by a wide range of Carboniferous stratigraphy subcropping Permian sediments (Leeder and Hardman, 1990). The subcrop pattern indicates a strong northwest-southeast tectonic influence during this inversion. This inversion event was followed by deposition of Upper Carboniferous red beds, which pass up into sandstones of the Permian Rotliegend Group.

BASIN DEVELOPMENT

Permian Salt Basins

In the North Sea, two main areas of Permian subsidence formed the Northern and Southern Permian basins, separated by the Mid North Sea–Ringkobing Fyn system of highs (Figure 2) (Glennie, 1990). A small Permian basin also occurs in the Moray Firth. Although the trends of Permian sedimentary facies bear some relation to those of the underlying basement structures (Glennie, 1990; Taylor, 1990), there is no published evidence of direct fault control of Permian sedimentation in the southern North Sea. This contrasts with the Central Graben to the northeast, where various workers have attributed up to 1100 m of topographic relief to Permian extension (Hodgson et al., 1992). Permian facies boundaries in the southern North Sea may be associated with inherited Carboniferous topography (Bailey et al., 1993).

The early basin sediments are called the *Rotliegend,* an old German miners' term for the red beds that underlie the Zechstein (Glennie, 1990). Rotliegend sedimentation

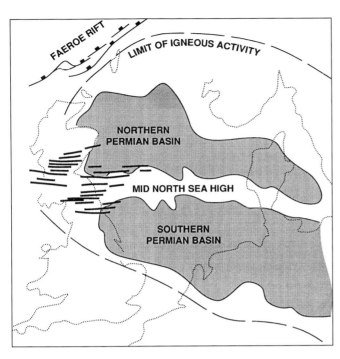

Figure 2—Simplified map showing the Northern and Southern Permian basins, the approximate limit of Carboniferous–Permian igneous activity (dashed lines), and the trend of dikes (heavy black lines) in the western North Sea. Modified from Ziegler (1990).

occurred during earliest–Late Permian time; Zechstein deposition took place entirely within the Tartarian, the youngest stage of the Permian. Early Permian volcanism (early Rotliegend) is most evident in northern Germany and Poland and within the Horn-Bamble-Oslo grabens. The thickest Rotliegend sequences are preserved in northern Germany in areas of the most widespread lower Rotliegend volcanism, suggesting that thermal subsidence was important (Glennie, 1990).

Early Rotliegend igneous activity includes the intrusion of the Whin sill in northern England and some of the dike swarms in the Midland Valley of Scotland. In these areas, volcanism followed Westphalian inversion and predated much of the Permian–Triassic extension, suggesting that the region was underlain by hot asthenosphere, possibly the edge of a northwest European hot spot. The trend of the dike swarms in the Midland Valley of Scotland and adjacent regions suggests local northeast-southwest extension. If the region were underlain by a hot spot, only small amounts of extension could lead to the upwelling of asthenospheric melt. Some of the latest Carboniferous regional uplift could be due to thermal doming, whereas mantle cooling could contribute to much of the subsequent Permian subsidence.

Several models may explain the origin of the Southern Permian Basin.

1. **Rifting model**—Badley et al. (1988) and Smith et al. (1993) argued that Early Permian extension initiated the north-striking Viking and Central graben systems and that the Southern Permian Basin was related to this stretching. This model is considered unlikely because evidence for important Early Permian extension is lacking in the Southern Permian Basin. The Upper Carboniferous Barren Red Beds and the Rotliegend infill a topography resulting from Carboniferous inversion tectonics. Thickness changes are a result of infilled topography rather than graben development. Permian rifting, however, occurred in the Northern Permian Basin and west of Shetland. Some minor rifting events could have occurred to the north and east of the Southern Permian Basin.

2. **Flexural basin model**—The Southern Permian Basin, its continuation into Poland, and its offset to the Peri-Caspian Basin lie on the foreland of the Variscan mountain belt. Lithospheric thickening could have produced a flexural foreland basin and could account for some of the subsidence. However, the edge of the Southern Permian Basin lies 200 km north of the main thrust front, and the basin formation coincided with mountain belt erosion and uplift rather than lithospheric thickening. This model is therefore unlikely.

3. **Thermal subsidence model**—Thermal subsidence following Carboniferous rifting, and more importantly, following Late Carboniferous volcanism, could account for the large subsidence basin. The probable dimensions of the Late Carboniferous thermal dome were probably close to that of the subsequent Permian subsidence basin.

Similarly, various models can be proposed for the origin of the Mid North Sea High. This high postdates Variscan inversion and foreland basin development. It overlies a region of intense Late Carboniferous intrusive activity and may represent a zone of thickened crust less liable to subsequent thermal subsidence.

In the Late Permian, there was a glacioeustatic rise in sea level related to the melting of Carboniferous–Permian ice of Gondwana. Rifting in the Faeroes–East Greenland region established a seaway to the Arctic Ocean. The resulting marine transgression from the north resulted in the establishment of a Zechstein Sea across northern and central Europe. In the North Sea, the Zechstein basins are separated by remnants of the Mid North Sea High. The Zechstein sequences are underlain by the organic-rich shales of the Kupferschiefer, which form a distinctive time marker and imply that the Rotliegend basins had subsided below sea level before the Zechstein transgression (Ziegler, 1988).

The Zechstein contains a number of carbonate-evaporite cycles reflecting glacioeustatic sea level fluctuations and is one of the world's saline giants. The southern margin of the Zechstein Sea was formed by the Brabant massif, while in the north, embayments occupied the southern end of the Viking Graben and the Moray Firth.

Four main evaporitic cycles and a partial fifth cycle are recorded from the North Sea. Each cycle contains a general basinward transition from shales or continental sandstones at the basin margins into and overlapped by

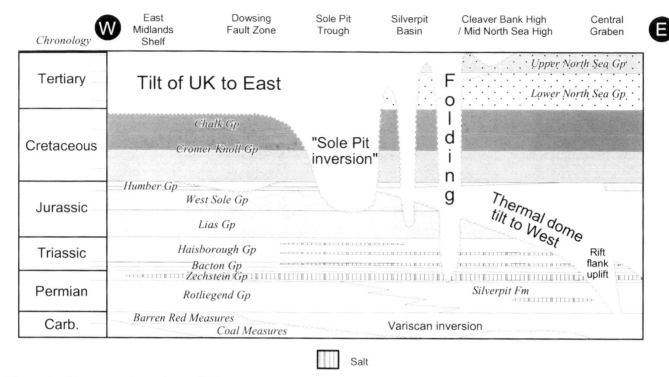

Figure 3—Chronostratigraphic profile from the East Midlands Shelf to the Central Graben, illustrating preserved stratigraphy and the impact of basinwide tilt events. Horizontal scale arbitrary; vertical scale in proportion to time.

carbonates, anhydrites, and halites (Taylor, 1990). The resultant pattern is partially a product of the sea shrinking under the influence of evaporation following a single flooding episode, and partially the effects of regional subsidence. There are stratigraphic variations where different amounts of subsidence occurred between or during cycles. Successive cycles are increasingly evaporitic; significant carbonates are confined to the first three cycles.

Triassic Basin Development

Rocks younger than Permian in the southern North Sea Basin have been progressively removed by Late Jurassic erosion (Figure 3). However, the isopach map of the Bacton Group illustrates the pattern of Early Triassic basin subsidence (Figure 4). The distribution of Bacton Group sediments is similar to that of the underlying Permian deposits (Alberts and Underhill, 1991; Cameron et al., 1992). The general pattern of deposition reflects an east-west trending sag basin. The east-west trend is interrupted by a high zone extending southward from the Mid North Sea High through the Cleaver Bank High onto the Hewett shelf (Figure 4). This high zone may have been produced by rift flank uplift associated with Early Triassic rifting in the Central Graben (Hodgson et al., 1992). Smaller scale variations in Lower Triassic sedimentary thickness, on the order of 300 m over tens of kilometers, have been documented in the central North Sea and attributed to early halokinesis (Hodgson et al., 1992; Penge et al., 1993). Such lateral variations in thickness of the Bacton Group, however, have not yet been documented in the southern North Sea.

The east-west trend continued to control sediment deposition during the Middle Triassic (Dowsing Dolomitic Formation), but the Late Triassic (Dudgeon Saliferous Formation) was marked by the first pronounced spatial influence of the Dowsing fault zone upon sedimentation (Cameron et al., 1992) (Figure 5). Subsidence due to extensional faulting soon obscured the effects of thermal subsidence that had characterized the Early Triassic; the full Triassic succession typically thickens by 100% across the Dowsing fault zone to reach about 1500 m in the Sole Pit Trough (Fisher and Mudge, 1990). Salt tectonics were also initiated in the Late Triassic. Salt swells of this age are recorded in the southeastern part of quadrant 49 (Pritchard, 1991), and extensional faulting decoupled from basement structures is seen in the postsalt section of block 48/11 (Arthur, 1993).

At the base of the Haisborough Group, the Dowsing Dolomitic Formation has a variably developed evaporite zone, the Röt Halite. The thickness of the salt varies, possibly owing to deposition in extension-generated depressions (Fisher and Mudge, 1990). Ziegler (1975) suggested that the salt may have originated from leached and reprecipitated Zechstein halites, but Holser (1979) cited the high bromide content of the halites as indicative of marine origin. Fisher and Mudge (1990) suggested increased marine influence eastward and hence a marine incursion from the east. However, the onset of extension during the Middle–Late Triassic may have exposed Zechstein salts to the seabed. Detailed section balancing and restoration in the southwestern part of the Sole Pit Basin (MPC unpublished data) suggest much higher extension of Lower Triassic Bacton Group sediments than

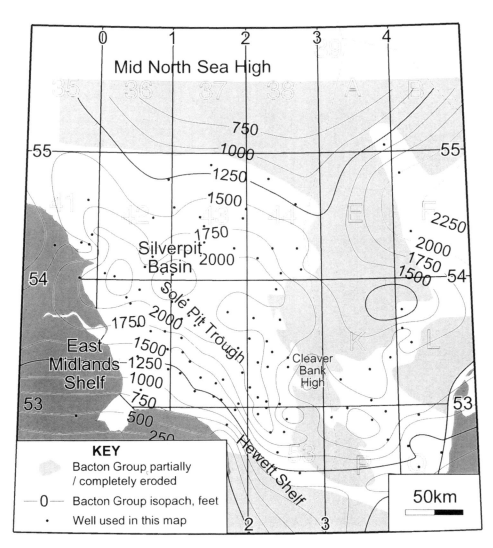

Figure 4—Lower Triassic (Bacton Group) isopach map, generated from arbitrarily chosen released wells (one from each license block where possible), with a fully preserved Bacton Group. Isopachs have been interpolated through the center of the basin where the Bacton Group was partially eroded during the Jurassic. The map therefore shows the shape of the Early Triassic basin prior to later erosion. The main depocenter is an east–west axis lying at 54°N; the Sole Pit Trough is apparent as a northwest-trending embayment in the southern side of the basin. The basin floor slope from the Mid North Sea High to the Silverpit Basin depocenter is 0.2°.

KEY

Bacton Group partially / completely eroded

—0— Bacton Group isopach, feet

• Well used in this map

50km

the overlying sequences. Permian salt may have been exposed subaerially in grabens during the early stages of Haisborough Group rifting and hence could have contributed to the saline conditions during Röt Halite deposition.

Jurassic–Early Cretaceous Basin Development

Subsidence in the Jurassic was fundamentally controlled by continued extension on the Dowsing fault zone, but sediment distribution was increasingly affected by salt tectonics.

The thickness of the Jurassic–Lower Cretaceous sequence increases gradually from the East Midlands shelf northeastward across the Dowsing fault zone into the Sole Pit Trough (see Figure 6 for location). The geometry of the Dowsing fault zone at the Rotliegend level was a series of terraces stepping down eastward and defined by northeastward-dipping extensional faults. Basinward thickening of the postsalt section was achieved without growth faulting because the Upper Permian salt provided a viscous layer that molded to the evolving terrace topography at the Rotliegend level, providing a relative-

ly smooth northeastward-dipping slope at the top of the salt (Figure 6). This pattern of sedimentary growth is unusual in that synkinematic sediments tend to thicken toward, rather than away from, synsedimentary basin margin faults (Jackson and McKenzie, 1983).

The postsalt sequence contains a major graben system (Dowsing graben system) that strikes southeastward from Flamborough Head for 200 km to the Hewett shelf (Figure 5). Segments of this graben system have been described onshore in northeastern England (Kirby et al., 1987) and offshore (Walker and Cooper, 1987; Cooke-Yarborough, 1991; Pritchard, 1991; Arthur, 1993).

Figure 5 illustrates the spatial relationships of the Dowsing graben system and both the lateral pinch-out of the Upper Permian salt and the underlying basement structures. From this map, it is clear that, although the Dowsing graben system is limited at each end by the Upper Permian facies boundary between salt and basin margin carbonates, the location and trend of the Dowsing graben system are closely controlled by basement structures. The first faulting within the Dowsing graben system occurred in the Late Triassic; restoration of the Mesozoic section locally shows loss of section in the Bacton Group, indicating locally high values of

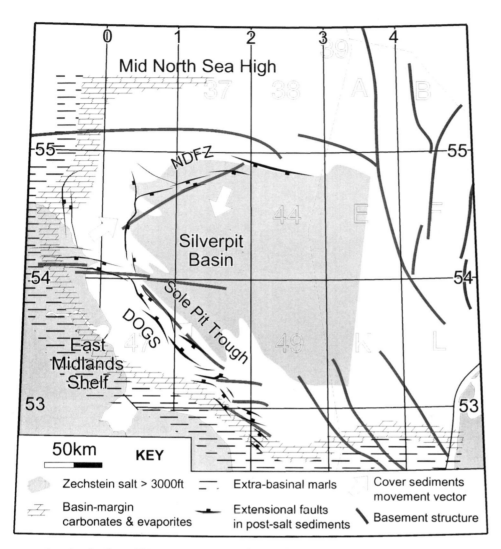

Figure 5—Map of the southern North Sea postsalt peripheral graben system, showing that the system is confined to the Permian salt basin, the margins of which are defined by the distribution of Zechstein basin margin deposits. Although the peripheral graben system is confined to the Permian salt basin, the locations and trends of the graben system elements within the basin are probably controlled by basement fault structures.

extension in the Late Triassic. However, the graben formation lasted through the Middle Jurassic and Early Cretaceous (Figure 6) (Arthur, 1993).

The heave across the Dowsing graben system in quadrant 48 is typically between 1 and 2 km. The postsalt sequence was not free to move southwestward because the salt pinches out within 20 km in that direction—and there are no structures in the pre- or postsalt sections that might accommodate observed extension in the Dowsing graben system. If proximity of the Dowsing graben system to the underlying Dowsing fault zone is considered (Figure 6), at the time of graben formation, the Dowsing graben system lay above the Dowsing fault zone scarp which rose several thousand feet above the Sole Pit Trough. The extension across the Dowsing graben system may be balanced northeastward by extension in the presalt across the Dowsing fault zone itself.

The Dowsing graben system and fault zone association resembles modeled basement–cover fault systems that include a viscous layer (Richard, 1991; Larroque, 1993; Oudmayer and de Jager, 1993). These experimental studies have indicated that where a viscous layer lies between basement and cover, fault systems in the cover tend to occur horizontally offset toward the basement

footwall some distance away from the basement fault crest. The magnitude of this offset is some function of the thickness ratio of viscous and cover layers and the change in surface slope associated with the stretching.

The northeastern limit of salt tectonics in the southern North Sea is marked by an arcuate graben system, similar in character to the Dowsing graben system, striking about 110 km through the northern parts of quadrants 42, 43, and 44 (Figures 5, 7). This graben system has been termed the North Dogger fault zone (Griffiths et al., 1995). Salt changes laterally into basin margin carbonates north of the North Dogger fault zone. Sections (Figure 7) demonstrate that extension occurred prior to the Cretaceous. Griffiths et al. (1995) dated the first faulting as Late Triassic, noting thickening of the Triton Anhydritic Formation toward the North Dogger fault zone.

To the northeast and east of the Silverpit Basin, Jurassic sediments have been eroded prior to Cretaceous sedimentation. The depth of erosion below the Cretaceous in this part of the basin is related to rift flank uplift on the edge of the Broad Fourteens and Central grabens, modified slightly by Middle Jurassic domal uplift centered on the central North Sea (Underhill and Partington, 1993). In combination with subsidence across the Dowsing fault zone, this uplift generated a basin slope from northeast to southwest (Sole Pit Trough). Erosion was responsible for loss of the Jurassic and much of the Triassic in the northern and eastern parts of the southern North Sea Basin.

Jenyon (1985) postulated an origin for the North Dogger fault zone associated with flow of salt updip and out of the basin, which might have dammed at the facies change to basin margin carbonates and actively risen as a series of "basin margin diapirs." However, well-documented examples of postsalt sequences sliding downslope in other tilted basins (Wu et al., 1990; Duval et al., 1992) provide analogs for this part of the southern North Sea Basin. We suggest that here the overburden slid downslope southwestward away from the North Dogger fault zone, toward the Sole Pit Trough. The North Dogger fault zone was a fault boundary between the autochthonous postsalt sequence on the Mid North Sea High and allochthonous postsalt sequence in the Silverpit Basin and was therefore similar in geometry to bergschrund faults, which define the upslope limits of mountain glaciers. In this interpretation of the North Dogger fault zone, salt plays a passive role in the evolution of the graben system, rising only where extensional geometries permitted reactive diapirism (Vendeville and Jackson, 1992). This model therefore disagrees with Jenyon's (1985) interpretation, but concurs with the interpretation of Allen et al. (1994), who suggested that Triassic extension across the North Dogger fault zone balanced shortening across salt swells within the basin. The heave across the North Dogger fault zone is incorporated into the structural model below.

The zone of thin-skinned rifting marked by the North Dogger fault zone may have continued toward the east into the Cleaver Bank High at the western edge of the Dutch Central Graben. In this region, however, there was considerable Late Jurassic–Early Cretaceous uplift at the edge of the rift basin, so evidence for thin-skinned extension in the postsalt cover has been obliterated.

The central part of the southern North Sea Basin—the Silver Pit, Sole Pit Trough, and western margin of the Cleaver Bank High—is characterized by a series of major salt swells and walls. However, much of the amplification of these structures occurred during the Cretaceous and Tertiary, and cutoff relationships below the Jurassic unconformity must be examined to reconstruct structures contemporaneous with the extensional faulting already described from the peripheral graben systems. When such structures are identified, they are invariably found to be asymmetric fold pairs overturned toward the southwest (Figure 8). The significance of this sense of overturn is that it suggests movement of the postsalt sequence to the southwest relative to the presalt section at a time when the basin floor was dipping in that direction and extensional faults were developing upslope. Examples of asymmetric fold structures can be found as far southwest as the eastern margin of the Sole Pit Trough, where a southwest-facing monocline more than 1200 m in amplitude is responsible for loss of most of the Jurassic and Triassic section below the unconformity (Figure 8).

PERMIAN–EARLY CRETACEOUS BASIN MODEL

A model integrating the previously described salt-related tectonics of the margins and center of the basin prior to the Late Jurassic unconformity can be constructed (Figure 9). This model presents structures of the postsalt section across the basin as a combination of both a basement-driven and gravity-driven linked system. Extension across the Dowsing graben system is essentially thick skinned; the extension in the carapace links through the salt to offset extension in the basement. Extension in the North Dogger fault zone is essentially thin skinned. In total, the strains should balance, that is, extension across the peripheral graben systems (eDOGS + eNDFZ) should be balanced by extension below the salt (eTOP ROT) and shortening across fold structures within the basin (cFOLDS):

$$eDOGS + eNDFZ = eTOP\ ROT + cFOLDS$$

This model has some similarities with published models accounting for salt tectonics in the Gulf of Mexico (Wu et al., 1990), offshore Brazil (Cobbold and Szatmari, 1991), and offshore Angola (Duval et al., 1992), except that, in the southern North Sea, (1) the basin is encratonic, and the postsalt section is not free to slide downslope almost indefinitely as on passive margins; and (2) basement tectonics account for some of the extension in the postsalt cover.

In the model presented here, salt plays a passive role by molding to the shape of structures controlled by deformation of the overlying sedimentary section. An alternative explanation of salt pillow growth and asymmetry has been given by Geil (1991), who suggested that thicker accumulation of postsalt sediment downslope toward the basin in the Danish sector would be sufficient to generate inherently asymmetric salt pillows. Although Geil's (1991) model and ours are not mutually exclusive, only the model discussed here accounts for both salt pillow geometry and the observed tectonics in the southern North Sea postsalt section.

LATE CRETACEOUS–TERTIARY BASIN EVOLUTION

The post-Early Cretaceous history of the southern North Sea Basin is characterized by inversion, in terms of both basin tilt and reverse slip on individual faults. The basin tilted to the northeast, a reversal of the Jurassic tilt direction. The driving mechanism for Late Cretaceous and Tertiary tilt is considered to be related to (1) thermal subsidence of the stretched lithosphere beneath the central North Sea and (2) reversal of movement on the Dowsing fault zone, causing uplift along the western flank of the Sole Pit Basin.

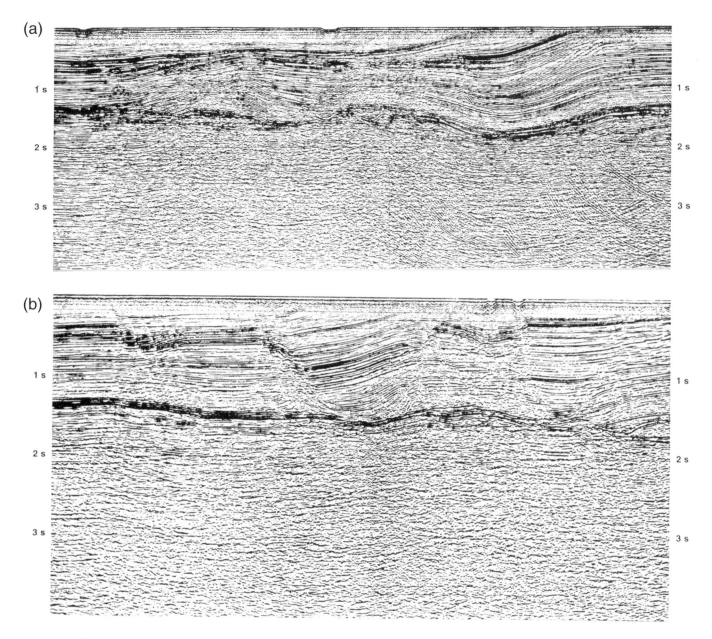

Figure 6—Seismic lines (this page) and interpretive drawings (opposite page) across the west margin of the southern North Sea Basin, showing the peripheral Dowsing graben system, (DOGS) and the Dowsing fault zone (DFZ). (a) Line SF90-DF9 (top). (b) Line SF90-DF6 (bottom). Note detachment between fault systems in the presalt and postsalt sections. In both examples, the northeastern ends of the lines lay on the west margin of the Sole Pit Trough and have been inverted during the Tertiary. The original geometry of the Sole Pit Trough is suggested by the northeastward thickening of the postsalt section, as shown in the inset (lower right).

Post-Early Cretaceous tilting has allowed accumulation of over 1 km of Tertiary sediments on the Cleaver Bank High. Similarly, on the western margin of the basin, apatite fission track data indicate over 1 km of Tertiary uplift (Green, 1989). Regional seismic lines show that an approximate return of Lower Permian Rotliegend stratigraphy to horizontal across the basin was a result of Late Cretaceous–Tertiary basin tilting (Figure 10).

Tectonic inversion can be more precisely dated along strike from the Dowsing fault zone on the Hewett shelf and Broad Fourteens Basin where unconformable sedi-

ments are preserved. In both areas, east-dipping faults with normal displacement at the Rotliegend level commonly pass up into westward-facing fault-bend folds in Upper Cretaceous–Tertiary sediments. Onlapping and unconformable beds from this area constrain several phases of inversion, from the Turonian to late Oligocene (Figure 11) (Van Wijhe, 1987; Badley et al., 1989; Van Hoorn, 1990). In view of the lateral continuity of the Dowsing fault zone, it is assumed that inversion of the Sole Pit occurred during one or several of the discrete phases that can be dated where suitable sediments are

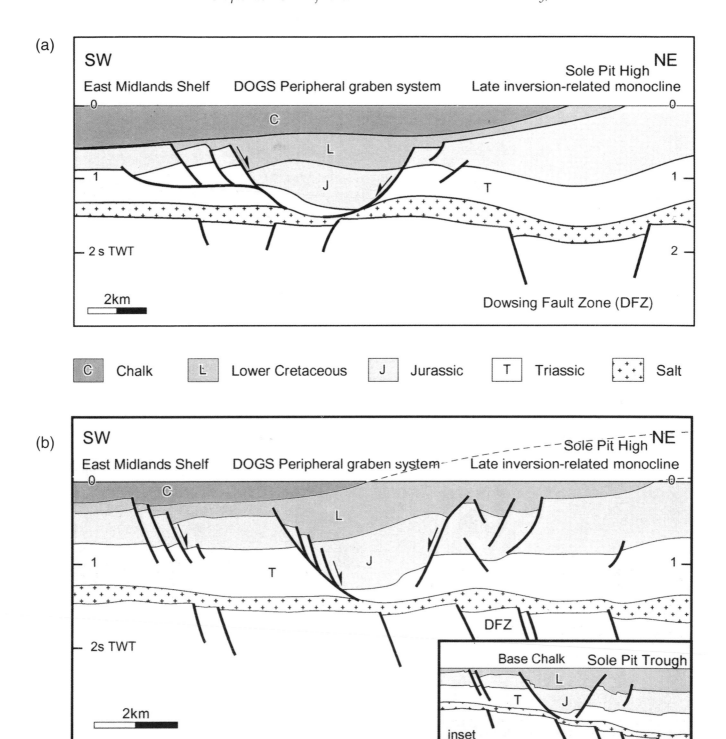

preserved. Although there are examples of compressional faulting at the Rotliegend level elsewhere in the basin (e.g., block 44/23; Ritchie and Pratsides, 1993), there is no evidence for basement uplifts of several thousand feet other than close to the Dowsing fault zone.

As was the case during extension, the geometry of inversion structures near the Dowsing fault zone was influenced by Permian salt. In the Hewitt shelf, individual inversion anticlines are between 2 and 10 km across. They are essentially fault-bend folds with long, gently northeast-dipping backlimbs and short, hooked south-

west-dipping limbs (Figure 11). The inversion is associated with back rotation of the fault blocks associated with reverse slip obliquely updip of the faults (Coward, 1995). During inversion and fault reactivation, the steeply dipping faults propagate upward through the postrift sequence, becoming flatter to die out in fault tips beneath the steep limb of the fault-bend fold. However, where Permian salt is present, rotation of the basement fault blocks in the Dowsing fault zone is decoupled along the salt. The synrift and postrift sediments are uplifted in a large, broad anticline with a half-wavelength of more

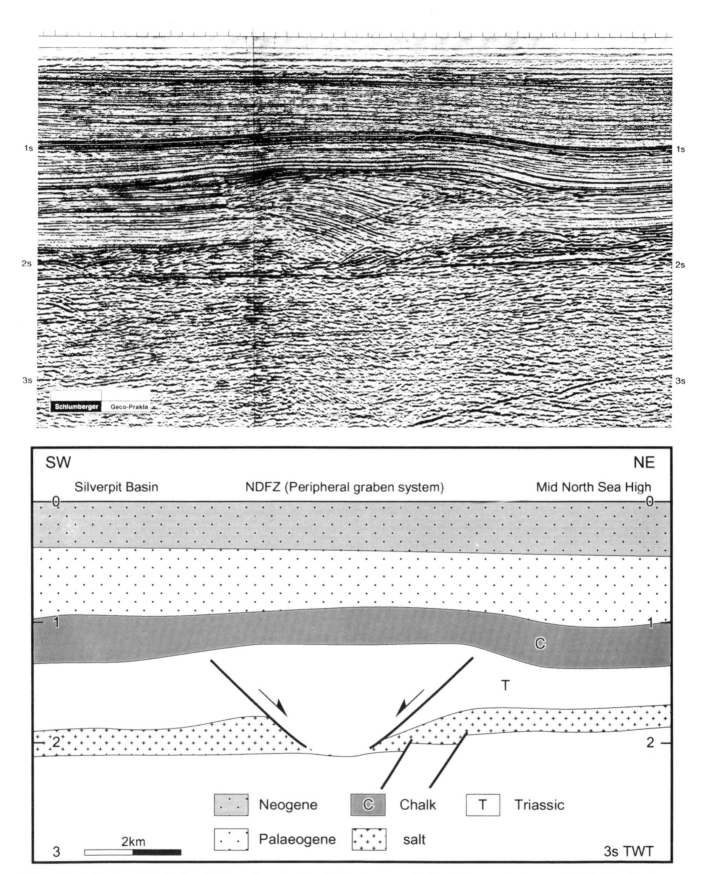

Figure 7—Seismic profile (top) and interpretive drawing (bottom) across the North Dogger fault zone (NDFZ) showing the preserved hanging-wall wedge of Triassic and possibly Jurassic sediments. Line NSP-85-20A. The overlying fold in the Cretaceous and Tertiary may be a forced (drape) structure formed by local salt migration in response to sedimentary loading.

than 50 km. Although the reverse component of movement on the Dowsing fault zone was about 1000 m, reverse displacement did not propagate up through the salt into the Dowsing graben system, where faulting that may be contemporaneous with Dowsing fault zone inversion is in fact extensional (Figure 12). Absence of shortening across or to the west of the Dowsing graben system, while shortening was occurring across the Dowsing fault zone, must indicate eastward translation of the Sole Pit Trough postsalt section relative to basement. Similarly, on the northeastern margin of the southern North Sea Basin, no significant reverse displacement is seen in the North Dogger fault zone.

In the center of the southern North Sea Basin, the Late Cretaceous and Tertiary periods were characterized by amplification or initiation of major fold structures (Figures 10, 13). Also, examples of collapse of major pre-Cretaceous folds are recorded by the major east-facing monocline in Tertiary sediments above the Swarte Bank Hinge and "rim synclines" in block 48/10 (see later discussion). Key features of this second generation of fold structures in the basin are axial symmetry and temporal coincidence of growth with periods of tectonic inversion. Concurrent with tectonic inversion and fold growth, a system of extensional faults developed in the Sole Pit Trough postsalt sequence, particularly in the southern area.

LATE CRETACEOUS–TERTIARY BASIN MODEL

As in the extensional model presented earlier, the presence of salt as an important decoupling horizon underpins the model for basin inversion. The model is based on the observations that Permian salt separates a basement that was shortened by rotation and reverse faulting from cover that was shortened by buckle folding. A simplified depiction of basin inversion is shown in Figure 14. By the end of the Early Cretaceous, the basin was an asymmetric trough; during progressive uplift, this trough became an asymmetric high, whose crest was the Sole Pit High.

Sedimentary sequences of this scale that are underlain by salt having a surface slope, as was generated here by inversion, might be expected to experience gravity spreading (Ramberg, 1981). Gravity spreading within sedimentary wedges has two well-documented effects: (1) extensional faulting in the thickest, highest parts of the wedge balanced by (2) shortening in the thinner, peripheral zones (Dewey, 1988). Gravity spreading in the postsalt section of the southern North Sea can therefore account for pulses of extensional faulting in the Sole Pit Trough and fold amplification to the northeast, which are contemporaneous with inversion movements on the Dowsing fault zone (Figures 13, 14).

This model identifies gravity spreading following inversion as an important driving mechanism for the development of structures in the postsalt section of the southern North Sea. However, the amount of extension in

the Late Cretaceous–Tertiary grabens (eGRABEN) cannot balance the shortening shown by the compressional pillows. A considerable amount of shortening in the cover (cTOP ROT) must be balanced by shortening in the basement (cFOLDS), presumably by inversion of the Dowsing fault zone. Hence,

$$cFOLDS = eGRABEN + cTOP\ ROT$$

The southern North Sea is very different from the gravity gliding zones in the Gulf of Mexico or offshore Angola because there is postsalt deformation of the basement as well as the postsalt cover. The structures in the cover must be balanced with structures in the basement. Thus, the salt structures in the southern North Sea must be classified as thick-skinned and thin-skinned, extensional and contractional structures (Figure 15).

DISCUSSION

Buckle Folds in the Southern North Sea Basin: Salt Swells and Pillows

Several analyses of salt structures in the southern North Sea and nearby basins have concluded that buoyant rise of salt, driven by buoyancy instability alone or differential sedimentary loading, accounts for fold structures in the overlying sediments (Jenyon, 1988b; Geil, 1991). Although these processes are not excluded from the model discussed here, the overburden fold growth mechanism implied by these authors is one of vertical shear in the fold limbs. The model discussed here, however, requires that these structures are buckle folds rather than shear folds.

For all examples of folds appearing on the database available to this study (Figure 1), rudimentary depth conversion was carried out, defining the postsalt sequence as a single layer; depths were calculated from time thicknesses using time–depth relationships recorded in released offset wells. The longest fold structures within the basin will have been sampled several times because they appear on consecutive serial sections. However, such resampling was felt to be valuable because the folds are characteristically noncylindrical. Fold initial wavelengths (λ_D) and layer thicknesses (t) were measured from the depth-converted sections and plotted on a graph of wavelength versus thickness, having corrected measurements where the sections were not normal to fold axes (Figure 16). The data demonstrate that a reasonably linear relationship exists between λ_D and t. A least-squares regression passing through the origin gives

$$\lambda_D/t = 5.4 \quad (r^2 = 0.73)$$

which lies between 5 and 10, the range of λ_D/t ratios identified by Price and Cosgrove (1990) as characteristic of geologic buckle folding.

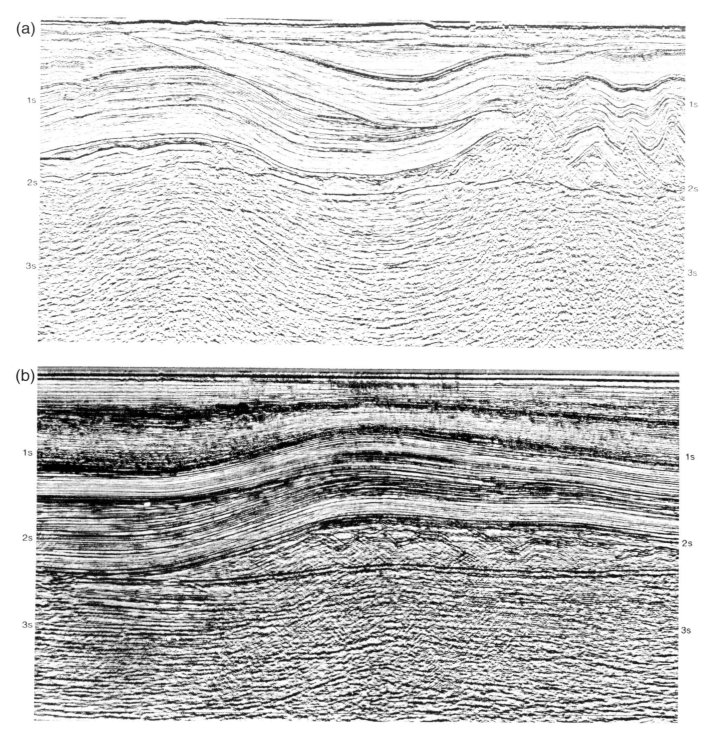

Figure 8—Seismic profiles (this page) and interpretive drawings (opposite page) showing fold structures preserved beneath the Upper Jurassic unconformity; note the consistent sense of facing toward the southwest. (a) From line SNSTI-UK-87-10 (top). (b) From line SNST-83-10 (bottom).

Note that this is an analysis of only the first increment of folding during which the dominant fold wavelength was established in any given area. During much of the subsequent amplification history of folds in the southern North Sea, buckling forces were supplemented by salt movement in response to differential sedimentary load-ing. Evidence for differential loading is the form of Tertiary growth sequences in synclinal troughs adjacent to major anticlines in the center of the salt basin (e.g., block 49/1). Such halokinesis would assist fold amplifi-cation, reducing the magnitude of work against gravity required of the buckling forces.

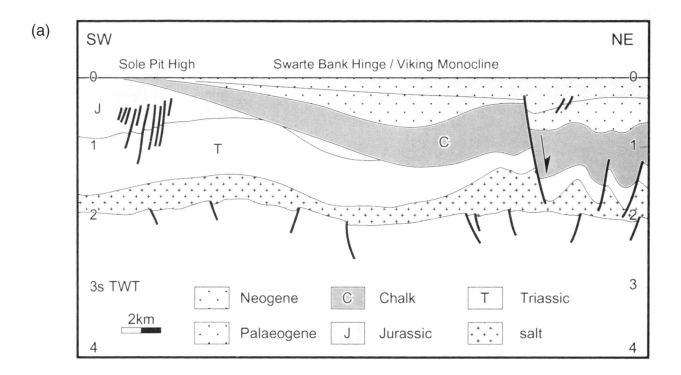

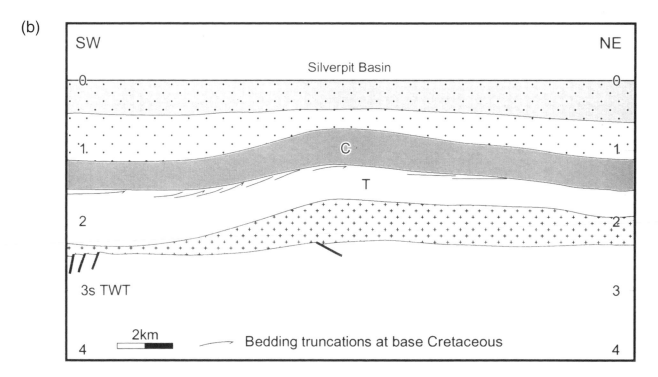

Line–Length Balance Between Extensional and Compressional Structures

One test of this model is to compare the shortening accommodated by fold structures within the basin with the amount of extension recorded in the peripheral graben systems. Several sections across the peripheral graben systems and the foldbelt within the basin were depth converted and line lengths were measured; the results are collated with previously published measurements in Figure 17. These results show that extension across the Dowsing graben system relative to the presalt section is variable along strike, but on the order of 1.5 km, whereas extension across the North Dogger fault zone (NDFZ) graben system on the other side of the basin is about 1 km. The shortening measured across the fold

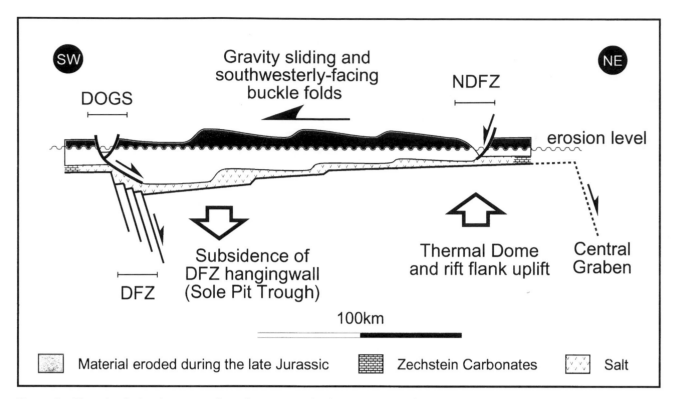

Figure 9—Hypothesis for the generation of structures in the postsalt section during basin extension and tilting. Note how extension across the peripheral graben systems (Figures 5, 7) may have been balanced by extension in the presalt and shortening across the folds that now subcrop the Cretaceous (e. g., Figure 8). Also note the sense of overturn of the folds within the basin. The postsalt sediments are pinned to the basement by marginal carbonates.

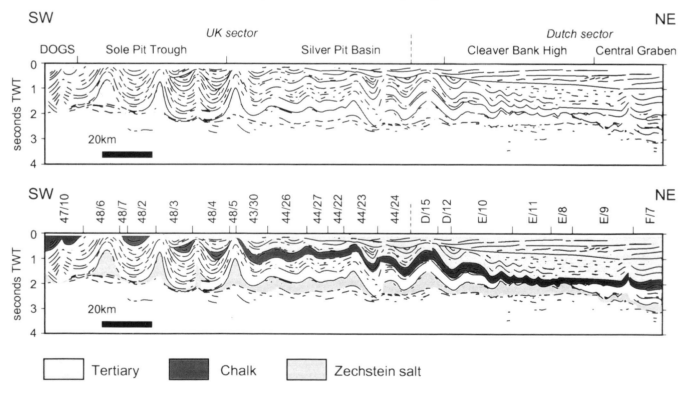

Figure 10—Drawing of line SNST-83-10 (top) annotated with structural domains. Vertical scale is in two-way time; true vertical exaggeration is about 5×. The same drawing (bottom) showing license blocks and Cretaceous chalk (shaded) defining a Triassic and Jurassic wedge whose geometry records Late Jurassic erosion. The chalk is overlain by a second wedge of Tertiary sediments recording reversal of the Jurassic basin tilt.

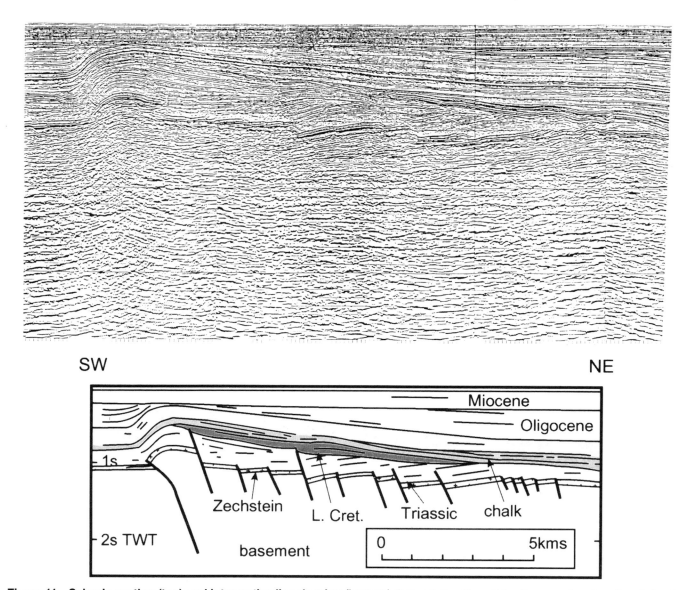

SW

NE

Figure 11—Seismic section (top) and interpretive line drawing (bottom) through the Hewett fault zone, showing the style of inversion where Permian salt is absent. Data from Badley et al. (1989).

structures in the basin was found to be about 2 km in both of the regional lines examined here. The total measured extension was thus slightly greater than the measured shortening; this might be explained by underestimation of net extension in the presalt section on the regional lines. The approximate balance suggested by these figures shows the importance of shortening across salt-cored structures in the southern North Sea.

Comparison of extension across the peripheral graben system and shortening in the basin may seem invalid because these tectonics mostly occurred before and after the Late Cretaceous, respectively. However, extension on faults in the presalt section was reversed during Cretaceous and Tertiary basin inversion. Heave on these faults had previously been part of the line–length balance with the peripheral graben system. Some of the shortening accommodated by fold growth therefore represents a balance of extension that had been "buffered" in the presalt section since the Jurassic.

Morphology of Salt Structures Controlled by Buckle Folding

The model described in this paper identifies buckle folds as the main control on location and morphology of salt structures in the southern North Sea. The depiction of buckle fold evolution in Figure 18 is based on observations made in the southern North Sea Basin and incorporates second-order structural styles noted on several different structures (subdivision into stages A–D is merely for clarity). As buckle folds grow, salt in the anticlinal cores is molded first into swells, then into pillows as fold amplitude increases (Figure 18A, B). The noncylindrical character of swells and pillows observed in the southern North Sea is also a feature of experimental studies of fold growth (Dubey and Cobbold, 1977).

The buckle folds change to active and then passive diapirs by two processes. First, where the cover to the salt is relatively thin, the crest of the fold may be eroded down

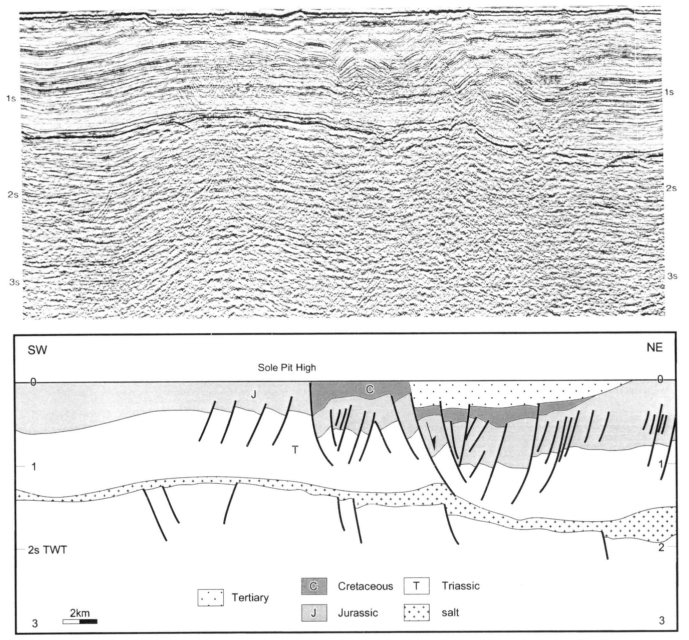

Figure 12—Seismic section (top) and interpretive line drawing (bottom) showing an example of an extensional system in the Sole Pit Trough, line SNSTI-UK-87-10. Some of these normal faults have Tertiary sediments preserved in the hanging wall. We suggest that this zone represents collapse of the Sole Pit High due to gravity spreading.

close to the level of the Zechstein salt. Second, as buckle fold amplification in the southern North Sea increases, gravity-induced collapse structures in the limbs become increasingly common, weakening the crestal region of the fold.

The collapse structures are planar extensional faults, dipping away from the anticlinal crests then detaching upon Middle Triassic Röt Halite; note the angular rather than listric geometry of these faults (Figure 18). This style of deformation is confined to the central and northern parts of the Sole Pit Trough (e.g., block 48/6); illustrated examples are provided by Jenyon (1988b) and Owen and Taylor (1983). Localization of these structures reflects spa-

tial coincidence of large fold structures, substantial depositional Röt Halite thickness, and preservation of the Röt Halite below the Upper Jurassic unconformity. Although these gravity-induced normal faults may have displacements of up to 100 m, in the context of the model discussed here, they are not a direct reflection of regional stresses because they are only present in association with gravity-induced buckle folds.

Fold limbs will collapse back to horizontal following invasion of the fold hinge by an active and then passive downbuilding diapir. Rim synclines within horizontal strata are records of this process (Figure 18D). Neogene examples of this process are rare in the southern North

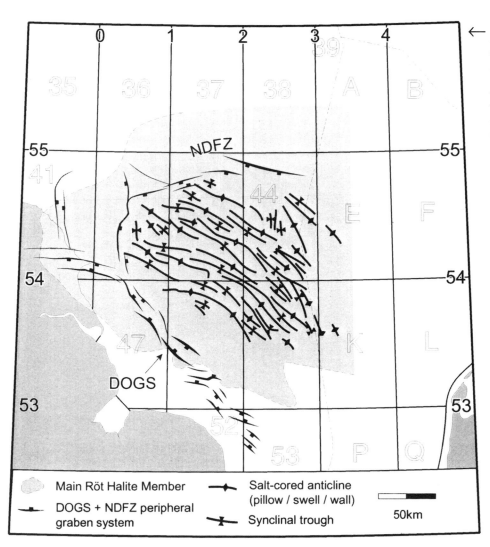

← **Figure 13—Map of major salt-cored anticlines and synclinal troughs of Tertiary age in the southern North Sea. Also shown is the peripheral graben system, which was formed by the end of the Early Cretaceous and is not contemporaneous with the folds on this figure.**

(below)

Figure 14—Hypothesis for tectonics within the postsalt section during basin inversion. Inset (lower right) shows the potential geometry of the Triassic–Jurassic sedimentary wedge (overlain by chalk) following inversion and tilt to the northeast, which induced gravity spreading (main figure).

Main Röt Halite Member

DOGS + NDFZ peripheral graben system

Salt-cored anticline (pillow / swell / wall)

Synclinal trough

50km

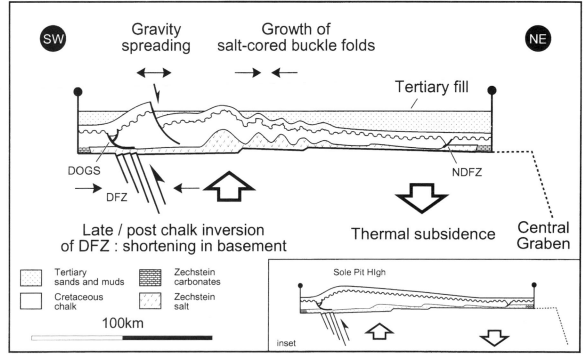

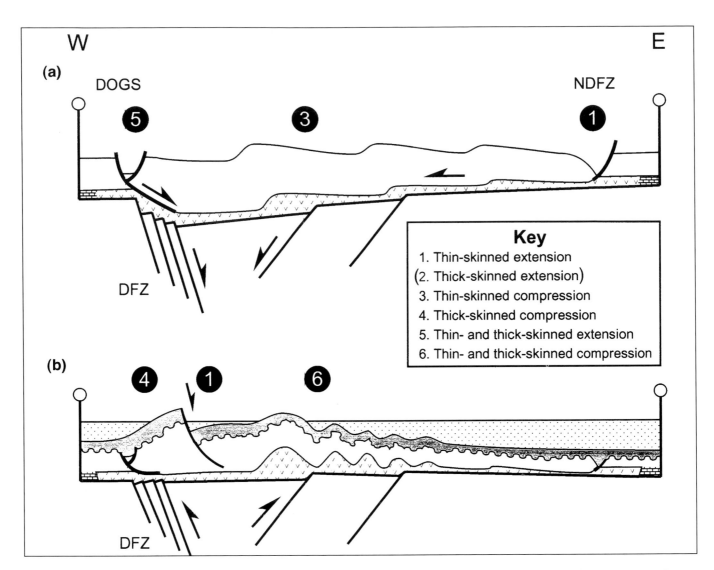

Figure 15—The driving mechanisms for tectonics in the cover to Zechstein salt in the southern North Sea. All the mechanisms listed are present except for simple thick-skinned extension (2). There is always an element of detachment tectonics along the salt.

Sea; the best-developed rim syncline in the basin overlies a pre-Cretaceous fold and can be found in block 48/10. Note that the presence of a rim syncline flanking a salt wall depends on the altitude attained by the fold crest relative to the level of erosion, which is arbitrary. It follows that this process can also account for salt wall emplacement in instances where no record of uplift is preserved in the adjacent strata, such as in the Outer Silverpit salt wall, blocks 48/1 and 48/2 (Jenyon, 1988a).

The morphologic evolution of salt structures described here, while based on observations made in the southern North Sea belt of compressional tectonics, agrees with several conclusions arising from experimental studies of salt structure evolution, which have tended to consider extensional environments (Vendeville and Jackson, 1992). Perhaps the most fundamental point of agreement is that overburden folding and faulting are the primary controls on salt structure initiation and location, rather than vice versa (Vendeville and Jackson, 1992).

Salt Structures Elsewhere in the North Sea

The models just described can be applied to several other regions of the North Sea underlain by Permian salt. Gravity-driven raft tectonics, with sliding toward the deeper parts of the rift basins, have been described by Penge et al. (1993) for the east Central Graben and by Thomas and Coward (in press) for the southern Viking Graben. Both thin-skinned and thick-skinned structures of several generations characterize salt tectonics of the North Sea.

CONCLUSIONS

The evolution of the southern North Sea Basin from the Permian onward has been described in terms of structural models, with Permian salt forming a key detachment horizon upon which the younger sediments (1)

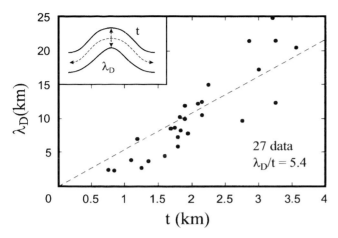

Figure 16—Graph of fold wavelength λ_D against postsalt layer thickness t. Note that λ_D represents the initial rather than finite fold wavelength (see inset). The least-squares regression has been constrained to pass through the origin.

show zones of shortening or extension offset from shortening or extension in the basement and (2) slid under the influence of gravity during various phases of basin tilt. The southern North Sea Basin therefore contains both thick-skinned and thin-skinned tectonics. Structures in the postsalt cover must be balanced with structures in the basement. Furthermore, thin-skinned structures require balancing across the whole basin. Hence, the structural analysis must be carried out on a basin scale. A detailed analysis of individual blocks will give an incomplete and probably erroneous picture.

The history of thick-skinned tectonics affecting the overburden includes the following events:

1. Rifting occurred along the Dowsing fault zone during the Late Triassic with renewed rifting during the Middle–Late Jurassic and Early Cretaceous. The Dowsing graben system was offset from the Dowsing fault zone.

2. Reactivation and inversion of the Dowsing fault zone occurred by block-fault rotation in the basement, decoupled from the postsalt cover. This basement reactivation may have involved oblique to strike-slip displacements. The Rotliegend was essentially restored to horizontal. However, in the postsalt section, the inversion produced a large-wavelength forced fold above uplifted synrift sediments, together with shortening of the postsalt section in a series of buckle folds in the center of the Sole Pit Basin.

The history of gravity gliding of the cover to the salt can be summarized as follows:

1. Prior to the Cretaceous, tilt toward the Dowsing fault zone caused gravity sliding toward the deepest part of the basin (Sole Pit Trough), creating a peripheral graben system around the basin whose extension was balanced by folding within the basin and extension in the presalt section.

2. A reversal of basin tilt during the Tertiary uplifted the thick sedimentary wedge in the Sole Pit Trough, which experienced gravity spreading. This late extension in the postsalt section, together with loss of fault heave due to inversion in the presalt section, was balanced by growth, or amplification, of fold structures within the basin.

Within this structural framework, salt structures in the southern North Sea Basin seem to fall into two broad categories—major pillows or swells that lie in the cores of large buckle fold structures and smaller diapirs that may intrude the crests of salt-cored fold structures or the peripheral graben system. Focusing on temporal evolution of salt structures, morphologic change from pillow through swell to residual wall has been described, noting the occurrence of parasitic structures in thinner, higher salt horizons.

a) Line	Top Trias.	Top Perm.	e
1.	13.6	15.6	2.0
2.	14.7	16.5	1.8
3.	16.3	17.4	1.1
4.	11.5	14.1	2.6
5.	13.9	13.9	0.9
6.	-	-	2.3
7.	36.6	34.7	-1.9
8.	84.6	82.8	-1.8

Figure 17—(a) Table comparing line lengths of top Rotliegend with top Triassic reflectors on depth-converted sections incorporating previously published data; e = top Rotliegend – top Triassic. Some measurements are restricted to parts rather than entire lengths of seismic lines. All lengths are in kilometers and adjusted to give northwest-southeast trending components. Lines 4 and 5 from Arthur (1993); line 6 from Allen et al. (1994). (b) Location map of measured sections.

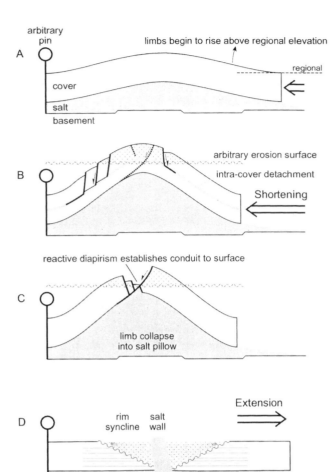

Figure 18—Summary of fold evolution and stages of salt structure evolution based on examples in the southern North Sea. Secondary parasitic structures are also shown.

Acknowledgments   We thank NOPEC/GECO-PRAKLA for permission to reproduce seismic data and our colleagues at Amerada Hess and Imperial College.

REFERENCES CITED

Alberts, M., and J. R. Underhill, 1991, The effect of Tertiary structuration on Permian gas prospectivity, Cleaver Bank area, southern North Sea, U.K, in A. M. Spencer, ed., Generation, accumulation and production of Europe's hydrocarbons: Special Publication, European Association of Petroleum Geology, v. 1, p. 161–173.

Arthur, T. J., 1993, Mesozoic structural evolution of the U.K. southern North Sea: insights from analysis of fault systems, in J. R. Parker, ed., Petroleum geology of northwest Europe: Proceedings of the Fourth Conference, Geological Society of London, p. 1269–1279.

Badley, M. E., J.D. Price, C. Rambech Dahl, and T. Agdestein, 1988, The structural evolution of the Northern Viking Graben and its bearing upon extensional modes of basin formation: Journal of the Geological Society of London, v. 145, p. 455–472.

Badley, M. E., J. D. Price, and L. C. Backshall, 1989, Inversion, reactivated faults and related structures: seismic examples from the southern North Sea, in M. A. Cooper and G. D. Williams, eds., Inversion tectonics: Special Publication, Geological Society of London, v. 44, p. 201–219.

Bailey, J. B., P. Arbin, O. Daffinoti, P. Gibson, and J. S. Ritchie, 1993, Permo-Carboniferous plays of the Silver Pit basin, in J. R. Parker, ed., Petroleum geology of northwest Europe: Proceedings of the Fourth Conference, Geological Society of London, p. 707–715.

Blundell, D. J., R. W. Hobbs, S. L. Klemperer, R. Scott-Robinson, R. E. Long, T. E. West, and E. Duin, 1991, Crustal structure of the central and southern North Sea from BIRPS deep seismic reflection profiling: Journal of Geological Society of London., v. 148, p. 445–457.

Cameron, T. J. D., A. Crosby, P. S. Balson, D. H. Jeffrey, G. K. Lott, J. Bulat, and D. J. Harrison, 1992, United Kingdom offshore regional report: the geology of the southern North Sea: HMSO for the British Geological Survey, London, 152 p.

Christian, H. E., 1969, Some observations on the initiation of salt structures of the southern British North Sea, in P. Hepple, ed., The exploration for petroleum in Europe and North Africa: London, Institute of Petroleum, p. 231–250.

Cobbold, P. R., and P. Szatmari, 1991, Radial gravitational gliding on passive margins: Tectonophysics, v. 188, p. 249–289.

Cooke-Yarborough, P., 1991, The Hewett field, blocks 48/28-29-30, 52/4a-5a, U.K. North Sea, in I. L. Abbotts, ed., United Kingdom oil and gas fields, 25 years commemorative volume: Geological Society of London, Memoir 14, p. 433–442.

Coward, M. P., 1993, The effect of late Caledonian and Variscan continental escape tectonics on basement structure, Paleozoic basin kinematics and subsequent Mesozoic basin development in NW Europe, in J. R. Parker, ed., Petroleum geology of northwest Europe: Proceedings of the Fourth Conference, Geological Society of London, p. 1095–1108.

Coward, M. P., 1995, Balanced sections through inverted basins, in P. G. Buchanan and D. A. Nieuwland, eds., Modern developments in structural interpretation, validation, and modelling: Geological Society of London, Special Publication, No. 99.

Dewey, J. F., 1988, Extensional collapse of orogens: Tectonics, v. 7, no. 6, p. 123–139.

Dubey, A. K., and P. R. Cobbold, 1977, Non-cylindrical, flexural slip folds in nature and experiment: Tectonophysics, v. 38, p. 223–239.

Duval, B., C. Cramez, and M. P. A. Jackson, 1992, Raft tectonics in the Kwanza Basin, Angola: Marine Petroleum Geology, v. 9, p. 389–405.

Fisher, M. J., and D. C. Mudge, 1990, Triassic, in K. W. Glennie, ed., Introduction to the petroleum geology of the North Sea, 3rd edition: Oxford, Blackwell Scientific Publications, p. 191–218.

Geil, K., 1991, The development of salt structures in Denmark and adjacent areas: the role of basin floor dip and differential pressure: First Break, v. 9, no. 10, p. 467.

Glennie, K. W., 1990, Rotliegend sediment distribution: a result of Late Carboniferous movements, in R. F. P. Hardman and J. Brooks, eds., Tectonic events responsible for Britain's oil and gas reserves: Special Publication, Geological Society of London, v. 55, p. 127–138.

Green, P. F., 1989, Thermal and tectonic history of the East Midlands shelf (onshore U.K.) and surrounding regions assessed by apatite fission track analysis: Journal of the Geological Society of London, v. 146, p. 755–773.

Griffiths, P. A., M. R. Allen, J. Craig, W. R. Fitches, and R. J. Whittington, 1995, Distinction between fault and salt control of Mesozoic sedimentation on the southern margin of the Mid-North Sea High, in S. A. R. Boldy, ed., Permian and Triassic rifting in northwest Europe: Geological Society of London, Special Publication No. 91, p. 145–160.

Hodgson, N. A., J. Farnsworth, and A. J. Fraser, 1992, Salt-related tectonics, sedimentation and hydrocarbon plays in the Central Graben, North Sea, UKCS, in R. F. P. Hardman, ed., Exploration Britain: geological insights for the next decade: Special Publication, Geological Society of London, v. 67, p. 31–63.

Holser, W. T., 1979. Rotliegend evaporites, Lower Permian of north-west Europe: geochemical confirmation of the non-marine origin: Erdöl und Kohle, Erdgas, Petrochemie, v. 32, p. 159–162.

Hospers, J., J. S. Rathore, F. Jianhua, E. G. Finnstrom, and J. Holthe, 1988, Salt tectonics in the Norwegian-Danish Basin: Tectonophysics, v. 149, p. 35–60.

Jackson, J., and D. McKenzie, 1983, The geometrical evolution of normal fault systems. Journal of Structural Geology, v. 5, no. 5, p. 471–482.

Jackson, M. P. A., and C. J. Talbot, 1986, External shapes, strain rates and dynamics of salt structures: GSA Bulletin, v. 97, p. 305–323.

Jenyon, M. K., 1985, Basin-edge diapirism and updip salt flow in Zechstein of southern North Sea: AAPG Bulletin, v. 69, no. 1, p. 53–64.

Jenyon, M. K., 1988a, Fault–salt wall relationships, southern North Sea: Oil and Gas Journal, Sept. 5, p. 76–81.

Jenyon, M. K., 1988b, Overburden deformation related to the pre-piercement development of salt structures in the North Sea: Journal of the Geological Society of London, v. 145, p. 445–454.

Jenyon, M. K., and P. M. Cresswell, 1987, The southern Zechstein salt basin of the British North Sea, as observed in regional seismic traverses, in J. Brooks and K. Glennie, eds., Petroleum geology of northwest Europe: London, Graham and Trotman, p. 171–180.

Kirby, G. A., K. Smith, N. J. P. Smith, and P. W. Swallow, 1987, Oil and gas generation in eastern England, in J. Brooks and K. Glennie, eds., Petroleum geology of northwest Europe: London, Graham and Trotman, p. 277–292.

Larroque, J. M., 1993, Decoupling of basement and overburden deformation by a viscous layer (salt): normal faulting and reactivation (abs.): AAPG, International Hedberg Research Conference Abstracts, Sept. 13–17, Bath, U.K., p. 108.

Leeder, M. R., and M. Hardman, 1990, Carboniferous geology of the southern North Sea Basin and controls on hydrocarbon prospectivity, in R. F. P. Hardman and J. Brooks, eds., Tectonic events responsible for Britain's oil and gas reserves: Special Publication, Geological Society of London, v. 55, p. 87–105.

Oudmayer, B. C., and J. de Jager, 1993, Fault reactivation and oblique slip in the southern North Sea, in J. R. Parker, ed., Petroleum geology of northwest Europe: Proceedings of the Fourth conference: Geological Society of London, p. 1281–1290.

Owen, P. F., and N. G. Taylor, 1983, A salt pillow structure in the southern North Sea, in A. W. Bally, ed., Seismic expression of structural styles: AAPG Studies in Geology 15, p. 2.3.2.7–2.3.2.10.

Penge, J., B. Taylor, J. A. Huckerby, and J. W. Munns, 1993, Extension and salt tectonics in the east Central Graben, in J. R. Parker, eds., Petroleum geology of northwest Europe: Proceedings of the Fourth Conference: Geological Society of London, p. 1197–1209.

Price, N. J., and J. W. Cosgrove, 1990, Analysis of geological structures: Cambridge, Cambridge University Press, 502 p.

Pritchard, M. J., 1991, The V-fields, blocks 49/16, 49/21, 48/20a, 48/25b, U.K. North Sea, in I. L. Abbotts, ed., United Kingdom oil and gas fields, 25 years commemorative volume: Geological Society of London Memoir 14, p. 497–502.

Ramberg, H., 1981, The role of gravity in orogenic belts, in K. McClay and N. J. Price, eds., Thrust and nappe tectonics: Special Publication, Geological Society of London, v. 9, p. 125–140.

Richard, P., 1991, Experiments on faulting in a two-layer cover sequence overlying a reactivated basement fault with oblique slip: Journal of Structural Geology, v. 13, no. 4, p. 459–469.

Ritchie, J. S., and P. Pratsides, 1993, The Caister fields, Block 44/23a, U.K. North Sea, in J. R. Parker, ed., Petroleum geology of northwest Europe: Proceedings of the Fourth Conference, Geological Society of London, p. 759–769.

Smith, R. I., N. Hodgson, and M. Fulton, 1993. Salt control on Triassic reservoir distribution, UKCS central North Sea, in J. R. Parker, ed., Petroleum geology of northwest Europe: Proceedings of the Fourth Conference, Geological Society of London, p. 547–57.

Taylor, J. C. M., 1990, Upper Permian–Zechstein, in K. W. Glennie, ed., Introduction to the petroleum geology of the North Sea, 3rd edition: Oxford, Blackwell Scientific Publications, p. 153–190.

Thomas, D. W., and M. P. Coward, in press, Mesozoic regional tectonics and South Viking Graben formation: evidence for localised thin-skinned detachments during rift development and inversion: Marine and Petroleum Geology.

Underhill, J. R., and M. A. Partington, 1993, Jurassic thermal doming and deflation in the North Sea: implications of the sequence stratigraphic evidence, in J. R. Parker, ed., Petroleum geology of northwest Europe: Proceedings of the Fourth Conference, Geological Society of London, p. 337–345.

Van Hoorn, B., 1990, Tectonic events responsible for Britain's oil and gas reserves: a summary, in R. F. P. Hardman and J. Brooks, eds., Tectonic events responsible for Britain's oil and gas reserves: Special Publication, Geological Society of London, v. 55, p. 393–395.

Van Wijhe, D. H., 1987, The structural evolution of the Broad Fourteens Basin, in J. Brooks and K. Glennie, eds., Petroleum geology of northwest Europe: London, Graham and Trotman, p. 315–323.

Vendeville, B. C., 1987, Champs de failles et tectonique en extension: modélisation expérimentale. Mémoires et Documents du Centre Armoricain d'Etude Structurale des Socles, Rennes, 15, 392 p.

Vendeville, B. C., and M. P. A. Jackson, 1992, The rise of diapirs during thin-skinned extension: Marine Petroleum Geology, v. 9, p. 331–353.

Walker, I. M., and W. G. Cooper, 1987, The structural and stratigraphic evolution of the northeast margin of the Sole Pit Basin, in J. Brooks and K. Glennie, eds., Petroleum geology of northwest Europe: London, Graham and Trotman, p. 263–275.

Wu, S., A. W. Bally, and C. Cramez, 1990, Allochthonous salt, structure, and stratigraphy of the northeastern Gulf of Mexico, part II: structure: Marine Petroleum Geology, v. 7, p. 334–370.

Ziegler, P. A., 1975. Geological evolution of the North Sea and its tectonic framework: AAPG Bulletin, v. 59, p. 1073–1097.

Ziegler, P. A., 1988. Evolution of the Arctic–North Atlantic and the western Tethys: AAPG Memoir 43, 198 p.

Ziegler, P. A., 1990, Geological atlas of western and central Europe, 2nd edition: The Hague, Shell International Petroleum Maatschappij B.V., 238 p.

Hooper, R. J., and C. More, 1995, Evaluation of salt-related overburden structures in the U.K. southern North Sea, *in* M. P. A. Jackson, D. G. Roberts, and S. Snelson, eds., Salt tectonics: a global perspective: AAPG Memoir 65, p. 251–259.

Chapter 11

Evaluation of Some Salt-Related Overburden Structures in the U.K. Southern North Sea

Robert J. Hooper
Conoco Inc., Geoscience Resources–Houston
Houston, Texas, U.S.A.

Colin More
Conoco (U.K.) Limited
Aberdeen, U.K.

Abstract

Interpretation of recently acquired high-resolution three-dimensional seismic data has been combined with two-dimensional cross-section restorations and new insights into salt tectonics derived from scaled physical and numerical models. This prompted a reevaluation of the development of salt-related structures in our areas of interest in the U.K. Southern Gas Basin. Salt-related structures in the overburden comprise a series of broadly northwest-trending grabens and associated salt diapirs and walls. These structures are considered to be caused by thin-skinned gravity-driven deformation that triggered and controlled the growth of grabens and diapirs and the later inversion of selected grabens. Additional structures were created by bending and by vertical movements associated with extensionally driven diapiric collapse. The development of the overburden structure was not driven by salt movement; salt structures developed as a simple "reaction" to the thin-skinned extension and subsequent contraction of the overburden.

INTRODUCTION

Much new information is now available from scaled physical and numerical models designed to investigate the development of structures associated with salt tectonics. The following papers and references therein provide a broad spectrum of current research: Vendeville and Jackson (1992), Davison et al. (1993), Nalpas and Brun (1993), Schultz-Ela et al. (1993), Weston et al. (1993), Jackson and Vendeville (1994), and Jackson et al. (1994). Of particular significance to the Southern Gas Basin are data from experiments designed to simulate diapirism and graben growth.

Experimental data reveal that in areas of originally thick salt, thin-skinned extension is an important process in triggering and controlling diapirism and graben growth (Jackson and Vendeville, 1994). Diapiric piercement proceeds through a predictable series of modes: reactive, active, and passive (Vendeville and Jackson, 1992). It must be appreciated that the mode of piercement can change along the trend of a graben–diapir system and can also change through time. Diapiric development typically occurs in two stages: rise (or piercement) and fall (collapse) (Vendeville and Jackson, 1992). During exten-

sion, if salt supply is unrestricted, a small, shallow crestal graben will develop over an essentially triangular diapir. The graben flanks will be downfaulted by a series of stairstep faults, and a broad anticlinal flexure will develop over the diapir. This situation can be maintained as long as extension continues and the supply of salt remains unrestricted. Continued extension simply causes an increase in the width of the diapir and graben. If for some reason the salt supply becomes restricted as extension continues, the diapir can no longer grow and begins to fall. The diapir can then no longer support the graben floor, and the graben collapses into the falling diapir. The change from diapiric rise to diapiric collapse is often marked by fundamental changes in sediment distribution patterns.

In this chapter, we apply the concepts derived from experimental data to one of our areas of interest in the Southern Gas Basin where we have recently acquired and processed a three-dimensional seismic survey. Our studies suggest that the general salt tectonic history here can be explained satisfactorily by concepts derived from experimental data. We discuss this history through the restoration of a key line from the data volume, supplemented by additional data where needed.

STUDY AREA

Our study area in the Southern Gas Basin contains the standard southern North Sea stratigraphy (Figure 1a). Faulted Paleozoic basement (Rotliegendes and older formations) is separated from a faulted overburden sequence (Bacton Group to Chalk) by the salt-dominated Zechstein Group. The basement–overburden package is overlain by a relatively undeformed Tertiary–Recent section. The area has been affected by both Mesozoic extension and Late Cretaceous–Tertiary inversion (e.g., Walker and Cooper, 1987; Arthur, 1993; Oudmayer and de Jager, 1993). Structure in the overburden is dominated by a series of northwest-trending grabens and associated salt diapirs and walls (Figure 1a, b). Basement structure is dominated by northwest-striking faults and fault zones that in places define graben and horst structures.

RESTORATIONS

The seismic data and the depth-converted geoseismic section used as input for the restorations are presented in Figure 1. The geoseismic section was restored in five stages, from the Lower Triassic to the present, using the computer program GEOSEC (Figure 2). The geoseismic section was balanced by area and line length using a combination of oblique simple shear and rigid body rotation and translation. The restored sections were fully decompacted at each stage using the algorithms built into GEOSEC. The restoration panel (Figure 2) reveals that through time the overburden becomes progressively longer by ~15% of the original section length. Restoration of currently mapped basement faults indicates that basement extension was of a similar value. It was impossible, however, from a restoration standpoint alone to address accurately the development of faulting in the basement, and thus basement does not appear on the incremental restorations (Figure 2). Nor was it possible to track accurately the salt budget through time, and we made no attempt to do so. Analysis of our three-dimensional data volume, however, suggests that there was no significant horizontal motion of the various parts of the overburden with respect to one another through time. It is thus possible, at least in a conceptual sense, to restore the overburden through time using standard two-dimensional techniques.

DEVELOPMENTAL HISTORY

The developmental history is discussed in four stages (Figure 2) that correspond to the principal seismic markers picked in the overburden (Figure 1a).

Top of Triassic

The timing of the beginning of extension is, in our opinion, debatable. A widely held view seems to be that because the Lower Triassic Bacton Group is of fairly uni-

form thickness, extension began sometime after its deposition. Extension of the overburden must have occurred during deposition of the Upper Triassic Haisborough Group to account for the observed thickness variations (Figures 1b, 2). Marked thickness variations exist, however, in the Lower Triassic section in several subgraben areas. It has been impossible to determine unequivocally whether the thinning is entirely postdepositional structural thinning or is, in part, related to early salt tectonics. The thinning may suggest that extension began earlier than generally acknowledged (during the deposition of the Bunter Shale Formation?) and thus may imply that cumulative extension over the region is slightly greater than current estimates.

Initial extension created a series of long, linear half-grabens. Long-wavelength footwall uplifts were created (Figure 2) as salt migrated away from downdropping hanging walls and moved in under the adjacent footwalls. Two graben systems can be defined in Figure 1. The major graben system, on the right, is bounded by faults that ultimately sole out in the Zechstein Group; faults bounding the graben to the left sole out within the Haisborough Group. The graben–diapir systems, established early in the extension process, controlled subsequent sedimentation throughout the extensional phase.

Base of Cretaceous Unconformity

Continued post-Triassic extension caused a general increase in half-graben width. Throughout this widening phase, the subgraben diapirs continued to build reactively (Figure 2). In several cases, a shift in half-graben polarity occurred as faults antithetic to the original half-graben-bounding faults became the dominant active graben-bounding faults (Figure 2).

Restorations indicate that the total amount of extension in the section is about the same within the overburden and the basement. The extension, however, is not partitioned in the same manner between the basement and the lower and upper overburden. Extension in the overburden was decoupled from the basement by several detachments (Figure 3). The principal lower detachment was in the Zechstein Group. A locally important higher level detachment formed in the evaporites within the Haisborough Group (principally the Röt Halite). This resulted in the extension being balanced throughout a two-layer overburden sequence (Figure 4). The right-hand graben in Figure 1 penetrates the whole overburden and is bounded by faults that sole out in the Zechstein Group. The left-hand graben in Figure 1, however, is restricted to the upper part of the overburden, its graben-bounding faults soling into the Upper Triassic detachment.

Correlating faults between horizons is complicated by two factors: not only does the area have a protracted and complicated deformational history, involving extension, inversion, and salt tectonics, but there are also multiple detachment levels. Faults at the Top Triassic level, for example, sole out into any of four detachments: the

Figure 1—(a) Northeast-southwest trending depth-converted geoseismic section. (b) Nearby seismic section, through our area of interest in the Southern Gas Basin, highlighting the typical stratigraphy and structure. A faulted basement (Rotliegendes and older) is separated from a faulted overburden sequence (lower Bunter Shale to Chalk) by the salt-dominated Zechstein Group. The whole basement–overburden package is overlain by a relatively undeformed Tertiary–Recent section. Structure in the overburden is dominated by a series of northwest-trending grabens and associated salt diapirs and walls.

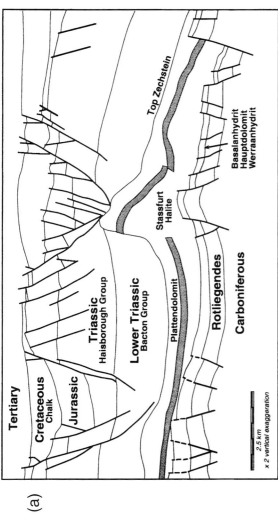

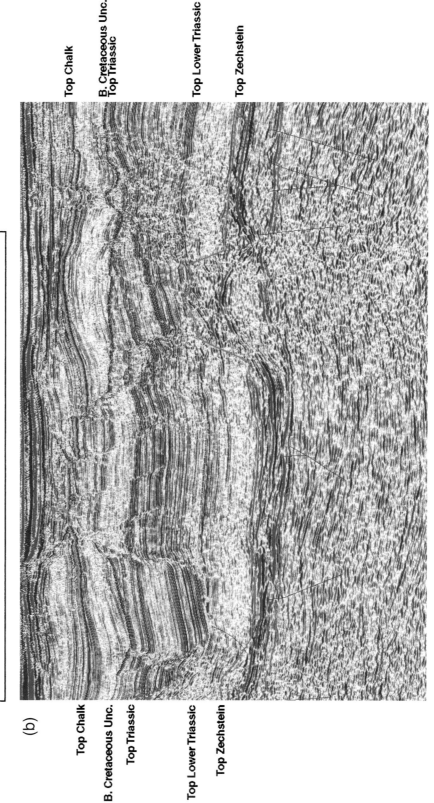

(a)

(b)

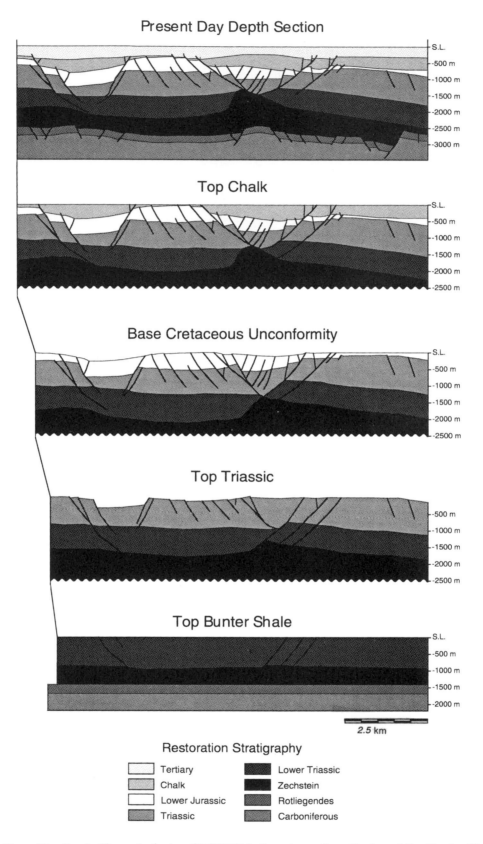

Figure 2—Restoration of the line in Figure 1a (using GEOSEC) in five stages from the top of the Bunter Shale to the present day. It was impossible from restoration alone to address the development of faulting in the basement or to track accurately the salt budget through time. The restoration panel is thus a schematic representation of deformation in the overburden above the Zechstein Group.

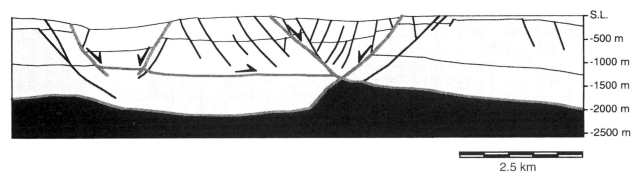

Figure 3—Restoration to the basal Cretaceous unconformity (from Figure 2) annotated to illustrate the partitioning of extension in the section. The total amount of extension is about the same throughout the overburden and the basement. However, the extension is not distributed in the same manner within the basement and the lower and upper overburden. The section acts as a four-layer system where a two-layer overburden is fully decoupled from the basement by the Zechstein Group.

Keuper Halite, the Muschelkalk Halite, the Röt Halite, and the Zechstein Group. Detailed inspection of our three-dimensional data has revealed that only a few major graben-bounding faults actually link through the overburden into the Zechstein Group; most overburden faults sole out within the overburden (e.g., Figure 3).

Top of Chalk

Diapiric rise in association with extension continued through the Jurassic. A fundamental change in the character of salt-related structures occurred after the base of Cretaceous unconformity was formed. Continued extension could no longer be balanced by salt migration, and diapiric collapse began (Figure 3). Grabens continued to widen during the collapse phase, with most graben growth being accommodated by slip on faults active during diapiric rise. Intergraben troughs became locally welded to the underlying basement as salt continued to migrate in an attempt to feed and support the still-extending graben systems. The magnitude of collapse varies along the trend of any given graben–diapir system ranging from relatively minor, accompanied by the development of small salt horns (Figure 5a), to almost complete collapse, with graben flanks and, in some places, floors welded to the underlying basement (Figure 5b).

Continued extension and associated diapiric collapse resulted in syn-Chalk depositional patterns that are fundamentally different from older ones. While pre-Chalk grabens continued to act as primary Chalk depocenters (Figures 1, 2), many graben flanks, areas that were relatively high in the pre-Chalk and had thin pre-Chalk stratigraphic sections, collapsed; they now have comparatively thick Chalk sections (Figures 1, 2). However, many pre-Chalk intergraben troughs, areas that were loci for pre-Chalk deposition, have comparatively thin Chalk sections (Figures 1, 2).

Present Day

Subsequent to, or late in the deposition of the Chalk, the Southern Gas Basin was affected by multiple phases of inversion. In our area, however, the effects of inversion are difficult to unravel from deformation associated with salt tectonics. Throughout most of our data volume, there are no structures unequivocally attributable to inversion tectonics. Only a couple of minor faults in our whole data set show thrust separation; there are no faults where a null point can be defined. In the absence of null points and new contractional faults, one way to quantify the effects of inversion is to look at the motion of blocks relative to their preinversion regional levels. In an area affected by salt tectonics, however, the difficulty comes in defining these regional levels. One potential effect of inversion may be an arching of the Tertiary basal unconformity that occurs locally above some, but not all, grabens (Figure 1, 2). The right-hand graben in Figure 1, the bounding faults of which sole out into the Zechstein Group, shows a prominent arching of the Tertiary basal unconformity. The same unconformity is not arched in the left-hand graben on the same section (Figure 1); that graben shows extension at the same time as the immediately adjacent graben was being contracted.

DISCUSSION AND SUMMARY

There are several key aspects of the general developmental history derived from our study area in the Southern Gas Basin that can be well explained using concepts derived from experimental data. We consider graben and diapiric growth to be triggered and controlled by brittle extension of the overburden. The graben–diapir systems were in place early in the developmental history, and they controlled sedimentation throughout the extensional phase. Extension led to reactive diapiric rise and a widening of both the grabens and underlying diapirs. Restorations demonstrate that the diapirs developed in a reactive mode throughout extension. Eventually the supply of salt to some diapirs became restricted, and continued extension led to diapiric collapse. The fundamental change in the depositional patterns of the Chalk (when compared with those in the underlying Jurassic and Triassic strata) is related to diapiric collapse. Pre-Chalk

Figure 4—Northeast-southwest trending seismic section to illustrate deformation partitioning within the overburden. Most of the faults in the prominent array visible between the basal Cretaceous unconformity and the top of the Lower Triassic sole out into the Röt Halite (upper Lower Triassic). Only one fault at the left end of the section propagates through the Lower Triassic section into the Zechstein Group.

grabens continued to act as Chalk depocenters. Many graben flanks that were pre-Chalk highs (i.e., relatively thin pre-Chalk section) became lows with a relatively thick Chalk section. Intergraben troughs that were pre-Chalk depocenters (i.e., thick pre-Chalk section) became highs and have only a thin Chalk section.

The restorations reveal that the total amount of extension in the section is approximately the same throughout the overburden and the basement. The extension, however, is not partitioned in the same manner within the basement and the overburden. Analysis of fault patterns from overburden and basement horizons suggests that the overburden behaved as a multilayered system, detaching on several Triassic halites (principally the Röt Halite), and was fully decoupled from the Rotliegendes and older formations by the Zechstein Group. Deformation of the overburden is thus considered to be thin skinned.

The cause of the deformation within the overburden cannot be solely determined from within the bounds of our survey area; the local survey area needs to be viewed within the regional context of the southern North Sea. We consider the early extensional and later contractional drive in the overburden to result principally from simple tilting as basement levels across the area changed in response to extension and later inversion of the Sole Pit and adjacent southern North Sea basins. Overburden structures resulted mainly from thin-skinned gravity-driven deformation that was responsible for triggering and controlling graben and diapir growth and for the selective later inversion of some grabens. Additional structures were created by bending and vertical movements associated with extensionally driven diapiric collapse. The development of the overburden structure was not driven by salt movement; salt structures developed as a simple "reaction" to the thin-skinned extension and subsequent contraction of the overburden.

Acknowledgments   *We would like to thank the management of Conoco (U.K.) Limited, Conoco Inc., and our partners in the study area for their support and approval to publish this paper. The interpretations and conclusions, however, are those of the authors and do not necessarily reflect the views of Conoco (U.K.) Limited, Conoco Inc., or our partners in the study area. The comments of two anonymous reviewers and an editor, D. G. Roberts, resulted in significant improvements to the organization and focus of the manuscript and are gratefully acknowledged.*

REFERENCES CITED

Arthur, T. J., 1993, Mesozoic structural evolution of the UK Southern North Sea: insights from the analysis of fault systems, *in* J. R. Parker, ed., Petroleum geology of northwest Europe: Proceedings of the 4th Conference, Geological Society of London, p. 1269–1279.

Davison, I., M. Insley, M. Harper, P. Weston, D. Blundell, K. McClay, and A. Quallington, 1993, Physical modelling of overburden deformation around salt diapirs: Tectonophysics, v. 228, p. 255–274.

Jackson, M. P. A., and B. C. Vendeville, 1994, Regional extension as a geologic trigger for diapirism: GSA Bulletin, v. 106, p. 57–73.

Jackson, M. P. A., B. C. Vendeville, and D. D. Schultz-Ela, 1994, Structural dynamics of salt systems: Annual Review of Earth and Planetary Sciences, v. 22, p. 93–117.

Nalpas, T., and J.-P. Brun, 1993, Salt flow and diapirism related to extension at a crustal scale: Tectonophysics, v. 228, p. 349–362.

Oudmayer, B. C., and J. de Jager, 1993, Fault reactivation and oblique-slip in the southern North Sea, *in* J. R. Parker, ed., Petroleum geology of northwest Europe: Proceedings of the 4th Conference, Geological Society of London, p. 1281–1290.

Schultz-Ela, D. D., M. P. A. Jackson, and B. C. Vendeville, 1993, Mechanics of active salt diapirism: Tectonophysics, v. 228, p. 275–312.

Vendeville, B. C., and M.P.A. Jackson, 1992, The rise and fall of diapirs during thin-skinned extension: The University of Texas at Austin, Bureau of Economic Geology Report of Investigations 209, 60 p.

Weston, P. J., I. Davison, and M. W. Insley, 1993, Physical modelling of North Sea salt diapirism: *in* J. R. Parker, ed., Petroleum geology of northwest Europe: Proceedings of the 4th Conference, Geological Society of London, p. 559–567.

Walker, I. M., and W. G. Cooper, 1987, The structural and stratigraphic evolution of the northeast margin of the Sole Pit Basin, *in* J. Brooks, and K. W. Glennie, eds., Petroleum geology of north west Europe: London, Graham and Trotman, p. 263–275.

(a)

Top Chalk

B. Cretaceous Unc.

Top Triassic

Top Lower Triassic

Top Zechstein

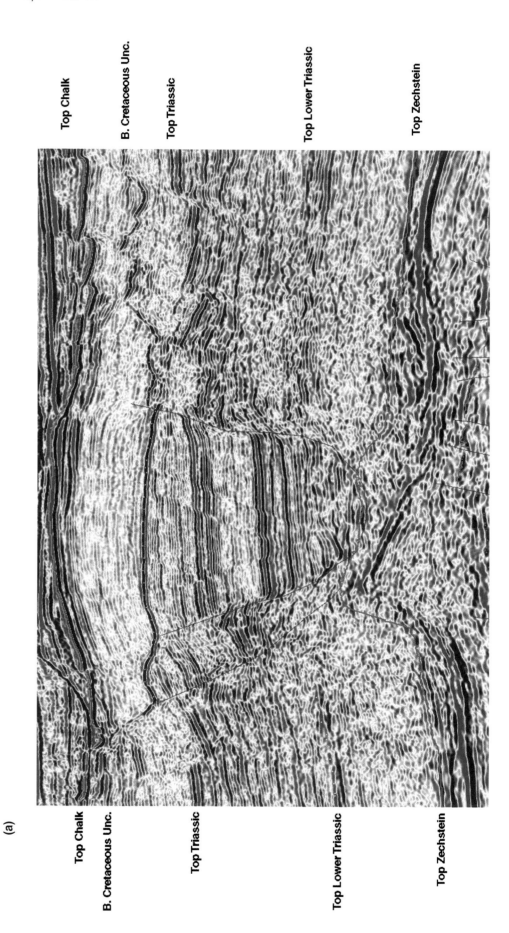

Top Chalk

B. Cretaceous Unc.

Top Triassic

Top Lower Triassic

Top Zechstein

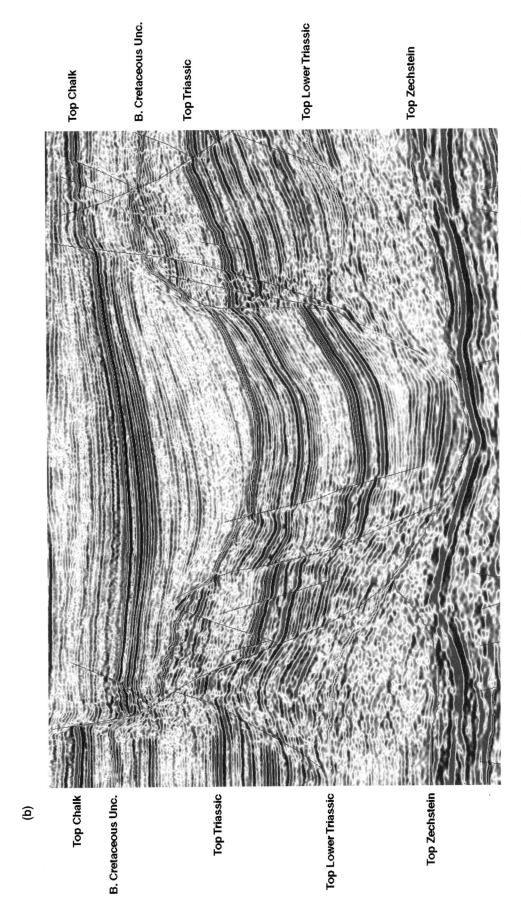

Top Chalk

B. Cretaceous Unc.

Top Triassic

Top Lower Triassic

Top Zechstein

(b)

Top Chalk

B. Cretaceous Unc.

Top Triassic

Top Lower Triassic

Top Zechstein

Figure 5—Northeast-southwest trending seismic sections illustrating the progressive development of diapiric collapse features. The magnitude of collapse varies along the trend of any given graben–diapir system, ranging from (a) *(opposite page)* relatively minor collapse, accompanied by the development of small salt horns, to (b) *(this page)* almost complete collapse, with graben flanks and, in some places, floors welded to the underlying basement.

Remmelts, G., 1995, Fault-related salt tectonics in the southern North Sea, The
Netherlands, *in* M. P. A. Jackson, D. G. Roberts, and S. Snelson, eds., Salt
tectonics: a global perspective: AAPG Memoir 65, p. 261–272.

Chapter 12

Fault-Related Salt Tectonics in the Southern North Sea, The Netherlands

G. Remmelts

Rijks Geologische Dienst
Haarlem, The Netherlands

Abstract

This chapter describes the relationship between basement faulting and salt flow in the southern North Sea in general and in the Dutch Central North Sea Graben area specifically. The research was executed within a national program on radioactive waste disposal in The Netherlands and is based on a 2000-km regional two-dimensional seismic survey.

Salt structures consist of Upper Permian Zechstein salt. The salt structures are, almost without exception, related to basement faults. This paper concerns the relative location of the basement faults and salt structures, the triggering of salt flow, and the rate of this flow.

The Netherlands sector of the continental shelf can be subdivided into a number of salt provinces on the basis of their developmental stage. This subdivision coincides with the structural units. The maturity of the salt structure is proportional to the throw of the basement fault. A relationship exists between basement faulting and increased salt flow, and the interference of fault systems in the basement is reflected in the geometry of the salt structures. The effect of basement faulting on salt flow appears to correspond with the results of previously published physical modeling. The buoyancy force related to the density inversion alone seems insufficient to pierce the overburden. Weakening of the overburden and enhancing the buoyancy forces by differential loading enable the salt to flow and breach the overburden.

INTRODUCTION

This paper is a result of a study on salt flow executed within a national research program on the feasibility of radioactive waste disposal in salt structures in The Netherlands (Remmelts and van Rees, 1992; Remmelts et al., 1993). The aim of this part of the research program was to enhance understanding of the relationship between the geologic setting, development, and architecture of salt structures. The study is based on released seismic and well data; 2000 km of the regional SNST83 survey by Nopec/Geco-Prakla (Figure 1), crossing a large number of salt structures, provided the framework for this research.

For many years, the mechanism of salt flow has been the subject of numerous papers. The relationship between salt flow and basement faulting has been described for different parts of the North Sea Basin: for Germany by Jaritz (1987) and Kockel (1990); for Denmark by Boldreel (1985), Koyi (1991), Koyi and Petersen (1993), and Koyi et al. (1993); for England by Jenyon (1986, 1988) and

Oudmayer and De Jager (1993); and for The Netherlands by Clark-Lowes et al. (1987). Pre-Zechstein rocks are considered "basement" in this paper. Salt flow and deformation in the southern North Sea are closely related. Basement faulting enables differential loading on the salt layer and causes weakening of the overburden of the salt, thus creating a situation favorable for salt flow (also described by Koyi et al., 1993). Salt structures tend to develop on the upthrown basement block, slightly offset but parallel to the basement fault. This link with basement faulting has implications for the location, orientation, geometry, and timing of flow of the salt structures with respect to the basement faults. In this chapter, we describe and illustrate this relationship for the southern North Sea in general and for the Dutch Central North Sea Graben more specifically. Observations made agree with the results of physical models presented by Koyi (1991) and Vendeville and Jackson (1992) and with numerical models indicating that the overburden is too strong to be breached by rock salt when salt is driven only by buoyancy forces (VUA, 1992a,b).

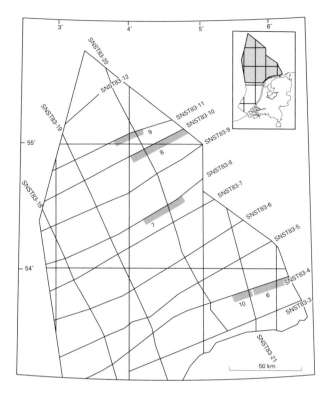

Figure 1—Location map of the SNST83 seismic survey. Shaded areas indicate the locations of the seismic sections and are accompanied by their figure numbers.

SOUTHERN NORTH SEA

Geologic Setting

The southern North Sea coincides largely with the western part of the intracratonic Southern Permian Basin (Figure 2) formed at the end of the Variscan orogeny (Ziegler, 1990). In the north, the Southern Permian Basin was separated from the Northern Permian Basin by the Mid North Sea High and the Ringkøbing-Fyn High. The Variscan structural grain has influenced the entire development of the basin, and many Variscan faults were rejuvenated during subsequent tectonic phases. The highs surrounding the basin were often re-uplifted to form new sediment sources. During the Late Permian, periodic isolation from influx of fresh seawater resulted in deposition of the thick sequences of evaporites belonging to the Zechstein Group in the basin center. The Zechstein Group generally comprises five main evaporite cycles. In the basin center in Germany, sixth and seventh cycles have been reported. In map view, these cycles are developed as clastic fringe facies and as carbonate and anhydrite platforms in shallow water along the basin margins and on the Mid North Sea High. The basin center was occupied by thick salt, mainly of the second and third cycle (Van Adrichem Boogaert and Kouwe, 1993). Salt deposited in these second and third cycles has a thickness of 400–800 m and 200–400 m, respectively, and is the most important unit for salt flow.

In Late Triassic–Jurassic time, the area was transected by rift systems. The North Sea rift system, which dominates the area, propagated southward in Triassic (Viking Graben) and Late Jurassic time (Dutch Central North Sea Graben). Thick-skinned extension in an essentially subsiding area characterizes the geologic setting of the southern North Sea. This structural style is one of the controlling factors in the evolution of salt structures. In this respect, the southern North Sea differs from areas where the geologic setting is dominated by high rates of sedimentation, progradation of sediments, and growth faulting (e.g., Gulf of Mexico) or from passive margins where gravity gliding plays a major role (e.g., Brazil and Angola).

Salt Structures: Orientation, Maturity, and Timing

Salt flow in the southern North Sea is limited to the center of the Southern Permian Basin and displays a systematic pattern in the location, orientation, and maturity (pillow or piercement) of the salt structures. This pattern bears a remarkable resemblance to the structural grain of the southern North Sea. It divides the southern North Sea into domains of salt structures having a specific dominant orientation and maturity (Figure 2). Three domains comprise elongated structures, each of different orientation. First, in the north, from the Central North Sea Graben to northern Germany, the salt structures trend north to north-northeast. In the Central North Sea Graben, the Glückstadt Graben, and to a lesser extent, the Horn Graben, the salt forms diapiric walls. Second, from English waters to onshore Netherlands, the salt structures are elongated northwest and comprise both pillows and walls. Third, in onshore Germany, an eastward trend is most conspicuous, and salt walls are generally lower than elsewhere; pillows are also present. Between these three areas, small salt structures are mainly present with no preferred orientation.

Triassic faults strike north to north-northeast and are mainly located in the northern part of the basin where they form the boundary of the Central North Sea Graben, the Horn Graben, and the Glückstadt Graben. In onshore Germany, this orientation is expressed as the Ems Low and the Weser Depression (Figure 2). The strongest faulting was along the margins of the Central North Sea Graben; during the main phase of Jurassic rifting, many Triassic faults in the Central North Sea Graben were rejuvenated. Farther south, extension was accommodated by transtensional slip in a northwest direction.

Not only do the location and orientation of basement faults and salt structures coincide but so does the timing of flow. Thus, the age of piercement of the salt structures (Jaritz, 1987) as well as the age of the basement faults is dominantly Triassic in the northern part of northern Germany (north-trending Glückstadt Graben and Gifthorn Trough). Conversely, in the south, a Jurassic age prevails (northwest to east oriented). In the Central North Sea Graben, the age of piercement is Late Jurassic, as is the main tectonic phase.

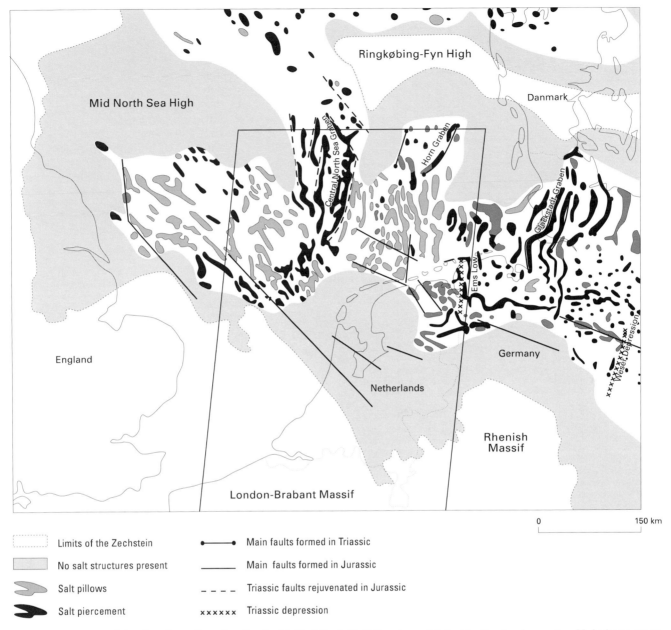

Figure 2—Outlines of the Southern Permian Basin (dotted line). Salt flow is restricted to the basin center. Main basement faults, salt pillows, and diapirs are indicated (after Heybroek et al., 1967). Note the coincidence of the location, orientation, and geometry of the faults and the salt structures. Rectangle shows the location of Figure 3.

CENTRAL NORTH SEA GRABEN AREA

Geologic Setting

The relationships described for the southern North Sea are treated here in more detail for the Central North Sea Graben area, comprising the graben itself and its adjacent platform areas (Figure 3). The Central North Sea Graben is bounded to the north by large faults with throws that diminish toward the south. The depth of the graben increases stepwise from the western flank (Intermediate Block and Step Graben) toward the central zone. In the east, this depth increase is more abrupt. The flanking plat-

forms comprise the Elbow Spit High and the Cleaver Bank High to the west and the Schill Grund High to the east (Heybroek, 1975; Van Wijhe, 1987). Toward the south, the boundaries of the Central Graben are poorly defined, and northwest-trending structures start to dominate. Here, the Terschelling Basin forms the southeastern extension of the Central Graben. The Terschelling Basin is bordered to the southeast by the higher Ameland Block.

The faults that delineate the Central North Sea Graben strike north in the north. Toward the south, they shift to north-northeast and interfere with northwest-oriented faults. The north-northeast strike is a continuation of the Horn Graben and can be traced south of the Central

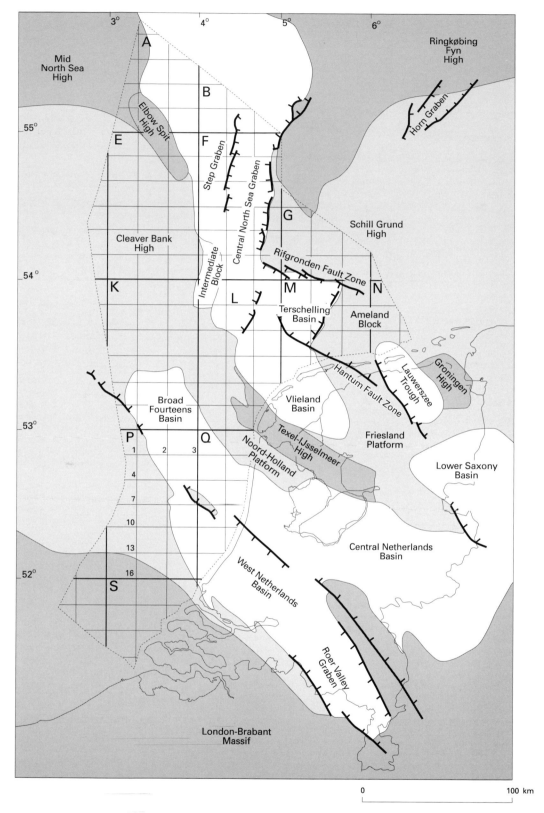

Figure 3—Structural units in The Netherlands and The Netherlands section of the continental shelf. The massifs and highs are indicated with dark shading. Subsided areas are white and intermediate areas are shown with light shading. The study area is located north of lat. 53°30'N. From north to south, there is a shift in the dominant structural trend from north to northwest. In the P quadrant, the numbering of the blocks is shown as an example for the other quadrants (see text).

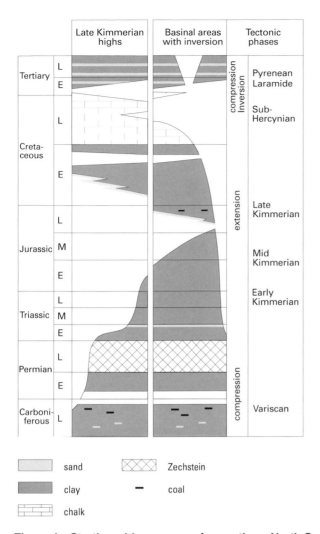

Late Kimmerian highs		Basinal areas with inversion	Tectonic phases

(columns from left to right: age, sub-stage)

Figure 4—Stratigraphic sequence for southern North Sea. The left column represents the platforms and the right column represents the basins. Hiatuses (shown in white) are the result of various tectonic phases.

Legend:
- sand
- clay
- chalk
- Zechstein
- coal

Late Cretaceous (Subhercynian phase, Santonian–Campanian), early Tertiary (Laramide phase, middle Paleocene) and late Eocene–early Oligocene (Pyrenean phase). The center of the Central Graben was inverted by reverse slip along preexisting normal faults. Meanwhile, the platforms flanking the Graben continued to subside, and northwest-oriented dextral strike-slip, accompanied by accelerated salt flow, took place during the Pyrenean phase (Van Hoorn, 1987; Oudmayer and De Jager, 1993). During Neogene and Quaternary time, the entire North Sea subsided.

A schematic overview of the stratigraphy and timing of the tectonic events is presented in Figure 4. The left column represents the platforms, and the right column represents the basins.

Because of the extreme volume of salt that has moved from its original position into the salt walls and the accompanying erosion or dissolution, it is impossible to calculate the original thickness of the Zechstein Group in the Central North Sea Graben. Regional, relatively undisturbed thicknesses on the neighboring platforms indicate an original salt thickness there of about 1200 m (Heybroek, 1975). The present-day thickness in the central part of the graben is commonly less than 500 m in withdrawal areas. These estimates are based on seismic data that, at the depth of these withdrawal basins, have lost significant resolution. Moreover, owing to lack of economic interest, the Zechstein Group has never been fully drilled here. The thickness (or height) of the salt in the salt structures is generally more than 2500 m. The greatest height is more than 5000 m in block F17, in a salt wall related to the western boundary fault between the Central Graben and the Intermediate Block (Figure 5).

On the Cleaver Bank High, the top of the Zechstein Group is at a depth between 2000 m in the north and 2500 m in the south. The regional thickness of the Zechstein averages about 1000 m, although in the south it decreases to about 500 m. During the erosion period, salt swells were truncated and large amounts of salt were dissolved (e.g., the Elbow Spit High). For many salt structures on the highs, the original salt was completely withdrawn during Tertiary time.

On the Schill Grund High and Ameland Block, the top of the Zechstein sequence generally consists of the Zechstein 3 Anhydrite (Main Anhydrite) or Carbonate (Platten Dolomite) members. On the Ameland Block, erosion has not removed all of the Triassic sequence, so the underlying Zechstein 4 Formation is preserved. The depth of the top of the Zechstein Group is about 3000 m and gradually decreases to 2500 m on the Ameland Block. The thickness of the Zechstein generally varies from about 200 m in the withdrawal areas to about 1250 m in the salt pillows. The thickness deviates from these figures at a few localities; the salt is almost absent in areas where severe erosion has taken place (block G13) and measures more than 2000 m in the thickest salt (block N4). In the Northwest German Basin, east of the Schill Grund High, the thickness of the Zechstein is 1000–1500 m, of which 75% consists of the Zechstein 2 cycle (Jaritz, 1987).

North Sea Graben (Heybroek, 1975; Ziegler, 1990). The northwest strike in the southern Central North Sea Graben and Terschelling Basin is related to basins farther south (Broad Fourteens, Central Netherlands, West Netherlands, and Lower Saxony basins). All these fault orientations are largely reflected in the location and geometry of the salt structures.

After deposition of the Zechstein evaporites, Triassic faulting created the initial Central North Sea Graben. The Early Jurassic was a period of relative tectonic quiescence. The late Kimmerian (Late Jurassic–Early Cretaceous) phase accentuated the graben structure and again uplifted the graben flanks, rejuvenating Triassic basement faults. Fault-controlled differential subsidence, characteristic of Jurassic–Early Cretaceous time, was replaced by a more regional subsidence during Cretaceous–Tertiary time (Ziegler, 1987, 1990). However, as a result of compressive stress, regional subsidence was later interrupted by three inversion phases (Van Wijhe, 1987) during the

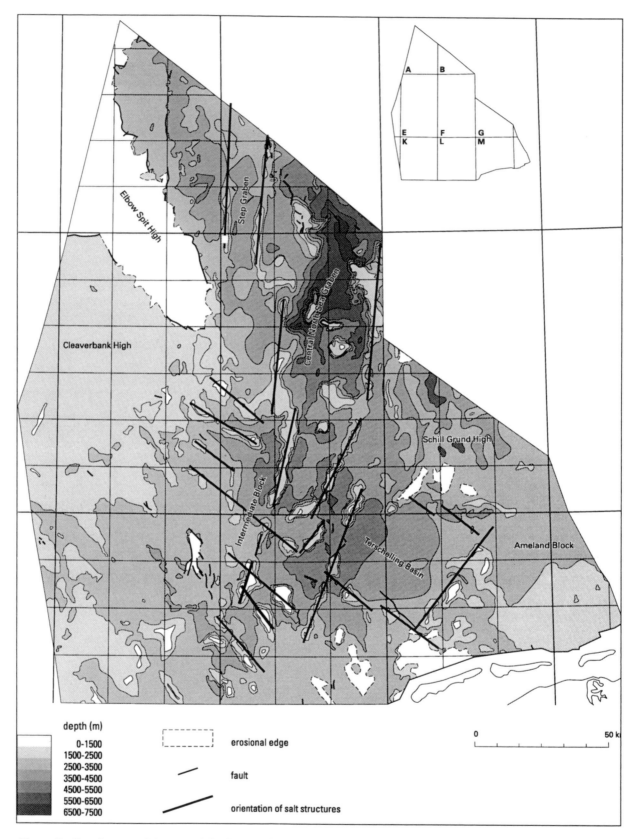

Figure 5—Depth map of the top of the Zechstein (after ECL, 1983). The sub-areas, structural trends, and orientations of the salt structures are indicated. Elongated salt walls in the north trend north, whereas smaller salt structures in the south trend northwest (after ECL, 1983).

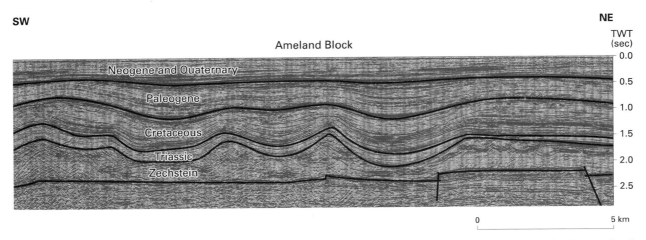

Figure 6—Seismic section SNST83-4 over the Ameland Block. Late Cretaceous salt flow related to the Subhercynian inversion phase has resulted in numerous internal unconformities in the Upper Cretaceous Chalk sequence.

Salt Provinces

The location, morphology, and geometry of the salt structures in the Central North Sea Graben area are shown on the top Zechstein depth map (Figure 5). The salt structures occur in two different settings: platforms and basins. Each is characterized by the morphology of the salt structures and their geometry and orientation.

On the platform areas, such as the Cleaver Bank High, the Schill Grund High, and the Ameland Block, salt flow was minor and resulted in salt pillows only. In the basinal areas, the Central Graben, the Step Graben, the Intermediate Block, and the Terschelling Basin, considerable salt flow has occurred, and salt domes and salt walls were formed.

Salt Structures on the Platform Areas

The platform areas are relatively stable, and faults are concentrated along their edges and in the northeastern corner of the Cleaver Bank High. The salt structures on the platform areas consist of pillows only (Figure 6). Only along the platform edges have some diapirs formed, and these are associated with boundary faults. The distance from the center of a salt structure to the center of the next structure (the wavelength) is ~7.5 km (ranging from 6 to 10 km) on the Schill Grund High and the Ameland Block. Variations also occur due to undulating salt ridges. On the Cleaver Bank High, the wavelength of the salt pillows is 5–7 km. Along the fringe of the platform areas, larger pillows have developed, whose wavelength (10–15 km) coincides with the spacing of faults in the basement (e.g., the Terschelling Basin).

The structural grain of the platform areas is closely related to salt structures there. Salt structures show a vague north or north-northeast trend on the Schill Grund High, whereas on the Cleaver Bank High, they have a northwest trend. Where only minor salt flow has taken place (e.g., on the northwestern and central parts of the Cleaver Bank High), basement faults of large throw are rare. Along the edges of the highs and in the northeastern corner of the Cleaver Bank High, larger salt structures exist over boundary faults having large throw. The absence of large fault slip retards differential loading because the overburden blankets the salt without being weakened by normal faulting. In consequence, triggering effects do not occur and the overburden is probably too strong to be breached by buoyancy forces only (Tsai et al., 1987; Koyi, 1991; VUA, 1992a,b).

In the northeastern corner of the Cleaver Bank High, flow was greatest. Northwest-trending salt walls developed parallel to the faults in the basement. This area shows interference of the north-northeast striking fault system related to the Central North Sea Graben and the northwest-oriented fault system related to the basins farther south.

The fault-bounded southern extension of the Elbow Spit High (Figure 7) is an example of salt flow from the downthrown block upward along the fault. The overburden at the downthrown side is much thicker due to greater subsidence of the basement. Buoyancy forces driving upward flow increased by differential loading, and basement faulting weakened the overburden.

Salt Structures in the Basinal Areas

The Central North Sea Graben is bounded by faults having very large throws, typically many hundreds of meters at the base Zechstein level. The larger throws result in differential loading of the salt on either side of these faults (Jenyon, 1986; Koyi et al., 1993).

The maturity and morphology of the salt structures seem to be closely related to the structural grain of the area and the intensity of the differential subsidence of the basement (Figures 3, 5). North-trending salt walls are developed in the northern part of the Central North Sea Graben and in the Step Graben. Toward the south, however, this orientation shifts to the north-northeast. The salt walls formed in the most pronounced part of the Central North Sea Graben parallel to the boundary faults and to the faults bisecting the Step Graben.

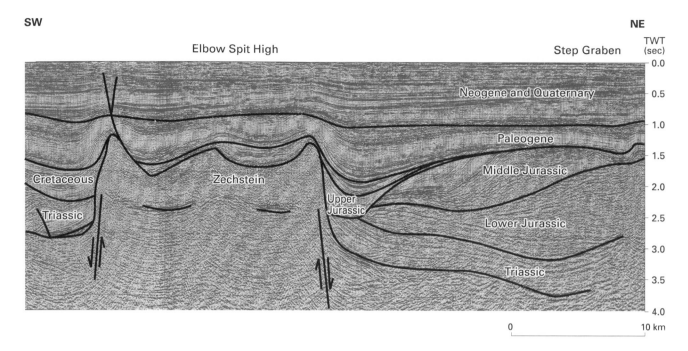

Figure 7—Seismic section SNST83-8 over the fault-bounded southern extension of the Elbow Spit High. This is a classic example of salt flow from the downthrown block along the fault toward the upthrown block.

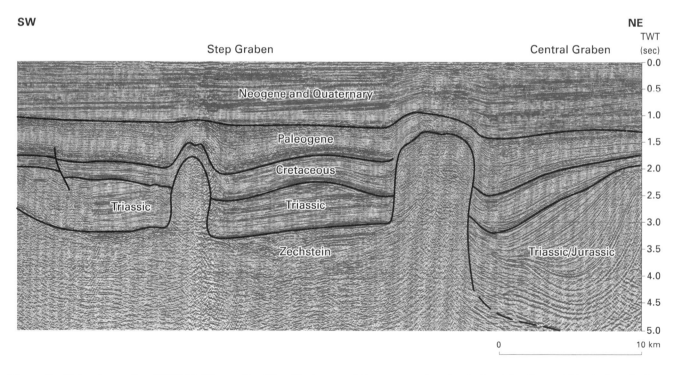

Figure 8—Seismic section SNST83-10 over the Step Graben and the western border fault of the Central North Sea Graben. Salt structures directly overlay the basement fault, forming an extensive salt wall. Secondary peripheral sinks migrate toward the salt structure on the northeastern side.

(a)

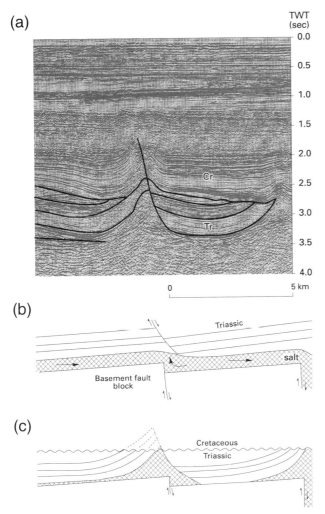

(b)

(c)

Figure 9—(a) Seismic section SNST83-11 over the Step Graben. (b) and (c) Schematic drawings of the evolution of such a structure. (b) Block faulting and accompanying rotation, which causes salt flow, mainly along the bedding plane partially up the fault plane. (c) Geometry of the overburden after salt flow and subsequent truncation by erosion.

In the southern end of the Central Graben and the Terschelling Basin, as well as in the northern end of the Central North Sea Graben (B quadrant and German waters), a northwest fault strike intersects with the north-northeast and north strikes. Alternating slip along these two fault systems seems to be the reason for the development of solitary salt domes instead of salt walls along the single fault trend. Intersection of faults results in a local maximum salt flow at that point. The salt structures tend to line up along both fault trends, but elongated structures are parallel to the dominant trend. Thus, south of the Central Graben, the trends of salt structures are mainly northwest and to a lesser extent north-northeast, whereas in the north, the orientation is more northward (Figure 5). That this phenomenon occurs both in the north and the south indicates that the intersection of fault trends has a greater influence than basement depth and the throw of the faults.

In the Step Graben, the two north-trending salt walls in the east (blocks B10, B13, B16, F1) are most prominent (Figures 5, 8). Toward the Elbow Spit High (A quadrant), the salt thins, and only a few isolated salt structures are present. The salt walls are clearly linked to large fault displacements in the basement. However, no salt wall is present above the fault zone separating the Elbow Spit High from the Step Graben. Chaotic structures in the sediments directly overlying the fault indicate salt flow. Severe erosion at the beginning of the Cretaceous has probably removed a significant amount of salt, which may have caused the salt wall to collapse. West of the boundary fault, another north-striking fault transects the Step Graben. This fault has a recent throw of 150–200 m and is overlain by a salt wall.

Figure 9 shows a section through this part of the Step Graben, where tilting of the overburden is the cumulative effect of salt tectonics. The angular discordance between the Triassic reflectors and the base Cretaceous reflector is much larger in the footwall than in the hanging wall. This can be explained by tilting of the basement blocks, which resulted in differential loading and subsequent updip salt flow. Depletion of the salt in the downthrown block has resulted in subsidence of the overlying sediments, which preserved them from Early Cretaceous erosion. Accumulation of salt in the upthrown block steepened the overburden, which was subsequently eroded. Apart from this salt flow along the top of basement, salt also flowed along the fault plane from the downthrown block upward. Diapiric piercement took place in Cretaceous time.

The boundary fault separating the Step Graben from the Central Graben shows some large salt structures (blocks B14, B17), but in contrast to the salt walls, they are irregularly aligned along the fault. This may be related to the interference of different fault trends. The salt wall above the boundary fault shows an enormous rim syncline at the side of the Central Graben (Figure 8). This Upper Jurassic depocenter indicates salt flow toward this structure. There is, however, no well data to confirm the precise age of the peripheral sink. In agreement with Trusheim (1960), the subsequent secondary peripheral sinks gradually move toward the dome. The diapiric stage seems to have been reached in Cretaceous time. The structure shows that high growth rates occurred during the Eocene–Oligocene. The seismic sections show that from Neogene time onward, no rim synclines formed. The slight doming of very young sediments indicates recent flow of the salt or differential compaction.

The main salt structures on the Intermediate Block are related to the boundary fault with the Central Graben. The smaller size of structures here compared with those along the boundary between the Step Graben and Central Graben is thought to result from more restricted differential loading on opposite sides of smaller "basement" faults. Because of erosion in the Intermediate Block, there is no evidence of the first phase of salt flow here. The diapiric stage was reached in Late Jurassic time with a period of major salt flow, which continued during the Early Cretaceous.

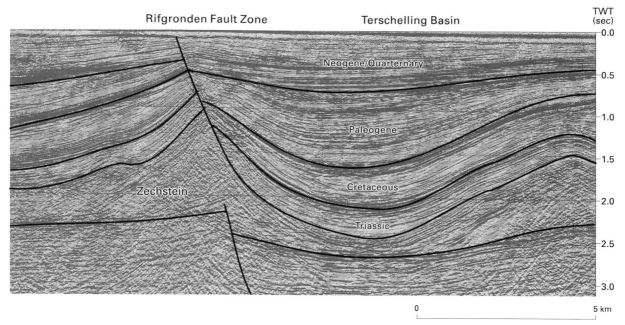

Figure 10—Seismic section SNST83-4 crossing the Rifgronden fault zone, which is accompanied by a salt structure. Listric faulting in the overburden can be traced up to ~150 m below the sea bed. The salt swell is on the upthrown block.

In large parts of the Terschelling Basin, Zechstein salt is no longer present. Salt structures initiated before Middle Jurassic were presumably truncated during the Kimmerian tectonism (Figure 4). Chaotic structures in the Mesozoic sediments indicate early salt flow and dissolution. Salt structures still present here are restricted to the fault-bounded edges of the Terschelling Basin. The first salt flow is clearly pre-Cretaceous, but no further constraints exist. The main periods of movement were in Late Cretaceous and especially Eocene–Oligocene time when the basement faults were rejuvenated.

Figure 10 shows the asymmetric depletion of the salt above the Rifgronden fault. The salt deposits on the upthrown block are less affected by salt flow, whereas salt in the downthrown block is almost the entire source for the structure. This implies that the asymmetric flow was generated by differential loading. Like most of the salt structures here, this particular structure probably developed before Cretaceous time.

Timing of Salt Flow

The timing of the initial salt flow is difficult to estimate because of the large hiatus in the sedimentary record, especially on the platform areas (Figure 4). In the Central Graben, subtle thickness changes in the Triassic of the F quadrant indicate salt flow then (Dronkert et al., 1989; Fontaine et al., 1993). Evidence is poor for the platform areas, although some rim synclines seem to show sediment thickness changes indicating Middle or Late Triassic salt flow. On the platform areas, the first definite record of salt flow is the beginning of the Cretaceous. In the basinal areas, no evidence for salt flow in Early Jurassic time has been found, but this may be due to the poor seismic qual-

ity at the Lower Jurassic level. During the Late Jurassic, most of the salt structures in the basinal areas reached their diapiric phase. Large secondary rim synclines filled with as much as 1000 m of Upper Jurassic strata record this phase. Breakthrough in Late Jurassic time is illustrated in Figure 7. Based on well data (F11-03, F8-02) and the seismic character of the Triassic sequence, a relatively constant thickness is assumed for the Lower Triassic. The Upper Jurassic at the eastern side of the salt structure (block F11, Step Graben), however, thickens toward the structure, indicating a secondary rim syncline.

Strong salt flow can be deduced for Late Cretaceous time. This acceleration of salt flow is thought to be related to Subhercynian inversion. Many internal unconformities in the Upper Cretaceous Chalk have recorded this inversion (e.g., G13, G17, M3, M5, N1) (Figure 6). Thin-skinned compression squeezed the salt structures, forcing them to flow. A comparable phase of strong salt flow coincided with a compressive phase during Eocene–Oligocene time. The salt flow had generally ceased at the beginning of the Miocene because the source layer became depleted (Figure 8). In certain cases, the Miocene reflector is almost undisturbed above the salt structures.

In Neogene time, salt flow had ceased in general. In a few salt structures, however, salt flow related to recent faulting (e.g., Figure 10) did continue. Undisturbed layers can only be found in the first 150 m below the seabed. The underlying basement fault has a small throw, possibly suggesting mainly strike-slip during the Tertiary. Some gravitationally driven faulting occurred along the flanks of salt structures, for which the salt is assumed to have acted as a passive detachment level.

In the southern part of the Central Graben (blocks L5, L6, L7), salt pillows probably formed in Late Jurassic time, whereas salt domes and walls pierced during the Early Cretaceous. Salt structures in this part of the Central Graben formed slightly later, which corresponds to the observation that structural development of the graben propagated southward (Ziegler, 1990).

MODEL

The observations described here are in agreement with results of various physical models (Koyi, 1991; Vendeville and Jackson, 1992) and numerical models (e.g., VUA, 1992a,b). Salt flow related to basement faulting occurred in the following sequence (see also Koyi et al., 1993). The initial condition was a basement covered by a sequence of salt having a sedimentary overburden (Figure 11a). If this overburden was fluid and heavier than the underlying salt (due to compaction of the sediments), an instability developed (Rayleigh-Taylor instability) in which buoyancy forced the salt to move upward. Physical models of this instability, using two fluids as equivalents for the salt and the overburden, have produced beautiful domal structures (e.g., Whitehead and Luther, 1975; Rönnlund, 1987; Koyi, 1991).

Recent numerical models, however, which model the overburden as a brittle (non-Newtonian) material, have shown that these buoyancy forces alone are not strong enough to break the strength of the overburden (VUA, 1992a,b). No salt flow occurs and the situation remains static. Basement faulting, however, affect this situation. Although the slip along the basement fault is partly accommodated by the salt sequence, the resulting differential subsidence affects the overburden. Local extension of the overburden over the fault trace weakens the overburden by fracturing and faulting (Figure 11b). Moreover, thicker sediment accumulation on the hanging-wall block leads to differential loading, which plays an important role in the flow of the salt. Once the overburden becomes weaker than the enhanced buoyancy force (and, if significant, the regional tectonic stresses), the salt is able to flow. Asymmetric depletion of the salt layer, with a maximum on the downthrown side, illustrates the importance of differential loading (Figure 11c). Strong salt flow in a compressive strike-slip setting implies the relevance of tectonic stresses in the process of salt flow. Thin-skinned shortening squeezes already existing salt structures, forcing upward flow (Nilsen et al., 1995; Rowan, 1995).

CONCLUSIONS

In the southern part of the North Sea, the salt structures follow the structural grain of the region and are directly related to faulting in the basement. Amplitudes of salt structures are directly proportional to the throw along the fault in the basement. On the relatively stable highs, small pillows are mainly found, while along the edges, larger pillows occur in combination with larger fault throws. In the basins, salt walls have developed in combination with long uninterrupted fault zones having large throw.

(a)

(b)

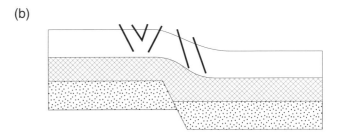

(c)

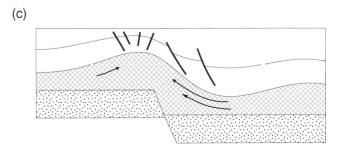

Figure 11—Model for salt flow related to basement faulting in the southern North Sea. (a) Density inversion is created when a sufficiently thick overburden overlies a salt sequence. However, buoyancy forces in the salt are not large enough to overcome the strength of the overburden. (b) Extensional basement faulting is accommodated by the salt layer, but the overburden is weakened by stretching and bending. (c) Differential loading and tectonic stresses in addition to buoyancy forces are greater than the strength of the overburden.

Periods of enhanced salt flow correspond to the timing of tectonic phases. Salt flow in the southern part of the Central North Sea Graben started slightly later than in the north. This is in agreement with the southward propagation of regional deformation. The increase of salt flow during tectonic phases is related mainly to differential loading (induced by fault slip). The accelerated growth rates during inversion phases are not reflected in significant throws along the basement faults. This may imply a direct role of compressive stresses on the flow of salt during inversion.

Salt flow induced only by buoyancy forces in a tectonically undisturbed area is probably prevented by the strength of the overburden. Basement faulting that has large slip deforms and weakens the overburden. In combination with increased buoyancy forces generated by differential loading, this enables salt to pierce the overburden above the basement fault (Figure 11). The position of the withdrawal areas of the salt structures may shift laterally due either to alternating normal or reversed slip along the underlying basement fault or to interference of two or more fault systems.

Acknowledgments   *This paper is based on results from research executed for a program on radioactive waste disposal on land (OPLA, Phase 1A). Nopec/Geco-Prakla is thanked for permission to use their SNST83 survey. Douwe van Rees and Ton Wildenborg are thanked for their valuable and constructive discussions. Critical remarks on the draft version by Hemin Koyi significantly improved the quality of the manuscript. Figures were drafted by Roel van de Kraan, Han Bruinenberg, and Christa van Houten.*

REFERENCES CITED

Boldreel, L. O., 1985, On the structural development of the salt dome province in northwest Jutland, Denmark, based on seismic studies: First Break. v. 3, p. 15–21.

Clark-Lowes, D. D. N, C. J. Kuzemko, and D. A. Scott, 1987, Structure and petroleum prospectivity of the Dutch Central Graben and neighboring platform areas, *in* J. Brooks and K. W. Glennie, eds., Proceedings of the Third Conference on Petroleum Geology of northwest Europe: London, Graham and Trotman, p. 337–356.

Dronkert, H., S. D. Nio, W. Kouwe, N. Van der Poel, and Y. Baumfalk, 1989, Exploration and production potential of the Buntsandstein: Non-Exclusive Study Intergeos BV, 3 vol., 450 p.

ECL, 1983, Offshore Netherlands, petroleum exploration appraisal, vol. 1. regional exploration review: Henley on Thames, Exploration Consultants Limited.

Fontaine, J. M., G. Guastella, P. Jouault, and P. de la Vega, 1993, F15-A: a Triassic gas field on the eastern limit of the Dutch Central Graben, *in* J. R. Parker, ed., Petroleum Geology of Northwest Europe: Proceedings of the Fourth Conference, Geological Society of London, p. 583–593.

Heybroek, P., 1975, On the structure of the Dutch part of the Central North Sea Graben, *in* A. W. Woodland, ed., Petroleum and continental shelf of northwest Europe: London, Applied Science Publishers, p. 339–351.

Heybroek, P., U. Haanstra, and D. A. Erdman, 1967, Observations on the geology of the North Sea area: Proceedings of the 7th World Petroleum Congress, v. 2, p. 905–916.

Jaritz, W., 1987, The origin and development of salt structures in northwest Germany, *in* I. Lerche and J. J. O'Brien, eds., Dynamical geology of salt and related structures: Orlando, Florida, Academic Press, p. 480–493.

Jenyon, M. K., 1986, Salt tectonics: New York, Elsevier, 191 p.

Jenyon, M. K., 1988, Fault-salt wall relationships, southern North Sea: Oil and Gas Journal, v. 88, p. 76–81.

Kockel, F., 1990, Morphology and genesis of northwest German salt structures: Proceedings of symposium on diapirism with special reference to Iran, Geologic Survey of Iran, v. 2, p. 225–249.

Koyi, H., 1991, Gravity overturns, extension, and basement fault activation: Journal of Petroleum Geology, v. 14, p. 117–142.

Koyi, H., and K. Petersen, 1993, The influence of basement faults on the development of salt structures in the Danish Basin: Marine and Petroleum Geology, v. 10, p. 81–94.

Koyi, H., M. K. Jenyon, and K. Petersen, 1993, The effect of basement faulting on diapirism: Journal of Petroleum Geology, v. 16, p. 285–312.

Nilsen, K. T, B. C. Vendeville, and J.-T. Johansen, 1995,

Influence of regional tectonics on halokinesis in the Nordkapp Basin, Barents Sea, *in* M. P. A. Jackson, D. G. Roberts, and S. Snelson, eds., Salt tectonics: a global perspective: AAPG Memoir 65, this volume.

Oudmayer, B. C., and J. De Jager, 1993, Fault reactivation and oblique-slip in the southern North Sea, *in* J. R. Parker, ed., Petroleum geology of northwest Europe: Proceedings of the Fourth conference, Geologic Society of London, p. 1281–1292.

Remmelts, G., and D. J. van Rees, 1992, Structureel geologische ontwikkeling van zoutstructuren in Noord Nederland en het Nederlandse deel van het Continentaal Plat: Rap. no 30.105/TRB2, project GEO-1A, OPLA Fase 1A, Rijks Geologische Dienst, 106 p.

Remmelts, G., E. Muyzert, D. J. van Rees, M. C. Geluk, C. C. de Ruyter, and A. F. B. Wildenborg, 1993, Evaluation of salt bodies and their overburden in The Netherlands for the disposal of radioactive waste, b. salt flow: Rijks Geologische Dienst, Report 30.012B/ERB, 87 p.

Rönnlund, P., 1987, Diapiric walls, initial edge effects and lateral boundaries: UUDMP research report no. 45, Uppsala University, Uppsala, Sweden, 31 p.

Rowan, M. G., 1995, Structural styles and evolution of allochthonous salt, central Louisiana outer shelf and upper slope, *in* M. P. A. Jackson, D. G. Roberts, and S. Snelson, eds., Salt tectonics: a global perspective: AAPG Memoir 65, this volume.

Trusheim, F., 1960, Mechanism of salt migration: AAPG Bulletin, v. 44, p. 1519–1540.

Tsai, F. C., J. E. O'Rourke, and W. Silva, 1987, Basement rock faulting as a primary mechanism for initiating major salt deformation features: Proceedings of the 28th Symposium on Rock Mechanics, Tucson, Arizona, p. 621–631.

Van Adrichem Boogaert, H. A., and W. F. P. Kouwe, eds., 1993, Stratigraphic nomenclature of The Netherlands, revision and update by RGD and NOGEPA: Mededelingen Rijks Geologische Dienst, 50 p.

Van Hoorn, B., 1987, Structural evolution, timing, and tectonic style of the Sole Pit inversion: Tectonophysics, v. 137, p. 239–284.

Van Wijhe, D. H., 1987, Structural evolution of inverted basins in the Dutch offshore: Tectonophysics, v. 137, p. 171–219.

Vendeville B. C., and M. P. A. Jackson, 1992, The rise of diapirs during thin-skinned extension: Marine and Petroleum Geology, v. 9, p. 331–353.

VUA, 1992a, Intraplaattektoniek, bekkenmodellering en zoutdiapirisme. Implementatietechnieken: Rapportage OPLA project februari 1992, OPLA92-56, Sectie Structurele Geologie/Tektoniek, Vrije Universiteit Amsterdam.

VUA, 1992b, Intraplaattektoniek, bekkenmodellering en zoutdiapirisme. Inhoudelijke deelstudie: Modellering Deel 1 en 2: Rapportage OPLA project augustus 1992, OPLA92-xx, Sectie Structurele Geologie/Tektoniek, Vrije Universiteit Amsterdam.

Whitehead, J. A., and D. S. Luther, 1975, Dynamics of laboratory diapir and plume models: Journal of Geophysical Research, v. 80, p. 705–717.

Ziegler, P. A., 1987, Late-Cretaceous intra-plate compressional deformations in the Alpine foreland—a geodynamic model: Tectonophysics, v. 137, p. 389–420.

Ziegler, P. A., 1990, Geological atlas of western and central Europe, 2nd edition: Shell International Petroleum Maatschappij, 239 p

Mohriak, W. U., J. M. Macedo, R. T. Castellani, H. D. Rangel, A. Z. N. Barros, M. A. L. Latgé, J. A. Ricci, A. M. P. Mizusaki, P. Szatmari, L. S. Demercian, J. G. Rizzo, and J. R. Aires, 1995, Salt tectonics and structural styles in the deep water province of the Cabo Frio region, Rio de Janeiro, Brazil, *in* M. P. A. Jackson, D. G. Roberts, and S. Snelson, eds., Salt tectonics: a global perspective: AAPG Memoir 65, p. 273–304.

Chapter 13

Salt Tectonics and Structural Styles in the Deep-Water Province of the Cabo Frio Region, Rio de Janeiro, Brazil

W. U. Mohriak

J. M. Macedo

R. T. Castellani

H. D. Rangel

A. Z. N. Barros

M. A. L. Latgé

J. A. Ricci

A. M. P. Mizusaki

P. Szatmari

L. S. Demercian

J. G. Rizzo

J. R. Aires

Petróleo Brasileiro S. A. Department of Exploration and Petróleo Brasileiro S. A. Research Center Rio de Janeiro, Brazil

Abstract

The Cabo Frio region, offshore Rio de Janeiro, lies between two of the most prolific Brazilian oil provinces, the Campos and Santos basins. Major geologic features have been identified using a multidisciplinary approach integrating seismic, gravity, petrographic, and borehole data. The Cabo Frio frontier region is characterized by marked changes in stratigraphy and structural style and is unique among the Brazilian marginal basins. Major geologic features include the deflection of the coastline and pre-Aptian hinge line from northeast to east; a large east-striking offshore graben related to salt tectonics; a northwest-trending lineament extending from oceanic crust to the continent; basement-involved landward-dipping (antithetic) normal faults in shallow water; a stable platform in the southern Campos Basin; a thick sequence of postbreakup intrusive and extrusive rocks; and, near the Santos Basin, a mobilized sequence of deep-water postrift strata affected by landward-dipping listric normal faults. These faults are unusual in salt-related passive margins in that they dip landward, apparently detach on the Aptian salt, and show large late Tertiary offsets. Locally, the older sequences do not show substantial growth in the downthrown blocks.

South of the Rio de Janeiro coast, a phenomenal landward-dipping fault system detaches blocks of the Albian platform to the north and, to the south, coincides with the depositional limit of the Albian platform. This deep-water fault system controls features that can be mapped for hundreds of kilometers along strike and forms an Albian stratigraphic gap tens of kilometers wide.

Two end-member processes of salt tectonics in the Cabo Frio region result in either synthetic or antithetic basal shear along the fault weld under the overburden: (1) thin-skinned processes, in which the listric faults were caused by salt flow in response to gravity forces related to massive clastic progradation from the continent; and (2) thick-skinned processes, in which faulting was indirectly triggered by diastrophic causes or disequilibrium in the basement topography. The structural styles in the Cabo Frio region are compared with analogs from other basins and with sandbox models.

INTRODUCTION

The Cabo Frio province is the frontier between the two most important oil provinces in offshore southeastern Brazil—the Campos and Santos basins (Figure 1). Both basins are genetically related to the breakup of Gond- wana and formation of the South Atlantic passive margin in the Early Cretaceous (Asmus and Ponte, 1973; Campos et al., 1974; Mohriak, 1988; Guardado et al., 1989). However, several outstanding differences in the temporal and geographic distribution of stratigraphic sequences, structural evolution, magmatism, and salt tectonics

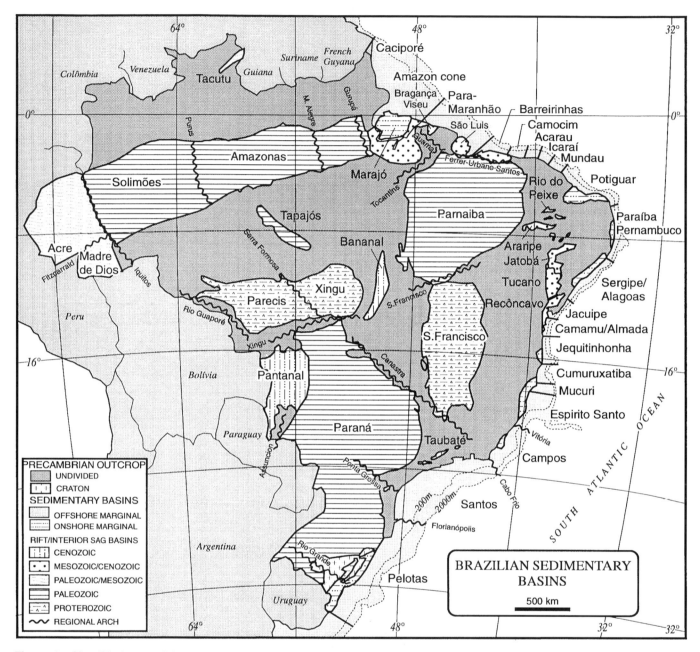

Figure 1—Simplified map of the onshore and offshore Brazilian sedimentary basins and intervening regional arches. The northwest-trending Cabo Frio Arch divides the Campos and Santos basins. Toward central Brazil, northwest-trending regional arches identified as Alto da Canastra and Serra Formosa constitute the Cabo Frio–Parecis lineament, which is punctuated by Mesozoic–Cenozoic alkaline intrusions.

distinguish this interbasinal region as having singular geological characteristics and exploratory plays compared with other South Atlantic passive margin basins (Mohriak et al., 1989).

A few regional studies have analyzed parts of the Cabo Frio province, particularly the Santos Basin (e.g., Ojeda et al., 1983; Ven, 1983; Andrade et al., 1983; Williams and Hubbard, 1984; Pereira et al., 1986). Most of these works have concentrated on sequence stratigraphy and structural interpretation of the shallow northern embayment of the Santos Basin without advancing toward the Cabo

Frio Arch, which was also the limit for similar research on the Campos Basin (e.g., Lobo and Ferradaes, 1983; Dias et al., 1987). A multidisciplinary basin analysis of the Cabo Frio province, conducted from 1989 to 1991, aimed at assessing the tectonic framework and identifying possible exploratory plays (Mohriak, 1991). The data set consisted of gravity, magnetic, and seismic data (about 15,000 km of regional profiles) and about 40 exploratory boreholes drilled by Petrobrás and several major oil companies operating under risk contract licenses in the Santos Basin. The basin analysis integrated small, local features in a regional context, resulting in a better understanding of the overall basin evolution and the identification of several exploratory leads.

In this paper, we analyze structural styles in the Cabo Frio province and contrast two end-member mechanisms of salt tectonics: *synthetic basal shear,* which has been described for several Atlantic-type marginal basins (e.g., Campos and Kwanza), and *antithetic basal shear,* which is proposed as a major mechanism of overburden extension associated with massive sediment progradation and salt withdrawal (Mohriak, 1991; Mohriak et al., 1992). Salt tectonics in the northern embayment of the Santos Basin is also compared with other interpretations of landward-dipping faults in the Campos Basin, the conjugate margins in West Africa, the Gulf of Mexico, and the North Atlantic. Sandbox models are also discussed and compared with regional seismic reflection profiles recently shot in the Santos Basin, which greatly illuminate spectacular structures in the deep-water salt diapir province. The Cabo Frio frontier region may provide a useful analog for similar phenomena in other salt basins worldwide, where thicker sedimentary sequences or inadequate geophysical coverage hinder visualization of deep autochthonous salt layers.

REGIONAL SETTING OF THE CABO FRIO PLATFORM

Onland, southeastern Brazil is characterized by major geotectonic features inherited from the late Precambrian–early Paleozoic Brazilian orogeny, which was associated with lithospheric convergence and collision between Africa and South America (Almeida, 1967; Hasui et al., 1975). This orogeny formed the northeast-trending Ribeira folded belt, which is cored by Precambrian metamorphic rocks and contains synorogenic batholiths and widespread postorogenic granites (DNPM, 1978; Radambrasil, 1983). The Paraíba River shear zone (PRSZ in Figure 2) is a late Precambrian northeast-trending ductile shear zone composed of several wrench faults and extending for more than 500 km from São Paulo to Rio de Janeiro. It runs adjacent to the uplifted maritime ranges of Serra do Mar and Serra da Mantiqueira (Campanha, 1981; Hasui and Oliveira, 1984).

In the Late Jurassic, Gondwana started to rift in the extreme south between Argentina and South Africa. In the Early Cretaceous, the rift propagated toward the Pelotas, Santos, and Campos basins (Larson and Ladd, 1973; Rabinowitz and LaBrecque, 1979). Rifting was accompanied by voluminous extrusion of tholeiitic basalts (Fodor et al., 1983; Rocha-Campos et al., 1988; Renne et al., 1992), both on land (Paraná Basin) and in the incipient marginal basins from Pelotas to Espírito Santo. The Campos and Santos basins were formed as rhombograbens that stretched simultaneously but discontinuously in the Early Cretaceous (Bacoccoli and Aranha, 1984). Propagation of the Santos rift was interrupted and transferred to the Campos rift by the east-trending Rio de Janeiro transfer zone south of Cabo Frio (Asmus, 1978; Asmus, 1982; Macedo, 1989). Old weakness zones dating from the Brazilian orogeny were reactivated during the breakup of Gondwana and during postbreakup tectonic episodes in southeastern Brazil, particularly in the early Tertiary. These zones formed small Cenozoic continental rift basins along the Paraíba River shear zone (Almeida, 1976; Asmus and Ferrari, 1978; Melo et al., 1985). Most rift phase normal faults in the Campos Basin follow the northeast-trending Precambrian structural grain (Cordani et al., 1984; Dias et al., 1987). In the southern part, however, the pre-Aptian limit (rift border), which corresponds to a hinge line rather than a fault zone, is deflected around the Cabo Frio Arch, trending eastward in the northern Santos Basin, running parallel to the coastline from Cabo Frio to Ilha Grande (Figure 2).

After continental breakup and emplacement of oceanic crust in the South Atlantic, two further episodes of magmatism affected the Cabo Frio region. First, extrusive basalts and sills (seen in boreholes drilled in the Santos Basin) were intercalated with Turonian–Campanian sediments, dating this magmatic event as middle Cretaceous (80–90 Ma; Mohriak, 1991). Second, in the Cabo Frio High, toward the southern Campos Basin, early Tertiary magmatism formed volcanic cones (Mohriak et al., 1989, 1990c). This magmatism probably began in the Paleocene but climaxed in the Eocene (at ~50 Ma; Mizusaki and Mohriak, 1992).

Geophysical maps (gravity and magnetic) show major northeast-trending lineaments in the continental crust. On oceanic crust, eastward-trending lineaments associated with fracture zones have been thoroughly described (e.g., Le Pichon and Hayes, 1971; Francheteau and Le Pichon, 1972; Rabinowitz and LaBrecque, 1979; Asmus and Guazelli, 1981). More recently, several authors have identified major northwest-trending features both onland and in oceanic crust adjacent to the Santos Basin (Bostrom, 1989; Mohriak, 1991; Souza, 1991). The crust was intruded along these lineaments via alkaline plugs and kimberlites. Southwest of the São Francisco craton in central Brazil, there is evidence for extensional processes affecting the middle Cretaceous volcaniclastic rocks of the Mata da Corda Formation (Bizzi et al., 1991).

STRATIGRAPHIC AND STRUCTURAL EVOLUTION

Rift and postrift subsidence in the Cabo Frio region formed a depositional prism typical of South Atlantic margins, where maximum sedimentary thickness reaches ~10 km in the northern Santos Basin (Figure 3). Three main tectono-stratigraphic sequences bounded by major unconformities of regional expression are identified in the margin: rift, transitional, and drift (Asmus and Ponte, 1973; Ojeda, 1982). These sequences are characterized by contrasting lithologic facies and structural styles (Figure 4).

The Neocomian–Barremian rift sequence (Lagoa Feia Formation in Campos and Guaratiba Formation in

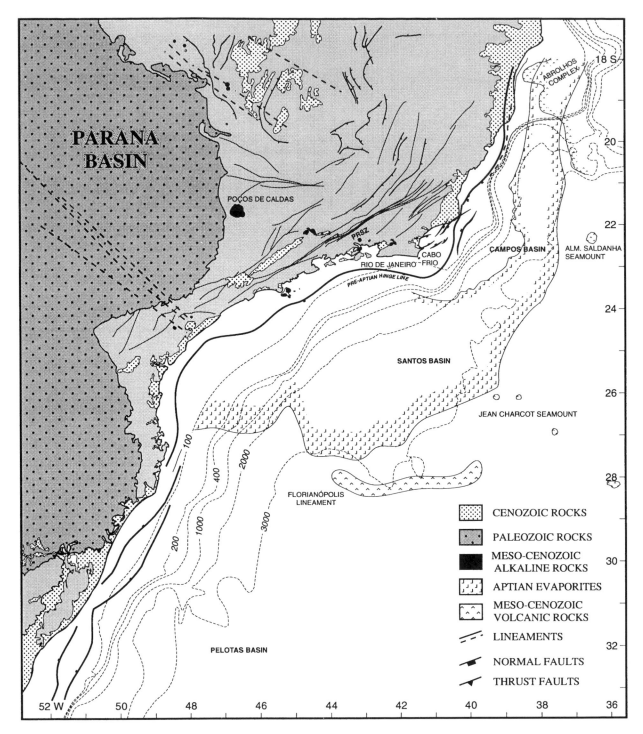

Figure 2—Simplified geologic map of southeastern Brazil showing the main tectonic features onland and offshore, north-ward from Pelotas to Espírito Santo basins. The Paleozoic–Mesozoic Paraná basin contains thick flows of Lower Cretaceous tholeiitic basalts. The northeast-trending Paraíba River shear zone (PRSZ) is associated with a Precambrian orogenic belt extending through the São Paulo and Rio de Janeiro states. The onshore Cenozoic basins along the shear zone are limited to the north by the east-trending Poços de Caldas–Cabo Frio magmatic lineament. The São Paulo Plateau, in the deep-water region of the Santos and Campos basins, is a diapir province containing salt stocks and walls and intensely deformed post-Aptian overburden. The Abrolhos volcanic complex (east of Espírito Santo Basin) and the Rio Grande Rise (south of the Florianópolis lineament) are aseismic ridges formed by volcanic rocks younger than the Neocomian basalts in the Paraná basin and the continental margin. The Florianópolis lineament (São Paulo Ridge) appar-ently coincides with the southernmost extension of the São Paulo Plateau. South of the plateau, the Pelotas Basin is char-acterized by proximal half-grabens bounded by landward-dipping faults. There is no evidence for Aptian salt deposition in the Pelotas Basin.

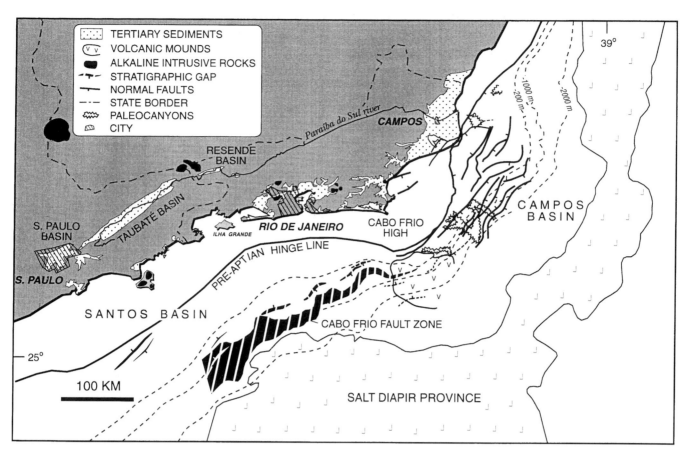

Figure 3—Map showing the major geologic features of the Cabo Frio frontier region. The Cabo Frio High, extending from the Campos to the Santos basins, has shallow basement and postbreakup magmatism. The pre-Aptian hinge line marks the western boundary of the offshore rift. It forms a salient around Cabo Frio, contrasting with reentrants in the Santos and Campos basins. The basement deepens abruptly from the southern Campos Basin to the northern embayment of the Santos Basin.

Santos) includes lacustrine sediments (shale, marl, limestone, coquina, and sandstone; e.g., Dias et al., 1987; Abrahão and Warme, 1990) overlying Lower Cretaceous tholeiitic basalts (Mizusaki et al., 1988). Characteristic of the Cabo Frio platform are basement-involved landward-dipping (antithetic) normal faults having small offsets. The faults extend from the pre-Aptian hinge line toward the shelf edge (Mohriak, 1991; Mohriak et al., 1992). In the central Campos Basin, a similar style can be observed: landward-dipping faults predominate in the proximal, less thinned continental crust, whereas from shelf to basin, toward the rift depocenter, faults change polarity and rift blocks are rotated landward by predominantly seaward-dipping faults having large offsets (Mohriak, 1988; Mohriak et al., 1990b). A major unconformity separates the rotated blocks (filled by Barremian lacustrine sediments) from overlying Aptian conglomerates (Ojeda, 1982; Guardado et al., 1989; Mohriak et al., 1990a).

The Aptian transitional phase (Ariri Formation in Santos and evaporitic Lagoa Feia Formation in Campos) is characterized by evaporites (mainly halite and anhydrite) deposited in an elongate gulf that extended from the southern Santos to the northern Sergipe-Alagoas Basin (Asmus and Ponte, 1973; Ojeda, 1982). The salt

basin, bounded on the west by the Aptian hinge line, is as wide as 300 km on the São Paulo plateau and is mostly underlain by rifted continental crust (Chang et al., 1992). The seaward limit of the salt basin approximately coincides with the possible boundary between the continental and oceanic crust (Lobo and Ferradaes, 1983; Chang and Kowsmann, 1987).

The western limit of the salt diapir province is closer to the Aptian hinge line in the northern Santos Basin than in the central Campos Basin (Mohriak, 1991; Szatmari and Demercian, 1991). The salt province (Figure 3) narrows to the north and south of the Campos Basin. The northern constriction (at 22° S) coincides with a marked change in regional trends from northeast to north-northeast, and the southern constriction (at 25° S) coincides with the Cabo Frio High (Figure 3). This results in a wide zone of extensional tectonics associated with salt mobilization in the central Campos Basin, where rates of overburden extension are probably higher than to the north and south (Szatmari and Demercian, 1991). The structural map for the horizon corresponding to the top of the Aptian salt (Figure 5) is marked by irregular clusters of salt diapirs and salt walls in the deep-water region of the Cabo Frio High.

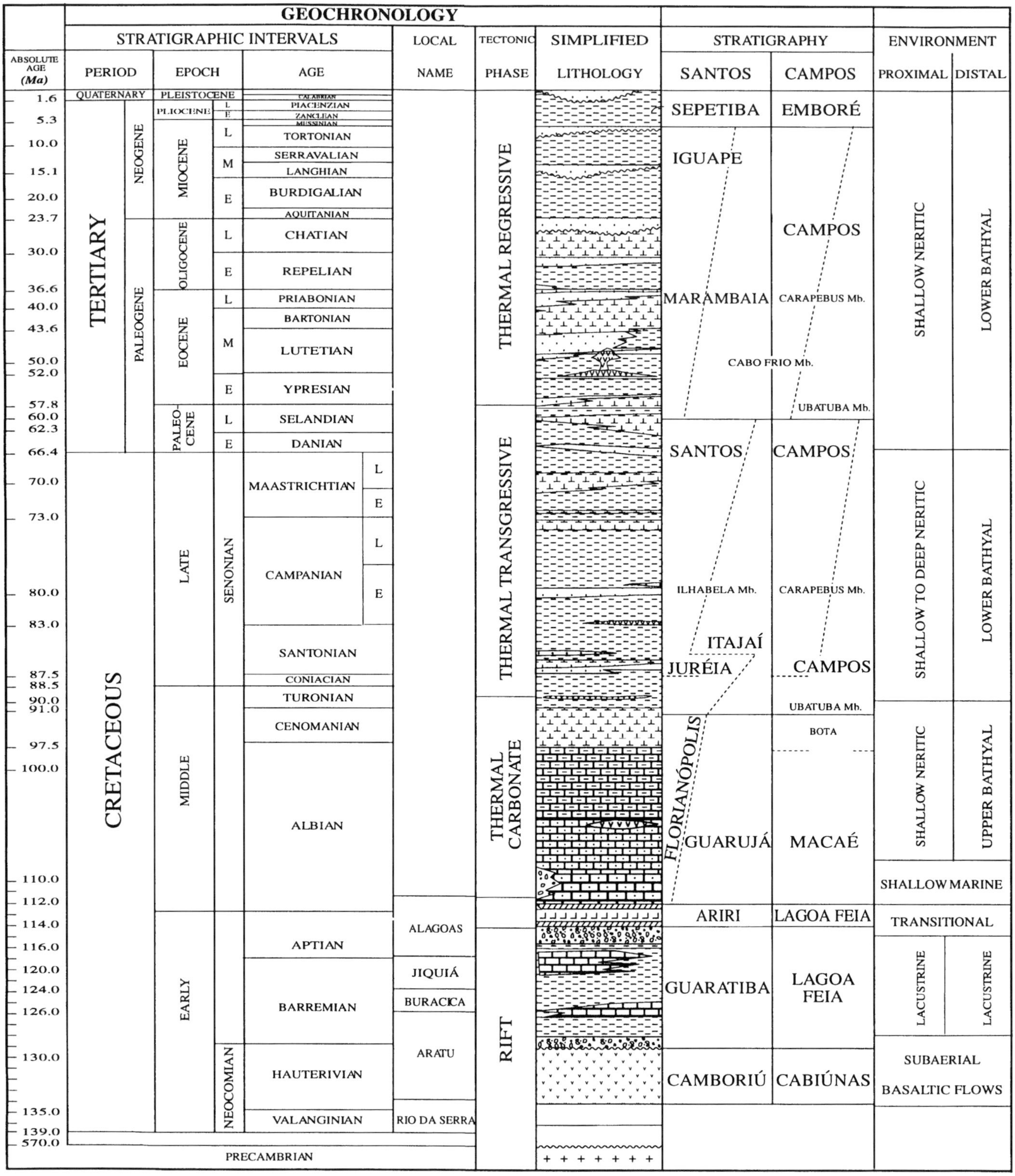

GEOCHRONOLOGY

ABSOLUTE AGE (Ma)	PERIOD	EPOCH	AGE	LOCAL NAME	TECTONIC PHASE	SIMPLIFIED LITHOLOGY	STRATIGRAPHY SANTOS	STRATIGRAPHY CAMPOS	ENVIRONMENT PROXIMAL	ENVIRONMENT DISTAL
1.6	Quaternary	Pleistocene	Calabrian				SEPETIBA	EMBORÉ		
5.3		Pliocene	Piacenzian / Zanclean / Messinian		THERMAL REGRESSIVE				SHALLOW NERITIC	LOWER BATHYAL
10.0	Neogene	Miocene	L Tortonian				IGUAPE			
15.1			Serravalian / Langhian					CAMPOS		
20.0			E Burdigalian							
23.7			Aquitanian							
30.0	Paleogene	Oligocene	L Chatian							
36.6			E Repelian				MARAMBAIA	CARAPEBUS Mb.		
40.0		Eocene	L Priabonian							
43.6			Bartonian					CABO FRIO Mb.		
50.0 / 52.0			M Lutetian							
57.8			E Ypresian					UBATUBA Mb.		
60.0 / 62.3		Paleocene	L Selandian		THERMAL TRANSGRESSIVE					
66.4			E Danian				SANTOS	CAMPOS	SHALLOW TO DEEP NERITIC	LOWER BATHYAL
70.0	Cretaceous	Late (Senonian)	Maastrichtian L/E							
73.0			Campanian L/E							
80.0							ILHABELA Mb.	CARAPEBUS Mb.		
83.0										
87.5 / 88.5			Santonian				ITAJAÍ / JURÉIA	CAMPOS		
90.0 / 91.0			Coniacian / Turonian					UBATUBA Mb.		
97.5		Middle	Cenomanian		THERMAL CARBONATE		FLORIANÓPOLIS	BOTA	SHALLOW NERITIC	UPPER BATHYAL
100.0			Albian				GUARUJÁ	MACAÉ		
110.0									SHALLOW MARINE	
112.0										
114.0		Early	Aptian	ALAGOAS			ARIRI	LAGOA FEIA	TRANSITIONAL	
116.0										
120.0				JIQUIÁ	RIFT		GUARATIBA	LAGOA FEIA	LACUSTRINE	LACUSTRINE
124.0 / 126.0			Barremian	BURACICA						
130.0			Hauterivian	ARATU			CAMBORIÚ	CABIÚNAS	SUBAERIAL BASALTIC FLOWS	
135.0		Neocomian								
139.0			Valanginian	RIO DA SERRA						
570.0			PRECAMBRIAN							

Figure 4—Generalized geochronostratigraphic column comparing the Campos and Santos basins. It shows absolute ages, stratigraphic units, main tectonic phases, and depositional environments on the Cabo Frio platform.

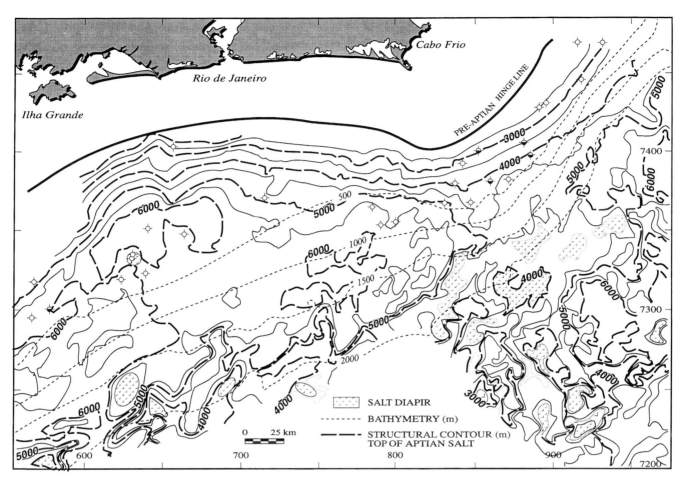

Figure 5—Simplified depth-converted structural map of the top of the Aptian salt. In deep water, salt stocks and salt walls predominate. Upper Tertiary–Quaternary strata thicken in local depocenters. Universal Transverse Mercator coordinate system is centered at Central Meridian 45° W.

Detailed paleontologic analysis and log interpretation indicate that the postrift sedimentary layers are separated by several geochronostratigraphic markers whose approximate absolute ages (based on several geologic time scales; e.g., Harland et al., 1982, and AAPG's 1983 time scale) point to major unconformities at 15, 49, 70–74, 80–90, and 117 Ma (Richter, 1987; Mohriak et al., 1990c; Rangel et al., 1990). These unconformities seem to be related not only to sea level fluctuations but also to major tectonomagmatic events in the South Atlantic (Mohriak, 1991).

The drift phase started in the Albian with a transitional to shallow marine platform that was rapidly flooded in the Cenomanian (Dias-Brito, 1982; Dias-Brito, 1987). The Albian–Cenomanian stratigraphy (Guarujá Formation in Santos, Macaé Formation in Campos) is characterized by shallow-water limestones in the Campos Basin, grading vertically and distally into calcilutites and marls (Azevedo et al., 1987b; Mohriak et al., 1989; Rangel et al., 1990), whereas Santos Basin stratigraphy is marked by major siliciclastic influx (Ven, 1983; Williams and Hubbard, 1984; Pereira et al., 1986; Viviers, 1986). The structural map of the top of the Albian–Cenomanian sequence (Figure 6) shows major listric faults trending subparallel to the pre-Aptian hinge line. South of the Rio

de Janeiro coastline, the absence of Albian–Cenomanian rocks along a northeast-trending zone is related to salt tectonics (Mohriak, 1991; Mohriak et al., 1992).

The geologic evolution of the Campos and Santos basins is marked by singular stratigraphic and structural contrasts during the middle–Late Cretaceous. In the Campos Basin, rising sea level and rapid thermal subsidence resulted in transgression from the late Albian to the Paleocene. Subsequent regression in the Tertiary resulted in a thick sequence of coarse-grained deltaic to turbidite siliciclastic rocks overlying a deep-water condensed section of pelitic rocks (Koutsoukos, 1984; Azevedo et al., 1987a; Milward et al, 1987). Conversely, continually reactivated uplift along the coastal ranges bordering the Santos Basin resulted in alluvial fan deposition and shedding of enormous volumes of siliciclastics into the basin since post-middle Cretaceous time. These sediments (Santos Formation) include continental red beds and shallow neritic coarse-grained siliciclastics that prograded from the shelf toward the upper slope (Ven, 1983; Williams and Hubbard, 1984; Pereira et al., 1986; Viviers, 1986).

This massive progradation resulted in unusual regressive systems tracts ranging in age from middle to Late Cretaceous, which contrast markedly with the global ten-

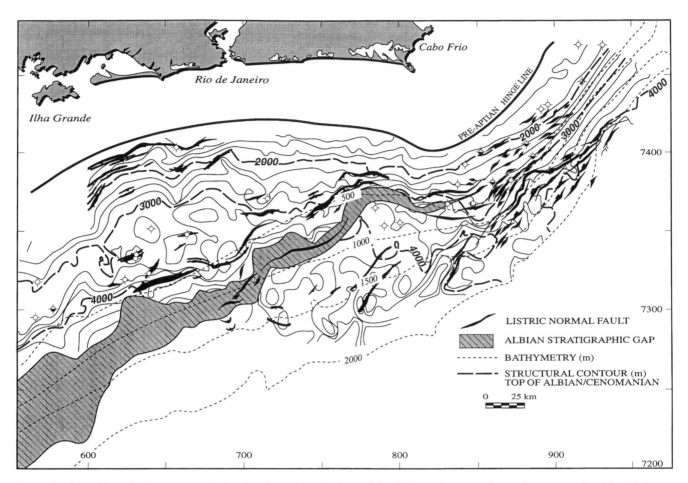

Figure 6—Simplified depth-converted structural map for the top of the Albian–Cenomanian carbonate rocks. The Albian stratigraphic gap, associated with salt tectonics, extends from the Cabo Frio High to the southwestern end of the northern Santos Basin south of Ilha Grande.

dency of rising sea level and marine transgression from Albian to Late Cretaceous (Vail et al., 1977). The Carapebus (Campos) and Ilhabela (Santos) members correspond to sandstone turbidites embedded in deep-water shales of the Campos and Itajaí formations, respectively (Figure 4). The middle–Upper Cretaceous sequence, which is very thick in the northern Santos Basin, thins toward the Cabo Frio Arch. In contrast, the upper Tertiary is characterized by thick sequences in the Campos Basin, but it rapidly thins south of the Cabo Frio Arch toward the Santos Basin (Mohriak, 1991). In both basins, large salt-withdrawal zones were initiated in the Albian but climaxed in the Tertiary of the Campos Basin. In the ultradeepest water (depths >2000 m) of the Cabo Frio region, salt tectonics has remained active, forming depocenters between salt walls and diapirs.

The predominant sediment isopachs in the sequences above the pre-Aptian unconformity (Figure 7) indicate a large sediment influx south of the Cabo Frio Arch in Cretaceous time. Conversely in the Campos Basin, the maximum sedimentary influx on the present-day platform was in the late Tertiary, indicating that diachronous depocenters migrated northward from the Santos to the Campos basins. This migration probably influenced the

contrasting thermal evolution of the rift phase sediments and controlled the styles of salt tectonics (Mohriak, 1991).

The stratigraphy of the Cabo Frio region is illustrated in Figure 8, which shows part of a regional seismic profile in the northern Santos Basin (see Figure 9 for location). This profile shows the main seismic reflectors and the stratigraphic sequences, particularly the prograding Upper Cretaceous siliciclastics above a transparent middle Cretaceous seismic facies. Overlying an Aptian salt pillow, Turonian siliciclastics and Albian–Cenomanian carbonates have been penetrated by an exploration well. The seismic data suggest the presence of pre-Aptian rift phase sedimentary rocks, but these have not been penetrated by drilling.

Figure 9 shows the major features of the Cabo Frio region and the location of a few regional seismic profiles. Figure 10 (66-RL-42) is an oblique regional seismic profile with its schematic interpretation, extending from the southern Campos Basin (Cabo Frio High) to the deepwater region of the Santos Basin. It illustrates the transition from shallow-water platform to deep basin with large salt diapirs. Figure 11 is a schematic seismic profile (231-RL-1332) that shows several volcanic mounds in the Cabo Frio platform, near the northeastern extremity of

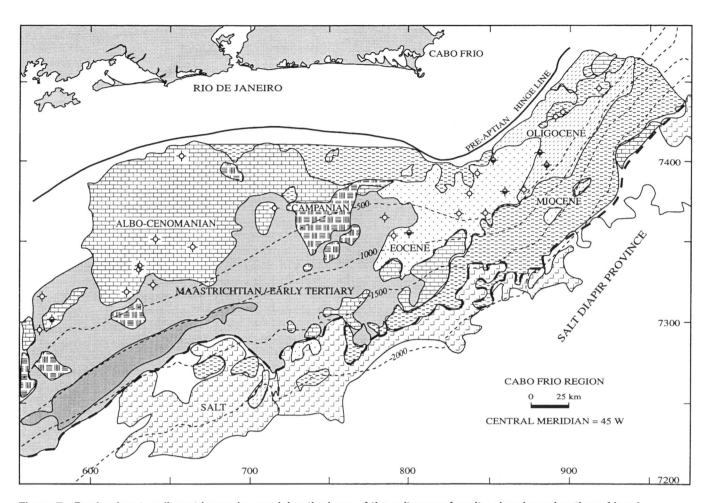

Figure 7—Predominant sediment isopachs overlying the base of the salt unconformity, showing migration of local depocenters in the Cabo Frio region. The general younging of postsalt depocenters indicates that sediment influx migrated from the Santos Basin in the Cretaceous to the Campos Basin in the Tertiary.

the profile. Originally, these features were interpreted as salt diapirs or turbidite mounds (J. C. Vieira, personal communication, 1989) or even as processing artifacts (L. R. Guardado, personal communication, 1989). Exploratory drilling recovered an Eocene sequence of volcaniclastic rocks, including tuff, hyaloclastite, volcanic breccia, basalt, and diabase (Mohriak et al., 1990c; Mizusaki and Mohriak, 1992).

This southwest-northeast transect crosses a large east-striking graben (related to salt tectonics) south of the Cabo Frio High (at shotpoint 10000) and also shows the gradual deepening of the volcanic basement from less than 3 sec (~4 km) on the platform (shot point 11000) to more than 5 sec (~8 km) in the Santos Basin (shot point 5000). The Upper Cretaceous–lower Tertiary sequence thickens toward the Santos Basin. Conversely, Miocene strata are very thick in the central and northern Campos Basin and thin rapidly south of the Cabo Frio High. Figure 11 also shows that the Albian strata disappear along the transect after entering a northeast-trending zone (see Figure 9). This zone, referred to here as the Albian gap, is interpreted to be associated with salt tectonics, as is discussed later.

The boundary between the Campos and Santos basins is characterized by the northwest-trending lineament from oceanic crust toward the Cabo Frio Arch, by contrasting depths to basement, by recurrent postbreakup magmatism, and by two end-member styles of salt tectonics. In the following sections we discuss (1) driving forces for triggering salt tectonics in the Cabo Frio region, (2) contrasting structural styles of salt tectonics in the Campos and Santos basins, and (3) the interaction of sedimentation and salt tectonics associated with the formation of seaward- and landward-dipping faults.

DRIVING MECHANISMS FOR SALT TECTONICS

In general, two mechanisms have been invoked to explain salt tectonics (Nettleton, 1943; Jenyon, 1986a): (1) basement-detached (adiastrophic) or gravity-induced processes, caused by buoyancy effects, basement inclination, or differential loading, which result in thin-skinned tectonics; and (2) basement-involved (diastrophic) or

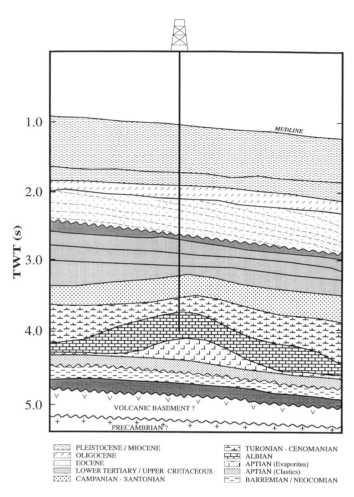

TWT (s)

PLEISTOCENE / MIOCENE
OLIGOCENE
EOCENE
LOWER TERTIARY / UPPER CRETACEOUS
CAMPANIAN - SANTONIAN

TURONIAN - CENOMANIAN
ALBIAN
APTIAN (Evaporites)
APTIAN (Clastics)
BARREMIAN / NEOCOMIAN

Figure 8—Interpretation of part of a regional seismic profile in the northern Santos Basin (228-RL-3707; see location in Figure 9), integrated with the main lithostratigraphic sequences sampled in a borehole drilled by Elf-Aquitaine in 1982. Lithologies below the Albian carbonate sequence are inferred. Length of section is about 10 km.

thick-skinned tectonic processes, caused by faults that affect basement, the salt layers, and the overburden. The interplay between these mechanisms results in coeval salt flow and overburden mobilization. Early work emphasized that salt mobilization is driven mainly by density differences between salt and the overlying sediments (e.g., Nettleton, 1943; Jenyon, 1988). More recent work, however, points to the great importance of differential loading, basement-detached extension (e.g., Schaller and Dauzacker, 1986; Vendeville et al., 1987; Vendeville and Jackson, 1992a; Jackson and Vendeville, 1994), and basement-involved extension (e.g., Jenyon, 1985, 1986b; Koyi and Petersen, 1993). These phenomena have attracted the attention of several researchers and are not completely understood. There is no consensus about a universal driving mechanism for triggering salt tectonics in all sedimentary basins. We believe that the best approach is to analyze geologic features within a regional context that may provide clues to the origin of specific structures.

In the Campos Basin, a simplistic model, in which halokinesis was driven solely by density contrasts and eastward salt flow due to simple, uniform basement tilting during thermal subsidence, is inadequate to explain recently recognized geologic features. These earlier models were based merely on shallow-water seismic profiles and borehole isopach maps made in the late 1970s and early 1980s. In contrast, seismic structural maps extending to deep water show that the volcanic basement is characterized by different steps and reversals of dip, which may form regional highs (external highs) in deep water. Moreover, the base of the evaporites is not simply a homocline dipping basinward. Rather, it is characterized by steps and locally by landward dips, indicating that differential basement subsidence was the norm rather than the exception throughout the basin. The rift phase was not controlled by randomly distributed seaward-dipping normal faults but instead was marked by distinct patterns in different tectonic compartments, particularly by the landward-dipping faults in the proximal region.

Significantly, the region most affected by salt mobilization in the Cabo Frio High has also been affected by one of the largest postbreakup volcanic episodes in southeastern Brazil (Mohriak et al, 1991; Mizusaki and Mohriak, 1992). Consequently, the area must have been considerably uplifted by magmatism in the continental crust. Regional seismic lines suggest that this magmatism and uplift may have been associated with the northwest-trending lineament that extends from the oceanic crust, close to the Jean Charcot seamount (Figure 2) toward the Cabo Frio Arch, as indicated by gravity maps (Mohriak, 1991; Souza, 1991). Here, fault offsets in the base of salt along the lineament probably control salt tectonics in deep water.

Several mechanisms have been suggested to explain salt mobilization in the Campos Basin. Although the Lower Cretaceous basalts and rift phase (Neocomian–Barremian) sediments are characteristically cut by basement-involved normal faults, the overburden is most often cut only by listric faults that detach on the ductile evaporite layers. These listric faults are caused by basinward salt flow on an inclined surface (usually the base of the evaporites; Vendeville and Cobbold, 1988), variations in bathymetry (Szatmari and Demercian, 1991), sedimentary load on the shelf (Figueiredo and Mohriak, 1984; Schaller and Dauzacker, 1986; Rizzo, 1987; Cobbold and Szatmari, 1991; Szatmari and Demercian, 1991), and gravity spreading of the overburden (Mohriak et al., 1992; Szatmari et al., 1993).

Triggering of salt tectonics by reactivation of basement-involved faults during thermal subsidence in sedimentary basins is still poorly understood. A systematic array of landward-dipping normal faults in the southern Campos Basin contrasts markedly with the listric normal faults observed elsewhere (Ojeda et al., 1983; Mohriak et al., 1989). These faults affect Albian–upper Miocene sedimentary layers, apparently detached on Aptian salt, and are characterized by a relatively uncommon geometry for

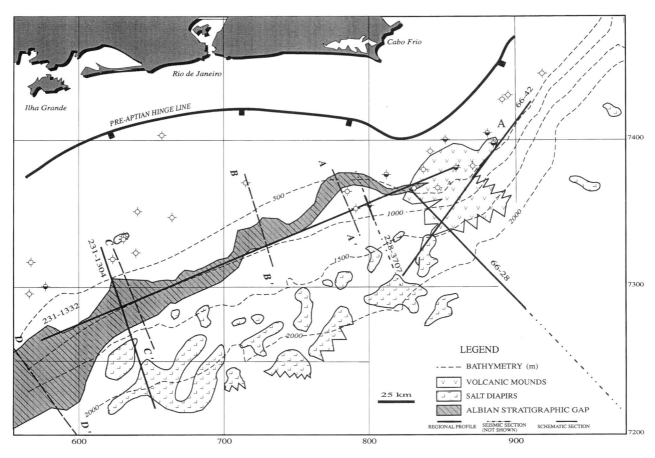

Figure 9—Regional map of the Cabo Frio province showing the major geologic features, regional seismic profiles, and geologic transects. Seismic line 66-RL-42 (northeast) runs along strike (subparallel to the pre-Aptian hinge line) in the Campos Basin and obliquely cuts the eastward regional trend in the northern Santos Basin. The oblique transect along seismic line 231-RL-1332 (center) extends from the Cabo Frio High to the northern Santos Basin south of Ilha Grande. Both regional profiles show Eocene volcanic mounds in the Cabo Frio High. Absence of Albian rocks is inferred inside the northeast-trending hatched area south of the Cabo Frio High. (Dashed lines indicate locations of geologic dip sections shown in Figure 19.)

salt-related faults: master faults dip landward and show large offsets in the younger overburden without substantial growth in the older overburden (Mohriak et al., 1989). Although inconclusive, sandbox models (executed at Petrobrás Research Center by J. R. Aires) indicate that eastward basement tilting and basinward salt flow cannot explain the observed features and suggest that postsalt tectonic activity might be involved (Rizzo et al., 1990). However, analogies with other sedimentary basins and physical modeling point to other possible mechanisms that can form landward-dipping faults by forced folding (Withjack et al., 1989; McClay, 1990) and by two-phase processes of overburden extension (Worrall and Snelson, 1989; Vendeville et al., 1992).

Mohriak (1991) and Mohriak et al. (1992) envisaged three mechanisms driving salt tectonics in the Cabo Frio region: (1) gravitational gliding above a detachment plane, (2) gravitational spreading of the overburden, and (3) basinward push of the salt mass by prograding sedimentary wedges. Demercian et al. (1993) pointed out that salt tectonics in the Campos and Santos basins is gravitationally induced and is independent of any basement-

involved tectonic activity. Szatmari et al. (1993) physically modeled the landward-dipping faults in the Santos Basin by gravitational spreading of the overburden above a flat basement. Nonetheless, there is growing evidence that basement-involved faults may be important in affecting postsalt layers in several basins of offshore Brazil, particularly in the northeast (e. g., Sergipe-Alagoas Basin). In the Santos Basin, south of Ilha Grande (Figure 5), an important landward-dipping basement-involved fault has a large offset at the base of the salt (Mohriak, 1991; Mohriak et al., 1992).

The southernmost part of the Campos Basin is characterized by salt tectonic structures that are markedly different from the more common structural styles observed in the central part of the basin (Mohriak et al., 1989, 1992; Mohriak, 1991). The following facts are relevant to the system of landward-dipping faults in the Cabo Frio region:

1. The landward-dipping faults are associated with the Cabo Frio High, where the orientation of the pre-Aptian hinge line deflects from north-northeast (Campos Basin) to east (Santos Basin).

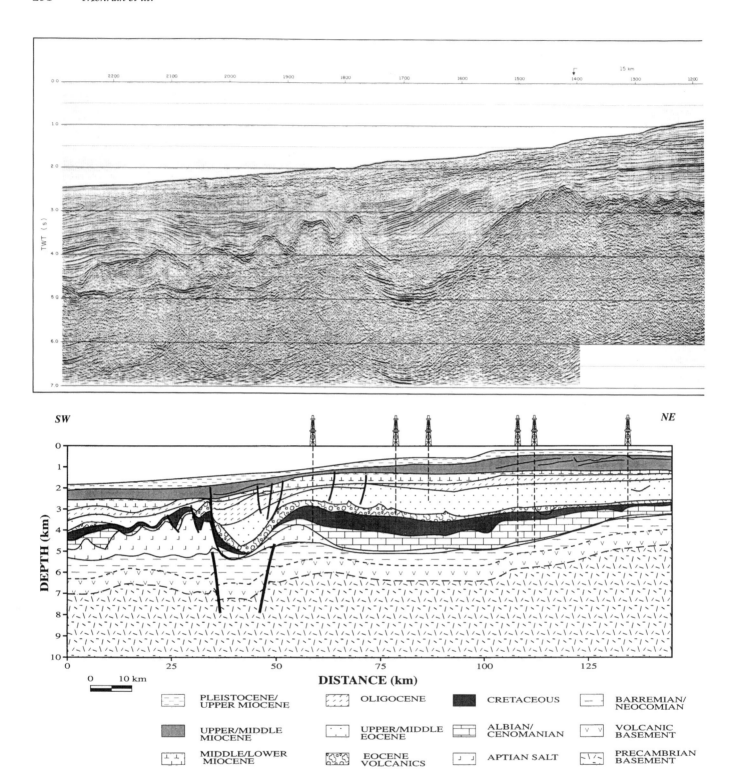

Figure 10—Uninterpreted (top, across spread) and interpreted (bottom, this page) oblique regional seismic profile 66-RL-42 showing the transition from a stable platform in the southern Campos Basin to a deep-water province in the northern Santos Basin. Postrift Eocene volcaniclastic mounds in the Cabo Frio Arch (center) appear to be isolated or amalgamated. The Tertiary sequence is very thick in the salt-controlled eastward-striking trough (center left). Lower Cretaceous tholeiitic basalts (undrilled thus far in the deep water here) probably constitute the economic basement underlying the Neocomian clastic rocks. The Albian carbonate sequence (brick pattern) has a reduced thickness (probably condensed) in the deep-water salt diapir province at the southwestern extremity of the profile. The westernmost borehole in the Cabo Frio High, located 15 km updip and projected onto the profile, intersected Tertiary clastic rocks, Eocene volcanic rocks, Late Cretaceous clastic rocks, Albian–Cenomanian carbonates, and Aptian salt and limestone.

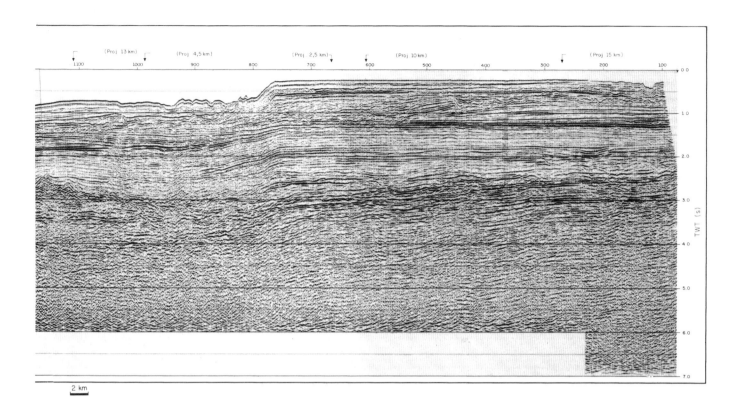

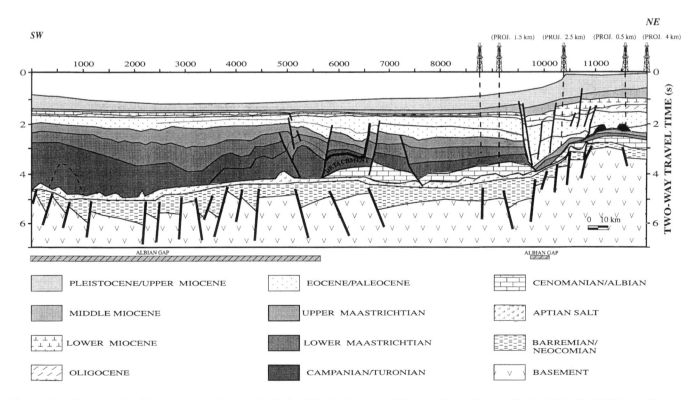

Figure 11—Schematic oblique transect from the Cabo Frio Arch toward the northern Santos Basin (231-RL-1332; see Figure 9 for location). The Cretaceous strata thin markedly toward the Cabo Frio Arch. The Albian sequence disappears southwestward in a detachment-controlled stratigraphic gap associated with salt tectonics.

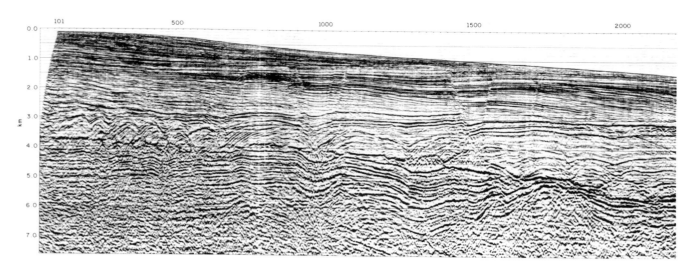

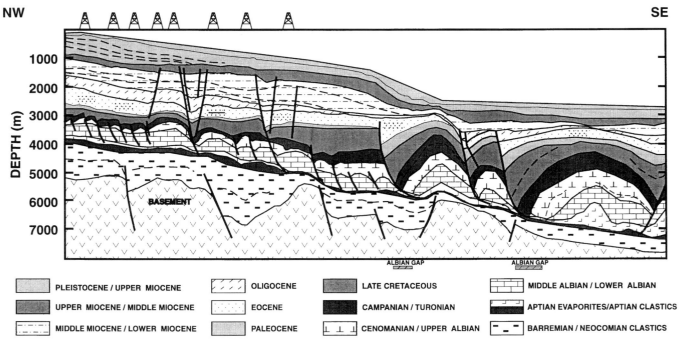

PLEISTOCENE / UPPER MIOCENE	OLIGOCENE	LATE CRETACEOUS	MIDDLE ALBIAN / LOWER ALBIAN
UPPER MIOCENE / MIDDLE MIOCENE	EOCENE	CAMPANIAN / TURONIAN	APTIAN EVAPORITES/APTIAN CLASTICS
MIDDLE MIOCENE / LOWER MIOCENE	PALEOCENE	CENOMANIAN / UPPER ALBIAN	BARREMIAN / NEOCOMIAN CLASTICS

Figure 12—Uninterpreted (top, across spread) and schematically interpreted (bottom) depth-migrated regional seismic section 214-RL-175, showing structural styles associated with salt tectonics in the central to southern Campos Basin. This dip-oriented section has a total length of about 100 km, with vertical exaggeration of about 5×. It shows various structural domains from platform to lower slope, particularly extensional features in the proximal region. Seaward-dipping (synthetic) faults are associated with salt tectonics and gravity spreading and gliding, forming salt rollers and extensional turtle structures of increasing amplitude seaward. Albian rafts moved basinward above a detachment along the salt weld.

2. Faulting was associated with periods of increased sediment progradation.
3. The base of the salt is apparently offset or changes dip abruptly locally.
4. The faults detached on the Aptian evaporites.
5. After being triggered, the faults evolved by salt tectonics alone.
6. The faults were active up to the late Miocene–Pliocene.
7. The Eocene–Miocene section is most thickened.

8. The older sequences are characterized by sub-parallel reflectors, indicating that the landward-dipping faults were virtually inactive then.
9. When faulting started, postbreakup magmatism was active nearby, particularly in the Eocene.
10. Postbreakup basement-involved normal faulting occurred in the westernmost Campos Basin, near Cape São Tomé (Dias et al., 1987), and in the Barra de São João graben, north of Cabo Frio (Mohriak and Barros, 1990).

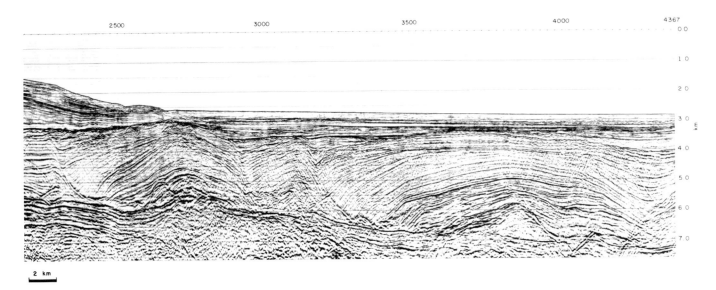

CONTRASTING STYLES OF SALT TECTONICS

Here we analyze salt tectonics in the Cabo Frio region and contrast the transition from the structural styles in the Campos and Santos basins. Figure 12 is a regional depth-migrated seismic line with a schematic geologic interpretation showing typical structural styles associated with salt tectonics in the central to southern part of the deep-water Campos Basin. This region is characterized by major thin-skinned extension of overburden by listric normal faulting (Figueiredo and Mohriak, 1984; Guardado et al., 1989; Cobbold and Szatmari, 1991). Four tectonic compartments divide the dip section. First, a proximal zone is characterized by extension; small seaward-dipping listric normal faults affect only relatively old (Albian–Santonian) stratigraphic sequences. Rollovers and anticlines are small, and most fault blocks of Albian limestones are still in contact with one another. Because sedimentation occurred during listric faulting, the sediment thickness increases markedly in hanging walls. Second, an intermediate part of the transect comprises a relatively stable block containing fewer faults. Third, a distal region contains large seaward- and landward-dipping faults bounding rollovers and anticlines, with amplitudes and wavelengths that increase basinward. These structures have not yet been drilled, but seismic interpretation and regional data indicate that Albian limestones overlie the Aptian salt cores. Finally, beyond this seismic section in ultradeep water, there is a contractional domain characterized by anticlines of large amplitude and wavelength, eroded anticlinal crests, thrust faults, and salt diapirs that may coalesce into elongated salt walls (Lobo and Ferradaes, 1983; Cobbold and Szatmari, 1991; Mohriak, 1991; Szatmari and Demercian, 1991; Demercian et al., 1992; Cobbold et al., 1995).

Most salt diapirs in the Campos Basin are associated with thin-skinned extension of the overburden. Reactive diapirs rose to fill newly created space associated with overburden stretching and thinning (Vendeville and Jackson, 1992a; Jackson and Vendeville, 1994) or grew by downbuilding rather than actively piercing the overburden by forceful intrusion (upbuilding). Continued extension may result in the fall of these diapirs (Vendeville and Jackson, 1992b). In the Campos Basin, evacuation grabens caused by salt withdrawal were rapidly filled by post-Albian sediments, particularly in the Tertiary. Evacuation graben systems are normally bounded by an asymmetric pair of faults; the tectonically thickened strata get younger basinward. In ultradeep water, depocenters are associated with pelagic sediment downbuilding around salt diapirs (Cobbold et al., 1995).

This style of salt tectonics is typical of passive margin basins in the South Atlantic, particularly the Campos Basin (Figueredo and Mohriak, 1984; Shaller and Dauzacker, 1986; Guardado et al., 1989; Mohriak et al., 1990b), Gabon Basin (Teisserenc and Villemin, 1989), Cabinda Basin (Brice et al., 1982), and Kwanza Basin (Duval et al., 1992; Lundin, 1992). Salt tectonics was mainly controlled by overburden extension associated with seaward-dipping faults. The overburden blocks moved downdip along the fault weld or detachment plane, and most listric faults flattened seaward. Hanging-wall strata rotated landward to accommodate the extension, and subordinate landward-dipping faults formed at the basinward side to compensate for the extension. Most experimental and conceptual cross sections in the literature (e.g., Brice et al., 1982; Vendeville et al., 1987; Vendeville and Cobbold, 1988; Cobbold et al., 1989) show the development of these growth faults with basinward (synthetic) dips. Several theoretical and physical models explain why it is mechanically easier to separate the overburden blocks that move basinward along a décollement by synthetic normal faults (e.g., Vendeville et al., 1987; Vendeville and Cobbold, 1988; Cobbold et al., 1989). If the overburden is greatly stretched, the hanging-wall blocks become detached from one another and displaced basinward as rafts, resulting in allochthonous blocks embedded in younger rocks (Duval et al., 1992).

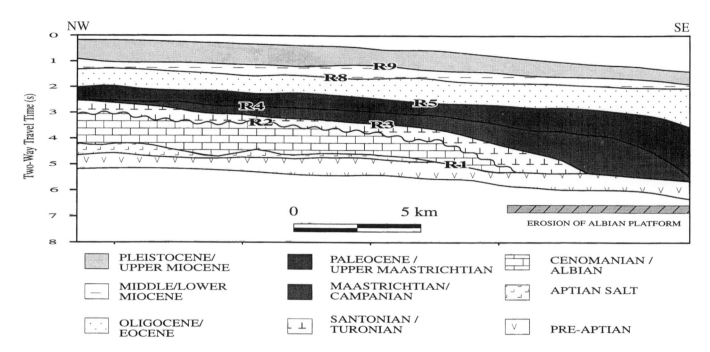

Figure 13—Schematic interpretation of a seismic profile in the northern Santos Basin, illustrating the pinch-out of the Albian–Cenomanian limestones and thickening of the overlying Turonian–Maastrichtian sequence (modified from Ven, 1983). The middle Cretaceous layers dip strongly southeastward. The reason for this enhanced dip is not readily seen in the shelf area, and multiple hypotheses are tenable. Earlier interpretations suggested stratigraphic control on thickening of the strata caused by eustatic sea level rise in the middle–Late Cretaceous. The Albian platform pinch-out was ascribed to submarine erosion by contour currents, resulting in the Albian gap in deep water. Stratigraphic sequence boundaries are indicated by regional markers R1–R9. Length of section is about 125 km.

A large graben related to salt tectonics occurs south of the Cabo Frio High, running subparallel to the eastward deflection of the pre-Aptian hinge line (Figures 3, 9). An exploratory borehole in this graben drilled through a thick section of Eocene–Upper Cretaceous volcaniclastics and siliciclastics. Southwest of here, landward-dipping faults dominate the salt tectonics, and Albian stratigraphic gaps are recognized, particularly in deep water (Figure 6).

Before discussing the salt tectonics in the Cabo Frio region, let us place earlier concepts in historical perspective. Figure 13 is a schematic interpretation of a regional seismic section (84-RL-65) in the northern Santos Basin. This is a typical example of the seismic profiles shot in the late 1970s and early 1980s which do not extend much beyond the shelf break of the Santos Basin and commonly stop short of the diapir province. The profile (Figure 13) shows small salt pillows underlying the Albian carbonate rocks in the platform. Nearer the shelf edge, reflectors arch slightly toward the end of the line. Based on seismic interpretation and geologic analogs, various interpretations were initially proposed for the dipping reflectors associated with the Albian limestone platform and middle Cretaceous sequence. The reflectors related to the Albian platform were interpreted to disappear by downlap onto the salt layer or the pre-Aptian unconformity (Andrade et al., 1983; Pereira et al., 1986) and by erosional processes associated with deep-water contour currents (Ven, 1983). The thickened Cretaceous strata were

thought to be associated (by Petrobrás and other oil companies) with rising eustatic sea level. Some authors also suggested that these features were associated with salt tectonics (e.g., Andrade et al., 1983; Ven, 1983; Williams and Hubbard, 1984), but no genetic mechanism had been proposed. As we will show, the early interpretations were handicapped by their restricted study area.

If we extend the survey to deep water (Figure 14), the Cretaceous seaward-dipping reflectors are seen to be related to a major landward-dipping fault tens of kilometers beyond the shelf break (compare with Figure 13, which corresponds to the first 20 km of Figure 14). Although very distant, this fault indicates that the dipping reflectors are likely to be rotated hanging-wall strata that were deposited subhorizontally updip of the landward-dipping fault, here bounding a salt diapir. The salt mass, affected by episodes of massive clastic progradation, moved basinward along a detachment plane at the base of the salt layer or at the pre-Aptian unconformity (Mohriak, 1991; Mohriak et al., 1992). The presence of a relict salt diapir (near shotpoint 2700 in Figure 14) remains as a witness to the mechanism associated with intense sedimentary progradation and salt tectonics. Collapse of the salt layer due to lateral and seaward salt flow and dissolution formed a salt weld and stratigraphic gap along the fault zone.

Salt tectonics controlled the deposition of a northeast-trending elongated sedimentary wedge that thickens basinward, resulting in accumulation of several kilometers of middle Cretaceous–lower Tertiary sediments (Mohriak et al., 1992). The post-Aptian strata dip relatively steeply as apparent downlaps onto the Cabo Frio salt weld (shotpoints 3500–2300, Figure 14) and become younger basinward. Near the footwall, Maastrichtian sediments rest directly on salt or subsalt strata above the pre-Aptian unconformity in deep water, with a tectonic jump across the salt weld of about 45 m.y., from 70 to 115 Ma.

This model emphasizes that block rotation caused by salt tectonics is not simply associated with growth faulting related to purely downslope gravitational gliding of overburden. Load-induced growth faulting associated with landward-dipping faults and gravity spreading above a mobile substratum is the second end-member process of salt tectonics in the Cabo Frio region, as opposed to the structural styles observed in the central Campos Basin (Figure 12). Figure 15 illustrates conceptual models for these end-members. The top shows the undeformed salt layer with vertical lines representing passive markers. During salt flow, the passive markers are deformed as shown in the early stage of flow. If salt flows basinward (to the right), we define synthetic and antithetic faults as overburden faults that dip basinward and landward, respectively. Slip along the listric fault is accommodated on top, within, or at the base of the salt layer (B. Vendeville, personal communication, 1994). However, if the salt mass is totally evacuated, the fault plane grounds onto the base of the salt layer, resulting in a salt weld.

While salt is present, deposition of overburden is controlled by the listric fault, which rotates and offsets overburden. If fault dips and salt flow are both seaward, synthetic basal shear is accommodated along the fault beneath the rotated overburden (left column, Figure 15). The fault offset creates an Albian gap, which is normally filled by siliciclastics during faulting. Hanging-wall blocks typically move basinward, and relict salt rollers occupy footwalls. Conversely, if the fault dips opposite to the seaward salt flow, antithetic basal shear is accommodated along the fault plane beneath the overburden (right column, Figure 15). Here the footwall (*not* the hanging wall) moves basinward because of progradation and concomitant salt flow, resulting in thick sediment accumulation in the basinward side of the hanging wall. In both models, continued salt flow and overburden extension may result in large stratigraphic gaps (bottom row, Figure 15).

In the Cabo Frio province, the basinward expansion of the growth sequence is mostly controlled by lateral salt withdrawal and downdip slip of both the salt mass and detached blocks of overburden, although stratigraphic control is also present. A major stratigraphic gap along the salt weld (Figures 3, 6, 9) starts at the landward limit of the gap, where Albian rocks overlie the salt weld, and increases basinward, reaching a width of almost 50 km in some places. For comparison, the onshore part of the Sergipe-Alagoas Basin has a maximum width of about 50 km, and the total width of the largest Cenozoic rift in onshore southeastern Brazil, the Taubaté Basin, is about 20 km.

GENETIC MODELS FOR THE CABO FRIO FAULT SYSTEM

Landward-dipping faults are less common than seaward-dipping listric faults, but they have been identified in several Atlantic marginal basins, such as the South Gabon (Rosendahl et al., 1991; Liro and Coen, 1995) and the Carolina Trough (Dillon et al., 1982). They also occur in the Gulf of Mexico (e.g., Larberg, 1983; Worrall and Snelson, 1989) and in the Haltenbanken area of Norway (Withjack et al., 1989). Although genetic mechanisms for landward-dipping faults elsewhere have been only touched on in the literature (e. g., Cloos, 1968; Dillon et al., 1982; Larberg, 1983; Withjack et al., 1989; Rosendahl et al., 1991), their importance as a fundamental process in the tectonic evolution of the Cabo Frio province has been discussed in recent work (e.g., Mohriak et al., 1989, 1990a, 1992; Rizzo et al., 1990; Mohriak, 1991).

Salt tectonics was recognized as fundamentally important in geologic sections published by Andrade et al. (1983), Williams and Hubbard (1984), Pereira et al. (1986), and Cobbold and Szatmari (1991). Notably, most of these studies used seismic profile WS-03, a regional transect shot in the Santos Basin by The University of Texas at Austin. This profile (e.g., Cobbold and Szatmari, 1991) shows typical features associated with landward-dipping listric faults in water that is 1000–2000 m deep.

The Cabo Frio system of landward-dipping faults (Figure 6) points to the great importance of exogenous factors such as tectonic uplift in source areas of the Santos Basin during the middle Cretaceous (Serra do Mar uplift). Conversely, middle Cretaceous progradation in the area of the present-day Campos platform was comparatively minor (Guardado et al., 1989; Mohriak et al., 1990b; Mohriak, 1991). In the Santos Basin, middle–Upper Cretaceous episodes of massive clastic progradation allowed salt tectonics to create accommodation space efficiently. The antithetic basal shear mechanism (Figure 15) assumes seaward salt mobilization over a subhorizontal detachment, creating troughs in the hanging-wall block. Prograding sediments accommodated in these troughs imposed an additional load that pushed both the footwall salt and the detachment fault basinward and rotated strata along the landward flank of the salt mass.

However, other hypotheses to explain the stratigraphic gap and landward-dipping faults in the Cabo Frio region can also be put forward. The genetic mechanisms for this spectacular fault system are not trivial, and some features in the seismic sections might mislead an interpreter even if analyzed within a basin-wide framework. The end-member models proposed for the structural styles in the Cabo Frio region are amalgams of several working hypotheses that must be clearly distinguished (M. P. A. Jackson, personal communication, 1994). They are discussed here.

Hypothesis 1—*The systematic array of landward-dipping faults is induced by reactivation of basement-involved landward-dipping faults.* Several lines of evidence suggest

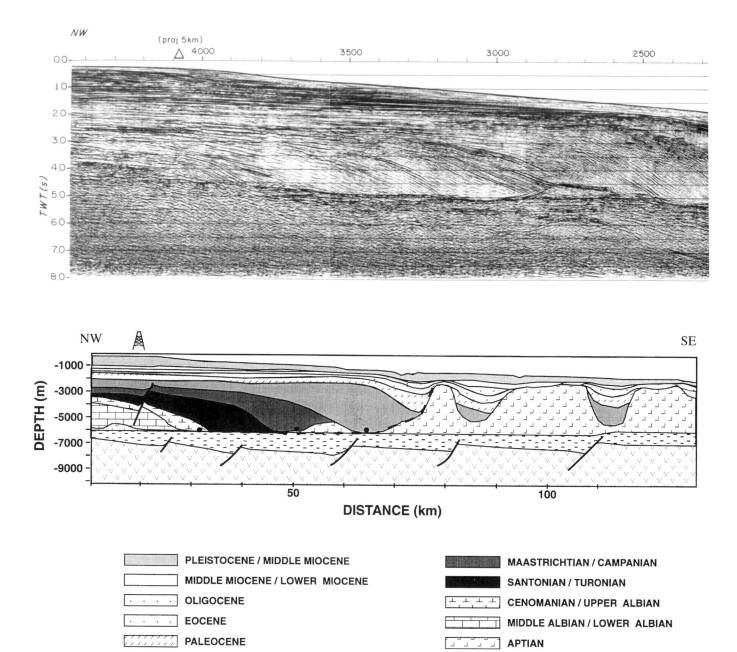

PLEISTOCENE / MIDDLE MIOCENE

MIDDLE MIOCENE / LOWER MIOCENE

OLIGOCENE

EOCENE

PALEOCENE

MAASTRICHTIAN

● ● ● SALT WELD

MAASTRICHTIAN / CAMPANIAN

SANTONIAN / TURONIAN

CENOMANIAN / UPPER ALBIAN

MIDDLE ALBIAN / LOWER ALBIAN

APTIAN

BARREMIAN / NEOCOMIAN

BASEMENT

Figure 14—Uninterpreted (time-migrated) (top, across spread) and schematically interpreted (depth converted using approximate interval velocities) (bottom) seismic section 231-RL-1304 in the northern Santos Basin (see Figure 9 for location). This profile crosses a major landward-dipping fault in deep water near shotpoint 2000. Expanding the area of study from that in Figure 13, the apparent downlap in the platform is seen to be associated with salt tectonics presently beyond the shelf edge. A possible depositional limit for the Albian platform (near shotpoint 3900) was probably controlled by a very thick salt basin having a depocenter south of the Albian platform. Post-Aptian sediments show an apparent downlap on the salt or presalt strata, forming pseudo-clinoforms that become younger basinward, as a result of several episodes of massive clastic progradation in the middle–Late Cretaceous. Coeval with progradation, salt withdrawal was associated with the Cabo Frio system of landward-dipping faults that detach on the base of the Aptian salt (Mohriak, 1991; Mohriak et al., 1992). The vertical tectonic gap resulting from this process may reach the 45-Ma level; rocks as young as 70 Ma are deposited on layers dated 115 Ma (shot point 2000). The salt weld is interpreted as a major detachment above which the overburden subsided while the salt mass moved basinward and laterally during progradation.

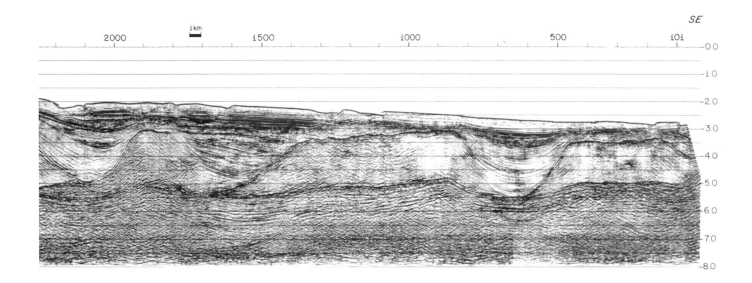

basement faulting after breakup in the Cabo Frio region (Mohriak et al., 1989, 1990c; Mohriak and Barros, 1990). However, only a few seismic profiles actually depict their possible connection with thin-skinned salt tectonics. Nonetheless, salt is an almost perfect lubricant, so small fault offsets may be absorbed into thick salt. Overburden extension can also obscure direct links between faults in the basement and the overburden. South of Ilha Grande, a large basement-involved fault cuts overburden and might have triggered salt mobilization. Also, Rosendahl et al. (1991) suggested basement-involved faulting to explain salt tectonic antithetic faults in South Gabon. Sandbox experiments for landward-dipping faults (Rizzo et al., 1990; Szatmari et al., 1993) have provided dubious and contradictory evidence.

Hypothesis 2—The Albian gap is an erosional limit or primary pinch-out of the Albian platform. The apparent pinch-out of the Albian platform limestones may be due to submarine erosion and drowning of a low-energy, non-rimmed carbonate ramp following coastal subsidence or deltaic encroachment by a rapid eustatic rise of sea level (Ven, 1983). This hypothesis is tenable in the southwest end of the studied area, but northward, the Cabo Frio fault system clearly offsets blocks of Albian carbonates. Nearer the Campos Basin, remnants of the Albian platform occur in deep water south of the fault zone.

Hypothesis 3—The apparent downlap of overburden strata on the base of the salt was caused by salt dissolution. The edge dissolution hypothesis (Jenyon, 1986a, his figures 6-26, 6-29) was proposed to explain seismic features that are more common in the North Sea than in Atlantic-type margins. We think that feather-edge salt dissolution may have played a role in forming the structures in the Cabo Frio province, but its main drawback is timing. This dissolution hypothesis assumes that apparent downlaps were formed by later collapse of overburden, whereas off

Cabo Frio, postsalt strata thickened during salt flow and sedimentary troughs grew younger basinward due to different episodes of clastic progradation.

Hypothesis 4—The apparent downlap of overburden strata on the base of the salt was caused by prograding wedges filling a starved basin. This hypothesis assumes that a large salt mass extended basinward of the prograding wedges (e. g., Andrade et al., 1983; Pereira et al., 1986). It implies that the dipping strata are genuinely downlapping, nonrotated, sigmoidal clinoforms. However, a deep-water diapir would probably have dissolved during the time required for the youngest strata to onlap its flank. Another drawback is that sigmoidal clinoforms would form subhorizontal bottom sets distally. However, in the Cabo Frio fault zone, individual reflectors increase their dip nearer the detachment surface (salt weld), suggesting that they actually correspond to pseudo-clinoforms (Wu et al., 1990).

Hypothesis 5—The Albian gap is a basal fault linked to a genuine landward-dipping fault. This hypothesis assumes that overburden extension occurs by blocks rotating seaward and moving upslope. Under gravity spreading, this is inherently improbable, particularly because there is no seismic evidence for contraction updip of the fault system (M. P. A. Jackson, personal communication, 1994). However, physical modeling also suggests that overburden can translate seaward even where landward-dipping faults are involved (Cloos, 1968; McClay, 1989, his figure 3). Unlike hypothesis 7 (see later), salt is not necessarily involved in this model. However, the necessary décollement must be subhorizontal (McClay, 1990) because even small basinward tilts promote basinward-dipping faults.

Hypothesis 6—The Albian gap is a basal salt weld linked to a landward-dipping pseudo-fault weld. The Albian gap and the apparent downlap were caused by prograding wedges loading a thick salt basin that was progressively evacuated by flow of salt both seaward and along strike

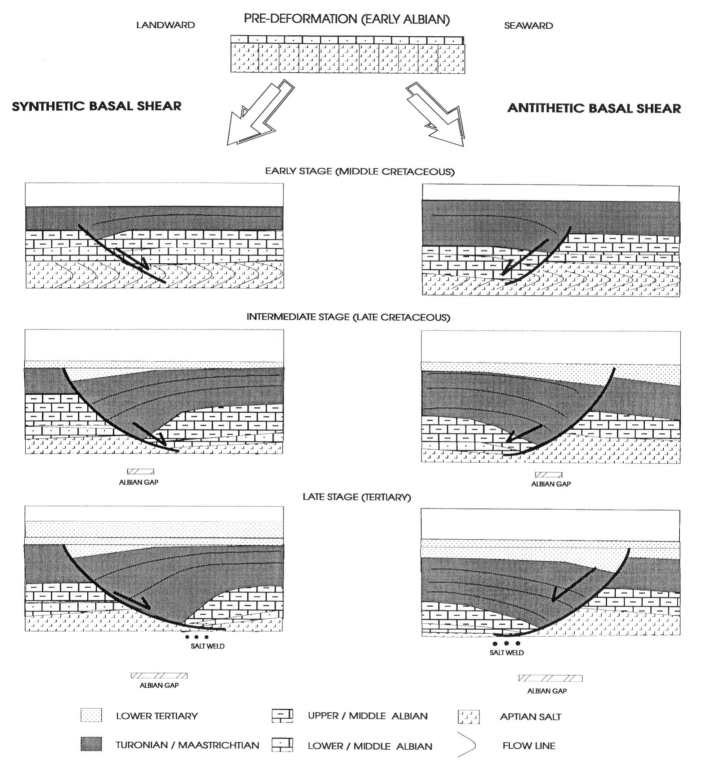

Figure 15—Two end-member models of salt tectonics in Atlantic-type passive margins. If both the listric fault plane and the salt flow vector pointed in the same direction (basinward), synthetic basal shear (left column) was accommodated between the overburden and the base of the salt mass. If the salt was totally evacuated, the fault plane grounded along a salt weld. Alternatively, if the fault dip was landward and the salt flow basinward, antithetic basal shear (right column) was accommodated between the rotating overburden and the base of the salt layer. Prograding sedimentary wedges were continuously rotated by the listric fault, which was overrun by younger sediments that accumulated in depocenters adjoining the salt mass that moved basinward, resulting in widening of the stratigraphic gap.

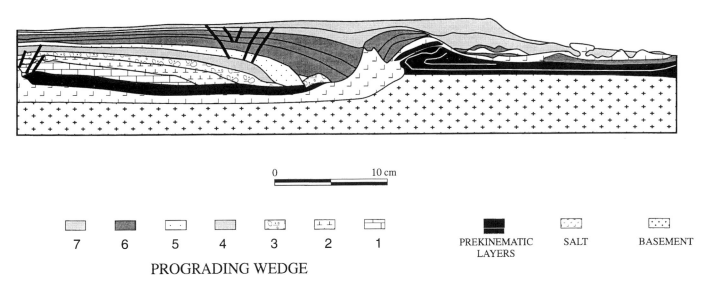

0   10 cm

| 7 | 6 | 5 | 4 | 3 | 2 | 1 | PREKINEMATIC LAYERS | SALT | BASEMENT |

PROGRADING WEDGE

Figure 16—Schematic drawing of a sandbox model in which silicone simulated salt and dry sand simulated a prograding wedge with seven episodes of progradation above continuous prekinematic layers capping the evaporite basin. It reproduced structures associated with sediment progradation and salt mobilization, particularly the formation of salt tongues in the Gulf of Mexico (modified from Vendeville et al., 1992, The University of Texas at Austin, Applied Geodynamics Laboratory, experiment 146). It also reproduced some of the structural and stratigraphic features observed in the northern embayment of the Santos Basin. (Compare with Figure 14, but several important differences exist between these analogs.)

(mainly southward). Following this hypothesis, neither the salt basin nor the overburden were affected by significant horizontal slip along the detachment. The dipping reflectors were steepened by salt withdrawal, forming a salt weld. This hypothesis assumes that in some regions the Albian platform never extended beyond the Albian gap, except as a condensed section; the Albian gap would be a primary stratigraphic limit. A similar hypothesis was suggested by Jenyon (1986a, his figure 5-34).

This hypothesis implies that salt flow basinward could even rise through the stratigraphic sequence, forming salt tongues detached from the mother salt (Wu et al., 1990). This has been substantiated by physical modeling for structures in the Gulf of Mexico. Figure 16 shows the results of a sandbox experiment by Vendeville et al. (1992). The salt basin pinched out basinward below initially flat-topped, subhorizontal prekinematic layers (Figure 17). Sediment progradation then loaded the salt basin and displaced the salt seaward, rupturing the prekinematic layer and expelling salt to the surface. This forced salt to migrate basinward and upward through the stratigraphic sequence, forming cuspate, extrusive salt sheets that spread like salt glaciers in the abyssal plain (Vendeville et al., 1992). These features are remarkably similar to geologic structures observed in the Cabo Frio region (see Figure 14). However, this experiment simulated overburden progradation without overall extension or slip between overburden and salt weld. Thus, the detachment is merely a pseudo-fault (Vendeville et al., 1992; Schuster, 1995).

Because of the exploration implications, this possibility was heatedly discussed among the interpretation group, particularly the first three co-authors. Two alternative models having radically different assumptions were proposed for the Albian gap in Figure 14. The first

model (Figure 18A) envisions an initially continuous platform of shallow-water Albian limestone throughout the Cabo Frio region. Major extension moved the disrupted rafts of the Albian platform to deep water, above and between salt diapirs. The second model (Figure 18B) assumes that the pinch-out is the seaward stratigraphic limit of the Albian platform, subparallel to the pre-Aptian hinge line. South of Ilha Grande, the pinch-out coincided with a thick salt basin. This implies that south of the Albian stratigraphic gap, Albian strata were reduced to only a condensed section of calcilutites and marls originally deposited in substantial water depths. Despite some dissension, the second model (Figure 18B, hypothesis 6) prevailed because of an overwhelming weight of evidence. Accordingly, the structural and lithofacies maps portray a marked pinch-out of the Albian section in the southwest Cabo Frio province. The Cabo Frio fault system cuts obliquely through this deep-water facies in the south and disrupts shallow-water facies toward the northeast. The stratigraphic gap therefore widened diachronously along strike, across different stratigraphic facies.

The schematic sections in Figure 19 illustrate the progressive detachment and separation of two large blocks of the Albian platform by the landward-dipping faults south of the Cabo Frio High (northernmost cross section) and the enlargement of the stratigraphic gap toward the southernmost section (see Figure 9). Albian strata overlying salt pillows in deep water were detached by the Cabo Frio fault system south of the eastward-striking graben, forming rafts cored by salt relics. The stratigraphic gap widened from north to south (Figures 19B, C). In the southernmost section (Figure 19D), a major basement-involved landward-dipping fault was mapped near the seaward limit of the Albian sequence. This structure pro-

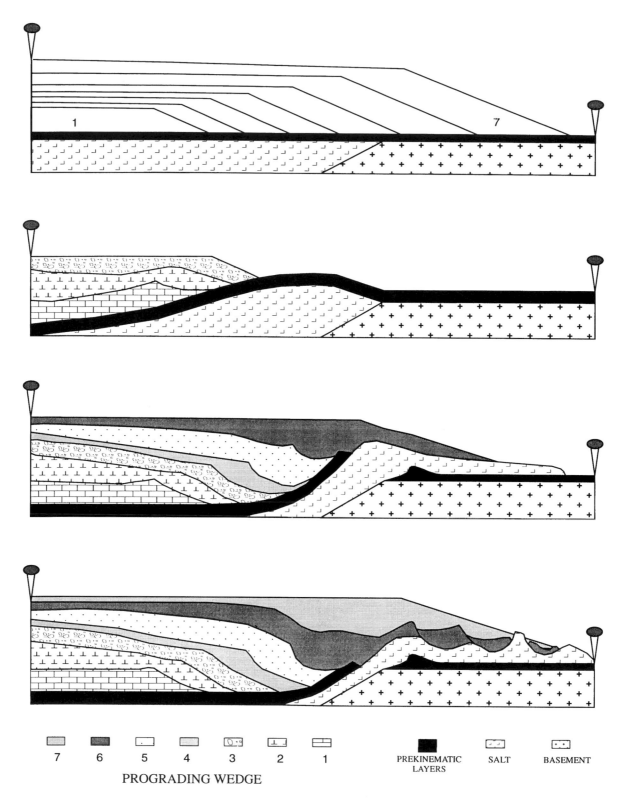

Figure 17—Schematic cross sections showing the sequence of events for the Gulf of Mexico physical model discussed in Figure 16 (modified from Vendeville et al., 1992, experiment 146). The salt was expelled by the clastic wedges that loaded the basin, resulting in the formation of salt tongues on the distal uplifted blocks. In this experiment, there was no slipping along the salt weld, which was formed by salt evacuation following loading by prograding wedges. For this experiment, the progradation episodes were much younger than salt deposition, simulating the stratigraphic development of the Gulf of Mexico (see Jenyon, 1986a, his figure 5-34). Conversely, the Cabo Frio fault-salt weld is conceptually interpreted as a mega-detachment caused by antithetic basal shear associated with basinward salt flow and massive clastic progradation.

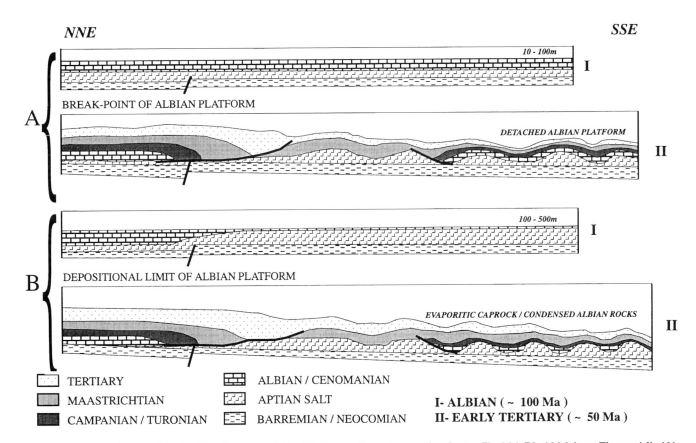

Figure 18—Alternative models for the absence of the Albian section along seismic profile 231-RL-1304 (see Figure 14). **(A)** The Aptian salt broke up the Albian carbonate platform, which originally consisted of shallow-water limestones. Remnants of these limestones should be present in deep water as detached blocks that slipped basinward. This model also implies major contraction downdip due to overburden extension updip. **(B)** The Aptian salt province south of Ilha Grande limited deposition of the Albian platform to the northern shallow water part of the margin. In deep water, the Albian sequence is condensed to thin calcilutites and marls capping the evaporites. South of the landward-dipping fault, a condensed section of Albian carbonate possibly slipped with the salt mass that flowed basinward.

vides evidence for postsalt basement tectonics in the Santos Basin. The reactivation of the antithetic fault resulted in a basement high with a salt diapir on top, flanked by a thick section of Cretaceous rocks.

Hypothesis 7—The Albian blocks in the footwall were displaced basinward as piggy-backs, possibly carried by salt. However, the hanging-wall blocks above the salt weld have not extended or moved significantly (Figure 14). In fact, the remnant salt weld actually resembles a pseudo-fault, as interpreted in the Gulf of Mexico sandbox model (Figure 16). Nonetheless, several lines of evidence suggest that the Cabo Frio structure may be associated with extensional tectonics involving a combination of overburden spreading, clastic progradation, and salt withdrawal. Perhaps we should quote the Bard: "What's in a name? That which we call a rose by any other word would smell as sweet."

Considering the Albian gap as a salt weld associated with a genuine landward-dipping fault, updip extension must be compensated somehow by contraction, diapirism, or gliding of blocks farther basinward. This compensation could involve the third dimension and may not be observable in the study area. However, Figure 20 shows an example of downdip contraction associated with massive clastic progradation south of the Cabo Frio Arch (see Figure 9 for location). Two small stratigraphic gaps are present: a proximal gap at the western end and a deep-water gap between shotpoints 500 and 600. The first gap, in the eastward-striking evacuation graben south of the Cabo Frio Arch, is clearly associated with offset of Albian blocks. Seaward of the graben, the well shown on Figure 8 recovered shallow-water limestone. South of the graben, beyond the shelf break, there are salt diapirs and steep reflectors apparently controlled by a major landward-dipping fault. Although they resemble the prograding wedges shown in Figure 14, there are major differences. In Figure 20, the salt did not move far basinward and the reflectors apparently do not downlap. Farther basinward, the seismic section suggests that Albian blocks may occur in deep water. Downslope, contractional structures are represented by possible thrusts and folds, the crests of which are eroded (Figure 20). This section is possibly a snapshot of another stratigraphic gap in the making. The progradation here is much younger than that in

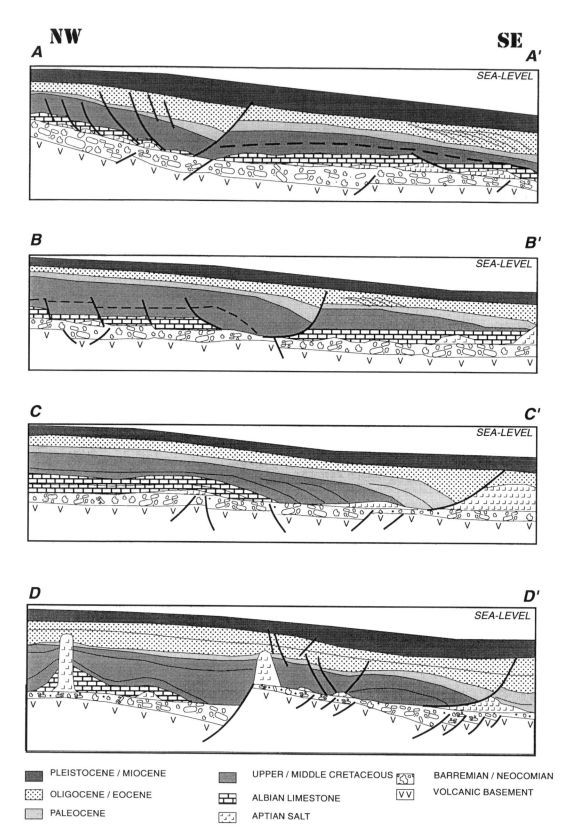

Figure 19—Sequential schematic cross sections from the Cabo Frio Arch (A–A') to the northern Santos Basin (D–D') (see Figure 9 for location). The stratigraphic gap progressively widens and the Cretaceous sequence thickens toward the southernmost profile (D–D'). The Albian gap in the northernmost section (A–A') is caused by overburden extension and basinward salt flow. Albian platform sediments have been drilled south of the eastward-striking salt-related graben (center left). The southernmost section (D–D') shows that the Albian gap is related to seaward salt flow and massive clastic progradation during the Cretaceous. The intermediate sections (B–B', C–C') illustrate pinch-out of the Albian platform.

Figure 14. Apparently, the prograding wedges had space problems and could not easily be accommodated by basinward flow of salt and overburden extension.

Figure 21 shows a model for the Cabo Frio fault zone based on antithetic basal shear between the overburden and the landward-facing wall of a diapir that moved basinward ahead of the prograding wedges. The detachment resulted from rotation of the upper part of the fault, which was successively pushed down and overrun by progressively younger prograding layers, creating a stratigraphic gap.

Another controversial subject in the Cabo Frio region is the depositional environment of the Aptian evaporites. Was the initial salt province a continuous gulf or did it comprise several shallow epineritic seas? Was the salt layer deposited in deep ocean or shallow water? Did the salt flow into a deep basin from an adjacent lagoonal environment of salt deposition? It is difficult to reconcile a deep-water origin for the evaporites (e.g., Schmalz, 1969) with the fact that, wherever drilled in Brazil, sediments below the salt are continental lacustrine and those above the salt are typically shallow-water limestones or coarse-grained siliciclastics. We infer that the salt was deposited in shallow water, although the basin may have shallowed in response to both climatic changes and tectonic uplift (M. J. Pereira, personal communication, 1994). Could salt have been introduced into the basin from magmatic fluids emanating from the mantle (Momenzadeh, 1990)? Magmatic enrichment is not too far-fetched. The Sergipe-Alagoas Basin in northeastern Brazil was the only evaporitic basin considered to be nonvolcanic during its inception. However, even there, new deep seismic data indicate a massive magmatic effusion characterized by seaward-dipping reflectors in deep water (Mohriak et al., 1993).

CONCLUSIONS

The Cabo Frio region is an interbasinal province distinctly different from other marginal basins in southeastern Brazil. Both the adjoining Campos and Santos basins share a common origin resulting from the breakup of Gondwana. However, their distributions of stratigraphic sequences, structural evolution, magmatism, and salt tectonics are distinctly different in style and age.

The postsalt depositional systems show conspicuous contrasts in lithologic facies and depocenters in the Cabo Frio region. From Albian to Miocene time, progradational depocenters migrated northward from the Santos Basin to the Campos Basin. This differential loading of evaporites may have produced the major differences in structural style. Postbreakup magmatism along the Cabo Frio Arch, which is possibly associated with a major northwest-trending lineament, may have lifted the basement and controlled salt tectonics. Postbreakup reactivation of basement faults, affecting Upper Cretaceous–Tertiary sediments, may also have triggered salt tectonics, particularly in the western Campos Basin and south of Ilha Grande in the Santos Basin.

The Albian stratigraphic gap is a large northeast-trending zone in which Albian strata are thin or missing, whereas post-Albian strata are thickened and apparently downlap onto the base of the salt. These features are interpreted to have been caused by salt flow and massive clastic progradation associated with a landward-dipping fault that detached within or at the base of the Aptian evaporites.

For the deep-water Cabo Frio province, two end-member models of salt tectonics are suggested. In reality, several mechanisms probably combined to form the structures. Several working hypotheses and sandbox models have been used to examine the genesis of these salt structures. The first end-member is associated with synthetic basal shear and is recognized in passive margins worldwide, especially in the South Atlantic. Seaward-dipping listric normal faults are usually formed during overburden extension, gravity spreading, and gravity sliding. Landward-dipping faults are subordinate. Sediment wedges in each rollover expand on the seaward side of each seaward-dipping growth fault because hanging walls move seaward and leave relict structures in the footwalls. Severe extension may completely separate the blocks, forming rafts.

The second end-member is very important in the Cabo Frio region. The system of landward-dipping faults is identified from the southern Campos Basin to the northern Santos Basin and even farther south. This structural style is associated with antithetic basal shear and results from overburden spreading and massive clastic progradation, which pushes the salt mass seaward. Sediment wedges in the rollover expand on the landward side of each landward-dipping fault. Reactivation of antithetic basement-involved normal faults, as observed south of Ilha Grande in the Santos Basin, might also trigger this peculiar style of salt tectonics.

Sediment progradation against a landward-dipping listric fault along the landward side of a salt mass results in basinward expansion and rotation of the hanging-wall strata as salt flows seaward. The hanging-wall block is essentially fixed after the rotated sedimentary layers ground onto the detachment plane following displacement of the salt layer laterally. Rotation of strata by landward-dipping faults results in apparent downlap onto the salt weld. Thereafter, vertical subsidence by salt evacuation ceases, and horizontal progradation above the fixed hanging-wall block predominates. This may form troughs filled by younger sedimentary wedges. These wedges load and continually push the salt mass seaward, carrying sediments as piggybacks between diapirs. Conversely, restrained salt flow results in contractional features recognized by folds, thrusts, uplift, and erosion.

In the northern Santos Basin, differential loading of Aptian salt by prograding middle Cretaceous–lower Tertiary clastic wedges also produced the phenomenal Cabo Frio stratigraphic gap of Albian age that extends for tens of kilometers downdip and hundreds of kilometers along strike. In a deep-water region south of Ilha Grande, absence or condensation of the Albian sequence is

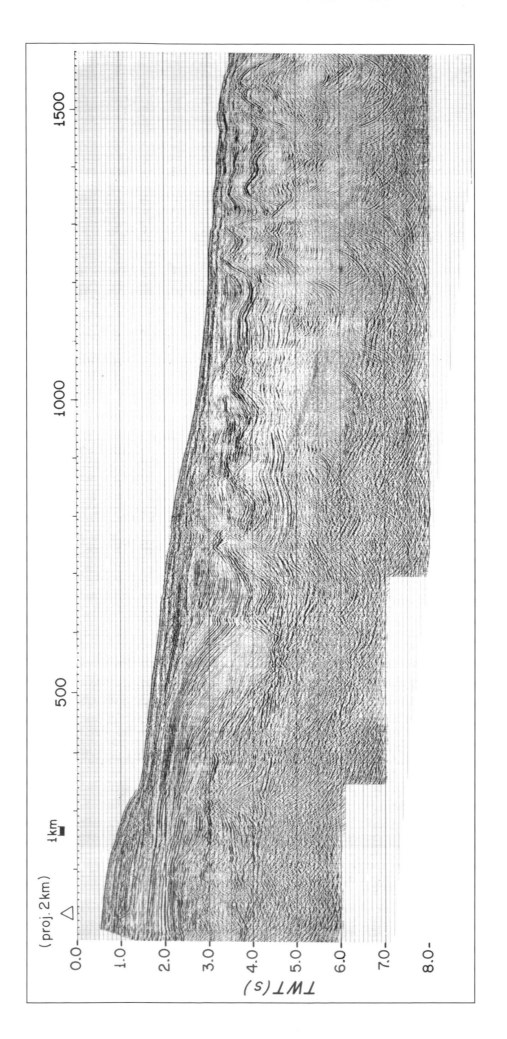

Figure 20 (top)

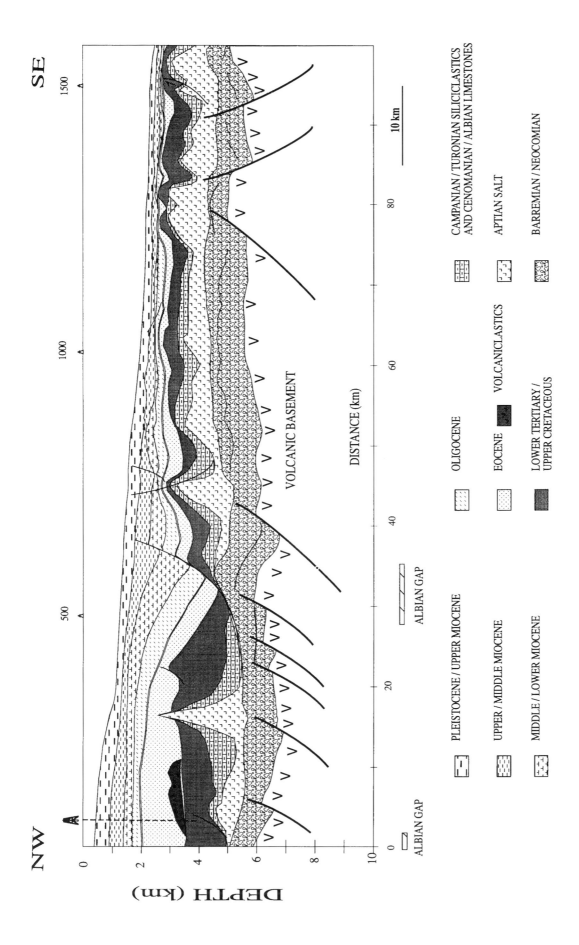

Figure 20—Uninterpreted (time migrated) (opposite page) and interpreted (depth converted using approximate interval velocities) (this page) seismic profile 66-RL-28 south of the Cabo Frio High (see Figure 9 for location). This regional dip line shows massive clastic progradation during the Tertiary and detachment of the Albian platform forming two Albian gaps. The proximal gap, extreme left near Cabo Frio, formed an eastward-striking salt-related graben. The distal gap (shotpoints 500–600) is bounded by a much larger landward-dipping fault associated with a prominent Miocene depocenter. Contractional structures in the deep-water overburden are folds, eroded rollovers and anticlines, and possible thrust faults bounding depocenters.

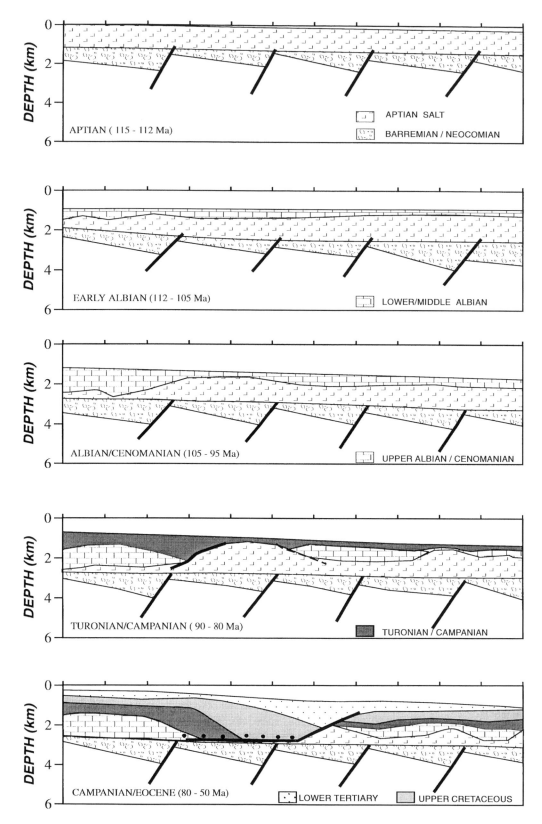

Figure 21—Sequential schematic cross sections showing the genetic mechanism proposed for the Cabo Frio–Santos Basin system of landward-dipping faults and associated stratigraphic gaps. Antithetic basal shear is inferred to have occurred between the Aptian salt mass and the overburden rocks that prograded massively in the Cretaceous. Progradation and gravity spreading of the overburden pushed the salt mass basinward and formed a landward-dipping fault that was subsequently overrun by younger prograding wedges, resulting in a mega-detachment along the salt weld.

inferred seaward of the Albian gap. The immense scale of the Cabo Frio fault zone and its stratigraphic gap is possibly unique in salt basins worldwide; it is one of the largest structures on the South Atlantic passive margin. Concepts of the role of the Cabo Frio fault zone as a detachment and its bearing on hydrocarbon migration will be tested by exploratory drilling in the near future.

Finally, we conclude that exploration interpretation in complex regions can be optimized by widening the area of study to the scale of a whole basin or several basins. Some areas require integration of several tools to allow better geologic characterization. In the Cabo Frio region, for example, it is possible to recognize the northwest trend of the postbreakup volcanism only if we extrapolate the gravity maps to regions underlain by oceanic crust. Again, structures near the shelf edge are actually controlled by salt tectonics presently tens of kilometers away in deep water. Finally, it is important to emphasize that although we have only touched on the profound implications of extending geologic interpretation to deep water, this has greatly increased our understanding of salt tectonics and perhaps provided a reference model to explain similar features in other evaporite basins.

Acknowledgments *The Cabo Frio Project was conducted at the Petrobrás Department of Exploration from 1989 to 1991. This work results from many technical discussions about this frontier region. Specific contributions were given by the following explorationists: project coordination, regional geologic evaluation, and tectonics by Webster U. Mohriak; seismic mapping and geologic integration (Santos Basin) by Juliano M. Macedo and Ricardo T. Castellani; seismic mapping and geologic integration (Campos Basin) by A. Z. N. Barros, A. Fujita, and J. A. Ricci; stratigraphy and geologic integration (Campos Basin) by H. D. Rangel; paleontology, stratigraphy, and sedimentology by A. Richter and M. G. F. Costa; potential methods by M. A. L. Latgé and B. S. Gomes; tectonics and physical modeling by P. Szatmari, L. S. Demercian, J. G. Rizzo, and J. R. Aires; igneous petrography and geochronology by A. M. P. Mizusaki; log interpretation by M. S. Scutta; and geochemistry by M. R. Mello, G. Henz, and A. L. Soldan.*

We gratefully acknowledge A. M. F. de Figueiredo, R. S. Carvalho, and L. G. F. Aranha for constant support throughout this work. We also thank Orlando A. Lima and the SEDESE group for draftsmanship and artwork.

Authorization to present this work at the AAPG Hedberg International Research Conference on Salt Tectonics and permission to publish were given by the Petrobrás Board of Directors and by the Department of Exploration managers, C. F. Lucchesi and S. Possato. We are grateful to D. Abrahão and L. R. Guardado for many constructive suggestions. We thank J. J. Moraes, Jr., for reading the first draft of this paper before submission to AAPG.

We appreciate the critical reading and many suggestions by M. P. A. Jackson, B. Vendeville, and S. Seni, which greatly improved the original manuscript. We express our gratitude to M. P. A. Jackson for his constant cooperation, enlightening suggestions, and insightful comments.

REFERENCES CITED

Abrahão, D., and J. E. Warme, 1990, Lacustrine and associated deposits in a rifted continental margin—Lower Cretaceous Lagoa Feia Formation, Campos Basin, offshore Brazil, *in* B. J. Katz, Lacustrine basin exploration, case studies and modern analogs: AAPG Memoir 50, p. 287–305.

Almeida, F. F. M., 1967, Origem e evolução da plataforma brasileira: DNPM/DGM, Boletim 241, Rio de Janeiro.

Almeida, F. F. M., 1976, The system of continental rifts bordering the Santos Basin, Brazil: An. Acad. bras. Ciên., Suppl. 48, p. 15–26.

Andrade, V. F., J. C. Vieira, C. F. Luchesi, F. J. Feijó, G. G. Herter, and S. Shimabukuro, 1983, Regional depositional and tectonic characterization of Brazilian marginal basins: Petrobrás Internal Report, Exploration Department, Rio de Janeiro, 47 p.

Asmus, H. E., 1978, Hipótese sobre a origem dos sistemas de zonas de fraturas oceânicas/alinhamentos continentais que ocorrem nas regiões sudeste e sul do Brasil: Série Projeto Remac, v. 4, p. 39–73.

Asmus, H. E., 1982, Significado geotectônico das feições estruturais das bacias marginais brasileiras e áreas adjacentes: Congr. Bras. Geol. 32, Salvador, v. 4, p. 1547–1557.

Asmus, H. E., and A. L. Ferrari, 1978, Hipótese sobre a causa do tectonismo Cenozóico na região sudeste do Brasil: Série Projeto Remac, v. 4, p. 75–88.

Asmus, H. E., and W. Guazelli, 1981, Descrição sumária das estruturas da margem continental brasileira e das áreas oceânicas e continentais, adjacentes: Série Projeto Remac, v. 9, p. 187–269.

Asmus, H. E., and F. C. Ponte, 1973, The Brazilian marginal basins, *in* A. E. Nair and F. G. Stehli, eds., The ocean basins and margins, v. 1, The South Atlantic: New York, Plenum Press, p. 87–132.

Azevedo, R. L. M., J. Gomide, and M. C. Viviers, 1987a, Geohistória da Bacia de Campos, Brasil: do Albiano ao Maastrichtiano: Revista Brasileira de Geociências, v. 17, no. 2, p. 139–146.

Azevedo, R. L. M., J. Gomide, M. C. Viviers, and A. T. Hashimoto, 1987b, Bioestratigrafia do Cretáceo marinho da Bacia de Campos, Brasil: Revista Brasileira de Geociências, v. 17, no. 2, p. 147–153.

Bacoccoli, G., and L. G. F. Aranha, 1984, Evolução estrutural fanerozóica do Brasil meridional: Relatório Interno Petrobrás, Depex, Rio de Janeiro, 153 p.

Bizzi, L. A., C. B. Smith, H. O. A. Meyer, R. Armstrong, and M. J. De Wit, 1991, Mesozoic kimberlites and related alkalic rocks in south-western São Francisco craton, Brazil: a case for local mantle reservoirs and their interaction: Fifth International Kimberlite Conference, Araxá, June 1991, Extended Abstracts, p. 17–19.

Bostrom, R. C., 1989, Subsurface exploration via satellite: structure visible in Seasat images of North Sea, Atlantic continental margin, and Australia: AAPG Bulletin, v. 73, no. 9, p. 1053–1064.

Brice, S. E., M. D. Cochran, G. Pardo, and A. D. Edwards, 1982, Tectonics and sedimentation of the South Atlantic rift sequence: Cabinda, Angola, *in* J. S. Watkins and C. L. Drake, eds., Studies in continental margin geology: AAPG Memoir 34, p. 5–18.

Campanha, G. A. da C., 1981, O lineamento de Além-Paraíba na área de Três Rios (RJ): Revista Brasileira de Geociências, v. 11, no. 3, p. 159–171.

Campos, C. W. M., F. C. Ponte, and K. Miura, 1974, Geology of the Brazilian continental margin, *in* C. A. Burk and C. L. Drake, eds., The geology of the continental margins: New York, Springer-Verlag, p. 447–461.

Chang, H. K., and R. O. Kowsmann, 1987, Interpretação genética das seqüências estratigráficas das bacias da margem continental brasileira: Revista Brasileira de Geociências, v. 17, no. 2, p. 74–80.

Chang, H. K., R. O. Kowsmann, A. M .F. Figueiredo, and A. Bender, 1992, Tectonics and stratigraphy of the East Brazil rift system: an overview: Tectonophysics, v. 213, p. 97–138.

Cloos, E., 1968. Experimental analysis of Gulf Coast fracture patterns: AAPG Bulletin, v. 52, no. 3, p. 420–444.

Cobbold, P. R., and P. Szatmari, 1991, Radial gravitational gliding on passive margins: Tectonophysics, v. 188, p. 249–289.

Cobbold, P. R., E. A. Rossello, and B. Vendeville, 1989, Some experiments on interacting sedimentation and deformation above salt horizons: Bulletin Soc. Géol. France, v. 8, t. V, no. 3, p. 453–460.

Cobbold, P. R., P. Szatmari, L. S. Demercian, D. Coelho, and E. A. Rossello, 1995, Seismic and experimental evidence for thin-skinned horizontal shortening by convergent radial gliding on evaporites, deep-water Santos Basin, Brazil, *in* M. P. A. Jackson, D. G. Roberts, and S. Snelson, eds., Salt tectonics: a global perspective: AAPG Memoir 65, this volume.

Cordani, U. G., et al., 1984, Estudo preliminar de integração do Precambriano com eventos tectônicos das bacias sedimentares brasileiras: Ciência Técnica Petróleo, Exploração de Petróleo, v. 15, Cenpes/Petrobrás, 70 p.

Demercian, L. S., P. Szatmari, P. R. Cobbold, and D. F. Coelho, 1992, Halocinese na Bacia de Campos: Anais do XXXVII Congresso Brasileiro de Geologia, São Paulo, SP, v. 1, p. 560–561.

Demercian, L. S., P. Szatmari, and P. R. Cobbold, 1993, Style and pattern of salt diapirs due to thin-skinned gravitational gliding, Campos and Santos basins, offshore Brazil: Tectonophysics, v. 228, p. 393–433.

Dias, J. L., J. C. Vieira, A. J. Catto, J. Q. Oliveira, W. Guazelli, L. A. F. Trindade, R. O. Kowsmann, C. H. Kiang, U. T. Mello, A. M. P. Mizusaki, and J. A. Moura, 1987, Estudo Regional da Formação Lagoa Feia: Relatório Interno Petrobrás, Depex, Rio de Janeiro, 143 p.

Dias-Brito, D., 1982, Evolução paleoecológica da Bacia de Campos durante a deposição dos calcilutitos, margas e folhelhos da Formação Macaé (Albiano e Cenomaniano?): Boletim Técnico Petrobrás, v. 25, no. 2, p. 11–24.

Dias-Brito, D., 1987, A Bacia de Campos no Mesocretáceo: uma contribuição à paleoceanografia do Atlântico Sul primitivo: Revista Brasileira de Geociências, v. 17, no. 2, p. 162–167.

Dillon, W. P., P. Popenoe, J. A. Grow, K. D. Klitgord, B. Swift, C. K. Paull, and K. V. Cashman, 1982, Growth faulting and salt diapirism: their relationship and control in the Carolina Trough, Eastern North America, *in* J. S. Watkins and C. L. Drake, eds., Studies in continental margin geology: AAPG Memoir 34, p. 21–46.

Departamento Nacional da Produçáo Mineral, 1978, Carta Geológica do Brasil ao milionésimo: Ministério das Minas e Energia, 1:1,000,000, 46 sheets.

Duval, B., C. Cramez, and M. P. A. Jackson, 1992, Raft tectonics in the Kwanza Basin, Angola: Marine and Petroleum Geology, v. 9, p. 389–404.

Figueiredo, A. M. F., and W. U. Mohriak, 1984, A tectônica salífera e as acumulações de petróleo na Bacia de Campos: Congresso Brasileiro de Geologia, 33, Rio de Janeiro, 1984, SBG, v. 3., p. 1380–1384.

Fodor, R. V., E. H. McKee, and H. E. Asmus, 1983, K-Ar ages and the opening of the South Atlantic Ocean: basaltic rocks from the Brazilian margin: Marine Geology, v. 54, p. M1–M8.

Francheteau, J., and X. Le Pichon, 1972, Marginal fracture zones as structural framework of continental margins of South Atlantic Ocean: AAPG Bulletin, v. 56, p. 991–1007.

Guardado, L. R., L. A. P. Gamboa, and C. F. Lucchesi, 1989, Petroleum geology of the Campos Basin, a model for a producing Atlantic-type basin, *in* J. D. Edwards and P.A. Santogrossi, eds., Divergent/passive margin basins: AAPG Memoir 48, p. 3–79.

Harland, W.B., A. V. Cox, P. G. Llewellyn, C. A. G. Pickton, D. G. Smith, and S. R. Walter, 1983, A geologic time scale: Cambridge, Cambridge University Press, 131 p.

Hasui, Y., C. D. R. Carneiro, and A. M. Coimbra, 1975, The Ribeira folded belt: Rev. Bras. Geoc., v. 5, no. 4, p. 257–266.

Hasui, Y., and M. A. F. Oliveira, 1984, Província Mantiqueira–Setor Central, *in* F. F. M. Almeida and Y. Hasui, eds., O Pré-Cambriano do Brasil, p. 308–344.

Jackson, M. P. A., and B. C. Vendeville, 1994, Regional extension as a geologic trigger for diapirism: GSA Bulletin, v. 106, p. 57–73.

Jenyon, M. K., 1985, Fault-associated salt flow and mass movement: Journal of the Geological Society of London, v. 142, p. 547–553.

Jenyon, M. K., 1986a, Salt tectonics: London, Elsevier Applied Science Publishers, 191 p.

Jenyon, M. K., 1986b, Some consequences of faulting in the presence of a salt interval: Journal of Petroleum Geology, v. 9, p. 29–52.

Jenyon, M. K., 1988, Some deformation effects in a clastic overburden resulting from salt mobility: Journal of Petroleum Geology, v. 11, no. 3, p. 309–324.

Koutsoukos, E. A. M., 1984, Evolução paleoecológica do Albiano ao Maastrichtiano na área noroeste da Bacia de Campos, Brasii: Congresso Brasileiro de Geologia, 33, Rio de Janeiro, 1984, v. 3, p. 685–698.

Koyi, H., and K. Petersen, 1993, Influence of basement faults on the development of salt structures in the Danish Basin: Marine and Petroleum Geology, v. 10, p. 82–94.

Larberg, G. M. B., 1983, Contra-regional faulting: salt withdrawal compensation, offshore Louisiana, Gulf of Mexico, *in* A. W. Bally, ed., Seismic expression of structural styles: AAPG Studies in Geology Series 15, v. 2, p. 2.3.2.42–2.3.2.43.

Larson, R. L., and J. W. Ladd, 1973, Evidence for the opening of the South Atlantic in the Early Cretaceous: Nature, v. 246, p. 209–212.

Le Pichon, X., and D. E. Hayes, 1971, Marginal offsets, fracture zones, and the early opening of the South Atlantic: Journal of Geophysical Research, v. 76, p. 6283–6293.

Liro, L. M., and R. Coen, 1995, Salt deformation history and postsalt structural trends, offshore southern Gabon, West Africa, *in* M. P. A. Jackson, D. G. Roberts, and S. Snelson, eds., Salt tectonics: a global perspective for exploration: AAPG Memoir 65, this volume.

Lobo, A. P., and J. O. Ferradaes, 1983, Reconhecimento preliminar do talude e sopé continentais da Bacia de Campos: Relatório Interno Petrobrás, Depex, Rio de Janeiro, 29 p.

Lundin, E. R., 1992, Thin-skinned extensional tectonics on a salt detachment, northern Kwanza Basin, Angola: Marine and Petroleum Geology, v. 9, p. 405–411.

Macedo, J. M., 1989, Evolução tectônica da Bacia de Santos e áreas continentais adjacentes: Boletim de Geociências da Petrobrás, Rio de Janeiro, v. 3, no. 3, p. 159–173.

McClay, K. R., 1989. Physical models of structural styles during extension, *in* A. J. Tankard and H. R. Balkwill, eds., Extensional tectonics and stratigraphy of the North Atlantic margins: AAPG Memoir 46, p. 95–110.

McClay, K. R., 1990, Extensional fault systems in sedimentary basins: a review of analogue model studies: Marine and Petroleum Geology, v. 7, p. 206–233.

Melo, M. S., C. Riccomini, Y. Hasui, F. F. M. Almeida, and A. M. Coimbra, 1985, Geologia e evolução do sistema de bacias tafrogênicas continentais do sudeste do Brasil: Rev. Bras. Geoc., v. 15, no. 3, p. 193–201.

Milward, R. L., J. Gomide, and M. C. Viviers, 1987, Geohistória da Bacia de Campos, Brasil: do Albiano ao Maastrichtiano: Revista Brasileira de Geociências, v. 17, no. 2, p. 139–146.

Mizusaki, A. M. P., A. Thomaz Filho, and J. G. Valença, 1988, Volcano-sedimentary sequence of Neocomian age in Campos Basin (Brazil): Rev. Bras. Geociênc., v. 18, p. 247–251.

Mizusaki, A. M. P., and W. U. Mohriak, 1992, Sequências vulcano-sedimentares na região da plataforma continental de Cabo Frio, RJ: Anais do XXXVII Congresso Brasileiro de Geologia, Resumos Expandidos, São Paulo, SP, v. 2, p. 468–469.

Mohriak, W. U., 1988, The tectonic evolution of the Campos Basin, offshore Brazil: Ph.D. dissertation, University of Oxford, U.K., 381 p.

Mohriak, W. U, 1991, Evolução tectono-sedimentar da área "offshore" de Cabo Frio, Rio de Janeiro: Petrobrás Internal Report, v. 1 (text), v. 2 (atlas), v. 3 (exploratory plays): Depex/Dirsul/Serab, Rio de Janeiro.

Mohriak, W. U., and A. Z. Barros, 1990, Novas evidências de tectonismo cenozóico na região sudeste do Brasil: o graben de Barra de são João na plataforma continental de Cabo Frio, Rio de Janeiro: Revista Brasileira de Geociências, v. 20, no. 1-4, p. 187–196.

Mohriak, W. U., A. Z. Barros, and A. M. Fujita, 1989, Geologia da plataforma continental de Cabo Frio, Rio de Janeiro: Primeiro Simpósio de Geologia do Sudeste, Rio de Janeiro, SBG, Boletim de Resumos, p. 21.

Mohriak, W., M. R. Mello, J. F. Dewey, and J. R. Maxwell, 1990a, Petroleum geology of the Campos Basin, offshore Brazil, *in* J. Brooks, ed., Classic petroleum provinces: Geological Society of London, p. 119–141.

Mohriak, W. U., M. R. Mello, G. D. Karner, J. F. Dewey and J. R. Maxwell, 1990b, Structural and stratigraphic evolution of the Campos Basin, offshore Brazil, *in* A .J. Tankard and H. R. Balkwill, eds., Extensional tectonics and stratigraphy of the North Atlantic margins: AAPG Memoir 46, p. 577–598.

Mohriak, W. U., A. Z. Barros, and A. Fujita, 1990c, Magmatismo e tectonismo cenozóicos na região de Cabo Frio, RJ: Congresso Brasileiro de Geologia, 37, Natal, 1990, v. 6, p. 2873–2885.

Mohriak, W. U., J. M. Macedo, and R. T. Tarabini, R. T., 1992, Estilos estruturais e tectônica de sal na região de Cabo Frio, RJ: Anais do XXXVII Congresso Brasileiro de Geologia, Resumos Expandidos, São Paulo, SP, v. 1, p. 336–337.

Mohriak, W. U., M. C. Barros, J. H. L. Rabelo, and R. D. Matos, 1993, Deep seismic survey of Brazilian passive basins: the northern and northeastern regions: Third International Congress of the Brazilian Geophysical Society, Rio de Janeiro, RJ, v. 2, p. 1134–1139.

Momenzadeh, M., 1990, Saline deposits and alkaline magmatism: a genetic model: Journal of Petroleum Geology, v. 13, no. 3, p. 341–356.

Nettleton, L. L., 1943, Recent experimental and geophysical evidence of mechanics of salt dome formation: AAPG Bulletin, v. 27, p. 51–63.

Ojeda, H. A. O., 1982, Structural framework, stratigraphy, and evolution of Brazilian marginal basins: AAPG Bulletin, v. 66, p. 732–749.

Ojeda, H. A. O., M. Carminatti, and J. B. Rodarte, 1983, Bacia de Campos: arcabouço estrutural regional e interpretação genética preliminar: Relatório Interno Petrobrás, Depex, Rio de Janeiro, 50 p.

Pereira, M. J., C. M. Barbosa, J. Agra, J. B., Gomes, L. G. F. Aranha, M. Saito, M. A. Ramos, M. D. Carvalho, M. Stamato, and O. Bagni, 1986, Estratigrafia da Bacia de Santos: análise das seqüências, sistemas deposicionais e revisão lito-estratigráfica: Congresso Brasileiro de Geologia, 34, Goiânia, 1986, v. 1, p. 65–79.

Rabinowitz, P. D, and J. LaBreque, 1979, The Mesozoic South Atlantic Ocean and evolution of its continental margins: Journal of Geophysical Research, v. 84, no. B11, p. 5973–6002.

Radambrasil, 1983, Mapa Geológico, Folhas SF-23/24, Rio de Janeiro/Vitória, scale 1:1,000,000, v. 32, p. 27–304.

Rangel, H. D., W. U. Mohriak, A. Richter, A. Z. Barros, and C. J. Appi, 1990, Evolução estrutural e estratigráfica da porção sul da bacia de Campos: Fourth Congresso Brasileiro de Petróleo, Rio de Janeiro, TT 207, p. 1–10.

Renne, P. R., M. Ernesto, I. G. Pacca, R. S. Coe, J. M. Glen, M. Prévot, and M. Perrin, 1992, The age of Paraná flood volcanism, rifting of Gondwanaland, and the Jurassic–Cretaceous boundary: Science, v. 258, p. 975–979.

Richter, A. J., 1987, Subafloramento das discordâncias Turoniana e Campaniana no sul da Bacia de Campos: Rev. Bras. Geoc., v. 17, no. 2, p. 173–176.

Rizzo, J. G., 1987, Falhas das seqüência rift e pós-rift na Bacia de Campos, Rio de Janeiro, Brasil–possibilidades de relacionamento. Master's thesis, Universidade Federal de Ouro Preto, 74 p.

Rizzo, J. G., W. U. Mohriak, J. G. Aires, and A. Z. N. Barros, 1990, Modelagem física de falhamentos antitéticos em águas profundas da região de Cabo Frio na Bacia de Campos, RJ: Congresso Brasileiro de Geologia, 37, Natal, 1990, v. 5, p. 2228–2249.

Rocha-Campos, A. C., U. G. Cordani, K. Kawashita, H. M. Sonoki, and I. K. Sonoki, 1988, Age of the Paraná flood volcanism, *in* E. M. Piccirillo and A .J. Melfi, eds., The Mesozoic flood volcanism of the Paraná basin: petrogenetic and geophysical aspects: Inst. Astron. Geofis., University of São Paulo, Brazil, p. 25–45.

Rosendahl, B. R., H. Groschel-Becker, J. Meyers, and K. Kaczmarick, 1991, Deep seismic reflection study of a passive margin, southeastern Gulf of Guinea: Geology, v. 19, p. 291–295.

Schaller, H., and M. V. Dauzacker, 1986, Tectônica gravitacional e sua aplicação na exploração de hidrocarbonetos: Boletim Técnico da Petrobrás, v. 29, no. 3, p. 143–206.

Schmalz, R. F., 1969, deep-water evaporite deposition: a genetic model: AAPG Bulletin, v. 53, p. 798–823.

Schuster, D. C., 1995, Deformation of allochthonous salt and evolution of related salt/structural systems, eastern Louisiana Gulf Coast, *in* M. P. A. Jackson, D. G. Roberts, and S. Snelson, eds., Salt tectonics: a global perspective: AAPG Memoir 65, this volume.

Souza, K. G., 1991, La marge continentale bresilienne sub-orientale et les domaines oceaniques adjacents: structure et evolution: Ph.D. dissertation, Laboratorie de Géodynamique Sous-Marine de Villefrance sur Mer, Université Pierre et Marie Curie, 230 p.

Szatmari, P., and L. S. Demercian, 1991, Halocinese na Bacia de Campos e área de Cabo Frio: Relatório Interno, Petrobrás/Cenpes/Divex, 17 p.

Szatmari, P., J. C. J. Conceição, M. C. Lana, E. J. Milani and A. P. Lobo, 1984, Mecanismo tectônico do rifteamento Sul-Atlântico: Congresso Brasileiro de Geologia, 33, Rio de Janeiro, 1984, p. 1589–1601.

Szatmari, P., M. C. M. Guerra, and M. A. Pequeno, 1993, Modelagem física da falha antitética de Santos: Petrobrás Internal Report, Cenpes, Rio de Janeiro, 9 p.

Teisserenc, P., and Villemin, J., 1989, Sedimentary basin of Gabon: geology and oil systems, *in* J. D. Edwards and P.A. Santogrossi, eds., Divergent/passive margin basins: AAPG Memoir 48, p. 117–199.

Vail, P. R., R. M. Mitchum, and S. Thompson, 1977, Seismic stratigraphy and global changes of sea level, part 4: global cycles of relative changes of sea level, *in* C. E. Payton, eds., Seismic stratigraphy—applications to hydrocarbon exploration: AAPG Memoir 26, p. 83–97.

Ven, P. H. V. D., 1983, Seismic stratigraphy and depositional systems of northeastern Santos Basin, offshore southeastern Brazil: Master's thesis, The University of Texas at Austin, 151 p.

Vendeville, B., and P. R. Cobbold, 1988, How normal faulting and sedimentation interact to produce listric fault profiles and stratigraphic wedges: Journal of Structural Geology, v. 10, no. 7, p. 649–659.

Vendeville, B. C., and M. P. A. Jackson, 1992a, The rise of diapirs during thin-skinned extension: Marine and Petroleum Geology, v. 9, p. 331–353.

Vendeville, B. C., and M. P. A. Jackson, 1992b, The fall of diapirs during thin-skinned extension: Marine and Petroleum Geology, v. 9, p. 354–371.

Vendeville, B., P. R. Cobbold, P. Davy, J. P. Brun, and P. Choukroune, 1987, Physical models of extensional tectonics at various scales, *in* M. P. Coward, J. F. Dewey, and P. Hancock, eds., Continental extensional tectonics: GSA Special Publication 28, p. 95–107.

Vendeville, B. C., M. P. A. Jackson, and D. D. Schultz-Ela, 1992, Applied Geodynamics Laboratory, second semi-annual progress report to industrial associates for 1991: Bureau of Economic Geology, The University of Texas at Austin, February, 48 p.

Viviers, M. C., 1986, Bioestratigrafia e evolu ão paleoambiental do Meso-Neocretáceo da Bacia de Santos, Brasil: Congresso Brasileiro de Geologia, 34, Goiânia, 1986, v. 1, p. 50–64.

Williams, B. G., and R. J. Hubbard, 1984, Seismic stratigraphy framework and depositional sequences in the Santos Basin, Brazil: Marine and Petroleum Geology, v. 1, p. 90–104.

Withjack, M. O., K. E. Meisling, and L. R. Russel, 1989, Forced folding and basement-detached normal faulting in the Haltenbanken area, offshore Norway, *in* A. J. Tankard and H. R. Balkwill, eds., Extensional tectonics and stratigraphy of the North Atlantic margins: AAPG Memoir 46, p. 567–575.

Worrall, D. M., and S. Snelson, 1989, Evolution of the northern Gulf of Mexico, with emphasis on Cenozoic growth faulting and the role of salt, *in* A. W. Bally and A. R. Palmer, eds., The geology of North America—an overview: GSA, The Geology of North America, v. A, p. 97–138.

Wu, S., A. W. Bally, and C. Cramez, 1990, Allochthonous salt, structure, and stratigraphy of the northeastern Gulf of Mexico, part II, structure: Marine and Petroleum Geology, v. 7, p. 334–370.

Cobbold, P. R., P. Szatmari, L. S. Demercian, D. Coelho, and E. A. Rossello, 1995,
Seismic and experimental evidence for thin-skinned horizontal shortening by
convergent radial gliding on evaporites, deep-water Santos Basin, Brazil, *in*
M. P. A. Jackson, D. G. Roberts, and S. Snelson, eds., Salt tectonics: a global perspective: AAPG Memoir 65, p. 305–321.

Chapter 14

Seismic and Experimental Evidence for Thin-Skinned Horizontal Shortening by Convergent Radial Gliding on Evaporites, Deep-Water Santos Basin, Brazil

Peter R. Cobbold

Géosciences-Rennes
Université de Rennes
Rennes, France

Peter Szatmari

L. Santiago Demercian

Petrobras Research Center
Cidade Universitária
Rio de Janeiro, Brazil

Dimas Coelho

Petrobras Exploration Department
Rio de Janeiro, Brazil

Eduardo A. Rossello

Géosciences-Rennes
Université de Rennes
Rennes, France

Present address:
Departamento de Ciencias Geológicas
Universidad de Buenos Aires
Buenos Aires, Argentina

Abstract

Thin-skinned gravitational gliding of sediments above a detachment layer of salt or shale is common on passive margins. Changes in surface slope result in a domain of extension upslope and a domain of contraction
downslope. Contractional domains tend to occur under present-day deep water and are thus not well understood.

In the deep-water Santos Basin, Brazil, a contractional domain contains a suite of salt-cored structures.
Angular folds (chevron and box folds), as well as concentric folds, are common in the upper part of the Aptian
evaporite sequence, which appears to comprise alternating layers. In general, angular and concentric folds form
by flexural slip during shortening of mechanically layered sequences. Their occurrence in the Santos Basin is evidence in favor of horizontal contraction. The lower part of the Aptian evaporite sequence appears to be mostly
rock salt. It has been squeezed out from under synclines into spaces created by growing anticlines. In places, the
layered evaporite sequence has been thickened or even repeated across thrust faults and ramp anticlines. An
overlying sequence of open-marine sediments has been condensed or eroded over anticlines but forms local
depocenters. These depocenters are asymmetric (of foreland style) next to isolated thrusts but symmetric in synclines or between thrusts of opposite vergence.

The structural styles have been reproduced in physical models, properly scaled for gravitational forces, in
which salt is represented by silicone putty and sediments are represented by sand. The models were shortened
horizontally by a screw jack. The experiments illustrate the importance of horizontal contraction and syntectonic
sedimentation in shaping salt-cored structures. They have been used to establish criteria that may be diagnostic
of contraction.

INTRODUCTION

Many of the world's hydrocarbon provinces are associated with evaporites. Impermeable layers of salt form
good hydraulic seals for reservoirs, and they guide the
migration of hydrocarbons. Because of their ductility and
low viscosity, salt layers also provide detachment layers
and form domes responsible for many structural and
stratigraphic traps.

As a result of intense exploration for hydrocarbons, it
has recently become apparent that thin-skinned deformation above salt layers is very common, even in stable tectonic settings, such as passive continental margins. Here,
surface slopes are generated by prograding sedimentation or by basinward tilting of the entire margin, as a
result of thermal subsidence or formation of new oceanic
crust. If there is a layer of salt or soft mud at depth, the
overlying sedimentary sequence can detach and glide

downslope under its own weight. A slope of only a few degrees is sufficient. The detachment may be at a depth of up to several kilometers, and the gliding may occur over several hundred kilometers. Resulting structures may be surprisingly large and structural styles complex (see Jackson and Talbot, 1991). This has led to difficulties with seismic interpretation, especially before the arrival of modern seismic acquisition and processing techniques. Physical modeling has been of great help in investigating the mechanical origin of structures and in elucidating their styles.

In a gliding thin-skinned sheet, a central plate may shift (translate) downslope with little internal deformation. However, as in plate tectonics, such a process will lead to divergence and convergence at the upper and lower margins, respectively, of the sliding plate. These margins may not be sharp everywhere. Instead, they may be domains of internal deformation (bulk strain)—extensional upslope and contractional downslope. The strain is expressed as characteristic structures, which may be salt cored. From physical modeling at the scale of an entire continental margin, Cobbold et al. (1989) have suggested that extensional and contractional domains may form where there are significant changes in surface slope, for example, at the upper and lower hinge lines of a margin or at the slope edges of prograding sedimentary wedges and deltas.

Because extensional domains tend to be in proximal basin positions, such as onshore or under shallow water, they have been better studied than contractional domains. Structures characteristic of extensional domains are salt rollers in the footwalls of listric normal growth faults, salt walls (triangular in cross section) between intersecting conjugate normal faults, turtle anticlines, and salt welds (see Jackson and Talbot, 1991).

Contractional domains due to gravitational gliding have been less well studied because they tend to occur in deeper water at the foot of a continental slope or a prograding delta front. Better known are contractional structures from orogenic areas, where fold and thrust belts may be underlain by salt (see Davis and Engelder, 1987; Jackson and Talbot, 1991; de Ruig, 1995; Harrison, 1995; Letouzey et al., 1995; Sans and Vergés, 1995). Nevertheless, contractional domains have been recognized along the passive margins of some continents, including the Atlantic margin of Africa (Lehner and de Ruiter, 1977), the southern margin of Australia (Wilcox et al., 1988), the southern margin of the Red Sea (Heaton et al., 1995), and the Gulf of Mexico (Weimer and Buffler, 1992).

A well-studied area for salt tectonics is the Gulf of Mexico. Here, Buffler et al. (1979) described the style of folding and thrusting in the Mexican Ridges and considered the possibility that they formed as a result of gravitational gliding over Miocene shale. Worrall and Snelson (1989) reviewed structural styles throughout the Gulf, including deep-water contractional ones. Weimer and Buffler (1992) compared the styles of folds and reverse faults in three gravity-induced contractional domains of different ages: the Mississippi Fan, Perdido, and Mexican

Ridges fold belts. In the first two, the structures are cored by salt; in the Mexican Ridges, by shale. Concentrating on the Mississippi Fan foldbelt, Weimer and Buffler (1992) identified reverse faults having sediment wedges in their footwalls and asymmetric salt bodies in their hanging walls.

On the Brazilian Atlantic margin, thin-skinned salt-cored structures overlie a detachment layer of Aptian evaporites. Almost all authors have attributed these structures to gravitational gliding (Ojeda, 1982; Petrobras, 1983; Schaller and Dauzacker, 1986; Chang et al., 1988, 1992; Guardado et al., 1989; Dias et al., 1990; Figueiredo and Martins, 1990; Mohriak et al., 1990a,b,c; Pereira and Macedo, 1990; Carminatti and Scarton, 1991; Cobbold and Szatmari, 1991; Demercian et al., 1993; Mohriak et al., 1995).

Contractional growth folds in the deep-water parts of the Campos and Santos basins were described by Cobbold and Szatmari (1991). In the Santos Basin, growth folds were identified on both dip-oriented and strike-oriented seismic lines, providing evidence for radially convergent gliding in variable directions perpendicular to the arcuate coastline. Salt-cored thrusts were identified on one regional seismic line across the Santos Basin, but this interpretation was debatable because of the poor quality of the unmigrated seismic image (Cobbold and Szatmari, 1991, their figure 9).

In a more recent paper, Demercian et al. (1993) used time-migrated seismic sections and a regional structure contour map on the top of the evaporite sequence to provide further evidence for horizontal contraction in both the Campos and Santos basins. In the deep-water Campos Basin, trains of contractional growth folds have two dominant wavelengths, each proportional to total stratigraphic thickness at the time of buckling (Demercian et al., 1993, their figure 5). Salt-cored domes with sharp, straight overhanging edges were attributed to folding and synsedimentary reverse faulting (Demercian et al., 1993, their figure 6). More complex structures were attributed to compressional inversion of triangular salt walls, formed earlier and farther upslope as a result of horizontal extension (Demercian et al., 1993, their figure 3). Finally, drawing on unpublished results from physical modeling, Demercian et al. (1993) interpreted regional seismic lines with poorer definition, identifying a suite of contractional structures: growth folds, depocenters, reverse faults with sediment wedges in their footwalls, and an allochthonous tongue at the edge of the salt.

The main purpose of the present paper is to provide seismic evidence for horizontal contraction in the deep-water Santos Basin, adding to the examples described by Cobbold and Szatmari (1991) and improving on them. A second purpose is to describe some experiments on horizontal contraction in layered physical models. The experiments help explain structural mechanisms and styles in areas of synsedimentary horizontal contraction. They also help to support some of the seismic interpretations of Cobbold and Szatmari (1991), Weimer and Buffler (1992), and Demercian et al. (1993).

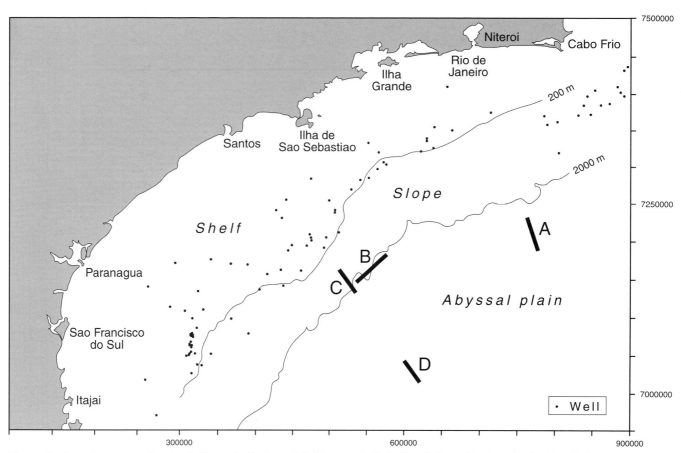

Figure 1—Location map of seismic lines A, B, C, and D (Figures 2, 3, 4, 5), offshore Santos Basin, Brazil. Contours show the approximate limits of the continental slope. Map coordinates are from the Brazilian National Grid (UTM projection).

SEISMIC EVIDENCE FOR HORIZONTAL SHORTENING IN THE DEEP-WATER SANTOS BASIN

In the deep-water Santos Basin, Cobbold and Szatmari (1991) identified growth folds on both dip-oriented and strike-oriented seismic lines. They attributed the strike-parallel contraction to radial gliding of overburden in directions perpendicular to the arcuate coastline, which leads to a convergence of displacement paths along strike. A regional structure contour map on the top of the evaporite sequence provided further evidence for a pattern of radially convergent gliding in the Santos Basin. Salt-cored thrusts were identified on one seismic dip line across the basin.

Here we reproduce four short seismic sections (Figures 2, 3, 4, 5) extracted from recent regional surveys. For proprietary reasons, the locations of these sections are approximate (Figure 1). Three lines run downdip, and the fourth runs along strike. They were chosen to complement the three sections previously published by Cobbold and Szatmari (1991, their figures 9, 10, 11). Thus, line B of the current paper (Figure 3) is the westward continuation of a previously published section (Cobbold and Szatmari, 1991, line C, their figure 11).

The seismic data were originally available as time-migrated versions, which we interpreted in the form of simple line drawings. For three of the lines, depth-migrated data subsequently became available and are reproduced here without interpretation (Figures 2, 3, 5).

In the Santos Basin, Aptian evaporites (Alagoas Stage, Ariri Formation) may have been up to several kilometers thick when deposited (Pereira and Macedo, 1990; Demercian et al., 1993). Anhydrite dominates the evaporitic interval toward the western edge of the basin, whereas halite is more common offshore. However, the nature of the evaporite sequence is not well understood under deep water. Few wells have been drilled to date in the Santos Basin and none are in deep water. From our observations of seismic stratigraphy and structural style, we believe that the lower part of the sequence is mainly halite. In contrast, in the upper part of the sequence, seismic reflectors are evenly spaced, suggesting either regular contrasts in seismic velocity or a limited seismic bandwidth (Figures 2, 3, 5). However, since the latter possibility can be discarded, we infer that the sequence is layered. Where the sequence is folded, chevron folds, concentric folds, or box folds are the most common styles (Figures 2, 3, 4, 5). Layers tend to be uniformly thick around the folds, so we infer that folding occurred after deposition of the evaporite sequence. It is well known from physical

(Text continues on p. 312)

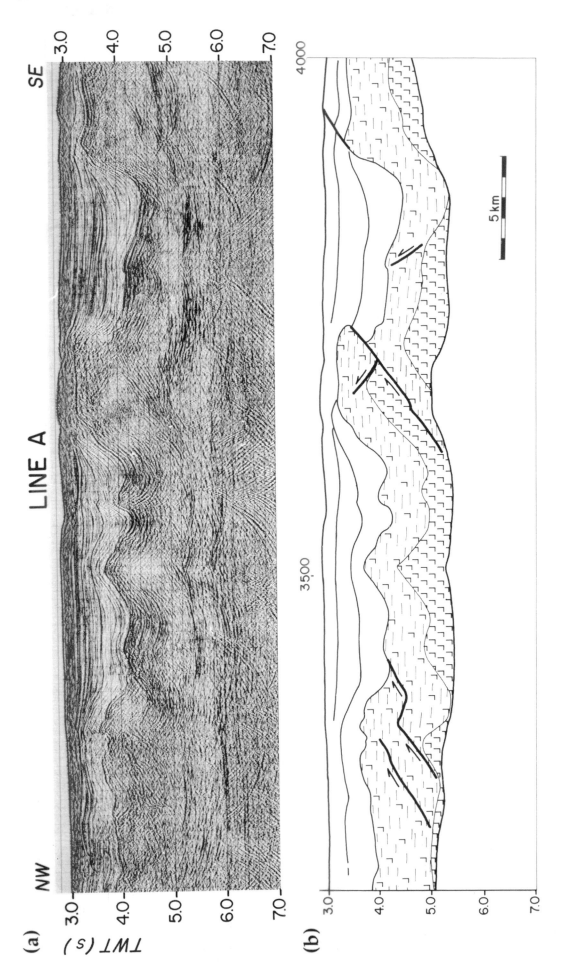

Figure 2—Seismic section A, a deep dip line trending southeastward (to the right) down the northeastern part of the basin (see Figure 1 for location). (a) Time-migrated seismic section. (b) Drawing interpreting the time-migrated section and showing major evaporite units (patterned), bedding (thin lines), reverse faults (heavy lines with arrows), and a normal fault (heavy line). Vertical scale is in seconds (two-way traveltime). Numbers along the top refer to shot points. Folds and faults occur above a detachment layer with little internal layering interpreted as rock salt (dense L pattern). Sequence with fine bedding and angular or concentric folds is interpreted as layered evaporite (alternating dashes and L pattern). Notice small, nearly isoclinal fold (center) and thrust faults with ramp anticlines and dominant southeastward vergence. Anticline crests are eroded. Overburden thickness ranges widely from <1 km to ~3 km. From these thickness variations in overlying open marine sediments (blank), we infer growth folding. Presalt basement (blank) shows mild velocity pull-ups under thick salt, suggesting that it has the highest seismic velocity. See Figure 2(c) on foldout for depth-migrated seismic section A without vertical exaggeration.

LINE B

1500

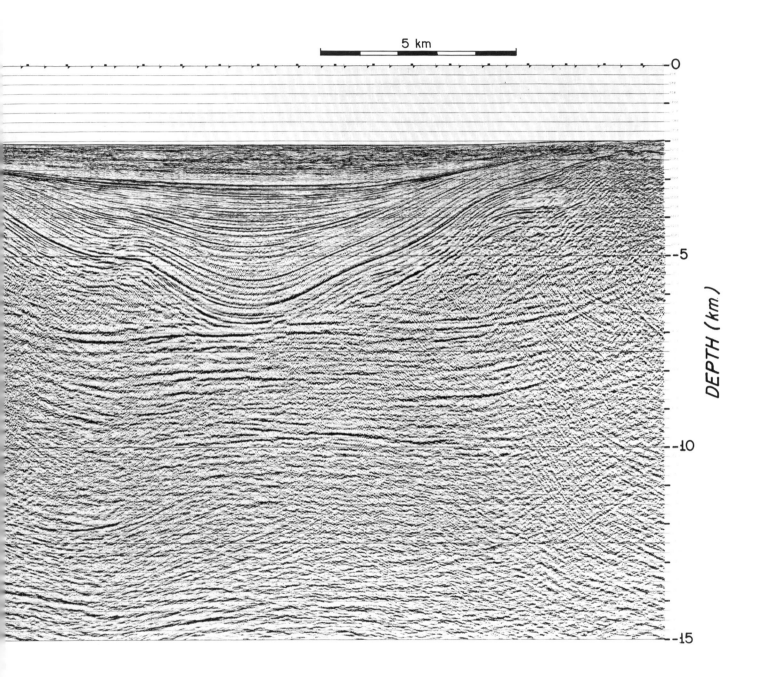

5 km

0

-5

-10

-15

DEPTH (km.)

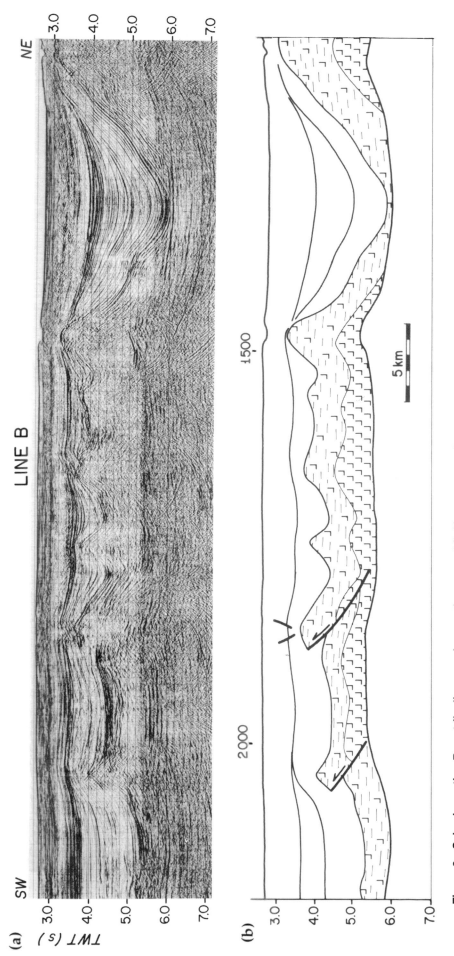

Figure 3—Seismic section B, a strike line, running southwestward (left) across the base of the slope in the center of the basin (see Figure 1 for location). (a) Time-migrated seismic section. (b) Drawing interpreting the time-migrated section and showing major evaporite units (patterned), bedding (thin lines), and reverse faults (heavy lines with arrows). Vertical scale is in seconds (two-way traveltime). Numbers along the top refer to shot points. Folds and faults overlie a detachment layer with little internal layering, interpreted as rock salt (dense L pattern). The overlying sequence with fine bedding and angular or concentric folds is interpreted as layered evaporite (alternating dashes and L pattern). From thickness variations in overlying open marine sediments (blank), we infer growth folding. Notice salt withdrawal beneath large growth syncline (right). Presalt basement (blank) shows mild velocity pull-ups under areas of thick salt, suggesting that it has the highest seismic velocity. However, some of these remain after depth migration, suggesting that they may not be wholly velocity effects. A prominent seafloor multiple occurs at about 5.7 seconds. See Figure 3(c) on foldout for depth-migrated seismic section B without vertical exaggeration.

modeling (Ghosh, 1968; Cobbold et al., 1971) and numerical modeling (DeBremacker and Becker, 1978) that chevron folds, concentric folds, and box folds form in a multilayered sequence with alternating mechanical stiffness as a result of contraction parallel to the layering. The occurrence of such fold styles in the Santos Basin is therefore evidence that the upper evaporite is a sequence of alternating layers (e.g., salt and shale, salt and anhydrite, or salt and carbonate) and that it has contracted horizontally.

The lower evaporite layer (probably halite) is thinner beneath synclines and thicker beneath anticlines. Hence, we infer that it has flowed in ductile fashion, having been squeezed out from under growing synclines into spaces made available in the cores of growing anticlines.

Above the evaporite, Albian marine carbonates (near the basin edge) are overlain by an open-marine megasequence of Upper Cretaceous–Tertiary sediments containing growth folds. From the stratigraphic relationships, we infer that smaller folds grew in the Albian but became buried and inactive during the Late Cretaceous, whereas larger folds, cored by evaporite, appeared in the Late Cretaceous and continued to grow during the Tertiary. Some appear to be still active today (Figure 3, far right).

In experiments on buckling in layered media, the characteristic or dominant wavelength of folds is proportional to layer thickness, especially the total thickness of the buckling system (Ghosh, 1968). However, if the total thickness increases in small increments, the characteristic wavelength does not increase with it, step by step (Cobbold et al., 1989). Instead, small folds formed in the first stages continue to grow, but at a steadily decreasing rate. Later, new folds appear that have a much longer characteristic wavelength. Relationships of this kind, observed for folds in the Santos Basin, reinforce the idea that they formed by horizontal contraction of an aggrading sequence between the Late Cretaceous and Tertiary.

Seismic lines A and B (Figures 2, 3) are orthogonal, yet both show evidence for horizontal contraction. In fact, most seismic lines we have examined in the deep-water Santos Basin, whatever their azimuth, show evidence for horizontal contraction during the Tertiary. The simplest way to explain this is by a mechanism of convergent radial gliding, in which all material particles glide downslope, with the bottom contours being concave seaward. The particle paths are then radially convergent, and particles converge along strike (for more details, see Cobbold and Szatmari, 1991).

In some areas, seismic definition is sufficient to infer thickening or repetition of the layered evaporites across thrust faults and ramp anticlines (Figures 2, 3). The overlying open-marine sequence is condensed over ramp anticlines but forms depocenters. These depocenters are asymmetric (of typical foreland style) next to isolated thrusts (Figure 3, left). They are symmetric where they fill synclines or are caught between thrusts of opposite vergence (Figure 2, right; compare Peel et al., 1995).

Normal faults of small offset are present above many salt-cored anticlines. We attribute these normal faults not to regional extension but to outer arc extension during anticlinal bending. The bending may result from bulk horizontal contraction, from active diapirism due to salt buoyancy (Schultz-Ela et al., 1993), or from both.

By crude balancing of depth-migrated dip lines in the lower contractional domain of the Santos Basin, we estimate the amount of downdip contraction to be about 50% (Figure 2) or more.

NEW EXPERIMENTS ON HORIZONTAL CONTRACTION

We have used physical modeling to investigate the styles and formation mechanisms of salt-cored structures during horizontal contraction of a layered sequence under gravitational loading. For the principles of the modeling technique and its applications to gravitational gliding and salt tectonics, we refer the reader to the abundant recent literature on the subject (Vendeville et al., 1987; Vendeville and Cobbold, 1987, 1988; Cobbold et al., 1989; Cobbold and Szatmari, 1991; Davy and Cobbold, 1991; Vendeville, 1991; Cobbold and Jackson, 1992; Vendeville and Jackson, 1992a,b; Childs et al., 1993; Gaullier et al., 1993; Nalpas and Brun, 1993; Schultz-Ela et al., 1993; Weijermars et al., 1993).

In their experiments on gravitational gliding at the scale of an entire passive margin, Cobbold et al. (1989) obtained contraction at the foot of a surface slope due to prograding sedimentation or to tilting of a model margin. The contractional structures were salt-cored growth folds and reverse faults with associated depocenters. However, the models were too small to allow detailed observation of these structures.

Here we describe two new experiments designed to investigate interacting sedimentation and contraction in layered sequences above salt. The experiments differ in the mechanical properties of the layering and the history of sedimentation.

As with previous experiments, we used dry quartz sand having a grain size of about 200 μm to represent clastic sediments and Silbione silicone putty with a viscosity of about 10^4 Pa s to represent salt. Rather than constructing very large models that could glide under their own weights, we chose to shorten models horizontally with a screw jack operating at a steady computer-controlled velocity. Nevertheless, to ensure that the ratio of gravitational forces to horizontal stresses would be realistic, we chose a screw jack velocity (1 cm/hr) that was small compared to the initial length of the model (50 cm for experiment 1 and 30 cm for experiment 2) and to the viscosity of the silicone (~10^4 Pa s). Thus, the model ratio of length was 10^{-5} (1 cm in the model representing 1 km in nature). The other model ratios were 10^{-10} for time (1 hr representing about 10^5 years), 10^{-5} for stress, 10^{-15} for viscosity, and 10^5 for velocity (1 cm/hr representing about 1 km/10^5 year).

(Text continues on p. 317)

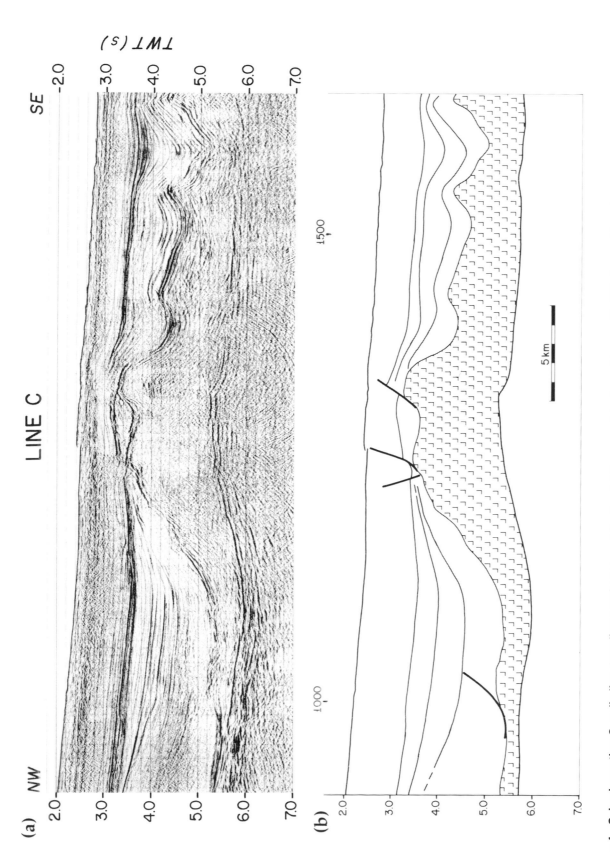

Figure 4—Seismic section C, a dip line, trending southeastward (to the right) down the lower slope in the center of the basin (see Figure 1 for location). (a) Time-migrated seismic section. (b) Drawing interpreting the time-migrated section and showing major evaporite units (patterned), bedding (thin lines), and normal faults (heavy lines). Vertical scale is in seconds (two-way travel-time). Numbers along the top refer to shot points. The evaporite sequence has not been differentiated here. Nevertheless, folds and faults occur above a detachment layer with little internal layering, interpreted as mainly rock salt, whereas the upper part of the evaporite sequence shows layering and angular or concentric folds on the right. From thickness variations in overlying open-marine sediments (blank), we infer growth folding. This section crosses the boundary between the extensional domain (left) and the contractional domain (right). Presalt basement (blank) shows mild velocity pull-ups under areas of thick salt, suggesting that it has the highest seismic velocity.

LINE D

5 km

4500

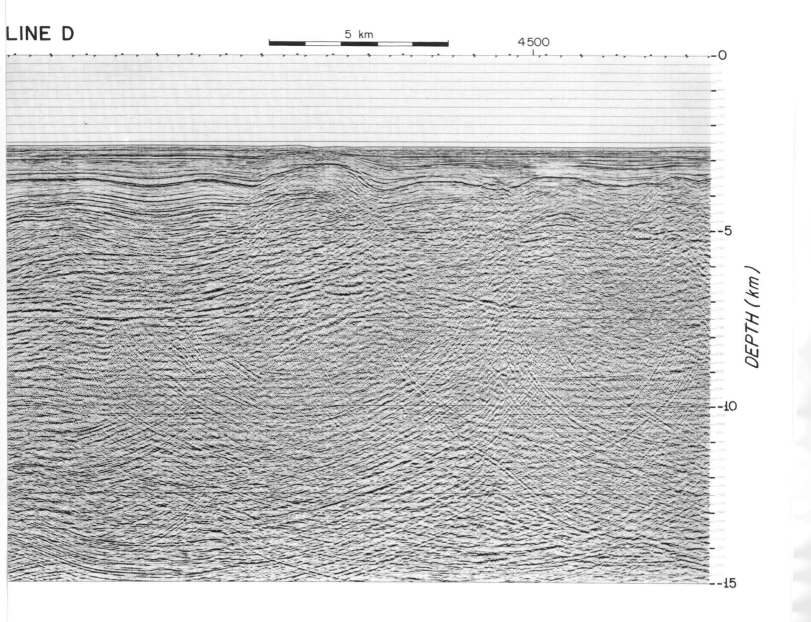

DEPTH (km)

0

-5

-10

-15

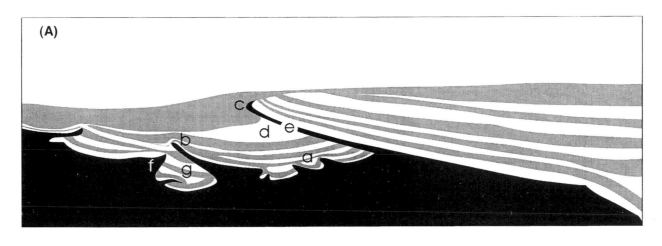

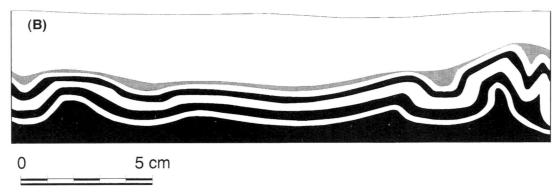

0   5 cm

Figure 6—Contrasting styles of contractional structures in two physical models having different kinds of layering and different histories of sedimentation. **(A)** Synsedimentary folds and thrusts at the end of experiment 1 after nearly 50% bulk shortening. A single thick layer of silicone putty (black) representing salt overlies a rigid, planar base (bottom of frame). Above the silicone are ten sand layers (alternating white and gray), deposited intermittently to represent synkinematic clastic sediments. A thick postkinematic layer (white at top) was deposited at the end of the experiment to preserve the surface topography before wetting and sectioning. Buried at the base of the sand are growth folds of short wavelength (a); these formed by early buckling where the sand sequence was thin. Also buried, but higher in the sequence, are reverse faults and thrusts (b). Only one thrust reached the surface (c), having been active throughout the experiment; its hanging wall was eroded. The footwalls of thrusts (d) thickened. Along the hanging walls of low-angle thrusts, progressive slip smeared out the silicone, forming thin tongues (e). Conversely, against high-angle reverse faults, the silicone formed asymmetric domes (f). Contractional depocenters (g) formed beneath thrusts of opposite vergence. **(B)** Flexural-slip folds at the end of experiment 2 after about 30% bulk shortening. The central part of the model underwent negligible deformation. The lowermost thick layer of silicone putty (black) represents salt overlying a rigid, flat base (bottom of frame). Five alternating layers of sand (white) and silicone putty (black), each of uniform thickness, represent a prekinematic layered sequence. Angular or concentric folds (left and right) formed by flexure of the sand layers and slip along the silicone layers. During deformation, synkinematic sand layers (gray) were deposited at regular intervals, filling topographic lows. Finally, a thick postkinematic layer (white at top) helped preserve the surface topography before wetting and sectioning.

Experiment 1: Synsedimentary Folds and Thrusts

In this experiment (Figure 6A), a thick layer of silicone putty represented salt above a horizontal, rigid basement. Successive sand layers representing clastic sediments were deposited above the silicone. The average rate of aggradation in the model was about 1 mm/hr (representing ~1 km/m.y. in nature), but in detail this varied in time and space. In time, layers of different colors were deposited episodically. In space, the thickness of each layer varied in two ways. First, from one end of the model to the other (right to left, Figure 6A), the thickness

decreased approximately linearly to simulate a deltaic wedge. Second, in areas of local relief due to growing structures, some sand was eroded from topographic highs, but more was deposited in topographic lows.

During sedimentation, the model was horizontally shortened and vertically thickened by a screw jack advancing at a computer-controlled velocity of 1 cm/hr. The experiment was stopped after 24 hr, when the final length was nearly half the initial length. A final layer of sand was added after deformation to preserve the surface topography. Then the accumulated sand was carefully wetted via the free surface, and the model was sectioned.

While the sand sequence was still thin, folds of short

wavelength formed by buckling (see also Cobbold et al., 1989). At each point along the buckling layer, the ratio of wavelength to thickness of the folds was about 5:1. In other words, folds were narrower where the sediment thickness was smaller (Figure 6A, left). As more sediment accumulated, the initial folds gradually stopped growing vertically and finally became buried and inactive.

Further shortening of the model was taken up by reverse faulting or thrusting. Most faults formed near the hinges of older buckles, and their vergence was inherited from limb dips. With further sedimentation, some of these faults became buried and inactive. The spacing of remaining active faults increased episodically in proportion to stratigraphic thickness, in a manner similar to that reported for normal faults in sandpacks extended horizontally (Vendeville et al., 1987). In the final stages of experiment 1, only one fault remained active. Its vergence may have been controlled by the average surface slope of the entire sediment wedge.

Where faults were widely spaced, asymmetric depocenters formed in their footwalls. The depocenters filled with sediment wedges, nearly triangular in cross section and similar to those of foreland basins at crustal scale. Other small basins developed more symmetrically, between thrusts of opposite vergence, becoming deep and narrow in the process. These depocenters were similar to the symmetric ramp basins described at crustal scale by Cobbold et al. (1993).

During progressive slip on faults, silicone domes or sheets became emplaced in their hanging walls. Along low-angle thrusts, the silicone formed thin tongues. These resulted not so much from injection as from smearing out of the silicone during fault slip. Against high-angle reverse faults, the silicone formed asymmetric triangular domes. Their nonoverturned flanks remained concordant with overlying sediments, whereas their overturned flanks glided into contact with footwall cutoffs in the sand. Hanging wall cutoffs were seldom resolvable because of the condensed nature of the hanging wall sequence or because of gravitational collapse or erosion at the thrust front.

In general, the structural styles in this model were similar to those described by Cobbold et al. (1989) for experiments on true gravitational gliding. However, in the new model, layers of sand were more numerous and hence thinner in relation to total sand thickness. As a result, small-scale folds and thrusts developed in the first stages of the experiment before becoming buried. These small structures did not form or were not resolved in the earlier experiments of Cobbold et al. (1989).

Experiment 2: Flexural-Slip Folds

In this experiment (Figure 6B), once again a relatively thick layer of silicone putty, representing salt, overlay a rigid basement. Next, alternating layers of silicone putty and sand, each of uniform thickness, represented a prekinematic multilayered evaporite sequence. The multilayer underwent horizontal shortening, provided by a screw-jack, advancing at a computer-controlled velocity of 1 cm/hr. At intervals during deformation, successive synkinematic layers of sand were deposited at the free surface, preferentially filling topographic lows. The experiment was stopped after 30% total shortening. A final thick layer of postkinematic sand was deposited, and the model was carefully wetted to strengthen the sand layers. Because of the impermeable silicone layers, the model was wetted through its sides. The multilayered model was then sectioned with a sharp knife.

During shortening, folds appeared in the multilayered sequence. The folds were angular or concentric due to flexure of the sand layers and slip along the silicone layers. The ratio of initial wavelength to total thickness of the multilayer was about 2.5:1. During shortening, the thick basal silicone layer became slightly squeezed out from under synclines but flowed into the spaces created by growing anticlines, forming concordant domes. Some folds reached high amplitudes and steep limb dips, but no faults appeared. Most of the contraction was accommodated by ductile flow in the silicone layers and by rotation and bending of the individual sand layers.

DISCUSSION

The two physical models show very different styles of contractional structures. In the multilayered model (experiment 2, Figure 6B), the only structures were angular or concentric folds with a wavelength to thickness ratio of about 2.5:1. Conversely, in the model with more uniform brittle overburden (experiment 1, Figure 6A), the oldest contractional structures were buckles formed with a wavelength to thickness ratio of about 5:1. However, these soon became inactive and were replaced by thrusts higher in the sequence. Clearly, a well-layered sequence favors angular folds, whereas a more uniform brittle sequence favors small folds when the sequence is thin and favors thrusts after the sequence thickens.

Similar in shape to the folds of experiment 2 are the folds with angular and concentric styles developed in the multilayered evaporite sequence of the deep-water Santos Basin (Figures 2–5). Thrusts are less common, but their ramps and flats are similar to those described for many foothill basins next to mountain belts. Possibly, such thrusts would be generated in a multilayered model if it were shortened by more than 30%. Angular folds, cored by salt, have also been reported for the Perdido foldbelt of the northwestern Gulf of Mexico (e.g., Weimer and Buffler, 1992).

Similar to the folds and thrusts of experiment 1 are the structures described by Cobbold and Szatmari (1991) and Demercian et al. (1993) for the Campos Basin. Here, the evaporite sequence appears to be less well layered, and few internal structures have been identified. In the overlying marine sediments, the basal folds have shapes and initial ratios of wavelength to thickness similar to those of experiment 1. There are also asymmetric salt domes with sharp, straight overhanging edges, which Demercian et al. (1993) interpreted as emplaced above reverse faults. Similar also to the structures of experiment 1 are some of those described by Weimer and Buffler et al. (1992) for the Mississippi Fan foldbelt and interpreted by them in terms of horizontal contraction.

In general, for a typical sedimentary sequence on a passive margin subject to gravitational gliding, we expect horizontal contraction to result in hybrid structural styles intermediate between those of experiments 1 and 2. In other words, combinations of growth folds and growth thrusts would form, cored by ductile salt or shale. An alternative interpretation for salt domes with overhanging edges is that they are due to passive piercement (downbuilding). In other words, simultaneous sediment aggradation and salt flow occur at comparable rates, but without horizontal contraction. The top of the salt remains at the free surface at all times (Jackson and Talbot, 1991).

However, the asymmetric domes in the Campos Basin show the following features, which we believe are suggestive, if not diagnostic, of horizontal contraction:

1. The overhanging edge of the salt is straight and sharp. Footwall cutoffs are also sharp.
2. Salt in the hanging wall is overlain by a concordant sedimentary sequence, which is a condensed equivalent of the thicker sequence found *beneath* the footwall.

The absence of hanging wall cutoffs may not be detrimental to the contractional hypothesis. Cutoffs may be unresolvable in condensed sequences, or they may be blurred by drape folding, gravitational collapse, or erosion at an emerging thrust front, as in experiment 1 (Figure 6A).

The following features may also be diagnostic of horizontal contraction:

3. Salt forms thin, gently dipping tongues where it has been smeared out along thrust faults.
4. Depocenters, triangular in section, are bounded on both sides by asymmetric salt domes or salt tongues verging basinward.
5. Angular and concentric folds occur within a layered sequence, in which mechanical contrasts can be inferred from well data or from the nature of seismic reflectors.
6. Fold trains having small characteristic wavelengths occur near the base of a sedimentary sequence and show growth over an early stratigraphic interval when the sequence was thin. In contrast, trains with longer wavelengths show growth over later intervals when the sequence was correspondingly thicker.

If all six of these features are found in one area, we believe that the cumulative evidence for horizontal contraction becomes strong. This is the situation in many deep-water areas of the Santos or Campos basins.

Salt-cored thrusts may be especially difficult to image seismically because ray paths tend to be complex beneath overhanging edges or thin tongues of high-velocity salt. Hence, we suspect that many such structures have gone unnoticed so far or have not been ascribed to horizontal contraction.

Acknowledgments   This paper is one outcome of a Petrobras project on salt tectonics (1987–1992) originally organized by Antônio M. F. de Figueiredo. It is published with the company's kind permission. We are grateful for the support of Petrobras Exploration Department (DEPEX) and especially of José J. de Moraes, Jr., head of the Santos Basin Sector. Experiment 1 was done at Géosciences-Rennes by PRC and EAR; we thank CNRS technician Jean-Jacques Kermarrec for his help with apparatus. Experiment 2 was done at the Petrobras Research Center (CENPES); we are grateful to Marta C. M. Guerra and Mônica A. Pequeno for their help in the laboratory. Figure 1 was drawn by Hongxing Ge, Bureau of Economic Geology, The University of Texas at Austin. Comments by two anonymous reviewers and Martin Jackson helped improve the paper.

REFERENCES CITED

Buffler, R. T., F. J. Shaub, J. S. Watkins, and J. L. Worzel, 1979, Anatomy of the Mexican Ridges, southwestern Gulf of Mexico: AAPG Memoir, v. 29, p. 319–327.

Carminatti, M., and J. C. Scarton, 1991, Seismic stratigraphic aspects of the Oligocene sequences in Campos Basin, offshore Brazil, *in* M. Link and P. Weimer, eds., Seismic facies and sedimentary processes of submarine fans and turbidite systems: Berlin, Springer-Verlag, p. 241–246.

Chang, H. K., R. O. Kowsmann, and A. M. F. Figueiredo, 1988, New concepts on the development of East Brazilian marginal basins: Episodes, v. 11, p. 194–202.

Chang, H. K., R. O. Kowsmann, A. M. F. Figueiredo, and A. A. Bender, 1992, Tectonics and stratigraphy of the East Brazil rift system: an overview: Tectonophysics, v. 213, p. 97–138.

Childs, C., S. J. Easton, B. C. Vendeville, M. P. A. Jackson, S. T. Lin, J. J. Walsh, and J. Watterson, 1993, Kinematic analysis of faults in a physical model of growth faulting above a viscous salt analogue: Tectonophysics, v. 228, p. 313–329.

Cobbold, P. R., J. W. Cosgrove, and J. M. Summers, 1971, Development of internal structures in deformed anisotropic rocks: Tectonophysics, v. 12, p. 23–53.

Cobbold, P. R., D. Gapais, C. Lecorre, E. A. Rossello, J. C. Thomas, J. J. Tondji Biyo, and M. de Urreiztieta, 1993, Sedimentary basins and crustal thickening: Sedimentary Geology, v. 86, p. 77–89.

Cobbold, P. R., and M. P. A. Jackson, 1992, Gum rosin (colophony): a suitable material for thermomechanical modeling of the lithosphere: Tectonophysics, v. 210, p. 255–271.

Cobbold, P. R., E. R. Rossello, and B. Vendeville, 1989, Some experiments on interacting sedimentation and deformation above salt horizons: Bulletin de la Société Géologique de France, v. 8, p. 453–460.

Cobbold, P. R., and P. Szatmari, 1991, Radial gravitational gliding on passive margins: Tectonophysics, v. 188, p. 249–289.

Davis, D. M., and T. Engelder, 1987, Thin-skinned deformation over salt, *in* I. Lerche and J. J. O'Brien, eds., Dynamical geology of salt and related structures: Orlando, Florida, Academic Press, p. 301–337.

Davy, P., and P. R. Cobbold, 1991, Experiments on shortening of a 4-layer model of the continental lithosphere: Tectonophysics, v. 188, p. 1–25.

DeBremacker, J. A., and E. B. Becker, 1978, Finite element models of folding: Tectonophysics, v. 50, p. 349–367.

Demercian, S., P. Szatmari, and P. R. Cobbold, 1993, Style and pattern of salt diapirs due to thin-skinned gravitational gliding, Campos and Santos basins, offshore Brazil: Tectonophysics, v. 228, p. 393–433.

de Ruig, M. J. , 1995, Extensional diapirism in the Eastern Prebetic fold belt, southeastern Spain: *in* M. P. A. Jackson, D. G. Roberts, and S. Snelson, eds., Salt tectonics: a global perspective: AAPG Memoir 65, this volume.

Dias, J. L., J. C. Scarton, F. R. Esteves, M. Carminatti, and L. R. Guardado, 1990, Aspectos da evolução tectono-sedimentar e a ocorrência de hidrocarbonetos na Bacia de Campos, *in* G. P. De Raja Gabaglia and E. J. Milani, eds., Origem e evolução de bacias sedimentares: Rio de Janeiro, Petrobras, p. 333–360.

Figueiredo, A. M. F. de, and C. C. Martins, 1990, 20 años de exploração da Bacia de Campos e o sucesso nas águas profundas: Boletim de Geociências da Petrobras, v. 4, p. 105–123.

Gaullier, V., J. P. Brun, G. Guérin, and H. Lecanu, 1993, Raft tectonics: the effects of residual topography below a salt décollement: Tectonophysics, v. 228, p. 363–381.

Ghosh, S. K., 1968, Experiments of buckling of multilayers which permit interlayer gliding: Tectonophysics, v. 6, p. 207–250.

Guardado, L. R., L. A. P. Gamboa, and C. F. Lucchesi, 1989, Petroleum geology of the Campos Basin, a model for a producing Atlantic-type basin, *in* J. D. Edwards and P. A. Santogrossi, eds., Divergent/passive margin basins: AAPG Memoir 48, p. 3–79.

Harrison, J. C., 1995, Tectonics and kinematics of a foreland folded belt influenced by salt, Arctic Canada, *in* M. P. A. Jackson, D. G. Roberts, and S. Snelson, eds., Salt tectonics: a global perspective: AAPG Memoir 65, this volume.

Heaton, R. C., M. P. A. Jackson, M. Bamahmoud, and A. S. O. Nani, 1995, Superposed Neogene extension, contraction, and salt canopy emplacement in the Yemeni Red Sea, *in* M. P. A. Jackson, D. G. Roberts, and S. Snelson, eds., Salt tectonics: a global perspective: AAPG Memoir 65, this volume.

Jackson, M. P. A., and C. J. Talbot, 1991, A glossary of salt tectonics: The University of Texas at Austin, Bureau of Economic Geology Geological Circular 91-4, 44 p.

Lehner, P., and P. A. C. de Ruiter, 1977, Structural history of Atlantic margin of Africa: AAPG Bulletin, v. 61, p. 963–981.

Letouzey, J., B. Colletta, R. Vially, and J. C. Chermette, 1995, Evolution of salt-related structures in compressional settings, *in* M. P. A. Jackson, D. G. Roberts, and S. Snelson, eds., Salt tectonics: a global perspective: AAPG Memoir 65, this volume.

Mohriak, W. U., R. Hobbs, and J. F. Dewey, 1990a, Basin-forming processes and the deep structure of the Campos Basin, offshore Brazil: Marine and Petroleum Geology, v. 7, p. 94–112.

Mohriak, W. U., M. R. Mello, J. F. Dewey, and J. R. Maxwell, 1990b, Petroleum geology of the Campos Basin, offshore Brazil, *in* J. Brooks, ed., Classic petroleum provinces: Geological Society of London, Special Publication, v. 50, p. 119–141.

Mohriak, W. U., M. R. Mello, G. D. Karner, J. F. Dewey, and J. R. Maxwell, 1990c, Structural and stratigraphic evolution of the Campos Basin, offshore Brazil, *in* A. J. Tankard and H. R. Balkwill, eds., Extensional tectonics of North Atlantic margins: AAPG Memoir, v. 46, p. 577–598.

Mohriak, W. U., J. M. Macedo, R. T. Castellani, H. D. Rangel, A. Z. N. Barros, M. A. L. Latgé, A. M. P. Mizusaki, P. Szatmari, L. S. Demercian, J. G. Rizzo, and J. R. Aires, 1995, Salt tectonics and structural styles in the deep water province of the Cabo Frio region, Rio De Janeiro, Brazil, *in* M. P. A. Jackson, D. G. Roberts, and S. Snelson, eds., Salt tectonics: a global perspective: AAPG Memoir 65, this volume.

Nalpas, T., and J. P. Brun, 1993, Salt flow and diapirism related to extension at crustal scale: Tectonophysics, v. 228, p. 349–362.

Ojeda, H. A., 1982, Structural framework, stratigraphy, and evolution of Brazilian marginal basins: AAPG Bulletin, v. 66, p. 732–749.

Peel, F. J., C. J. Travis, and J. R. Hossack, 1995, Genetic structural provinces and salt tectonics of the Cenozoic offshore U.S. Gulf of Mexico: a preliminary analysis, *in* M. P. A. Jackson, D. G. Roberts, and S. Snelson, eds., Salt tectonics: a global perspective: AAPG Memoir 65, this volume.

Pereira, M. J., and J. M. Macedo, 1990, A Bacia de Santos: perspectivas de uma nova província petrolífera na plataforma continental sudeste brasileira: Boletim de Geociências da Petrobras, v. 4, p. 3–11.

Petrobras, 1983, Campos and Espirito Santo basins, offshore Brazil, *in* A. W. Bally, ed., Seismic expression of structural styles: AAPG Studies in Geology 15, p. 2.3.51–2.3.58.

Sans, M., and J. Vergés, 1995, Fold development related to contractional salt tectonics, southeastern Pyrenean thrust front, Spain, *in* M. P. A. Jackson, D. G. Roberts, and S. Snelson, eds., Salt tectonics: a global perspective: AAPG Memoir 65, this volume.

Schaller, H., and M. V. Dauzacker, 1986, Tectônica gravitacional e sua aplicação na exploração de hidrocarbonetos: Boletim Técnico da Petrobras, v. 29, p. 193–206.

Schultz-Ela, D. D., M. P. A. Jackson, and B. C. Vendeville, 1993, Mechanics of active salt diapirism: Tectonophysics, v. 228, p. 275–312.

Vendeville, B. C., 1991, Mechanics generating normal fault curvature: a review illustrated by physical models, *in* A. M. Roberts, G. Yielding and B. Freeman, eds., The geometry of normal faults: Geological Society of London, Special Publication, v. 56, p. 241–249.

Vendeville, B. and P. R. Cobbold, 1987, Glissements gravitaires synsédimentaires et failles normales listriques: modèles expérimentaux: Comptes Rendus de l'Académie des Sciences de Paris, Séries II, v. 35, p. 1313–1320.

Vendeville, B., and P. R. Cobbold, 1988, How normal faulting and sedimentation interact to produce listric fault profiles and stratigraphic wedges: Journal of Structural Geology, v. 10, p. 649–659.

Vendeville, B., P. R. Cobbold, P. Davy, J. P. Brun, and P. Choukroune, 1987, Physical models of extensional tectonics at various scales, *in* M. P. Coward, J. F. Dewey and P. L. Hancock, eds., Continental extensional tectonics:

Geological Society of London, Special Publication No. 28, p. 95–107.

Vendeville, B. C., and M. P. A. Jackson, 1992a, The rise of diapirs during thin-skinned extension: Marine and Petroleum Geology, v. 9, p. 331–353.

Vendeville, B. C., and M. P. A. Jackson, 1992b, The fall of diapirs during thin-skinned extension: Marine and Petroleum Geology, v. 9, p. 354–371.

Weijermars, R., M. P. A. Jackson, and B. Vendeville, 1993, Rheological and tectonic modeling of salt provinces: Tectonophysics, v. 217, p. 143–174.

Weimer, P., and R. T. Buffler, 1992, Structural geology and evolution of the Mississippi Fan foldbelt, deep Gulf of Mexico: AAPG Bulletin, v. 76, p. 225–251.

Wilcox, J. B., et al., 1988, Rig seismic research cruises 10 and 11: geology of the central Australian Bight Region: Bureau of Mineral Resources, Geology and Geophysics, Report 286, 140 p.

Worrall, D. M., and S. Snelson, 1989, Evolution of the northern Gulf of Mexico, with emphasis on Cenozoic growth faulting and the role of salt, *in* A. W. Bally and A. R. Palmer, eds., The Geology of North America; an overview: GSA Decade of North American Geology, v. A, p. 97–138.

Liro, L. M., and R. Coen, 1995, Salt deformation history and postsalt structural trends, offshore southern Gabon, West Africa, *in* M. P. A. Jackson, D. G. Roberts, and S. Snelson, eds., Salt tectonics: a global perspective: AAPG Memoir 65, p. 323–331.

Chapter 15

Salt Deformation History and Postsalt Structural Trends, Offshore Southern Gabon, West Africa

L. M. Liro

Texaco Exploration and Production Technology Department
Houston, Texas, U.S.A.

Present address:

Texaco Central Exploration Department
Bellaire, Texas, U.S.A.

R. Coen

British Gas
Houston, Texas, U.S.A.

Present address:

AGIP Petroleum
Houston, Texas, U.S.A.

Abstract

Salt deformation in offshore southern Gabon is represented by mobilization of an Aptian salt layer in reaction to Tertiary clastic progradation. Seismic mapping of salt bodies and associated faulting has resulted in increased understanding of the types and distribution of these salt bodies, their associated faulting patterns, and some aspects of their origin.

Away from the Tertiary depocenter, the growth history of salt swells or pillows can be determined by examining onlapping and draping seismic reflectors. Significant Tertiary clastic progradation into the area mobilized the salt and resulted in a series of linear, deep salt walls and asymmetric, basinward-dipping salt rollers, commonly associated with significant up-to-basin listric faulting. The up-to-basin faulting dominates the southern Gabon subbasin. The expansion history of associated sediments suggests that these faults expanded episodically throughout the Tertiary, continuing to present-day bathymetric fault scarps. The bias toward up-to-basin faults, to the apparent exclusion of down-to-basin expansion faults, remains enigmatic.

INTRODUCTION AND PREVIOUS WORK

This chapter describes styles of salt deformation evident from seismic data from offshore southern Gabon, West Africa. After an overview of the regional geologic framework of southern Gabon, we document the types of salt structures and illustrate them by seismic examples and a map. Finally, we offer some interpretations of the origin and genesis of the various salt structures.

Most studies of salt tectonics of the Aptian salt basin of West Africa result from hydrocarbon exploration (e.g., Edwards and Bignell, 1988; Teisserenc and Villemin, 1989; Edwards et al., 1990; Traynor et al., 1992). Although salt deformation in the northern Gabon subbasin has been well studied (Teisserenc and Villemin, 1989) due to successful hydrocarbon exploration, the southern Gabon subbasin has received little attention. Thompson et al. (1989) drew comparisons between salt deformation in offshore Gabon and elsewhere. Our chapter will document the salt deformation styles observed on seismic data off the southern Gabon coast and attempt to interpret their genesis and effects on postsalt structural styles here.

REGIONAL FRAMEWORK

The Aptian salt basin of the West African margin from Gabon south through Angola (Figure 1) (Uchupi, 1992) originated in a Jurassic–Early Cretaceous rifting episode (Teisserenc and Villemin, 1989). Rifting was followed by the initial opening of the South Atlantic Ocean, separating the South American and African cratons. The basins along the West African margin thus show a rift–drift history consistent with passive margin development.

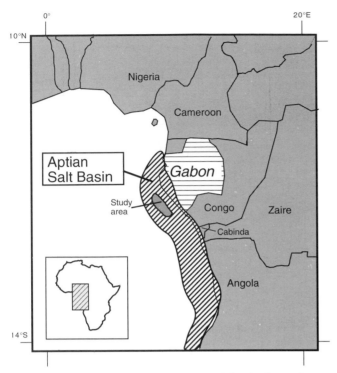

Figure 1—Extent of the study area and the Aptian salt basin on the West African margin. Adapted from Uchupi (1992).

Age, Ma		Formations		Tectono–stratigraphy	
66.4	Early Tertiary	Post – Madiela Units	Siliciclastic sequence	Postrift	
88.5	Senonian				
91	Turonian				
97.5	Cenoman.				
113	Albian	Madiéla	Carbonate sequence	Postrift	
119	Aptian	Ezanga	Evaporite sequence	Transitional	
		Gamba			
	Neocomian – Barremian	Dentale	Fluvial Lacustrine sequence	Synrift	
		Crabe			
		Melania			
		Lucina			
144		Kissenda			
	Pre– Jurassic	Basement			

Figure 2—Tectonostratigraphy of the southern Gabon subbasin. Adapted from Teisserenc and Villemin (1989).

Active rifting of the West African margin began in the latest Jurassic–Early Cretaceous. Synrift deposits accumulated from about the Neocomian through the Barremian (Figure 2). Initial deposits were Neocomian fluvial and fan delta coarse-grained siliciclastics. A major rifting event in the late Neocomian–Barremian created tilted fault blocks along the West African margin. Ensuing thermal subsidence, coupled with isolation from the marine waters of the opening South Atlantic, created a lacustrine environment.

Continuing rifting and subsidence were coupled with an arid climate and sporadic breaching of the Walvis Ridge off Namibia, which allowed marine incursions and evaporation. The primary result was deposition of a succession of evaporitic and dolomitic strata (the Ezanga Formation in Gabon) overlying the lacustrine sequence and constituting the Aptian salt basin of West Africa. The extent of the salt closely matches the maximum extent of the presalt lacustrine strata. The original thickness of the evaporite sequence is difficult to estimate because of subsequent halokinesis and possible dissolution, although previous workers have suggested that it did not exceed 800 m (Teisserenc and Villemin, 1989). We are not aware of any documented fault offsets in the base of the salt.

With the Albian onset of seafloor spreading in the South Atlantic, rifting ended and drift/subsidence began. Initial deposition during the drift phase was a broad, open marine carbonate platform (the Madiela Formation in Gabon), extending virtually uninterrupted from

Gabon to Angola under various names. In north Gabon, progradation is recorded in sediments of Santonian–early Eocene age. This episode is economically significant because much of the offshore Gabon oil and gas is in reservoirs of this age (Teisserenc and Villemin, 1989). A second major progradation began in the Miocene and continues to the present day (Teisserenc and Villemin, 1989).

There are three sedimentary basins in Gabon (Figure 3). The interior basin of onshore Gabon developed from the Early Cretaceous to Albian and did not receive sediments after that time. The coastal and offshore portions of Gabon are divided into the north subbasin and the south subbasin (Teisserenc and Villemin, 1989). Both have received sediments from Early Cretaceous to modern times, but they differ significantly in their salt tectonics (Teisserenc and Villemin, 1989). The southern subbasin (herein, the southern Gabon subbasin) is separated from the northern subbasin by the N'Komi fracture zone, a strike-slip transfer zone related to rifting. The N'Komi fracture zone clearly offsets the Madiela carbonate platform (Figure 3). This fracture zone also separates offshore Gabon into two basic families of salt deformation: dominantly simple diapirs to the north and more complex polygenetic salt structures to the south.

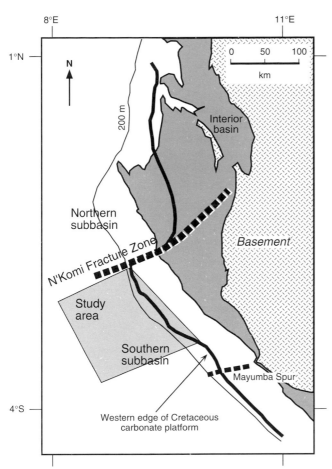

Figure 3—Basin distribution in Gabon. Edge of Cretaceous carbonate platform adapted from Teisserenc and Villemin, 1989.

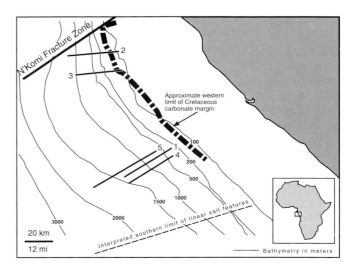

Figure 4—Structural limits and bathymetry in the study area. Seismic lines are approximately located.

SALT DEFORMATION

Study Area

Our study area (Figure 4) is within a seaward bulge of the present-day bathymetry between the N'Komi fracture zone and the interpreted southern limit of linear salt features. Because this bathymetric bulge coincides with a distinct region of salt deformation described here, we interpret that the seaward bulge is related to the distribution of salt features and their influence on sedimentary patterns there. A grid of two-dimensional seismic lines, on approximately 8-km spacing and oriented roughly perpendicular to the present shoreline, was used as the basis for this study.

Seismic Evidence of Salt Deformation

Seismic data (Figure 5) show ample evidence for salt mobilization and deformation. First, the suspected salt features must be differentiated from diapiric shale, which can have similar seismic appearance. In the overburden, reflection-free seismic zones cut or disrupt otherwise

coherent and continuous stratal reflections. Where there is reasonable seismic imaging above and below these disrupted zones, seismic stacking velocities consistently approach the seismic velocity of rock salt, 4400 m/sec. There is no seismic evidence for mobile shale masses here.

Where salt is deformed, overlying sequences thicken and thin. Faulting is associated with deformed salt, particularly where there has been significant rise of salt. This listric faulting is typically up-to-basin (Figure 5), an unusual and enigmatic structural relationship characteristic of Gabon.

Salt-Related Structures

Generally, two basic families of salt structures are observed: simple salt pillows in deep water and a complex suite of deeply buried stocks, walls, salt welds, and salt rollers associated with up-to-basin faults. The following salt-deformational styles are observed:

1. Salt pillows (Figure 6) are present throughout the northern parts of the southern Gabon subbasin and in deeper water. These salt pillows penetrate little of the stratigraphic section above them; more commonly, units immediately above them are thinned. From the age of the strata affected (late Mesozoic), this style of salt deformation is interpreted as relatively early and mild. This deformation, where observed on crossing and orthogonal seismic lines, induces simple four-way anticlinal closure in immediately overlying strata. These are growth structures: their closure and relief generally diminish upward to the point where Tertiary and younger strata are unaffected. Drape and onlap of strata deposited at and subsequent to salt mobilization identify the time of salt movement. Although these salt mobilization events appear to occur at different times depending on location in the basin, none appear to be younger than early Tertiary.

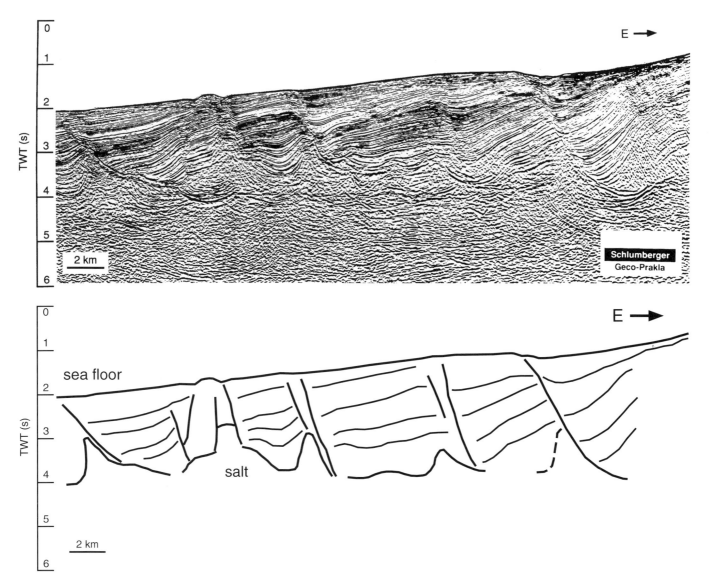

Figure 5—Uninterpreted (top) and interpreted (bottom) seismic profile 1, showing evidence of salt deformation. See Figure 4 for location.

2. Salt stocks (Figure 6) are present in deep water, typically seaward of the salt swells. The stocks clearly penetrate section on seismic profiles. Their apparent symmetry distinguishes them from the asymmetric salt structures discussed below.

3. Symmetric salt walls (Figure 7) form thick masses with extensive bases. The walls typically culminate upward into several individual spires or diapir-like forms, commonly with extensive and persistent normal faults at or along one side of their crests. Although the crests are typically symmetric in dip cross sections, their dominant faults most commonly dip landward.

4. Perhaps the most striking salt structures here are linear trends of **asymmetric salt rollers** (Figure 8). These may have risen through the sedimentary section because of withdrawal of salt from depth or from preexisting salt structures. These salt rollers (previously described by Teisserenc and Villemin, 1989) are basinward of the

Cretaceous carbonate platform. The salt rollers are invariably on the footwall of significant listric faults, and most of these faults are up-to-basin. The rollers are thus generally asymmetric and basinward tilting. Typically, several salt rollers are regularly spaced across a dip line. Salt flow and associated faulting typically persist through the stratigraphic section, in places to the present water bottom (Figure 5). The asymmetric salt rollers are overlain by half-grabens, all having the same basic form and sense of fault block rotation.

5. Salt welds (Figure 9) occur in the study area. Salt tectonism often results in the evacuation of salt from its original stratigraphic position. Salt welds, as described by Jackson and Cramez (1989), result in the juxtaposition or tectonic discordance of strata. In southern Gabon, welding juxtaposes postsalt drift-phase strata against presalt rift-phase strata. In Figure 9, a salt weld is shown immediately basinward of the Cretaceous platform edge.

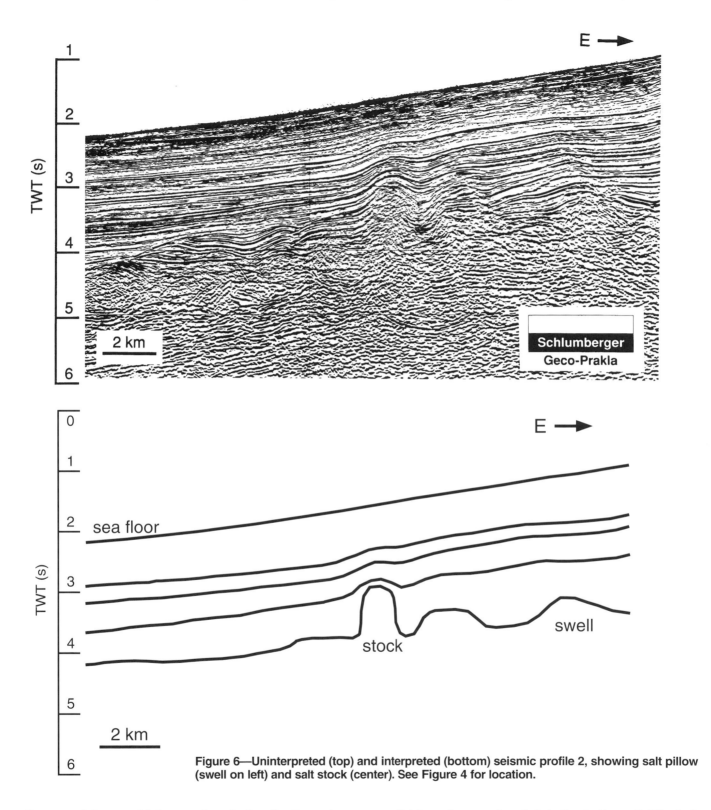

Figure 6—Uninterpreted (top) and interpreted (bottom) seismic profile 2, showing salt pillow (swell on left) and salt stock (center). See Figure 4 for location.

Because of the very high acoustic velocity of salt, precise seismic documentation of salt welds is difficult. Indirect evidence for thin to evacuated salt is the apparent volume of salt in diapiric structures and the geometry of associated listric faults.

6. Up-to-basin faulting is the dominant style of faulting in the southern Gabon subbasin (Figure 5). The faults are listric and appear to sole into the Aptian salt. The fault trends are interpreted as approximately linear. Sediments thicken toward the fault hanging walls (likely related to salt withdrawal). Fault offset persists above the salt contact. Major up-to-basin faults are all observed to be associated with salt rollers, and thickened salt is always present on their footwalls.

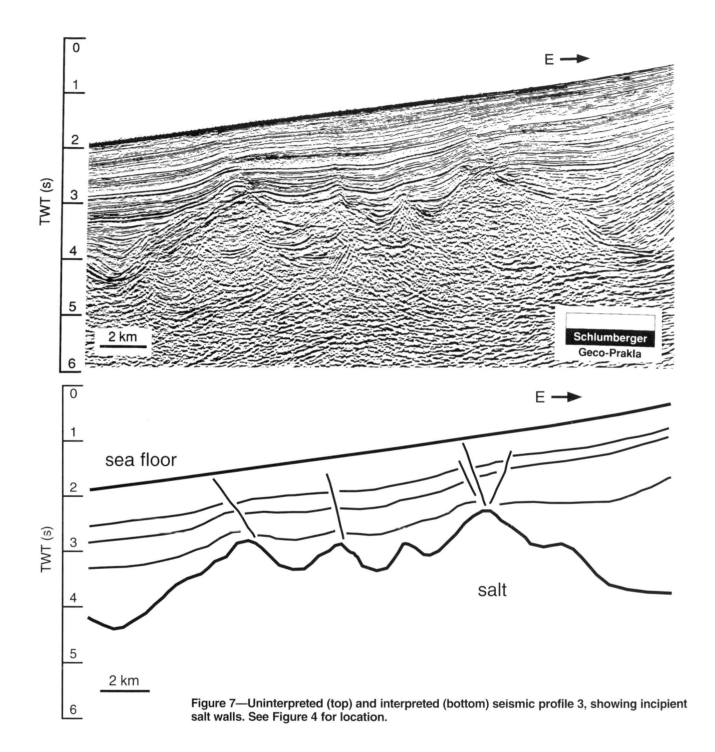

Figure 7—Uninterpreted (top) and interpreted (bottom) seismic profile 3, showing incipient salt walls. See Figure 4 for location.

Characterization of Salt Tectonic Processes

The inherent instability of prograding clastic strata, deposited as a basinward-thinning wedge over a mobile salt layer, can result in salt mobilization and a variety of salt structures (Jackson et al., 1988). In the southern Gabon subbasin, we think that the salt pillows in deep water, where the influence of Tertiary clastic loading is minimized, represent the initial style of salt deformation. With onset of significant clastic input in the Late Cretaceous and particularly in the later Tertiary, the salt bodies

deformed further with significant salt rise and creation of fault zones. On some seismic lines, less well developed salt wedges are seen as asymmetric, basinward-leaning, deep salt swells. Map correlation of the associated up-to-basin faults (Figure 10) suggests that the original configuration of the salt may have been linear, deep-seated salt walls. Growth history on the faults indicates that significant stratigraphic expansion (and hence, slip) is relatively young (Tertiary) (Figure 8) and occurred more than once.

In the southernmost area, present-day bathymetric expression is associated with listric faulting. This indi-

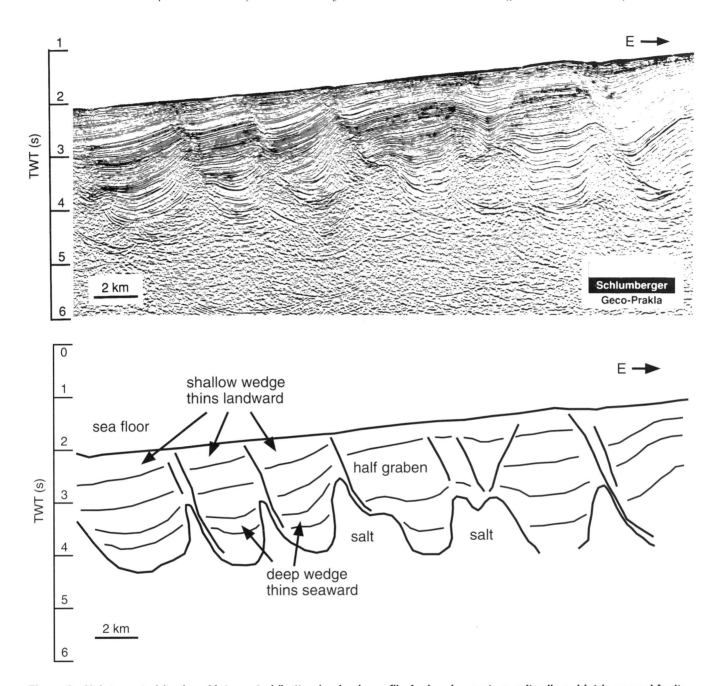

Figure 8—Uninterpreted (top) and interpreted (bottom) seismic profile 4, showing mature salt rollers. Listric normal faults have up-to-basin offset and mobilized salt in their footwalls. Marked intervals demonstrate relative time thickness changes associated with the faulting. See Figure 4 for location.

cates that the processes leading to salt accumulation in the salt rollers are continuing to the present day.

In the southern Gabon subbasin, fault zones in the overburden are listric and dominantly up-to-basin. With basinward tilting and thermal subsidence, down-to-basin faulting would normally be expected, as documented by Duval et al. (1992) for the Kwanza Basin of Angola. Why up-to-basin faulting dominates is problematic. Perhaps this style of faulting is related to tectonics unique to the southern Gabon tectonic corridor defined by bounding regional strike-slip elements; we have not observed it as

the dominant fault style elsewhere along the Aptian salt basin. Duval et al. (1992) and Lundin (1992) documented up-to-basin faulting in the Kwanza Basin of Angola, but these faults are genetically related to complementary down-to-basin faulting in the raft tectonics that characterize this region.

Polyphase salt deformation can result not only in the creation of salt features but also in their destruction. We note that there are salt features in the study area that may be the result of deflation, rather than inflation, of salt structures (Figure 8, right-center).

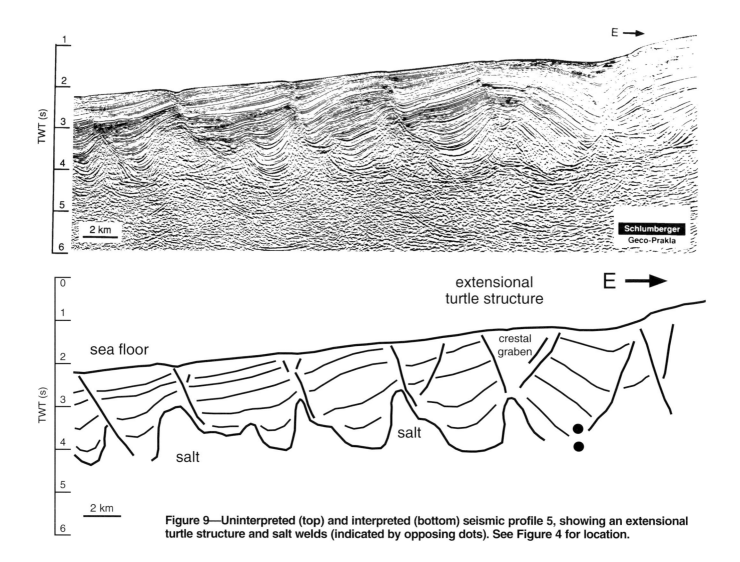

Figure 9—Uninterpreted (top) and interpreted (bottom) seismic profile 5, showing an extensional turtle structure and salt welds (indicated by opposing dots). See Figure 4 for location.

CONCLUSIONS

- All major structural trends in the overburden are influenced by salt tectonics, mostly related to mobilization of the Aptian salt.
- A variety of salt structures in the southern Gabon subbasin can be grouped into two basic families: (1) simple salt pillows in deep water and (2) a complex suite of deeply buried stocks and walls, salt welds, and salt rollers associated with up-to-basin faults basinward of the Cretaceous carbonate platform.
- Salt deformation was repetitive and appears to have responded primarily to the contemporary depositional conditions and basin elements. The Aptian salt was initially deformed in the Late Cretaceous. Although we have no firm estimates of the original thickness of the Aptian salt in the study area, it has been reported in the literature that it may have been as thick as 800 m nearby. Initial deformation of the salt layer likely resulted in salt pillows basinward of the Cretaceous car-

bonate platform, with successively younger deformation reaching deeper waters. Alternatively, where linear salt walls are present (Figure 10), incipient salt walls probably formed initially rather than pillows. During increased clastic input in the Tertiary, asymmetric salt rollers formed, intimately associated with normal faulting. The faulting is dominantly up-to-basin, and stratal packages thicken against the faults. This style of deformation continues to the present. In any region with polyphase salt activity, older positive salt features may have been cannibalized by later salt mobilization.

- Up-to-basin listric faulting in strike-parallel trends is the dominant fault pattern in the southern subbasin. Why these faults are prominently up-to-basin is problematic. We do not think that the southern Gabon subbasin is fundamentally different in its tectonics from adjacent parts of the West African margin. We suggest that this structural fabric may be related to depositional overstepping of earlier salt features by Tertiary regressive

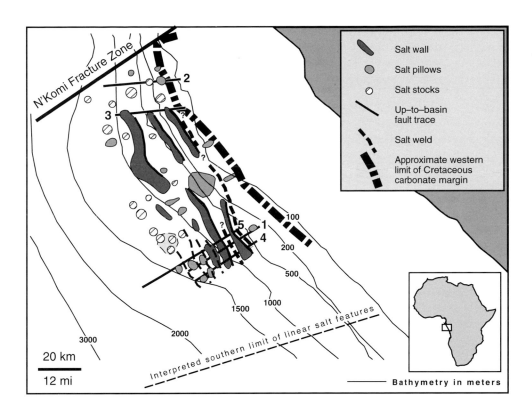

Figure 10—Distribution of major salt structures and tectonic trends in relation to the seismic lines.

sequences. However, neither a tectonic nor depositional scenario seems to adequately explain the dominance of this style of faulting in the southern Gabon subbasin.

- Subsalt imaging with conventional seismic data is commonly inadequate (Edwards et al., 1990) to evaluate presalt elements; these basement structures may play a significant role in the distribution of original salt and its later mobilization.

Acknowledgments   The authors wish to thank Martin Jackson, Mike Hudec, and Steve Henry for their comments and suggestions, which improved the final version of this paper. We thank the management of Texaco and British Gas for permission to publish this paper. The senior author wishes to thank coworkers from Texaco's Central Exploration Department for their comments and Henry Cedillo for photography. The seismic sections illustrated come from a variety of nonexclusive proprietary surveys owned by GECO-PRAKLA (UK) Limited. Permission to use these data is gratefully acknowledged.

REFERENCES CITED

Duval, B., C. Cramez, and M. P. A. Jackson, 1992, Raft tectonics in the Kwanza Basin, Angola: Marine and Petroleum Geology v. 9, p. 389–404.

Edwards, A., and R. Bignell, 1988, West Africa, part 1: Hydrocarbon potential of West African salt basin: Oil and Gas Journal, v. 86, no. 50, p. 71–74.

Edwards, A. D., D. J. Staughton, and P. M. Traynor, 1990, Problems of exploration data quality and interpretation in the Gabon and Congo areas of the West Africa salt basin: Society of Exploration Geophysicists, Technical Program Expanded Abstracts, v. 1, p. 206–208.

Jackson, M. P. A., and C. Cramez, 1989, Seismic recognition of salt welds in salt tectonic regimes: SEPM Gulf Coast Section, 10th Annual Research Conference, Program and Extended Abstracts, Houston, Texas, p. 66–71.

Jackson, M. P. A., C. J. Talbot, and R. R. Cornelius, 1988, Centrifuge modeling of the effects of aggradation and progradation on syndepositional salt structures: The University of Texas at Austin, Bureau of Economic Geology Report of Investigations No. 173, 93 p.

Lundin, E. R., 1992, Thin-skinned extensional tectonics on a salt detachment, northern Kwanza Basin, Angola: Marine and Petroleum Geology, v. 9, p. 405–411.

Teisserenc, P., and J. Villemin, 1989, Sedimentary basin of Gabon—geology and oil systems, in J. D. Edwards and P. A. Santogrossi, eds., Divergent/passive margins: AAPG Memoir 48, p. 117–199.

Thompson, D. T., I. Lerche, and J. J. O'Brien, 1989, Salt basins of western Europe and Gabon, West Africa: dynamical aspects and hydrocarbon production: SEPM Gulf Coast Section, 10th Annual Research Conference, Proceedings, Houston, Texas, p. 122–130.

Traynor, P., M. Doherty, G. Grant, J. McKenna, P. Philip, N. Wilson, and A. Edwards, 1992, Regional seismic interpretation offshore Gabon: results and implications for stratigraphic development (abs.): First Stratigraphy and Paleogeography of West Africa Sedimentary Basins Colloquium Proceedings, p. 177.

Uchupi, E., 1992, Angola Basin: geohistory and construction of the continental rise, in C. W. Poag and P. C. de Graciansky, eds., Geologic evolution of Atlantic continental rises: New York, Van Nostrand Reinhold, p. 77–99.

Heaton, R. C., M. P. A. Jackson, M. Bamahmoud, and A. S. O. Nani, 1995, Superposed
Neogene extension, contraction, and salt canopy emplacement in the Yemeni Red
Sea, *in* M. P. A. Jackson, D. G. Roberts, and S. Snelson, eds., Salt tectonics: a global
perspective: AAPG Memoir 65, p. 333–351.

Chapter 16

Superposed Neogene Extension, Contraction, and Salt Canopy Emplacement in the Yemeni Red Sea

R. C. Heaton

BP Exploration
Uxbridge, Middlesex, U.K.

Present address:

Cairn Energy PLC
Edinburgh, U.K.

M. P. A. Jackson

Bureau of Economic Geology
The University of Texas at Austin
Austin, Texas, U.S.A.

M. Bamahmoud

A. S. O. Nani

Yemeni Ministry of Oil and Mineral Resources
Sana'a, Yemen

Abstract

Although the Neogene Red Sea basin has been intensively examined as the type example of a young, narrow ocean, salt tectonics there has been neglected. The Yemeni part of the Red Sea exhibits a wide array of salt tectonic features within a small area. Above the rift section, a middle Miocene evaporite layer, originally 1.5–2 km thick, is the source for autochthonous and allochthonous salt structures. In the middle–late Miocene, evaporite-clastic overburden, halite, and anhydrite layers 50–350 m thick assisted deformation by providing several levels for décollement. Four southward-narrowing tectonic zones trend subparallel to the basin axis. Areas of extension in the easternmost Roller Zone, severe shortening in the central Canopy Zone, and mild shortening in the western Anticline Zone all narrow then pinch out at roughly the same latitude. This convergence suggests that extension and contraction are linked by various salt layers and by transfer structures transecting the tectonic zones. Extension, contraction, and coeval salt canopy emplacement were superposed, mostly between 8 and 5 Ma. The presence of allochthonous salt sheets casts doubt on previous estimates of salt 5 km thick in the southern Red Sea. Fault scarp asperities in the basin floor may have acted as buttresses against which contraction was initiated. The wide variety of salt structures may be due to the weakness and anisotropy of the partially evaporitic overburden and to high geothermal gradients (up to 77°C/km). These factors enhanced the deformation driven by gravity spreading and sedimentary differential loading.

INTRODUCTION

The Red Sea, and especially the Gulf of Suez, has been much examined as the premier showcase for the evolution of a rift, incipient divergent margin, mid-ocean ridge system, and associated hydrocarbon resources and metal deposits. Thematic issues of *Tectonophysics* (1991, v. 198, nos. 1–4, p. 129–468) and the *Journal of Petroleum Geology* (1992, v. 15, no. 2) contain papers detailing research in these fields. Almost all papers on the area show that the Red Sea is a major evaporite basin. Yet publications concentrating on the salt tectonics of the Red Sea are conspicuous by their paucity (El Anbaawy et al., 1992).

We wish to redress this imbalance by describing the complex mix of salt tectonics in the Neogene section across the entire width of the eastern flank of the Red Sea basin in Yemen. We also synthesize these data and deduce a structural evolution by superposed deformation.

The study area lies between latitudes 14°30'N and 16°25'N and comprises an offshore zone in the Yemeni Red Sea and the adjacent 30–50-km-wide Tihama coastal plain. The western edge of the area is the active seafloor spreading center dividing the Yemeni and Eritrean parts of the basin. The eastern boundary is the abrupt Great Escarpment of the mainly volcanic mountains, which rise

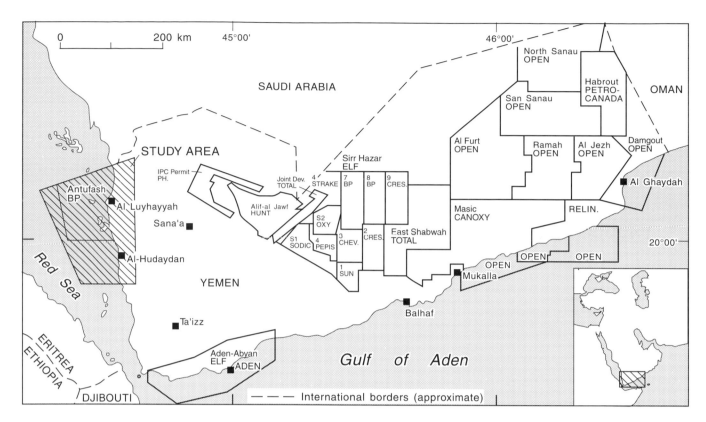

Figure 1—Location map of study area (hatched) in relationship to the Red Sea, Gulf of Aden, and Yemen hydrocarbon license blocks. All international borders shown in this and succeeding maps are highly approximate.

to 3660 m and form the southeastern shoulder of the Red Sea rift (Davison et al., 1994a) (Figure 1). Water depths range mostly from 15–100 m, plunging to more than 1000 m in the central trough of the Red Sea. Numerous islands and shoals represent almost emergent salt structures on the shelf or volcanoes in the south and west.

Geologically, the southern Red Sea basin is dominated by the Oligocene–Recent divergent continental margin and volcanic shoulders along the Great Escarpment. These overlie thinned continental crust of Precambrian–Mesozoic age.

DATA SOURCES

Our study is based on approximately 8000 km of offshore and onshore seismic lines acquired by Shell, Hunt, and British Petroleum in six surveys between 1974 and 1991 (Figure 2). Recent reprocessing of much of the data has significantly improved imaging, especially of allochthonous salt. Data from all 11 hydrocarbon exploration wells drilled between 1961 and 1992 (Figure 3) have been used. In the middle–upper Miocene section, diagnostic biostratigraphic data are absent because the depositional environments were either hypersaline or fluvial.

We used 10,000 km of aeromagnetic data in the northern offshore area primarily to detect volcanic rocks and to assess depth to basement rather than to map the salt

masses. Gravity data acquired by Shell during a 1974 seismic survey were used in conjunction with the aeromagnetic data to determine the underlying structural grain of the area. Gravity profile data were used locally to corroborate the distribution of salt masses interpreted on seismic lines.

Landsat images color coded by water depth helped bathymetric interpretation. On most lines, shallowly buried salt structures form shoals crowned by carbonate reefs, so the Landsat images proved valuable in interpolating shallow salt structures. Onshore, the salt diapirs lie within a few meters of the surface. One quarried diapir (Al Salif) has elevated Recent beach deposits at a rate of ~1 cm/year. Another quarried diapir (Jabal Al Milh) has stretched an 80–100-year-old mosque, which forms part of its overburden, by as much as 14% (Figure 3) (Davison et al, 1994b).

TECTONIC SETTING

The early synrift section of Oligocene–early Miocene age (~30–17 Ma) is poorly known because it is only partially penetrated by one well: Zeidieh-1 (Figure 3). During the Oligocene, the Red Sea area was probably centered over the Afar plume triple junction of the African and Arabian plates (Menzies et al., 1992). Work in onshore Yemen (Menzies et al., 1992; Davison et al., 1994a) sug-

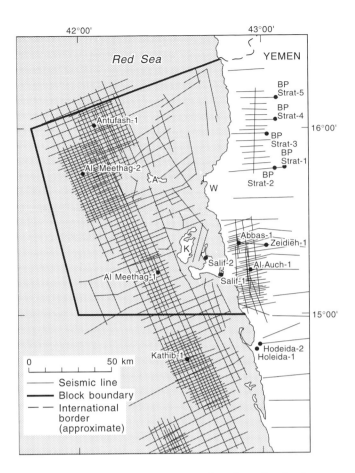

Figure 2—Map of the approximate location of seismic lines and exploration wells in the study area. A = Antufash Island, K = Kamaran Island, W = Wadi Mawr Delta.

gests that, beginning 30 Ma, major magmatism had already emitted 2–3 km of extrusions (Yemen Volcanic Group) before significant uplift and extension began 22 Ma (Figure 4). Domino-style faulting has extended the crust to a β (crustal stretch) value of 1.6–1.8 along the coastal plain (Davison et al., 1994a).

Rifting was followed in the middle Miocene by accumulation of the main evaporite group. Lack of biostratigraphic or isotopic data from this interval precludes accurate dating. Even the plate tectonic status of the salt is uncertain. Hughes and Beydoun (1992) interpret it as synrift, whereas Davison et al. (1994a) regard it as postrift. Onshore, the evaporites are clearly postrift. Offshore, in contrast, neither the evaporite's exact age nor the structure and composition of the underlying crust are known, and the salt may be affected by crustal extension (as documented here). Accordingly, we call the main salt *transitional*.

For similar convenience, we define *postrift* as the interval between the deposition of the main salt (middle Miocene, ~11 Ma) and the end of the Miocene (6–5 Ma). At the latter time, the African and Arabian plates finally

separated in the southern Red Sea (Sultan et al., 1992), initiating an indisputable drift phase.

The nature of the crust beneath the continental shelf of the Red Sea has been much disputed. One group (e.g., Girdler and Underwood, 1985; Sultan et al., 1992) proposes shelves underlain by oceanic crust. Conversely, a second group postulates a thinned and intruded continental crust below the shelf (e.g., Cochran, 1983; Davison et al., 1994a), a hypothesis supported by gravity modeling (Makris et al., 1991) and by observed maturity profiles and modeled basal heat flows from wells in the Red Sea (Barnard, 1992).

STRATIGRAPHY

As indicated by a simplified basin chronostratigraphy (Figure 5), an Archean and Pan-African granitoid and metamorphic basement is overlain by a thin Permian sedimentary succession (Akbra Shale, not shown). The basement, in turn, is overlain by the prerift sedimentary succession comprising the Mesozoic fill of a slowly sagging basin (e.g., Kholan, Amran, and Tawilah groups) in the lower parts and local volcanic rocks in the upper parts (Davison et al., 1994a).

By analogy elsewhere in the basin (e.g., Eritrea), we infer that the synrift succession in the area probably comprises thick Oligocene volcanic rocks overlain by a mixed continental and marine succession containing local volcanic rocks (Crossley et al., 1992; Mitchell et al., 1992).

In the transitional phase, large volumes of massive salt with subordinate anhydrite and marginal clastics were deposited in a restricted basin, possibly confined by continental fragments such as the Danakil Alps of Eritrea. Onshore wells penetrate both allochthonous and autochthonous salt. This salt has been described as up to 5 km thick locally and 2–4 km thick over a wide "Massive Salt Zone" in depositionally massive and halokinetically thickened forms (Mitchell et al., 1992). This impressive thickness is probably illusory because, as speculated by Mitchell et al. and as shown by our data, the apparently thickest salt comprises allochthonous sheets overlying younger sediments.

Above the main evaporite group, the postrift phase is represented by a thick cyclic succession of mixed evaporites (halite and anhydrite) and clastics of middle–late Miocene age, which increase in evaporite content from east to west (Figure 5). Most evaporite units are 0.5–1 km thick, and some are more than 3 km thick (Figure 3). Considerable well data are available for this interval (Hughes and Beydoun, 1992). However, because the sedimentary facies are mainly either subaerial or hypersaline, they are poor in age-diagnostic fauna.

When the two continental masses split at the end of the Miocene, sedimentation changed. A widespread open-marine carbonate platform, having biostratigraphic affinities to the Indian Ocean, formed offshore during the Pliocene–Recent. Fluvial and alluvial sedimentation continued onshore to the present day.

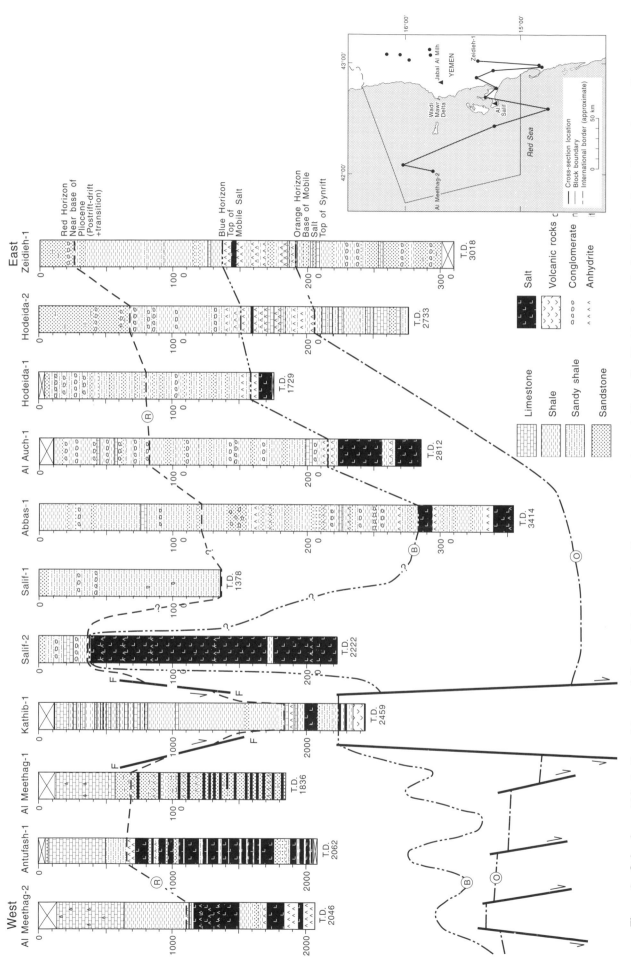

Figure 3—Seismic and biostratigraphic correlation of all deep exploration wells, of which only the easternmost reach the base of the main evaporite (Orange horizon), here predominantly anhydritic. The proportion of halite in the overlying Blue to Red interbedded clastic—evaporite interval increases basinward. Above the Red horizon, fluviodeltaic facies in the onshore wells grade into a carbonate shelf offshore. On inset location map, black dots = wells; black triangles = emergent salt diapirs (referred to in the text).

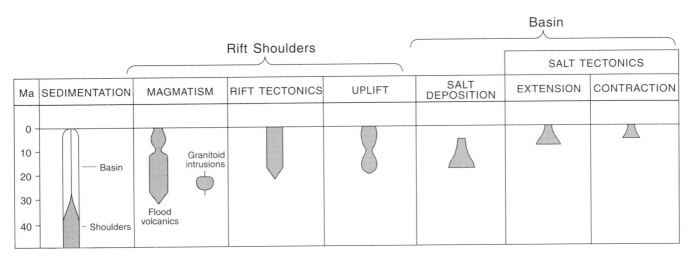

Figure 4—Schematic chronology of tectonic, magmatic, and sedimentary events in the Red Sea rift of Yemen. (Extended from Davison et al., 1994a.)

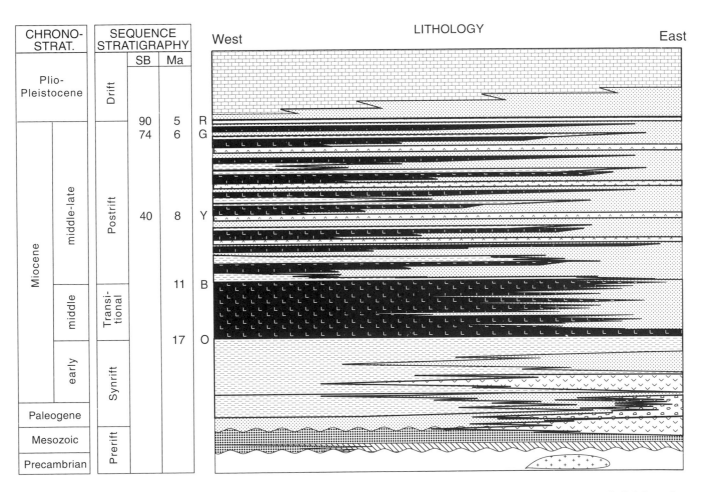

Figure 5—Chronostratigraphy of the study area. Although lack of biostratigraphic fauna prevents accurate subdivision of the middle–late Miocene, this interval contains increasingly abundant evaporites to the west; interfingering subaerial deposits increase landward to the east. Mobile salt at the base of this interval is probably middle Miocene in age and represents a transition from synrift to postrift. Numbers denote sequence boundaries (SB) or age (Ma). Letters O, B, Y, G, and R denote Orange, Blue, Yellow, Green, and Red horizons, respectively. See Figure 3 for key to lithologic patterns.

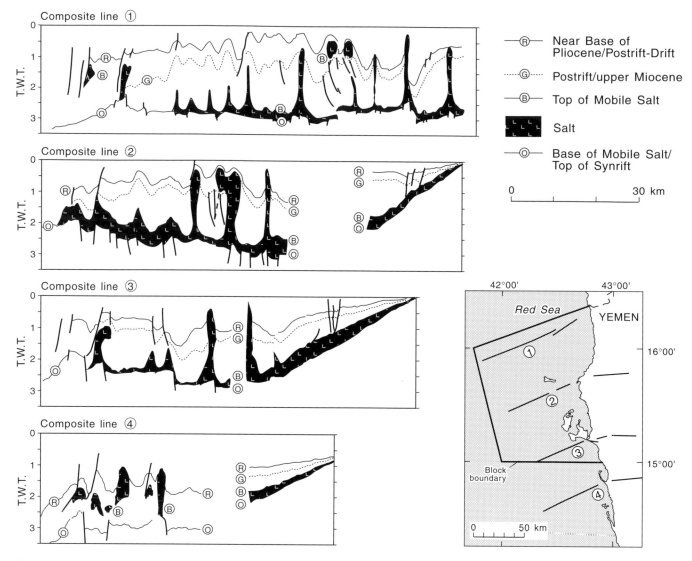

Figure 6—Four east-west composite geoseismic dip sections showing the overall form and variation in salt structures: relatively undeformed onshore salt wedge, major diapirs and detached salt bodies, salt-cored anticlines, and shelf edge slumps in extreme west. (Vertical exaggeration is roughly 62 times.) Inset map shows location of four composite sections.

MAPPING RESULTS

Four key horizons have been regionally interpreted on the seismic data (Figure 5):

- Orange: base mobile salt/top synrift, ~17 Ma
- Blue: top mobile salt/top transitional, ~11 Ma [Yellow: intra-postrift, sequence boundary (SB) 40, ~8 Ma]
- Green: intra-postrift, SB 74, ~6 Ma
- Red: postrift to drift, SB 90, ~5 Ma

In addition, the bracketed Yellow horizon was mapped locally where recognizable. In the western-central area, where seismic and well data allow, the postrift section penetrated by wells, which is less than half the full section, has been divided into 10 "sequences" represent-

ing major flooding–drying cycles. These sequence boundaries are included in the above list of key horizons. Isopachs of these sequences provide evidence of the timing and distribution of halokinetic pulses.

An initial examination of four composite regional dip sections (Figure 6) indicates that the area can be divided into four geographic zones on the basis of structural style. These zones trend parallel to the basin axis and narrow southward as the Red Sea narrows (Figure 7). From east to west, these are the Roller, Canopy, Anticline, and Outer Shelf zones.

Roller Zone

The Roller Zone occupies the largely onshore area in the east. The zone greatly resembles the updip zones of divergent margin salt basins such as the West African

Canopy Zone

The Canopy Zone is characterized by a postrift section roughly 4–5 km thick. It contains a wide variety of salt structures including salt pillows, stocks, detached stocks, walls, sheets, tongues, and canopies (Figures 9, 10). Primary salt and fault welds separate these structures. Seismic resolution at depth is poor, and the deep structure is correspondingly speculative. However, the base of the main salt (Orange horizon) appears offset by block faulting in many places (Figures 6, 9, 10, 11). If so, rifting continued offshore during or after deposition of the salt.

Anticlines underlie many of the salt sheets (Figure 11). They resemble contractional fault-bend folds and fault-propagation folds above the hanging wall ramps of thrust complexes. In places, apparent hanging wall cutoffs are imaged; so are thrust fault offsets, although the seismic data are not clear enough to resolve whether they are duplexes or imbricate systems.

Most salt sheets have a fairly flat crest and an irregular base (Figure 11), unlike those shown in Figures 9 and 10. Figure 12 shows the distribution of these salt structures where they rise to shallow levels (near the top of the Miocene or above). The salt sheets, tongues, walls, and salt-cored anticlines trend roughly northwestward, parallel to the coastline except in mid-latitudes, where the salt structures form irregular polygonal ridges surrounding withdrawal basins of younger sediments (Figure 12). Figure 13 shows the distribution of the deeper faults and salt cutoffs affecting the Green intra-postrift horizon.

The Canopy Zone has several tectonic subzones subparallel to the basin axis (Figure 14):

1. Subzones of extension and salt withdrawal containing normal faults and half-grabens, both landward and basinward verging
2. Subzones of contraction containing thrust faults, ramp anticlines, and highly irregular dips
3. Subzones containing simple passive diapirs and adjoining symmetric withdrawal basins
4. Subzones displaying major transfer-like offsets and random alignment of structures.

Anticline Zone

The Anticline Zone contains northwest-trending, relatively gentle sinusoidal buckles (Figure 15). These detachment folds are most likely cored by the main salt unit. Salt flow in this zone merely accommodated and responded to mild buckling involving only a few percent shortening. In the north (Figure 15), folding began just before deposition of the Green horizon (~6 Ma). Elsewhere, reflectors onlapping anticlines indicate growth folding from deposition of the Yellow horizon (~8 Ma) to post Red horizon (5–0 Ma). Most of the anticlines verge slightly seaward. They are locally box folds, as is typical of well-layered sequences decoupling over evaporites (compare Cobbold et al., 1995). As in the Canopy Zone, data could be interpreted to suggest that some of the

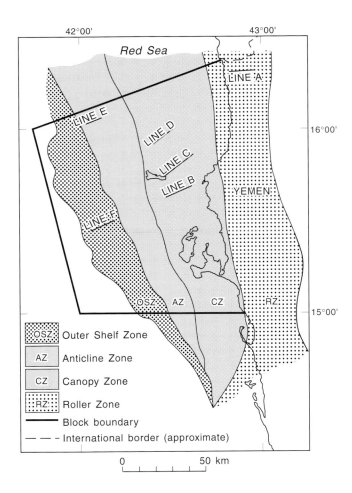

Figure 7—Map showing the distribution of the four main salt tectonic zones and the location of seismic lines A–F (in Figures 8–11, 15, 16).

basins. Features typical of the Roller Zone are illustrated in Figures 6 and 8. The zone is a sedimentary wedge thinning from ~4 km to ~300 m.

In the east, the wedge is unfaulted at seismic resolution. The main basal evaporite unit here is restricted to two thin (together ~100 m thick) beds of anhydrite and subordinate halite (~30 m thick).

In the west, listric normal growth faults detach on the main evaporite. Rotated fault blocks overlie salt rollers. Most faults dip basinward. Based on the oldest reflectors that diverge toward the faults, extension began just before deposition of the Yellow horizon (~8 Ma) and ended variably at around 5–6 Ma. Where penetrated by wells, the evaporites comprise two thick autochthonous units of dominantly halite at least 500 m thick. Both halite units acted as décollements. They are overlain by mudstone and sandstone red beds of fluvial-alluvial origin.

The boundary between the unfaulted and faulted zones is defined by the easternmost listric fault. In the northernmost part of the basin (Figure 8), this fault overlies a hinge point marking the basinward steepening of the base of the salt (Orange horizon).

340   *Heaton et al.*

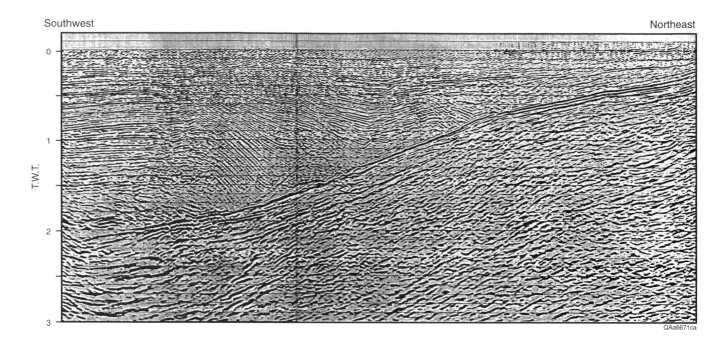

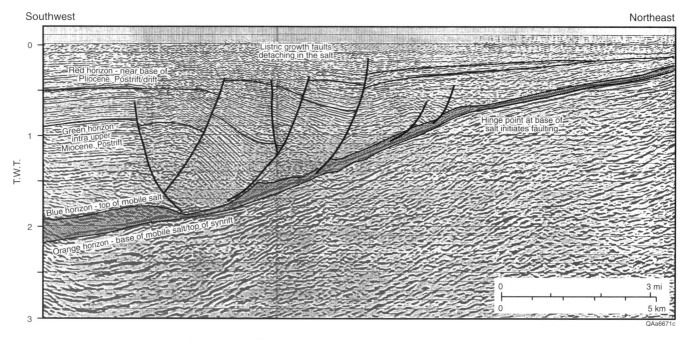

Figure 8—Uninterpreted (top) and interpreted (bottom) migrated seismic line A through the Roller Zone in the northern onshore region. A homoclinal eastern platform passes over a hinge point into a zone of listric, dominantly seaward-dipping, extensional faults detaching on the mobile salt, causing rollover anticlines in the hanging walls to rotate clockwise. See Figure 7 for location.

anticlines overlie landward-dipping normal faults offsetting the base of the main evaporite (Figure 15).

The eastern margin of the Anticline Zone is an asymmetric salt wall. In the north, transfer zones are indicated by consistent offset of parallel anticlinal axial traces. Overall, the number of anticlines decreases southward as the zone narrows, eventually dying out near the Kathib well (Figures 3, 14).

Outer Shelf Zone

The Outer Shelf Zone in the west is separated from the Anticline Zone by relay normal faults. The structural regime here is extensional (Figure 16). The role of salt is restricted to providing several décollements. Wells indicate that salt dominates the entire middle–late Miocene overburden in this zone (Figure 3). Listric faults detach

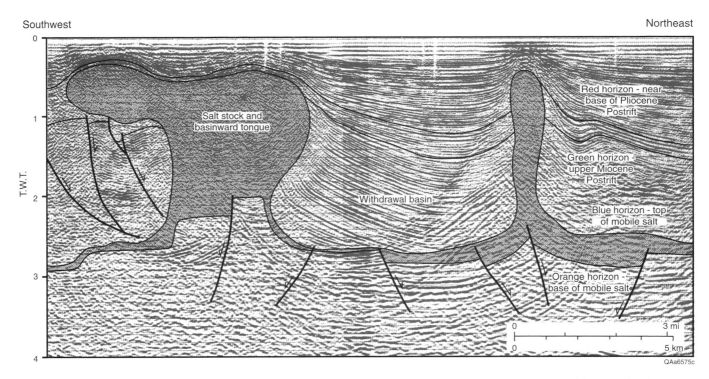

Figure 9—Interpreted migrated seismic line B through the offshore Canopy Zone. A broad salt stock with an allochthonous tongue is separated from a neighboring stock by a withdrawal basin. Velocity pull-up beneath the diapirs is mild due to the high proportion of salt in the Blue to Red overburden interval, which increases the latter's seismic velocity. See Figure 7 for location.

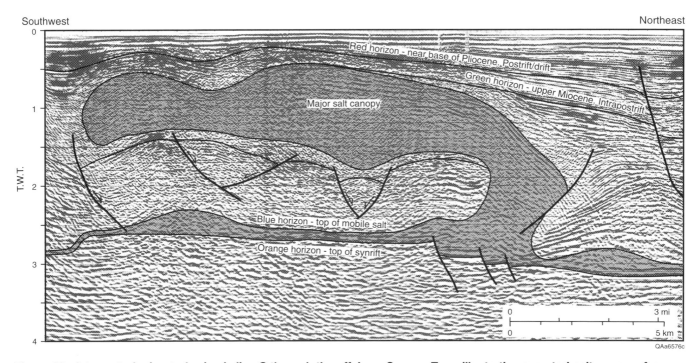

Figure 10—Interpreted migrated seismic line C through the offshore Canopy Zone illustrating a rooted salt canopy. A syncline in the base of the canopy may indicate an additional feeder stem off section or it could be the result of the subsalt graben below it. This salt structure forms part of a polygonal salt wall surrounding a withdrawal basin (Figure 12). See Figure 7 for location.

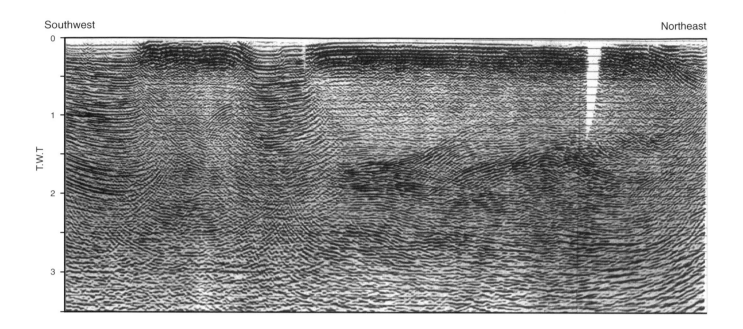

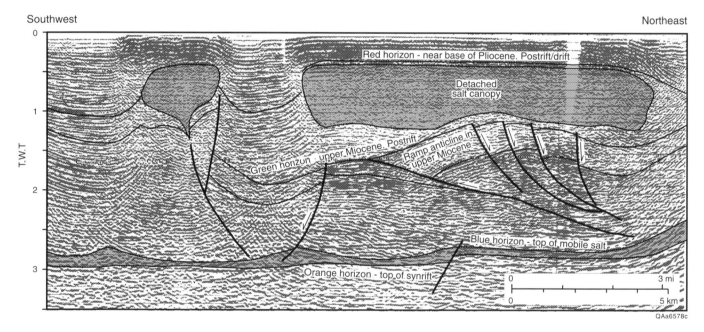

Figure 11—Uninterpreted (top) and interpreted (bottom) migrated seismic line D through the offshore Canopy Zone illustrating a detached (in this plane of section) salt allochthon overlying a faulted anticline, interpreted as a southwest-verging ramp anticline. (The horizon locally picked between the Blue and Green is the Yellow horizon.) See Figure 7 for location.

along several salt layers above the main evaporites; some hanging walls are deformed by extensional fault-bend folding.

The western limit of this zone is conjectural where the slope declines into the Red Sea abyss (Figure 16). The Pliocene–Pleistocene unit is probably interbedded with igneous rocks of the active spreading center, but the salt tectonic setting here is entirely speculative.

INTERPRETATION OF HALOKINETIC EVOLUTION

Toward the southern straits of the Red Sea (Bab el Mandab) (Figure 14), the Roller, Canopy, and Anticline zones narrow and pinch out in concert. The Anticline Zone reduces from three or four anticlines in the north to one or two opposite Kamaran Island, and then to a

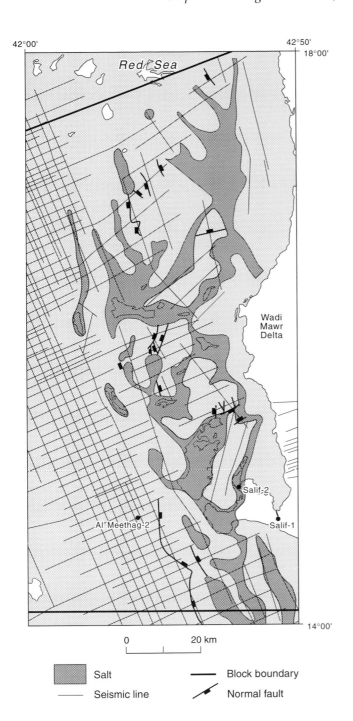

Figure 12—Map of the Canopy Zone showing salt structures and faults intersecting the shallow Red horizon. In
the north and south, the salt walls are subparallel to the
coastline. In the mid-latitudes, the walls define polygonal
upwellings surrounding withdrawal basins (see Figure 3).

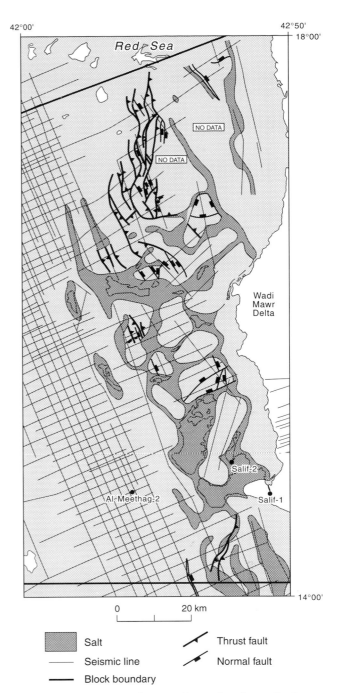

Figure 13—Map of the Canopy Zone showing salt structures and faults intersecting the Green horizon. The thrust
faults are associated with ramp anticlines. The thrusts
have the same preferred orientation as the salt walls in
the north and south but are discordant to salt contacts in
the mid-latitude polygon area.

narrow horst, before pinching out altogether. Likewise in
the Roller Zone, extensional rollers die out opposite the
point where the contractional parts of the Canopy and
Anticline zones pinch out. This approximate parity suggests that the extensional rollers, the contractional zone
beneath the canopy, and the anticlines are laterally linked
by salt layers and transfer structures transecting the tec-

tonic zones. The Red Sea basin also pinches out southward toward the straits, where regional extension trans-
fers from the Red Sea to the Afar triple junction west of
the long-lived Danakil horst of Eritrea (Abouzakhm et al.,
1991; Mitchell et al., 1992). Figure 17 schematically sum-
marizes the following interpretation of the salt tectonic
evolution.

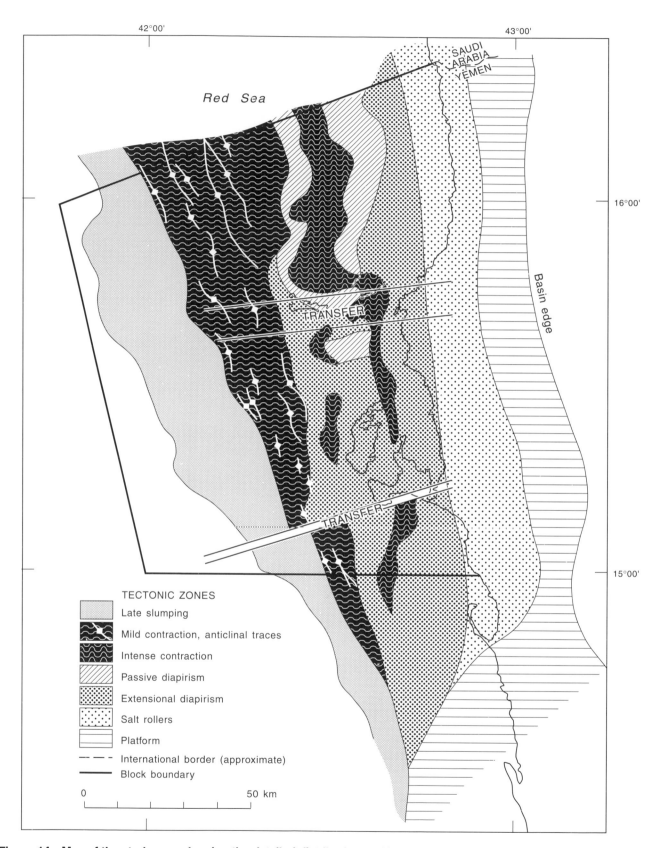

Figure 14—Map of the study area showing the detailed distribution and interaction of the salt tectonic zones. All the deformation zones pinch out southward, implying seaward translation of a laterally linked system of extension and contraction. The zone labeled "Extensional diapirism" comprises basins bounded by normal faults; these basins could also have been formed solely or partially by salt withdrawal.

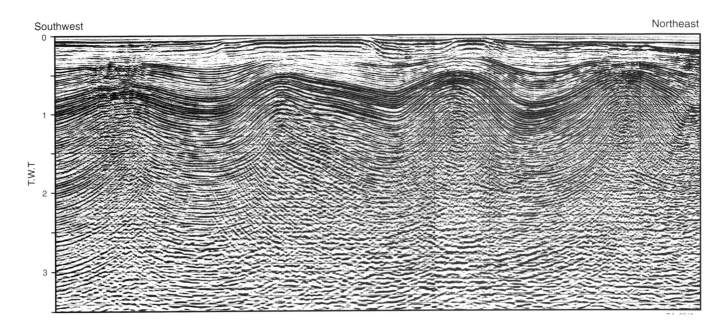

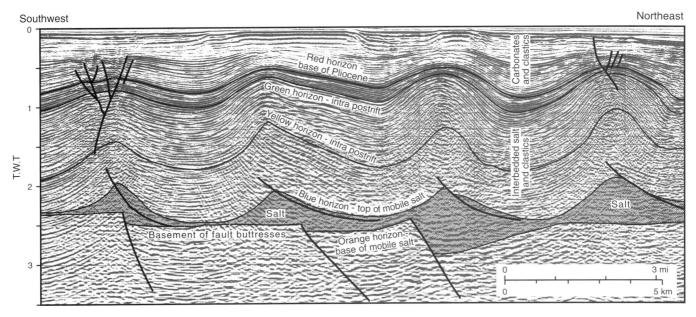

Figure 15—Uninterpreted (top) and interpreted (bottom) migrated seismic line E through the Anticlinal Zone illustrating the generally sinusoidal, parallel fold style and their weak southwest vergence. Inferred landward-dipping normal faults in the base of the salt may have localized the folds by buttressing seaward translation of the overburden. The folded overburden is much thicker than the salt, suggesting that the salt merely responded to regional contraction by flowing from below synclines into the lower pressure anticlinal cores. See Figure 7 for location.

Middle Miocene: Pre–SB 40 (Orange to Yellow Horizons)

Late Eocene–Oligocene rifting initiated the Red Sea basin. This faulting and subsidence determined the boundaries and tectonic elements of the basin where the main salt was deposited during the middle Miocene. Structure at the base of the salt (Figure 18) shows a depocenter centered on the Canopy Zone. Moreover, the structural maturity of salt structures (degree of allochthon

development) increases inward from the east and west. These observations independently suggest that when salt was accumulating, the basin depocenter coincided with the present Canopy Zone. Seismic lines have been depth converted and reconstructed using *Restore* © (output not shown) and LOCACE software (Figure 19). These restorations indicate a maximum prekinematic salt thickness of 1.5–2 km; the salt wedged out eastward onshore and westward onto the outer shelf high.

In the middle Miocene, clastics began to be shed into

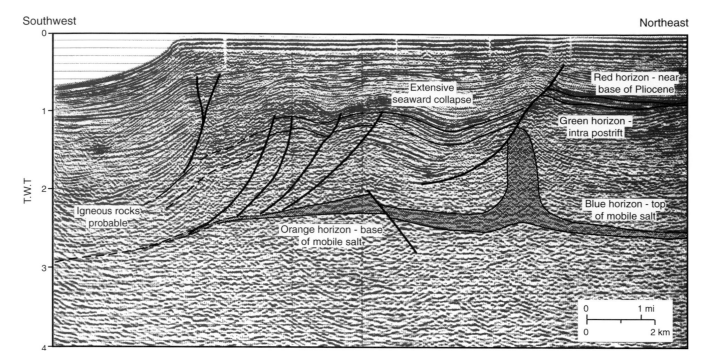

Figure 16—Interpreted migrated seismic line F through the Outer Shelf Zone illustrating the dominantly extensional gravitational collapse at the shelf break. Faults detach both at the base of the salt and within the salt-bearing overburden. See Figure 7 for location.

the basin from the rising rift shoulders. As the base of salt tilted basinward, the eastern part of the Roller Zone began to extend by rotational listric faults detaching over the salt.

The landward limit of visible extensional faulting coincides with the hinge point where the base of the salt steepens basinward. This coincidence suggests that extension was caused by gravity sliding due to increased basinward dip at the hinge. We speculate that this hinge point may separate domains characterized by different amounts of rifting and postrift subsidence due to cooling. However, this hinge point is absent in the southern part of the area. Another explanation of the abrupt landward limit of extension is the facies change in the main evaporite unit: the proportion and thickness of halite markedly increase seaward of the hinge point from ~10% (30 m) to ~75% (530 m) (compare Zeidieh-1 and Al Auch-1 wells in Figure 3). This halite would have promoted decoupling and extension on even a gentle slope. Downdip strata would have torn away from the updip part of the wedge, like strata in the Mexia-Talco fault zone (East Texas basin) at the updip pinch-out of the salt (Jackson, 1982).

Farther west in the Canopy Zone, diapirs were initiated below thin overburdens (<500 m thick). Two mechanisms could have triggered diapirism. The first is thin-skinned contraction induced by gravity gliding in the Roller Zone farther landward. However, numerous seismic examples and modeling (reviewed by Jackson and Vendeville, 1994) indicate that diapirs are initiated more easily by extension than by buckling. Initiation from buckles typically requires considerable erosion of anticlinal crests, and such erosion is unlikely in the deepest part

of the salt basin. More likely, the diapirs were initiated reactively by tectonic differential loading caused by stretching of the overburden during regional extension, as documented in many of the world's salt basins (Jackson and Vendeville, 1994). That episode of early extension is compatible with the structural restoration (Figure 19), but cannot be proven by this means.

The map patterns of diapiric walls suggest that more than one process initiated the walls. In the north and south, the walls trend subparallel to the coastline. They could therefore have been initiated by strike-parallel extensional faults in the overburden or by differential loading induced by prograding sediments. Seismic data are inadequate to allow us to eliminate either process. The mid-latitude polygonal pattern of withdrawal basins could have arisen by randomly superposed lobes of sediment. Alternatively, multidirectional extension of lobate, formerly flat-topped salt sheets and their overburdens could have caused reactive diapirs to rise as polygonal ridges below and through intersecting graben systems (Jackson and Vendeville, 1995). Divergent extension is likely off a coastal promontory, just as convergent extension is likely off a coastal embayment (Cobbold and Szatmari, 1991). The Wadi Mawr Delta, which debouches into the area of polygonal walls (Figure 12), represents just such a coastal promontory. This delta could have caused either multidirectional extension or randomly superposed lobes of sediment.

Extension in the Roller Zone and inferred extension in the Canopy Zone were accommodated by gentle buckling farther west in the Anticline Zone (Figures 17, 19).

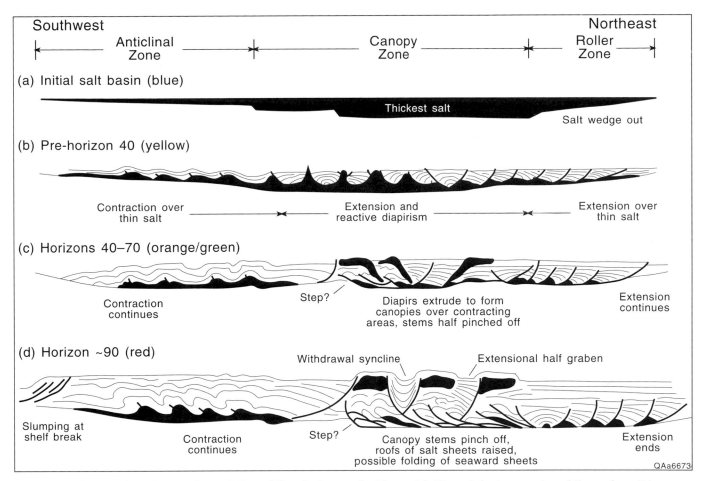

Figure 17—Schematic salt tectonic evolution of the study area (not to scale). The original geometry of the main salt layer (black) was inferred from salt tectonic styles and from the basement data in Figure 18. Half-grabens in (d) could have been created by either extension or salt withdrawal.

Middle Late Miocene: SB 74–90 (Green to Red Horizons)

The original salt thickness was greatly reduced by export into the rising stocks and salt sheets in the Canopy Zone. As the source layer thinned, faults offsetting the base of the salt would have increasingly affected salt flow by impinging on the subsiding top of salt (Figure 19). Any fault scarps facing landward would have acted as buttresses to the seaward translation of the overburden, initiating contraction in the Canopy Zone.

Being the weakest parts of the system, the salt structures were probably the first to shorten in the Canopy Zone. We expect that eventually most of the walls (including segments of polygonal ridges outlining withdrawal basins) subparallel to the coast were laterally squeezed. Their salt was expelled upward to form extrusive sheets and canopies. Thereafter, contraction was accommodated by thrust faulting and folding. Interbedded salt layers in the overburden would have influenced the location of thrust flats and enhanced flexural folding. Eventually, thrust slices in the Canopy Zone stacked up. Contraction thus propagated seaward, and folds in the Anticlinal Zone continued to tighten and increase in amplitude.

End of Miocene: SB 90 (Red Horizon)

Extension in the Roller Zone all but ended by deposition of the Red Horizon. Only local extension and salt withdrawal is observed in the Canopy Zone. However, contraction in the Canopy Zone continued to pinch canopy stems and encouraged lateral spreading and inflation of the salt sheets and canopies. By analogy with the Gulf of Mexico (Fletcher et al., 1995; Talbot, 1995), the allochthonous salt sheets were probably emplaced by extrusion rather than by sill injection. Nearly all the sheets and canopies have flat roofs but irregular bases. This difference suggests that the original salt glaciers extruded (1) during pulsed sedimentation, which caused primary ramps and flats in the base of salt (Jackson and Talbot, 1991), or (2) over a faulted surface, rather like a flat-topped salt glacier surmounting an irregular, stepped bedrock (Talbot, 1979).

Compared with the world average of 27°C/km, geothermal gradients in the southern Red Sea are extremely high: up to 77°C/km in the study area (Al Meethag-2 well; Barnard, 1992) and 180°C/km in the axial trough. Heat flow increases southward, so that the study area (and opposing margin) have the highest average heat

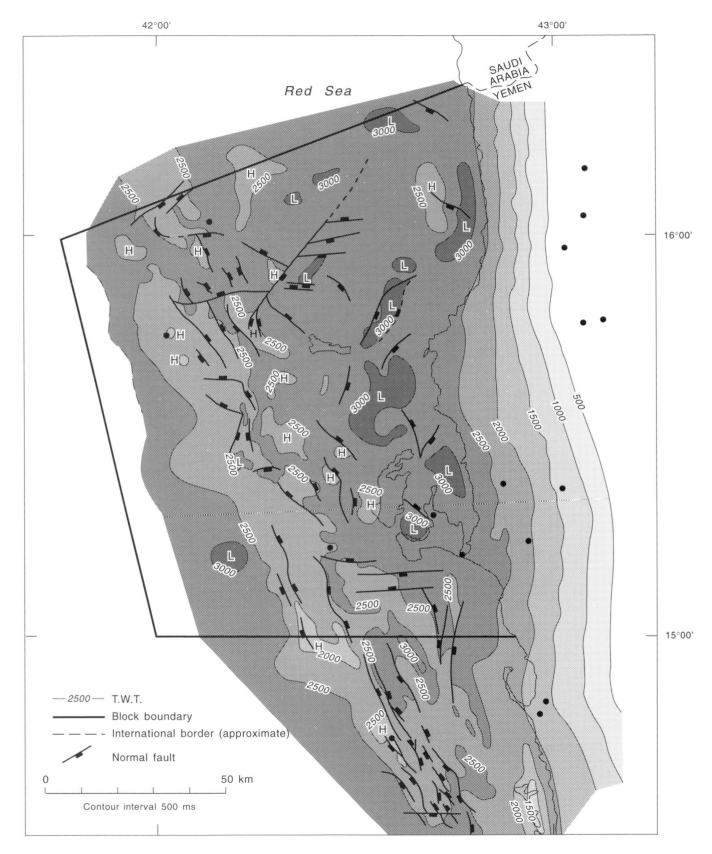

Figure 18—A time-structure map of the base of the salt (Orange horizon) illustrating the following: (1) the basin widens northward, (2) the postsalt depocenter is in the Canopy Zone, and (3) a probable horst forms the eastern boundary of the Outer Shelf Zone. Most faults appear to dip landward.

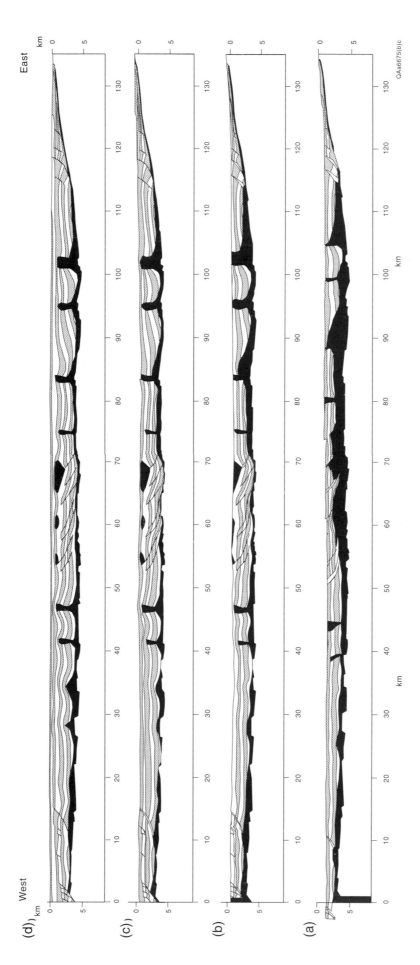

West

East

(d)

(c)

(b)

(a)

km

km

km

km

QAa6675(b)c

Figure 19—LOCACE structural restoration of a depth-converted composite regional dip section. The restoration is area balanced but not decompacted because of the abundance of noncompacting evaporites in the offshore overburden. The reconstruction begins when some diapirs were already emergent during the early extension. Restored stages from bottom to top are at approximately (a) 7 Ma, (b) 6 Ma, (c) 5 Ma, and (d) 0 Ma. The salt depocenter in the Canopy Zone and outer Roller Zone shows clearly. (Restoration by J. Hossack.)

flow: >55°C/km. Geotherms were even higher during earlier phases of rifting (Barnard, 1992). Creating an oil window <1 km thick, this heat has matured even postsalt source beds as young as 5 Ma in withdrawal basins between diapirs (Mitchell et al., 1992).

Salt was weakened by this heat. High temperatures lower the viscosity of salt and thereby increase its strain rate (van Keken et al., 1993). Moreover, the overburden above the main salt unit would also be considerably weakened by the presence of interlayered evaporites, which would have promoted slip between the stronger clastic units, as in the Great Kavir of Iran (Jackson et al., 1990). Thus, the wide range of salt structures in a small area of the Red Sea may have been partially induced by unusually weak salt and overburden.

CONCLUSIONS

1. Four structural zones parallel to the Red Sea coastline converge and pinch out southward in concert. This relationship suggests that the zones are laterally linked by various salt layers and transfer structures transecting the tectonic zones.
2. The easternmost *Roller Zone* is entirely extensional and comprises a sedimentary wedge thinning landward from ~4 km to ~0.3 km. In the east, the wedge is undeformed; the main evaporite unit contains only two thin beds of anhydrite (together ~100 m thick) and 30 m of halite. Farther west, however, the main evaporite comprises two 500-m-thick halite layers. Here, listric seaward-dipping normal faults sole into salt rollers and detach on the main salt. The landward limit of extension coincides with a steepening dip of the base of the salt or with an increase in thickness of the basal halite.
3. The *Canopy Zone* contains salt pillows, stocks, strike-parallel walls, sheets, and canopies. The canopies are shallowly buried onshore and form many salt islands and shoals crowned by reefs. Because of inadequate imaging, earlier estimates of salt thickness (5 km) confused these allochthonous canopies, which overlie younger sediments, with salt massifs, which consist entirely of salt. Below and between the canopies, the Canopy Zone comprises (1) subsalt contractional thrust faults, fault-bend anticlines, and strata of irregular dip; (2) symmetric and asymmetric withdrawal basins; and (3) diapiric walls.
4. The *Anticline Zone* has as many as four coast-parallel, seaward-verging, sinusoidal or box anticlines, apparently decoupling on the main salt.
5. The extensional *Outer Shelf Zone* is separated from the Anticlinal Zone by relay normal faults. In the overburden, 50–350-m-thick halite and anhydrite layers provide detachments for large gravity slumps down the steep continental slope.
6. The wide variety of salt structures in a small area probably results from the high geothermal gradient, which weakens the source layer, and from the intercalated salt layers within the overburden, which allow flexural slip and provide décollements.
7. The main evaporite originally wedged out to the east and west from a depocenter 1.5–2 km thick in the Canopy Zone. The overburden detached extensionally in the Roller Zone between 8 and 5 Ma. In the Canopy Zone, coeval extension of the overburden is inferred to have initiated reactive salt walls. The only direct evidence for this inference is the observed contraction in the Anticline Zone, which also began 8–6 Ma. As salt in the source layer became depleted by extension and diapiric withdrawal, hypothetical fault steps in the top of the basement could have buttressed seaward translation, thereby compressing the Canopy Zone. As diapiric stems were pinched by adjoining thrusting and folding, salt was squeezed upward, enhancing the growth of allochthonous salt sheets. These sheets coalesced to form canopies then inflated after burial. Contraction ended at ~5 Ma.

Acknowledgments We thank the Yemeni Ministry of Oil and Mineral Resources and BP Exploration Operating Company (Yemen), Ltd., for permission to publish the seismic and well data. Peter Shiner, Ernie Robertson, Dave Barr, and Trevor Witton carried out much of the seismic interpretation, and Jake Hossack did the LOCACE reconstruction. Illustrations were drawn by Tari Weaver and Jana S. Robinson, under the direction of Richard L. Dillon (BEG Cartography) from original figures prepared by BP Exploration's XFI Cartographic Office for the presentation of this paper in Bath. The original manuscript was stylistically improved by Amanda Masterson. Ian Davison provided invaluable refereeing comments and also supplied a preprint of his 1994a article. This paper is published by permission of the Director, Bureau of Economic Geology.

REFERENCES CITED

Abouzakhm, A. G., R. B., Allen, and A. H. Sikander, 1991, Characteristics of the Bab Al Mandab–northern Afar area of the southern Red Sea: AAPG Bulletin, v. 75, p. 1400.

Barnard, P. C., 1992, Thermal maturity and source rock occurrence in the Red Sea and Gulf of Aden: Journal of Petroleum Geology, v. 15, p. 173–186.

Cobbold, P. R., and P. Szatmari, 1991, Radial gravitational gliding on passive margins: Tectonophysics, v. 187, p. 249–289.

Cobbold, P. R., P. Szatmari, L. S. Demercian, D. Coelho, and E. A. Rossello, 1995, Seismic and experimental evidence for thin-skinned horizontal shortening by convergent radial gliding on evaporites, deep water Santos Basin, Brazil, *in* M. P. A. Jackson, D. G. Roberts, and S. Snelson, eds., Salt tectonics: a global perspective: AAPG Memoir 65, this volume.

Cochran, J. R., 1983, A model for the development of the Red Sea: AAPG Bulletin, v. 67, p. 41–69.

Crossley, R., C. Watkins, M. Raven, D. Cripps, A. Carnell, and D. Williams, 1992, The sedimentary evolution of the Red Sea and Gulf of Aden: Journal of Petroleum Geology, v. 15, p. 157–172.

Davison, I., S. Al-Kadasi, A. K. Al-Subbary, J. Baker, S. Blakey, D. Bosence, C. Dart, R. Heaton, K. McClay, M. Menzies, G. Nichols, L. Owen, and A. Yelland, 1994a, Geological evolution of the southeastern Red Sea Rift margin, Republic of Yemen: GSA Bulletin, v. 106, p. 1474–1493.

Davison, I., D. Bosence, I. Alsop, and M. Al-Aawah, 1994b, Deformation around exhumed Miocene salt diapirs Al Salif and Jabal al Milh, Tihama Plain, NW Aden, *in* I. Alsop, D. Blundell, and I. Davison, convenors, Salt tectonics: Geological Society of London Meeting, Sept. 14–15, Programme and Abstracts, p. 49–50.

El Anbaawy, M. I. H., M. A. H. Al-Aawah, K. A. Al-Thour, and M. Tucker, 1992, Miocene evaporites of the Red Sea Rift, Yemen Republic: sedimentology of the Salif halite: Sedimentary Geology, v. 81, p. 61–71.

Fletcher, R. C., M. R. Hudec, and I. Watson, 1995, Salt glacier and composite sediment–salt glacier models for the emplacement and early burial of allochthonous salt sheets, *in* M. P. A. Jackson, D. G. Roberts, and S. Snelson, eds., Salt tectonics: a global perspective: AAPG Memoir 65, this volume.

Girdler, R. W., and M. Underwood, 1985, The evolution of early oceanic lithosphere in the southern Red Sea: Tectonophysics, v. 116, p. 95–108.

Hughes, G. W., and Z. R. Beydoun, 1992, The Red Sea–Gulf of Aden: biostratigraphy, lithostratigraphy, and palaeoenvironments: Journal of Petroleum Geology, v. 15, p. 135–156.

Jackson, M. P. A., 1982, Fault tectonics of the East Texas basin: The University of Texas at Austin, Bureau of Economic Geology Geological Circular 82-4, 31 p.

Jackson, M. P. A., and C. J. Talbot, 1991, A glossary of salt tectonics: The University of Texas at Austin, Bureau of Economic Geology Geological Circular 91-4, 44 p.

Jackson, M. P. A., and B. C. Vendeville, 1994, Regional extension as a geologic trigger for diapirism: GSA Bulletin, v. 106, p. 57–73.

Jackson, M. P. A., and B. C. Vendeville, 1995, Origin of mini-basins by multidirectional extension above a spreading lobe of allochthonous salt (abs.): AAPG International Conference, Nice, Official Program, p. 33A.

Jackson, M. P. A., R. R. Cornelius, C. H. Craig, A. Gansser, J. Stöcklin, and C. J. Talbot, 1990, Salt diapirs of the Great Kavir, central Iran: GSA Memoir 177, 139 p.

Makris, J., C. H. Henke, F. Eglof, and T. Akamaluk, 1991, The gravity field of the Red Sea and East Africa, *in* J. Makris and R. Rihm, eds., Red Sea—birth and early history of a new oceanic basin: Tectonophysics, v. 198, p. 369–381.

Menzies, M. A., J. Baker, D. Bosence, C. Dart, I. Davison, A. Huford, M. Al Kadasi, K. McClay, G. Nichols, A. Al Subbary, and A. Yelland, 1992, The timing of crustal magmatism, uplift, and crustal extension—preliminary observations from Yemen, *in* B. C. Storey, T. Alabaster, and R. J. Pankhurst, eds., Magmatism and the causes of continental break-up: Geological Society of London, Special Publication 68, p. 293–304.

Mitchell, D. J. W., R. B. Allen, W. Salama, and A. Abouzakm, 1992, Tectonostratigraphic framework and hydrocarbon potential of the Red Sea: Journal of Petroleum Geology, v. 15, p. 187–210.

Sultan, M., R. Becker, R. E. Arvidson, P. Shore, R. J. Stern, Z. El Alfy, and E. A. Guinness, 1992, Nature of the Red Sea crust: a controversy revisited: Geology, v. 20, p. 593–596.

Talbot, C. J., 1979, Fold trains in a glacier of salt in southern Iran: Journal of Structural Geology, v. 1, p. 5–18.

Talbot, C. J., 1995, Molding of salt diapirs by stiff overburden, *in* M. P. A. Jackson, D. G. Roberts, and S. Snelson, eds., Salt tectonics: a global perspective: AAPG Memoir 65, this volume.

van Keken, P. E., C. J. Spiers, A. P. van den Berg, and E. J. Muyzert, 1993, The effective viscosity of rocksalt: implementation of steady-state creep laws in numerical models of salt diapirism: Tectonophysics, v. 225, p. 457–476.

de Ruig, M. J., 1995, Extensional diapirism in the eastern Prebetic foldbelt, southeastern Spain, *in* M. P. A. Jackson, D. G. Roberts, and S. Snelson, eds., Salt tectonics: a global perspective: AAPG Memoir 65, p. 353–367.

Chapter 17

Extensional Diapirism in the Eastern Prebetic Foldbelt, Southeastern Spain

Menno J. de Ruig

Shell U.K. Exploration and Production
London, U.K.

Abstract

The influence of tectonic stress on the initiation and development of evaporite diapirs is of great importance in the interpretation of diapiric structures and associated sediments. The eastern Prebetic foldbelt in southeastern Spain contains many diapirs that provide excellent examples of tectonically controlled diapirism. These diapirs are mainly composed of Triassic evaporite and shale that have pierced their overburden along extensional faults and in releasing oversteps along strike-slip faults. They occur as highly elongated diapiric walls or as large, fault-bounded bodies that can cover several tens of square kilometers at the center of grabens. Most of these diapirs reached the surface in Neogene times, constituting local depocenters for Miocene sediments. Outcrop and well data suggest that their source layer consists mainly of interbedded shale and anhydrite, which are denser than their carbonate overburden and thus preclude piercement diapirism driven by buoyancy. The external geometry of these diapirs and a variety of kinematic indicators in surrounding overburden suggest that their location and initiation was primarily controlled by (trans)tensional faulting. It is therefore concluded that the Prebetic diapirs formed in response to thin-skinned extension of their overburden, which induced differential loading and viscous flow of the Triassic evaporites. Regional paleostress analysis and chronostratigraphic correlation suggest that diapirism was triggered by rifting in the adjacent Western Mediterranean Basin.

INTRODUCTION

Evaporite diapirs occur in a wide range of tectonic settings, and various mechanisms have been proposed to explain their formation. Diapirism is commonly attributed to the buoyant rise of low-density evaporites through denser overburden, but several studies have shown that the buoyancy model often fails to explain the initiation, shape, and distribution of diapirs (e.g., Bishop, 1978; Jackson and Talbot, 1986; Vendeville and Jackson, 1992). Differential loading or other mechanisms inducing differential stress are therefore thought to be a prerequisite for the formation of diapirs (e.g., Jenyon, 1986). Diapirism can also be initiated or accelerated by tectonic forces that stretch, wrench, or compress sedimentary basins (Jackson and Talbot, 1986; Stephenson et al., 1992; Koyi et al., 1993; Jackson and Vendeville, 1994). The influence of tectonic stress on evaporite diapirism is of great importance for the interpretation of diapirs in deformed continental margins. In such a setting, their emplacement can be related to both tensile stresses during rifting and compressive stresses during the postrift evolution.

The eastern Prebetic foldbelt (Figure 1) contains many diapirs composed mainly of Triassic evaporite and mudstone. The concept of diapirism has been used to explain a variety of stratigraphic and structural complications in this area. Diapirism is believed to be responsible for lateral changes in sediment thickness and facies (e.g., Leclerc and Azéma, 1976) and for folds and faults that deviate from the regional structural trend (e.g., Rodriguez Estrella, 1977; Moseley et al., 1981). In some studies, diapirism is seen as a process that was essentially independent from the regional tectonic forces (Polvêche, 1961; Rodriguez Estrella, 1977), although most authors recognize that the process is tectonically controlled. Moseley et al. (1981) suggested that some Prebetic diapirs have risen along conjugate sets of wrench faults, which were formed during regional NNW-SSE compression. Navarro et al. (1960) explained the formation of the diapir of Altea as a result of isostatic readjustment following the Pyrenean tectonic phase, whereas further diapiric rise was accelerated by the Betic fold phase. According to Rodriguez Estrella (1977) and Azéma (1977), the location of the diapirs is controlled by basement faults. García-Rodrigo (1960, 1965) and Rondeel and Van de Gaag (1986) suggested that the diapiric rise of low-density Triassic evaporites was facilitated by extensional faults.

Evidently, there is little agreement about if and how the diapirs were controlled by regional tectonics. Yet the tectonic control and emplacement mechanism of these diapirs are crucial to understand the tectonic evolution of the eastern Prebetic. Furthermore, the good exposure and

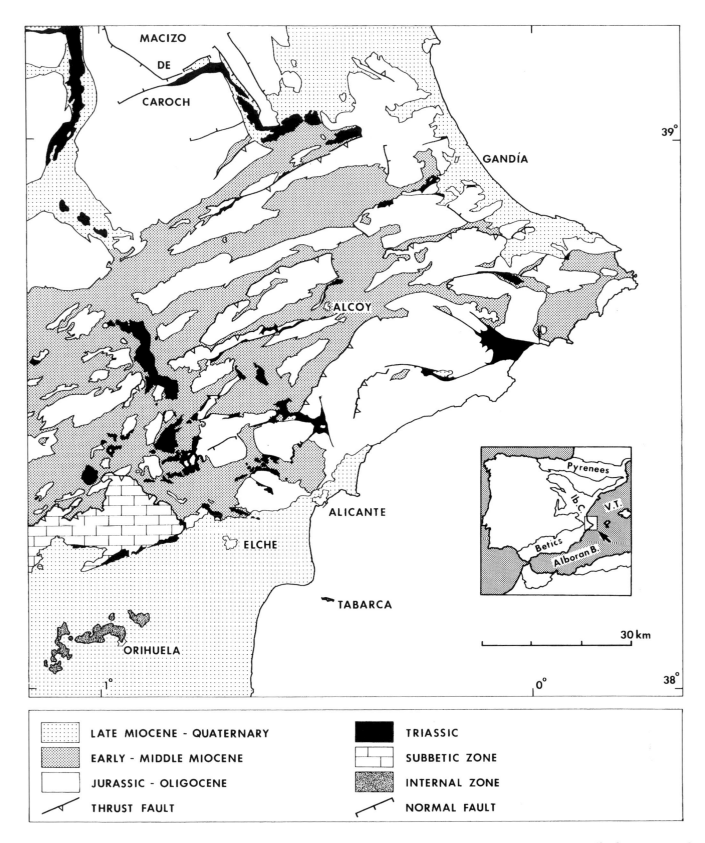

Figure 1—Simplified geologic map of the eastern Prebetic fold belt. The majority of Triassic outcrops are diapirs composed of Middle–Upper Triassic evaporites and shales. Inset is tectonic sketch map of the Iberian Peninsula, showing Alpine orogens (Ib.C. = Iberian chain) and Mediterranean basins (V.T. = Valencia trough); arrow points to eastern Prebetic promontory, enlarged on the main map.

accessibility of the Prebetic diapirs make them an excellent example for studying the relationship between faulting and diapir evolution in foldbelts. In this chapter, the distribution, structure, and age of the Prebetic diapirs is discussed. It is argued that diapirism was initiated by thin-skinned extensional faulting and by the resulting differential loading of the ductile Triassic rocks.

TECTONIC SETTING

The diapirs discussed here are in the eastern foreland of the Betic Cordilleras in southern Spain (Figure 1). The Betic orogen is traditionally subdivided into an Internal Zone in the south and an External Zone in the north. The Internal Zone consists of an allochthonous stack of thrust sheets composed mainly of Triassic and older metamorphic rocks (Egeler and Simon, 1969; Torres-Roldán, 1979; De Jong, 1990). The External Zone represents the foreland fold and thrust belt of the orogen and consists of a deformed sedimentary wedge of Mesozoic and Cenozoic carbonates that is detached from its Paleozoic basement along Triassic evaporites and shales. On the basis of both tectonic and stratigraphic criteria, the External Zone is subdivided into a Prebetic zone in the north, characterized by shallow-marine sediments, and a Subbetic zone in the south, where pelagic facies prevail. This facies distribution is thought to reflect the Mesozoic–Paleogene paleogeography of the former rifted continental margin of southern Iberia, in which the Prebetic zone represents the former shelf and the Subbetic the former basinal area (Azéma et al., 1979; García-Hernandez et al., 1980). During the Miocene orogeny, the External Zone developed into the present foreland fold and thrust belt as a result of the overthrusting of the Internal Zone onto the southern Iberian margin.

The eastern part of the Prebetic foldbelt, or Prebetic of Alicante (De Ruig, 1992), is located at the junction of several major tectonic units of the Western Mediterranean (Figure 1). Here, the Betic orogen joins the southern end of the Iberian chain, a Mesozoic intracontinental rift basin that was inverted in Oligocene–Miocene time. Offshore, the promontory of Alicante is bounded by the extensional basin margins of the Valencia Trough and the Alboran–North Algerian Basin. Crustal extension and the local formation of oceanic crust in these basins were largely coeval with crustal shortening in the adjacent fold and thrust belts (Rehault et al., 1984; Torres et al., 1993). This interaction between compressional and tensional stresses is also evident in the geologic record of the eastern Prebetic. Late Oligocene–early Miocene rifting in the Western Mediterranean Basin affected areas well within the Iberian plate, as demonstrated by onshore rift grabens and attenuated crust along the eastern margin of Iberia, including the eastern Prebetic (Zeyen et al., 1985). Within the eastern Prebetic, these rift structures have been severely overprinted by subsequent early–middle Miocene compressional and transpressional deformation.

REGIONAL STRUCTURE AND DISTRIBUTION OF DIAPIRS

The large-scale structure of the Prebetic of Alicante can be characterized as a deformed sedimentary wedge overlying thinned continental crust. The present-day thickness of the wedge varies from 2 km in the north to about 6 km in the south. Its basal detachment is probably in a relatively thin unit of evaporite and clay and interbedded carbonate directly below Middle–Upper Triassic Muschelkalk carbonates (Simon, 1987). The latter represents the lowest stratigraphic level exposed (Figure 2).

The structure of the post-Triassic cover is dominated by a series of east-northeast trending Betic folds and thrusts formed during several phases of compression in the early–middle Miocene (de Ruig et al., 1987; Ott d'Estevou et al., 1988). The anticlines are mainly composed of Mesozoic and Paleogene carbonates and supposedly cored by Triassic evaporites. They are separated by broad synclinal basins filled by synorogenic Miocene marl and limestone. The degree of deformation and structural complexity decreases from south to north. Particularly in the southern Prebetic, folding is strongly disharmonic, and local deviations from the regional fold trend occur along deep-seated wrench faults (de Ruig et al., 1987; de Ruig, 1990).

Toward the north, the Prebetic folds become more open and abut against a platform that is essentially unaffected by folding. This area is known as the Valencian domain (Baena Perez and Jerez Mir, 1982) or Macizo de Caroch (Figure 1) and is composed of a subhorizontal platform of Jurassic and Cretaceous sediments. Two nearly orthogonal sets of normal faults, striking east-northeast and north-northwest, divide the platform into rectangular blocks in a horst and graben geometry. The throw along these faults varies from several tens of meters to more than 700 m. The larger faults form the boundaries of narrow, elongate grabens that have been intruded by Triassic rocks (e.g., Triassic of Ayora-Cofrentes, Bicorp-Quesa, and Navarrés) (Figure 3). The association of steep, rectilinear normal faults with Triassic rocks has resulted in rectangular outcrop patterns, such as the Z-shaped diapir of Navarrés (Figure 4). Normal faulting here is thought to be related to early Neogene rifting in the adjacent Valencia Trough (Roca and Guimerà, 1992).

Figure 4 shows a balanced cross section through the graben of Bicorp-Quesa, which illustrates the typical geometry of the so-called diapiric Triassic rift valleys (Moissenet, 1989) of the Macizo de Caroch. The diapirs are flanked by half-grabens filled with Miocene continental deposits. These graben fills overlie tilted fault blocks of Jurassic–Cretaceous strata and generally show an open synclinal structure. Near the contact with Triassic rocks, the Miocene strata are commonly vertical to overturned. Near Cofrentes, the Triassic and Miocene rocks have been intruded by Pliocene peralkaline dikes associated with lava flows and small cinder cones. These rocks are basanites and nephelenite basalts and yield K/Ar ages of about 2.2 Ma (Sáenz and López, 1975). They form part of a

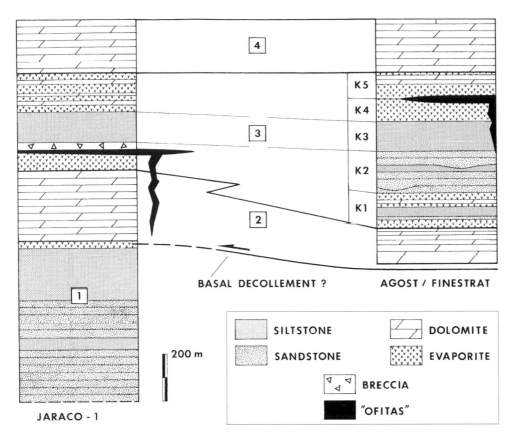

Figure 2—Schematic Triassic–Lower Jurassic stratigraphy of the Jaraco-1 well (Lanaja et al., 1987) and the Agost and Finestrat diapirs. Legend: (1) Lower–Middle Triassic Buntsandstein red beds, (2) Middle–Upper Triassic Muschelkalk carbonate, (3) Upper Triassic Keuper diapiric sequence (K1–K5 refer to formations discussed in the text), (4) Lower Jurassic dolomite.

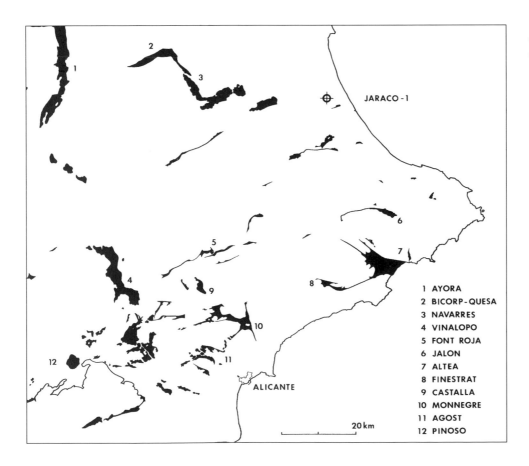

Figure 3—Distribution of Triassic diapiric outcrops in the eastern Prebetic fold belt. Numbers 1–12 give names and locations of major diapirs referred to in the text.

1 AYORA
2 BICORP-QUESA
3 NAVARRES
4 VINALOPO
5 FONT ROJA
6 JALON
7 ALTEA
8 FINESTRAT
9 CASTALLA
10 MONNEGRE
11 AGOST
12 PINOSO

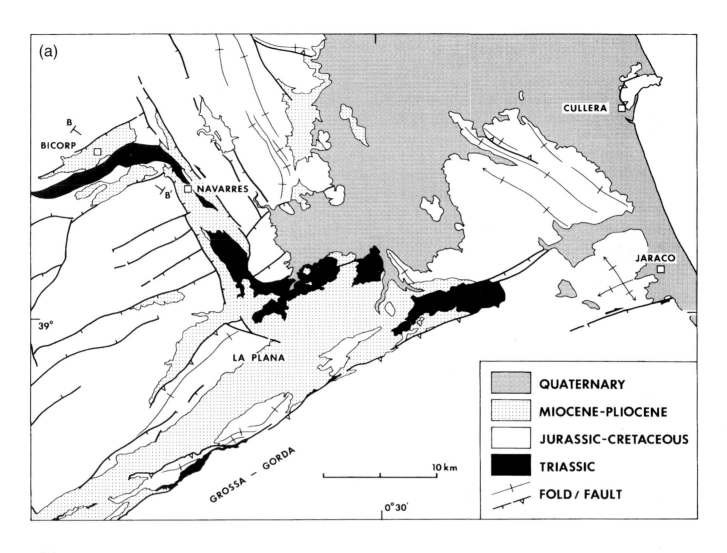

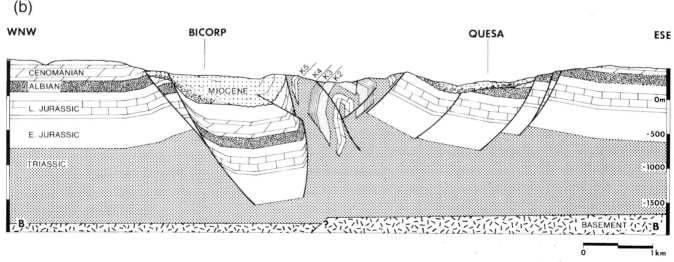

Figure 4—(a) Simplified geologic map of the Navarrés diapir and surroundings and (b) balanced cross section B–B' through the Bicorp-Quesa branch of the diapir (location of section shown on map). In the Triassic diapiric anticline, K2–K5 denote the main formations of Keuper stratigraphy (see text). Extension of the post-Triassic cover amounts to 2.1 km (local extension factor of 29%).

larger volcanic province that extends eastward to the Valencia Trough (Martí et al., 1992).

Southward in the Prebetic of Alicante, folds and thrusts are cut by complex fault zones associated with large diapirs of Triassic evaporite and shale similar to those of the Macizo de Caroch (Figure 3). These diapirs typically form topographic depressions. They occur as either discontinuous planar bodies along high-angle faults or large, fault-bounded bodies that can cover several tens of square kilometers. The latter are generally elongate in plan view, and their longest axis is typically perpendicular to the local fold trend.

INTERNAL STRUCTURE AND STRATIGRAPHY

The diapirs of the Prebetic zone are composed mainly of Upper Triassic continental deposits of the Keuper facies (Ortí Cabo , 1973). Although complex deformation commonly hampers stratigraphic analysis, the Keuper deposits can be confidently subdivided into five formations (Figure 2). The basal Keuper is known as the Jarafuel Formation (K1 on Figure 2) (Ortí Cabo, 1973), which is a sequence of gypsiferous mudstone containing many intercalations of sandstone and lacustrine carbonates. This sequence is covered by a thick series of red or green floodplain mudstones and siltstones containing intercalations of fine-grained fluvial sandstones (Manuel and Cofrentes formations; K2 and K3). The thick gypsum and mudstone sequences of the overlying Quesa Formation (K4) are commonly exposed as homogeneous masses rich in small red euhedral quartz crystals. Tholeiitic diabase sills ("ofitas") of presumed Jurassic age are common in this formation and more widespread than the local Pliocene intrusions at Cofrentes. The upper part of the Keuper sequence is built up of white and black marbled gypsum and thinly bedded dolomite interbeds, indicating a marine influx (Ayora Formation, K5). In some of the diapirs, tectonically isolated slabs of older Middle–Upper Triassic Muschelkalk carbonates are found, which stratigraphically precede the Keuper series.

Significantly, almost all Triassic evaporites found are sulfates. Salt is absent in outcrop and in most exploration wells in the region. In the wells that did encounter salt, halite is a minor constituent, rarely exceeding 50 m in thickness (Lanaja et al., 1987; Bartrina et al., 1990). On the basis of well data and composite field sections, the total original stratigraphic thickness of the diapiric sequence is estimated to vary between 500 and 1000 m.

The internal structure of the diapirs usually appears chaotic and shows evidence for polyphase deformation. Most Triassic outcrops are composed of what Moseley et al. (1981, p. 248) described as "large-scale melange structures of gypsiferous ferruginous mudstone." These melange structures usually line major fault planes either within the diapirs or at their margins. The bright red and green gypsiferous mudstones commonly show a coarse, vertical to steeply dipping tectonic banding in which layers of gypsum and mudstone alternate with breccias containing fragments of all other Triassic rock types (mainly dolomite and sandstone). Some fragments are composed of fine-grained, banded gypsum containing flow folds, which indicate an earlier stage of ductile deformation. Cap rock usually forms small, isolated outcrops crowning small hills within the diapirs. The outcrops consist of a largely homogeneous mass of gypsum and red mudstone having residual insoluble matter and fragments of the more brittle Triassic rocks. The cap rock commonly has a crude, subhorizontal, meter-scale pedogenic layering that crosscuts the bedding. Gypsum here commonly forms "swallow-tail" crystals that indicate neocrystallization.

Most of the larger diapirs contain isolated fault blocks of strongly tilted to overturned Upper Cretaceous or lower Tertiary limestones; Jurassic rocks are notably absent in the diapirs. The fault blocks vary in size from several hundred meters to over 1 km and are generally found near the margins of the diapirs. Because the fault blocks consist of the younger levels of the overburden, they cannot have been dragged upward from the subsurface by the rising diapir. Instead, they appear to be downthrown along normal faults surrounding the diapir or have been sheared off its margins by strike-slip movement. An example of the latter is in the Finestrat diapir. Close to its northern boundary, an 800-m-long lensoid slab of steeply dipping Cenomanian–Senonian limestone is exposed. This limestone raft cannot have moved upward from the subsurface with the rising diapir because the adjacent country rock is of Aptian–Albian age. Instead, the raft has probably been emplaced laterally by drag along a major dextral strike-slip fault that constitutes the northern margin of the diapir (Figure 5). Elsewhere, the fault blocks may represent remnants of a tectonically thinned roof, shouldered aside by the diapir in the late stages of piercement. This mechanism has been clearly illustrated by the experiments of Vendeville and Jackson (1992), where a reactive diapir rises along extensional faults then actively pierces its thinned overburden, rotating and dragging small blocks next to the emergent diapir.

There are several indications that the chaotic nature of the diapirs is the result of deformation of originally more coherent diapiric folds. In areas that have only been weakly deformed during Neogene shortening, such as the Macizo de Caroch, the diapirs still clearly show an anticlinal internal structure in cross section (Figure 4). According to Holzhaus (1989), the structure of the Ayora diapir can be characterized as a mushroom fold (Jackson and Talbot, 1989). The large slabs of overturned Triassic strata in some of the larger diapirs of the Internal Prebetic (e.g., Agost and Finestrat) may be remnants of comparable structures (Figure 5). An extreme example of the latter is found in the Triassic of Agost, where gently dipping sandstone and siltstone beds cover several square kilometers. Numerous younging criteria (e.g., cross-bedding and small channels) demonstrate that these beds are completely upside down (De Haan and Van der Wal, 1985), locally forming antiformal synclines and synformal anticlines.

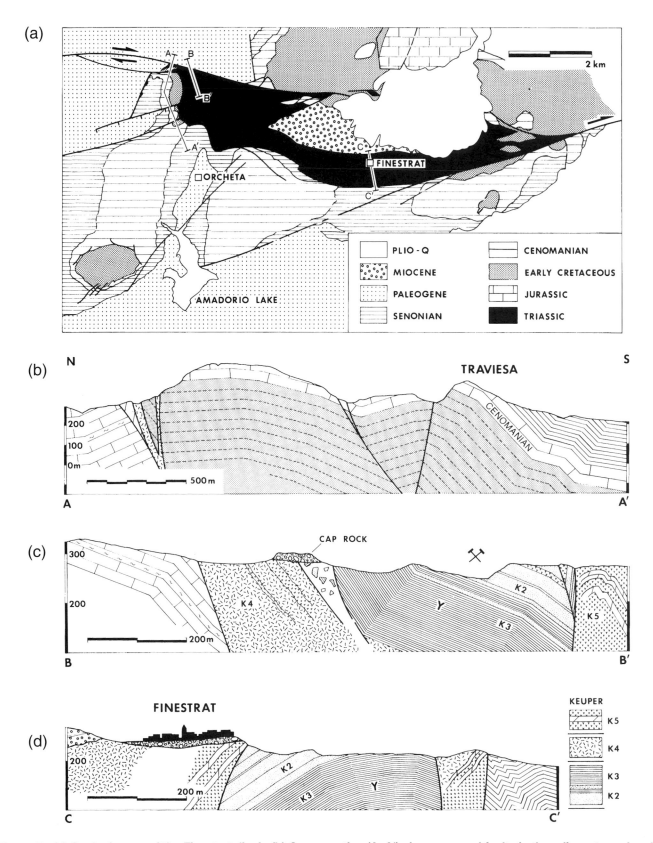

Figure 5—(a) Geologic map of the Finestrat diapir. (b) Cross section (A–A') shows normal faults in the adjacent overburden that are genetically related to the formation of the diapir. (c) and (d) Two cross sections (B–B', C–C' through part of the diapir show overturned slabs of Triassic rocks that are interpreted as remnants of mushroom folds (Y denotes inverted younging direction). The external shape of this diapir and its location between two overlapping dextral strike-slip faults suggest it has intruded a pull-apart structure.

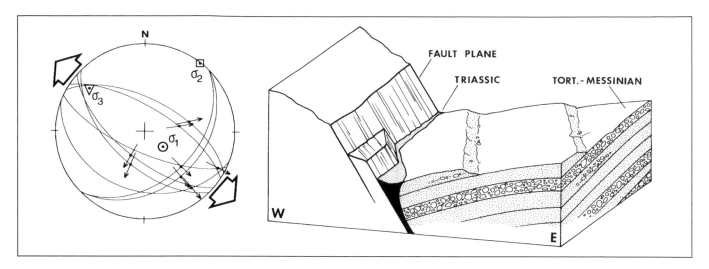

Figure 6—Schematic block diagram of a large normal fault injected by Triassic evaporites and mudstones. The fault juxtaposes a footwall of Eocene carbonates against Tortonian–Messinian red beds along the Font Rotja diapiric trend. Diagram on left shows the stress tensor inferred from fault-slip analysis (Schmidt lower hemisphere projection), indicating northwest-southeast directed tension, approximately perpendicular to the strike of the main fault (Font Rotja stress tensor in Table 1).

The domains of inverted strata are commonly bounded by steeply dipping fault zones containing banded gypsum and gypsum breccias. The fault zones are usually parallel or subparallel to the main trend of the diapir or to the nearest contact with the post-Triassic cover (also see Moseley et al., 1981). The comparison with less disturbed diapiric anticlines of the Valencian domain suggests that the domains of inverted strata represent the lower limbs of mushroom folds, which have subsequently been faulted and intruded by the gypsiferous mudstones.

On the basis of these observations, the movement history of the diapir can be divided into at least two stages. First, Triassic rocks rose diapirically by ductile flow, forming large anticlinal structures and mushroom folds. Ductile flow is indicated by the common occurrence of flow-folded gypsum presently found as fragments in gypsum breccias. Also, the negligible amount of deformation of the overturned sandstones (often thin, brittle layers) suggests that they were cushioned by the surrounding ductile evaporites during relatively slow diapiric rise. The second stage in the structural evolution of the diapirs involved brittle faulting, as indicated by the brecciation of banded gypsum and truncation of older fold structures. During this stage, the diapirs were probably at or near the surface.

EXTERNAL STRUCTURE AND TECTONIC CONTROLS

The external shape and piercement contact of a diapir and its relationship to the adjacent cover are important indicators of the tectonic controls on diapir emplacement. Although it is obvious that the Triassic rocks have moved relatively upward, the diapirs are generally not associated with reverse faults. On the contrary, the external geometry of the diapirs and a variety of offset criteria in the surrounding cover (e.g., slickensides and drag) indicate that most Triassic diapirs have intruded along extensional and transtensional faults.

The boundary faults or fault zones are commonly steep to subvertical and truncate all other structures. The fabric of these faults mainly depends on the local rock type of the adjacent cover. Competent carbonates are brecciated in places but usually form discrete fault planes. Such fault planes commonly have mineralized slickenside surfaces that indicate the slip direction (Figure 6). Marls or other incompetent rocks in contact with the Triassic are strongly sheared up to 50 m from the contact. In plan view, the diapir/country rock contact is either linear, where major faults form the boundary, or sinuous and highly irregular. Based on their external shape and boundary faults, the diapirs can be subdivided roughly for descriptive purposes into three types: graben, pull-apart, and fault injection. The distinction among these diapir geometries is loose, and hybrid forms occur.

Graben-type diapirs occupy the center of Miocene grabens and are generally 2–8 km wide. Examples of graben-type diapirs are the Triassic outcrops of the Rio Vinalopó, the Macizo de Caroch, and Castalla (Figures 4, 7). They are bounded by north-northwest and east-northeast striking normal or oblique-slip faults, which are typically perpendicular to the local fold trend. The boundary faults are commonly associated with small intrusions of red Triassic evaporite and mudstone. However, larger diapiric outcrops are in the center of the grabens, surrounded by steeply dipping to vertical

Table 1—Paleostress Tensors from Diapir Fault-Slip Data, Prebetic Diapirs

Diapir	n [a]	σ_1 [b]	σ_2 [b]	σ_3 [b]	ϕ [c]	$\Delta(°)$ [d]	Rock Age
Navarrés	7	287/70	111/19	021/01	0.28	13.8	Triassic–Miocene
Castalla	10	050/79	312/02	222/11	0.53	11.1	Miocene
Finestrat	11	124/78	274/10	005/06	0.85	14.9	Cretaceous
Font Rotja	8	131/68	038/01	308/22	0.23	14.1	Triassic–Miocene
Jalon	10	162/79	321/11	052/04	0.51	11.8	Cretaceous

[a] n = number of faults defining the tensor.

[b] Azimuth/dip of principal stress axes.

[c] Stress ratio $\phi = (\sigma_2 - \sigma_3)/(\sigma_1 - \sigma_3)$ defines the shape of the stress ellipsoid.

[d] Δ = average angle between measured fault striations and theoretical shear stress.

Miocene deposits. Away from the central diapir, these deposits rapidly assume shallow to subhorizontal dips, largely concealing the internal structure of the graben.

Pull-apart diapirs have a typical rhomb-shaped geometry and are bounded by strike-slip faults at their northern and southern boundaries (Figure 5). They are mainly found in the southern part of the Prebetic of Alicante (Jijona fold zone of de Ruig et al., 1987; de Ruig, 1990), such as the diapirs of Finestrat, Jalon, and part of Monnegre. These diapirs are interpreted to have intruded pull-apart basins at right-lateral oversteps between dextral strike-slip faults that strike east-northeast or east-southeast (see Moseley et al., 1981).

Diapiric fault-plane injections form narrow (0.1–100 m), discontinuous strings of gypsiferous mudstone that have intruded high-angle faults. They occur along a variety of structural trends, but predominate along major east-northeast striking faults, commonly subparallel to the axial plane of large east-northeast trending folds. Slickensides and other slip criteria show diverse directions, indicating a complex movement history. For individual faults, it is often impossible to determine when the diapiric injections were emplaced. Based on regional data and comparison with the larger diapirs just described, it is thought that transtensional movement facilitated the emplacement of Triassic evaporites. This interpretation is supported by the fact that thrusts and reverse faults having the same strike but lesser dip are generally not associated with diapiric injections, suggesting that such compressional faults hamper diapiric rise.

These observations strongly suggest that the external shape of the Prebetic diapirs is controlled by extensional faults. Furthermore, section balancing of structures such as the Navarrés diapir indicates that the diapirs are located where there has been significant net extension of the overburden (e.g., Figure 4). This interpretation is also supported by paleostress determinations (de Ruig, 1992). Fault-slip data derived from striated fault planes were collected from several diapiric zones and analyzed using a computer-based inversion process described by Michael (1984). The fault populations from these diapir boundary zones generally yield tensile stress tensors with a subhorizontal σ_3 axis and a subvertical σ_1 axis (Table 1). The long axes of these diapirs were perpendicular to the local direction of σ_3 at the time of their emplacement.

EMPLACEMENT HISTORY AND AGE OF DIAPIRISM

Structural and stratigraphic evidence indicates that the diapirs in the Prebetic of Alicante were emplaced episodically. Most diapirs reached the surface in Neogene time, but early piercement diapirism is generally thought to have started in Cretaceous time because Cretaceous and Paleogene sediments contain reworked Triassic rock fragments (Foucault 1966; Leclerc, 1971; Rodriguez Estrella, 1977). According to Leclerc and Azéma (1976), small amounts of Triassic mudstone, gypsum, and quartz crystals have been redeposited in the upper Albian–Cenomanian limestones north of Agost. Wilke (1988) reported small, red, idiomorphic quartz crystals in upper Berriasian–lower Valanginian deposits near Castalla, which are derived from Upper Triassic rocks. Similar observations were made by Rodriguez Estrella (1977), Llavador et al. (1983), and Geel et al. (1992). However, all these observations concern small amounts of reworked Triassic rock, sometimes limited to a single quartz crystal. There is no evidence that large diapiric bodies surfaced during Cretaceous or Paleogene time. Leclerc and Azéma (1976) suggested that the depositional hiatuses, thickness variations, and mass flows in the Upper Cretaceous sediments north of Agost are related to the diapiric rise of Triassic evaporites, which generated local submarine highs with unstable slopes. Similar suggestions were made to explain the thickness variations and unconformities in Upper Cretaceous and Paleogene series near the diapir of Finestrat (Polvêche, 1961) and Cabezón de Oro (Rodriguez Estrella, 1977). However, analysis of the Upper Cretaceous–Paleogene stratigraphy has shown that the unconformities and abrupt thickness variations result from regional block faulting rather than from local diapirism (de Ruig et al., 1991; de Ruig, 1992). Although the Triassic evaporites may have been mobilized during these tectonic phases, there is no direct causal relationship between the pattern of thickness variations in Cretaceous or Paleogene deposits and the present-day distribution of large Triassic outcrops.

The occurrence of reworked Triassic rock fragments in pre-Neogene sediments appears to correlate with periods of tensional stress, such as the Cenomanian and Early

Cretaceous rifting phases (see de Ruig, 1992). It is therefore suggested that the reworked Triassic rock fragments were injected along growth faults that formed during extension. Larger Triassic rock masses might have reached higher stratigraphic levels (e.g., Aptian–Albian) as diapiric sills or dikes but did not reach the surface.

The oldest sediments known to unconformably overlie large Triassic evaporite diapirs are of early Neogene age. In the Macizo de Caroch, the Bicorp-Quesa and Navarrés diapirs are flanked and partially covered by alluvial clastic deposits and lacustrine limestones, marls, and organic-rich shales. Most of these sediments are of middle–late Miocene age, but the base of the series contains early Burdigalian microvertebrate remains (Santisteban et al., 1989). In the southern Prebetic of Alicante, Triassic rocks are locally covered by thick anoxic calcarenites, which appear to be spatially associated with the graben-type diapirs. Near Tibi at the junction of the Monnegre and Castalla diapirs, a 600-m-thick sequence of lower Miocene hydrogen-sulfide-bearing calcarenites unconformably overlies Triassic rocks. The base of the sequence contains reworked fragments of mudstone, gypsum, and idiomorphic quartz crystals. Toward the edge of the diapir, the sulfurous limestones laterally grade into shallower marine coralline algal limestones, which rapidly thin toward the faulted boundaries of the diapir.

The facies distribution of the Miocene limestones thus mimics the outline of the diapir and indicates that the diapir formed a small, fault-bounded basin in which a thick sequence of calcarenite could accumulate while the boundaries of the basin were fringed by coralline algal reefs. The lower part of the Miocene sequence at Tibi has a $^{87}Sr/^{86}Sr$ age of about 23 Ma (Aquitanian) (Beets and de Ruig, 1992). The youngest pre-Neogene rocks pierced by the diapir are of Eocene–Oligocene age. These data indicate that the first large graben-type diapirs, such as Monnegre and Navarrés, surfaced in uppermost Oligocene–earliest Miocene time. The diapiric graben fill of sulfurous limestones and coralline algal reefs at Monnegre is strongly similar to the anoxic Casablanca limestones described by Soler et al. (1983), which constitute the early synrift graben fill in the Valencia Trough. This correspondence in tectonic setting, stress regime, and sedimentary facies suggests that the lower Miocene Prebetic diapiric grabens formed in response to a phase of regional extension induced by rifting in the adjacent Valencia Trough.

The middle Miocene sedimentary record shows evidence of repeated burial and resurfacing of Triassic rocks, indicating that diapirism continued during subsequent orogenic shortening. Reworked Triassic rock fragments occur in virtually all Miocene stratigraphic units, and the larger diapirs in the Prebetic of Alicante are covered by a patchwork of sediments, ranging in age from early Miocene to Recent. The repeated surfacing of Triassic rocks was due to creation of new diapirs and reactivation of preexisting ones. Reactivation is well documented at the diapirs of Vinalopó, Monnegre, and Castalla. The lower Miocene sulfurous calcarenites that cover the Monnegre and Castalla diapirs are cut by north-northwest striking normal faults along which the Triassic has resurfaced (Figure 7). These faults are partially covered by middle Miocene marls. Along the southern edge of the Monnegre diapir, both the lower and middle Miocene deposits that overlie the diapir are cut by east-striking dextral strike-slip faults associated with diapiric intrusions. These faults are covered by an unconformable sequence of lower Tortonian marls and conglomerates.

Newly formed diapirs of middle–late Miocene age are generally related to strike-slip faults. Most of the pull-apart diapirs were probably formed during this stage because the oldest deposits covering the rhomb-shaped diapirs of Finestrat and Jalon are of late Miocene age (Cater, 1987; de Ruig et al., 1987). Although these diapirs are contemporaneous with regional contraction, their emplacement appears to be controlled by local extension between strike-slip faults.

EMPLACEMENT MECHANISM

Several different mechanisms have been proposed for the initiation and emplacement of the Prebetic diapirs. Most studies emphasize the assumed buoyancy of the Triassic rocks. Because of the interaction of Western Mediterranean rifting and Alpine orogeny, the eastern Prebetic foldbelt has been subjected to both extension and contraction during the Neogene epoch, and both have been advocated as the primary tectonic control on diapir shape and location.

Buoyant Rise

It is generally thought that the Triassic gypsiferous mudstones are of low density compared to the predominantly limestone overburden; the resulting buoyancy would favor diapiric rise (e.g., Azéma, 1977; Moseley et al., 1981). There are, however, strong indications that no significant density inversion existed when diapirism was initiated.

Most of the gypsum now in the Triassic outcrops was probably deposited as such (Murray, 1964). However, gypsum generally dehydrates to anhydrite below depths of about 800 m (MacDonald, 1953; Murray, 1964), assuming an open system with respect to fluids. Indeed, although most exposed Triassic evaporites comprise gypsum, the sole Triassic evaporite encountered in exploration wells below 800 m depth in the Prebetic of Alicante is anhydrite, except for the offshore Calpe-1 well (see Lanaja et al., 1987). The Upper Triassic evaporites reached this burial depth in latest Jurassic or Early Cretaceous time in most parts of the Prebetic of Alicante. Based on the regional sediment thickness distribution, most of the Triassic gypsum was probably converted to anhydrite by the middle Cretaceous. Marfil Pérez (1970) provided evidence that most of the gypsum found in outcrop is after

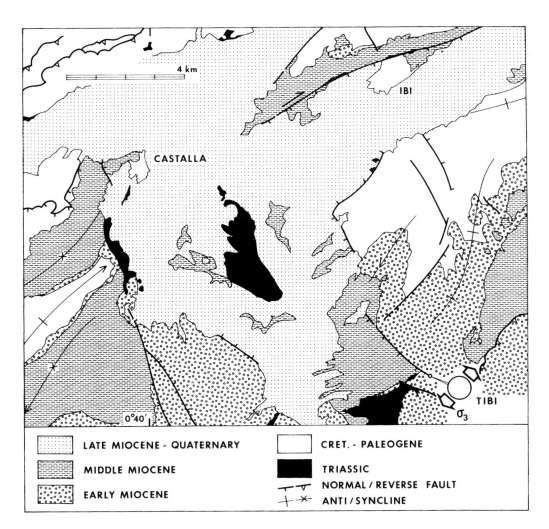

Figure 7—Simplified geologic map of the Castalla diapir. Note that the diapiric graben is roughly perpendicular to the local fold trend and to the direction of σ_3 near the town of Tibi (stress direction indicated by arrows) (see Table 1).

	LATE MIOCENE - QUATERNARY		CRET. - PALEOGENE
	MIDDLE MIOCENE		TRIASSIC
	EARLY MIOCENE	⊤⊽	NORMAL / REVERSE FAULT
		╬⁎	ANTI / SYNCLINE

anhydrite, because anhydrite is the main inclusion in idiomorphic quartz crystals embedded in the Triassic gypsum layers. Anhydrite is much denser (2900 kg/m³) than gypsum (2300 kg/m³) and even denser than most well-cemented carbonates (2600–2800 kg/m³). It is thus unlikely that a density inversion could have existed between the Triassic evaporites and their overburden. Consequently, the diapiric structure of the Triassic evaporites is primarily the result of their low viscosity, rather than their buoyancy.

Some anhydrite-bearing diapirs contain a core of rock salt, which is commonly not exposed in outcrop (e.g., Van Berkel, 1986). Buried rock salt is potentially buoyant because it is much less dense than most compacted sedimentary rocks. In the Jumilla region, directly west of the Prebetic of Alicante, rock salt is actively exploited at the surface in the Cerro de la Sal of Pinoso (Figure 3) and the La Rosa diapir of Jumilla (Mancheño Jiménez and Rodriguez Estrella, 1985; Rondeel and Van de Gaag, 1986). In contrast to the elongate depressions of the Alicante region, the diapirs of Jumilla form isolated, circular to elliptical domes rising as much as 300 m above the surrounding plain. This morphologic contrast is probably related to the presence of rock salt in Jumilla because the diapirs are in an otherwise identical geologic setting.

None of the Triassic outcrops in the Prebetic of Alicante contain rock salt, nor is it reported in most of the exploration wells (Lanaja et al., 1987). Even if one were to assume the presence of hidden or subsequently dissolved bodies of rock salt, which decrease the overall density of the Triassic series, the buoyancy mechanism does not explain the initiation and structure of the Prebetic diapirs. The Triassic evaporites in the Prebetic of Alicante did not start rising before Cretaceous time, and large-scale piercement diapirism was delayed until the Miocene epoch, about 185 m.y. after evaporite deposition, when the relatively dense carbonate cover was 2–4 km thick. Both experimental and theoretical studies have shown that buoyancy alone is neither necessary nor sufficient to trigger diapirism, particularly below a strong, thick overburden (e.g., Jenyon, 1986; Vendeville and Jackson, 1992). This delay clearly must have been caused by other factors than a density inversion, such as regional deformation that fractured and weakened the overburden.

Compression

Many of the Prebetic diapirs were active during the Alpine orogeny, as shown by the repeated piercement and reburial by Neogene synorogenic sediments. Some diapirs (e.g., Finestrat and Jalón) were probably initiated in this overall compressional stress regime. To assess the role of lateral compression in the initiation of diapirs, it is important to distinguish between piercement diapirism and the flow of Triassic evaporites due to regional décollement tectonics. Several authors have used the term *diapiric fold* or *diapiric uplift* to refer to folds in the overburden that are supposedly cored by Triassic evaporites; these folds are thought to have been forced by the active diapiric rise of evaporites (e.g., Moseley, 1973; Martínez del Olmo et al., 1975; Rodriguez Estrella, 1977; Moseley et al., 1981; Cater, 1987). However, when they formed, no clear density inversion existed between the Triassic evaporites and their carbonate cover, which precludes uplift of large blocks of overburden. Even if a density inversion were present, the major faults adjacent to these "uplifts" would offer a much easier pathway for diapiric ascent. Furthermore, on the basis of tectonic stylolites and other small-scale structures, it can be shown that these folds were formed by buckling under lateral compression (de Ruig et al., 1987) rather than by vertical diapiric motion.

It is therefore debatable whether evaporite cushions under folds should be regarded as a diapiric feature, even though flow of evaporites is involved. Cushioning of folds by Triassic evaporites is directly related to their function as a basal décollement. During folding, the evaporites are squeezed into the cores of major box folds to accommodate space problems between the rigid basement and the overlying cover. This process has been discussed in detail by Laubscher (1975, 1977) in his analysis of box folds in the Jura foldbelt. There is no evidence that this process was associated with piercement of the cover, because thrusts and reverse faults associated with major Prebetic anticlines are usually unassociated with diapiric intrusions.

Preexisting diapirs, which extruded to the surface during late Oligocene–earliest Miocene rifting (e.g., Monnegre and Agost), were distorted and squeezed during ensuing early–middle Miocene Alpine compression. The repeated injection of Triassic rocks into Miocene sediments overlying the diapirs is likely to be related to lateral compression of these preexisting, shallowly buried diapirs. Although the boundary faults of these diapirs are generally oblique to the regional fold trend, folds within the Triassic diapiric sequence commonly parallel the regional trend. Much of the late-stage brittle faulting within the diapirs also aligns with regional contractional structures.

In contrast, the diapirs in the Prebetic of Alicante that were initiated during Neogene contraction are generally elongated perpendicular to adjacent folds and thrusts. Their shape and location were controlled by the local direction of extension, as shown by the structural geometry of the pull-apart diapirs (Figure 5). During regional folding, diapirs may therefore be initiated from local extensional fracturing of the overburden rather than from overall lateral compression.

Extension and Reactive Diapirism

Ample evidence suggests that the principal mechanism for the initiation of diapirs in the eastern Prebetic is extension. First, the external geometry of the Triassic diapirs and their relationship to faults and folds in the adjacent cover suggests that they occupy volumes where the post-Triassic cover was pulled apart. Second, most of the Triassic outcrops are associated with thick fills of Neogene sediment indicating that they constituted graben-like depressions even in their earliest stages. Third, the phases of major diapiric upwelling correlate with periods of regional extension.

Thin-skinned extension is a highly effective mechanism for triggering diapirism through even a strong carbonate overburden. Extensional faults both weaken and thin the overburden and differentially load the evaporites by their surface relief (e.g., Vendeville and Jackson, 1992; Jackson and Vendeville, 1994). A similar type of reactive diapirism has been suggested to explain the structure of the Triassic grabens in the Macizo de Caroch (Ruiz et al., 1979; Ríos et al., 1980; Moissenet, 1989; Roca and Guimerà, 1992) and the Triassic outcrop that borders the Miocene Abarán Basin (Van der Straaten, 1993) about 90 km west of Alicante.

The experimental and theoretical studies summarized by Jackson and Vendeville (1994) have shown that in a geologic system in which a pressurized fluid source layer underlies a strong, brittle overburden, diapirism is controlled more by the relative strength than by the relative density of the overburden and source layer. When such a strong overburden is breached by extensional faulting, the pressure forces on the source layer will drive upward diapiric flow regardless of the density ratio. If, as suggested by well data and field observations, the Prebetic diapirs were sourced by an anhydrite-rich Triassic sequence containing little or no halite, they must have risen through less dense overburden (Figure 8).

Deformation experiments of Müller et al. (1981) and observations of Jordan and Nüesch (1989) have shown that anhydrite can deform by ductile flow at relatively shallow depths (>1.6 km). Major extensional faults will cause differential loading of the ductile Triassic evaporites and may therefore initiate diapiric flow. The high-density anhydrite diapirs are unable to rise all the way to the surface, however, because the weight of the diapirs has to be sustained by the lithostatic pressure exerted by the lower density overburden (in equilibrium, overburden density × overburden height = diapir density × diapir height) (see Bishop, 1978; Vendeville and Jackson, 1992). For a 2-km-thick carbonate overburden, such a diapir will typically lie in a depression of 100–200 m, which is within the topographic range of the Prebetic diapiric grabens (Figure 8). Further diapiric rise may have been sustained by (1) late-stage compression during orogeny (squeezing) (Nilsen et al., 1995; Rowan, 1995) and (2) the volume increase due to alteration of anhydrite to gypsum by rehydration, which may amount to 62% (Jenyon, 1986).

(a)

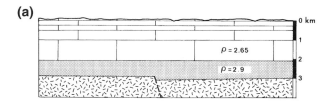

(b)

EXTENSION

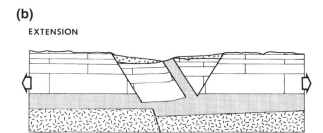

(c)

DIAPIRISM   ANHYDRITE ⇔ GYPSUM

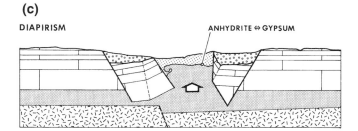

Figure 8—Model for extension-induced diapirism in the Prebetic of Alicante. (a) For the initial situation, an extreme case is taken in which the Triassic source layer has an anhydritic specific gravity (ρ = 2. 9) overlain by a less dense carbonate overburden (mean ρ = 2. 65). (b) Regional extension induces the formation of large normal faults, which cut the post-Triassic overburden, possibly nucleating on a basement fault. (c) Normal faulting induces differential loading of the evaporites, which flow toward the area of tectonically thinned overburden, resulting in the formation of a reactive diapir.

CONCLUSIONS

Diapirism of Triassic evaporites in the Prebetic foldbelt of Alicante was primarily controlled by regional tectonics. The relatively high density of the Triassic diapiric source layer and the late initiation of diapirism, occurring about 185 m.y. after evaporite deposition, indicate that buoyancy played a negligible role in diapir evolution. Major piercement diapirs were initiated by regional thin-skinned extensional faulting, which weakened the carbonate overburden and created the necessary space for diapiric ascent. Their emplacement was driven by tectonic differential loading and the resulting flow of the ductile evaporites. The oldest diapirs of considerable size reached the surface in late Oligocene–early Miocene time, triggered by Western Mediterranean rifting; they may have started rising during older phases of extension. Subsequent early–middle Miocene compression resulted in mainly brittle deformation and lateral squeezing of

these preexisting diapirs. The structural geometry and paleostress field of diapirs newly initiated during orogenic shortening suggest that their emplacement was driven by local (trans)tensional faulting rather than by regional compression.

In conclusion, the ability of the Triassic evaporites to generate fold structures in the post-Triassic cover by active diapirism has been strongly overestimated in most literature concerning the structure of the Prebetic of Alicante. During orogenic contraction, the evaporites provided a weak décollement layer allowing regional thin-skinned folding. Reactive piercement diapirs were formed by regional extension during rifting or localized transtensional faulting during orogenesis. Although the presence of Triassic evaporites has thus been of great importance in the deformation history of the region, their role was mainly passive.

Acknowledgments   *The author would particularly like to thank Harry Stel, Harm Rondeel (both of Vrije Universiteit, Amsterdam), and Otto Simon (University of Amsterdam) for their help and guidance during my research and the many fruitful discussions about the tectonics of southern Spain. Shell U.K. Exploration and Production is kindly acknowledged for the use of their facilities and permission to publish this paper. Thanks also to two anonymous referees who reviewed this paper and provided helpful suggestions for improvement.*

REFERENCES CITED

Azéma, J., 1977, Etude géologique des Zones Externes des Cordillères Bétiques aux confins des provinces d'Alicante et de Murcie (Espagne): Ph.D. dissertation, Université de Paris, Paris, 396 p.

Baena Perez, J., and L. Jerez Mir, 1982, Sintesis para un ensayo paleogeografico entre la Meseta y la Zona Betica (s. str.): Instituto Geologico y Minero de España, Madrid, 256 p.

Bartrina, T., E. Hernández, and A. Serrano, 1990, Estudio de subsuelo del Trias salino en la Depresión Intermedia, *in* F. Ortí Cabo and J. M. Salvany Duran, eds., Formaciones evaporíticas de la Cuenca del Ebro y cadenas periféricas, y de la zona de Levante: Publ. Departament de Geoquimica, Petrologia i Prospecció Geològica, Universidad de Barcelona, p. 232–238.

Beets, C. J., and M. J. de Ruig, 1992, [87]Sr/[86]Sr dating of coralline algal limestones and its implications for the tectono-stratigraphic evolution of the eastern Prebetic (Spain): Sedimentary Geology, v. 78, p. 233–250.

Bishop, R. S., 1978, Mechanism for emplacement of piercement diapirs: AAPG Bulletin, v. 62, p. 1561–1583.

Cater, J. M. L., 1987, Sedimentary evidence of the Neogene evolution of SE Spain: Journal of the Geologic Society, London, v. 144, p. 915-932.

De Haan, P., and J. Van der Wal, 1985, De geologie van het Agost gebied (provincie Alicante, zuid Spanje): Internal Report, Universiteit van Amsterdam, 115 p.

De Jong, K., 1990, Alpine tectonics and rotation pole evolution of Iberia, *in* G. Boillot and J. M. Fontboté, eds., Alpine evolution of Iberia and its continental margins: Tectonophysics, v. 184, p. 279–296.

De Lange, G. J., N. A. I. M. Boelrijk, G. Catalano, C. Corselli, G. P. Klinkhammer, J. J. Middelburg, D. W. Müller, W. J. Ullman, P. Van Gaans, and J. R. W. Woittiez, 1990, Sulfate-related equilibria in the hypersaline brines of the Tyro and Bannock basins, eastern Mediterranean: Marine Chemistry, v. 31, p. 89–112.

de Ruig, M. J., 1990, Fold trends and stress deviation in the Alicante fold belt, southeastern Spain, *in* G. Boillot and J. M. Fontboté, eds., Alpine evolution of Iberia and its continental margins: Tectonophysics, v. 184, p. 393–403.

de Ruig, M. J., 1992, Tectono-sedimentary evolution of the Prebetic fold belt of Alicante (SE Spain)—a study of stress fluctuations and foreland basin deformation: Ph.D. dissertation, Structural Geology and Tectonics Group, Vrije Universiteit, Amsterdam, 207 p.

de Ruig, M. J., R. M. Mier, and H. Stel, 1987, Interference of compressional and wrenching tectonics in the Alicante region, SE Spain: Geologie en Mijnbouw, v. 66, p. 201–212.

de Ruig, M. J., J. Smit, T. Geel, and H. Kooi, 1991, Effects of the Pyrenean collision on the Paleocene stratigraphic evolution of the southern Iberian margin (southeast Spain): GSA Bulletin, v. 103, p. 1504–1512.

Egeler, C. G., and O. J. Simon, 1969, Sur la tectonique de la Zone bétique (Cordillères bétiques, Espagne): Verhandelingen der Koninklijke Nederlandse Academie van Wetenschappen, v. 25, no. 3, 90 p.

Foucault, A., 1966, Le diapirisme des terrains triasiques au Secondaire et au Tertiaire dans le Subbétique du NE de la province de Grenade (Espagne méridionale): Bulletin de la Société Géologique de France, v. 7, p. 527–536.

García-Hernandez, M., A. C. López-Garrido, P. Rivas, C. Sanz de Galdeano, and A. J. Vera, 1980, Mesozoic palaeogeographic evolution of the external zones of the Betic Cordillera: Geologie en Mijnbouw, v. 59, p. 155–168.

García-Rodrigo, B., 1960, Sur la structure du Nord de la province d'Alicante: Bulletin de la Société géologique de France, v. 7, p. 273–277.

García-Rodrigo, B., 1965, Nuevas datos sobre el Paleogeno de la Zona Prebetica al norte de Alicante: Notas y communicaciones del Instituto Geologico y Minero de España, v. 79, p. 69–88.

Geel, T., T. B. Roep, W. Ten Kate, and J. Smit, 1992, Early–middle Miocene stratigraphic turning points in the Alicante region (SE Spain): reflections of western Mediterranean plate tectonic reorganizations: Sedimentary Geology, v. 75, p. 223–239.

Holzhaus, E., 1989, De geologie van het gebied tussen Jalance, Cofrentes en Casas del Río (prov. Valencia), zuidoost Spanje: Internal Report, Universiteit van Amsterdam, 100 p.

Jackson, M. P. A., and C. J. Talbot, 1986, External shapes, strain rates, and dynamics of salt structures: GSA Bulletin, v. 97, p. 305–323.

Jackson, M. P. A., and C. J. Talbot, 1989, Anatomy of mushroom-shaped diapirs: Journal of Structural Geology, v. 11, p. 211–230.

Jackson, M. P. A., and B. C. Vendeville, 1994, Regional extension as a geologic trigger for diapirism: GSA Bulletin, v. 106, p. 57–73.

Jenyon, M. K., 1986, Salt Tectonics: Essex, U.K., Elsevier Applied Science, 191 p.

Jordan, P., and R. Nüesch, 1989, Deformation structures in the Muschelkalk anhydrites of the Schafisheim well (Jura overthrust, northern Switzerland): Ecologae geologica Helvetica, v. 82, p. 429–454.

Koyi, H., M. K. Jenyon, and K. Petersen, 1993, The effect of basement faulting on diapirism: Journal of Petroleum Geology, v. 16, no. 3, p. 285–312.

Lanaja, J. M., R. Querol, and A. Navarro, A., 1987, Contribucion de la exploracion petrolifera al conocimiento de la geologia de España: Instituto Geologico y Minero de España, Madrid, 465 p.

Laubscher, H. P., 1975, Viscous components in Jura folding: Tectonophysics, v. 137, p. 239–254.

Laubscher, H. P., 1977, Fold development in the Jura: Tectonophysics, v. 37, p. 337–362.

Leclerc, J., 1971, Etude geologique du Massif du Maigmo et de ses abords (province d'Alicante, Espagne): Ph.D. dissertation, University of Paris, Paris, 128 p.

Leclerc, J., and J. Azéma, 1976, Le Crétace dans la region d'Agost (province d'Alicante–Espagne) et ses accidents sedimentaires: Cuadernos de Geología, Granada, v. 7, p. 35–51.

Llavador, F., J. A. Pina, and C. Auernheimer, 1983, Discriminacion geoquimica de algunas facies del Cretacico (Albense) en el sector oriental de la Zona Prebetica (provincia de Alicante): Mediterránea, Series Geologicos, v. 1, p. 31–69.

MacDonald, G. J. F., 1953, Anhydrite–gypsum equilibrium relations: American Journal of Science, v. 251, p. 884–898.

Mancheño Jiménez, M. A., and T. Rodriguez Estrella, 1985, Geologia de los diapiros triásicos en el noreste de la provincia de Murcia: Estudios Geologicos, v. 41, p. 189–200.

Marfil Pérez, R., 1970, Estudio petrogenético del Keuper en el sector meridional de la Cordillera Ibérica: Estudios Geologicos, v. 26, p. 113–161.

Martí, J., J. Mitjavila, E. Roca, and A. Aparicio, 1992, Cenozoic magmatism of the Valencia Trough (western Mediterranean): relationship between structural evolution and volcanism, *in* E. Banda and P. Santanach eds., Geology and geophysics of the Valencia Trough (western Mediterranean): Tectonophysics, v. 203, p. 145–165.

Martínez del Olmo, W., M. Benzaquen, I. Cabañas, and M. A. Uralde, 1975, Cartografía y Memoria de la Hoja 820 (Onteniente): Mapa Geologico de España, Instituto Geologico y Minero de España, Madrid, scale 1:50,000, 49 p.

Michael, A. J., 1984, Determination of stress from slip data, faults and folds: Journal of Geophysical Research, v. 89, p. 11517–11526.

Moissenet, E., 1989, Les fosses néogènes de la Chaîne ibérique: leur évolution dans le temps: Bulletin de la Société géologique de France, v. 5, no. 5, p. 919–926.

Moseley, F., 1973, Diapiric and gravity tectonics in the Pre-Betic (Sierra Bernia) of southeast Spain: Boletin del Instituto Geologico y Minero de España, v. 84, no. 3, p. 114–126.

Moseley, F., J. C. Cuttell, E. W. Lange, D. Stevens, and J. R. Warbrick, 1981, Alpine tectonics and diapiric structures in the Pre-Betic zone of southeast Spain: Journal of Structural Geology, v. 3, no. 3, p. 237–251.

Müller, W. H., S. M. Schmid, and U. Briegel, 1981, Deformation experiments on anhydrite rocks of different grain sizes: rheology and microfabric, *in* G. S. Lister, H.-J. Behr, K. Weber, and H. Zwart, eds., The effect of deformation on rocks: Tectonophysics, v. 78, p. 527–543.

Murray, R. C., 1964, Origin and diagenesis of gypsum and anhydrite: Journal of Sedimentary Petrology, v. 34, no. 3, p. 512–523.

Navarro, A., E. Trigueros, C. Villalón, and J. M. Ríos, 1960, Derniers progrès dans la connaissance de l'extrémité nord-est des Chaines subbétiques (région d'Altea-Benisa, province d'Alicante, Espagne), *in* M. Durand-Delga, ed., Livre a la mémoire du professeur Paul Fallot: Société Géologique de France, p. 143–153.

Nilsen, K. T., B. C. Vendeville, and J.-T. Johansen, 1995, Influence of regional tectonics on halokinesis in the Nordkapp Basin, Barents Sea, *in* M. P. A. Jackson, D. G. Roberts, and S. Snelson, eds., Salt tectonics: a global perspective: AAPG Memoir 65, this volume.

Ortí Cabo, F., 1973, El Keuper del Levante Español— Litostratigrafia, petrologia y paleogeografia de la cuenca: Ph.D. dissertation, University of Barcelona, Barcelona, 174 p.

Ott d'Estevou, P., C. Montenat, F. Ladure, and L. Pierson d'Autrey, 1988, Evolution tectono-sédimentaire du prébétique oriental (Espagne) au Miocène: Comptes Rendus de l'Acadèmie des Sciences, Paris, sér. D, v. 307(2), p. 789–796.

Polvêche, J., 1961, Tectonique et Trias dans la région d'Alicante: Annales de la Société géologique du Nord, v. 82, p. 155–160.

Rehault, J.-P., G. Boillot, and A. Mauffret, 1984, The western Mediterranean basin geologic evolution: Marine Geology, v. 55, p. 447–477.

Ríos, L. M., F. J. Beltrán, and M. A. Zapatero, 1980, Cartografía y Memoria de la Hoja 769 (Navarrés): Mapa Geologico de España, escala 1:50. 000, Instituto Geologico y Minero de España, Madrid, 28 p.

Roca, E., and J. Guimerà, 1992, The Neogene structure of the eastern Iberian margin: structural constraints on the crustal evolution of the Valencia Trough (western Mediterranean): Tectonophysics, v. 203, p. 203–218.

Rodriguez Estrella, T., 1977, Síntesis geológica del Prebético de la provincia de Alicante: Boletin del Instituto Geologico y Minero de España, v. 88, no. 3, p. 183–214, and v. 88, no. 4, p. 273–299.

Rondeel, H. E., and P. Van de Gaag, 1986, A two stage diapiric event in the eastern Prebetic: Estudios Geologicos, v. 42, p. 117–125.

Rowan, M. G., 1995, Structural styles and evolution of allochthonous salt, central Louisiana outer shelf and upper slope, *in* M. P. A. Jackson, D. G. Roberts, and S. Snelson, eds., Salt tectonics: a global perspective: AAPG Memoir 65, this volume.

Ruiz, V., A. Núñez, I. Colodrón, I. Cabañas, and M. A. Uralde, 1979, Cartografia y Memoria de la Hoja 768 (Ayora): Mapa Geologico de España, Instituto Geologico y Minero de España, Madrid, scale 1:50,000, 34 p.

Sáenz Ridruejo, C., and J. M. López Marinas, 1975, La edad del vulcanismo de Cofrentes (Valencia): Tecniterrae, v. 6, p. 8–14.

Santisteban, C. de, F. Ruíz-Sánchez, and D. Bello, 1989, Los depósitos lacustres del Terciario de Bicorp (Valencia): Acta Geológica Hispánica, v. 24, no. 3-4, p. 299–307.

Simon, O. J., 1987, On the Triassic of the Betic Cordilleras (southern Spain): Cuadernos de Geología Ibérica, v. 11, p. 385–402.

Soler, R., W. Martínez del Olmo, A. G. Megías, and J. A. Abeger Monteagudo, 1983, Rasgos básicos del Neógeno del Mediterráneo Español: Mediterraneo, Series Geologicos, v. 1, p. 71–82.

Stephenson, R. A., J. Van Berkel, and S. Cloetingh, 1992, Relationship between salt diapirism and the tectonic history of the Sverdrup Basin, Arctic Canada: Canadian Journal of Earth Science, v. 29, p. 2695–2705.

Torres, J., C. Bois, and J. Burrus, 1993, Initiation and evolution of the Valencia Trough (western Mediterranean): constraints from deep seismic profiling and subsidence analysis: Tectonophysics, v. 228, p. 57–80.

Torres-Roldán, R. L., 1979, The tectonic subdivision of the Betic zone (Betic Cordilleras, southern Spain): its significance and one possible geotectonic scenario for the westernmost Alpine belt: American Journal of Science, v. 279, p. 19–51.

Van Berkel, J. T., 1986, A structural study of evaporite diapirs, folds and faults, Axel Heiberg Island, Canadian Arctic Islands: Ph.D. dissertation, Universiteit van Amsterdam, GUA Papers of Geology, Amsterdam, 149 p.

Van der Straaten, H. C., 1993, Neogene strike-slip faulting in southeastern Spain: the deformation of the pull-apart basin of Abarán: Geologie en Mijnbouw, v. 71, no. 3, p. 205–225.

Vendeville, B. C., and M. P. A. Jackson, 1992, The rise of diapirs during thin-skinned extension: Marine and Petroleum Geology, v. 9, p. 331–353.

Wilke, H.-G., 1988, Stratigraphie und Sedimentologie der Kreide im Nordwesten der Provinz Alicante (SE-Spanien): Berliner Geowissenchaftlichen Abhandlungen, v. 95, 72 p.

Zeyen, H., E. Banda, J. Gallart, and J. Ansorge, 1985, A wide angle seismic reconnaissance survey of the crust and upper mantle in the Celtiberian Chain of eastern Spain: Earth and Planetary Science Letters, v. 75, p. 393–402.

Sans, M., and J. Vergés, 1995, Fold development related to contractional salt
tectonics: southeastern Pyrenean thrust front, Spain, *in* M. P. A. Jackson,
D. G. Roberts, and S. Snelson, eds., Salt tectonics: a global perspective: AAPG
Memoir 65, p. 369–378.

Chapter 18

Fold Development Related to Contractional Salt Tectonics: Southeastern Pyrenean Thrust Front, Spain

M. Sans

*Departament de Geologia Dinàmica,
    Geofísica i Paleontologia*
Universitat de Barcelona
Barcelona, Spain

Present address:

Súria K, S. A. Sales y Potasas
Barcelona, Spain

J. Vergés

*Departament de Geologia Dinàmica,
    Geofísica i Paleontologia*
Universitat de Barcelona
Barcelona, Spain

Abstract

In the outermost region of the southeastern Pyrenees, a suite of thrusted folds detach above an upper Eocene salt that was about 300 m thick before deformation. The foremost anticlines formed during the Oligocene and represent a small amount of shortening. In map view, they display a relay pattern slightly oblique to the margin of the salt layer, where deformation stops. The three-dimensional edge effects caused by the pinch-out of the Cardona salt play an important role in the development of the frontal structures in the southeastern Pyrenees.

Fold evolution has been reconstructed by interpreting variations along the strike of the folds as an indicator of deformation sequence. Where the sedimentary pile contains an upper detachment, thrusts developed fishtail geometries in which thrusts of alternating vergence were stacked up. Where an upper detachment is lacking, a thrusted anticline formed, into whose core salt migrated during the early phases of folding. Whether or not an upper detachment is present, anticlines continued to amplify during and after thrusting. Folding blocked further slip on some thrusts and promoted the development of pop-up structures.

INTRODUCTION

Many external regions of fold and thrust belts are deformed by detachment above evaporites, mainly salt. The external region of the southern Pyrenees is one such belt, and here the upper Eocene Cardona salt has acted as an effective detachment. Contractional structures extend in a wide area of the southeastern foreland. Narrow cross-sectional taper, wide deformed areas, and wide synclines separating narrow anticlines are some of the characteristics that differentiate fold belts developed above salt, such as the southeastern Pyrenees, from those lacking basal evaporitic detachments (Davis and Engelder, 1985). Unlike most models of fold development above salt (Wiltschko and Chapple, 1977; Davis and Engelder, 1985; Dahlstrom, 1990), the Cardona salt is wedge-shaped. This sedimentary geometry of the detachment level influences the location of the frontal structure (Vergés et al., 1992) and the vergence and along-strike changes of the folds.

The deformation front of the southeastern Pyrenees is well exposed and relatively simple (one main basal detachment). In addition, shortening has been slight, so that anticlines are preserved without further transport above thrusts. This area is thus ideal to study fold development above salt. The aims of this paper are to (1) document the geometry of these frontal structures and their variations along strike based on field data, potash wells, and new exploration seismic profiles; (2) establish the relationship between these structures and the salt layer thickness changes; (3) describe the influence of multiple detachment levels within the overburden on the tectonic style; and finally (4) present a hypothesis for fold development using observed differences along strike as clues to their structural evolution.

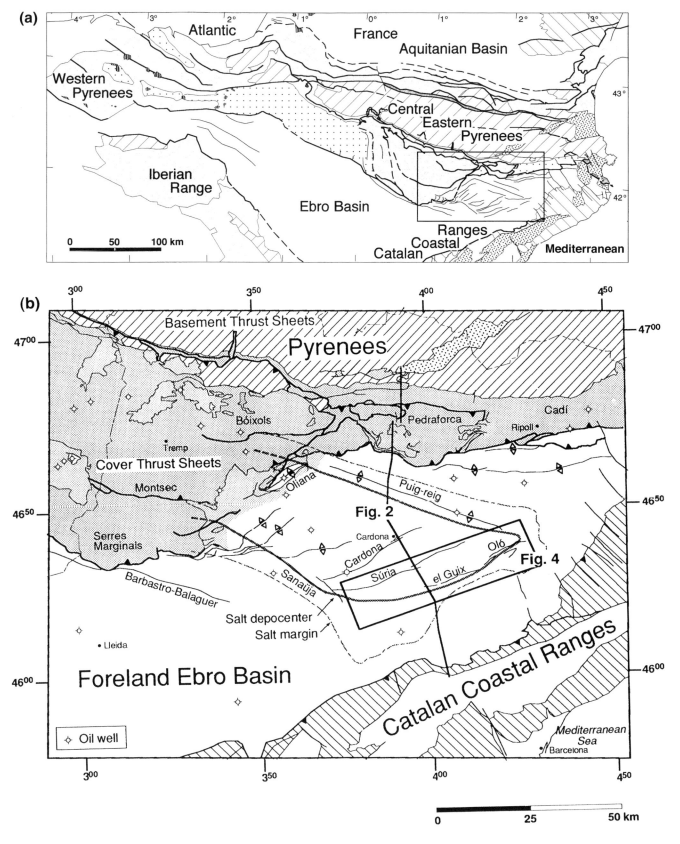

Figure 1—Structural map of the Pyrenees and the foreland Ebro Basin. (a) Location of the southeastern Pyrenees (rectangle). (b) The Ebro Basin is limited by the Pyrenees to the north and the Catalan Coastal Ranges to the south (shaded regions). The thin dashed line shows the extent of the Cardona salt at depth. Rectangle shows the study area.

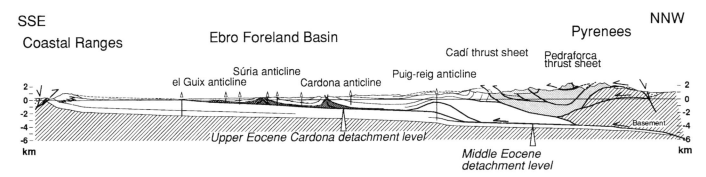

Figure 2—Geologic section across the eastern Pyrenees and Ebro foreland basin, without vertical exaggeration. The basal thrust of the Pyrenean thrust system has a staircase geometry. Thrust flats are located within the Eocene foreland evaporitic levels. See Figure 1 for location.

GEOLOGIC SETTING

The study area corresponds to the most external region of the southeastern Pyrenean fold and thrust belt (Figures 1, 2). The mountain belt formed during the continental collision of the Iberian and European plates from Late Cretaceous to Miocene time (Puigdefàbregas et al., 1992). The Ebro Basin is the latest foreland basin formed by lithospheric flexure due to the load of the Pyrenean thrust sheets. Cover thrust sheets piled up in the external zones of the chain, whereas the basement thrust sheets formed a south-directed antiformal stack in the inner part of the southern orogen (Figure 2). The Ebro foreland basin, triangular in map view, is limited by the Pyrenees to the north, the Catalan Coastal ranges to the southeast, and the Iberian Range to the south and southwest.

The stratigraphy of the eastern Pyrenean foreland basin is summarized in Figure 3. Sedimentation changed from marine to continental during emplacement of the thrust sheets (Puigdefàbregas et al., 1986; Vergés and Burbank, 1996). The marine infill of the basin ended after deposition of the Cardona salt in early Priabonian time. The Cardona Formation, which is the most effective detachment level in the basin, consists from bottom to top of anhydrite (5 m), a lower salt member represented by massive halite (200 m), and an upper evaporitic member of halite and interbedded potassium salt (50 m) (Pueyo, 1975). The transition from marine to continental sediments above takes place in a 60-m-thick lutitic unit containing interbedded gypsum and halite (gray lutites of Sáez, 1987). Continental deposits above the Cardona salt are represented by alluvial and fluvial sediments prograding over a lacustrine system (Figure 3b).

The lacustrine deposits are represented by the Barbastro and Castelltallat formations of late Eocene–Oligocene age (Sáez and Riba, 1986; Sáez, 1987; Anadón et al., 1993). The Barbastro Formation consists of 30 m of gypsum and interbedded lutites; the Castelltallat Formation is represented by 100–200 m of marl and interbedded limestones. These lacustrine deposits are interbedded with and grade southward and northward into alluvial and fluvial sediments (Súria, Solsona, and Artés formations). The Solsona and Artés formations are

fine- to coarse-grained red sediments interpreted as alluvial fan deposits. The Artés Formation originates in the Catalan Coastal ranges. The Solsona Formation was shed from the Pyrenees and grades southward into the Súria Formation sandstones, which have been interpreted as terminal alluvial fan deposits (Sáez, 1987). The uppermost deposits in the eastern Ebro foreland basin can be assigned to the upper part of the lower Oligocene (Vergés and Burbank, 1996).

A striking feature of the eastern Ebro Basin is the three sets of folds of different trends: northwest-trending folds in the north, northeast-trending folds in the center, and west-northwest trending folds in the southwest. Each set decouples over an Eocene–Oligocene foreland evaporitic layer (Vergés et al., 1992). The northern boundary of the central folded region is the Puig-reig anticline (Figures 1, 2), where the salt pinches out northward. This anticline represents the surface expression of a footwall ramp that allows the basal Pyrenean thrust to climb from the lower detachment level to the Cardona salt (Figure 2). The front of the thrust system can be followed to the west along the northeast-trending Santa Maria d'Oló, El Guix, and Súria anticlines (study area); the north-northwest trending Sanaüja anticline; and the west-northwest trending Barbastro-Balaguer anticline. The Sanaüja and Barbastro-Balaguer anticlines have been detached along upper décollements (Barbastro evaporites, which reach a thickness of more than 800 m outside the study area). The Sanaüja anticline (Figure 1) parallels the southern boundary of the Cardona salt and is the surface expression of a ramp that climbs from the Cardona salt to the Barbastro gypsum and salt (Vergés et al., 1992; Gil and Jurado, 1994). Thus, the northern, southern, and southwestern boundaries of the Cardona salt have all played a dominant tectonic role in creating marginal anticlines that represent edge effects of the Pyrenean foreland.

STRUCTURE

The southernmost frontal structures of the Pyrenean fold and thrust system are a set of northeast-trending anticlines cut by faults. These folds are 30 km long and

(a)

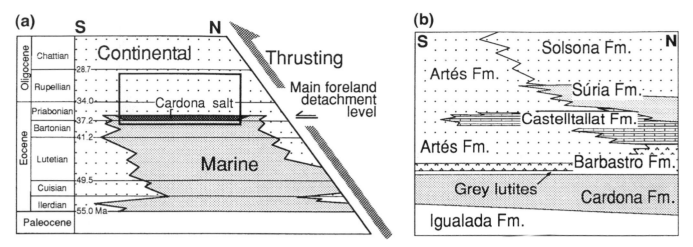

(b)

Figure 3—Sedimentary evolution during thrusting in the southeastern Pyrenean foreland basin. (a) Sedimentation changes from marine to continental after the Cardona salt accumulated in Priabonian times. Box locates figure on the right.
(b) Above the Cardona salt (main detachment) are lacustrine gypsum (chevron pattern), lacustrine limestones (brick pattern), and alluvial and fluvial deposits (stippled pattern).

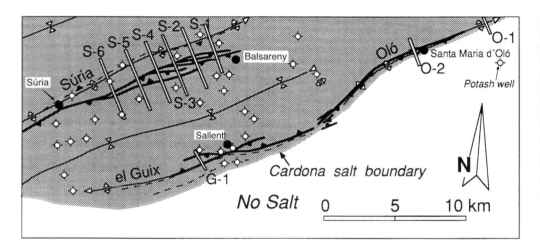

Figure 4—Structural map of the frontal structures of the eastern Pyrenean fold and thrust belt. Abundant potash exploration wells constrain the subsurface structure. Dashed lines indicate subsurface structures. Shaded area represents the subsurface extent of Cardona salt. G-1, O-1, O-2, and S-1 to S-6 are lines of cross sections shown in Figures 5–7.

display a relay geometry (Figure 4). From east to west, the north-vergent Santa Maria d'Oló anticline merges with the south-vergent El Guix anticline. In the north-northwest, the Súria anticline displays a complex double vergence.

Santa Maria d'Oló–El Guix Anticline

The Santa Maria d'Oló structure represents the eastern segment of an approximately 35-km-long anticline (Santa Maria d'Oló–El Guix anticline). It is differentiated from the El Guix anticline because it has the opposite vergence. This northward vergence can be observed in the northeastern segment of the Santa Maria d'Oló anticline (Figure 5a), where the structure is formed by a simple, slightly asymmetric fold with an interlimb angle of 97°. The anticline opens and ends toward the east with a gentle plunge of 2–3°. The central segment of the north-vergent Santa Maria d'Oló anticline is modified by thrusting (Figure 5b). The thrust dips steeply between the two

limbs. The anticlinal crest is detached by a pair of thrusts (Figure 5b). The fold opens with increasing depth. Salt is asymmetrically distributed under the Santa Maria d'Oló fold.

The Cardona salt upper contact is located at different elevations in both limbs; the northern limb becomes horizontal more rapidly and overlies more salt than does the southern. This geometry reflects the change in original sedimentary salt thickness toward the south. Several potash exploration wells delimit the change in salt thickness along the strike of the anticlines. An eastward decrease in salt thickness strongly suggests that the salt layer initially thinned toward the eastern margin as well as the southern margin of the evaporitic basin. To the southwest, the Santa Maria d'Oló anticline plunges southwestward and terminates in a small splay of north-directed thrusts (Figure 4) aligned with the El Guix south-directed thrust. Between the two structures, there is little deformation and it is difficult to analyze their age relationships. One locality shows that the hanging-wall anti-

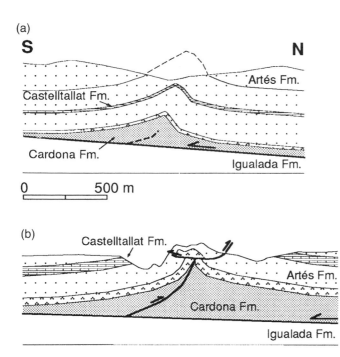

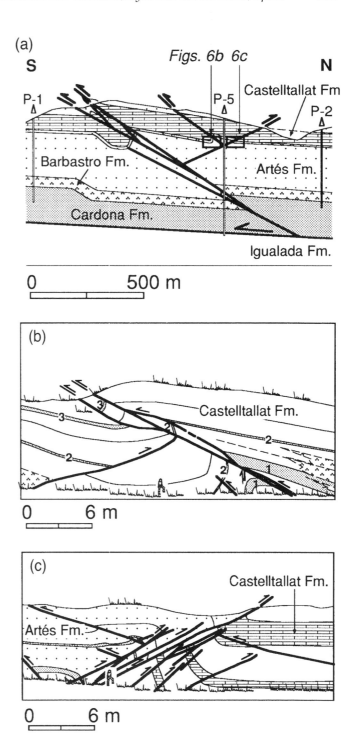

Figure 5—North-south sections of the Santa Maria d'Oló anticline, without vertical exaggeration. (a) Eastern section (O-1 in Figure 4) near the Cardona salt pinch-out verges northward. (b) Western section (O-2 in Figure 4); the pop-up structure in the anticline crest is related to the folding, steepening, and kinematic blocking of the main north-directed backthrust.

cline of the Santa Maria d'Oló thrust is cut by a set of small south-directed thrusts. The Santa Maria d'Oló structure is most shortened in the central portion. The backthrust displacement here is more than 100 m (just west of Santa Maria d'Oló) (Figure 4).

The south-verging El Guix anticline extends for more than 20 km west of the Santa Maria d'Oló anticline (Figure 4). Its long, north-dipping limb and short, sub-vertical southern limb have an interlimb angle of about 110° (Figure 6a). The anticline is cut by a set of thrusts with opposing vergence; their geometry is constrained by field exposures and potash exploration wells. Both groups of thrusts dip from 27° to 40°. Good exposures along the Sallent potash railway reveal the minor structures associated with thrusting and folding and their age relationships. Anticlines and thrusts are clearly related in the El Guix anticline. Thrusts cut through different segments of the fold. Figure 6b shows that the crest of the anticline is displaced upsection from the footwall (anticline hinge in bed 1) to the hanging wall of the thrust (anticline hinge in bed 3). This fold is geometrically similar to a fault-propagation fold, although the tip of the thrust has been removed by erosion. Pop-up structures are common on different scales (Figures 6a, c).

North- and south-directed thrusts intersect one another, suggesting that both groups of thrusts were developed at the same time. Cleavage is well developed near the thrust faults. This cleavage, which formed by pressure solution normal to the bedding, is clearly folded and cut

Figure 6—(a) Section G-1 through the central part of the faulted south-directed El Guix anticline (see Figure 4 for location). Minor structures (shown in part b) include conjugate thrust faults of similar dip. (b) Some fault-propagation folds are also related to the south-directed thrusts. Numbers 1, 2, and 3 represent displaced members. (c) North-directed backthrust and pop-up structures on various scales.

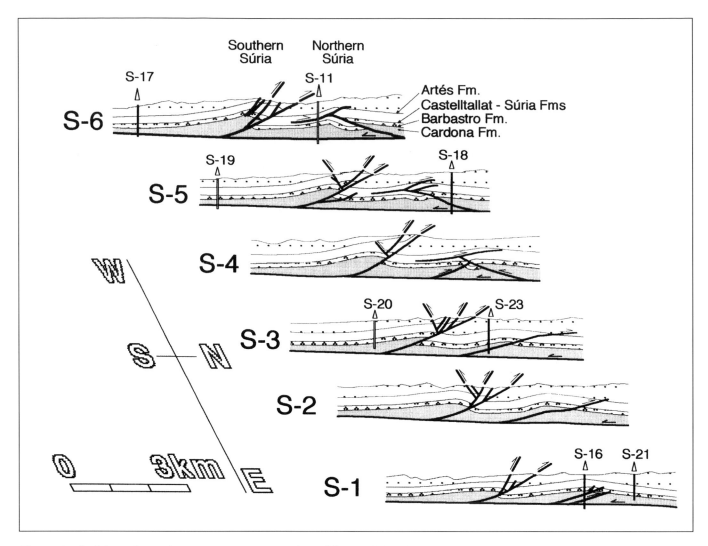

Figure 7—Serial sections of the eastern segment of the Súria anticline (see Figure 4 for location). The most distinct feature of the Súria northern anticline is an upper detachment within the alternating limestones and marls of the Castelltallat Formation. Sections are constrained by seismic sections and the potash exploration wells shown.

by thrusts at low angles, indicating early localized deformation prior to folding and thrusting. The proximity of the cleavage to thrusts suggests that areas of high strain were the locus of folding and thrusting. The total shortening of the El Guix anticline progressively increases from 0% in the terminations to 15% in the central segment. Displacement of the south-directed thrusts is greater than in the north-directed ones. The major south-directed thrust has 180 m of displacement.

Súria Anticline

The Súria anticline is northwest of the Santa Maria d'Oló and El Guix anticlines and separated from them by a broad syncline (Figures 2, 4). The termination of the El Guix anticline toward the southwest means that the Súria anticline becomes the frontal structure of the fold and thrust belt in the west, following the curve of the Cardona salt pinch-out. The Súria anticline is a complex structure

represented at the surface by two structures of opposite vergence—a south-verging anticline in the north and a north-directed thrust in the south (Figure 7). Both structures are mined for potash 450–850 m below the surface.

The northern anticline can be mapped for at least 35 km along strike (Figures 1, 4). Its structural style changes along strike, as seen in cross sections of the eastern and central segments (Figure 7). In the east, the northern anticline is symmetric and cored by small north-directed thrusts (section S-1, Figure 7). Conversely, in the central section, the northern fold is south-vergent and cored by a complex array of thrusts (sections S-6, S-5). The geometric arrangement of these thrusts in the northern Súria anticline (sections S-6, S-5) forms a fishtail structure (Figure 8) in the sense of Harrison and Bally (1988), where thrusts of alternating vergences are stacked up. The flat segment of the fishtail upper thrust follows an intermediate detachment level in the interbedded limestones and marls of the Castelltallat Formation (sections S-4, S-5, S-6). The lateral

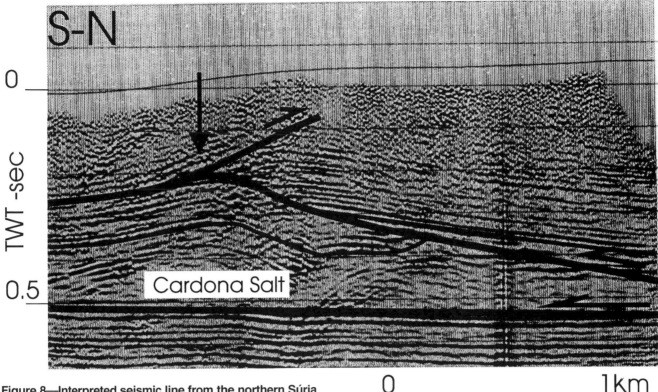

Figure 8—Interpreted seismic line from the northern Súria anticline near cross section S-6 (see Figure 4). Fishtail upper thrust separates divergent reflectors (arrow).

0   1km

change of this fishtail structure is documented in Figure 7. The easternmost north-directed thrust in the core of the symmetric anticline (sections S-1, S-2) is intersected by a south-directed thrust to the west (section S-3), which roots in an intermediate detachment level in the Castell-tallat Formation. From this level, a new north-directed thrust propagates upward (sections S-4, S-5, S-6). This fishtail structure decouples the structure of the upper part of the saline formation from the shallower structure exposed at the surface. The crest of the anticline at the surface corresponds at depth to the superposition of two blind synclines: one in the footwall of the south-directed thrust and the other in the footwall of the north-directed thrust. The surface anticline is located in the hanging wall of the south-directed thrust (Figures 7, 8) and verges southward.

The southern structure of the Súria anticline is a north-directed backthrust with related imbricates in its hanging wall (Figure 7). Some imbricates are south-directed and some north-directed. The thrust faults dip at 30°–50°. In the hanging wall of this backthrust, a well-developed fault-bend anticline has an interlimb angle of 150°. In the footwall, a smooth syncline shows an increase in limb dip near the thrust fault in the lower layers (Barbastro Formation). This dip increase can be related to drag folding and to a small mirror image of the hanging wall due to the pressure exerted by the salts under the thrust. As in the El Guix anticline, a well-developed cleavage normal to bedding is near the thrust zones. Shortening related to

the Súria anticline increases from east to west, as does displacement of the major backthrust. The minimum displacement was calculated to be 350 m in the central segment of the thrust (in Súria) (Figure 4).

FOLD DEVELOPMENT

Assuming that folds develop by lateral propagation, serial sections cut across strike represent different stages of fold evolution. In the study region, we propose a fold evolution based on (1) well-constrained serial sections of salt-detached anticlines and (2) cross-cutting relationships among structures on different scales.

To describe fold development, the deformed cover can be grouped into two units according to their mechanical behavior in response to nearly horizontal compression: the lower ductile Cardona salt and the upper brittle overburden (Figure 9). Prior to deformation, the reconstructed ductile salt thickened toward the north (center of the basin), reaching a maximum of 300 m (Pueyo, 1975). The salt was weak and therefore highly mobile. The brittle overburden had nearly constant thickness throughout the study area, and its mechanical behavior was isotropic in the south and anisotropic in the north.

In the first stage of deformation, pressure-solution cleavage formed normal to bedding in localized areas (Figure 9a). As deformation increased, the overburden buckled and the ductile Cardona salt flowed into the core

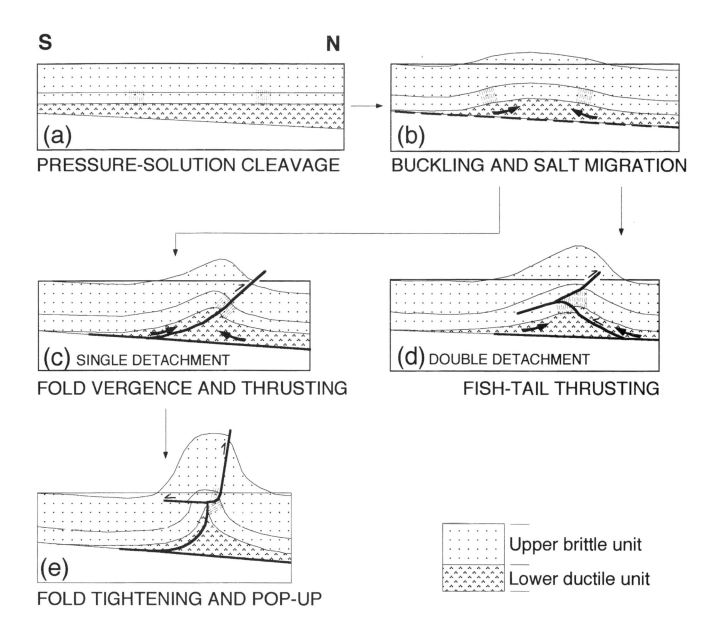

Figure 9—Proposed evolution for detachment folds based on the changes along strike of the anticlines studied. (a) Local formation of pressure-solution cleavage (fine dashed lines). (b) Folding and salt migration into the anticline. (c) and (d) Thrusting develops after the initiation of folding, creating either (c) a single thrust where the brittle upper unit is isotropic or (d) a fishtail geometry where there is an upper detachment (e.g., Súria anticline). (e) Tightening of the fold and blocking of the thrust promote the formation of a pop-up structure in the anticlinal crest.

of the anticlines (Figure 9b). These anticlines were initially symmetric (Súria anticline, Figure 7, section S-1) then became more asymmetric (Santa Maria d'Oló, Figure 5a). Although both northwest and southeast vergences were theoretically equally possible in the detached fold and thrust belt, southeast vergences were favored because of the greater salt thickness in the northern limbs of the anticlines. This greater northern thickness reflected the initial salt geometry. Assuming that the basal detachment was subparallel to the hinges of the synclines and that both the basal detachment and the hinges of the synclines sub-

sided during folding, the southern syncline of a fold was welded or nearly welded and thus pinned earlier than the northern synclines. Greater pinning of the southern synclines caused anticlines to be asymmetric. By analogy, the greatest thickness of salt below the anticlines favored an increase in shearing toward the crest of the fold; this may also have contributed to the asymmetry of the anticline, as interpreted in the Zagros Mountains of Iran (Colman-Sadd, 1978).

After buckling, thrusting formed a single thrust or a narrow brittle thrust zone in the early cleavage zones

(Figure 9c) of all three anticlines (Súria, Guix, and Santa Maria d'Oló). The superposition of thrusting and the early cleavage zone suggest that deformation was localized in time. Northern vergences of the folds were related to faulting at the salt pinch-out. Increased basal friction due to salt reduction favored the development of north-directed backthrusts subparallel to the bedding, which modified the previous vergence of their related folds. Salt continued to migrate into the cores of the anticlines during thrusting. Finally, tightening of the anticlinal crest folded and steepened the thrust (Figure 9e). This blocked further slip on the thrust and promoted a new, shallower thrust with opposite vergence (Santa Maria d'Oló, Figure 5b). The initiation of an opposed vergence thrust fault upward in the section created a pop-up geometry.

The anisotropic mechanical behavior of the brittle overburden in the northern area modified fold development, although the final resulting geometry can be similar. An upper detachment level (sections S-4, S-5, S-6, Figure 7) produced a ramp-flat thrust geometry. Flats are located in the Cardona salt (lower detachment) and in the overburden (upper detachment). Both flats are linked by a ramp. Amplification of the anticline folded and blocked the upper detachment, which retarded further slip on the thrust fault and favored the development of an opposite vergence thrust. Branched thrusts formed typical fishtail geometries (Figure 9d) in the sense of Harrison and Bally (1988). The difference between pop-up structures in the isotropic overburden (Figure 9e) and fishtail structures (Figure 9d) in the anisotropic overburden is easily determined by cross-cutting relationships between thrust and strata. Fishtail-type upper thrusts have long hanging-wall flats, whereas pop-up structures have hanging-wall and footwall ramps.

DISCUSSION AND CONCLUSIONS

The geometries and proposed evolution described here for detachment folds in the southernmost anticlines of the southeastern Pyrenees show some similarities to other hypotheses proposed for fold and thrust belts above salt. Synclines are broad, and they separate narrow, complex anticlines (Davis and Engelder, 1985). Anticlines are genetically associated with salt migration and underlain at depth by one or more thrusts. Hinterland- and foreland-directed thrusts are equally developed across the fold belt, which results in pop-up structures. The lack of dominant vergence reflects the vertical arrangement of the thrusts, which forms fishtail structures, such as in the Parry Islands foldbelt (Harrison and Bally, 1988; Harrison, 1991, 1995). The kink-fold geometry and the restriction of deformation to narrow bands through time are predicted by numerical (Higgs, 1993) and physical models (Vendeville, 1991; Letouzey et al., 1995).

Davis and Engelder (1985) also described the formation of structures of anomalous trend at the edges of the evaporitic basins. In the study area, the geometry of the salt layer plays an important role not only in localizing the frontal structures but also in influencing fold vergence. The southern Pyrenean anticlines are clearly asymmetric, although they were apparently symmetric earlier on, as seen in preserved incipient examples. The increase in thickness of the salt layer to the north (Figure 3b) ensures a greater salt thickness under the northern limb, which induces asymmetry and south vergence. This is a clear difference from symmetric folds described above tabular salt layers (Wiltschko and Chapple, 1977; Davis and Engelder, 1985; Dahlstrom, 1990).

Sections S-1 through S-6 in Figure 7 show that the salt area increases on average from east to west. This fact suggests salt flow parallel to the fold axes. Shortening in the Súria anticline also increases from east to west. Two hypotheses can correlate salt flow and shortening. Either salt flowed to areas of maximum shortening, where anticlines were forced to rise the most, or, because salt was originally thickest in the west, this allowed maximum decoupling and shortening there. Of the two hypotheses, the former seems to be supported by the seismic sections that show no dip of the base of the salt and thus imply a constant thickness through the area.

Salt migration from the synclines into the anticlines continued during folding. There is no evidence of a reversal in the salt flow out of the anticlines, even where the interlimb angle is less than 60° (Figure 5b). These Pyrenean examples contrast with examples from the Jura and Zagros, where salt has been expelled. This also differs from the salt expulsion observed in physical models (Vendeville, 1991) and calculated by Wiltschko and Chapple (1977) to be initiated after the interlimb angle reaches 60°. The lack of salt in the uppermost and tightest part of the anticline and in the thrust suggests that the fold tightened by a detachment of the brittle overburden without salt involvement.

In the study area, two distinct final geometries resulted from folding and thrusting of an isotropic overburden in the north and an anisotropic overburden in the south. In both areas, the synchronism of folding and thrusting once the anticline had been initiated is proved by the folding and blocking of the thrust faults. Pinning of the thrust faults during continuous folding triggered the formation of new thrust faults, generally of opposite vergence. Pop-up structures formed where the brittle upper unit was homogeneous, whereas fishtail structures formed where there was an upper detachment.

Acknowledgments *We would like to thank Súria K, S. A. Sales y Potasas, for permission to study the potash exploration seismic sections and wells and for permission to publish the seismic section included in this paper. We would also like to thank Pere Santanach, the two reviewers Patricia Dickerson and Steve Laubach, and Martin Jackson for their comments and suggestions. This project has been partially financed by DGICYT PB 91-0805 and PB 91-0252.*

REFERENCES CITED

Anadón, P., L. Cabrera, B. Colldeforns, and A. Sáez, 1993, Los sistemas lacustres del Eoceno superior y Oligoceno del sector oriental de la Cuenca del Ebro: Acta Geológica Hispánica, v. 24, p. 205–230.

Colman-Sadd, S. P., 1978, Fold development in Zagros simply folded belt, southwest Iran: GSA Bulletin, v. 62, p. 984–1003.

Dahlstrom, D. A., 1990, Geometric constraints derived from the law of conservation of volume and applied to evolutionary models for detachment folding: AAPG Bulletin, v. 74, p. 336–344.

Davis, D. M., and T. Engelder, 1985, The role of salt in fold-and-thrust belts: Tectonophysics, v. 119, p. 67–88.

Gil, J. A., and M. J. Jurado, 1994, Evolución tectonostratigráfica de la terminación occidental del anticlinal de Barbastro-Balaguer: II Congreso del Grupo Español del Terciario, p. 113–116.

Harrison, J. C., 1991, Melville Island's salt-based fold belt (Arctic Canada): Ph.D. dissertation, Rice University, Houston, Texas, 449 p.

Harrison, J. C., 1995, Tectonics and kinematics of a foreland folded belt influenced by salt, Arctic Canada, in M. P. A. Jackson, D. G. Roberts, and S. Snelson, eds., Salt tectonics: a global perspective: AAPG Memoir 65, this volume.

Harrison, J. C., and A. W. Bally, 1988, Cross-sections of the Parry Island fold belt on Melville Island, Canadian Arctic Islands: implications for the timing and kinematic history of some thin-skinned décollement systems: Canadian Petroleum Geology Bulletin, v. 36, p. 311–332.

Higgs, N. G., 1993, Finite element models of structural style in salt-floored fold-and-thrust belts (abs.): AAPG, International Hedberg Research Conference Abstracts, Sept. 13–17, Bath, U.K., p. 81–83.

Letouzey, J., B. Colletta, R. Vially, and J. C. Chermette, 1995, Evolution of salt-related structures in compressional settings, in M. P. A. Jackson, D. G. Roberts, and S. Snelson, eds., Salt tectonics: a global perspective: AAPG Memoir 65, this volume.

Puigdefàbregas, C., J. A. Muñoz, and M. Marzo, 1986, Thrust belt development in the eastern Pyrenees and related depositional sequences in the southern foreland basin, in Ph. Allen and P. Homewood, Foreland basins: International Association of Sedimentologists Special Publication, v. 8, p. 229–246.

Puigdefàbregas, C., J. A. Muñoz, and J. Vergés, 1992, Thrusting and foreland basin evolution in the southern Pyrenees, in K. R. McClay, ed., Thrust tectonics: London, Chapman & Hall, p. 247–254.

Pueyo, J. J., 1975, Estudio petrológico y geoquímico de los yacimientos potásicos de Cardona, Súria, Sallent y Balsareny (Barcelona, España): Ph.D. dissertation, Universitat de Barcelona, Barcelona, Spain, 350 p.

Sáez, A., 1987, Estratigrafía y sedimentología de las formaciones lacustres del tránsito Eoceno-Oligoceno del NE de la Cuenca del Ebro: Ph.D. thesis, Universitat de Barcelona, Barcelona, Spain, 353 p.

Sáez, A., and O. Riba, 1986, Depósitos aluviales y lacustres Paleógenos del margen pirenaico catalán de la Cuenca del Ebro, in P. Anadón and L. Cabrera, Guia de las Excursiones del XI Congreso Español de Sedimentología, p. 6–29.

Vendeville, B. C., 1991, Thin-skinned compressional structures above frictional-plastic and viscous decollement layers: GSA Abstracts with Program, v. 23, p. A423.

Vergés J., and D. W. Burbank, 1996, Eocene–Oligocene thrusting and basin configuration in the eastern and central Pyrenees (Spain), in P. F. Friend and C. J. Dabrio, eds., Tertiary basins of Spain: Cambridge, U.K., Cambridge University Press, p. 120–133.

Vergés, J., J. A. Muñoz, and A. Martínez, 1992, South Pyrenean fold-and-thrust belt: role of foreland evaporitic levels in thrust geometry, in K. R. McClay, ed., Thrust tectonics: London, Chapman and Hall, p. 255–264.

Wiltschko, D. V., and W. M. Chapple, 1977, Flow of weak rocks in Appalachian Plateau folds: AAPG Bulletin, v. 61, p. 653–670.

Harrison, J. C., 1995, Tectonics and kinematics of a foreland folded belt influenced
by salt, Arctic Canada, *in* M. P. A. Jackson, D. G. Roberts, and S. Snelson, eds.,
Salt tectonics: a global perspective: AAPG Memoir 65, p. 379–412.

Chapter 19

Tectonics and Kinematics of a Foreland Folded Belt Influenced by Salt, Arctic Canada*

J. Christopher Harrison

Geological Survey of Canada
Calgary, Alberta
Canada

Abstract

The Ordovician (upper Arenig–Llanvirn) Bay Fiord Formation is one of three widespread evaporite units known to have profoundly influenced the style of contractional tectonics within the Innuitian orogen of Arctic Canada. In the western Arctic Islands, the salt-bearing Bay Fiord Formation has accommodated buckling and mostly subsurface thrusting in the west-trending Parry Islands foldbelt. A characteristic feature of this belt is a stratigraphic succession more than 10 km thick featuring three rigid and widespread sedimentary layers and two intervening ductile layers (lower salt and upper shale). The ductile strata have migrated to anticlinal welts during buckling. Other features of the foldbelt include (1) an extreme length of individual upright folds (up to 330 km), (2) extreme foldbelt width (up to 200 km) governed by the distribution of underlying salt, (3) an equal occurrence of stacked and duplexed forethrusts and backthrusts (some carrying salt upsection) within anticlines, (4) modest overall shortening (up to 11%), (5) a shallow dipping salt décollement system (0. 1°–0. 6°) that has also been folded in the hinterland and later extended, and (6) a complete absence of halokinetic piercing diapirs. The progression from simple thrust–fold structure on the foldbelt periphery to complex in the interior provides a viable kinematic model for this and other contractional salt provinces. One feature of this model is a single massive triangle zone structure (passive roof duplex) that may envelop the entire 200-km width of the foldbelt and underlie an area exceeding 52,000 km².

Geological Survey of Canada contribution number 57294.

INTRODUCTION

In salt diapir provinces, structural style is primarily governed by the ductility and low density of salt and a triggering mechanism such as extension faulting—all features that permit the salt to intrude or otherwise migrate to the top of the associated stratigraphic pile (Jackson and Vendeville, 1994). The problem with understanding the behavior of salt in compressional settings is that many salt-involved thrust belts are also associated with halokinetic piercement diapirs. The combined effects of halokinetic flow (which can produce forced folds in overburden) and lateral compression (which also folds the overburden) are not readily differentiated (see Nilsen et al., 1995). Classic examples of this problem are provided by the Zagros belt of Iran, the Eurekan orogen of the central Sverdrup Basin in Arctic Canada, and some portions of the Gulf of Mexico salt province.

In other belts, the involvement of salt is known or suspected from surface geology and understood to a lesser extent by subsurface data (scattered wells or inadequate seismic expression). Such is the case for the Salt Ranges of Pakistan, the Jura, the Pennsylvanian Appalachians, the Verkhoyansk Mountains (far eastern Russia), and the Franklin Mountains in Canada's Northwest Territories (Davis and Engelder, 1985, 1987). In contrast, the Parry Islands foldbelt of Canada's Arctic Islands (Figures 1, 2) is both superbly exposed and also remarkably imaged with a grid of regional seismic profiles acquired by industry (Harrison, 1994a). Information from wells provides ties to reflectors from the surface down to the top of the subsalt succession.

An important feature of the Parry Islands foldbelt is the widespread imaging of the para-autochthonous subsalt succession. This has provided a detailed view of variation in tectonized salt thickness, depth to the salt décollement, structure beneath the salt, and the relationship between subsalt faulting and faulting in overburden during separate phases of contraction and superimposed extension. In addition, area balancing of the salt layer between synclines on each of eight regional structural cross sections has permitted the mapping and contouring of pretectonic salt thickness.

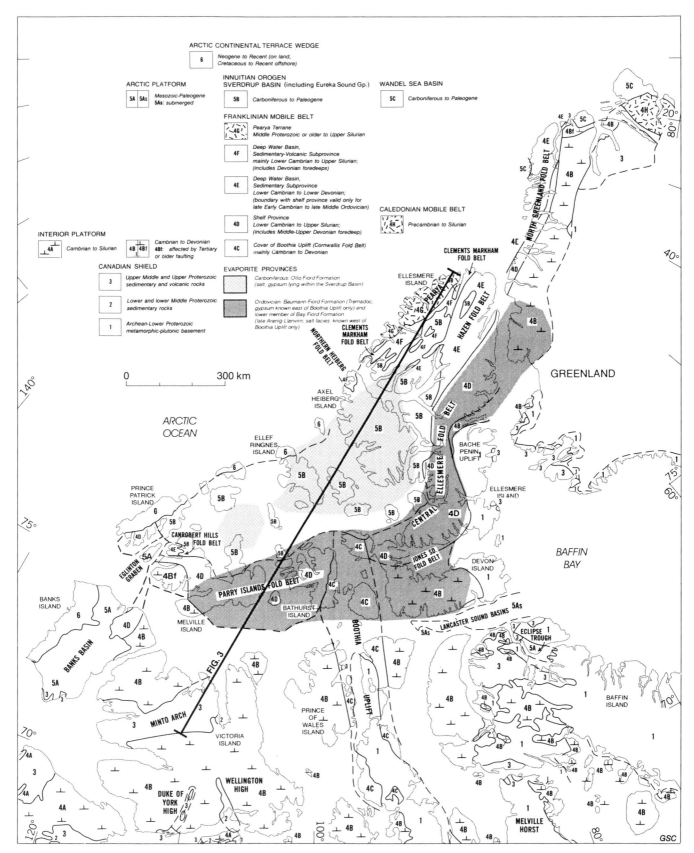

Figure 1—Geologic provinces and evaporitic basins of the Canadian Arctic Islands and northern Greenland (modified from Trettin, 1991c; Harrison, 1995).

Figure 2—Oblique aerial photograph of the arcuate, east-plunging foldbelt of eastern Melville Island expressed by various units of an extensive Middle–Upper Devonian clastic wedge (modified from Harrison, 1995). View is to the east; foreground width is about 11 km. Prominent fold closures in the foreground include the doubly plunging Robertson Point anticline (dark, on right) and the east-plunging termination of the Sabine Bay anticline (pale, on left). The foldbelt has been truncated by a widespread peneplain overlain unconformably by erosional remnants of Lower Cretaceous strata.

Large-scale syndepositional salt flow features, now identified in many of the larger salt diapir provinces, are absent from the Parry Islands foldbelt. Instead, salt has flowed laterally to resolve space problems created by postdepositional buckling of overburden. Upward transport of salt has been permitted by thrust ramps that rise off a flat-based throughgoing salt décollement. Forward-

and backward-vergent thrusts also accommodate space problems in evolving anticlines and are linked along intermediate detachments and stacked vertically within each anticline. A wide range of thrust structures that are not normally encountered in traditional foreland-vergent thrust belts are illustrated here.

An additional element of the Parry Islands foldbelt is a

382   *Harrison*

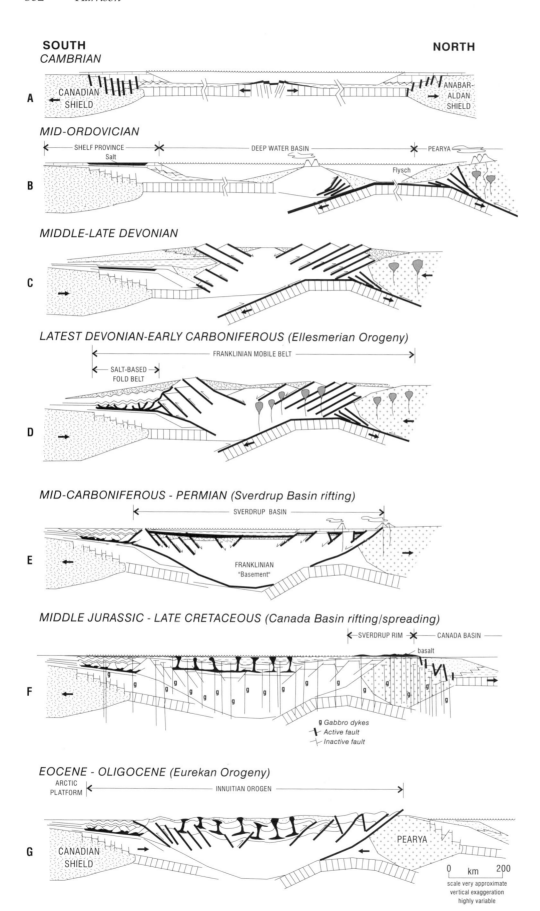

SOUTH / NORTH

Figure 3—Schematic cross sections illustrating the Phanerozoic tectonic evolution of the Arctic Islands region. See Figure 1 for location of sections. The salt-based foldbelt was formed in the latest Devonian–Early Carboniferous (D) and was only modestly affected by subsequent tectonics. This belt should not be confused with the diapirism and contraction affecting middle Carboniferous evaporites of the Sverdrup Basin (G).

basin-fill shale formation near the top of the succession that, like the salt at depth, has thickened within some anticlines to resolve space problems during buckling of shallow overburden. This second ductile unit elucidates the relative timing of fold evolution and the extent and activity of upper detachment surfaces as provided by seismic constraints and surface structural measurements. (In this account, the placement of bedding-parallel faults on the boundary of both ductile layers merely fulfills the geometric necessity of accounting for discrepancies in shortening of the deformed succession rather than being actual shear zones recognizable in outcrop, core, or cuttings.)

Arising from these observations and the restoration of a set of regional structural cross sections is a general kinematic model for the foldbelt that features a triangle zone structure (passive roof duplex) up to 200 km wide and potentially enveloping the entire 470-km length of the belt.

This paper illustrates and reviews the distinctive structural style in the western half of the Parry Islands foldbelt, with special emphasis on deformation kinematics and seismic expression of salt and other evaporites in the compressional setting. Many of the ideas and observations presented here have also been stated elsewhere; most notably in Fox (1985), Harrison and Bally (1988), Harrison et al. (1991), and Harrison (1995). Many of the illustrations have been modified or simplified from Geological Survey of Canada, Map 1844A and Bulletin 472 (Harrison, 1994a, 1995).

REGIONAL GEOLOGY

Geologic provinces of Canada's Arctic Islands are shown in Figure 1. Peneplained Archean and Proterozoic rocks of the Canadian shield are exposed on Banks, Victoria, Baffin, eastern Devon, and southeastern Ellesmere islands and in Greenland. Nonconformable flat-lying Cambrian–Devonian cover is widespread in these areas and is included in the Arctic platform. Farther north and west is the Innuitian orogen, which comprises Proterozoic–Tertiary strata variably affected by protracted collisional deformation in both the Paleozoic and Tertiary. Structural provinces of the Paleozoic orogen are assigned to the (1) Franklinian mobile belt, including foreland fold and thrust belts (central Ellesmere, Jones Sound, and Parry Islands belts); (2) hinterland fold and thrust belts, involving deep-water sediments and volcanic rocks (north Greenland, Hazen, Clements Markham, Northern Heiberg, and Canrobert Hills belts); and (3) accreted terranes (Pearya). The Franklinian mobile belt is overlain with pronounced angular unconformity by Carboniferous–Eocene strata of the Sverdrup Basin. Cretaceous and Paleogene strata are also preserved in rift-related basins peripheral to Baffin Bay (Eclipse trough and Lancaster Sound basins) and the Arctic Ocean (Banks Basin and Eglinton Graben) and in scattered erosional remnants on many islands. Eocene and all older rocks

were affected to varying degrees by the middle Tertiary Eurekan orogeny, the effects of which are most pronounced on Ellesmere and Axel Heiberg islands.

TECTONIC HISTORY

Trailing margin shelf, slope, and ocean basin depositional realms were already firmly established in the Arctic Islands region by the Early Cambrian (Figure 3A) following a period of Late Proterozoic(?) rifting between ancestral North America and a second continental landmass that may have included the Anabar-Aldan shield of eastern Siberia (Condie and Rosen, 1994).

Sediment deposition on the shelf and craton was more extensive in the Ordovician (Figure 3B) and included the widespread accumulation of evaporites in the Tremadoc (Baumann Fiord Formation of the eastern Arctic Islands) and Arenig-Llanvirn (Bay Fiord Formation). Coeval sediments deposited in the deep-water basin include chert and graptolitic shales. These are now exposed in the Canrobert and Hazen belts and, at greater distance, arc-related volcanic rocks of the Clements Markham belt. Forearc collisional deformation and plutonic activity are recorded for this time period in Pearya (Trettin, 1991a; Trettin et al., 1991).

Evidence for the progressive closure of the deep-water realm is provided by progradation of Silurian–Lower Devonian orogen-derived deep-sea fan deposits (flysch) transported from the Caledonian belt of eastern Greenland and by a Middle–Upper Devonian clastic wedge transported from both the Caledonides and the evolving Franklinian mobile belt (Figure 3C) (Embry, 1991a; Trettin et al., 1991; McNicoll et al., 1995).

The final phase in the evolution of the Paleozoic orogen occurred in the late Famennian–Viséan and produced foreland folds and thrusts on Ellesmere Island, the Parry Islands, and elsewhere (Figure 3D) (Harrison et al., 1991; Okulitch, 1991). This was the principal interval of deformation associated with the development of the salt-based foldbelt discussed in the present paper. Synorogenic clastics were also likely to have been deposited during this time (to depths of 1–8 km according to Embry, 1991a). However, these strata have been entirely stripped during subsequent tectonic events.

Rift-related structures of the embryonic Sverdrup Basin have developed across the peneplained roots of the Franklinian mobile belt (Figure 3E), including the northern margin of the salt-based foldbelt. This was probably in response to thermal collapse of the underlying orogen (Stephenson et al., 1987). Extensional growth faulting with redbed alluvial fan progradation, evaporite deposition, and scattered rift-marginal volcanism persisted episodically from the Early Carboniferous (late Viséan) to Late Permian. Nevertheless, distinct shelf and deep-water realms were already well established in the Sverdrup Basin by middle Carboniferous (Bashkirian) time (Davies and Nassichuk, 1991).

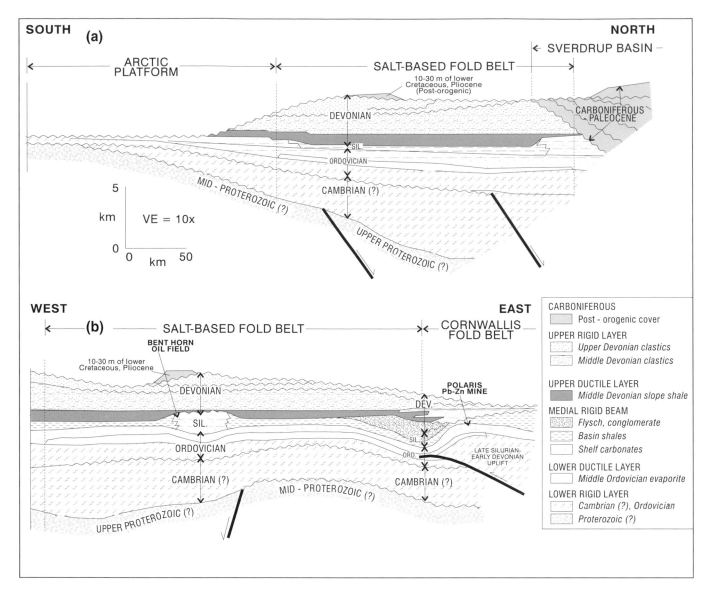

Figure 4—Stratigraphic relationships across the Parry Islands foldbelt in the western Canadian Arctic Islands. (a) North–south section from the Sverdrup Basin to the Arctic platform. (b) East–west section from the Cornwallis foldbelt to the Canrobert Hills foldbelt.

A repetitive pattern of passive subsidence, basinward sand sheet progradation, and shelfward marine transgression persisted in the Sverdrup Basin from the Triassic to Middle Jurassic. Episodic diapirism of Carboniferous evaporites was an important feature of basin evolution (Figure 3F) (Embry, 1991b). However, these events had little effect on the salt-based foldbelt. Renewed rifting centered over the future Canada Basin (part of the modern Arctic Ocean Basin) in the Middle Jurassic, producing shelf-parallel rift basins including Eglinton graben and Banks Basin, both located west of Melville Island (Figure 1). Gabbro dike swarms and general lithospheric extension were widespread in the Cretaceous. Related to this were coeval flood basalts on northern Ellesmere Island and on the submerged and now dormant Alpha ridge spreading center (Embry, 1991b).

Eocene to pre-Miocene intraplate convergence between Greenland and North America produced out-of-basin thrusts, folds, and regional uplift of the entire Sverdrup Basin and of Franklinian "basement" (Figure 3G). Currently manifested as mountainous topography and as pronounced shortening on Axel Heiberg and Ellesmere islands, these events are kinematically linked to coincident sea floor spreading in Baffin Bay and the North Atlantic (de Paor et al., 1989; Okulitch and Trettin, 1991). In contrast, the western Arctic Islands, including the salt-based foldbelt, presently feature a dissected and partially flooded, low-relief plateau (Figure 2) overlain unconformably by Cretaceous outliers and, to the northwest, by Neogene strata of the continental terrace wedge of the Arctic Ocean (Harrison, 1995; Trettin, 1991b).

ARCTIC EVAPORITES

Three major evaporite formations have played key roles in the structural style and evolution of the Innuitian orogen in Arctic Canada (Figure 1).

Otto Fiord Formation

The Early–middle Carboniferous (Serpukhovian–Moscovian) Otto Fiord Formation was deposited in a rift tectonic setting within the axial depression of the embryonic Sverdrup Basin (Figure 3E) (Davies and Nassichuk, 1991). Diapirs of salt and gypsum with intruded gabbro bodies are prominent in the central Sverdrup Basin and are known to have been active since the Early Triassic (Fox, 1983). On western Axel Heiberg Island, Otto Fiord evaporites have migrated upsection and now occur as bedding-parallel tongues and sheets in folded Lower Cretaceous strata. Diapirs have also penetrated through Tertiary anticlines producing a distinctive "wall-and-basin" style of compressive deformation (van Berkel et al., 1985; Thorsteinsson, 1971a). On eastern Axel Heiberg and northern Ellesmere islands, middle Tertiary thrusts have also placed the Otto Fiord Formation over Eocene and older strata (Thorsteinsson, 1971b, 1972).

Baumann Fiord Formation

On central Ellesmere Island, Early Ordovician (middle Tremadoc) evaporites of the Baumann Fiord Formation (mostly gypsum and restricted marine carbonates) were deposited in an extensive but shallow intrashelf basin (de Freitas et al., 1994) (Figure 1). These evaporites have facilitated both Tertiary and Paleozoic phases of contraction and represent a widespread ductile unit and intermediate detachment level for east-vergent thrusts rising from within or below the Lower Cambrian in the west to upper detachment levels in the Tertiary in the east (Harrison et al., 1994). Gypsum-bearing Tremadoc evaporites are also represented by the coeval but generally undeformed Poulsen Cliff and Nygaard Bay formations of Washington Land in northwestern Greenland (Higgins et al., 1991). The Baumann Fiord Formation is unknown in the western Arctic Islands.

Bay Fiord Formation

The third major evaporite lies in the lower part of the Ordovician (late Arenig–Llanvirn) Bay Fiord Formation (Trettin et al., 1991; Cape Webster Formation of Greenland, Higgins et al., 1991) and was deposited in a shallow intrashelf basin similar to that of the Tremadoc evaporites but more extensive (Figures 3b, 4). On Ellesmere Island, this formation is thin and contains gypsum and thin-bedded carbonates. The unit acts as an intermediate detachment level for east-vergent and eastward-climbing thrusts above the generally thicker Baumann Fiord Formation (Harrison, et al., 1994). The lower Bay Fiord Formation is substantially thicker and

continuously preserved throughout Bathurst Island and the eastern half of Melville Island (Figure 4), where at least three exploratory wells have intersected intervals of rock salt (halite) and other evaporites (Figure 5) (Mayr, 1980; Harrison, 1995). These rocks comprise the critical ductile unit responsible in many ways for the existence of the Parry Islands foldbelt and its associated subsurface thin-skinned fold and thrust system in the western Arctic Islands (Fox, 1983, 1985; Harrison and Bally, 1988).

Miscellaneous Evaporites

Other evaporitic strata, comprising mostly gypsum with variable carbonates, redbeds, and other clastic rocks, have only minor or local tectonic significance. Cited occurrences are found in the Cape Crauford (Ludlow), Sophia Lake (Lochkovian), Vendom Fiord (Pragian), medial Blue Fiord (Emsian), lower Bird Fiord (Emsian), Antoinette (Moscovian–Asselian), and Mount Bayley (Asselian) formations (Goodbody, 1989; Harrison et al., 1994; Thorsteinsson, 1974; Trettin et al., 1991).

TECTONO-STRATIGRAPHIC UNITS

Unlike many salt diapir provinces where structural style is overwhelmingly governed by the intrusion of rising salt masses, the salt-bearing Bay Fiord Formation in the Parry Islands foldbelt is only one, albeit significant, layer in a tectonic sandwich of five layers more than 10 km thick (Figures 4, 6) (Harrison, 1995). The layers include three relatively rigid units separated by two more distinctly ductile and weak units.

Lower Rigid Layer

The basal lower rigid layer (abbreviated LRL in the figures) (Figures 4, 6) comprises previously deformed Proterozoic(?) assemblages interpreted from seismic reflection profiles. It is overlain with a pronounced angular unconformity by 4–7 km of presumed Lower Cambrian through definite Lower Ordovician (middle Arenig) strata. Shelf dolostones are known in the upper part. The lower rigid layer mainly represents a flat-lying floor to deformation beneath the thin-skinned foldbelt. However, deep-seated folds (also of latest Devonian–Early Carboniferous age) locally affect this entire succession. In addition, thrusts and transpressional strike-slip faults are interpreted to pass through the lower rigid layer and emerge onto the décollement at the base of the overlying lower ductile layer.

Lower Ductile Layer

The lower ductile layer (abbreviated LDL in the figures), which marks the base of the foldbelt in many areas, correlates with the salt-bearing strata of the Middle Ordovician lower Bay Fiord Formation (Figures 4–7). This unit, defined seismically by bounding reflectors, currently ranges from less than 60 m to 2200 m thick and varies greatly in local thickness. The evaporites include interbedded rock salt, anhydrite, and dolostone, with lesser thicknesses of marl and limestone and traces of carnallite. Surface exposures and core collected from several anticlinal culminations feature thin boudinaged beds, minor folds, steep dips, tectonic brecciation, and coarse halite veins (up to 2 cm grains). A key structural element of the lower ductile layer is a bedding-parallel décollement identified throughout the foldbelt subsurface that carries up to 28 km of southward-directed slip as measured above the northern depositional limit of evaporites. In this paper, the base of the lower ductile layer is taken to be the décollement surface. However, it is equally likely that foreland-directed strain has been accommodated more or less evenly across other salt units within the lower ductile layer and that slip across the base of the lowest salt unit is much less than the measured total.

Medial Rigid Beam

The medial rigid beam (abbreviated MRB in the figures) represents a stiff panel of Middle Ordovician (Llanvirn) through Middle Devonian (late Emsian–early Eifelian) shelf carbonates, graptolitic shales, and flysch 1.3–3.2 km thick (Figure 4). The beam is bounded by the two ductile layers and includes four significant internal and bounding reflection surfaces that correlate with a similar number of formations intersected in drill holes and also mapped locally at the surface (Figure 6). The lower part of the beam features incipiently cleaved shelf dolostones and lesser limestone with intense calcite veining and tectonic brecciation, particularly in pop-up structures and in the footwall of major and minor faults. The upper part of the beam displays incipiently cleaved graptolitic shales, argillaceous limestone, chert, siliciclastic turbidites, and lesser shelf carbonates. The shelf carbonates grade into basin facies in the Silurian–Lower Devonian (upper) parts of the beam (Figures 4, 6, 7). Fissile strata are intensely kink banded, and evidence of bedding plane slip is common. Major and minor thrusts in this part of the beam lie preferentially along kink band boundaries. Larger structures in the beam include faulted buckles and thrust ramps that link lower and upper décollements within the bounding lower and upper ductile layers, respectively.

Upper Ductile Layer

Basin-fill shales of the Eifelian Cape de Bray Formation (up to 950 m), which lie at the base of a thick clastic succession, form the upper ductile layer (abbreviated UDL in the figures). Large-scale southwest-facing clinoforms in the upper part of the Cape de Bray Formation are a prominent primary depositional feature at the surface and on many seismic profiles (Figure 6). Elsewhere, the upper ductile layer is tectonically thickened up to 230% below some surface anticlines and in front of thrusts that emerge from the medial rigid beam. Other low-amplitude

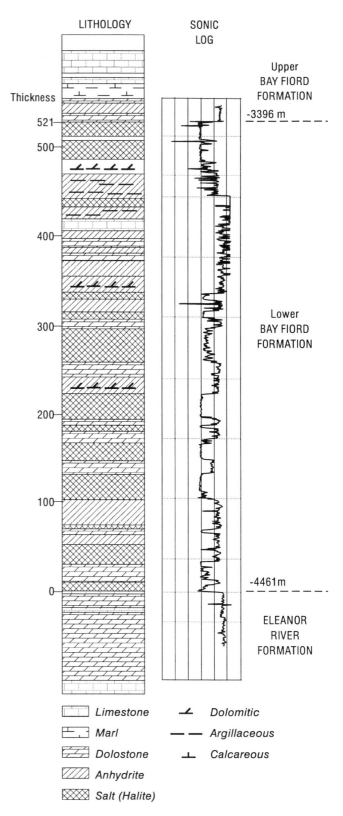

Figure 5—Sonic log and lithostratigraphy (from cuttings and well log analysis) of the lower Bay Fiord Formation in the Panarctic et al. Sabine Bay A-07 well (reproduced from Harrison, 1995). Proportions of component lithologies are calculated as 42% bedded rock salt, 30% dolostone, and 28% anhydrite.

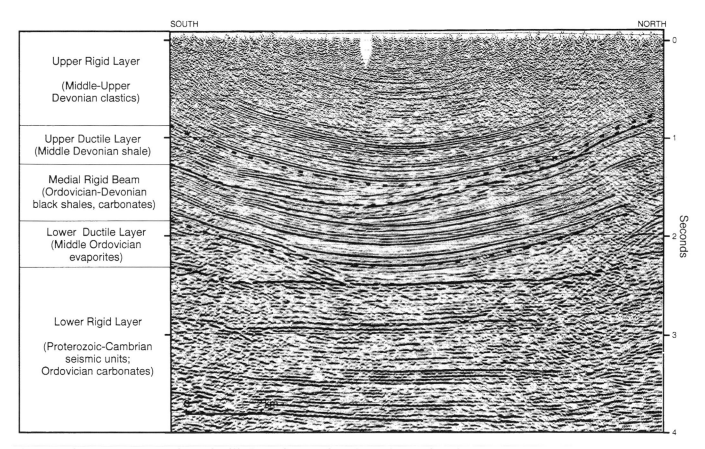

Figure 6—Seismic section of Cambrian(?)–Devonian stratigraphy and tectonic units, southeastern Melville Island. A regionally mappable facies change southward from shelf carbonates to condensed basinal shales is imaged on this profile in the upper part of the medial rigid beam. Vertical exaggeration is ~1.0 at 2 sec and ~1.5 at the surface (modified from Harrison, 1995). Additional stratigraphic data are provided in the legend of Figure 11 (second foldout sheet).

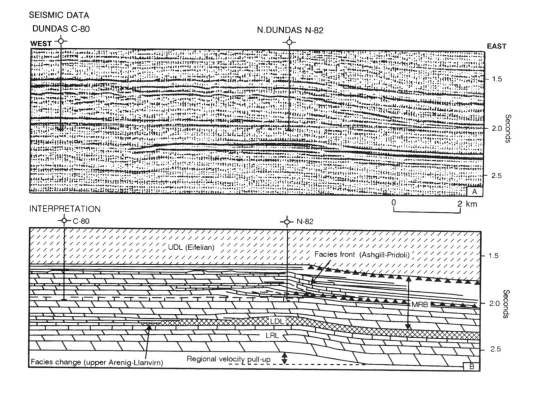

Figure 7—(A) Seismic section and (B) interpreted line drawing of the westward pinch-out of the evaporitic member of the Bay Fiord Formation (lower ductile layer) and other primary depositional features of southern Melville Island (modified from Harrison, 1995). See Figure 11 for legend and section E for location.

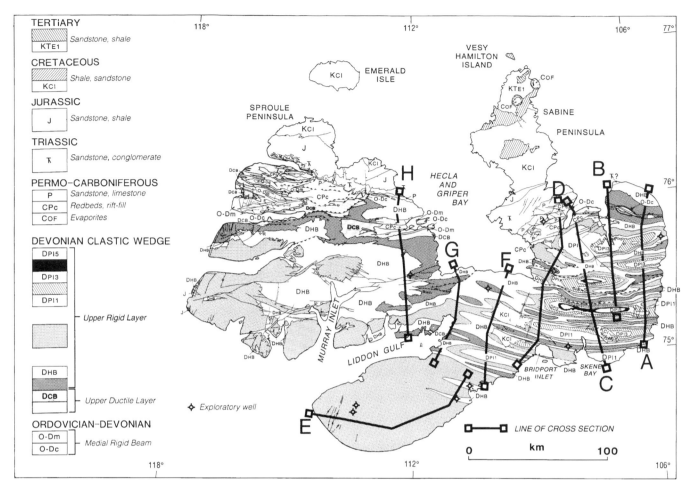

Figure 8—Surface geology of Melville Island, including the western half of the Parry Islands foldbelt (modified from Harrison, 1994a,b). Located exploratory wells are those used in cross-section construction (Figure 11).

anticlines feature compactional thinning of the upper ductile layer. Otherwise, the unit appears macroscopically undeformed beneath most synclines. Surface exposures of the upper ductile layer include isoclinal minor folds, thin boudinaged sandstone beds, and tightly spaced fractures in centimeter-scale kink bands and bedding-parallel slip surfaces. A regional décollement surface (or zone) may lie in or immediately above the upper ductile layer.

Upper Rigid Layer

The upper rigid layer (abbreviated URL in the figures) comprises four sandstone-dominated formations (max. 3.6 km preserved) in a widespread and widely exposed Middle–Upper Devonian (Eifelian–middle Famennian) foreland clastic wedge (Figures 2, 4, 6). These four formations are further divisible into eight map units (Figure 8). Although these strata are orogen derived, there is no local evidence of synorogenic sediment accumulation or thrust-associated growth faulting. The dominant structural features in the upper rigid layer are upright anticlines and synclines and scattered north- and south-vergent thrusts (Figures 2, 8). Noticeably rare in outcrop and

drill cores are minor structures such as cleavage, minor folds, kink bands, slip surfaces, and strain fabrics. In contrast, sedimentary structures are well preserved.

SURFACE STRUCTURE

Regional-scale geologic mapping and seismic interpretation point to a close correspondence between the distribution of long, evenly spaced arcuate folds at surface and the distribution of salt at depth (Figures 8, 9). Individual anticlines are up to 330 km long and typically have wavelengths of 12–17 km. Axial planes commonly dip between 80° and 90° with no preferred vergence. Fold amplitudes range from less than 350 m for the folds nearest the platform in the south to about 3200 m for larger folds in the center of the foldbelt. Depth of erosion is fairly uniform throughout the belt.

On eastern Melville Island, for example, the top of the upper ductile unit (upper Eifelian) is at or near the surface in each of seven parallel anticlines, and middle Famennian strata represent the youngest beds in each intervening syncline. Fold axes plunge variably from 0° to

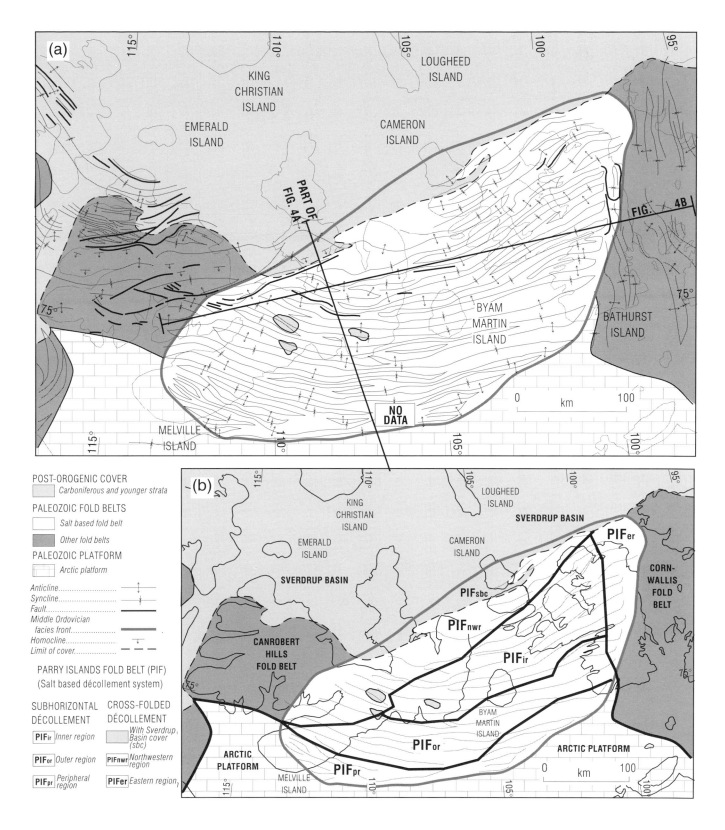

Figure 9—(a) Fold axial traces in the Franklinian mobile belt of the western Canadian Arctic Islands (modified from Harrison, 1995). All these folds are middle Paleozoic in age and have been projected to the surface where covered by postorogenic Carboniferous and younger strata. (b) Structural subprovinces of the Parry Islands foldbelt.

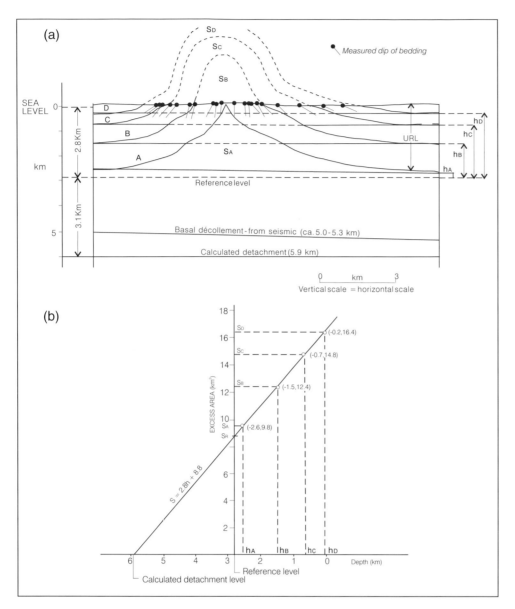

Figure 10—Near-surface structure and depth to detachment of Beverley Inlet anticline (section C, Figure 11). (a) Structural section based on surface geology and calculated depths to detachment. (b) Depths to the reference level plotted against excess area under four mapped geologic contacts (base of units A–D). Based on the method of Epard and Groshong (1993).

~6°. Nevertheless, the foldbelt above the upper ductile layer as a whole plunges at 0.7° to the west-southwest across Bathurst Island, passes through a plunge depression, then plunges 0.4° east-southeast on central Melville Island. The arcuate shape of the foldbelt, the termination of individual folds by sinistral shear against the eastern and southern limit of subsurface salt (Figure 9), and the orientation of subsurface thrusts define a subhorizontal and S5°W direction of overall compressive tectonic transport during folding (Harrison, 1995). The generally uniform depth of erosion across the surface foldbelt implies an equally uniform, shallow regional dip on the basal décollement under the foldbelt.

The southward direction of tectonic transport is oblique to the foreland and hinterland limits of salt and most other facies belts within the lower Paleozoic succession (Harrison, 1995). Isopach maps of Cambrian and Ordovician strata also indicate that the lower Paleozoic continental margin miogeocline faced a basinal realm to the northwest (Trettin et al., 1991). Similarly, seismic clinoforms and other paleocurrents in the Devonian clastic wedge imply southwestward delta progradation and basin filling (McNicoll et al., 1995). This direction is oblique to both the regional tectonic transport direction and lower Paleozoic facies belts. This obliquity implies that differential sediment accumulation is unlikely to have played a significant part in initiating foldbelt growth during deposition of the preserved foreland sediment wedge.

DEPTH TO DETACHMENT

Detailed geologic mapping along with bedding dips measured over one of the typical regional anticlines on eastern Melville Island (Figure 10) provided the data for a depth to detachment calculation following the procedure of Epard and Groshong (1993). This anticline features vertical bedding on both fold limbs in the immediate axial area. A significant space problem had apparently developed during folding, which must have been accommodated by a ductile response in strata located in the immediate subsurface. The reference level in Figure 10a ($h = 2.8$ km below sea level) provides a first estimate of the depth to detachment based on qualitative appraisal of the surface geology. To obtain the true depth to detachment, the depth to the reference level ($h_A - h_D$) is plotted against the excess area ($S_A - S_D$) under each of four horizons (A to D) in the upper rigid layer (Figure 10b). The line of best fit has a slope equal to the magnitude of shortening (2.8 km) and intersects the abscissa ($S = 0$) at the calculated depth to detachment (5.9 km below sea level). This is not significantly deeper than the 5.0–5.3 km depth to detachment obtained from the time to depth conversion of local seismic reflection profiles.

Using the continuous seismic reflectors from the base and top of the salt as reliable guides, the salt décollement is calculated to lie variably at or below 3.5 km on eastern Bathurst Island to an average maximum depth of 5.8 km below central Melville Island.

CROSS-SECTION CONSTRUCTION

The following account is illustrated with eight regional structure sections and one composite restored section (Figures 11, 12; simplified from Harrison, 1994a). Figure 8 shows the location of the sections. Examples of the migrated seismic data incorporated in the cross sections are given elsewhere in this paper. The method of cross section construction outlined here is similar to that described in detail by Harrison (1995). Bed lengths of six units in the lower rigid layer and four units in the medial rigid beam were measured and restored between terminal and intermediate pin lines generally located within each syncline. The lower ductile (salt) layer and upper ductile (shale) layer were each separately area balanced between the same pin lines and restored to lengths matching that of the balanced units in the medial rigid beam.

Because the deformed bed length of the upper rigid layer is apparently shorter than that of the thrust-faulted units in the beam, the upper rigid layer was also line-length balanced (Figure 12). The decision to line-length balance the upper rigid layer assumes that the magnitude of shortening in this layer is insignificant at the mesoscopic or microscopic scale, an assumption supported by available field evidence. The implication is that foreland-directed slip on the salt-based décollement is matched in part by hinterland-directed slip on an upper detachment located in or immediately above the upper ductile layer. An alternative model, requiring area balancing of the upper rigid layer, has been promoted by Harrison (1995) and assumes that a significant component of unrecognized strain in this layer can account for the shortfall in macroscopic shortening.

TECTONIC REGIONS OF THE SALT-BASED FOLDBELT

Tectonic provinces surrounding the Parry Islands foldbelt are shown in Figure 9 and include the following:

1. The flat-lying and generally undeformed lower Paleozoic strata of the *Arctic platform* located to the south (Figure 7; section E in Figure 11)
2. The postorogenic *Sverdrup Basin* to the north
3. The northward-trending *Boothia uplift* and its folded cover succession (the *Cornwallis foldbelt*) on and beyond easternmost Bathurst Island
4. The east-trending *Canrobert Hills foldbelt* with its Middle Cambrian detachment and interference cross folds along its eastern contact with the salt-based foldbelt.

The salt-based Parry Islands foldbelt is further subdivided into six distinct tectonic regions. The *peripheral* (PIF$_{pr}$ in Figure 9), *outer* (PIF$_{or}$), and *inner* (PIF$_{ir}$) regions share a common subhorizontal salt-involved décollement. In general, the scale and style of deformation increases progressively from the foreland edge of the peripheral region (section E) to the inner region (sections A–D). If fold and thrust belts expand outward toward the foreland through time, as proposed by Bally et al. (1966), then the progression from simple folds on the periphery to complex folds in the inner part of the salt-based foldbelt can provide a kinematic model for this evolution (Harrison and Bally, 1988).

The *northwestern cross-folded region* (PIF$_{nwr}$ in Figure 9) of central Melville Island represents a zone of at least one latest Devonian–Early Carboniferous phase of cross folding superposed on the salt-based décollement system (sections A–D, F, G). Nevertheless, slip on some deep-seated thrusts is carried forward and dissipated in the salt-based décollement system (sections C, D, G, H). All deep-seated structures here are kinematically linked to southward-directed shortening within the partly coeval and partly younger Canrobert Hills foldbelt (to the west).

The *eastern cross-folded region* (PIF$_{er}$ in Figure 9) represents a belt of north-trending deep-seated folds that have refolded the salt-based décollement system on Bathurst Island. This deep fold system is kinematically linked to westward-directed shortening documented in the adjacent Cornwallis foldbelt, which was likely most active in the middle Tertiary. Other more geographically restricted deformation in the Cornwallis foldbelt is dated by a sub-Emsian angular unconformity on Cornwallis Island and by uplift fringing Late Silurian–Early Devonian syntectonic sedimentary deposits (Figure 4b) (Thorsteinsson, 1986; Thorsteinsson and Uyeno, 1980). The tectonic history and styles of deformation in the eastern cross-folded region are beyond the scope of the present paper. Additional details are provided by Kerr (1974), Fox (1985), Okulitch et al. (1991), de Freitas et al. (1993), and Harrison et al. (1993).

The hinterland edge of the cross-folded salt-based décollement system is partly covered in the north by the postorogenic Sverdrup Basin (region PIF$_{sbc}$ in Figure 9). Carboniferous half-grabens here are kinematically linked to reactivated extensional slip on the Ordovician salt and collapse of deep-seated subsalt anticlinoria (section C, Figure 11).

Peripheral Region

On southern Melville Island, the seismically defined foreland limit of salt lies along the limit of compressive deformation (Figure 7; section E, Figure 11). The first signs of shortening northeast of the deformation front are open, short-wavelength (6–7 km), faulted buckle folds in the medial rigid beam (section E, Figure 13). Fold amplitude is greatest in the beam, progressively diminishes upsection, and is undetected in the clastic rocks of the upper rigid layer at the surface. Fault offsets are invariably present but small. Thrusts appear to have nucleated in the most competent medial part of the beam as a consequence of brittle failure as buckling continued. The faults then propagated from the point of failure toward bounding detachment surfaces above and below.

The base of the salt layer here is essentially horizontal (northeastward dip of only 0.1°). The overlying salt ranges from a deformed thickness of ~250 m below synclines to ~550 m in salt *welts* (compressively thickened, ductile cores of anticlines; Harrison and Bally, 1988). Area balancing and restoration calculations indicate that the original salt thickness here was ~300 m (Figure 14). Therefore, salt (and other evaporites) have been pulled (or sucked) along pressure gradients from synclines to adjacent anticlines. Potential internal geometries of these and other salt welts are shown in Figure 15.

Because horizontal shortening has occurred in all units above the lower ductile layer, a subhorizontal and foreland-vergent slip zone (or surface) is required in each of the five tectonic models near the base of the shortened units. Models A and B, although resembling the shape of welts, are actually rigidly deformed duplex anticlines (and therefore not true welts). In models C and D, the deformed strata are almost homogeneously ductile. The basal sliding zone is overlain by either a ductile box-fold with conjugate thrusts (C) or unfaulted similar folds with downward decreasing amplitudes (D). The final model is based on contrasting lithologies and deformation styles in the welt (Figure 15E). This model best fits the style of deformation and mix of rock types actually encountered in drill cores and outcrop. It is important to emphasize that in each of the last three welt models, ductile rocks can be introduced from either or both adjacent synclines. However, net slip within the basal sliding zone must be in one direction only—toward the foreland.

The opposite direction of transport, from anticlines to synclines of shale in the upper ductile layer, is also implied by the cross section of the peripheral folds. However, other processes, such as differential sediment compaction over anticlines, may also have operated.

Outer Region

The boundary between the peripheral and outer regions of the foldbelt (Figure 9) is drawn on the foreland side of the southernmost anticline that has a surface-mappable expression in the Devonian clastic rocks of the upper rigid layer. On southern Melville Island, this line follows the hinge of the syncline east of Hearne Point anticline (section E, Figure 11). Absolute and percentage shortening in the medial rigid beam from the foreland limit of deformation to this point is ~1 km and 3.5%. In contrast, accumulated visible shortening in the bounding upper and lower rigid layers is insignificant.

The outer region of the foldbelt on southern Melville Island includes a salt-based décollement with a measured northward dip of only 0.1° to 0.4° which underlies four long wavelength anticlines on sections E and F. Similar salt-based tectonic structures, cross-folded at depth, are on other parts of sections F and G. Figure 9 reveals that several of these anticlines exceed 300 km in length, pass eastward into the peripheral region of the foldbelt, and gradationally terminate near the southern depositional limit of salt.

Anticlines here have recognizable surface expression in the upper rigid layer and have more complexly organized internal thrust faults in the medial rigid beam (than in the peripheral region). Above the basal décollement surface, the tip of other thrusts is exclusively beneath each surface anticline. Vertical limits to faulting are strictly defined by the salt-based décollement and an upper detachment lying within or immediately above the upper ductile layer.

The lateral migration of salt into welts coring each anticline was accompanied by salt migration to parasitic welts under parasitic buckle folds (Figure 16). The larger welts have backthrusts that carry salt over encapsulated fragments of the medial rigid beam (Figure 16). Salt thickness ranges from 1200 m in these welts to 220 m under synclines. Restored thickness of the salt is 400–550 m (Figure 14).

Shale in the upper ductile layer shows the most pronounced thickness increase in front of thrust faults emerging upsection from the beam. These local changes in thickness could be accounted for by brittle intraformational imbrication, but the deformation mechanism is uncertain.

Inner Region

The boundary between the outer and inner regions of the foldbelt (Figure 9) is drawn on the foreland side of the southernmost anticline cored by a shale welt in the upper ductile layer above a similar salt welt in the lower ductile layer. In the upper rigid layer a shale welt under a surface anticline is indicated by steep to vertical bedding on one or both fold limbs near the fold hinge (Figure 10). On Melville Island, the boundary of the inner region of the foldbelt follows the surface trace of the first syncline southwest of Robertson Point anticline on section A

(Text continues on p. 398)

SECTION D
SOUTHWEST

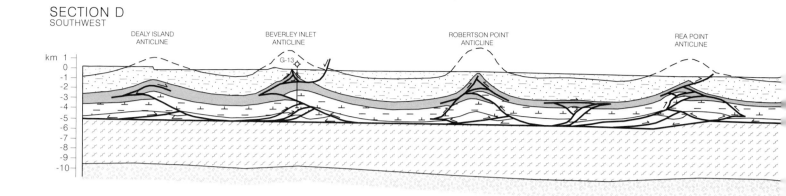

DEALY ISLAND
ANTICLINE

BEVERLEY INLET
ANTICLINE

ROBERTSON POINT
ANTICLINE

REA POINT
ANTICLINE

G-13

km 1
0
-1
-2
-3
-4
-5
-6
-7
-8
-9
-10

SECTION E

WEST

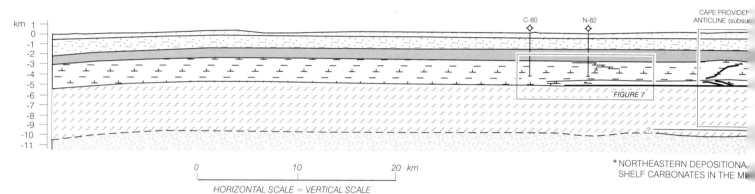

C-80

N-82

CAPE PROVIDEN
ANTICLINE (subsu

km 1
0
-1
-2
-3
-4
-5
-6
-7
-8
-9
-10
-11

FIGURE 7

0   10   20 km

HORIZONTAL SCALE = VERTICAL SCALE

* NORTHEASTERN DEPOSITIONA
SHELF CARBONATES IN THE M

SECTION F

SOUTHWEST

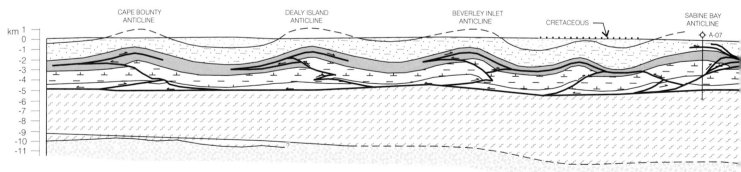

CAPE BOUNTY
ANTICLINE

DEALY ISLAND
ANTICLINE

BEVERLEY INLET
ANTICLINE

CRETACEOUS

SABINE BAY
ANTICLINE

A-07

km 1
0
-1
-2
-3
-4
-5
-6
-7
-8
-9
-10
-11

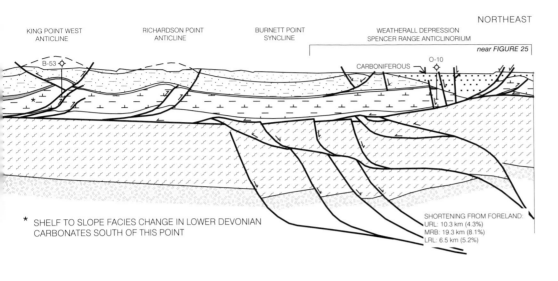

NORTHEAST

KING POINT WEST
ANTICLINE

RICHARDSON POINT
ANTICLINE

BURNETT POINT
SYNCLINE

WEATHERALL DEPRESSION
SPENCER RANGE ANTICLINORIUM

near FIGURE 25

B-53

CARBONIFEROUS

O-10

* SHELF TO SLOPE FACIES CHANGE IN LOWER DEVONIAN
CARBONATES SOUTH OF THIS POINT

SHORTENING FROM FORELAND:
URL: 10.3 km (4.3%)
MRB: 19.3 km (8.1%)
LRL: 6.5 km (5.2%)

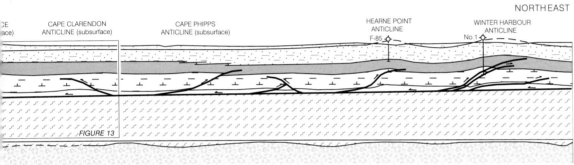

NORTHEAST

CE
ace)

CAPE CLARENDON
ANTICLINE (subsurface)

CAPE PHIPPS
ANTICLINE (subsurface)

HEARNE POINT
ANTICLINE
F-85

WINTER HARBOUR
ANTICLINE
No.1

FIGURE 13

LIMIT OF SILURIAN
DIAL RIGID BEAM

SHORTENING FROM FORELAND:
URL: 1.1 km (1.3%)
MRB: 2.3 km (2.8%)
LRL: NIL

NORTHEAST

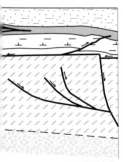

SHORTENING FROM FORELAND:
URL: 4.6 km (2.8%)
MRB: 10.6 km (6.2%)
LRL: 0.2 km (0.2%)

Figure 11 *(continued)*—Sections D–F

SECTION G

SOUTHWEST

NORTHEA

CAPE BOUNTY
ANTICLINE

DEALY ISLAND
ANTICLINE

MECHAM RIVER WEST
ANTICLINE

NIAS POINT
ANTICLINORIUM

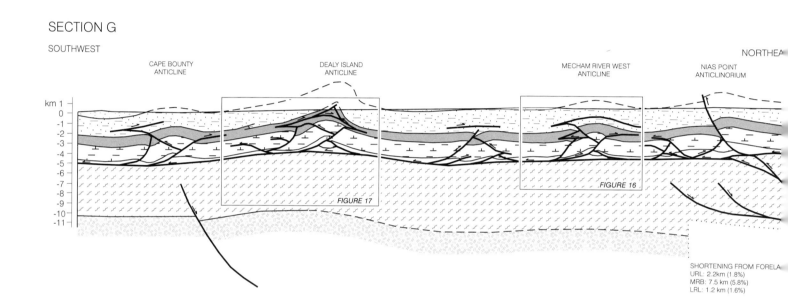

FIGURE 17

FIGURE 16

SHORTENING FROM FORELA
URL: 2.2km (1.8%)
MRB: 7.5 km (5.8%)
LRL: 1.2 km (1.6%)

SECTION H

SOUTH

near FIGURE 24

NIAS POINT
ANTICLINORIUM

APOLLO
ANTICLINE

C-73

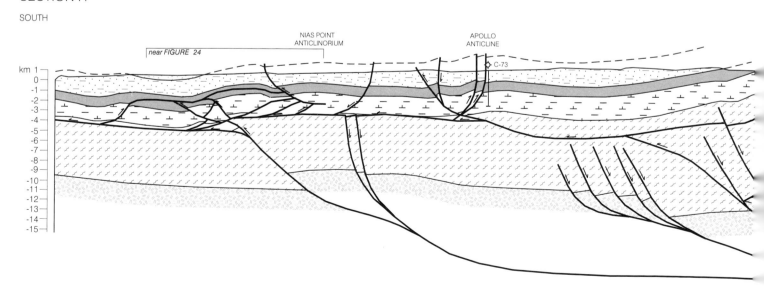

Figure 11 *(continued)*—Sections G and H and legend.

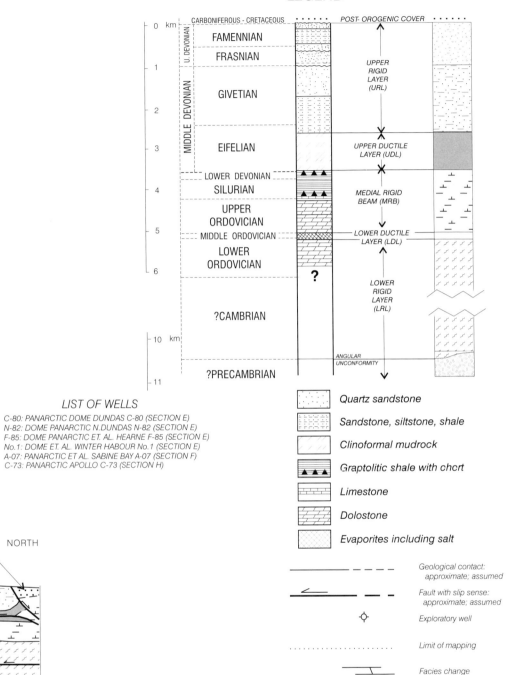

LEGEND

	CARBONIFEROUS - CRETACEOUS	POST- OROGENIC COVER

U. DEVONIAN	FAMENNIAN	
	FRASNIAN	UPPER RIGID LAYER (URL)
MIDDLE DEVONIAN	GIVETIAN	
	EIFELIAN	UPPER DUCTILE LAYER (UDL)
	LOWER DEVONIAN	
	SILURIAN	MEDIAL RIGID BEAM (MRB)
	UPPER ORDOVICIAN	
	MIDDLE ORDOVICIAN	LOWER DUCTILE LAYER (LDL)
	LOWER ORDOVICIAN	
		LOWER RIGID LAYER (LRL)
	?CAMBRIAN	
	?PRECAMBRIAN	ANGULAR UNCONFORMITY

Quartz sandstone

Sandstone, siltstone, shale

Clinoformal mudrock

Graptolitic shale with chert

Limestone

Dolostone

Evaporites including salt

Geological contact: approximate; assumed

Fault with slip sense: approximate; assumed

Exploratory well

Limit of mapping

Facies change

LIST OF WELLS

C-80: PANARCTIC DOME DUNDAS C-80 (SECTION E)
N-82: DOME PANARCTIC N.DUNDAS N-82 (SECTION E)
F-85: DOME PANARCTIC ET. AL. HEARNE F-85 (SECTION E)
No.1: DOME ET. AL. WINTER HABOUR No.1 (SECTION E)
A-07: PANARCTIC ET AL. SABINE BAY A-07 (SECTION F)
C-73: PANARCTIC APOLLO C-73 (SECTION H)

NORTH

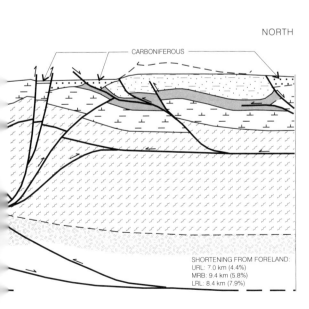

SHORTENING FROM FORELAND:
URL: 7.0 km (4.4%)
MRB: 9.4 km (5.8%)
LRL: 8.4 km (7.9%)

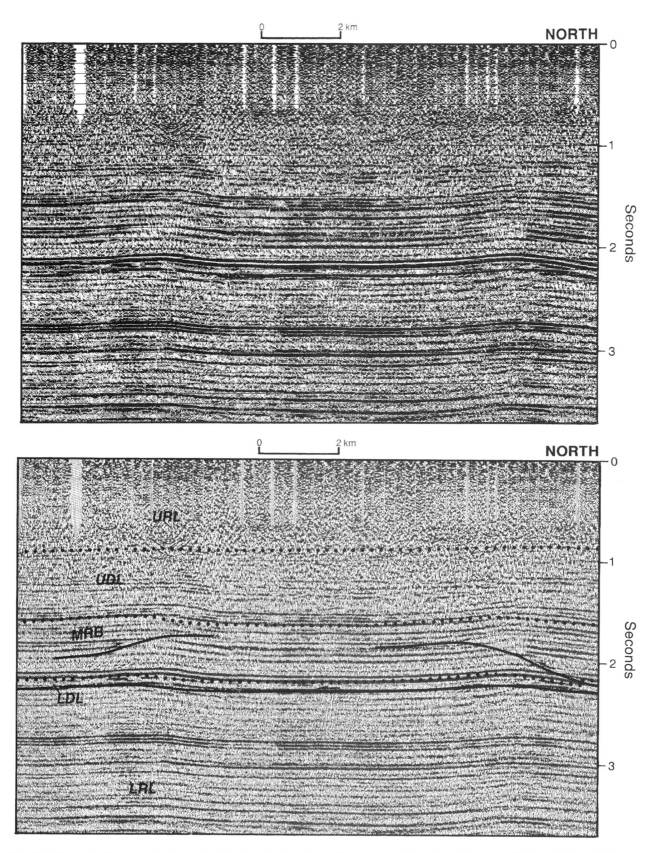

Figure 13—Uninterpreted (top) and interpreted (bottom) seismic profiles across Cape Providence anticline (left) and Cape Clarendon anticline (modified from Harrison, 1995). See section E in Figure 11 for location. Component rock types of each tectonic layer are described in the text and in the legend of Figure 11 (second foldout). Maximum stratigraphic separation on thrusts in the medial rigid beam is 25–30 msec (80–100 m).

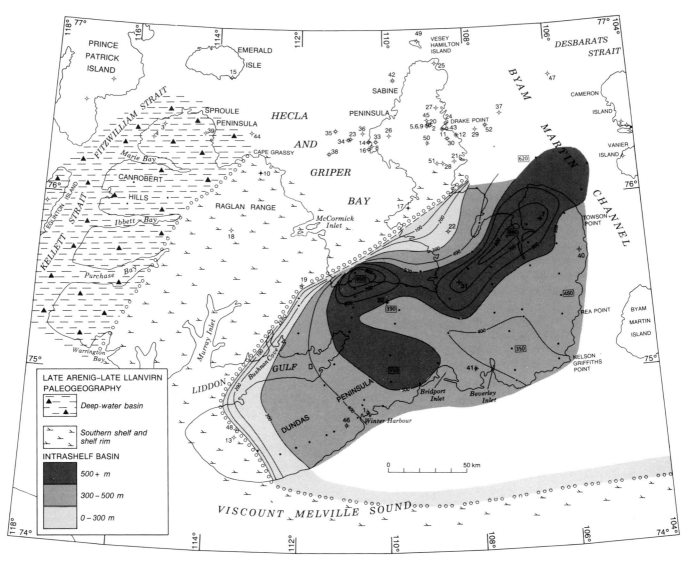

Figure 14—Paleogeography of Melville Island in late Arenig–late Llanvirn time and isopachs of the halite facies of the intrashelf lower Bay Fiord Formation. Facies limits are delineated by open circles. Boxed spot thicknesses and contours in meters are based on 68 data points (black dots) and an interval velocity of 5.3 km/sec. Data were obtained by area balancing and palinspastic restoration of the salt in sections A–H in Figure 11.

(Figure 11) and southwest of Beverley Inlet anticline on sections C and D. The total and percentage accumulated shortening from the foreland to the identified boundary on section D is 5.5 km and 6% at the level of the medial rigid beam. Visible shortening in the upper rigid layer is about half this amount.

Seismic examples of shale welts and related deformation in the beam and salt layer are provided by Figures 17, 18, and 19 and sections A–D in Figure 11. In general, migration of shale to anticlines (like that of salt) is a response to a space problem during buckling of the overlying rigid layer. However, hinge thickening of shale can be more subtle. Faulting in the medial rigid beam is more complex within the inner region anticlines. Parasitic thrusts and parasitic salt welts are particularly abundant on section A and on sections in Figures 17, 18, and 19. In some cases, salt welts and minor thrust anticlines extend

under adjacent synclines. In Figure 19, large pop-up structures and extreme variations in thickness of both ductile layers underlie a relatively gentle surface syncline. Outside the region of deep-seated folding, the salt lies on a décollement dipping northward at 0.4°–0.6°. Above this level, the salt layer has a tectonized thickness range of 60–2200 m. Locally, salt migration appears to have completely evacuated the lower ductile layer so that the medial rigid beam has welded onto the lower rigid layer beneath several synclines (Figure 19 and sections A and B). Restored depositional thicknesses of the salt layer range from 350 to >800 m (Figure 14).

The northern halves of sections A–D (and Figure 6) also show depositional thinning of the ductile shale unit due to shale onlap on a shallow marine carbonate bank that existed only here. Associated folds are transected by one or more thrusts that pass through a depositionally

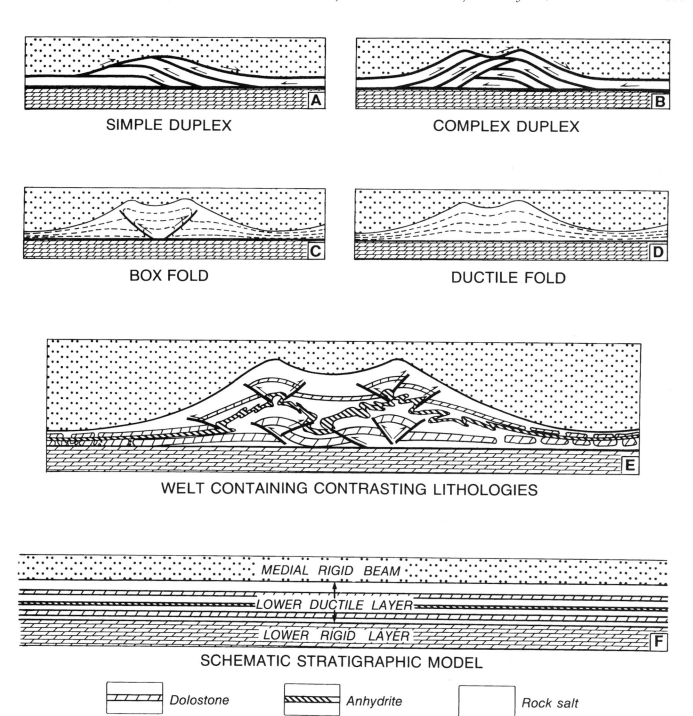

SIMPLE DUPLEX

COMPLEX DUPLEX

BOX FOLD

DUCTILE FOLD

WELT CONTAINING CONTRASTING LITHOLOGIES

MEDIAL RIGID BEAM

LOWER DUCTILE LAYER

LOWER RIGID LAYER

SCHEMATIC STRATIGRAPHIC MODEL

Dolostone   Anhydrite   Rock salt

Figure 15—Possible internal geometries of salt welts and related structures (modified from Harrison, 1995). No scale is indicated, but vertical and horizontal axes are of equal scale. Welts typically range in width from 2 to 15 km.

thickened medial rigid beam and terminate at the surface in the upper rigid layer.

The nature and variety of thrust structures in the beam are governed partially by mechanisms of displacement transfer on overlapping subsurface thrusts (Figures 20, 21) and partially by foreland- and hinterland-vergent thrusts being equally common in this salt-based décollement system. Adding to the complexity is the occurrence of foreland- and hinterland-vergent duplex structures,

lying either in the footwalls or hanging walls of larger thrusts, and "fishtail" structures comprising stacked thrusts of opposing vergence that are linked to one another along intermediate detachment surfaces (Figure 22).

Cross-Folded Region

Deep-seated folds involving the entire Cambrian–Devonian succession are seen on all sections A–D and

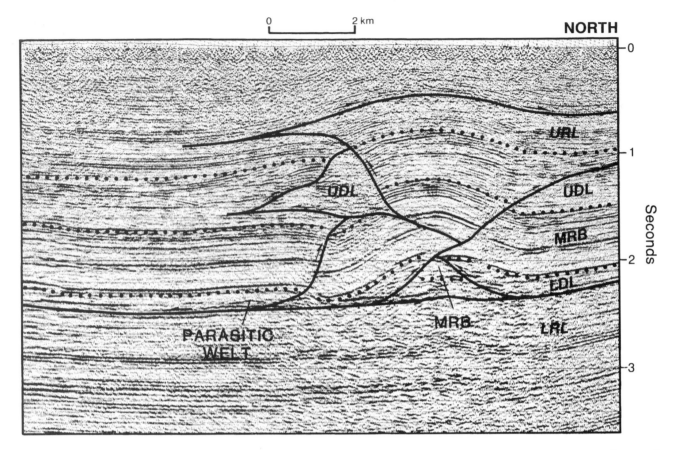

Figure 16—Interpreted seismic section showing Mecham River West anticline (modified from Harrison, 1995). See section G in Figure 11 for location. Component rock types of each tectonic layer (and in Figures 17, 18, 19, 24, 25) are described in the text and in the legend of Figure 11 (second foldout).

F–H. A seismic example of deep-seated flexure is provided by a regional anticline in Figure 17. Only a small, upper part of the entire structure is visible. Here hinges of the shallow and deep anticlines nearly coincide. This is strictly fortuitous: structure contours drawn on the subsalt reflection indicate a 25°–35° obliquity between linear, west-northwest trending salt-detached folds and west-trending, deep-seated periclinal culminations and depressions (Figure 23). Their age is uncertain, but these deep-seated periclines are probably Early Carboniferous. They have folded the uppermost Devonian–Lower Carboniferous salt-based décollement system and refolded all units above the base of the salt and below unconformable middle Carboniferous and Lower Cretaceous outliers.

Conversely, other deep-seated structures are associated with southward-transported thrusts that rise from Middle(?) Proterozoic and Middle Cambrian detachment surfaces, and they flatten into the overlying salt (Figure 24). These faults and their associated hanging wall anticlines must be coeval with the salt-based décollement system because some slip on the deep-seated thrusts has been accommodated within the salt and dissipated by shortening above the salt. The salt has largely decoupled strata above and below. The salt is thickest under synclines and in front of each deep-seated thrust, and it has migrated in such a way that extreme structural relief below the salt is decreased above it.

Region of Superposed Extension

The location of Carboniferous half-grabens associated with the embryonic Sverdrup Basin are also kinematically linked to preexisting deep-seated thrust structures. Southward-directed extension above the Ordovician salt (lower ductile layer) created the half-graben accommodation space (sections C and D; Figure 25). This extension was apparently matched by an equal northward-directed subsalt extension on reactivated subsalt thrusts that extend mainly to the north into Paleozoic "basement" under the Sverdrup Basin. The restoration of the sub-Carboniferous angular unconformity to its previous horizontal state (Figure 26, top) also necessitates the preexistence of a massive anticlinal duplex and an inclined, tectonized salt layer on a regional southward-dipping anticlinal forelimb. The salt in these circumstances has not promoted the growth of diapirs. Rather, the salt has tended to migrate during extension such that deep-seated fold amplitude and related faulting are decreased above the salt.

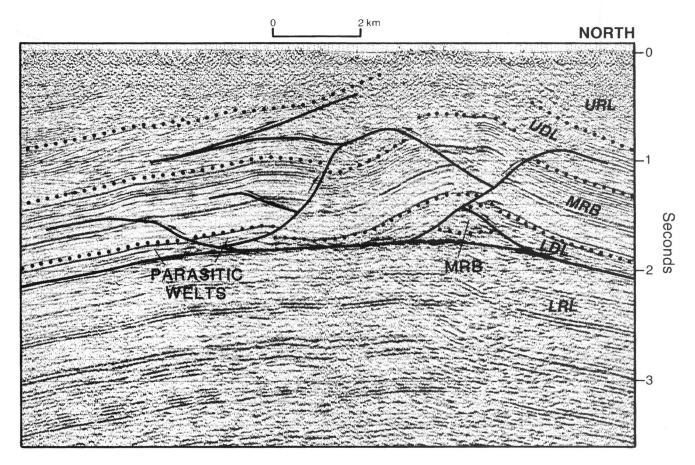

Figure 17—Interpreted seismic section showing Dealy Island anticline (modified from Harrison, 1995). See section G in Figure 11 for location.

HINTERLAND LIMIT OF THE SALT-BASED FOLDBELT

The hinterland depositional limit of salt is indicated by gradual pinch-out of the lower ductile layer in Figure 24 and on section H (Figure 11). Reduced wavelength and amplitude of salt-detached folds are clearly indicated here. Well data and seismic images also delineate a salt limit on or near the extreme northern end of sections C and D and in Figure 25. From the foreland, the maximum absolute and percentage shortening across the foldbelt in the medial rigid beam is 28 km and 11% at the northern end of section A (~25 km south of the salt limit). In contrast, the beam has been shortened 19 km at the northern end of section D and only 3 km at the salt limit on section H. The westward decrease in total shortening, apparent in all cross sections, is attributed to a westward narrowing of the foldbelt and the obliquity between salt-based folds and the geometry of the Ordovician salt basin. A similar pattern of eastward decreasing shortening is expected on cross sections that might be drawn for Bathurst Island where the salt-based foldbelt impinges on the deep structures of the Cornwallis foldbelt and Boothia uplift.

DISCUSSION

A fundamental consequence of the weakness of salt in some salt-detached thrust foldbelts is the absence of a preferred sense of thrust transport (Davis and Engelder, 1987). Therefore, without some knowledge of the regional geology, the geometry of individual anticlines and thrusts reveals nothing about the location of the tectonic hinterland. Other clues must be found, such as the location of deep-seated thrust folds linked to the salt décollement, the basinward increase in thickness of units in the deformed miogeoclinal wedge (not fully reliable), or the direction of transport of orogen-derived clastics (also not fully reliable). The mix of forethrusts and backthrusts is also apparent within individual anticlines. In these instances, the vertical stacking of opposed thrusts can be considered an example of thrust wedging in miniature (Figure 22). Each anticlinal salt welt can also be thought of as one or more intercutaneous wedges or mini-triangle zone structures (Figure 15a,b). In these cases, slip in one direction on the base of salt is matched by opposing slip on thrusts on the top of salt.

This introduces the concept of an *intercutaneous wedge geometry* for the entire salt-based décollement system

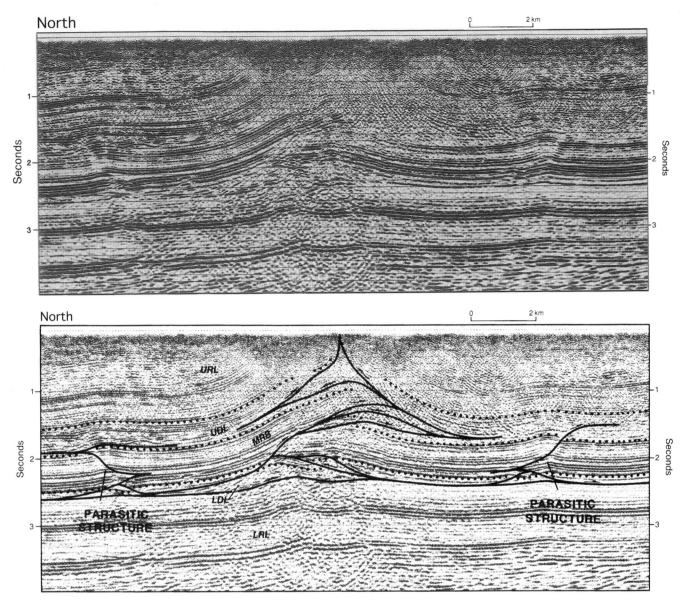

Figure 18—Uninterpreted (top) and interpreted (bottom) seismic profiles of Beverley Inlet anticline (modified from Harrison, 1995). See section C in Figure 11 for location.

(Harrison, 1994c, 1995). Apparent shortening in the medial rigid beam is greater than that observed and measured in the bounding upper and lower rigid layers (Figure 12). The lack of shortening in the lower rigid layer is accounted for by accumulated slip on the salt décollement, amounting to as much as 28 km near the hinterland edge of salt at the northern end of section A (Figure 11). For the upper rigid layer, shortening mostly by folding is significant but is still less than half that measured in the medial rigid beam. This discrepancy is seen on each cross section and is also apparent on each seismic example where unbroken reflectors can be traced over the hinge of each anticline (Figures 13, 16). Geologic constraints also do not permit significant displacement on mapped thrusts at the surface in the upper rigid layer (Figure 8) (Harrison, 1995).

It is beyond the scope of the present paper to discuss all the implications or alternative models to account for the various shortening problems arising from the construction of the cross sections. Most of this has been discussed at length in Harrison (1995) and references contained therein. In general, however, mesoscopic and microscopic structures are not sufficiently abundant in the clastic units of the upper rigid layer to account for the shortfall in required shortening, and there is no evidence that these clastic formations were deposited during or after the thin-skinned deformation.

Figure 27 entertains the notion that a regional upper décollement in or above the upper ductile layer carries the slip that cannot be accounted for by alternative deformation processes in the upper rigid layer. In this hypothesis, south-directed slip on the salt décollement is

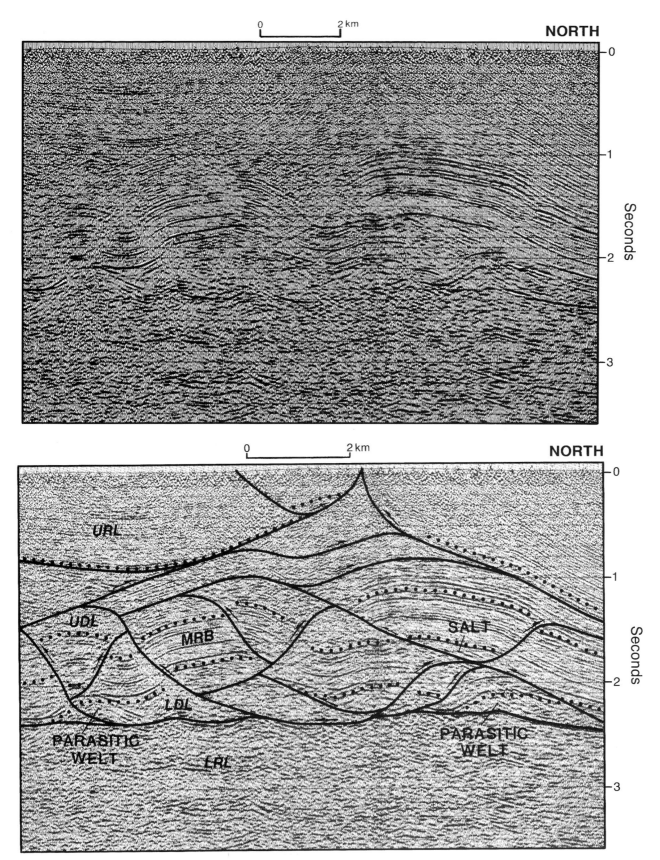

Figure 19—Uninterpreted (top) and interpreted (bottom) seismic profiles of King Point East anticline (modified from Harrison, 1995). See section A in Figure 11 for location.

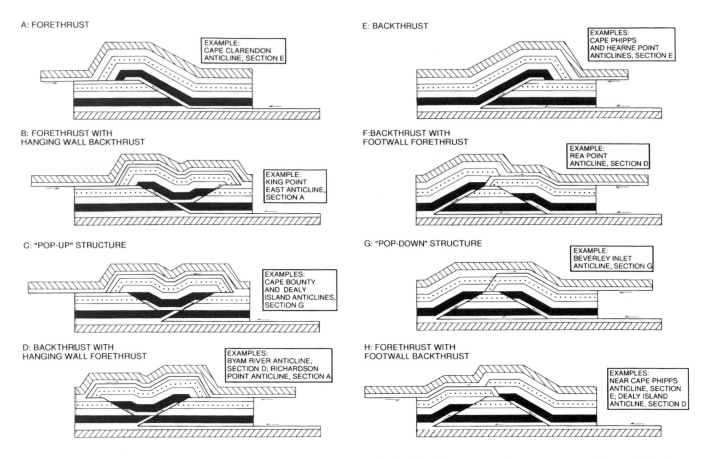

Figure 20—Schematic cross sections showing eight types of vertically linked thrust structures over a weak basal detachment (lower white layer). Examples from the salt-based foldbelt on Melville Island are listed for each. In these schematic sections, variation in salt thickness and related influences on structural style in overlying strata have been eliminated (modified from Harrison, 1995).

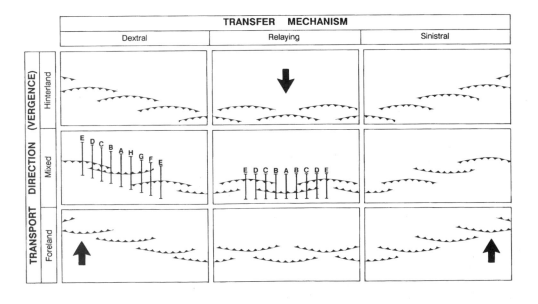

Figure 21—Mechanisms of displacement transfer between thrusts over a basal décollement. All nine types of transfer systems could exist in a single tectonic reentrant bounded to the left and right by salients. If one turns the figure upside down, the thrust systems are seen to envelop a salient with reentrants to the left and right. The main direction of tectonic transport is indicated by arrows. Lettered cross-section lines refer to fundamental structures shown in Figure 20 (reproduced from Harrison, 1995).

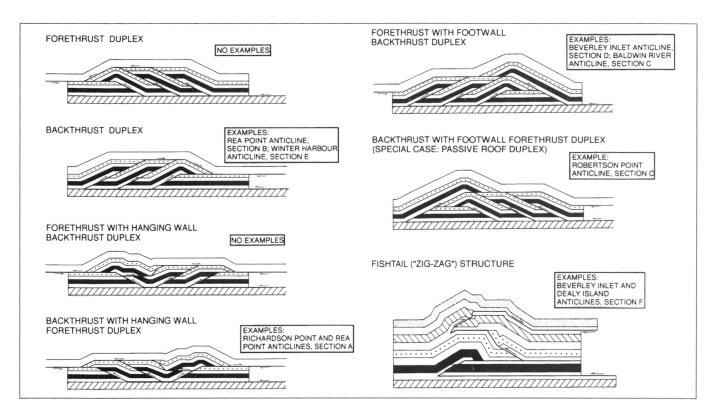

Figure 22—Schematic cross sections showing seven types of thrust systems with a foreland- and hinterland-vergent hanging wall and footwall duplex developed on a weak, basal decoupling surface. Examples from the salt-based foldbelt on Melville Island are listed for each. As in Figure 20, variation in salt thickness has been eliminated (modified from Harrison, 1995).

matched by an equal amount of shortening by folding and thrusting in the beam, by folding with minor thrusting in the upper rigid layer, and by a component of north-directed slip on the upper detachment surface. The latter carries the maximum amount of slip in the north, where it exits the section at the sub-Carboniferous erosion surface. This general kinematic hypothesis for the foldbelt is analogous to the triangle zones of the Rocky Mountain foothills (Gordy et al., 1977; McMechan, 1985) but at a scale that dwarfs these classic examples. If true, the Parry Islands triangle zone could be up to 200 km wide, enveloping the entire 470-km length of the foldbelt and underlying an area of more than 52,000 km². Development of such an enormous zone can be explained by the efficient lubricating effects of salt.

Although a unique upper decoupling surface has not been identified, shale in the upper ductile layer, like the salt layer at depth, is penetratively deformed and clearly weak enough to carry a large displacement subhorizontal fault. The structure sections and seismic examples indicate several potential upper detachment surfaces within or immediately above the upper ductile layer. Other faults (on sections A–D, G, H) extend from the beam right through the upper rigid layer to the surface and must therefore offset and postdate the hypothesized upper detachment. Therefore, any slip on the upper detachment must have been early in thrust belt evolution, that is, before the transecting faults appeared and before hinter-

land folding of the upper detachment made continued slip on this surface a mechanical improbability.

In this hypothesis, the upper detachment is active early when most folds are growing but still embryonic and immature. During this interval, the upper and lower rigid layers are pinned on the foreland side of the orogen and ride passively above and below the migrating nascent anticlines in the beam (as the skin of the snake folds and unfolds as it swallows an egg). Slip on the upper detachment is then terminated by several impediments, including (1) progressive steepening of rigid beam fold limbs in the inner part of the foldbelt, (2) growth of shale welts and high-amplitude anticlines in the upper rigid layer vertically above the hinge of matching anticlines in the beam, and (3) thrusting that propagated from the beam and transected both the upper ductile and upper rigid layers.

IMPLICATIONS FOR EXPLORATION

Implications for hydrocarbon exploration are many but can be reduced to a few key observations. Rocks of the Devonian clastic wedge lie within the oil window (Gentzis and Goodarzi, 1993) but tend to be poor potential reservoirs due to reduced porosity or lack of seal. The Ordovician–Lower Devonian strata of the medial rigid beam are overmature except locally in the hinterland

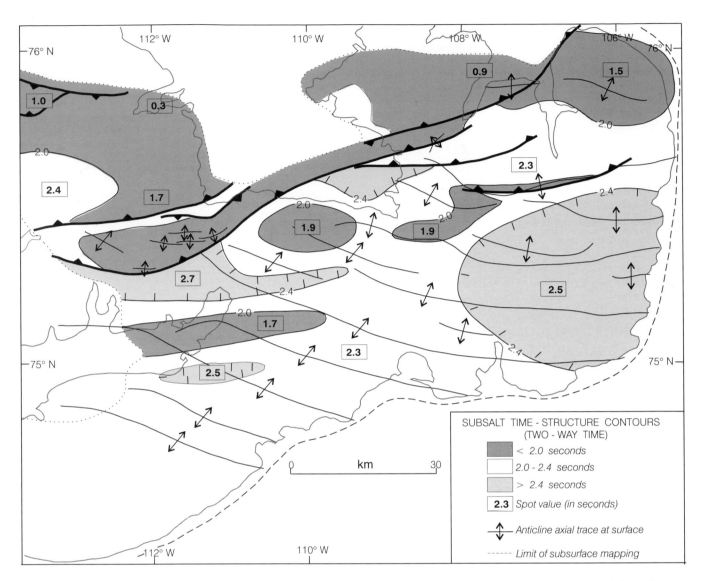

Figure 23—Map of axial traces of anticlines at the surface and selected two-way time contours on the base of the salt reflection. Posted times of 1.0, 2.0, 2.4, and 2.7 sec correspond to depths of about 2.5, 5.0, 5.7, and 6.2 km, respectively.

where Devonian overburden is currently thin. Strata in the beam may also be thermally mature in extensive underexplored areas of the foldbelt periphery and in the foreland beneath the offshore channels. These latter areas represent the best prospects for new exploration because hydrocarbons generated during thrust belt evolution in the latest Devonian–Early Carboniferous are most likely to have been driven into structural and stratigraphic traps located to the south. At least 13 major fold closures remain untested within the peripheral region of the foldbelt (Harrison, 1995). However, the high cost of frontier drilling and the apparently low porosity of the Paleozoic succession have, since the early 1980s, tended to focus exploration on targets located in karsted Devonian limestones near the producing Bent Horn oil field of Cameron Island (above the hinterland limit of Ordovician salt) and in Mesozoic strata of the Sverdrup Basin.

CONCLUSIONS

The salt-based foldbelt of Canada's Parry Islands provides useful insight into the distinctive tectonic style of compressional orogens involving salt. These insights arise from a well-exposed surface geology, high-quality seismic reflection profiles, and some well control. The reflection data permit identification of both top-of-salt and base-of-salt reflectors and a characteristic seismic stratigraphy above the salt.

Bedded salt, with its unusual combination of low shear strength and low density, provides the structural geologist with a wide array of unusual salt tectonic structures. The weakness and ductility of salt in the Parry Islands foldbelt are directly responsible for many characteristic features, including extreme foldbelt width, subhorizontal salt décollement, and equal occurrence of forethrusts and backthrusts. Indeed, many of the key features

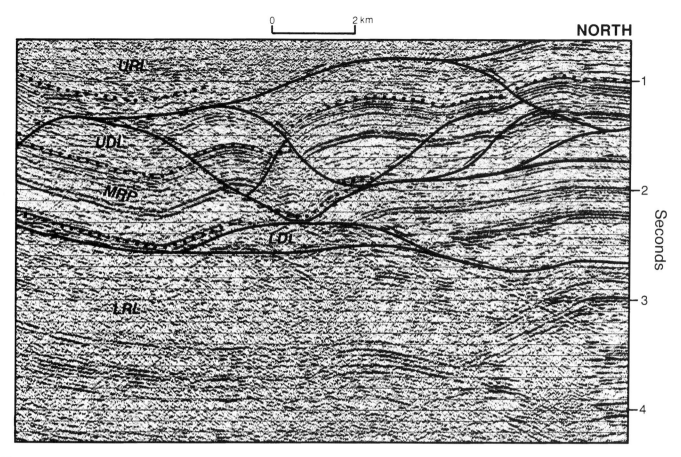

Figure 24—Interpreted seismic section showing Nias Point anticlinorium (modified from Harrison, 1995). Located near section H in Figure 11.

noted by Davis and Engelder (1987) as typical of salt-based foldbelts are borne out by the present case study. For example, foldbelt limits controlled by the depositional limit of salt, recognized as typical by Davis and Engelder (1987), are also indicated here. In addition, the en echelon fold terminations near the salt limit in the Parry Islands belt imply that the direction of tectonic transport was oblique to the edge of the salt basin. The salt edge can be seen to have induced a component of sinistral drag along the eastern and southern deformation front.

The correspondence between extreme foldbelt width and underlying salt distribution led to a second presumption by Davis and Engelder (1987)—that the hinterland dip on the salt décollement must be less than 1° in salt-based foldbelts. This relationship is also supported on Melville Island. The subsalt reflection, assumed to be the lower limit of foreland-directed slip on the basal décollement, is calculated to be 0.1°–0.6°. However, this surface has also been folded to the east and northwest by partly younger deep-seated folds oblique to the salt-based folds. Nevertheless, the direction of tectonic transport need not be (and is not) perpendicular to any specific preexisting basin anisotropy (e.g., hinterland or foreland limits of salt, preexisting shelf edge, or direction of foreland clastic wedge progradation).

Jackson and Vendeville (1994) have suggested that halokinetic structures will form in any basin containing a reasonable thickness of salt, given the weakness of salt and sufficient regional extension. The complete absence of intrusive salt in the Parry Islands foldbelt suggests either that the salt layer has been too thin or stronger than normal (due to an abundance of various nonevaporite interbeds) or that extension has been insufficient to promote salt piercement. The array of observed contractional structures owe their existence to a weak but not necessarily low-density salt-bearing layer and to external tectonic stresses that have been transmitted through competent overburden to produce buckling and thrust faulting. In these circumstances, the salt plays a relatively passive role. It is pulled from regions of high pressure under synclines to regions of low pressure under anticlines to correct space problems generated by buckling. It is equally apparent that salt has migrated during superposed deep-seated compression and superposed extension such that the amplitude and complexity of deep-seated structure are decreased above the salt.

Salt is only effectively moved upsection in the hanging wall of thrust ramps and is apparently unable to migrate without conformable overburden in the Parry Islands foldbelt. In contrast, salt migration in diapiric provinces can produce a wide assortment of intrusive salt-cored

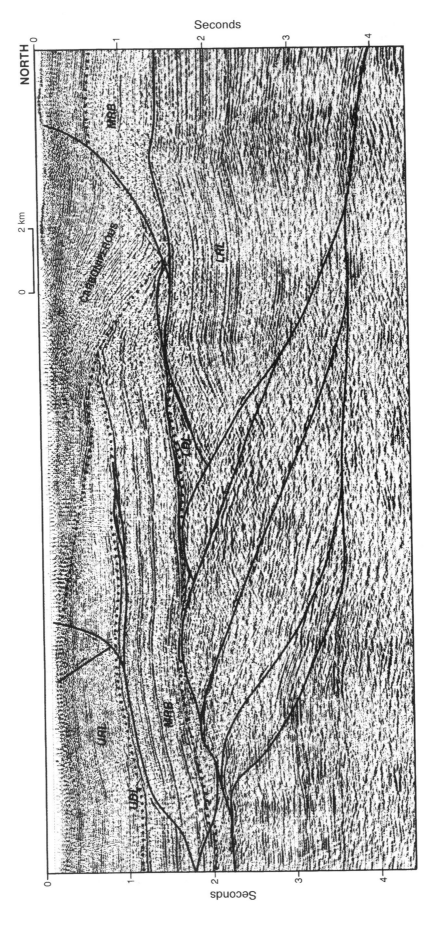

Figure 25. Interpreted seismic section showing Weatherall depression and Spencer Range anticlinorium (reproduced from Harrison, 1995). Located near section D in Figure 11. See Figure 26.

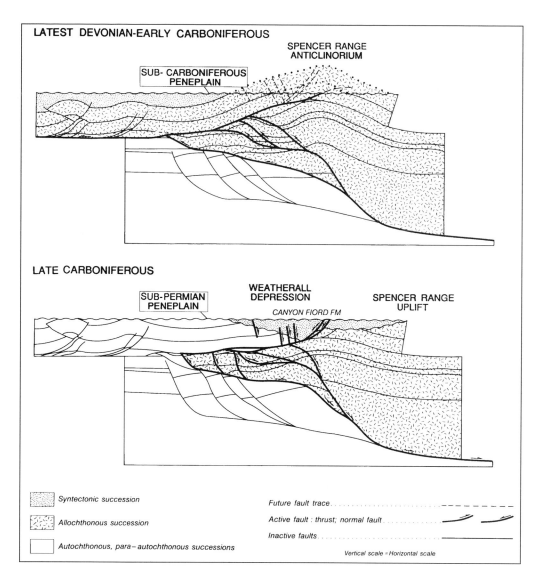

Figure 26—Kinematic model for Spencer Range anticlinorium for (a) latest Devonian–Early Carboniferous, evolving to (b) Late Carboniferous (modified from Harrison, 1995).

masses and a related array of syndepositional structures in overburden (e.g., crestal thinning, rim synclines, and turtle structures), all of which are unknown in this study. It could be argued that the Parry Islands belt represents an immature salt tectonic province. However, the range of salt welt and overburden fold and thrust structures (from simple in the foldbelt periphery to complex in the interior) presents a kinematic progression of deformation from embryonic to fully mature, a potential model for this and other contractional salt provinces that complements the Jackson and Talbot (1991) model for the evolution of salt diapirs (from salt rollers to salt walls to salt canopies) in extensional salt provinces.

The existence of two ductile layers and three rigid layers in the Parry Islands foldbelt presents additional opportunities to study the contrasting response of shale and salt to compressive deformation. Salt has clearly migrated from synclines to anticlines, but the behavior of shale is less clear. Where shale is thickened in front of thrusts emerging from underlying competent strata, the thickening process may be entirely related to intraformational thrust imbrication. However, shale in hinge area

welts has apparently migrated in a more ductile fashion in response to buckling in overburden and related space problems and pressure gradients. Nevertheless, thinning of ductile shale within synclines remains to be proven.

An equal distribution of forethrusts and backthrusts is characteristic of the Parry Islands foldbelt and of others described by Davis and Engelder (1987). The present study recognizes that thrusts of opposing vergence can be stacked in individual anticlines and otherwise arranged in a variety of complicated configurations not normally seen in traditional foreland-vergent thrust belts. This arrangement of faults provides for the study and understanding of triangle zones at scales ranging from those in individual salt welts to others in regional-scale anticlines. Most intriguing is the possibility that the entire 200-km width and 52,000-km^2 area of the salt-based foldbelt might be a single massive triangle zone (passive roof duplex), whose foreland-directed slip on the salt detachment is balanced by shallow folding and thrusting and a component of hinterland-directed slip on an upper detachment. Although admittedly speculative, this hypothetical structure is consistent with the ductility and

410   *Harrison*

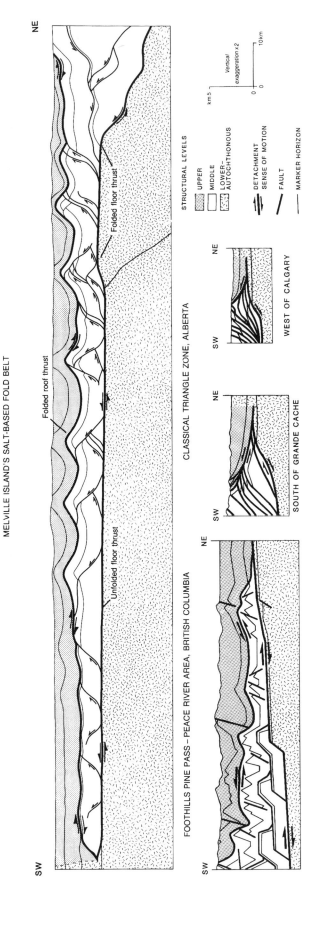

MELVILLE ISLAND'S SALT-BASED FOLD BELT

FOOTHILLS PINE PASS – PEACE RIVER AREA, BRITISH COLUMBIA

CLASSICAL TRIANGLE ZONE, ALBERTA

WEST OF CALGARY

SOUTH OF GRANDE CACHE

STRUCTURAL LEVELS

UPPER
MIDDLE
LOWER–AUTOCHTHONOUS

DETACHMENT
SENSE OF MOTION
FAULT
MARKER HORIZON

Vertical exaggeration x2

Figure 27—Cross sections showing possible structural geometry of the Parry Islands foldbelt on Melville Island compared with similar triangle zone (passive roof duplex) geometries of the Pine Pass–Peace River area in British Columbia (after McMechan, 1985) and the classical triangle zone in Alberta (after Gordy et al., 1977). (Figure modified from Harrison, 1994c, 1995.)

extreme weakness of both salt and shale layers and effectively accounts for all available data concerning shortening mechanisms and shortening discrepancies in the foldbelt. At the very least, this hypothesis should cause us to reevaluate the existence and mechanical limits of the size of triangle zones in much the same way that the mechanical feasibility of large thrust sheets challenged geoscientists last century.

Acknowledgments   The author thanks Panarctic Oils Limited, Texaco Canada Resources, and Shell Canada Limited for providing original seismic records and for permission to publish selected examples of these data. Research funding has been provided by the Polar Continental Shelf Project and the Geological Survey of Canada. The author is also grateful to the Geological Survey of Canada for allowing the reproduction of many illustrations in this paper. Text figures have benefited from drafting by Carmen Lee and Peter Neelands. The scientific value of the paper has been greatly improved by the critical review comments of David J. Anastasio, Gloria Eisenstadt, Martin P. A. Jackson, and Andrew V. Okulitch. Many thanks to Kathy Walker of Editorial Technologies of Renton, WA, for editorial improvements.

REFERENCES CITED

Bally, A. W., P. L. Gordey, and G. A. Stewart, 1966, Structure, seismic data, and orogenic evolution of the southern Canadian Rocky Mountains: Bulletin of Canadian Petroleum Geology, v. 14, p. 337–381.

Condie, K. C., and O. M. Rosen, 1994, Laurentia–Siberia connection revisited: Geology, v. 22, no. 2, p. 168–170.

Davies, G. R., and W. W. Nassichuk, 1991, Carboniferous to Permian geology of the Sverdrup Basin, Canadian Arctic Archipelago, *in* H. P. Trettin, ed., Innuitian orogen and Arctic platform: Canada and Greenland: Geological Survey of Canada, Geology of Canada, no. 3, p. 343–367.

Davis, D. M., and T. Engelder, 1985, The role of salt in fold-and-thrust belts: Tectonophysics, v. 119, p. 67–88.

Davis, D. M., and T. Engelder, 1987, Thin-skinned deformation over salt, *in* I. Lerche and J. J. O'Brien, eds., Dynamical geology of salt-related structures: Orlando, Florida, Academic Press, p. 301–337.

de Freitas, T. A., J. C. Harrison, and R. Thorsteinsson, 1993, New field observations on the geology of Bathurst Island, Arctic Canada, part A, stratigraphy and sedimentology of the Phanerozoic succession: Geological Survey of Canada Paper 93-1B, p. 1–11.

de Freitas, T. A., J. C. Harrison, and U. Mayr, 1994, Lower Paleozoic T-R sequence stratigraphy, central Canadian Arctic, *in* D. K. Thurston and K. Fujita, eds., 1992 Proceedings of the International Conference on Arctic Margins, Anchorage, Alaska: Minerals Management Service Report 94-0040, p. 123–128.

de Paor, D. G., D. C. Bradley, G. Eisenstadt, and S. M. Phillips, 1989, The Arctic Eurekan orogen: A most unusual fold-and-thrust belt: GSA Bulletin, v. 101, p. 952–967.

Embry, A. F., 1991a, Middle–Upper Devonian clastic wedge of the Arctic Islands, *in* H. P. Trettin, ed., Innuitian orogen and Arctic platform: Canada and Greenland: Geological Survey of Canada, Geology of Canada, no. 3, p. 263–279.

Embry, A. F., 1991b, Mesozoic history of the Arctic Islands, *in* H. P. Trettin, ed., Innuitian orogen and Arctic platform: Canada and Greenland: Geological Survey of Canada, Geology of Canada, no. 3, p. 371–434.

Epard, J.-L., and R. H. Groshong, Jr., 1993, Excess area and depth to detachment: AAPG, Bulletin, v. 77, p. 1291–1302.

Fox, F. G., 1983, Structure sections across Parry Islands foldbelt and Vesey Hamilton salt wall, Arctic archipelago, *in* A. W. Bally, ed., Seismic expression of structure styles, vol. 3: AAPG Studies in Geology Series 15, p. 3.4.1.54–3.4.1.72.

Fox, F. G., 1985, Structural geology of the Parry Islands foldbelt: Canadian Petroleum Geology Bulletin, v. 33, no. 3, p. 306–340.

Gentzis, T., and F. Goodarzi, 1993, Regional thermal maturity in the Franklinian mobile belt, Melville Island, Arctic Canada: Marine and Petroleum Geology, v. 10, p. 215–230.

Goodbody, Q. H., 1989, Stratigraphy of the Lower to Middle Devonian Bird Fiord Formation, Canadian Arctic archipelago: Canadian Petroleum Geology Bulletin, v. 37, p. 48–82.

Gordy, P. L., F. R. Frey, and D. K. Norris, 1977, Geological guide for the Canadian Society of Petroleum Geologists, and 1977 Waterton–Glacier Park Field Conference: Calgary, Alberta, Canadian Society of Petroleum Geologists, 93 p.

Harrison, J. C., 1994a, Melville Island and adjacent smaller islands, Canadian Arctic archipelago, District of Franklin, Northwest Territories: Geological Survey of Canada, Map 1844A, scale 1:250,000 (12 sheets).

Harrison, J. C., 1994b, A summary of the structural geology of Melville Island, Canadian Arctic archipelago, *in* R. L. Christie and N. J. McMillan, eds., The geology of Melville Island, Arctic Canada: Geological Survey of Canada Bulletin, v. 450, p. 257–283.

Harrison, J. C., 1994c, Negligible-taper triangle zone model for the salt-based foldbelt in Canada's western Arctic Islands, *in* D. K. Thurston and K. Fujita, eds., 1992 Proceedings of the International Conference on Arctic Margins, Anchorage, Alaska: Minerals Management Service Report 94-0040, p. 143–148.

Harrison, J. C., 1995, Melville Island's salt-based foldbelt: Geological Survey of Canada Bulletin, v. 472, 331 p.

Harrison, J. C., and A. W. Bally, 1988, Cross-sections of the Parry Islands foldbelt on Melville Island: Canadian Petroleum Geology Bulletin, v. 36, p. 311–332.

Harrison, J. C., F. G. Fox, and A. V. Okulitch, 1991, Late Devonian–Early Carboniferous deformation of the Parry Islands and Canrobert Hills foldbelts, Bathurst and Melville islands, *in* H. P. Trettin, ed., Innuitian orogen and Arctic platform: Canada and Greenland: Geological Survey of Canada, Geology of Canada, no. 3, p. 321–333.

Harrison, J. C., T. A. de Freitas, and R. Thorsteinsson, 1993, New field observations on the geology of Bathurst Island, Arctic Canada, part B, structure and tectonic history: Geological Survey of Canada Paper 93-1B, p. 12–21.

Harrison, J. C., R. Thorsteinsson, and T. A. de Freitas, 1994, Phanerozoic geology of Strathcona Fiord map area (NTS 49E and part of 39F), District of Franklin, Northwest Territories: Geological Survey of Canada Open File 2881, scale 1:125,000, 1 sheet.

Higgins, A. K., J. R. Ineson, J. S. Peel, F. Surlyk, and M. Snderholm, 1991, Lower Paleozoic Franklinian Basin of North Greenland, *in* J. S. Peel and M. Snderholm, eds., Sedimentary basins of North Greenland: Grønlands Geologiske Underslgelse, Rapport Nr. 160, p. 71–140.

Jackson, M. P. A., and C. J. Talbot, 1991, A glossary of salt tectonics: The University of Texas at Austin, Bureau of Economic Geology Geological Circular 91-4, 44 p.

Jackson, M. P. A., and B. C. Vendeville, 1994, Regional extension as a trigger for diapirism: GSA Bulletin, v. 106, p. 57–73.

Kerr, J. W., 1974, Geology of the Bathurst Island group and Byam Martin Island, Arctic Canada (Operation Bathurst Island): Geological Survey of Canada Memoir 378, 152 p.

Mayr, U., 1980, Stratigraphy and correlation of lower Paleozoic formations, subsurface of Bathurst Island and adjacent smaller islands, Canadian Arctic archipelago: Geological Survey of Canada Bulletin, v. 306, 52 p.

McMechan, M. E., 1985, Low-taper triangle zone geometry: an interpretation for the Rocky Mountain foothills, Pine Pass–Peace River area, British Columbia: Canadian Petroleum Geology Bulletin, v. 33, p. 31–38.

McNicoll, V. J., J. C. Harrison, and R. Thorsteinsson, 1995, Provenance of the Devonian clastic wedge of Arctic Canada: constraints imposed by detrital zircon ages, *in* S. Dorobek and G. Ross, eds., Stratigraphic development in foreland basins: SEPM Special Publication, p. 77–93.

Okulitch, A. V., 1991, Late Devonian–Early Carboniferous deformation of the central Ellesmere and Jones Sound foldbelts, *in* H. P. Trettin, ed., Innuitian orogen and Arctic platform: Canada and Greenland: Geological Survey of Canada, Geology of Canada, no. 3, p. 318–320.

Okulitch, A. V., and H. P. Trettin, 1991, Late Cretaceous–early Tertiary deformation, Arctic Islands, *in* H. P. Trettin, ed., Innuitian orogen and Arctic platform: Canada and Greenland: Geological Survey of Canada, Geology of Canada, no. 3, p. 469–489.

Okulitch, A. V., J. J. Packard, and A. I. Zolnai, 1991, Late Silurian–Early Devonian deformation of the Boothia uplift, *in* H. P. Trettin, ed., Innuitian orogen and Arctic platform: Canada and Greenland: Geological Survey of Canada, Geology of Canada, no. 3, p. 302–307.

Stephenson, R. A., A. F. Embry, S. M. Nakiboglu, and M. A. Hastaoglu, 1987, Rift-initiated Permian to Early Cretaceous subsidence of the Sverdrup basin, *in* C. Beaumont and A. J. Tankard, eds., Sedimentary basins and basin-forming mechanisms: Canadian Society of Petroleum Geologists Memoir 12, p. 213–231.

Thorsteinsson, R., 1971a, Geology, Strand Bay map area (NTS 59H), District of Franklin, Northwest Territories: Geological Survey of Canada Map 1301A, scale 1:250,000, 1 sheet.

Thorsteinsson, R., 1971b, Geology, Eureka Sound North map area (NTS 49G), District of Franklin, Northwest Territories: Geological Survey of Canada Map 1302A, scale 1:250,000, 1 sheet.

Thorsteinsson, R., 1972, Geology, Otto Fiord map area (NTS 340C), District of Franklin, Northwest Territories: Geological Survey of Canada Map 1309A, scale 1:250,000, 1 sheet.

Thorsteinsson, R., 1974, Carboniferous and Permian stratigraphy of Axel Heiberg Island and western Ellesmere Island, Canadian Arctic archipelago: Geological Survey of Canada Bulletin, v. 224, 115 p.

Thorsteinsson, R., 1986, Geology of Cornwallis Island and neighboring smaller islands, District of Franklin, Northwest Territories: Geological Survey of Canada Map 1626A, scale 1:250,000 (1 sheet).

Thorsteinsson, R., and T. T. Uyeno, 1980, Stratigraphy and conodonts of Upper Silurian and Lower Devonian rocks in the environs of the Boothia uplift, Canadian Arctic archipelago, part 1, contributions to stratigraphy by R. Thorsteinsson with contributions by T. T. Uyeno: Geological Survey of Canada Bulletin, v. 292, 75 p.

Trettin, H. P., 1991a, The Proterozoic to Late Silurian record of Pearya, *in* H. P. Trettin, ed., Innuitian orogen and Arctic platform: Canada and Greenland: Geological Survey of Canada, Geology of Canada, no. 3, p. 239–259.

Trettin, H. P., 1991b, Middle and late Tertiary tectonic and physiographic developments, *in* H. P. Trettin, ed., Innuitian orogen and Arctic platform: Canada and Greenland: Geological Survey of Canada, Geology of Canada, no. 3, p. 493–496.

Trettin, H. P., 1991c, Tectonic framework, *in* H. P. Trettin, ed., Innuitian orogen and Arctic platform: Canada and Greenland: Geological Survey of Canada, Geology of Canada, no. 3, p. 59–66.

Trettin, H. P., U. Mayr, G. D. F. Long, and J. J. Packard, 1991, Cambrian to Early Devonian deposition, Arctic Islands, *in* H. P. Trettin, ed., Innuitian orogen and Arctic platform: Canada and Greenland: Geological Survey of Canada, Geology of Canada, no. 3, p. 163–237.

van Berkel, J. T., W. M. Schwerdtner, and J. G. Torrance, 1985, Origin of wall-and-basin structure—an intriguing tectonic-halokinetic feature on west-central Axel Heiberg Island, Sverdrup basin, Canadian Arctic archipelago: Canadian Petroleum Geology Bulletin, v. 32, p. 343–358.

Nilsen, K. T., B. C. Vendeville, J.-T. Johansen, 1995, Influence of regional tectonics
on halokinesis in the Nordkapp Basin, Barents Sea, *in* M. P. A. Jackson, D. G.
Roberts, and S. Snelson, eds., Salt tectonics: a global perspective: AAPG
Memoir 65, p. 413–436.

Influence of Regional Tectonics on Halokinesis in the Nordkapp Basin, Barents Sea

Kåre T. Nilsen

Norwegian Petroleum Directorate
Harstad, Norway

Bruno C. Vendeville

Bureau of Economic Geology
The University of Texas at Austin
Austin, Texas, U.S.A.

Jan-Terje Johansen

Norwegian Petroleum Directorate
Harstad, Norway

Abstract

Seismic analysis of salt structures in the Nordkapp Basin, a deep salt basin in the southern Barents Sea, combined with experimental modeling suggests that regional tectonics closely controlled diapiric growth. Diapirs formed in the Early Triassic during basement-involved regional extension. The diapirs then rose rapidly by passive growth and exhausted their source layer. Regional extension in the Middle–Late Triassic triggered down-to-the-basin gravity gliding, which laterally shortened the diapirs. This squeezed salt out of diapir stems, forcing diapirs to rise, extrude, and form diapir overhangs. After burial under more than 1000 m of Upper Triassic–Lower Cretaceous sediments, the diapirs were rejuvenated by a Late Cretaceous episode of regional extension and gravity gliding, which deformed their thick roofs. After extension, diapirs stopped rising and were buried under 1500 m of lower Tertiary sediments. Regional compression of the Barents Sea region in the middle Tertiary caused one more episode of diapiric rise. Diapirs in the Nordkapp Basin are now extinct.

INTRODUCTION

The Nordkapp Basin (Figure 1) is a deep, narrow salt basin in the southern Barents Sea. The southwestern part of the Nordkapp Basin is a narrow, northeast-trending subbasin 150 km long and 25–50 km wide; about 17 salt domes are located along the basin's axis (Figure 2a). The northeastern part (Figure 2a) is a wider, east-trending subbasin about 200 km long and 50–70 km wide. More than 16 salt domes occur west of the 32°E meridian (no exploration has been conducted east of here). Exploration in the Nordkapp Basin started in the 1980s but remained limited until the early 1990s. Only one well has been drilled so far, although another well is scheduled for 1996. Published work on the Nordkapp Basin (e.g., Gabrielsen et al., 1990) remained scarce until the 1992 Norwegian Petroleum Society (NPF) special publication *Structural and Tectonic Modelling and Its Applications to Petroleum Geology* (e.g., Dengo and Røssland, 1992; Gabrielsen et al., 1992; Jensen and Sørensen, 1992). Recently, the oil industry showed renewed interest in the basin during the 14th

concession round in 1993 (Jensen et al., 1993). Furthermore, recent improvements in the interpretation of the basin's structural history and discovery of traces of hydrocarbons in wells outside the basin suggest that the Nordkapp Basin could soon become a promising exploration target.

As the quality of seismic data and structural interpretations has improved, the inferred geometry of salt structures in the basin has progressed from wide salt stocks with vertical flanks to more complex shapes with broad diapir overhangs above narrow stems (Koyi et al., 1993; Talbot et al., 1993; Yu and Lerche, 1993). These refined interpretations have reduced the estimates of the total volume of salt remaining in the basin (Koyi et al., 1993). Also, new seismic data with higher resolution have imaged numerous normal faults in the subsalt basement on the margins (Gabrielsen et al., 1992) and in the basin itself (Koyi et al., 1993).

Since 1993, the Norwegian Petroleum Directorate (Harstad, Norway) and the Applied Geodynamics Laboratory (Bureau of Economic Geology, The University

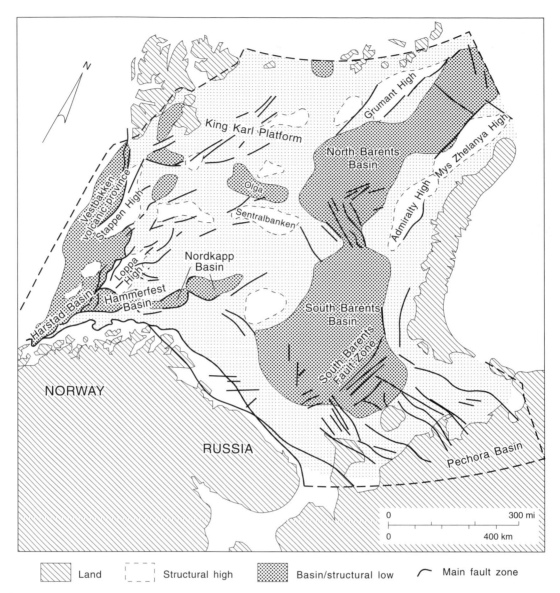

Figure 1—Main structural elements in the Barents Sea area and location of the Nordkapp Basin. Modified from Johansen et al. (1993).

of Texas at Austin) have conducted a detailed structural study of the Nordkapp Basin applying concepts derived from scaled tectonic experiments (e.g., Nilsen et al., 1994; Vendeville et al., 1994). This combination has helped to refine the structural interpretation of salt tectonics. In this chapter, we focus on how regional tectonics in the Barents Sea has tightly controlled the initiation, growth, and reactivation of salt structures in the Nordkapp Basin (Figure 3). Salt tectonics in the Nordkapp Basin is characterized by short periods of diapiric growth, commonly corresponding to regional tectonic phases, interspersed between longer intervals of diapiric inactivity during regional tectonic lulls. We show that successive phases of regional extension (1) formed the initial basin in which rock salt was deposited during the late Paleozoic, (2) initiated Early Triassic diapiric rise by fracturing the salt overburden, and (3) triggered gravity gliding during

Middle–Late Triassic and Late Cretaceous time, thereby forcing the diapirs to rise after source layer depletion, even after they had been capped by thick sedimentary roofs. Middle Tertiary regional shortening also reactivated the diapirs.

LATE PALEOZOIC: FORMATION OF THE NORDKAPP BASIN

Seismic sections along the margins of the basin (Figure 4) (e.g., Gabrielsen et al., 1992) clearly show the offsets of subsalt strata across basement faults. In contrast, in the deeper, central part of the basin, subsalt faults are commonly obscured by seismic noise because of the overlying diapirs and folded or faulted suprasalt strata (Figure 5). Free-air gravity data help delineate the basin's

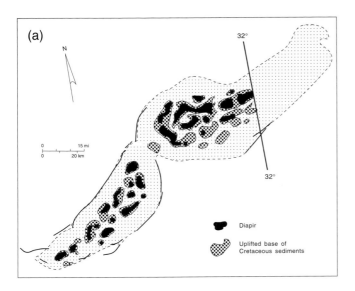

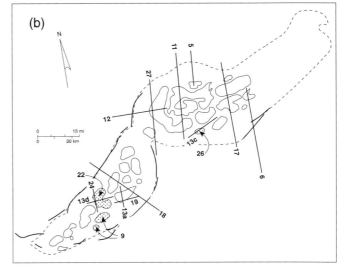

(above)

Figure 2—(a) Simplified structural map of the Nordkapp Basin showing salt diapirs and main fault zones. The map shows no diapirs east of longitude 32°E because of the absence of exploration data here, not because of the lack of diapirs. Black = subcrop of diapirs at or near the Pliocene–Pleistocene erosion surface; dark stippled = base of Cretaceous sediments uplifted above regional datum during diapiric rise. (b) Location map for section lines and detail maps (stippled diapirs) in Figures 5–27. Numbers on the map correspond to figure numbers.

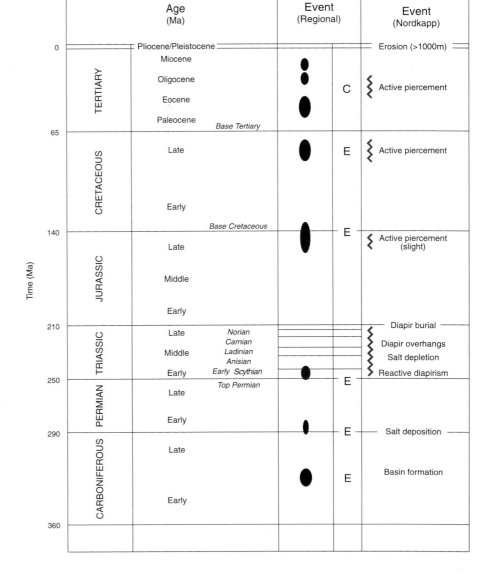

Figure 3—Time chart showing the connection between diapiric history in the Nordkapp Basin and regional tectonic events in the Barents Sea area. E and C denote episodes of regional extension and compression, respectively, illustrated by black ellipses.

Figure 4—Salt rollers and normal faults on the margins of the southwestern Nordkapp Basin. Line tracings from (a) northwest-southeast and (b) west-east seismic sections by Gabrielsen et al. (1992). Sections also show faults in the subsalt basement; normal faults in the overburden remained active until Cretaceous time. Cross pattern = subsalt basement; black = salt; white = over-burden; dotted line = base of Anisian (Middle Triassic); thick line = base of Cretaceous.

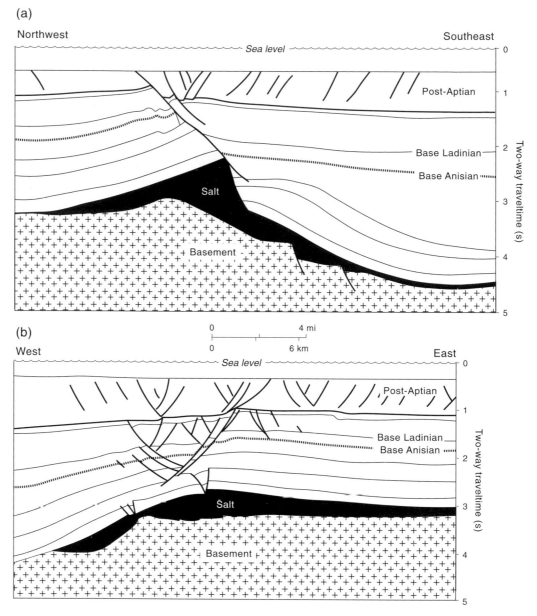

geometry in the subsalt basement as a 25- to 70-km-wide, 350-km-long northeast-trending basin made of two segments separated by an intrabasin ridge (Dengo and Røssland, 1992; Gudlaugson et al., 1994). The initial basement faulting is attributed to a phase of regional extension during the middle Carboniferous that created the basic architecture of grabens, half-grabens, and basement highs in the Barents Sea (Stemmerik and Worsley, 1989; Dengo and Røssland, 1992; Gabrielsen et al., 1992; Jensen and Sørensen, 1992; Gudlaugson et al., 1994; Johansen et al., 1994). By correlation with regional tectonics on the Finnmark platform, the tectonic event responsible for the formation of the basin can be dated as Serphukopian (Early–middle Carboniferous). The structural grain inherited from the Caledonian orogeny probably partially controlled the deformation style of the presalt units (Kjøde et al., 1978; Dore, 1991; Karpaz et al., 1993; Gudlaugson et al., 1994; Lippard, 1994; Nilsen, 1994). Seismic sections across the Nordkapp Basin show asymmetric half-grabens and grabens whose polarity shifts rapidly along the basin's strike (Bergendahl, 1989; Gabrielsen et al., 1990, 1992; Dengo and Røssland, 1992; Gudlaugson et al., 1994). In the southwestern Nordkapp Basin, the dominant fault vergence shifts from northwest-facing faults in the south (Måsøy fault complex) to southeast-facing faults in the north (Nordsel High along the Nysleppen fault complex). In the intrabasin high, basement faults and grabens are more symmetric. The northeastern subbasin shows similar shifts in dominant fault polarity along the basin strike.

After rifting, a thick, Upper Carboniferous–Lower Permian layer of evaporites was deposited in the basin and partially overstepped the basin margins (Gerard and Buhrig, 1990; Bruce and Toomy, 1993; Nilsen et al., 1993).

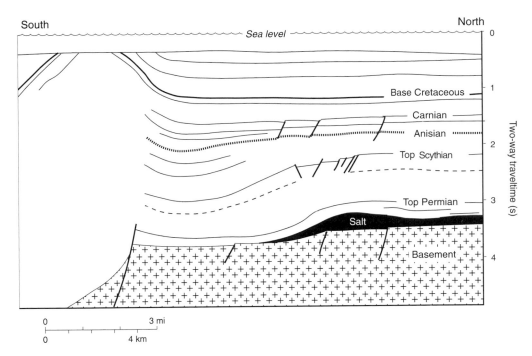

Figure 5—Line drawing from seismic section (Norwegian Petroleum Directorate line 302230-86) along the northern margin of the Nordkapp Basin. (See Figure 2b for location.) Section shows normal faults formed by Early Triassic regional extension (top of Scythian). Similar normal faults are suspected to have initiated diapirs in the basin but cannot be imaged in seismic reflection (e.g., left side of diapir). Also note basinward stratigraphic thickening of the Middle–Upper Triassic (top of Scythian–Anisian) associated with rapid passive diapiric growth. Black = salt; dotted line = base of Anisian (Middle Triassic).

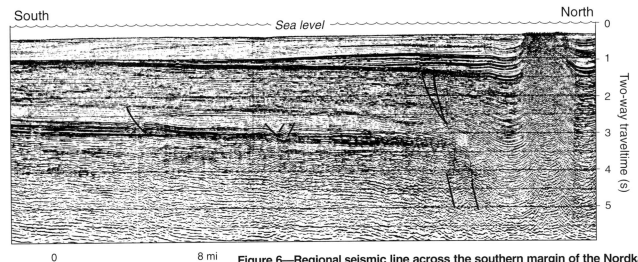

Figure 6—Regional seismic line across the southern margin of the Nordkapp Basin (Norwegian Petroleum Directorate line 3145-85). (See Figure 2b for location.) The line shows Early Triassic normal faults (center and left side) detaching above thin salt.

Gudlaugson et al. (1994) have proposed a more regional model in which the entire Barents shelf subsided and evaporites were deposited in the deep parts of the region. The thickest halite sections were deposited only in the deepest depocenters, such as the Nordkapp Basin. In the Nordkapp Basin itself, the estimates of initial thickness vary from 2.0–2.5 km in the southwestern subbasin to 4.0–5.0 km in the northeastern subbasin (Bergendhal, 1989; Jensen and Sørensen, 1992). The difference in salt thickness between the two subbasins can be attributed to greater basement subsidence in the northeastern than in the southwestern subbasin.

By Late Permian time, the evaporites were overlain by a carbonate section and a Lower Triassic siliciclastic section of relatively uniform thickness (Figures 5, 6), indicating that no significant subsidence occurred in the Nordkapp Basin (Gabrielsen et al., 1992; Jensen and Sørensen, 1992). The evaporitic source layer has long since been depleted by salt flow toward the diapirs, and the presalt, Permian, and Lower Triassic strata are commonly subparallel and tilted toward the basin axis. Sagging of the overburden was caused mainly by salt withdrawal, but faulting and rotation of the subsalt strata suggest that there has been a component of basement subsidence and deepening of the basin during later regional extension.

END OF EARLY TRIASSIC: REGIONAL EXTENSION AND DIAPIRIC INITIATION

Although most authors (e.g., Dengo and Røssland, 1992; Gabrielsen et al., 1992; Jensen and Sørensen, 1992; Koyi et al., 1993; Willoughby and Øverli, 1994) agree on an Early Triassic age for the onset of salt flow, its trigger continues to be debated. Dengo and Røssland (1992) proposed that massive deposition of sediment originating from the Ural Mountains caused differential loading and triggered diapiric rise. However, evidence of Early Triassic collapse of the Urals foreland makes it doubtful that the Urals could have supplied enough sediment to span the rapidly subsiding basin (Dore, 1991). Instead, both seismic and well data point toward the Norwegian mainland, rather than the Ural Mountains, as the main sediment source (Dore, 1991). Our seismic interpretation agrees with the hypothesis of Gabrielsen et al. (1992) that there was no large-scale differential subsidence of the Nordkapp Basin at that time, as indicated by the lack of any significant thickness change in the Lower Triassic sequence (Figures 4, 5, 6).

We agree with the hypothesis of Gabrielsen et al. (1992), Jensen and Sørensen (1992), and Koyi et al. (1993) that diapirs were triggered by normal faulting and thinning of the sediment overburden probably during a phase of basement-involved regional extension that affected the Nordkapp Basin in late Scythian (late Smithian–early Spathian) time. Seismic sections (especially Figure 6) clearly indicate Early Triassic faulting of the Permian and Lower Triassic overburden, and faulting and basinward tilting of the subsalt basement (Figures 4, 5, 6). Moreover, the simultaneous initiation of diapirs in both the northeastern and the southwestern subbasins supports this hypothesis and suggests that the trigger for diapirism was regional tectonics rather than sedimentation. This extensional episode can be correlated with the Variscan–Uralian tectonic phase that affected the Barents Sea area (Dore, 1991; Dengo and Røssland, 1992). Extension triggered diapiric reactive rise (Figure 7), a process described by Vendeville and Jackson (1992a) and Jackson and Vendeville (1994). Normal faulting locally thinned the brittle overburden and created the space necessary for salt to flow upward (Figures 7a, b). Diapirs eventually pierced the thinned graben floors above them (Figure 7c).

Structural clues for this early phase of reactive diapirism are not always visible. Because of subsequent diapiric growth by downbuilding and formation of broad diapir overhangs, normal faults associated with Early Triassic reactive rise, typically located near or against the base of diapir stems, are commonly obscured on seismic reflection data. But evidence for normal faulting can still be found on sections where the structures did not later fully evolve into tall, passive diapirs (right side of Figure 5 and left side of Figure 6). Furthermore, serial seismic sections (Figures 8, 9) illustrate how diapirs connect along strike to salt ridges or pillows overlain by extensional

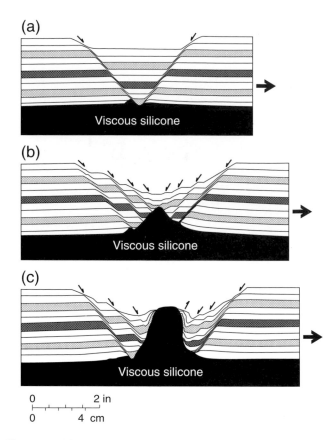

Figure 7—Line drawings of model sections of extensional diapirism. (a) and (b) During the reactive stage, the diapir rose by filling the space created by fault block displacement in the overlying graben. (c) Later, the diapir pierced the thin overburden and rose actively. After Jackson and Vendeville (1994) and Vendeville and Jackson (1992a).

grabens. This geometry is identical to that of scaled models of extensional diapirs that formed first by reactive rise, then grew passively (i.e., by downbuilding) during sedimentation (Vendeville and Jackson, 1992a,b).

Although Koyi et al. (1993) have proposed that at least some diapirs were originally salt pillows, the relatively uniform thickness of Permian and Lower Triassic strata across the basin suggests that diapirism was triggered by extension rather than by spontaneous rise from salt pillows (as Rayleigh–Taylor instabilities). This is also confirmed by the common lack of rim synclines around many of the Nordkapp Basin diapirs. On most seismic sections, there is no significant change in stratal thickness before active piercement and passive diapiric rise.

Regional extension during the Early Triassic took place at crustal scale and hence deformed basement, salt, and overburden. Therefore, in the Nordkapp Basin, as in the North Sea (e.g., Jenyon, 1986), it is tempting to regard basement faults as triggers for the formation of salt diapirs above. Koyi et al. (1993) suggested that reactivation of basement faults in the Nordkapp Basin has controlled deformation of the salt and its overburden and has localized diapirs above or near the basement faults.

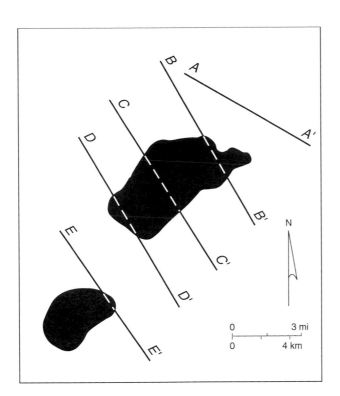

Figure 8—Map (from Geoteam, line NBGS-90) of a salt structure in southwestern Nordkapp Basin. (See Figure 2b for location.) The diapir (black) is shown, along with the location of sections in Figure 9.

However, seismic sections do not show any clear correlation between the location of basement faults and that of the diapirs, suggesting that basement faults and diapirs are not geometrically and genetically related. This conclusion is reinforced by experimental and theoretical work by Vendeville et al. (1993a, 1995). Their results (Figure 10) indicate that the low viscosity of salt rock, combined with the low rate at which crustal extension commonly occurs, means that a thick salt layer acts as a perfect lubricant between its basement and its cover. Deformation localized by faults in the basement is diffused by the thick, weak overlying salt layer rather than being directly transmitted upward through the salt into the overlying sediments.

Because of the large initial salt thickness in the Nordkapp Basin, estimated to be 2–4 km, the source layer acted as a cushion that effectively insulated the sediment overburden from the faulted subsalt basement. Both basement and overburden hence deformed independently (until the source layer had been thinned by salt withdrawal). During thick-skinned extension, diapiric initiation was controlled by overburden thinning (Vendeville et al., 1993a, 1995). The location of the diapirs therefore indicates where overburden grabens, not basement faults, formed in the Early Triassic. As for thin-skinned extension, the location of initial normal faults in the overburden is controlled by various geologic flaws, such as lateral changes in overburden facies or thickness. L. N. Jensen (personal communication, 1994) has suggested

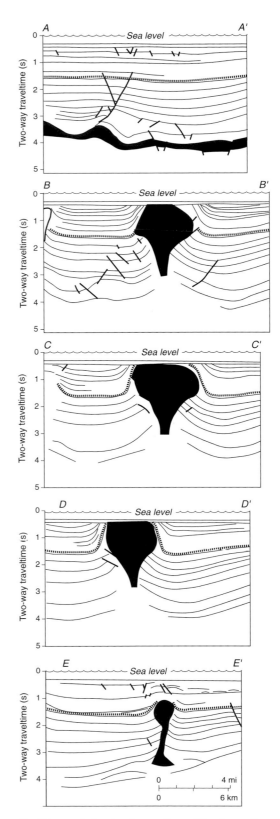

Figure 9—Five serial seismic sections (Geoteam line NBGS-90) cutting across the salt structure in southwestern Nordkapp Basin. (See Figure 8 for locations.) Along strike, the tall diapir in sections B–B' through E–E' becomes a deeply buried salt roller or pillow (section A–A') overlain by an extensional graben.

(a)

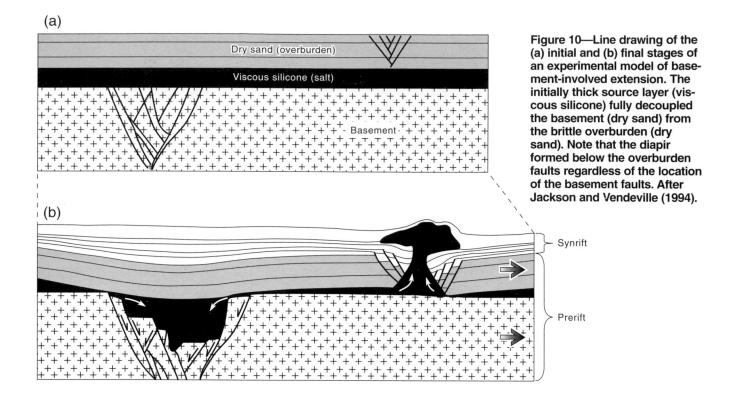

Dry sand (overburden)

Viscous silicone (salt)

Basement

(b)

Synrift

Prerift

Figure 10—Line drawing of the (a) initial and (b) final stages of an experimental model of basement-involved extension. The initially thick source layer (viscous silicone) fully decoupled the basement (dry sand) from the brittle overburden (dry sand). Note that the diapir formed below the overburden faults regardless of the location of the basement faults. After Jackson and Vendeville (1994).

that asymmetric loading of the salt layer in the Nordkapp Basin by sediments coming from the southeast induced a gentle flexure of the overburden, which in turn forced the Early Triassic overburden faults to initiate preferentially in the center of the basin.

MIDDLE–LATE TRIASSIC: DIAPIRIC GROWTH BY DOWNBUILDING AND GRAVITY GLIDING

Reactive diapirs eventually grew tall enough to actively pierce their thinned roof and emerge at the sea floor (Figure 11). Subsequently, diapirs grew passively by downbuilding, a process described by Nelson (1989, 1991) and Jackson and Talbot (1991), in which diapirs rise by maintaining their crest at or close to the sea floor while sediments accumulate in depotroughs between diapirs. Seismic sections in the Nordkapp Basin indicate that passive growth of diapirs was rapid and salt flow was vigorous. Middle Triassic strata thickened tremendously from the basin margins toward the diapirs (e.g., left side of Figure 11) and formed locally thick wedges that abut against the diapir flank with the apparent geometry of salt rollers. Many of the salt depotroughs are asymmetric. Until Ladinian time, many diapirs in the center of the basin grew with steep or vertical flanks (Figure 11). Some diapirs widened upward during passive growth. This geometry suggests that the rate of net diapiric rise (i.e., vertical salt flow) up the diapir conduit was at least as high as the rate of sediment aggradation in the rapidly

subsiding adjacent depotroughs (Vendeville and Jackson, 1991; Vendeville et al., 1993b; Talbot, 1995).

The Anisian marked a significant change in structural style. Although many diapirs continued to grow passively, there is clear evidence that the source layer was thinned or depleted. A major unconformity at the base of the Anisian on seismic sections in the northwestern part of the basin (Figure 12) corresponds to the top Lower Triassic unconformity described by Koyi et al. (1993, their figure 2). Sediments below the unconformity have been strongly deformed and flexed (Figure 12) while the center of the depotroughs was rapidly sinking into the thick salt layer. Sediments on the sides of the depotroughs, being closer to the rising diapirs, were dragged upward by vigorous salt flow and were later truncated by the unconformity. In contrast, sediments above the basal Anisian unconformity (Figure 12) remained largely undeformed, flat-lying, and of uniform thickness. This indicates that, by Anisian time, most overburden depotroughs had grounded onto the subsalt and the source layer had been virtually depleted. Paradoxically, although source layer depletion had restricted or cut off the salt supply, diapirs continued to rise passively after the unconformity episode until Ladinian–Carnian time (Middle Triassic). Seismic sections show that diapirs rose fast enough to even overflow their margins and produce salt overhangs (Figure 13) when the rate of sediment aggradation decreased in Carnian time. The precise geometry and extent of diapir overhangs in the Nordkapp Basin are still debated (Koyi et al., 1992, 1993). Diapir overhangs seem to be larger and more numerous in the northeastern sub-basin.

South   North

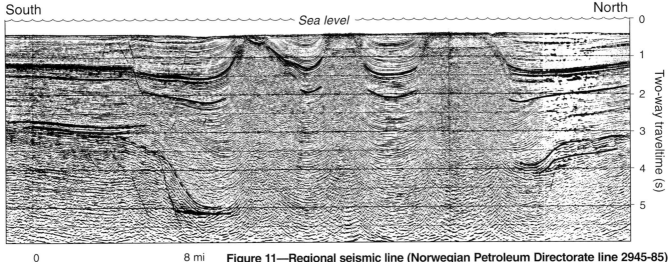

Figure 11—Regional seismic line (Norwegian Petroleum Directorate line 2945-85) across the northeastern Nordkapp Basin. (See Figure 2b for location.) Section shows three tall passive diapirs that rapidly rose during the Middle–Late Triassic. Also note the extreme thickening of the overburden section between the margins and the center of the basin.

Diapiric rise after depletion of its source layer raises both geometrical and mechanical questions. First, there should have been no more salt available to fill the increasing volume of the rising diapirs after Anisian time. Some extra salt could have been squeezed out from under the sides of the depotroughs by flattening the flexed-up base and inverting the depotroughs into turtle structures (as on the Angolan margin; Duval et al., 1992; Lundin, 1992). However, seismic sections in the Nordkapp Basin show no signs of such inversion (e.g., Figures 9, 11, 12, 13). Instead, sediments in the depotroughs seem to have acted as rigid overburden blocks. Second, the mechanical force that drove diapirism during reactive and passive rise could no longer act after the source layer was depleted. As long as the source layer was thick, the increasing weight of sediments in the depotroughs pressurized the salt and forced it to flow laterally from the source layer into the rising passive diapirs. Because the base of the overburden rested directly on the subsalt basement after source layer depletion, thickening of the overburden by sediment aggradation could no longer increase the pressure in the source layer. Therefore, salt flow should have stopped and diapirs should have remained static.

How a diapir can rise without available salt supply from the source layer is illustrated in Figure 14. The basic principle is similar to the mechanism that causes diapiric fall during extension (Vendeville and Jackson, 1992b), but it involves horizontal shortening of the diapir rather than widening. To rise without changing its width, a diapir would have to import more salt to fill the volume increase of the growing diapir (Figure 14b). However, a diapir can rise even without extra salt available if the increase in height is compensated by a decrease in diapir width (Figure 14c). This mechanism is made possible by the overburden blocks behaving far more rigidly than salt.

The applicability of this mechanism was tested using the experimental model shown in Figure 15 (Vendeville, 1993). The model comprised an initially thick source layer of viscous silicone (representing salt) overlain by a brittle sand overburden (representing sediments). The model was first regionally extended during deposition. Extension formed two reactive diapiric ridges that rose below faulted grabens (structures G1 and G2, Figure 15a). Later, the southern segments of the diapiric ridges actively pierced the thin overlying graben floors, emerged at the model surface (structures D1 and D2, Figure 15a), and then grew passively during ongoing extension and deposition. The northern part of the ridges remained in the reactive stage throughout the extensional phase (G1 and G2, Figure 15a).

The experiment was stopped after passive diapirism had largely depleted the source layer and diapiric rise had slowed down. The model was then buried and regionally shortened (Figure 15b). Because viscous silicone is far weaker than sand, the shortening was accommodated by reducing the width of the diapirs, leaving the adjacent overburden blocks virtually undeformed. In the southern part of the model (Figures 15b, c) where diapirs were already at or near the model surface, regional shortening squeezed the silicone out of the diapir stems and forced the diapir to rise. In the northern part of the model (Figures 15b, d) where diapirs were still overlain by a thick roof, shortening flexed the model surface and reactivated some of the old reactive normal faults in reverse slip (Figure 15d). Of particular interest is the absence of obvious signs of shortening in sections where diapirs remained at or near the surface (Figure 15c). Because all the shortening was accommodated by diapiric narrowing rather than by folding or reverse faulting of the adjacent overburden, the diapir in Figure 15c would not be con-

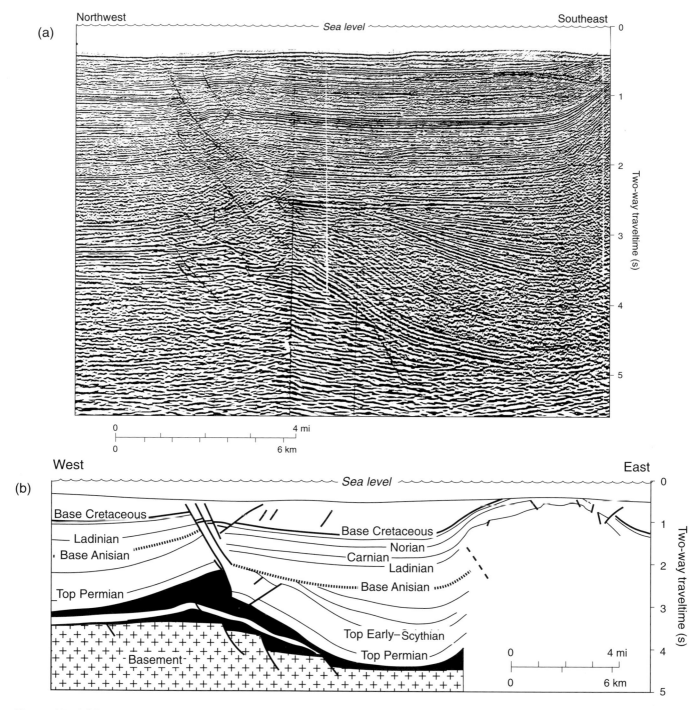

Figure 12—(a) Largely uninterpreted seismic section (Norwegian Petroleum Directorate line D8-85), which forms center-left part of (b) interpreted line drawing showing the change in style of overburden deformation above the Anisian unconformity (hatched line on drawing). (See Figure 2b for location.) Sediments below the Anisian unconformity have been flexed down by salt withdrawal during vigorous diapiric rise. In contrast, sediments above the unconformity are subhorizontal except for drag against the diapir flank and normal faulting on the basin margin.

ventionally interpreted as being driven by shortening. The only indications of diapiric shortening are (1) the evidence for continued diapiric rise after source layer depletion (as in Figure 12) and (2) the reverse faults or folds along strike, where the diapiric ridge had always remained deeply buried.

What caused shortening of the diapirs in the Middle Triassic? We propose that normal displacement along basement faults during an episode of regional extension deepened the basin floor, steepened the basal slope, and triggered down-to-the-basin gravity gliding of sediments above the thinned source layer (Figure 16). This

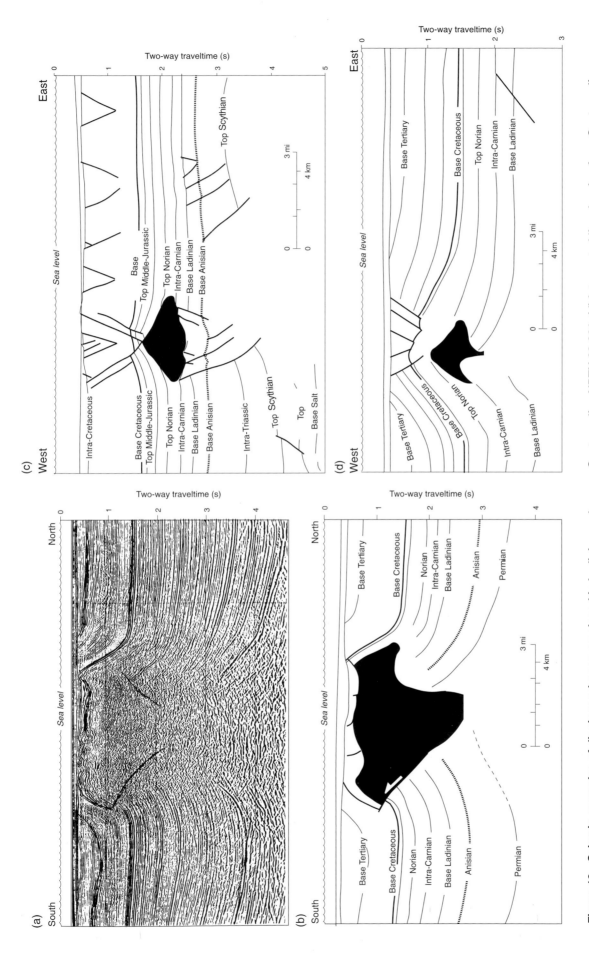

Figure 13—Seismic examples of diapir overhangs emplaced in Ladinian and Carnian time. (a) Uninterpreted seismic section and (b) interpreted line drawing from Geoteam line NBGS-90-209. (c) Interpreted line drawing from Geoteam line NPTN-92-201. (d) Interpreted line drawing from Geoteam line NPTS-92-203. (See Figure 2b for locations.)

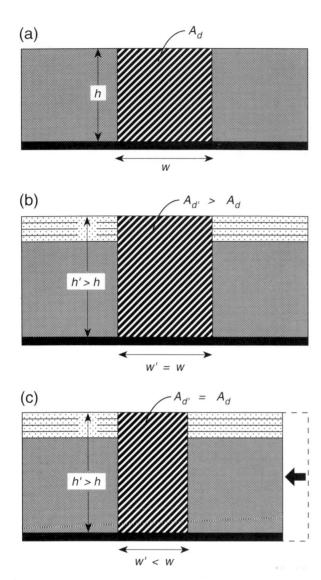

Figure 14—Salt volume problem associated with diapiric growth above a thinned source layer. Black = source layer; striped = diapiric salt. (a) Initial stage with diapiric height = *h*, width = *w*, and salt area in section = A_d. (b) If the diapir grows vertically while maintaining a constant width, *w*, the salt volume A_d must increase to $A_{d'}$ as *h* increases to *h'*. (c) The diapir can grow without a change in salt area, A_d, if its width, *w*, decreases to *w'*.

hypothesis is supported by coeval normal faulting and growth of salt rollers along the basin margins (Figure 4) (Gabrielsen et al., 1992). Steepening of the basin slope caused the overburden blocks to glide downslope under the effect of gravity, squeezing the salt diapirs in the basin center and opening extensional gaps upslope along the basin margin. Because the timing of diapiric growth in Middle–Late Triassic time remains strikingly similar across the Nordkapp Basin (e.g., diapirs formed overhangs in Carnian time in both the northeast and southwest subbasins), we advocate that gravity gliding was triggered by regional tectonics that steepened the slope, rather than by more local influences. Thus, during

Middle–Late Triassic time, regional extension tightly controlled diapiric growth, albeit indirectly by triggering gravity gliding. Without regional extension and deepening of the basin, diapirs in the Nordkapp Basin would have stopped rising as early as the onset of Anisian time when salt withdrawal had already depleted the source layer.

By Late Triassic, the diapirs stopped rising and were buried. Although widespread block faulting occurred throughout the western Barents Sea in Late Jurassic–Early Cretaceous time (Brekke and Riis, 1987; Gabrielsen et al., 1990, 1992; Dengo and Røssland, 1992), it does not seem to have significantly affected the Nordkapp Basin, except for the southernmost part (Krokan, 1988; Bergendahl, 1989). In the rest of the basin, the only visible effect of this block faulting was a minor pulse of active diapiric rise recorded by a slight thinning of Upper Jurassic strata against the diapir flanks. The diapirs were then buried under 1000–1500 m of flat-lying Cretaceous sediments. That diapirs were then extinct is clearly demonstrated by the fact that sedimentary wedges prograded without interruption across the basin above the diapir crests (Figures 6, 17). If the diapirs had been active, they would have deformed the sea floor topography, which would have affected or retarded progradation of the wedges. Such wedges were later deformed by diapiric rejuvenation during Late Cretaceous–middle Tertiary time.

LATE CRETACEOUS: DIAPIRIC REACTIVATION BY GRAVITY GLIDING

Another episode of diapiric rise occurred during the Late Cretaceous. Although many structural clues in the upper part of the sedimentary cover were later removed by Pliocene–Pleistocene erosion (1000–1200 m of sediments removed) (Nyland et al., 1992; Riis, 1992; Vågnes et al., 1992), the Cretaceous episode of active diapirism is clearly visible on most seismic sections in the Nordkapp Basin (Figures 6, 11, 13, 17, 18). Active diapirs rose above the regional datum by arching their thick roofs. Most sections show diapir roofs containing grabens at their crest and monoclinal folds above diapir shoulders. A few sections (Figure 19) show synthetic reverse faults on the side of the uplifted roof that are similar to those formed in experimental models of active diapirism (Vendeville and Jackson, 1992a; Schultz-Ela et al., 1993). In the southwestern subbasin, the diapir crests were eroded. In addition, mound-like and fan-like deposits can be interpreted as the result of mass flows triggered by subsidence. Their age has been interpreted by Henriksen (1990) and Henriksen and Vorren (in press) as early Tertiary, but correlation with nearby wells suggests that their age is Late Cretaceous. The timing for diapiric reactivation also neatly coincides with Late Cretaceous rifting of the North Atlantic (Faleide et al., 1993; Hinz et al., 1993, Knott et al., 1993). This tectonic episode is known to have affected the western Barents Sea, producing associated salt tectonics in the Tromsø and Sørvestsnaget basins (Faleide et al., 1993).

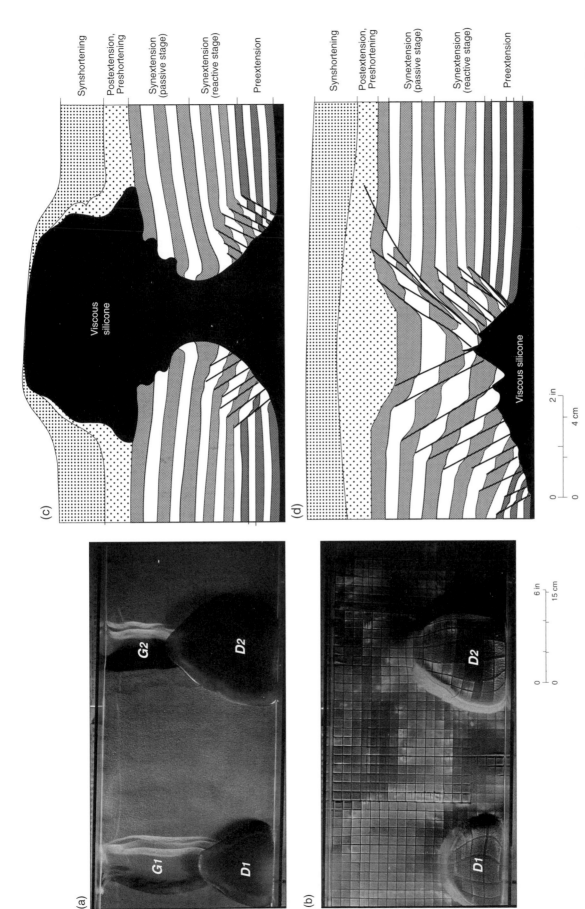

Figure 15—Experimental model of early extension and diapirism followed by regional shortening. (a) Overhead view of model following early extension. Extension produced two grabens (G1 and G2) below which diapirs rose reactively. Rounded diapirs (D1 and D2) later actively pierced the southern part of the model grabens, then grew passively. (b) Overhead view of model following burial and late shortening. Grid lines are passive markers printed on the model surface. The old passive diapirs (D1 and D2) were reactivated during shortening and rose actively, while stretching and distorting their thin roof. Northern part of the model shows little evidence of shortening. (c) Cross section of the southern part of the model, where diapirs had emerged before shortening. Section shows the old normal faults formed during the reactive stage, stratigraphic thickening during the passive stage, and bending of the diapir roof during late shortening and vertical rise. There are no folds or reverse faults in the adjacent overburden. (d) Cross section of the northern part of the model, where diapirs had remained in the reactive stage before shortening and always had a thick roof. Late shortening deformed the structure by anticlinally warping the model surface and reactivating some of the old normal faults as reverse faults.

(a)

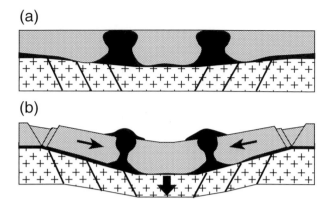

(b)

Figure 16—Schematic diagrams illustrating how basement subsidence during crustal extension deepens the basin, steepens the basin slopes, and triggers down-to-the-basin gravity gliding. This induces shortening of the diapirs in the basin center and normal faulting on the margins.

Following the hypothesis of Rønnevik (1982) that the Late Cretaceous regional extension had reactivated structures in the Nordkapp Basin, we propose that regional extension triggered down-to-the-basin gravity gliding and reactivated the diapirs. The mechanism causing diapiric rise is similar to that of the Middle–Late Triassic episode illustrated in Figures 14 and 16, except that diapirs were overlain by a thick roof prior to the Late Cretaceous gravity gliding and renewed rise. Unlike spontaneous active diapirism, which is driven by pressure differences in a thick source layer and can easily be

prevented by a thick roof (Vendeville and Jackson, 1992; Schultz-Ela et al., 1993), diapiric reactivation by local shortening after source layer depletion was driven by the slope-parallel component of the weight of the entire overburden blocks between the basin margin and the diapirs in the basin center. This process pressurizes the diapir and can easily overcome the strength of even a thick overburden roof. Our hypothesis that Late Cretaceous diapiric reactivation was driven by gravity gliding is supported by coeval extensional faulting along the basin margins, which created the Nysleppen, Måsøy, and Thor Iversen fault complexes (Gabrielsen et al., 1990). Gabrielsen et al. (1990) also described reverse faults of Late Cretaceous age above a salt pillow (Figure 4b, bottom left) and suggested that the margins of the basin were compressed. However, we have interpreted such apparent reverse faults as Early Triassic normal faults that were later rotated passively during Late Cretaceous subsidence and basement tilting.

We conducted a second set of experiments (Figure 20) to test the applicability of this mechanism and to investigate the deformation style above circular and linear salt structures during late shortening. For simplicity, diapirs were triggered purely by sedimentary differential loading. An initially thick source layer of viscous silicone was overlain by a thin sand layer of uniform thickness. Deformation started after we vacuumed six holes in the sand layer (two linear and four circular; see location and shape in Figure 20a). The viscous silicone rose diapirically through the holes, forming passive diapirs that maintained their crest at the model surface during subsequent deposition of new sand layers. The diapirs rose passively until source layer depletion caused them to stop growing. We buried the then-inactive diapirs under a 2-cm-thick

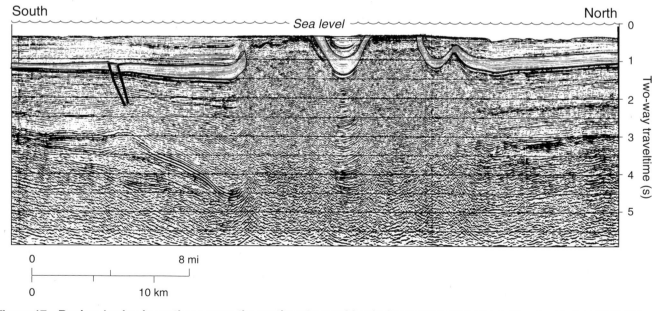

Figure 17—Regional seismic section across the northeastern subbasin (Norwegian Petroleum Directorate line 3115-85) showing a Lower Cretaceous sediment wedge (gray; also visible in Figure 6) that prograded southward across the basin above the then inactive diapirs. The wedge was later deformed during Late Cretaceous and middle Tertiary diapiric reactivation. (See Figure 2b for location.)

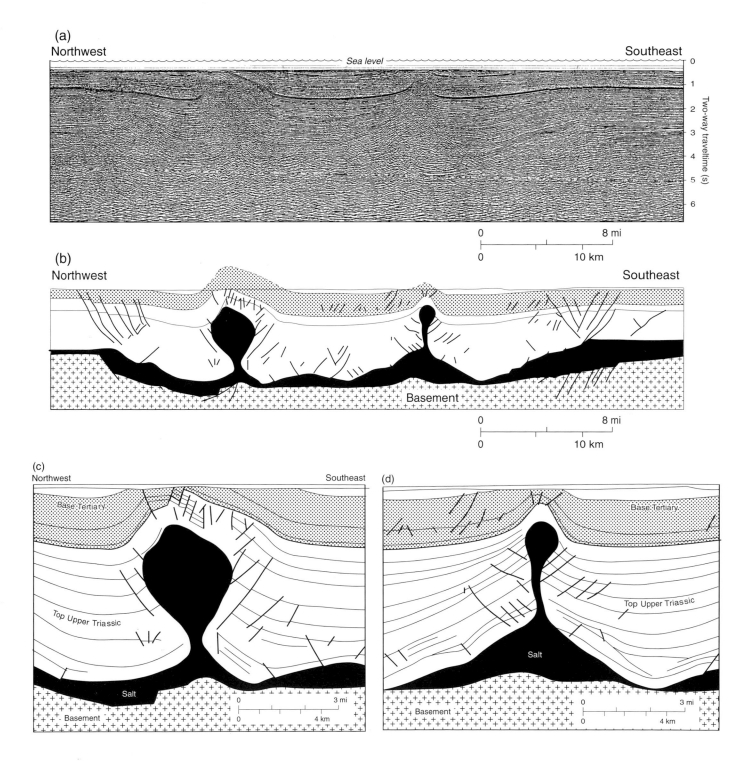

Figure 18—(a) Regional time-migrated seismic line across the southwestern subbasin. (b) Line drawing of depth-converted section. (c) and (d) Details of diapirs in this section, also depth converted. No vertical exaggeration. Sections show that Jurassic and Cretaceous sediments (top stippled layer) of regionally uniform thickness were deformed by active rise in the Late Cretaceous and middle Tertiary. (See Figure 2b for location.)

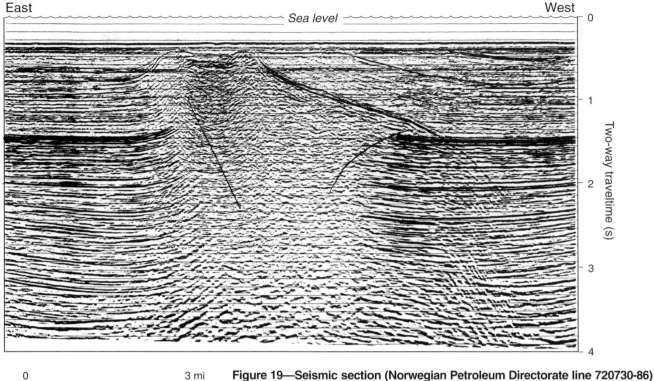

East   West

—— *Sea level* ——

Two-way traveltime (s)

0   3 mi

0   4 km

Figure 19—Seismic section (Norwegian Petroleum Directorate line 720730-86) across a diapir reactivated during Late Cretaceous and middle Tertiary time. Note the reverse fault induced by active diapiric rise. (See Figure 2b for location.)

sand layer (representing the 1000–1500 m of uniformly thick Cretaceous sediments in the Nordkapp Basin; lightly stippled in Figure 20c,d) and regionally shortened the entire model.

An overhead view of the model after shortening (Figure 20b) shows that most diapirs were reactivated during shortening. Diapirs rose by lifting their thin roofs, forming crestal grabens (diapirs 5, 6, 7 in Figure 20b) and flexing the sediments above the diapir shoulders. The traces of the crestal normal faults are not consistently oriented with respect to the direction of regional shortening. Instead, fault traces appear to follow the shape of the diapir in planform, indicating that the horizontal component of stretching of the diapir roof caused by diapir rise far outweighed the horizontal component due to regional shortening. In other words, the height of most diapirs increased much faster than their width decreased. Two squeezed diapirs (diapirs 3 and 4 in Figure 20b) even emerged at the surface and formed new overhangs. We attribute the rapid diapiric rise to the fact that they were initially tall and narrow prior to regional shortening. Tall, narrow diapirs rise vigorously in response to a small amount of shortening (Figure 21, left side). Their deformation style is thus closer to that of structures formed purely by vertical diapiric rise than that of structures formed by regional shortening. In contrast, initially low, wide diapirs (Figure 21, right side) do not rise as much, and their deformation style is closer to that typical of regional shortening.

Because most diapirs in the Nordkapp Basin had initially narrow stems, the deformation style of their roofs is similar to that above vertically rising diapirs (Figures 22, 23). Around some of the model diapirs (see large north-trending fault trace east of diapirs 4 and 5 in Figure 20b), the flexure above the diapir shoulders was tight enough to initiate shallow reverse faults. Sections in the model (Figure 20c) indicate that such shallow reverse faults are similar to those above active diapirs (Vendeville and Jackson, 1992a; Schultz-Ela et al., 1993) and are not caused directly by regional shortening. These shallow reverse faults commonly root at depth onto the top of the diapir or at the tip of diapir overhangs. The model showed only one large, through-going reverse fault (southwest corner of Figure 20 B, near diapir 7) that can unmistakably be interpreted as a direct result of regional shortening. In the rest of the model, regional shortening was entirely accommodated by laterally squeezing the diapirs, forcing them to rise.

Model sections (Figures 20c, d) and seismic examples from the Nordkapp Basin (Figures 22, 23, 24) illustrate the difficulty of finding evidence of shortening in areas where most of the deformation was accommodated by narrowing the diapirs. Sections cutting across diapirs (Figures 20c, 23a, b) typically show no signs of shortening and display only active diapirs. The only clues that could lead an interpreter to suspect that shortening has occurred are mechanical. That diapirs would have actively risen again after depleting their source layer, would

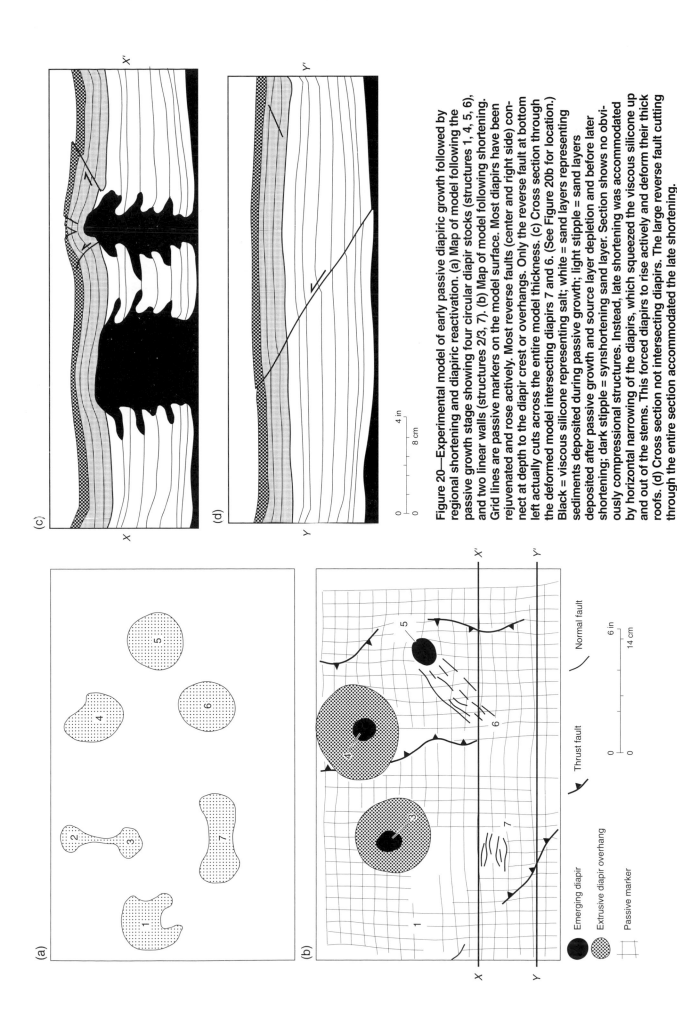

Figure 20—Experimental model of early passive diapiric growth followed by regional shortening and diapiric reactivation. (a) Map of model following the passive growth stage showing four circular diapir stocks (structures 1, 4, 5, 6), and two linear walls (structures 2/3, 7). (b) Map of model following shortening. Grid lines are passive markers on the model surface. Most diapirs have been rejuvenated and rose actively. Most reverse faults (center and right side) connect at depth to the diapir crest or overhangs. Only the reverse fault at bottom left actually cuts across the entire model thickness. (c) Cross section through the deformed model intersecting diapirs 7 and 6. (See Figure 20b for location.) Black = viscous silicone representing salt; light stipple = sand layers representing sediments deposited during passive growth; dark stipple = sand layers deposited after passive growth and source layer depletion and before later shortening; dark stipple = synshortening sand layer. Section shows no obviously compressional structures. Instead, late shortening was accommodated by horizontal narrowing of the diapirs, which squeezed the viscous silicone up and out of the stems. This forced diapirs to rise actively and deform their thick roofs. (d) Cross section not intersecting diapirs. The large reverse fault cutting through the entire section accommodated the late shortening.

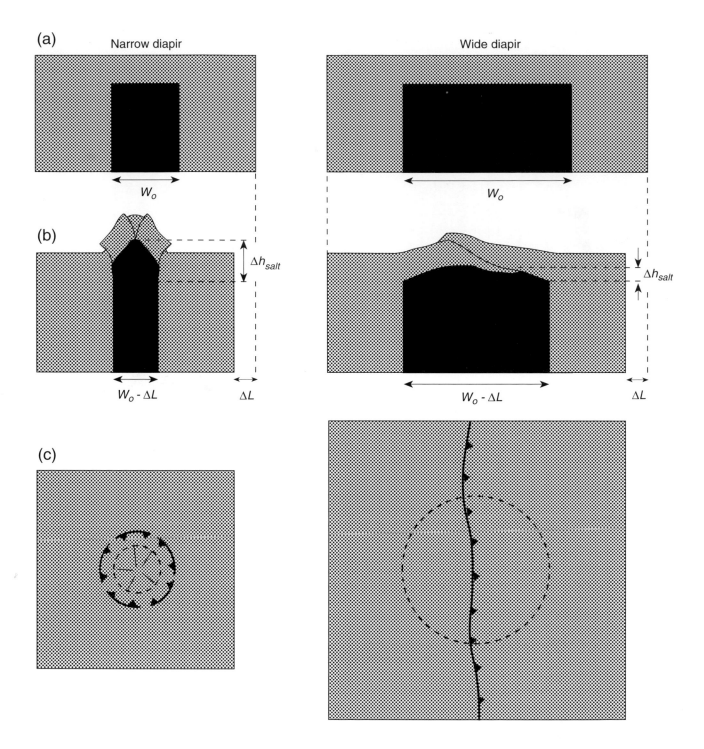

Figure 21—Conceptual diagram illustrating how (a) initially tall and narrow diapirs (left column) (b) rise higher and faster than initially wide and stocky diapirs (right column) for the same amount of horizontal shortening, ΔL. The increase in diapiric height, Δh_{salt}, depends on the initial diapiric width, W_o, and on the component of horizontal shortening, ΔL. Because the vertical displacement, Δh_{salt}, during reactivation of narrow diapirs largely exceeds the component of horizontal shortening, ΔL, the deformation style is that of a vertically rising active diapir. (c) Normal and reverse faults above narrow diapirs follow the diapiric shape regardless of the direction of regional shortening (left side). In contrast, the component of horizontal shortening, ΔL, during reactivation of wide diapirs exceeds the vertical displacement, Δh_{salt}. The deformation style is thus closer to that of regional compression, and the diapir forms only a slight topographic bulge that does not significantly disturb the local stress field. Reverse faults do not necessarily follow the diapiric shape but can form perpendicular to the direction of regional shortening (right side).

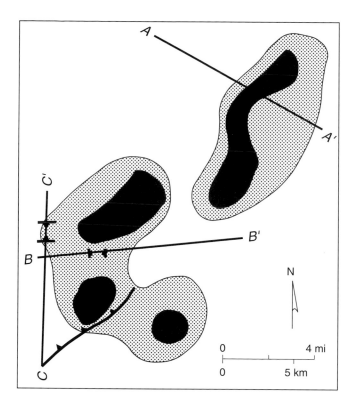

Figure 22—Map of salt diapirs reactivated by shortening in southwestern Nordkapp Basin. (See Figure 2b for location.) Locations of the seismic sections in Figure 23 are shown. Black = diapirs close to the Pliocene–Pleistocene erosion surface; stippled = base of the Cretaceous sediments uplifted above regional datum.

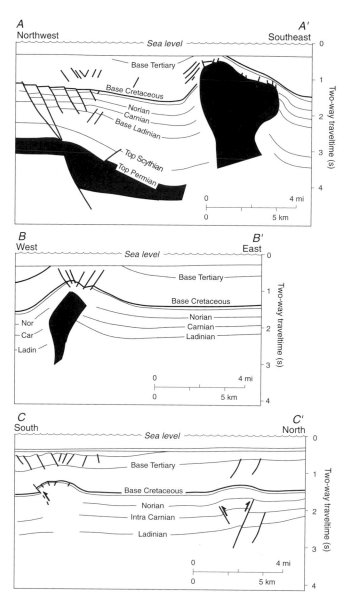

Figure 23—Interpreted line drawings of seismic sections located in Figure 22. Sections A–A' (Geoteam line NBGS-90-423) and B–B' (Geoteam line NBGS-90-203) cut across preexisting diapirs, thus no obviously contractional structure is visible. Instead, shortening was accommodated mostly by narrowing of the diapir stems. Note the nearly complete pinch-off of the diapir stem on section B–B'. Section C–C' (Geoteam line NPTS-92-301) is away from the diapir stems and shows reverse faults and folds caused by Late Cretaceous and middle Tertiary shortening that was accommodated elsewhere by diapiric narrowing.

have stopped growing for a long time, and would then have been buried under a thick roof implies that there was an external trigger. Another clue can be found by looking for contractional structures in sections along strike that do not cut across diapirs.

In the experiment, the section in Figure 20d shows a large reverse fault that cuts across the entire overburden section and therefore must have been induced by regional shortening rather than by vertical diapiric rise. In the seismic example, only sections away from the diapir stems (Figures 23c, 24) show folding and reverse faulting associated with shortening of the central part of the basin. The last clue for late-stage diapiric shortening is the shape of the diapir stem in planform. Assuming that diapirs had circular planforms before shortening, their planforms should have become elliptical after diapiric narrowing (Figure 25), with the short axis parallel to the direction of maximum shortening. The Nordkapp Basin comprises many salt diapirs whose stems are elliptical (Figure 26). Although the elliptical shape might have been partially inherited from older reactive salt ridges during Late Triassic extension, it also reflects later shortening.

The Late Cretaceous pulse of active diapiric rise ended, and diapirs were buried under 1500 m of lower Tertiary sediments.

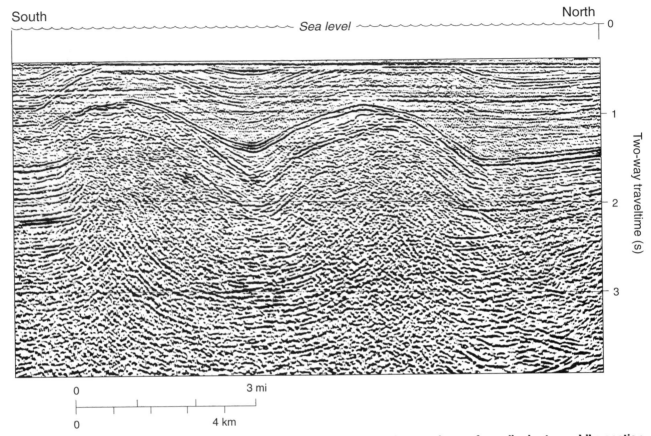

Figure 24—Seismic section (Norwegian Petroleum Directorate line 2700-87) located away from diapir stems. Like section C–C' in Figure 23, this section shows reverse faults and folds caused by Late Cretaceous and middle Tertiary shortening. (See Figure 2b for location.)

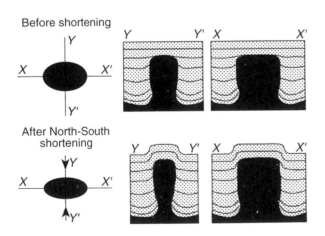

Figure 25—Conceptual diagram showing the change in planform (left) and vertical sections (right) of a diapir reactivated by late shortening. The initially elliptical diapiric planform is deformed into an even tighter ellipse after north-south shortening. Sections normal to the direction of shortening (section X–X') show no change in diapiric width. On section Y–Y', parallel to the direction of regional shortening, the diapiric width has decreased. (See Figure 26 for examples of this concept.)

MIDDLE TERTIARY: DIAPIRIC REACTIVATION BY REGIONAL SHORTENING

In the middle Tertiary (Eocene–Oligocene), a new regional tectonic event (described by Brekke and Riis, 1987; Gabrielsen and Faerseth, 1988; Gabrielsen et al., 1990; Dengo and Røssland, 1992; Riis and Fjeldskar, 1992; Sales, 1992; Nilsen, 1994; Reemst et al., 1994) triggered the last phase of diapiric rise in the Nordkapp Basin. Unlike the Middle–Late Triassic and Late Cretaceous phases, no normal faulting along the basin margin was associated with the middle Tertiary diapiric rise. Instead, because the base of the Cretaceous in some sections is slightly warped above the regional datum along the basin margin (Figure 27), we infer that middle Tertiary diapiric growth was not induced by gravity gliding triggered by regional extension but instead by regional contraction in the Nordkapp Basin. At the scale of the diapirs, deformation was very similar to that during the Late Cretaceous. Diapirs responded to horizontal shortening by rising vertically and lifting and fracturing their roofs. This last episode of horizontal shortening pinched off the stem of many diapirs, giving them a tear-drop geometry.

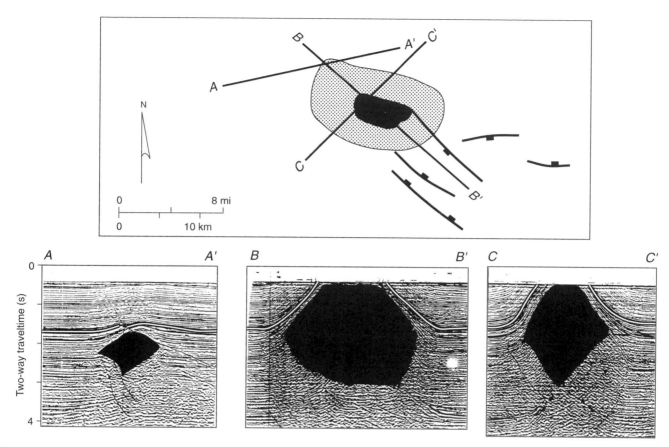

Figure 26—Seismic examples from the Nordkapp Basin illustrating the concept shown in Figure 25. (See Figure 2b for locations.) Section A–A' is Geoteam line NPTN-92-201. The diapir stem is much narrower in section C–C' (Geoteam line NPTN-92-407) than in B–B' (Geoteam line NPTN-92-210).

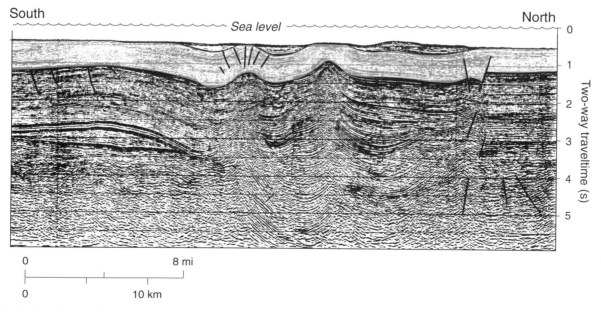

Figure 27—Regional seismic line across the northeastern subbasin (Norwegian Petroleum Directorate line 2845-85) showing Cretaceous sediments (gray) deformed during diapiric rejuvenation in the Late Cretaceous and middle Tertiary (open folds in the center of basin). In contrast, the broad fold on the south margin of the basin (left side) was directly caused by middle Tertiary regional compression. (See Figure 2b for location.)

In models simulating a fluid overburden, tear-drop diapirs are interpreted as very mature diapirs that formed by Rayleigh–Taylor instability in which salt rises buoyantly and eventually pinches off the initially wide stems by deforming the adjacent viscous overburden. In the examples illustrated here, stem pinch-off was neither induced by buoyancy forces, nor did it occur by deforming the adjacent overburden. Salt was expelled upward under the lateral pressure of the rigid, undeformed overburden blocks. The driving force was not imposed by the salt diapir itself but was either the slope-parallel component of the weight of the overburden gliding downslope or simply horizontal regional compression. Because many of the depotroughs had acquired a concave-upward base during downbuilding in Triassic time, the diapirs were usually wider at their base than higher up in the section. When shortening in Late Cretaceous and middle Triassic time closed the diapirs at their narrowest width, it commonly left a small, triangular pedestal of remaining salt at the diapir base (Figures 18c, d). Koyi et al. (1992) attributed this apparent geometry to a velocity pull-up effect caused by overlying diapir overhangs. However, the presence of such triangular pockets of salt at the base of diapirs devoid of overhangs (e.g., Figure 18d) indicates that they are real structures and not mere seismic artifacts.

No diapiric activity occurred after the middle Tertiary pulse. The Barents Sea strata were subsequently uplifted during the Pliocene–Pleistocene and were deeply eroded. Erosion is thought to have removed more than 1000 m of sediments in the Nordkapp Basin (Eidvin and Riis, 1989; Riis and Jensen, 1992; Vågnes et al., 1992; Reemst et al., 1994). The area then subsided and was overlain by thin (100–200 m) Pleistocene sediments; the sea floor is now below 300–400 m of water. The diapirs now have negligible bathymetric expression.

CONCLUSIONS

The evolution of salt structures was once commonly regarded as depending mainly on the intrinsic properties of salt and overburden rocks, such as density contrast, viscosity difference, and initial thicknesses. The geometry and growth rate of salt diapirs were believed to be controlled much more by these parameters than by external processes such as regional tectonics. Our observations in the Nordkapp Basin suggest that halokinesis can be intimately and primarily controlled by regional tectonics. In the Nordkapp Basin, diapirs did not rise continually through geologic time. Rather, they grew by short pulses separated by long periods of inactivity. Each burst of diapiric rise can be chronologically and genetically tied to an episode of regional tectonics in the Barents Sea area. Early Triassic regional extension triggered the onset of diapirs by fracturing and thinning their overburden. Middle–Late Triassic regional extension and basin subsidence allowed the diapirs to keep growing even after they had depleted the source layer. Another episode of

gravity gliding during Late Cretaceous basement-involved extension reactivated the diapirs. Finally, middle Tertiary compression triggered the last episode of diapiric rise in the basin.

In this chapter, we have presented a new process of diapiric reactivation in which regional extension deepens the basin floor, steepens the basal slope, and triggers gravity gliding. This has induced normal faulting on the basin margin and lateral contraction and renewed diapiric growth in the basin center. We think this process is also applicable to diapirs in other salt basins. For example, diapirs at the base of the slope in the Gulf of Mexico have pinched-off stems and had their last surge of vertical rise during an episode of local shortening. We have also described a new mode of diapir stem pinch-off, a feature encountered on many seismic sections. This process operates when local shortening squeezes initially wider, pre-existing diapirs and pinches off their stems, without requiring any significant strain of the rigid adjacent overburden blocks.

Acknowledgments   *We thank Geoteam, Geco-Prakla and their partners (Esso, Norsk Hydro, and Statoil) , and the Norwegian Petroleum Directorate for allowing us to use and reproduce seismic data. The experimental modeling was conducted at the Applied Geodynamics Laboratory, Bureau of Economic Geology, The University of Texas at Austin, with funds from the Texas Advanced Research Program and the following companies: Agip S.p.A, Amoco Production Company, Anadarko, ARCO Exploration and Production Technology, BP Exploration Inc., Chevron Petroleum Technology Company, Conoco Inc., Exxon Production Research Company, Louisiana Land and Exploration, Marathon Oil Company, Mobil Research and Development Corporation, Petroleo Brasileiro S.A., Phillips Petroleum Company, Société National Elf-Aquitaine Production, Statoil, Texaco Inc., and Total Minatome Corporation. We thank Bobby Duncan, Martin Jackson, Tucker Hentz, Jake Hossack, Lars Jensen, Hemin Koyi, and Dave Roberts for comments on the manuscript, and Michele Bailey for drawing the figures. Special thanks to Beth Dishman and Dottie Johnson for their patience in handling countless phone calls between Norway and Texas during this research.*

REFERENCES CITED

Bergendhal, E., 1989, Halokinetisk utvikling av Nordkapp-bassengets sørvestre segment: Master's thesis, University of Oslo, Norway, 120 p.

Brekke, H., and F. Riis, 1987, Tectonics and basin evolution of the Norwegian shelf between 62°–72°: Norsk Geologisk Tidsskrift, v. 67, p. 295–322.

Bruce, J. R., and D. F. Toomy, 1993, Late Paleozoic biotherm occurrences of the Finnmark shelf, Norwegian Barents Sea: analogues and regional significance, *in* T. Vorren, ed., Arctic geology and petroleum potential: London, Elsevier, p. 377–392.

Dengo, C. A., and K. G. Røssland, 1992, Extensional tectonic history of the western Barents Sea, *in* R. M. Larsen et al., eds., Structural and tectonic modelling and its application to petroleum geology: Norwegian Petroleum Society Special Publication, p. 91–107.

Dore, A. G., 1991, The structural foundation and evolution of Mesozoic seaways between Europe and Arctic: Paleogeography, Paleoclimatology, Paleoecology, v. 87, p. 441–492.

Duval, B., C. Cramez, and M. P. A. Jackson, 1992, Raft tectonics in the Kwanza basin: Marine and Petroleum Geology, v. 9, p. 389–404.

Eidvin, T., and F. Riis, 1989, Nye dateringer av de tre vestlige borehullen i Barentshavet. Resultater og konsekvenser for den tertiaere hevningen: Norwegian Petroleum Directorate Bulletin No. 27, 44 p.

Faleide, J. I., S. T. Gudlaugson, and G. Jacquart, 1984, Evolution of the western Barents Sea: Marine and Petroleum Geology, v. 1, p. 120–150.

Faleide, J. I., E. Vågnes, and S. T. Gudlaugson, 1993, Late Mesozoic–Cenozoic evolution of the southwestern Barents Sea, *in* J. R. Parker, ed., Petroleum geology of northwest Europe: Proceedings of the Fourth Conference, Geological Society of London, p. 933–950.

Gabrielsen, R. H., and R. R. Faerseth, 1988, Cretaceous and Tertiary reactivation of master fault zones of the Barents Sea: Norsk Polar Institute Report No. 46, p. 93–97.

Gabrielsen, R. H., R. R. Faerseth, L. N. Jensen, J. E. Kalheim, and F. Riis, 1990, Structural elements of the Norwegian continental shelf, part I, the Barents Sea region: Norwegian Petroleum Directorate Bulletin No. 6, 33 p.

Gabrielsen, R. H., O. S. Kløvjan, and T. Stølan, 1992, Interaction between halokinesis and faulting: structuring of the margins of the Nordkapp Basin, Barents Sea region, *in* R. M. Larsen et al., eds., Structural and tectonic modelling and its application to petroleum geology: Norwegian Petroleum Society Special Publication, p. 121–131.

Gerard, J., and C. Buhrig, 1990, Seismic facies of the Barents shelf: analysis and interpretation: Marine and Petroleum Geology, v. 7, p. 234–252.

Gudlaugson, S. T., J. I. Faleide, S. E. Johansen, and A. Breivik, 1994, Late Paleozoic structural development of the southwestern Barents Sea, *in* S. E. Johansen, Geological evolution of the Barents Sea with special emphasis on late Paleozoic development: Ph.D. dissertation, University of Oslo, Norway, 46 p.

Henriksen, S., 1990, Den Kenozoiske utvikling av det sørvestlige Nordkappbasenget: Master's thesis, Institute for Biologi og Geologi, Uitø, 168 p.

Henriksen, S., and T. O. Vorren, in press, Early Cenozoic sequence stratigraphy and salt tectonics in the Nordkapp Basin: Norsk Geologisk Tidsskrift

Hinz, K., O. Eldholm, M. Block, and J. Skogseid, 1993, Evolution of North Atlantic volcanic continental margins, *in* J. R. Parker, ed., Petroleum geology of northwest Europe: Proceedings of the Fourth Conference, Geological Society of London, p. 901–914.

Jackson, M. P. A., and C. J. Talbot, 1991, A glossary of salt tectonics: The University of Texas at Austin, Bureau of Economic Geology Geological Circular 91-4, 44 p.

Jackson, M. P. A., and B. C. Vendeville, 1994, Regional extension as a geologic trigger for diapirism: GSA Bulletin, v. 106, p. 57–73.

Jensen, L. N., and K. Sørensen, 1992, The tectonic framework and halokinesis of the Nordkapp Basin, Norwegian Barents Sea, *in* R. M. Larsen et al., eds., Structural and tectonic modelling and its application to petroleum geology: Norwegian Petroleum Society Special Publication, p. 109–120.

Jensen, L. N., K. Sørensen, K. Kåsli, O. Riise, and J. R. T. Granli, 1993, Regional assessment and palinspastic restoration of halokinesis in the Nordkapp Basin, Barents Sea (abs.): AAPG, International Hedberg Research Conference Abstracts, Sept. 13–17, Bath, U.K., p. 99.

Jenyon, M. K., 1986, Salt tectonics: London, Elsevier, 191 p.

Johansen, S. F., B. K. Ostisky, Ø. Birkeland, Y. F. Federovsky, V. N. Martirosjan, B. Christensen, S. I. Cheredeev, E. A. Ignatenko, and L. S. Margulis, 1993, Hydrocarbon potential in the Barents Sea region: play distribution and potential, *in* T. O. Vorren et al., eds., Arctic geology and petroleum potential: Elsevier, Norwegian Petroleum Society, p. 273–320.

Johansen, S. E., T. Henningsen, E. Rundhoude, B. M. Sather, C. Fichler, and H. G. Rueslåtten, 1994, Continuation of the Caledonides north of Norway: seismic reflectors within the basement beneath the southern Barents Sea: Marine and Petroleum Geology, v. 11, p. 190–201.

Karpaz, M. R., O. Roberts, V. Moralev, and E. Terekhov, 1993, Regional lineament framework of eastern Finnmark, Norway, western Kola Peninsula, Russia and southern Barents Sea: Proceedings of the Ninth Thematic Conference on Geological Remote Sensing, Pasadena, California, p. 733–750.

Kjøde, J., K. M. Storetvedt, O. Roberts, and A. Gidskehaüg, 1978, Paleomagnetic evidence for large-scale dextral movement along the Trollfjord-Komagelu fault, Finnmark north Norway: Physics of Earth and Planetary Interiors, v. 16, p. 132–144.

Knott, S. D., M. T. Burchell, E. J. Jolley, and A. J. Fraser, 1993, Mesozoic to Cenozoic plate reconstruction of the North Atlantic and hydrocarbon plays of the Atlantic margins, *in* J. R. Parker, ed., Petroleum geology of northwest Europe: Porceedings of the Fourth Conference, Geological Society of London, p. 953–974.

Koyi, H., S. Nybakken, K. Hogstad, and B. Tørrudbakken, 1992, Determining geometry of salt diapirs by depth modeling of seismic velocity pull-up: Oil and Gas Journal, October 5, p. 97–101.

Koyi, H., C. J. Talbot, and B. Tørrudbakken, 1993, Salt diapirs of the southwest Nordkapp Basin: analogue modelling: Tectonophysics, v. 228, p. 167–187.

Krokan, B., 1988, Et gravimetrisk studium av Nordkappbassenget: Master's thesis, Institute for Geology, University of Oslo, Norway, 174 p.

Lippard, S., 1994, Permo-Carboniferous basaltic dykes and fault systems on Magerøya; some new observations and implications for rifting in the southern Barents Sea (abs.): Tectonic and structural geology studies group meeting on reactivation of offshore and onshore basement structures, Norwegian Geological Society, Tromsø, Norway, Nov.

Lundin, E. R., 1992, Thin-skinned extensional tectonics on a salt detachment, northern Kwanza basin, Angola: Marine and Petroleum Geology, v. 9, p. 405–411.

Nelson, T. H., 1989, Style of salt diapirs as a function of the stage of evolution and the nature of encasing sediment, *in* Gulf of Mexico salt tectonics, associated processes and exploration potential: SEPM Gulf Coast Section, 10th Annual Research Conference, Program and Extended Abstracts, Houston, Texas, p. 109–110.

Nelson, T. H., 1991, Salt tectonics and listric normal faulting, *in* A. Salvador, ed., The Gulf of Mexico basin: GSA Decade of North American Geology, v. J, p. 73–89.

Nilsen, K. T., 1994, Comments on the tectono-stratigraphic evolution of the Barents Sea (abs.): Tectonic and structural geology studies group meeting on reactivation of offshore and onshore basement structures: Norwegian Geological Society, Tromsø, Norway, Nov.

Nilsen, K. T., F. Henriksen, and G. B. Larssen, 1993, Exploration of the Late Paleozoic carbonates in the Southern Barents Sea—a seismic stratigraphic study, *in* T. Vorren, ed., Arctic geology and petroleum potential: Elsevier, Norwegian Petroleum Society, p. 393–403.

Nilsen, K. T., B. C. Vendeville, and J. T. Johansen, 1994, An example of salt tectonics controlled by regional tectonics: the Nordkapp Basin, Norway: AAPG Annual Convention Official Program, Denver, Colorado, v. 3, p. 225.

Nyland, B., L. N. Jensen, J. I. Skagen, O. Skarpnes, and T. Vorren, 1992, Tertiary uplift and erosion in the Barents Sea: magnitude, timing and consequences, *in* R. M. Larsen et al., eds., Structural and tectonic modelling and its application to petroleum geology: Norwegian Petroleum Society Special Publication, p. 153–162.

Reemst, P., S. Cloething, and S. Fanavoll, 1994, Tectono-stratigraphic modelling of Cenozoic uplift and erosion in the southwestern Barents Sea: Marine and Petroleum Geology, v. 11, p. 478–490.

Riis, F., 1992, Dating and measuring of erosion, uplift and subsidence in Norway and the Norwegian shelf in glacial periods, *in* L. N. Jensen et al., eds., Post-Cretaceous uplift and sedimentation along the western Fennoscandian shield: Norsk Geologisk Tidsskrift, v. 72, no. 3, p. 325–331.

Riis, F., and W. Fjeldskaar, 1992, On the magnitude of the late Tertiary and Quaternary erosion and its significance for the uplift of Scandinavia and the Barents Sea, *in* R. M. Larsen et al., eds., Structural and tectonic modelling and its application to petroleum geology: Norwegian Petroleum Society Special Publication, p. 163–185.

Riis, F., and L. N. Jensen, 1992, Measuring uplift and erosion—proposal for a terminology, *in* L. N. Jensen et al., eds., Post-Cretaceous uplift and sedimentation along the western Fennoscandian shield: Norsk Geologisk Tidsskrift, v. 72, no. 3, p. 320–324.

Rønnevik, H. C., 1982, Structural and stratigraphic evolution of the Barents Sea: Norwegian Petroleum Directorate Bulletin No. 1, 77 p.

Sales, J. K., 1992, Uplift and subsidence of northwestern Europe: possible causes and influence on hydrocarbon productivity: Norsk Geologisk Tidsskrift, v. 72, no. 3, p. 253–258.

Schultz-Ela, D. D., M. P. A. Jackson, and B. C. Vendeville, 1993, Mechanics of active salt diapirism: Tectonophysics, v. 228, p. 275–312.

Stemmerik, L., and D. Worsley, 1989, Late Paleozoic sequence correlations, North Greenland, Svalbard and the Barents shelf, *in* J. D. Collision, ed., Correlation in hydrocarbon exploration: Graham and Trotman, Norwegian Petroleum Society, p. 303–331.

Talbot, C. J., 1995, Molding of salt diapirs by stiff overburden, *in* M. P. A. Jackson, D. G. Roberts, and S. Snelson, eds., Salt tectonics: a global perspective: AAPG Memoir 65, this volume.

Talbot, C. J., H. Koyi, and J. Clark, 1993, Multiphase halokinesis in the Nordkapp Basin, *in* T. O. Vorren et al., eds., Arctic geology and petroleum potential: Elsevier, Norwegian Petroleum Society, p. 665–668.

Vågnes, E., J. I. Faleide, S. T. Gudlaugson, 1992, Glacial erosion and tectonic uplift in the Barents Sea, *in* L. N. Jensen et al., eds., Post-Cretaceous uplift and sedimentation along the western Fennoscandian shield: Norsk Geologisk Tidsskrift, no. 3, p. 333–338.

Vendeville, B. C., 1993, Thin-skinned inversion tectonics of salt diapirs: GSA Abstracts with Programs, v. 25, no. 6, p. A408.

Vendeville, B. C., and M. P. A. Jackson, 1991, Deposition, extension, and the shape of downbuilding diapirs: AAPG Bulletin, v. 75, p. 687–688.

Vendeville, B. C., and M. P. A. Jackson, 1992a, The rise of diapirs during thin-skinned extension: Marine and Petroleum Geology, v. 9, p. 331–353.

Vendeville, B. C., and M. P. A. Jackson, 1992b, The fall of diapirs during thin-skinned extension: Marine and Petroleum Geology, v. 9, p. 354–371.

Vendeville, B. C., M. P. A. Jackson, and H. Ge, 1993a, Detached salt tectonics during basement-involved extension: AAPG Annual Convention, Official Program, New Orleans, Louisiana, p. 195.

Vendeville, B. C., M. P. A. Jackson, and R. Weijermars, 1993b, Rates of salt flow in passive diapirs and their source layers, *in* Rates of geologic processes: SEPM Gulf Coast Section, 10th Annual Research Conference, Program and Extended Abstracts, Houston, Texas, p. 269–276.

Vendeville, B. C., K. T. Nilsen, and J. T. Johansen, 1994, Using concepts derived from tectonic experiments to improve the structural interpretation of salt structures: the Nordkapp Basin example: Geological Society of London, meeting on modern developments in structural interpretation, validation and modelling, London, Feb. 21–23.

Vendeville, B. C., H. Ge, and M. P. A. Jackson, 1995, Models of salt tectonics during basement extension: Petroleum Geoscience, v. 1, p. 179–183.

Willoughby, J., and P. E. Øverli, 1994, Salt diapirism and the geological history of the western Nordkapp Basin: an exploration perspective (abs.): Tectonic and structural geology studies group meeting on reactivation of offshore and onshore basement structures, Norwegian Geological Society, Tromsø, Norway, Nov.

Yu, Z., and I. Lerche, 1993, Salt dynamics: simulation of mushroom cap on a salt diapir in the Barents Sea, Norway, *in* T. O. Vorren et al., eds., Arctic geology and petroleum potential: Elsevier, Norwegian Petroleum Society, p. 669–680.

Koyi, H., C. J. Talbot, and B. O. Tørudbakken, Salt tectonics in the northeastern
Nordkapp Basin, southwestern Barents Sea, in M. P. A. Jackson, D. G. Roberts,
and S. Snelson, eds., Salt tectonics: a global perspective: AAPG Memoir 65,
p. 437–447.

Chapter 21

Salt Tectonics in the Northeastern Nordkapp Basin, Southwestern Barents Sea

Hemin Koyi

Christopher J. Talbot

Hans Ramberg Tectonic Laboratory
Institute of Earth Sciences
Uppsala, Sweden

Bjørn O. Tørudbakken
Saga Petroleum a.s.
Sandvika, Norway

Abstract

Salt structures in the northeastern Nordkapp subbasin are interpreted on reflection seismic profiles. Thickness variations indicate localized accumulation of the mother salt in Late Carboniferous–Early Permian time. Rapid sedimentation in the Early Triassic accompanied rise of salt into asymmetric salt pillows during regional extension. These pillows domed the prekinematic Permian sediments and became diapiric during the late Early–Middle Triassic, perhaps as a result of thin-skinned normal faulting decoupled by the salt from old basement faults reactivated by thick-skinned regional (northwest-southeast) extension.

Variations in size, maturity, and evolution history of individual salt structures can be attributed to local differences in thickness of the initial salt layer and its burial history. Salt structures form three rows concentric to the basin margins and cover ~ 20% of the basin area. Some salt stocks appear to overlie basement faults. Asymmetric primary, secondary, and in places tertiary, peripheral sinks indicate that salt was withdrawn mainly from the basin side of most diapirs throughout Triassic downbuilding.

The ratio of net salt rise rate to net aggradation rate ($\overset{\bullet}{R} / \overset{\bullet}{A}$) increased slowly from <1 to >1 during Middle Triassic time and increased markedly during slow sedimentation in the Late Triassic and Jurassic. By Jurassic time, more than 18 enormous salt fountains extruded downslope and spread a partial salt canopy in the central and northern parts of the northeastern subbasin. Larger and more widely spaced salt extrusions in the northeastern subbasin spread significantly farther than their equivalents in the southwestern subbasin, where Triassic subsidence or downbuilding was slower. Salt extrusion (and perhaps dissolution) ceased during Cretaceous burial but probably resumed locally in the late Tertiary. Salt loss during Cretaceous–Tertiary reactivation of salt rise reduced the area of the salt canopy. Surviving remnants of the salt canopy may still trap any pre-Jurassic hydrocarbons despite hydrocarbon venting throughout the Arctic during Tertiary uplift.

INTRODUCTION

The geology of the southern Barents Sea has been described in several publications (Faleide et al., 1984; Rønnevik and Jacobsen, 1984; Berglund et al., 1986; Jensen and Sørensen, 1988, 1992; Bergendahl, 1989; Gabrielsen et al., 1990, 1992; Koyi et al., 1992, 1993b; Talbot et al., 1992). The Nordkapp Basin lies in the southwestern Barents Sea (Figure 1) and consists of two subbasins separated by an interbasinal ridge, the Nordkapp High (Dengo and Røssland, 1989; Gabrielsen et al., 1990). The southwestern subbasin is 150 km long in a northeast-trending half-graben having an average width of 40 km (Figure 1). The northeastern subbasin, the subject of this chapter, is an east-northeast striking symmetric graben with an average width of 60 km and a length of ~200 km.

Although salt structures occur elsewhere in the Norwegian sector of the Barents Sea, the Nordkapp Basin contains most of the seismically defined salt structures in the region (Figure 1). Most previous publications on the salt structures in the Nordkapp Basin (Bergendahl, 1989; Jensen and Sørensen, 1988, 1992; Gabrielsen et al., 1990, 1992) adopted a traditional interpretation in which the flanks of salt stocks and walls were assumed to be vertical below well-defined top salt seismic reflections. The traditional interpretations implied a 2-km-thick mother salt in the southwestern subbasin and a significantly thicker mother salt (4–5 km) in the northeastern Nordkapp subbasin (Jensen and Sørensen 1992).

Koyi et al. (1993b) used a dynamically scaled model to interpret the geometry and evolution history of the salt diapirs in the southwestern subbasin. Here, reflection

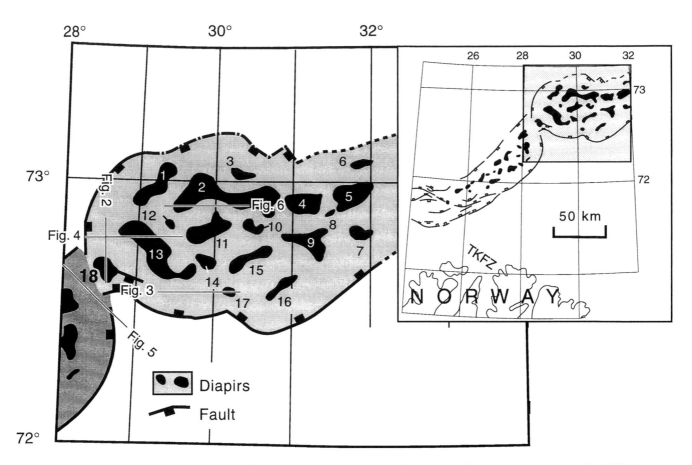

Figure 1—Map of the Nordkapp Basin showing salt structures subcropping on the Quaternary unconformity. TKFZ on the inset map is the Trollfjord–Komagelv fault zone.

seismic data alone are used to study the geometry and evolution of the larger salt diapirs in the northeastern subbasin, which spread farther than their equivalents in the southwestern subbasin and formed a partial salt canopy.

By 1990, the Nordkapp Basin was covered by a 5 × 5 km grid of generally good quality reflection seismic lines and locally denser coverage shot for the Norwegian Petroleum Directorate. The seismic lines were tied to the few wells drilled on the basin margins. Our interpretation of the salt structures in the northeastern subbasin was based on the reflection seismic data, which clearly resolve the crests of the closely spaced diapir overhangs, the intradome areas, and some dome margins. Subsalt reflections are obscured beneath the broad salt overhangs. Constraint on the geometry of the salt overhangs comes from the structural relationships of their tops and flanks and from basin history (Figures 2, 3, 4, 5, 6, 7).

GEOLOGY OF NORDKAPP BASIN

The Nordkapp Basin formed during late Paleozoic rifting in the southwestern Barents Sea. The general north-northeast strikes of the normal faults bounding large parts of the Nordkapp Basin suggest inversion of under-

lying Caledonian reverse faults during Devonian–Carboniferous transtension followed by repeated reactivation (Gabrielsen et al., 1990; Gudlaugsson et al. 1990; Jensen and Sørensen, 1992). Crustal extension between Greenland and Norway (Dengo and Røssland, 1989) was accompanied by dextral movement along the northwest-striking Trollfjord-Komagelv fault zone in northern Finnmark (Jensen and Sørensen, 1988, 1992). From the Late Carboniferous to Early Permian, the site of the future Nordkapp Basin was a stable carbonate platform (Riis et al., 1986; Jensen and Sørensen, 1988) of regional extent (Gabrielsen et al., 1990). A substantial salt sequence accumulated in the basin and on parts of its flanks, whereas the surrounding platforms were either exposed or were depositional sites of anhydrite or carbonates.

By analogy with genetic relationships with the late Paleozoic carbonate sequences of the Sverdrup Basin and Spitzbergen, the salt in the northeastern subbasin was deposited during the Late Carboniferous–Asselian (Jensen and Sørensen, 1988, 1992). Fine-grained clastic sediments were deposited in the Nordkapp Basin in the Early Triassic. Block faulting dominated during the Middle Jurassic–Early Cretaceous (Gabrielsen et al., 1990, 1992), but Late Cretaceous regional extension that affected the western Barents Sea (Faleide et al., 1984) reactivated the structures in the Nordkapp Basin. Regional com-

(a)

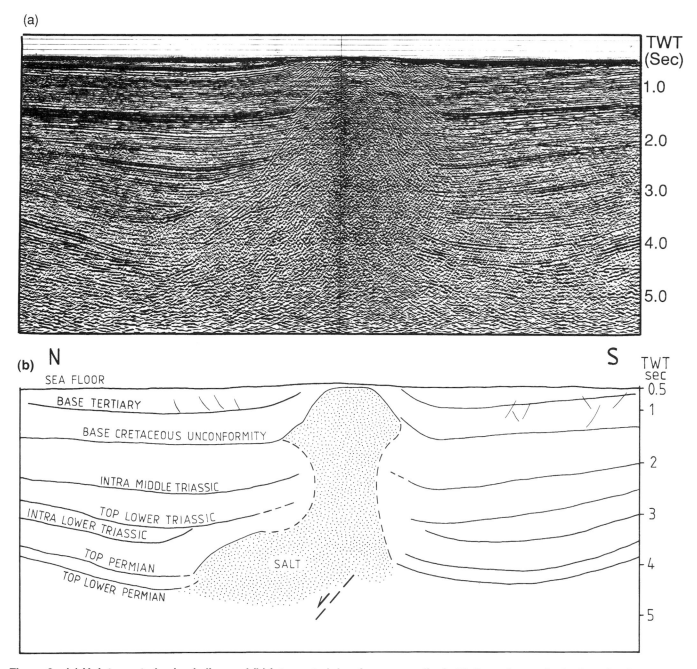

Figure 2—(a) Uninterpreted seismic line and (b) interpreted drawing across diapir 18, the only one in the junction between the northeastern and southwestern subbasins. Truncation of the intra–Lower Triassic reflectors (north of the diapir) by the top Lower Triassic reflectors implies major salt withdrawal from a half-graben to feed a growing pillow or salt roller. The thickened top Lower–Upper Triassic sequences to the north of the diapir indicate the formation of secondary periphery sinks when the structure became diapiric. The dashed lines in (b) indicate the limit of the diapir where the reflectors are not detected beneath the diapir overhang.

pression occurred in early Tertiary time (Jensen and Sørensen, 1992). Most of the hydrocarbons in the region were vented when major uplift and erosion took place during the Neogene (Berglund et al., 1986; Nøttvedt et al., 1988; Vorren et al., 1988, 1990; Nyland et al., 1992; Richardsen et al., 1993).

There is little change in thickness or seismic facies across the boundaries of what must have been a cryptic

Nordkapp Basin in the Permian and earliest Triassic prekinematic sequence (see Figure 2). The Nordkapp Basin was reestablished by fault reactivation and differential subsidence of fine-grained Lower–Middle Triassic clastic sediments (Gabrielsen et al., 1992). In general, Triassic sequences are thicker in the northeastern subbasin, implying faster subsidence than in the southwestern subbasin.

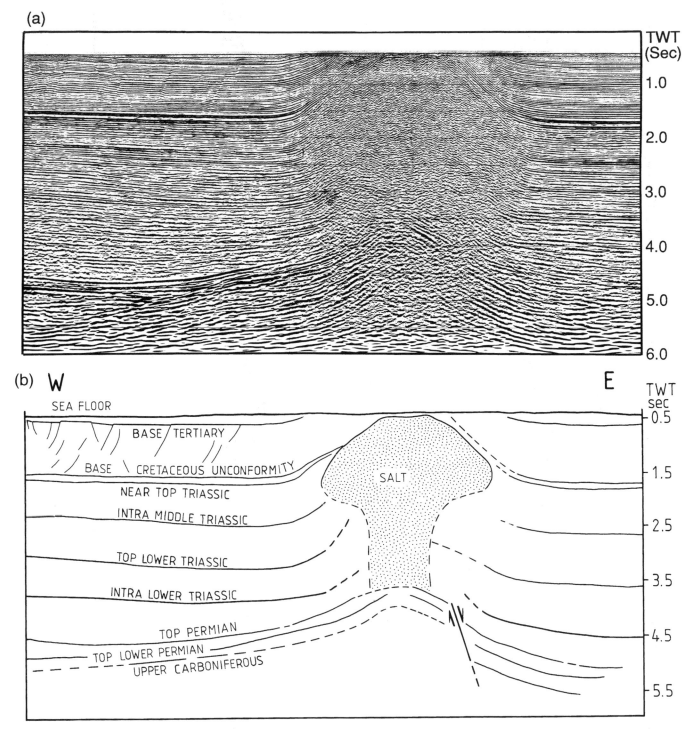

Figure 3—(a) Uninterpreted east-west seismic line and (b) interpreted drawing across diapir 14 (see Figure 1 for location). The intra-Lower Triassic units thin toward the structure, suggesting pillow formation during the Early Triassic.

SALT IN NORDKAPP BASIN

Age and Thickness

Salt in the northeastern subbasin was deposited in the last phase of rifting (Riis, 1990) during the Carboniferous–Early Permian (Jensen and Sørensen, 1988, 1992).

The mother salt in the northeastern subbasin was possibly deposited with nonuniform thickness on a faulted basement (Figure 5). Rim synclines around the margins of the basin indicate salt rise and piercement during the late Early Triassic. Within the basin, thicker Lower Triassic sediments around the salt diapirs indicate salt withdrawal and probable basin subsidence.

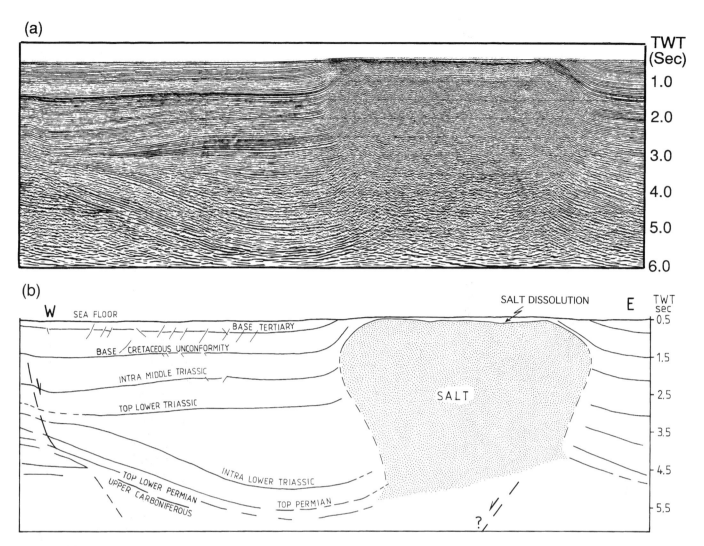

Figure 4—(a) Uninterpreted east-west seismic line and **(b)** interpreted drawing cutting obliquely across diapir 13. Diapir 13 (33 × 8 km) is the only salt structure in the basin that is elongate in the northwest direction parallel to the fault-defined southwestern margin of the basin (see Figure 1 for location). Compare the thick Lower–Middle(?) Triassic sediments on the western side of the diapir with their thinner equivalents, which thicken away from the eastern side of the diapir. This thicker pile is attributed to salt withdrawal and subsidence along a basement fault assumed to underlie the diapir. The Upper and Middle Triassic and Cretaceous sediments thicken away from the western side of the diapir, indicating later reactivation of the marginal fault (extreme left). The extensional faults closely spaced in the Upper Cretaceous and Tertiary units (west of the diapir) are due to the later regional uplift and erosion. The Cretaceous and younger units thin against the salt overhang. The salt shows dissolution effects beneath the major unconformity over the crest of the diapir.

Based on direct calculation of the volume of the salt structures and indirect calculation of the excess volume of sediments in the secondary rim synclines, Jensen and Sørensen (1988, 1992) estimated the original maximum thickness in the northeastern subbasin to be 4–5 km. This exceptional thickness for the northeastern subbasin salt is probably an overestimate due to the assumption that the salt diapirs were columnar (Talbot et al., 1992; Koyi, 1994).

Salt Structures

Eighteen major individual salt structures (numbered on Figure 1) are imaged in the northeastern subbasin. Compared with the 21 stocklike diapirs in the southwest-

ern subbasin, the diapirs in the northeastern subbasin are wall-like and concentric to the basin margins. Salt structures subcropping on the base of the Quaternary have larger areas and are farther apart in the northeastern subbasin than in the southwestern subbasin. Within the northeastern subbasin, diapirs along the northern and northwestern margin have larger subcrops than those near the southern and southeastern margins (Figure 1). This could be because (1) the mother salt was initially thicker along the northern and northwestern flanks of the basin, (2) the salt structures there were more mature, or (3) the salt was driven to the north and northwest by extra loading during Early Triassic progradation from the south and southeast (Rønnevik et al., 1982; Rønnevik and

Jacobsen, 1984; Jensen and Sørensen, 1992). Most diapirs in the Nordkapp Basin spread wide overhangs over narrower stems throughout the Late Triassic–Jurassic. However, the spreading was more pronounced than in the southwestern subbasin and formed a partial canopy in the northeastern subbasin.

Salt Flow and Sedimentation

The 1600–2200 m of prekinematic carbonates and rapidly deposited shaly sequences (Jensen and Sørensen, 1992) appear to have been sufficiently strong to suppress halokinesis of salt already rendered buoyant by a density inversion. Salt flow began in the northeastern subbasin during the Early–Middle Triassic when the basin evolved as a secondary rim syncline around closely spaced diapirs (Jensen and Sørensen 1988, 1992; Gabrielsen et al. 1990). Jensen and Sørensen (1988, 1992) suggested that most diapirs pierced without passing through a pillow stage. Although evidence for or against a pillow stage is generally obscured by broad salt overhangs (Figures 4, 6), there is clear evidence for at least two of the oldest structures (e.g., Figure 2) having initiated as either slightly asymmetric salt pillows or asymmetric salt rollers. Thus, Upper Permian–lowermost Triassic prekinematic layers are truncated by upper Lower Triassic syndiapiric sediments on the basin side of the diapir (Figure 2), but are only thinned on the other side.

The asymmetry in the Early Triassic primary rim syncline could be due to asymmetric salt withdrawal from a pillow or to a salt roller created by thin-skinned extension in the cover as it slid down into the basin. Jensen and Sørensen (1992) suggested that salt started to flow due to differential loading by subsiding overburden stretched during thick-skinned regional extension. Salt movement varied locally during the Early Triassic, but all salt structures were diapiric by the Middle Triassic. Slip of the marginal basement faults and subsidence of overburden units basinward (Figure 4) during Triassic may have remobilized the remaining salt on the shelf and the basin margins into salt pillows. The resulting pillows (Figure 5) are the only hydrocarbon targets drilled in the basin to date (1993).

Thin clastic sequences suggest that sedimentation rates slowed in the Late Triassic–Jurassic. Sequence thickness changes indicate that most of the salt diapirs were spreading at or just below the surface in the Late Triassic–Jurassic, resulting in upward-widening diapirs (Figures 4, 6). The salt diapirs of the northeastern subbasin continued to rise during the Middle Jurassic–Early Cretaceous, indicating continued salt supply from below. Salt spreading with or without dissolution of large volumes of salt ended abruptly during rapid filling of the basin by thick Cretaceous clastics. Roofs of uncompacted, less-dense overburden were either lifted or shed by salt flow. The surface expression of the extrusive salt mounds persisted longest at the western end of the northeastern subbasin. Cretaceous sedimentation rates were greater than the rate of diapiric rise but insufficient to preclude diapiric rise. On the contrary, Cretaceous sedimentary

onlap reactivated salt diapirs, which resumed rising during deposition. The Cretaceous–Tertiary overburden enhanced differential loading, so salt flow in the overhangs reversed from spreading downward and outward to draining upward and inward. A regional compression of unknown magnitude affected the Nordkapp Basin in early Tertiary time (Jensen and Sørensen, 1992). Jensen and Sørensen (1992) suggested that the combination of this gravitational instability and regional compression reactivated the Nordkapp diapirs (Nilsen et al., 1995). Tertiary uplift and erosion have removed much of the evidence for the salt fountains and their inversion during the compression.

Burial of salt structures continued until Pliocene–Pleistocene uplift and erosion. Closely spaced normal faults in the Cretaceous–Tertiary sequence indicate post–early Tertiary regional extension (Figures 2, 4), which is attributed to opening of the Norwegian Sea at 58 Ma (Berglund et al., 1986; Nøttvedt et al., 1988; Vorren et al., 1988, 1990; Nyland et al., 1992; Richardsen et al., 1993).

FAULTS

Using paleotectonic maps, Dengo and Røssland (1989) suggested three phases of extension in the southern Barents Sea. Episodes of regional extension are recorded by normal faults in middle Carboniferous, Upper Triassic, Upper Jurassic–Lower Cretaceous, and Tertiary strata (Figure 4). Any extension during Permian–Middle Triassic time is not detectable on the seismic data. The Ordovician(?)–Carboniferous basement was broken into blocks rotated by the early rifting (Figure 5). The rifting seems to have decreased during the Late Carboniferous, and most of the movement has been taken up by the basin margin faults that are still active (Figure 4).

Extensional faults in the overburden can be classified into two major groups. The first group is related to reactivation of deeper faults in the basement (with or without decoupling by the salt layer); most are along the basin margins (Figure 5). The second group comprises closely spaced listric normal faults mainly within the Cretaceous–Tertiary units. Although a few of these displace sub-Jurassic reflectors, most sole to Upper Jurassic units (Figures 2, 3, 4, 5). These faults aided venting of hydrocarbons during Pliocene–Pleistocene decompressional uplift, except perhaps for hydrocarbons beneath the Late Triassic–Jurassic salt overhangs, which provided particularly impermeable traps.

Although thin-skinned extension faulted the Cretaceous–Tertiary sediments, older basin margin faults below the salt reactivated as a result of thick-skinned regional extension (Figure 5). Apart from the marginal faults that propagate upward through younger units, any interbasinal basement faults are obscured by the broad diapir overhangs (Figures 4, 5). Figure 5 represents a seismic line that runs northwestward across the junction of the two subbasins and shows indications of early basin formation. Unobscured by salt overhangs, basement blocks (including Devonian sequences) appear to have

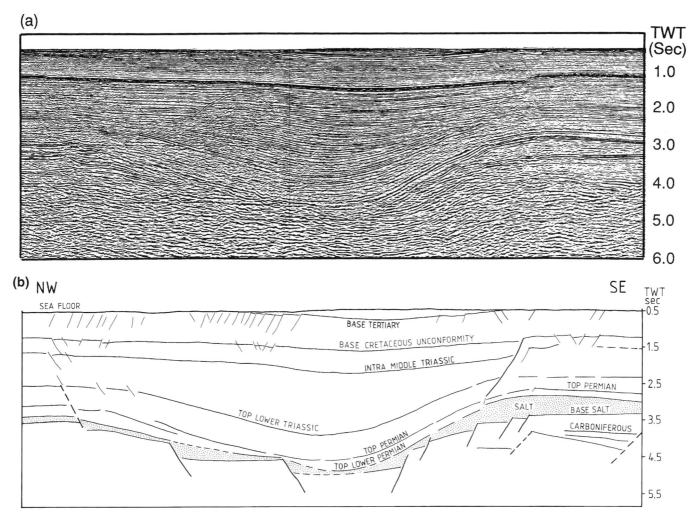

Figure 5—(a) Uninterpreted northwest-southeast seismic line and (b) interpreted drawing across the junction between the two subbasins. Unobscured by salt overhangs, this line shows rotated Carboniferous(?) basement blocks. Only the marginal basement faults have propagated into the cover units. Salt flow along the basin margin onto the shelf areas has contributed to collapse of the overlying sediments across the basin margin.

rotated along relatively minor faults within the basin and beyond (Figure 5). Basement faults acted as hinges for Triassic and later sediment accumulation, whereas differential loading led to subsidence and faulting of the overburden units that eventually triggered salt flow. The salt layer may have mechanically decoupled the overburden from the basement. However, slip along the basement faults during regional extension resulted in stretching and faulting of the overburden units, no matter what their location was relative to basement faults. Changes in spacing, dip, and levels of reflectors in the sediments indicate possible basement faults associated with some of the diapirs. Lower–Middle Triassic sediments that are thicker and deeper on the west side of diapir 13 (Figure 4) suggest a deeper secondary rim syncline to the west and a possible fault in the basement beneath this diapir. Lower Cretaceous–middle Tertiary sediments also thicken west of diapir 13 (Figure 4), indicating that reactivation of this marginal fault continued through Cretaceous–early Tertiary regional extension.

SALT FOUNTAINS, EXTRUSION, AND DISSOLUTION

Many of the seismic lines across salt diapirs in the Nordkapp Basin show that the Lower Cretaceous sediments thinned as they onlapped spreading salt overhangs (Figures 3, 4, 6). This implies deposition of these sediments as the diapiric overhangs were extruding. Jurassic–Lower Cretaceous sediments abut against steep faces of salt bulbs that faced upslope out of the basin. In contrast, they onlap the wider distal downslope salt glaciers (Figures 4, 6). The sediments thin toward and over the wider overhangs, implying vertical flow of salt or thickening of the overhangs during their deposition. Permian reflectors below some of the salt overhangs suggest that the stems of these diapirs are significantly narrower than their asymmetric overhangs (Figures 3, 4) (Koyi et al., 1992).

Salt diapirs rise and spread only if they have sufficient

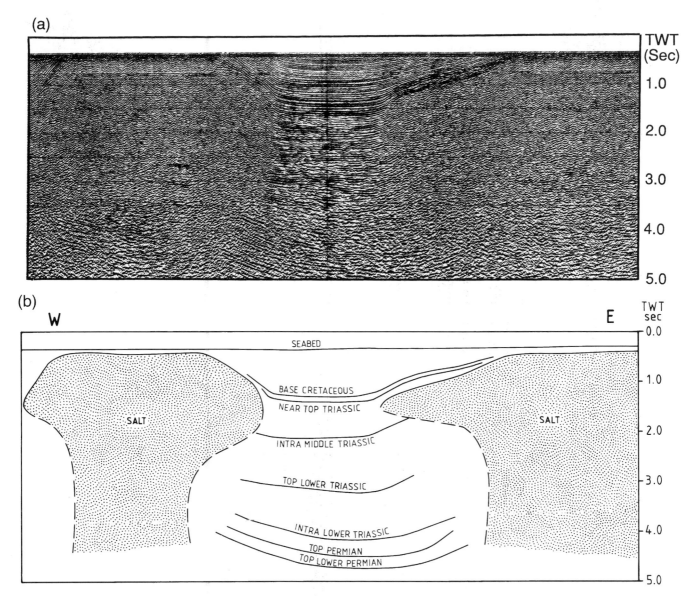

Figure 6—(a) Uninterpreted east-west seismic line and (b) interpreted drawing across two diapirs in the northern part of the basin. The line is subparallel to diapir A (on the right), which is elongate east-west and hence shows a wider stem. This diapir also shows a wide overhang onto which Cretaceous and younger sediments onlapped and thinned during their deposition. The thin Jurassic units are discordant to shallow levels of the overhang, implying that they onlapped a diapiric fountain.

time and supply of salt (Talbot 1993; but compare Nilsen et al., 1995). Most diapirs in the Nordkapp Basin had time to spread broad overhangs and form a discontinuous salt canopy during slow sedimentation in Late Triassic–Jurassic time. Their asymmetry (Figure 4, 6) suggests that salt spread down into a topographic basin influenced by asymmetric salt withdrawal due to faulting, basement slope, sediment compaction, or progradation of their overburden.

Little evidence for episodes of salt extrusion has survived subsequent reactivation and Tertiary erosion. Cretaceous and even lower Tertiary units might never have been deposited over diapir crests if they emerged as fountains above the surface, as seems likely. Voluminous

salt dissolution was also likely during the Late Triassic–Jurassic salt extrusion. Although some of the deeper salt diapirs may never have reached uncompacted levels where they were vulnerable to subsurface dissolution, most were sufficiently shallow for episodic salt dissolution to have occurred throughout the major phases of erosion in the last 58 Ma. Evidence of recent salt dissolution below the seafloor is seen in at least one place (Figure 4).

The crest of many of the salt structures was planed flat by the latest glacial erosion. This unconformity is still planar over structures that have been inactive since the last regional erosion, but it bulges over others, indicating current salt rise or differential compaction (Figure 2).

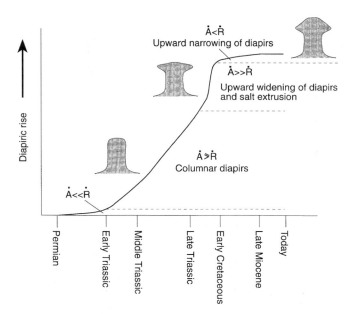

Figure 7—A qualitative curve plotting the history of the rate of sedimentation, $\dot{A}$, and the diapiric rise, $\dot{R}$, in the Nordkapp Basin.

DISCUSSION AND SUMMARY

During slow sedimentation in the Late Triassic–Jurassic, many of the salt diapirs in the Nordkapp Basin spread broad salt sheets that obscure underlying seismic reflectors (Gabrielsen et al., 1992; Koyi et al., 1992, 1993b; Talbot et al., 1992). Jensen and Sørensen's (1992) assumption that the diapirs have columnar geometry may have led them to overestimate the volume and initial thickness of the mother salt. A columnar profile beneath the crest of a diapir suggests 171% (45.4 km^2) more salt than does the interpretation of the same diapir with a significant overhang (Koyi, 1994). A typical axisymmetric columnar diapir would have a volume of 243 km^3, which implies 390% (193 km^3) more salt than that calculated for the same axisymmetric diapir with an overhang consistent with the velocity pull-up of subsalt horizons (Koyi et al., 1992). Distributing these two volumes of salt over a map area of 400 km^2 yields original salt thicknesses of 600 m and 120 m, respectively. Although these figures are not directly applicable to the Nordkapp Basin, they show the strong dependence of volume estimates and initial thickness on the three-dimensional geometry of salt structures.

Rates of diapiric rise and sedimentation varied throughout the history of the Nordkapp Basin. After the prekinematic sequence was deformed by salt pillows or rollers and pierced by active or reactive diapirs in the Early Triassic, the contacts of downbuilding salt structures were molded by their overburden (Vendeville and Jackson, 1991; Talbot, 1995). Figure 7 is a qualitative chart of the net aggradation rate ($\dot{A}$) relative to the rate of increase in diapir relief ($\dot{R}$).

At the onset of diapirism, the sedimentation rate was greater than the rate of diapiric rise ($\dot{R}$) (Figure 7). During the late Early Triassic and Middle Triassic, salt rise increased until it was equal to the accumulation rate, resulting in vertical contacts and columnar stocks or walls. In the Late Triassic, salt rise increased relative to clastic aggradation so that the contacts flared upward and outward. Salt rise outpaced the aggradation rate in Jurassic time, resulting in extrusion of fountains that spread salt overhangs (Figure 7). The Cretaceous sedimentation rate increased significantly relative to diapiric rise and buried diapirs, although they still probably continued to rise. Diapir crests tapered inward and upward at this stage. The Tertiary sedimentation rate also appears to have been higher than diapiric rise, resulting in further narrowing of the diapir crests. However, episodes of uplift and erosion during the late Tertiary partially exhumed diapirs so that they rose actively and faster. Because extension could cause diapiric fall (Vendeville and Jackson, 1992b), diapiric rise rates in the Nordkapp Basin might have decreased relative to sediment accumulation rates during episodes of lateral extension.

Lateral extension can be thick- or thin-skinned and can stretch and fault overburden units with or without involvement of the basement. Overburden faulting is a crucial factor in diapiric rise (Vendeville and Jackson, 1992a). Diapirs infill postsalt faults in the overburden that generally strike at high angles to the inferred direction of extension. Accordingly, the present locations of diapirs in the Nordkapp Basin indicate where the overburden initially faulted. As a result, we expect diapirs that rose along such faults in the overburden to be elongate at high angles to the extension direction but subparallel to basement faults.

In plan view, the majority of the diapirs in this basin are elongate parallel to the basin margins, whatever their orientation, rather than striking only at high angles to the inferred northeastward direction of regional extension (Figure 1). We consider that this pattern is more consistent with diapirs initiating as a result of basement fault reactivation during regional extension and differential loading that faulted their overburden than with diapirs rising along overburden faults created by thin-skinned extension alone.

In summary, the seismic data show the following:

1. In map view, salt structures in the northeastern Nordkapp Basin are generally elongate parallel to the basin margins.
2. The salt structures and their peripheral sedimentary fills had strongly asymmetric profiles from initiation, through downbuilding and extrusion, to their reactivation after burial.
3. Both sediment accumulation and salt rise or spreading were generally faster on the basinward side of most structures.
4. At the time of salt intrusion, the faults in the overburden probably had trends similar to the faults in the basement, although their locations might have differed because of decoupling along the salt layer.
5. Broad basinward overhangs could form significant traps for any hydrocarbons in reservoir rocks predating salt extrusion.

Acknowledgments   *Thanks are due to Steven Seni, Stein Nybakken, and two anonymous reviewers for reading and commenting on this manuscript and to Norwegian Petroleum Directorate, Saga Petroleum a.s., Geco-Prakla, BP Norge, and Enterprise for Oil for kindly releasing the pre-1990 seismic data. Saga Petroleum A.S. gave permission to publish interpretations and results.*

REFERENCES CITED

Bergendahl, E., 1989, Halokinetisk utvikling av Nordkapp bassengets sørvestre segment: Cand. Scient Oppgave, Institutt for Geologi, Universitet i Oslo, 120 p.

Berglund, L. T., G. Augustson, R. Færseth, and H. Ramberg-Moe, 1986, The evolution of the Hammerfest Basin, *in* A. M. Spencer, ed., Habitat of hydrocarbons on the Norwegian continental margin: Norsk Petroleum Forening, London, Graham and Trotman, p. 319–338.

Dengo, C. A., and K. G. Røssland, 1989, Extensional tectonic history of the western Barents Sea: Norsk Petroleum Forening meeting, Stavanger, Norway, p. 15A.

Faleide, J. I., S. T. Gudlaugson, and G. Jacquart, 1984, Evolution of the western Barents Sea: Marine and Petroleum Geology, v. 1, p. 123–150.

Gabrielsen, R. H., R. B. Færseth, L. N. Jensen, J. E. Kalheim, and F. Riis, 1990, Structural elements of the Norwegian continental shelf, part I: the Barents Sea region: NDP Bulletin, no. 6, 33 p.

Gabrielsen, R. H., O. S. Kløvjan, A. Rasmussen, and T. Stølan, 1992, Interaction between halokinesis and faulting: structuring of the margins of the Nordkapp Basin, Barents Sea region, *in* R. M. Larsen, H. Brekke, B. T. Larsen and E. Talleraas, eds., Structural and tectonic modeling and its application to petroleum geology: Norsk Petroleum Forening Special Publication 1, Amsterdam, Elsevier, p. 121–131.

Gudlaugsson, S. T., J. I. Faleide, and O. E. Baglo, 1990, Restoration of the eroded section in the western Barents Sea (abs.): International Conference on Arctic Geology and Petroleum Potential, Abstracts, Tromsø, Norway, p. 28.

Jackson, M. P. A., and R. R. Cornelius, 1987, Stepwise centrifuge modeling of the effects of differential sediment loading on the formation of salt structures, *in* I. Lerche and J. J. O'Brien, eds., Dynamical geology of salt and related structures: Orlando, Florida, Academic Press, p. 163–259.

Jackson, M. P. A., and C. J. Talbot, 1986, External shapes, strain rates, and dynamics of salt structures: GSA Bulletin, v. 97, p. 305–323.

Jackson, M. P. A., and C. J. Talbot, 1989, Anatomy of mushroom shaped diapirs: Journal of Structural Geology, v. 11, p. 211–230.

Jensen, L. N., and K. Sørensen, 1988, Geology and salt tectonics of the Nordkapp Basin, Barents Sea (abs.), *in* K. Binzer, I. Marcussen, and P. Konradi, eds., Nordiske Geologiske Vintermøde, København, Abstracts: Danmarks Geologiske Undergørelse, v. 18, p. 187–188.

Jensen, L. N., and K. Sørensen, 1989, Tectonic framework and halokinesis of the Nordkapp Basin, Barents Sea: Norsk Petroleum Forening meeting, Stavanger, Norway, p. 16.

Jensen, L. N., and K. Sørensen, 1992, Tectonic framework and halokinesis of the Nordkapp Basin, Barents Sea, *in* R. M.

Larsen, H. Brekke, B. T. Larsen, and E. Talleraas, eds., Structural and tectonic modeling and its application to petroleum geology: Norsk Petroleum Forening Special Publication 1, Amsterdam, Elsevier, p. 109–120.

Koyi, H., 1991, Gravity overturns, extension, and basement fault activation: Journal of Petroleum Geology, v. 14, p. 117–142.

Koyi, H., 1994, Estimation of salt thickness and restoration of cross-sections with diapiric structures: a few critical comments on two powerful methods: Journal of Structural Geology, v. 16, p. 1121–1128.

Koyi, H., O. J. Aasen, S. Nybakken, K. Hogstad, and B. O. Tørudbakken, 1992, Determining geometry of salt diapirs by depth modeling of seismic velocity pull-up: Oil and Gas Journal, v. 90, no. 40, p. 97–100.

Koyi, H., M. K. Jenyon, and K. Petersen, 1993a, The effect of basement faults on diapirism: Journal of Petroleum Geology, v. 16, p. 285–312.

Koyi, H., C. J. Talbot, and B. O. Tørudbakken, 1993b, Salt diapirs of the southwest Nordkapp Basin: analogue modelling: Tectonophysics, v. 228, p. 167–187.

Nardin, T. R., and K. G. Røssland, 1990, Restoration of the eroded section in the western Barents Sea (abs.): International Conference on Arctic Geology and Petroleum Potential, Abstracts, Tromsø, Norway, p. 30A.

Nilsen, K. T, B. C. Vendeville, and J.-T. Johansen, 1995, Influence of regional tectonics on halokinesis in the Nordkapp Basin, Barents Sea, *in* M. P. A. Jackson, D. G. Roberts, and S. Snelson, eds., Salt tectonics: a global perspective: AAPG Memoir 65, this volume.

Nøttvedt, A., T. Berglund, E. Rasmussen, and R. Steel, 1988, Some aspects of Tertiary tectonics and sediments along the western Barents shelf, *in* A. C. Morton and L. M. Parson, eds., Early Tertiary volcanism and the opening of the northeast Atlantic: Geological Society of London, Special Publication 39, p. 421–425.

Nyland, B., L. N. Jensen, J. Skagen, O. Skarpnes, and T. Vorren, 1992, Tertiary uplift and erosion in the Barents Sea: magnitude, timing, and consequences, *in* R. M. Larsen, H. Brekke, B. T. Larsen, and E. Talleraas, eds., Structural and tectonic modeling and its application to petroleum geology: Norsk Petroleum Forening Special Publication 1, Amsterdam, Elsevier, p. 153–162.

Richardsen, G., T. O. Vorren, and O. B. Tørudbakken, 1993, Post-Early Cretaceous uplift and erosion in the southern Barents Sea: a discussion based on analysis of seismic interval velocities: Norsk Geologisk Tidskrift, v. 7, p. 3–22.

Riis, F., 1990, Regional geology of the Norwegian Barents Sea (abs.): International Conference on Arctic Geology and Petroleum Potential, Abstracts, Tromsø, Norway, p. 18.

Riis, F., J. Vollset, and M. Sand, 1986, Tectonic development of the western margin of the Barents Sea and adjacent areas, *in* M. T. Halbouty, ed., Future petroleum provinces of the world: AAPG Memoir 40, p. 661–667.

Rønnevik, H. C., and H. P. Jacobsen, 1984, Structural highs and basins in the western Barents Sea, *in* A. M. Spencer, ed., Petroleum geology of the north European margin: Norsk Petroleum Forening, London, Graham and Trotman, p. 19–32.

Rønnevik, H. C., B. Beskow, and H. P. Jacobsen, 1982, Structural and stratigraphic evolution of the Barents Sea: Norsk Petroleum Forening Contribution no. 1, ONS-82.

Talbot, C. J., 1993, Spreading of salt structures in the Gulf of Mexico: Tectonophysics, v. 228, p. 151–166.

Talbot, C. J., 1995, Molding of salt diapirs by stiff overburden, *in* M. P. A. Jackson, D. G. Roberts, and S. Snelson, eds., Salt tectonics: a global perspective: AAPG Memoir 65, this volume.

Talbot, C. J., and R. J. Jarvis, 1984, Age, budget, and dynamics of salt diapirs: Journal of Structural Geology, v. 6, p. 521–533.

Talbot, C. J., H. Koyi, and J. Clark, 1992, Multiple halokinesis in Nordkapp Basin, *in* T. A. Vorren, ed., Proceedings Tromsö symposium on Arctic geology and petroleum potential: Norwegian Petroleum Society, Trondheim, Norway, p. 663–666.

Vendeville, B. C., and M. P. A. Jackson, 1991, Deposition, extension, and the shape of downbuilding diapirs (abs.): AAPG Bulletin, v. 75, p. 687–688.

Vendeville, B. C., and M. P. A. Jackson, 1992a, The rise of diapirs during thin-skinned extension: Marine and Petroleum Geology, v. 9, p. 331–353.

Vendeville, B. C., and M. P. A. Jackson, 1992b, The fall of diapirs during thin-skinned extension: Marine and Petroleum Geology, v. 9, p. 354–371.

Vorren, T. O., E. Lebesbye, E. Henriksen, S.-M. Knutsen, and G. Richardsen, 1988, Cenozoic erosion og sedimentasjon i det sørlige Barentshav (abs.): XVIII Nordisk Geologiska Vintermøte, København, p. 424.

Vorren, T. O., G. Richardsen, S.-M. Knutsen, and E. Henriksen, 1990, The western Barents Sea during the Cenozoic, *in* U. Bleil and J. Theide, eds., Geological history of the polar oceans: Arctic versus Antarctic: NATO ASI Series C 308, Dordrecht, The Netherlands, Kluwer Academic Publishers, p. 95–118.

Index